Environmental Geology

PRINCIPLES AND PRACTICE

Environmental Geology
PRINCIPLES AND PRACTICE

Fred G. Bell
Department of Geology and Applied Geology
University of Natal, Durban, South Africa

**Blackwell
Science**

© 1998 by
Blackwell Science Ltd
Editorial Offices:
Osney Mead, Oxford OX2 0EL
25 John Street, London WC1N 2BL
23 Ainslie Place, Edinburgh
  EH3 6AJ
350 Main Street, Malden
  MA 02148 5018, USA
54 University Street, Carlton
  Victoria 3053, Australia
10, rue Casimir Delavigne
  75006 Paris, France

Other Editorial Offices:
Blackwell Wissenschafts-Verlag
  GmbH
Kurfürstendamm 57
10707 Berlin, Germany

Blackwell Science KK
MG Kodenmacho Building
7–10 Kodenmacho Nihombashi
Chuo-ku, Tokyo 104, Japan

First published 1998

Set by Excel Typesetters Co.,
Hong Kong
Printed and bound in Great Britain
at the University Press, Cambridge

A catalogue record for this title
is available from the British Library

ISBN 0-86542-875-1

Library of Congress
Cataloging-in-publication Data

Bell, F.G.
  (Frederic Gladstone)
    Environmental geology/Fred G.
Bell.
      p.   cm.
    Includes bibliographical
references and index.
    ISBN 0-86542-875-1
    1. Environmental geology.
  I. Title.
QE38.B455   1998   97-51996
550—dc21
                              CIP

The Blackwell Science logo is a
trade mark of Blackwell Science Ltd,
registered at the United Kingdom
Trade Marks Registry

DISTRIBUTORS

Marston Book Services Ltd
PO Box 269
Abingdon, Oxon OX14 4YN
(*Orders*: Tel: 01235 465500
            Fax: 01235 465555)

USA
Blackwell Science, Inc.
Commerce Place
350 Main Street
Malden, MA 02148 5018
(*Orders*: Tel: 800 759 6102
            781 388 8250
        Fax: 781 388 8255)

Canada
Login Brothers Book Company
324 Saulteaux Crescent
Winnipeg, Manitoba R3J 3T2
(*Orders*: Tel: 204 224 4068)

Australia
Blackwell Science Pty Ltd
54 University Street
Carlton, Victoria 3053
(*Orders*: Tel: 3 9347 0300
            Fax: 3 9347 5001)

For further information on
Blackwell Science, visit our website:
www.blackwell-science.com

Contents

Preface, ix

1 Introduction, 1
1.1 Planning and geology, 1
1.2 Conservation, restoration and reclamation of land, 5
1.3 Geological hazards and planning, 6
1.4 Risk assessment, 8
1.5 Hazard maps, 10
1.6 Morphological maps, 11
1.7 Engineering geomorphological maps, 17
1.8 Environmental geological maps, 19
1.9 Engineering geological maps, 23
1.10 Geographical information systems, 29

2 Volcanic activity, 34
2.1 Introduction, 34
2.2 Volcanic form and structure, 34
2.3 Types of central eruption, 38
2.4 Volcanic products: volatiles, 40
2.5 Volcanic products: pyroclasts, 41
2.6 Volcanic products: lava flows, 44
2.7 Mudflows or lahars, 46
2.8 Volcanic hazard and prediction, 48
     2.8.1 Methods of prediction, 48
     2.8.2 Assessment of volcanic hazard and risk, 50
     2.8.3 Dealing with volcanic activity, 56

3 Earthquake activity, 58
3.1 Introduction, 58
3.2 Intensity and magnitude of earthquakes, 60
3.3 Ground conditions and seismicity, 63
3.4 Effects of earthquakes, 66
3.5 Methods of seismic investigation, 70
     3.5.1 Earthquake prediction, 70
     3.5.2 Assessment of movements along faults, 72
     3.5.3 Aseismic investigation, 72
     3.5.4 Accelerographs, 73
3.6 Seismic hazard and risk, 74

3.7 Seismic zoning, 76
3.8 Induced seismicity, 78

4 Mass movements, 83
4.1 Soil creep and valley bulging, 83
4.2 Causes of landslides, 84
4.3 Classification of landslides, 88
     4.3.1 Falls, 88
     4.3.2 Slides, 90
     4.3.3 Flows, 91
4.4 Landslides in soils, 93
4.5 Landslides in rock masses, 94
4.6 Monitoring slopes, 97
     4.6.1 Monitoring movement, 97
     4.6.2 Monitoring groundwater, 98
     4.6.3 Monitoring acoustic emissions or noise, 99
4.7 Landslide hazard, investigation and mapping, 99
4.8 Methods of slope control and stabilization, 106
     4.8.1 Rockfall treatment, 106
     4.8.2 Alteration of slope geometry, 107
     4.8.3 Reinforcement of slopes, 107
     4.8.4 Restraining structures, 108
     4.8.5 Drainage, 110

5 River activity, 114
5.1 The development of drainage systems, 114
5.2 Fluvial processes, 120
     5.2.1 River flow, 120
     5.2.2 River erosion, 122
     5.2.3 River transport, 123
     5.2.4 Deposition of sediments, 125
5.3 Floods, 126
5.4 Hazard zoning, warning systems and adjustments, 131
5.5 The design flood and flood control, 134

6 Marine activity, 139
6.1 Waves, 139
     6.1.1 Force and height of waves, 140

v

6.1.2 Wave refraction, 141
6.2   Tides, 142
6.3   Beach zones, 143
6.4   Coastal erosion, 144
6.5   Beaches and longshore drift, 147
6.6   Storm surges and marine inundation, 151
6.7   Protective barriers, 153
6.8   Stabilization of longshore drift, 157
6.9   Tsunamis, 159

7   Arid and semi-arid lands, 163
7.1   Introduction, 163
7.2   Wind action, 163
7.3   Desert dunes, 166
7.4   Stream action in arid and semi-arid regions, 169
7.5   Flooding and sediment problems, 173
7.6   Movement of dust and sand, 174
7.7   Sabkha soil conditions, 180
7.8   Salt weathering, 183
7.9   Desertification, 188
7.10   Irrigation, 190

8   Glacial and periglacial terrains, 199
8.1   Introduction, 199
8.2   Glacial erosion, 199
8.3   Glacial deposits: tills and moraines, 203
8.4   Basic properties of tills, 209
8.5   Compressibility and strength of tills, 212
8.6   Fluvio-glacial deposits; stratified drift, 213
8.7   Other glacial effects, 219
8.8   Glacial hazards, 220
8.9   Frozen ground phenomena in periglacial environments, 221
8.10   Construction in permafrost regions, 227
8.11   Frost heave, 228

9   Water resources, 233
9.1   The hydrological cycle, 233
9.2   Reservoirs, 234
9.3   Dam sites, 239
9.4   Geology and dam sites, 243
9.5   Groundwater, 246
9.6   A note on basement aquifers, 248
9.7   Springs, 249
9.8   Water budget studies, 250
9.9   Hydrogeological properties, 251
9.10   Groundwater exploration, 254
9.11   Assessment of permeability and flow in the field, 257
9.12   Water quality and uses, 260

9.13   Wells, 265
9.14   Safe yield, 270
9.15   Artificial recharge, 271
9.16   Groundwater pollution, 272
9.16.1 Landfill and groundwater pollution, 274
9.16.2 Saline intrusion, 275
9.16.3 Nitrate pollution, 277
9.16.4 Other causes of groundwater pollution, 277
9.17   Groundwater monitoring, 279
9.18   Conjunctive use, 281

10   Soil resources, 285
10.1   Origin of soil, 285
10.2   Soil horizons, 286
10.3   Soil fertility, 287
10.4   Pedological soil types, 288
10.5   Basic properties of soil, 292
10.5.1 Particle size distribution, 293
10.5.2 Consistency limits, 294
10.6   Soil classification, 298
10.7   Shear strength of soil, 300
10.8   Consolidation, 301
10.9   Soil erosion, 307
10.10 Estimation of soil loss, 311
10.10.1 The Universal Soil Loss Equation, 311
10.10.2 Soil erosion by wind, 313
10.11 Erosion control and conservation practices, 313
10.11.1 Conservation measures for water erosion, 314
10.11.2 Conservation measures for wind erosion, 316
10.12 Assessment of soil erosion, 316
10.13 Soil surveys and mapping, 323
10.14 Soils as a construction material, 324
10.14.1 Gravels and sands, 324
10.14.2 Clay deposits, refractory materials and bricks, 326
10.14.3 Fills and embankments, 328

11   Problem soils, 335
11.1   Quicksands, 335
11.2   Expansive clays, 338
11.3   Dispersive soils, 341
11.4   Collapsible soils, 345
11.5   Quickclays, 348
11.6   Soils of arid regions, 350
11.7   Tropical soils, 352
11.8   Peat, 357

12 Rock masses, their character, problems and uses, 362

12.1 Rock types, 362
  12.1.1 Igneous rocks, 362
  12.1.2 Metamorphic rocks, 362
  12.1.3 Sedimentary rocks, 365
  12.1.4 Deformation and strength of rocks, 367
12.2 Description of rocks and rock masses, 368
12.3 Discontinuities, 375
  12.3.1 Description of discontinuities in rock masses, 376
  12.3.2 Strength of discontinuous rock masses and its assessment, 378
  12.3.3 Discontinuities and rock quality indices, 379
  12.3.4 Discontinuity surveys, 381
12.4 Weathering, 384
  12.4.1 Mechanical weathering, 385
  12.4.2 Chemical and biological weathering, 386
  12.4.3 Engineering classification of weathering, 387
12.5 Igneous and metamorphic rocks, 388
12.6 Mudrocks, 395
12.7 Carbonate rocks, 399
12.8 Evaporitic rocks, 404
12.9 Building, roofing and facing stones, 406
12.10 Crushed rock, 408
  12.10.1 Concrete aggregate, 408
  12.10.2 Road aggregate, 410
12.11 Lime, cement and plaster, 411

13 The impact of mining on the environment, 415

13.1 Introduction, 415
13.2 Metalliferous mining and subsidence, 415
  13.2.1 Mining methods, 415
  13.2.2 Stabilization of old workings, 417
  13.2.3 Mining and sinkhole development, 417
13.3 Pillar workings, 418
13.4 Old mine workings and hazard zoning, 421
13.5 Measures to reduce or avoid subsidence effects due to old mine workings, 423
13.6 Old mine shafts, 424
13.7 Longwall mining and subsidence, 426
13.8 Measures to mitigate the effects of subsidence due to longwall mining, 430
13.9 Surface mining, 431
  13.9.1 Strip mining and opencasting, 431

13.9.2 Open pit mining, 433
  13.9.3 Blasting and vibrations, 434
  13.9.4 Dredge mining, 435
13.10 Subsidence associated with fluid abstraction, 437
13.11 Waste materials from mining, 441
  13.11.1 Spoil heaps and their restoration, 441
  13.11.2 Waste disposal in tailings dams, 444
13.12 The problem of acid mine drainage, 446
13.13 Waste waters and effluents from coal mining, 450
13.14 Heap leaching, 452
13.15 Other impacts, 453

14 Waste and its disposal, 459

14.1 Introduction, 459
14.2 Domestic refuse and sanitary landfills, 459
  14.2.1 Design considerations for a landfill, 461
  14.2.2 Degradation of waste in landfills, 466
  14.2.3 Attenuation of leachate, 466
  14.2.4 Surface and groundwater pollution, 468
  14.2.5 Landfill and gas formation, 470
14.3 Hazardous wastes, 471
14.4 Radioactive waste, 474
14.5 Contaminated land, 478
14.6 Remediation of contaminated land, 482
  14.6.1 Soil remediation, 483
  14.6.2 Groundwater remediation, 484

15 Environmental geology and health, 487

15.1 Introduction, 487
15.2 The occurrence of elements, 490
15.3 Geochemical surveys and maps, 496
15.4 Some trace elements and health, 498
  15.4.1 Arsenic, 498
  15.4.2 Cadmium, 500
  15.4.3 Mercury, 501
  15.4.4 Lead, 501
  15.4.5 Iodine, 503
  15.4.6 Fluorine, 506
  15.4.7 Selenium, 507
  15.4.8 Zinc, 508
  15.4.9 Molybdenum, 508
  15.4.10 Some other trace elements and health, 509
15.5 Mineral dusts and health, 511

15.6 Radon, 512
15.7 Environmental geology and chronic
      disease, 516

**16 Land evaluation and site
      assessment, 524**
16.1 Introduction, 524
16.2 Remote sensing, 525
      16.2.1 Infrared line scanning, 525
      16.2.2 Side-looking airborne radar, 526
      16.2.3 Satellite imagery, 527
16.3 Aerial photographs and photogeology, 528
      16.3.1 Types of aerial photograph, 528
      16.3.2 Photogeology, 530
16.4 Terrain evaluation, 531
16.5 Land capability studies, 532
16.6 Site investigation, 535

16.6.1 Desk study and preliminary
        reconnaissance, 535
16.6.2 Site exploration, 536
16.6.3 Subsurface exploration in soils, 537
16.6.4 Sampling in soils, 541
16.6.5 Subsurface exploration in rocks,
        542
16.7 Geophysical exploration, 546
      16.7.1 Seismic methods, 546
      16.7.2 Resistivity methods, 547
      16.7.3 Electromagnetic methods, 551
      16.7.4 Magnetic and gravity methods, 552
      16.7.5 Drill-hole logging techniques, 554
      16.7.6 Cross-hole seismic methods, 556
16.8 *In situ* testing, 556

Index, 563

Preface

Environmental geology has been defined by the American Geological Institute as the application of geological principles to the problems created by the occupancy and exploitation by man of the physical environment. However, environmental geology is not just the impact of man on the geological environment, it also involves the impact of the geological environment upon man. The origin of natural geohazards such as volcanic eruptions, earthquakes, landslides and floods generally is independent of man but can have a devastating impact upon society. Conversely, man's activities in the form of agriculture, mining and industry can have a notable effect upon the environment. Both these aspects are the concern of environmental geology. As such, environmental geology is of fundamental importance in relation to man's planning and development of the environment. The latter must take due account of geohazards and must seek to reduce the number of adverse environmental impacts of society on nature. Hence, environmental geology needs to be intimately involved in the planning process, providing basic information necessary to develop acceptable conditions in which people can live. Furthermore, the increasing public awareness of the importance of the environment requires a deeper understanding of the geological processes at play within the environment. It has been said that environmental geology is synonymous with urban geology—it is not. Urban geology is only part of environmental geology. The rural environment is just as important and must not be allowed to suffer at the hands of man. Areas of great scenic beauty and scientific interest must be protected, and, for example, the onslaught of tourism controlled. The greatest problem, of course, is man himself and his inability or unwillingness to control his growth in numbers, which places an ever increasing strain on the environment.

The author, with a prejudiced and British perspective, tended to regard environmental geology as part of engineering geology, even though some would claim that the former has its origins in the last century. Nonetheless, it has to be conceded that environmental geology has come to maturity in the recent past and that it now should be regarded as a subject in its own right, although remaining closely linked with engineering geology. Accordingly, the text should be of value to both students of environmental geology and of engineering geology. However, the book is not aimed solely at undergraduates and postgraduates in these disciplines, it also should be of use to those following courses in environmental science and geography. In addition, it is hoped that the book will find a home on the shelves of those engaged in the environmental professions, which includes not just those mentioned above, but also planners and civil engineers.

No book, because of the limitations of the author and the constraints on size, can be exhaustive in its treatment of its subject. Also, the manner and depth of treatment of the topics contained reflect the interests, bias and views of the author. Be that as it may, it is the author's opinion, to which he is entitled, that the text is as well balanced as others bearing this title. He also believes that it does have a more technical bias than its competitors and so should be of more use to the professional. As with the author's previous texts, this also contains numerous references, thereby allowing the interested reader to pursue topics to greater depth. Furthermore, it does have an international flavour which should give it global appeal.

Once again, the author must acknowledge his great debt to Roma Brackley who typed the manuscript and saved him much sweat, tears and toil. The author must also compliment Lorna Hind, Senior Production Editor at Blackwell Science, whose efforts turned a manuscript into a book in a remarkably short time.

F.G. Bell
Durban, May 1998

1   Introduction

1.1 Planning and geology

Environmental geology is intimately involved with land-use planning, and land-use planning inevitably will become more important as the world population continues to expand. This is even more important in the context of urban planning, because it is estimated that, by the end of the century, almost half the people in the world will be living in towns and cities (De Mulder, 1996). Most of this urbanization is occurring in developing countries. Obviously, the megacities of the future must be well planned if their inhabitants are to lead worthwhile lives. An outline of many of the geological and geotechnical factors involved in urban development has been summarized by Marker (1996). These include the behaviour of and problems associated with soils and rocks, the problem of groundwater, the problems associated with geohazards, ground evaluation and investigation, types and methods of construction, monitoring and remedial methods. Most of these topics are dealt with in this text and are more fully discussed in Bell (1978, 1980, 1992, 1993).

The objective of land-use planning is to improve the quality of life and the general welfare of the community. This is brought about by producing better environments in which people live. Therefore, land-use planning provides a system through which communities can address their development, and should involve environmental management. Land-use planning, or town and country planning, is not only concerned with creating decent living conditions, but needs to ensure that these function efficiently. However, this is by no means easy to achieve, because of the changing character of land use, especially of urban areas, the subjective nature of what is desirable and the conflicting expectations of people. Consequently, planning involves some degree of arbitration and of compromise, and planning action is an evolv-ing process. Although the policy which develops from planning embodies a particular course of action, planning proposals often are controversial in that they may offend one or more sections of the community. Hence, in the last analysis, planning policies are the prerogatives of government, because legislation is necessary to put them into effect. In other words, the state regulates the use of land through its planning laws, so that this valuable resource is not damaged unnecessarily, and maximum flexibility is maintained for future generations to exercise their options.

Worth (1987) maintained that the underlying rationale of land-use planning is that human activities and land uses are spatially distributed at different locations. Some locations are very specifically determined, for example, by a mineral resource, whereas other locations are almost a random selection, many land uses being able to be located in different places with more or less similar suitability. Therefore, land-use planning must attempt to establish acceptable criteria of location for each activity, as well as seeking an optional location for a particular activity in relation to others. Accordingly, the location problem for each activity becomes one of resolving the competing claims of different locations in order to achieve an overall optimization. Of course, one of the problems is that the same area of land may be suitable for different uses.

Land-use planning must also represent an attempt to reduce the number of conflicts and adverse environmental impacts both in relation to society and nature. In the first instance, land-use planning involves the collection and evaluation of relevant data from which plans can be formulated. The policies which result depend on economic, sociological and political influences, in addition to the perception of the problem. As mentioned above, land-use planning is a political process, with decisions generally being taken at various levels of government, depend-

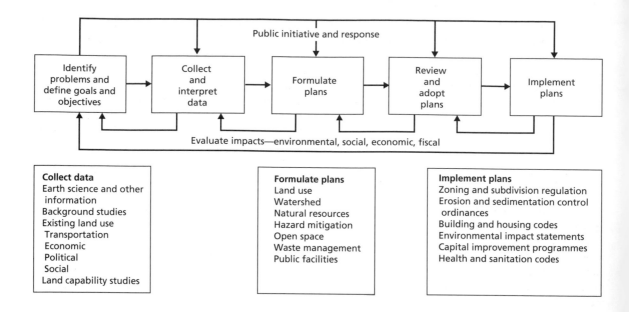

Fig. 1.1 Diagram of the land-use planning process.

ing on the project, after receiving advice from professional officers. As Worth (1987) pointed out, it is the planner who makes the recommendations in respect of planning proposals, but, with regard to a matter involving a specialist topic, the recommendation can only be as good as the advice received from the specialist. However, sufficient geological data should be provided to planners so that, ideally, they can develop the environment in harmony with nature. As indicated in Fig. 1.1, geological information is required at all levels of planning and development, from the initial identification of a social need to the construction stage. Even after construction, further involvement may be necessary in the form of advice on hazard monitoring, maintenance or remedial works.

Legget (1987) stressed the importance of geology in planning physical facilities and individual structures, and in the wise and best use of land. He pointedly noted the obvious: that land is the surface expression of the underlying geology, so that land-use planning can only be carried out with satisfaction if there is a proper understanding of the geology concerned. Legget further stated that the development of land must be planned with the full realization of the natural forces which have brought it to its present

state, taking into account the dynamic character of nature, so that the development does not upset the delicate balance any more than is essential. Geology must therefore be the starting point of all planning. Legget also strongly suggested that, before land is used for development, it is desirable to know whether any useful materials can be obtained from it. Whenever there is a possibility that materials at a particular location may be needed in the future, Legget advocated the concept of sequential land planning.

Because land use inevitably involves the different development of particular areas, some type of land classification(s) constitutes the basis on which land-use planning is carried out. However, land should also be graded according to its potential uses and capabilities. In other words, indices are required to assess the environmental status of natural resources and their potential. Such indices should establish limits, trends and thresholds, as well as providing insight that offers some measure of success of national and local programmes dealing with environmental problems.

Over recent years, public concern regarding the alteration and degradation of the environment has caused governmental and planning authorities to become more aware of the adverse effects of indiscriminate development. As a result, laws have been passed to help to protect the environment from damage. Most policies which deal with land use are concerned with either those processes which represent threats to life, health or property including, for instance, hazardous events, pollution of air or water and the exploitation, protection or conservation of natural resources, or the restoration of despoiled areas.

Many older industrial and urban areas are undergoing redevelopment; therefore, there is a need to plan the new environment so that land with difficult ground conditions is avoided by building development or the cost of building on it is fully appreciated. Hence, planners require relevant information. In many urban areas, data from site investigations are available, and Forster and Culshaw (1990) have described how these can be made use of to prepare engineering geological maps for planners.

Laws are now in force in many countries which demand the investigation and evaluation of the consequences to the environment of all large projects. An environmental evaluation can be defined as an activity designed to identify and predict the impact on human well-being of legislative proposals, policies, programmes, projects and operational procedures, and to interpret and communicate information about impacts. An environmental evaluation is a multidisciplinary process with a social component, as well as a technical component. Environmental evaluation now forms an established part of good development planning, and was formulated initially in the USA by way of the National Environmental Policy Act (1969). This Act requires the preparation of a formal statement on the consequences of any major federal activity on the environment. In other words, an environmental impact statement is required, and this is accompanied by other relevant documents which are necessary to the decision-making process. An environmental impact statement therefore usually involves a description of the proposed scheme, its impact on the environment, particularly noting any adverse effects, and alternatives to the proposed scheme. The aim of the environmental impact process is to improve the effectiveness of existing planning policy. It should bring together the requirements of the development and the constraints of the environment, so that conflicts can be minimized. Thus, the primary purpose of environmental evaluation is to aid decision making by national or local government by providing objective information relating to the

effects on the environment of a project or course of action. A number of associated objectives follow from this. First, an environmental evaluation must provide objective information on which decisions can be made. It therefore involves data collection relevant to environmental impact prediction. The information must be comprehensive, so that, after analysis and interpretation, it not only informs, but helps direct development planning. Environmental evaluation must be relevant and must involve comparative analyses of reasonable alternatives. Plans must be analysed, so as to ensure that the benefits of a project are fully appreciated. The negative effects, as well as the positive effects, of development proposals must be conveyed to the decision makers. Solutions should be proposed to any problems which may arise as a result of interactions between the environment and the project. Geologists invariably are involved in environmental impact investigations.

Checklists frequently are used in environmental evaluation (Fig. 4.11). These are comprehensive lists related to specific environmental parameters and specific human actions. They facilitate data gathering and presentation, help to order thinking and alert against omission of possible environmental impacts. The matrix approach greatly expands the scope of checklists (see Fig. 1.3). Such an approach may have one data set related to environmental parameters and one to human actions, with the two sets forming a cross-tabulation. For example, Leopold *et al.* (1971) devised a matrix system of analysis, which acts as a super-checklist for those involved in the preparation of environmental impact statements, and also tends to make assessment more objective. Their approach involved the quantification of data, the establishment of cause and effect relationships and the weighting of impacts. The matrix approach can highlight areas of particular concern, or those where further investigation is necessary. The method is adaptable. All cells in the matrix which represent a possible impact are evaluated individually, being scored in terms of the magnitude of the impact and its social significance. This helps in the assessment of the risk or uncertainty related to a particular impact.

One of the aspects of planning which intimately involves geology is the control or prevention of geological hazards that militate against the interests of humans (Bell *et al.*, 1987). The development of planning policies for dealing with hazards requires an

Table 1.1 Data required to reduce losses from geological hazards. (After Hays & Shearer, 1981.)

Reduction decisions	Technical information needed about the hazards from earthquakes, floods, ground failures and volcanic eruptions
Avoidance	Where has the hazard occurred in the past? Where is it occurring now? Where is it predicted to occur in the future? What is the frequency of occurrence?
Land-use zoning	Where has the hazard occurred in the past? Where is it occurring now? Where is it predicted to occur in the future? What is the frequency of occurrence? What is the physical cause? What are the physical effects of the hazard? How do the physical effects vary within an area? What zoning within the area will lead to reduced losses to certain types of construction?
Engineering design	Where has the hazard occurred in the past? Where is it occurring now? Where is it predicted to occur in the future? What is the frequency of occurrence? What is the physical cause? What are the physical effects of the hazard? How do the physical effects vary within an area? What engineering design methods and techniques will improve the capability of the site and the structure to withstand the physical effects of a hazard in accordance with the level of acceptable risk?
Distribution of losses	Where has the hazard occurred in the past? Where is it occurring now? Where is it predicted to occur in the future? What is the frequency of occurrence? What is the physical cause? What are the physical effects of the hazard? How do the physical effects vary within an area? What zoning has been implemented in the area? What engineering design methods and techniques have been adopted in the area to improve the capability of the structure to withstand the physical effects of a hazard in accordance with the level of acceptable risk? What annual loss is expected in the area? What is the maximum probable annual loss?

assessment of the severity, extent and frequency of the hazard in order to evaluate the degree of risk. Therefore, land-use planning for the prevention and mitigation of geological hazards should be based on criteria which establish the nature and degree of the risks present and their potential impact. Both the probable intensity and frequency of the hazard(s) and the susceptibility (or probability) of damage to human activities in the face of such a hazard are integral components of risk assessment. Vulnerability analyses, comprising risk identification and evaluation, should be carried out in order to make rational decisions on how best the effects of potentially disastrous events can be reduced or overcome through systems of per-

manent controls on land development. The geological data needed in planning and decision making, as summarized by Hays and Shearer (1981), are given in Table 1.1. Just as important is the quality of information transmission, and how readily it can be understood by politicians and planners, because this influences their reactions.

Once the risk has been assessed, methods whereby the risk can be reduced must be investigated and evaluated in terms of public costs and benefits. The risks associated with geological hazards may be reduced, for instance, by control measures carried out against the hazard-producing agent, by predisaster community preparedness, monitoring, surveillance and warning systems which allow evacuation, by restrictions on the development of land and by the use of appropriate building codes, together with the structural reinforcement of property. In addition, the character of the ground conditions can affect both the viability and implementation of planning proposals. The incorporation of geological information into planning processes should mean that proposals can be formulated which do not conflict with the ground conditions present.

1.2 Conservation, restoration and reclamation of land

Conservation is concerned with safeguarding natural phenomena and with the preservation and improvement of the quality of the environment. As such, it is closely associated with the geology, as the geology and environment are intimately interrelated. However, the interests of conservation frequently are in conflict with social and economic pressures for development, and so one of the roles of planning authorities is to balance the demands of urban, industrial and infrastructural development with the need to preserve the countryside and areas of scientific and cultural interest. Hence, the geologist may find himself in conflict with the conservationist, especially in the case of the mineral extraction industry. On the other hand, geological knowledge should be made use of in conservation programmes, because these should seek to make the wisest use of natural resources, and, in this regard, there is no conflict with the geologist.

Conservation is not simply preservation; it seeks to improve existing conditions rather than simply maintaining the status quo. Hence, conservation involves the reconciliation of differing views, so that the best compromise can be reached. Obviously, in this context, land use is important. In some instances, the same land can be used simultaneously to cater for several needs, whereas, in others, the uses to which it can be put are consecutive rather than concurrent, and the final effect can be either to restore the land to its original use or to a new use which forms an acceptable part of the landscape. Thus, a sequence of events must be planned to ensure the greatest efficiency in the use of resources, and the achievement of the most acceptable final state. Hence, in such situations, the geologist must be involved with planning from the

onset, otherwise natural resources may be sterilized as a consequence of premature, alternative development. Classic examples include the development of urban areas over valuable surface mineral deposits, which could have been extracted before the areas were developed.

Obviously, societies make demands upon land, which frequently mean that it is degraded. One of the major causes of the degradation of land is waste disposal from various sources, such as industry, mining and domestic. Not only has this led to the degradation of land, but it frequently has been responsible for the contamination of ground and the pollution of groundwater. Such sites require restoration. In fact, an appreciable contribution to the general economy can be made by bringing this derelict land back into worthwhile use. Eventually, such land must be restored. Furthermore, derelict land has a blighting effect on the surrounding area, which makes its restoration highly desirable. Whatever the ultimate use to which the land is put after restoration, it is imperative that it should fit the needs of the surrounding area and be compatible with other forms of land use that occur in the neighbourhood. Accordingly, planning of the eventual land use must take into account the overall plans for the area, and must endeavour to include the ecological integration of the restored area into the surrounding landscape. Restoring a site which represents an intrusion in the landscape to a condition which is well integrated into its surroundings upgrades the character of the environment far beyond the confines of the site.

Two of the major causes of the dereliction of land are the surface extraction of minerals and the disposal of waste products from mineral workings in the form of spoil heaps and tailings lagoons (Bell, 1996). Mineral extraction may also cause water pollution. In practice, the restoration and conservation of derelict sites involve the use of planning controls. Hence, conditions are attached to planning permissions granting mineral developments, and these apply not only to the form of development permitted, but also to the ultimate state of the land after extraction ceases. For example, the after-use of some sites where minerals have been worked has caused problems, notably when pollution has resulted from their utilization for the disposal of toxic wastes. On the other hand, the use of such sites as recreational areas has provided additional amenities for the communities they serve. Again this emphasizes the need for thorough planning.

A preliminary reconnaissance of a derelict site is desirable to determine the sequence of work for the site survey and investigation. The exact boundaries of the site and the various physical features it contains are recorded during the survey, and boreholes may be sunk to assess its geological character. These data allow plans to be drawn and the restoration project to be designed.

The reclamation of land involves its upgrading to a use which is considered to be more beneficial to the community. The reclamation of swamplands and marshlands by drainage, so that they can be used for agricultural or other purposes, illustrates this. Some impressive examples of reclamation are provided by the various schemes undertaken in the Netherlands, where land has been reclaimed from the sea; the Zuider Zee Scheme and the Delta Scheme offer noteworthy examples of the ingenuity of humans.

1.3 Geological hazards and planning

As far as hazards are concerned, it has been estimated that natural hazards cost the global economy over 50 000 million dollars per year. Two-thirds of this sum is accounted for by damage, and the remainder represents the cost of predicting, preventing and mitigating disasters. Man-made hazards, such as groundwater pollution, subsidence and damage caused by expansive soils, add to this sum.

A natural hazard has been defined by the United Nations Educational, Scientific and Cultural Organization (UNESCO) (see Varnes, 1984) as the probability of occurrence, within a specified period of time and within a given area, of a potentially damaging phenomenon. However, geological hazards are not all natural; some are influenced by or brought about by humans. Hazards pose a threat to humans, property and the environment, and so are of greater significance when they occur in highly populated areas. Moreover, as the global population rises, so the significance of hazard is likely to increase. For instance, according to the United Nations Disaster Relief Organization (UNDRO, 1991), the number of people affected by natural disasters increases by some 6% annually. In developed societies, hazards can cause great damage to property with associated high costs. In developing areas, loss of life and injury often

are of far more consequence. This is not to say that developing countries do not suffer heavy economic losses due to natural disasters. For example, Mora (1995) indicated that the Limon-Telire earthquake of 1991 cost Costa Rica between 5% and 8.5% of its gross national product that year. Indeed, estimates made by the Comisión Económica Para América Latina (CEPAL, 1990) suggested that, between 1960 and 1985, the countries of Central America lost, on average, 2.7% of their gross national product annually as a result of natural disasters. This compares with 1% for a developed country such as the USA over the same period. Furthermore, of the 3 million people killed by natural disasters between 1960 and 1990 (excluding the former communist bloc), only 0.7% were killed in developed countries, over three-quarters were killed in developing countries and the rest were killed in countries with average gross national products. Unfortunately, however, development in many of these countries is placing increasing numbers of people and property at risk as a consequence of unplanned occupancy and the use of marginal and high-risk zones. In particular, overcultivation and deforestation aggravate the problem. However, the associated problems of soil erosion and excessive run-off have not been confined to the developing world; such problems occurred, often with serious effects, in the USA in the 1930s. As geological hazards are widespread and fairly common events, they obviously give cause for concern. Many such geohazards impose notable constraints on development or exert a substantial influence on society. Hence, it is important that geological hazards are understood, in order that their occurrence and behaviour can be predicted and that measures can be taken to reduce their impact.

As individual types of geohazard tend to occur again and again, the frequency of a hazard event can be regarded as the number of events of a given magnitude which take place in a particular period of time. As such, a recurrence interval for such an event can be determined in terms of the average length of time between events of a certain size. Generally, however, the most catastrophic events are highly infrequent, so that their impact has the greatest overall significance. Similarly, frequent events normally are too small to have significant aggregate effect.

A hazard involves a degree of risk, the elements at risk being life, property, possessions and the environment. Risk involves the quantification of the probability that a hazard will be harmful, and the tolerable degree of risk depends upon what is being risked, life being much more important than property. The risk to society can be regarded as the magnitude of a hazard multiplied by the probability of its occurrence. If there are no mitigation measures, no warning systems and no evacuation plans for an area which is subjected to a recurring hazard, such an area has the highest vulnerability. The vulnerability V was defined by UNESCO as the degree of loss to a given element or set of elements at risk resulting from the occurrence of a natural phenomenon of a given magnitude. It is expressed on a scale from zero (no damage) to unity (total loss). The factors which usually influence loss are the population distribution, social infrastructure and socio-cultural and socio-political differentiation and diversity.

The effects of a disaster may be lessened by a reduction in vulnerability. Short-term forecasts a few days ahead of an event may be possible, and complement relief and rehabilitation planning. In addition, it is possible to reduce the risk of disaster by a combination of preventative and mitigating measures (see below). To do this successfully, the patterns of behaviour of the geological phenomenon posing the hazard need to be understood and the areas at risk identified. The level of potential risk may be decreased and the consequence of disastrous events mitigated by introducing regulatory measures or other inducements into the physical planning process. The impact of disasters may be reduced further by incorporating into building codes and other regulations appropriate measures so that structures will withstand or accommodate potentially devastating events.

UNESCO went on to define the specific risk, elements at risk and total risk. The specific risk R refers to the expected degree of loss associated with a particular hazard. It may be expressed as the product of the hazard H and the vulnerability ($R = H \times V$). The elements at risk E refer to the population, property and economic activity at risk in a given area. Lastly, the total risk R_T refers to the expected number of lives lost, injuries, property damage and disruption to economic activity brought about by a given hazard. It can be regarded as the product of the specific risk and the elements at risk ($R_T = R \times E$ or $R_T = E \times H \times V$). The net impact of a hazard can be looked upon in terms of the benefits derived from inhabiting a zone in which a hazard occurs minus the costs involved in mitigation measures. Mitigation measures may be structural or technological, on the one hand, or regulatory on the other (Table 1.2).

The response to geohazards by risk evaluation, leading to land-use planning and appropriate engineering design of structures, is lacking in many parts of the world; yet most geohazards are amenable to avoidance or prevention measures. In addition, the causes of geological hazards often are reasonably well understood, and therefore most can be identified and even predicted with a greater or lesser degree of accuracy. Nonetheless, geological hazards can represent obstacles which impede economic development, especially in developing countries. Consequently, measures should be included in planning processes in order to avoid severe problems attributable to hazards which lead to economic and social disruption. However, in developing countries, the capacity may not exist to effect planning processes or to establish mitigating measures, even though, in some instances, economic losses due to natural disasters could be reduced with only a small investment.

Geological hazards vary in nature, and can be complex. One type of hazard, for example, an earthquake, can be responsible for the generation of others, such as liquefaction of sandy soils, landslides or tsunamis. Certain hazards, such as earthquakes and landslides, are rapid onset hazards, and so give rise to sudden impacts. Others, such as soil erosion and subsidence due to the abstraction of groundwater, may take place gradually over an appreciable period of time. Furthermore, the effects of natural geohazards may be difficult to separate from the impact of humans. For example, although soil erosion is a natural process, it frequently has been accentuated by the interference of humans. Indeed, the modification of nature by humans often increases the frequency and severity of natural geohazards and, at the same time, these increase the threats to human occupancy. Other geohazards primarily are attributable to the activity of humans, such as the pollution of groundwater and subsidence due to mining.

As the world population increases, so the risks attributable to natural geohazards increase. Nevertheless, people continue to live in areas of known geohazard, despite the known levels of risk. The reasons for this are many and various, but, in doing so, a

Table 1.2 Non-structural and structural methods of disaster mitigation. (After Alexander, 1993.)

Non-structural methods		Coordinator(s)
(a) Short term		Police and firemen
Emergency plans:	Civil and military forces	Red Cross and charities
		Volunteer groups
		Medical service
Evacuation plans:	Routes and reception centres for the general public, for vulnerable groups: the very young, elderly, sick or handicapped	
Prediction of impact:	Monitoring equipment, forecasting methods and models	
Warning processes:	General and specialized warning (e.g. ethnic)	
(b) Long term		
Building codes and construction norms		
Hazard microzonation:	Selected risks	
	All risks	
Land-use control:	Regulations, prohibitions, moratoria, compulsory purchase	
Probabilistic risk analysis		
Insurance		
Taxation		
Education and training		
Structural methods		
Retrofitting of existing structures		
Reinforcement of new structures:	Design features, overdesign	
Safety features:	Structural safeguards, fail-safe design	
Engineering phenomenology		
Probabilistic prediction of impact strength		

demand for protection arises in developed countries. In poor countries, people remain in hazard areas because there is nowhere else to go or no other place to offer them a living. They therefore are obliged to accept the risks of hazards and their consequences. In such situations, a sense of resignation or fatalism frequently develops, and may be manipulated by unscrupulous political and economic interests. The extent to which protection measures can be adopted depends not just upon the type of geohazard and the technology available to combat the hazard, but also on the ability and willingness to pay for these measures. Permanent evacuation of areas of high hazard, even if feasible, necessitates the compensation of the people involved. Again, this can only be accomplished in societies which are rich enough to pay, and where the authorities are willing to do so.

Many losses attributable to geohazards may be compensated for by way of government grants or loans. Such aid frequently is linked to a defined hazard zone, such as the 100-year flood, which, in the USA, defines the area which is subject to flood damage compensation. If a community wishes to qualify for federal aid, it must join the Federal Flood Insurance Program. However, financial assistance in the form of government payouts does not help to reduce the hazard. Furthermore, the likelihood of financial aid in an emergency may encourage development in hazard-prone areas, which is contrary to the principles embodied in planning in relation to hazard.

1.4 Risk assessment

Risk, as mentioned above, should take into account the magnitude of the hazard and the probability of its occurrence. Risk arises out of uncertainty, as a result of insufficient information being available about a hazard and an incomplete understanding of the mechanisms involved. The uncertainties prevent accurate predictions being made of hazard occurrence. Risk analysis involves the identification of the degree of risk, its estimation and evaluation. An objective assessment of risk is obtained from a statistical assessment of instrumental data and/or data gathered from past events (Wu *et al.*, 1996). The risk to society can be quantified in terms of the number of deaths attributable to a particular hazard in a given period of time and the resultant damage to property. The costs involved can then be compared with the costs of hazard mitigation. Unfortunately, not all risks and benefits are readily amenable to measurement and assessment in financial terms. Even so, risk assessment is an important objective in decision making by urban planners, because it involves the vulnerability of people and the urban infrastructure on the basis of the probability of occurrence of an event. The aims of risk analysis are to improve the planning process, as well as to reduce vulnerability and mitigate damage.

The occurrence of a given risk in a particular period of time can be expressed in terms of the probability as follows:

$$R = \Sigma P(E)Ct \qquad (1.1)$$

where R is the level of risk per unit time, E is the event expressed in terms of the probability P, C refers to the consequences of the event and t is a unit of time. Risk analysis should include the magnitude of the hazard, as well as its probability. The magnitude of the hazard is related to the size of the population at risk. The assessment should mention all the assumptions and conditions on which it is based; if these are not constant, the conclusions should vary according to the degree of uncertainty, a range of values representing the probability distribution. The confidence limits of the predictions should be provided in the assessment.

Jefferies *et al.* (1996) referred to the use of the Bayesian approach in the assessment of risk probability. The Bayesian approach differs from the traditional use of statistics in assessing probability. In the conventional statistical assessment, the probability expresses the relative frequency of occurrence of an event based upon data collected from various sources. In the Bayesian approach, the probability expresses the current judgement of likely events, and does not require occurrence data to estimate the probability, although, if available, such data can be used. In other words, in the Bayesian approach, the probability quantifies judgement. A number of computer programs are available to carry out such analyses, and are referred to by Jefferies *et al.* They used the Bayesian approach to illustrate how the risk of contamination of groundwater from a landfill could be assessed.

The acceptable level of risk is inversely proportional to the number of people exposed to a hazard. Acceptable risk, however, is complicated by cultural factors and political attitudes. It is the last two factors which are likely to determine public policy in relation to hazard mitigation and regulatory control.

Although it can be assumed that there is a point at which the cost involved in the reduction of risk equals the savings earned by the reduction of risk, society tends to set arbitrary tolerance levels based largely on the perception of risk and the priorities for its management. Some hazards are far more emotive than others, notably the disposal of radioactive waste, and therefore may attract more funds from politically sensitive authorities to deal with associated problems. In essence, something is safe if its risks are judged to be acceptable, but there is little consensus as to what is acceptable.

The Health and Safety Executive (1988) in Britain defined a concept of tolerability. This they regarded as a willingness to live with a risk in order to obtain certain benefits in the confidence that the risk was being properly controlled. To tolerate a risk does not mean that it can be regarded as negligible or that it can be ignored. A risk needs to be kept under review, and further reduced if and when necessary. The use of tolerability to guide public policy involves both technical and social considerations.

A number of problems are inherent in risk management. For instance, the degree of risk does not increase linearly with the time of exposure to a hazard. Moreover, with time, the response to risk can change, so that mitigation and risk reduction can change. As Alexander (1993) pointed out, the dichotomy between actual and perceived risk does not help attempts to reduce risk and promotes a conflict of objectives. Nonetheless, risk management has to attempt to determine the level of risk which is acceptable. This has been referred to as risk balancing. Public and private resources may then be allocated to meet a level of safety which is acceptable to the public. Cost–benefit analyses can be made use of in order to develop a rational economic means for risk reduction expenditure.

Risk management requires a value system against which decisions are made, that is, the risk basis. In order to effect risk control, certain actions need to be taken. Risk assessment and control are made in terms of either monetary value or the potential loss of life. According to Jefferies *et al.* (1996), despite efforts to assign a monetary value to life in order to develop a single value system, it is desirable to use separate health and monetary value systems in terms of law. Risk management decisions which do not involve health risk can use a monetary value for the assessment of costs and benefits. If a health risk is used as a value system, the average incremental mortality rate of a population affected by a hazard may be used as the value criterion. This mortality rate is the expected additional deaths per year divided by the total affected population. Other health risk measures can be used, such as the average loss of life expectancy.

As far as risk is concerned, people more readily accept risks which are associated with voluntary activities than those which may be imposed upon them. In addition, people are more conscious of risk immediately after a disastrous hazard, and call for greater protection, than during periods of quiescence. Mora (1995), for instance, mentioned that, after the seismic events of 1990–1992 in Costa Rica, there was a greater awareness of geohazards and their consequences. Sadly, however, within a short while, the interest had declined.

1.5 Hazard maps

Any spatial aspect of hazard can be mapped, providing that there is sufficient information on its distribution. Hence, when hazard and risk assessments are made over a large area, the results can be expressed in the form of hazard and risk maps. An ideal hazard map should provide information relating to the spatial and temporal probabilities of the hazard mapped (Wu *et al.*, 1996). In addition to data gathered from surveys, hazard maps frequently are compiled from historical data related to past hazards which have occurred in the area concerned. The

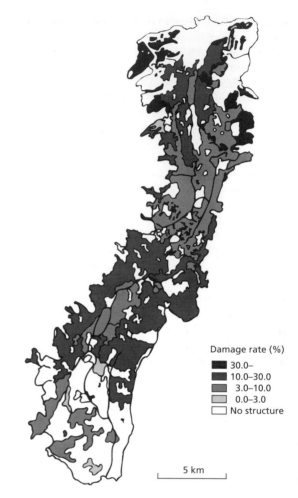

Damage rate (%)
- 30.0–
- 10.0–30.0
- 3.0–10.0
- 0.0–3.0
- No structure

5 km

Fig. 1.2 Distribution of structural damage resulting from a local earthquake in Quito, Ecuador. (After Villacis *et al.*, 1997.)

concept is based on the view that past hazards provide some guide to the nature of future hazards. Hazard zoning maps usually provide some indication of the degree of risk involved with a particular geological hazard (Blikra, 1990; Seeley & West, 1990). The hazard often is expressed in qualitative terms as high, medium and low. These terms must be adequately described so that their meaning is understood. The variation in intensity of a hazard from one location to another can be depicted by risk mapping. The latter attempts to quantify the hazard in terms of potential victims or damage (Fig. 1.2). Risk mapping therefore attempts to estimate the location, probability and relative severity of future, probable hazardous events, so that potential losses can be estimated, mitigated or avoided. Specific risk zoning maps divide a region into zones indicating exposure to a specific hazard. Vulnerability maps, for example, for the recognition of areas which offered potential risk to groundwater supply if developed, were described by Fobe and Goosens (1990). A more sophisticated development of the concept and use of hazard maps has been provided by Soule (1980). He outlined a method of mapping areas prone to geological hazards by using map units based primarily on the nature of the potential hazards associated with them. The resultant maps, together with their explanation, are combined with a land-use/geological hazard area matrix, which provides some idea of the problems which may arise in the area represented by the individual map. For instance, the matrix indicates the effects of any changes in slope or mechanical properties of rocks or soils, and attempts to evaluate the severity of hazard for various land uses. As an illustration of this method, Soule used a landslide hazard map of the Crested Butte–Gunnison area, Colorado (Fig. 1.3a). This map attempts to show which factors within individual map units have the most significance as far as potential hazard is concerned. The accompanying matrix outlines the problems likely to be encountered as a result of human activity (Fig. 1.3b).

However, hazard maps, like other maps, do have disadvantages. For instance, they are highly generalized, and represent a static view of reality. They need to be updated periodically as new data become available. The hazard map of Mount St Helens had to be completely redrawn after the eruption of 1980. As noted above, catastrophic events occur infrequently, and so information may be completely lacking in an area in which such an event subsequently occurs.

Microzonation has been used to depict the spatial variation of risk in relation to particular areas (Fig. 3.13). Figure 3.13 distinguishes three zones of ground response to seismic events. Ideally, land use should adjust to the recommendations suggested by such microzoning, so that the impacts of the hazard event(s) depicted have a reduced or minimum influence on people, buildings and infrastructure.

Foster (1980) distinguished single hazard–single purpose, single hazard–multiple purpose and multiple hazard–multiple purpose maps. The first type of

map obviously is appropriate where one type of hazard occurs and the effects are relatively straightforward. Single hazard–multiple purpose maps are prepared where the hazard is likely to affect more than one activity, and show the varying intensity of, and the response to, the hazard (Fig. 3.13). The most useful map as far as planners are concerned is the multiple hazard–multiple purpose map. Such a map can be used for risk analysis and assessment of the spatial variation in potential loss and damage. Both the hazard and its consequences should be quantified, and the relative significance of each hazard should be compared. The type and amount of data involved can be processed by a geographical information system (GIS).

1.6 Morphological maps

One of the important ways in which the geologist can be of service is by producing maps to aid the planner and others who are concerned with the development of land. A variety of maps can be produced, including morphological maps, engineering geomorphological maps, environmental geological maps (EGMs) and engineering geological maps. The distinction between these different types of map, particularly the last three, is not always clear-cut. The important thing, however, is that such maps are produced; their label is less important. As Varnes (1974) pointed out, maps represent a means of storing and transmitting information, in particular of conveying specific information about the spatial distribution of given factors or conditions. Other methods of using and portraying geological information for planning purposes are in terms of terrain evaluation (see Chapter 16) and, of increasing importance, via GISs.

The classical method of landform mapping is through surveyed contours. Waters (1958), however, devised a technique that was further refined by Savigear (1965), which defined the geometry of the ground surface in greater detail than normally found on contour maps. They proposed that the ground surface consisted of planes, which intersected in convex and concave, angular or curved 'discontinuities'. An angular discontinuity was defined as a break of slope and a curved discontinuity as a change of slope. A morphological map therefore is divided into slope units, which are delineated by breaks of slope, so defining the pattern of the ground (the distinction

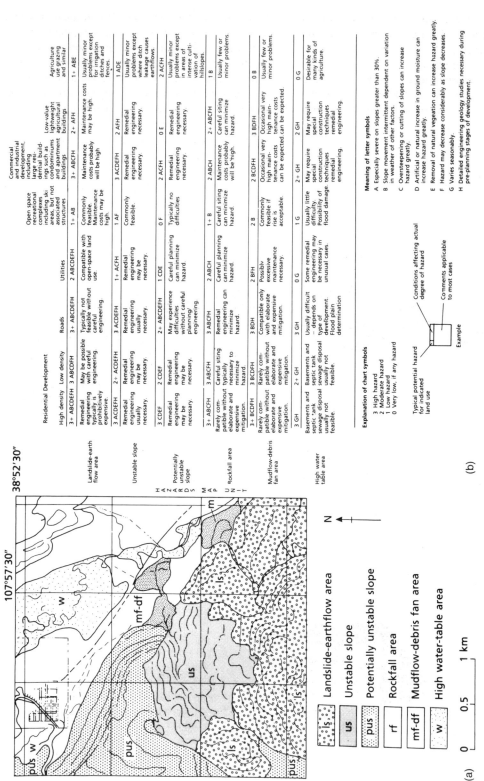

Fig. 1.3 (a) An example of engineering geological hazard mapping in the Crested Butte–Gunnison area, Colorado. (After Soule, 1980.) (b) The matrix is formed so as to indicate to the map user that several geological and geology-related factors should be considered when contemplating the land use in a given type of mapped area. This matrix can also serve to recommend additional types of engineering geological studies that may be needed for a site. Thus the map can be used to model or anticipate the kinds of problems that a land-use planner or land developer may have to overcome before a particular activity is permitted or undertaken.

between breaks and changes of slope provides a more precise appreciation of landform than is possible from reading contours).

When available, aerial photographs should be used for preliminary morphological mapping, because they furnish an idea of the terrain and may be employed to locate boundaries between the morpho-

logical units. These are recorded on the photographs and then transferred, by plotter, to the base map. The best scale for the field map depends on the objectives of the survey. Whatever the scale, however, some units will be recognized which have boundaries that are too close together to be represented separately. If small features that are regarded as important cannot be incorporated on the base map, they should be mapped on a larger scale. For example, Savigear (1965) maintained that certain features, such as cliff units, should always be represented in morphological mapping. Most standard geomorphological features

Fig. 1.4 (a) Morphological map of the Haven and Culverhole cliff landslips, Devon, England. (b) Slope categories of the Haven and Culverhole cliffs. (After Pitts, 1979.)

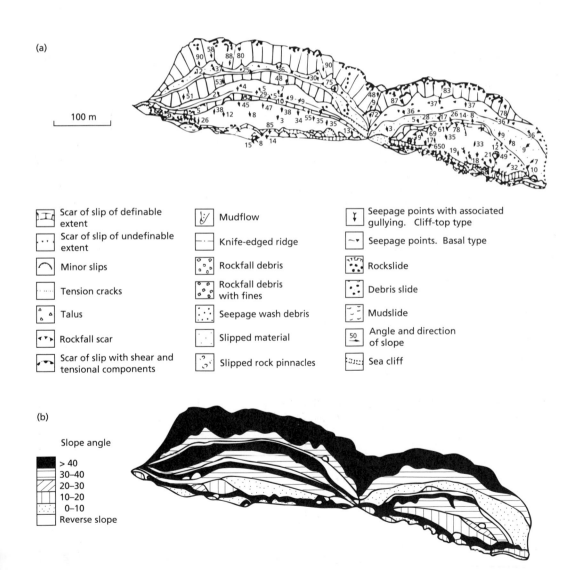

(a)

100 m

Scar of slip of definable extent

Scar of slip of undefinable extent

Minor slips

Tension cracks

Talus

Rockfall scar

Scar of slip with shear and tensional components

Mudflow

Knife-edged ridge

Rockfall debris

Rockfall debris with fines

Seepage wash debris

Slipped material

Slipped rock pinnacles

Seepage points with associated gullying.   Cliff-top type

Seepage points. Basal type

Rockslide

Debris slide

Mudslide

Angle and direction of slope

Sea cliff

(b)

Slope angle

> 40
30–40
20–30
10–20
0–10
Reverse slope

Slope category	Slope terminology	Gradient (%)	Slope length to slope height ratio
0–0°30′	Plain		
0°30′–2°	Slightly sloping	0–3.5	∞–28.6
2–5°	Gently inclined	3.5–8.7	28.6–11.4
5–15°	Strongly inclined	8.7–26.8	11.4–3.7
15–35°	Steep	26.8–70	3.7–1.4
(25–35°)	(Very steep)		
35–55°	Precipitous	70–143	1.4–0.7
55–90°	Vertical	143–∞	0.7–0
Over 90°	Overhanging		

Table 1.3 Slope angle classification for detailed geomorphological maps. (After Demek, 1972.)

can be represented on a base map which has a scale of 1 : 10 000. However, not only does the clear representation of all morphological information on one map provide difficult cartographic problems, it also gives rise to difficulties in interpretation and use, thus limiting the value of the map. This problem can, to some extent, be overcome by using overlays to show some special aspect of the land surface.

Convex and concave boundaries are distinguished in morphological mapping, and measurements can be made of slope steepness and, if present, slope curvature. A knowledge of slope angles is needed for the study of present-day processes and to understand the development of relief. Steepness can be shown by an arrow lying normal to the slope, pointing downhill, and the angle of the slope is marked in degrees on the arrow (Fig. 1.4a). Special symbols are used for very steep slopes, such as cliffs. Differences in slope steepness can be emphasized by shading or colours according to defined slope classes (Fig. 1.4b). Demek (1972) suggested several categories of slope (Table 1.3). Slope category maps depict the average inclination over an area, and make it easier to perceive the distribution of steep slopes, planation surfaces and valley asymmetry than is possible from contour maps (Kertesz, 1979). Slope steepness is of considerable importance in land management, for example, it frequently is a restricting factor in route selection, urban development and type of agriculture (Table 1.4). A morphogenetic map indicates the origin of the landforms recognized in the field, while a process map identifies the spatial pattern of geomorphological processes which are currently active (Fig. 1.5). Morphological mapping may prove to be useful as a quick reconnaissance exercise prior to a site investigation,

Table 1.4 Critical slope steepness for certain activities. (After Cooke & Doornkamp, 1990.)

Steepness (%)	Critical for
1	International airport runways
2	Main-line passenger and freight rail transport
	Maximum for loaded commercial vehicles without speed reduction
	Local aerodrome runways
	Free ploughing and cultivation
	Below 2%, flooding and drainage problems in site development
4	Major roads
5	Agricultural machinery for weeding, seeding
	Soil erosion begins to become a problem
	Land development (construction) difficult above 5%
8	Housing, roads
	Excessive slope for general development
	Intensive camp and picnic areas
9	Absolute maximum for railways
10	Heavy agricultural machinery
	Large-scale industrial site development
15	Site development
	Standard wheeled tractor
20	Two-way ploughing
	Combine harvesting
	Housing site development
25	Crop rotations
	Loading trailers
	Recreational paths and trails

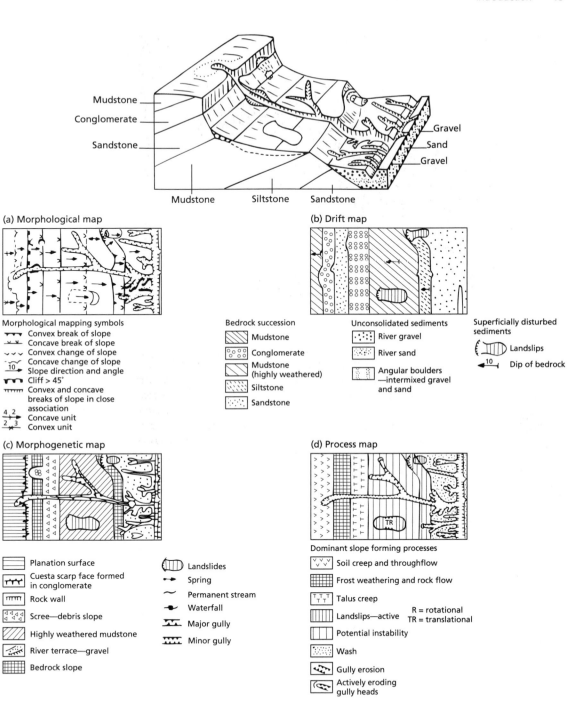

Mudstone

Conglomerate

Sandstone

Gravel

Sand

Gravel

Mudstone   Siltstone   Sandstone

(a) Morphological map

Morphological mapping symbols
⊤⊤⊤ Convex break of slope
⊥⊥⊥ Concave break of slope
⌄⌄⌄ Convex change of slope
⌒⌒ Concave change of slope
10↗ Slope direction and angle
⊓⊓ Cliff > 45°
⊤⊤⊤⊤⊤ Convex and concave
breaks of slope in close
association
4 2
⎯→ Concave unit
2 3
⎯→ Convex unit

(b) Drift map

Bedrock succession
▨ Mudstone
⬚ Conglomerate
◩ Mudstone
(highly weathered)
▧ Siltstone
▦ Sandstone

Unconsolidated sediments
▫ River gravel
∴ River sand
▦ Angular boulders
—intermixed gravel
and sand

Superficially disturbed
sediments
⬭ Landslips
10◄— Dip of bedrock

(c) Morphogenetic map

Planation surface
⊤⊤⊤ Cuesta scarp face formed
in conglomerate
⊓⊓⊓ Rock wall
◁◁◁ Scree—debris slope
▨ Highly weathered mudstone
〰 River terrace—gravel
▦ Bedrock slope

⬭ Landslides
↦ Spring
∼ Permanent stream
➤ Waterfall
⏚ Major gully
〰 Minor gully

(d) Process map

Dominant slope forming processes
⌄⌄⌄ Soil creep and throughflow
▦ Frost weathering and rock flow
⊤⊤⊤ Talus creep
▥ Landslips—active   R = rotational
TR = translational
▯ Potential instability
∴ Wash
⌁ Gully erosion
⌁ Actively eroding
gully heads

Fig. 1.5 Schematic diagram to illustrate the principles and products of geomorphological mapping. (a) The morphological map emphasizes slope form and steepness. (b) The materials or drift map indicates the nature of surface materials. (c) The morphogenetic map indicates the origin of the landforms. (d) The process map identifies the distribution of currently active geomorphological processes. (After Savigear, 1965.)

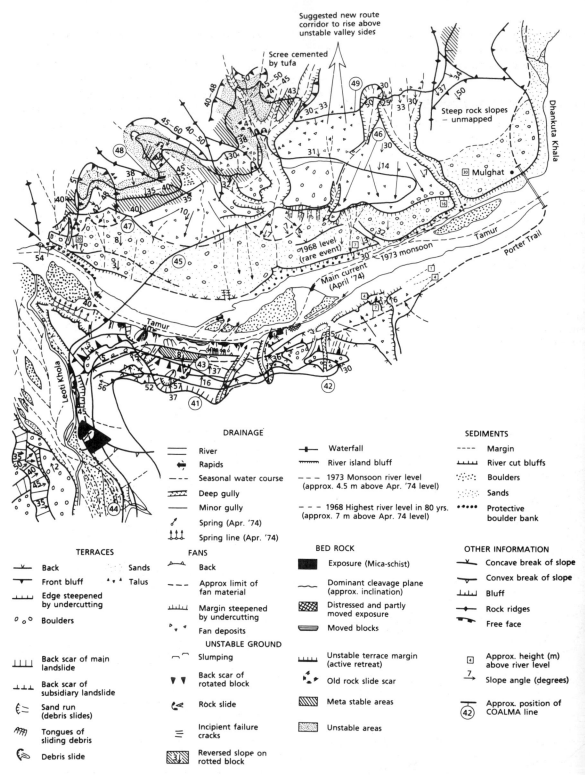

Fig. 1.6 Geomorphological map of the site and situation of a proposed bridge crossing on the Tamur River, eastern Nepal. Scale approximately 1 : 12 500. (After Brunsden *et al.*, 1975b.)

or as a more extensive undertaking where difficult or inaccessible terrain is concerned, and therefore restricts the use of some site investigation techniques.

1.7 Engineering geomorphological maps

Mapping the surface form is the first step in geomorphological mapping. The next is to make interpretations regarding the forms and to ascribe an origin to them. This must be carried out in relation to the geological materials which compose each feature, and in relation to the past and present processes operating in the area concerned. As such, geomorphological maps provide a comprehensive, integrated statement of landform and drainage (Fig. 1.6). Consequently, they contain much information of potential value as far as land-use planning and construction projects are concerned.

Brunsden *et al.* (1975a) maintained that the recording of geomorphological data on maps for planning and engineering purposes was of value, because the surface form and aerial pattern of geomorphological processes often influence the choice of a site. In other words, geomorphological maps provide a rapid appreciation of the nature of the ground, and thereby help in the design of more detailed investigations, as well as focusing attention on problem areas. Engineering geomorphological mapping involves the recognition of landforms, together with their delimitation in terms of size and shape. Many preliminary data can be obtained from aerial photographs. The principal object during a reconnaissance survey is the classification of every component of the land surface in relation to its origin, present evolution and likely material properties, based on techniques of landform interpretation (Fig. 1.7). A field survey provides additional data, which enable the preliminary views on

Fig. 1.7 Rationalization of borehole and trial pit information through a geomorphological and geological interpretation of landforms, Suez, Egypt. (After Doornkamp *et al.*, 1979.)

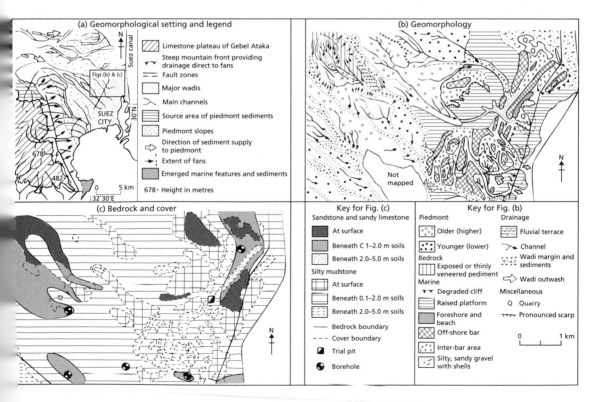

the causative processes to be revised, if necessary. Further precision can be afforded to geomorphological interpretations by obtaining details from climatic, hydrological or other records, and by analysis of the stability of landforms. Moreover, an understanding of the past and present development of an area is likely to aid in the prediction of its behaviour during and after any construction operations. Engineering geomorphological maps therefore should show how surface expression will influence a project, and should provide an indication of the general environmental relationship of the area concerned.

The aims of such a geomorphological survey have been summarized by Brunsden *et al.* (1975b) as follows:

1 The identification of the general characteristics of the terrain of an area, thereby providing a basis for the evaluation of alternative locations and avoidance of the worst hazard areas.

2 The identification of factors outside the site which may influence it, such as mass movement.

3 The provision of a synopsis of geomorphological development of the area which includes:

(a) a description of the extent and degree of weathering;

(b) a classification of slopes based on their steepness, material composition, mode of development and stability;

(c) a description of the location, pattern and magnitude of the surface and subsurface drainage features (including karst development);

(d) a definition of the shape and extent of geomorphological units, such as fans, scree slopes, terraces, etc.;

(e) the recognition of specific hazards, such as flooding and landslides.

4 The location of suitable supplies of construction materials.

Obtaining such information should facilitate the planning of any subsequent investigation. For instance, it should aid the location of boreholes, and these, it is hoped, will confirm what has been discovered by the geomorphological survey. Engineering geomorphological mapping therefore may help to reduce the cost of an investigation.

Doornkamp *et al.* (1979) pointed out that the recognition of the interrelationships between landforms on site and their individual or combined relationships to landforms beyond the site is fundamental. This is necessary in order to appreciate not

Table 1.5 Summary outline of working practice. (After Doornkamp *et al.*, 1979.)

Phase	Liaison with engineers	Desk studies (home based)	Field studies (based on site)

I	Brief received from client
	Discussions with senior
	    engineers and
	    engineering
	    geologist involved
	Brief re-examined
		Familiarization with project
		Examination of available
		    literature and maps
		Aerial photograph interpretation	

II	Continuing discussions with
	    engineer's field staff
			Field mapping
			Investigation of landforms,
			    materials and processes
			Review of trial pit and borehole
			    information (if available)
			Geomorphological map
			    compilation

III		Derivative maps compiled Data additional to initial brief compiled	

	Report with maps passed
	    to client
		Site investigation
		    suggestions defined	

only how the site conditions will affect the project, but, just as important, how the project will affect the site and the surrounding environment. In particular, a knowledge of ground hazards is required for good design of a project.

The general procedure in an engineering geomorphological investigation is summarized in Table 1.5. Phase I, which is carried out prior to the fieldwork, involves a familiarization with the project and the landscape. The amount of information which can be obtained from a literature survey varies with the location. In some developing countries, little or nothing may be available; even worthwhile topographical maps, which normally are a prerequisite of a geomorphological mapping programme, may not exist. Base maps can then be made from aerial photographs, which can be specially commissioned. A study of aerial photographs enables many of the significant landforms and their boundaries to be defined prior to the commencement of fieldwork. The scale of the photographs is usually 1:10000. Field mapping permits the correct identification of landforms, recognized on aerial photographs, as well as geomorphological processes, and indicates how they will affect the project. Mapping of a site can provide data on the nature of the surface materials.

The scale of an engineering geomorphological map is influenced by the project requirement, and the map should focus attention on the information relevant to the particular project. Maps produced for extended areas, such as needed for route selection, are drawn on a small scale. Small-scale maps have also been used for planning purposes, land-use evaluation, land reclamation, floodplain management, coastal conservation, etc. These general engineering geomorphological maps concentrate on portraying the form, origin, age and distribution of landforms, together with their formative processes, rock type and surface materials. In addition, if information is available, details of the actual frequency and magnitude of the processes can be shown by symbols, annotation, accompanying notes or successive maps of temporal change. Large-scale maps and plans of local surveys provide an accurate portrayal of surface form, drainage characteristics and the properties of surface materials, as well as an evaluation of currently active processes. If the maximum advantage is to be obtained from a geomorphological survey, derivative maps should be compiled from the geomorphological sheets. Such derivative maps generally are concerned with some aspect of ground conditions, for example, landslip areas, areas prone to flooding or over which sand dunes migrate.

1.8 Environmental geological maps

Environmental geological maps have been produced to meet the needs of planners (Robinson & Spieker, 1978; Culshaw *et al.*, 1990). It is important that geological information should be understood readily by the planner and those involved in development. Unfortunately, conventional geological maps often are inadequate for the needs of such individuals. For example, one of the shortcomings of conventional geological maps is that the boundaries are stratigraphical, and more than one type of rock may be included in a single mappable unit. Moreover, the geological map is lacking in quantitative information concerning facts, such as the physical properties of the rocks and soils, the amount and degree of weathering, the hydrogeological conditions, etc. Consequently, maps incorporating geological data are now being produced for planning and land-use purposes. Such maps essentially are simple, and provide some indication of those areas in which there are fewest geological constraints on development. Therefore they can be used by the planner or engineer at the feasibility stage of a project. The location of exploitable mineral resources is also of interest to planners. The obvious reason for presenting geological data in a manner that can be understood by planners, administrators and engineers is that they can then seek appropriate professional advice and, in this way, bring about safer and more cost-effective development and design of land, especially in relation to urban growth and redevelopment. In fact, two versions of an EGM may be produced, one for the specialist and one for the non-specialist.

Topics on EGMs vary, but may include solid geology, unconsolidated deposits, geotechnical properties of soils and rocks, landslides, depth to rock head, hydrogeology, mineral resources, shallow undermining and opencast workings, floodplain hazards, etc. Hence, each aspect of geology can be presented as a separate theme on a basic or element map. Derivative maps display two or more elements combined to show, for example, the foundation conditions, ease of excavation, aggregate potential, landslide susceptibility, subsidence potential, ground-water resources or capability for solid waste disposal. Environmental potential maps are compiled from basic data maps and derived maps. They present, in general terms, the constraints on development, such as areas where land is susceptible to landslipping or subsidence, or land likely to be subjected to flooding (Fig. 1.8). They can also present those resources with respect to minerals, groundwater or agricultural potential which might be used in development, or which should not be sterilized by building over or contamination from landfill sites. Again these maps should be readily understood by non-geologists.

Reports containing comprehensive suites of EGMs are now being produced. For example, in Britain, this began with a pilot study of the Glenrothes area of Fife, Scotland. This study produced a suite of 27 separate maps covering aspects such as the stratigraphy and lithology of bedrock, superficial deposits, rock head contours, engineering properties, mineral resources and workings, groundwater conditions and landslip potential (Nickless, 1982). The maps primarily were for the use of local and central planners, but were also useful sources of information for civil engineers, developers and mineral extraction companies. The main feature of the study was the presenta-

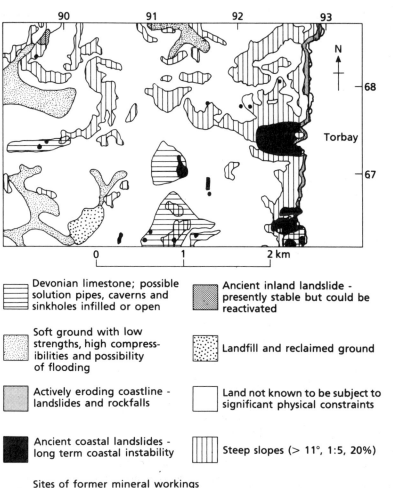

Devonian limestone; possible solution pipes, caverns and sinkholes infilled or open

Soft ground with low strengths, high compressibilities and possibility of flooding

Actively eroding coastline - landslides and rockfalls

Ancient coastal landslides - long term coastal instability

Sites of former mineral workings - steep rock faces or infill of varied properties

Ancient inland landslide - presently stable but could be reactivated

Landfill and reclaimed ground

Land not known to be subject to significant physical constraints

Steep slopes (> 11°, 1:5, 20%)

Fig. 1.8 Ground characteristics for planning and development for part of the Torbay area of southwest England. (After Culshaw *et al.*, 1990.)

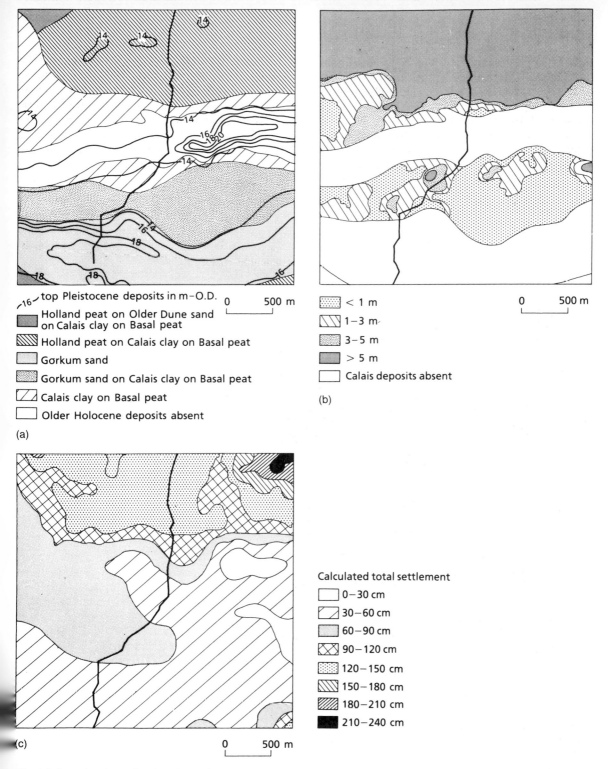

Fig. 1.9 Examples of sets of engineering geological maps of Leiden, the Netherlands: (a) multipurpose comprehensive map of older Holocene deposits; (b) isopach map of Calais deposits (the most notable formation in the Leiden area); (c) induced settlement map for an imaginary surface load of 100 kPa (After De Mulder & Hillen, 1990.)

tion of each element of the geology on a separate map in a manner that was easy for non-geologists to understand. Another feature was the interpretation of these elements so as to provide data other than mere outcrop distribution and lithology.

Similar developments have taken place in the Netherlands. For instance, De Mulder and Hillen (1990) described how sets of geological maps of certain areas have been produced to help planning. These sets include a location map, a geotechnical cross-section of the area, multipurpose comprehensive maps (e.g. sequence of soil layers, as well as the thickness and depth of pertinent individual layers (Fig. 1.9a,b)) and special maps (e.g. amount of settlement under load (Fig. 1.9c), foundation depth and hydrogeological environmental geology).

Wolden and Ericksen (1990) introduced the concept of geoplan maps, which outline the relevant geological factors likely to influence planning (Fig. 1.10). They maintained that such maps should help planners to avoid making poor decisions at an early stage, and should allow areas to be developed sequentially (e.g. mineral resources could be developed before an area is built over). Such maps would facilitate the development of a multiuse plan of a particu-

lar area and avoid one factor adversely affecting another.

Environmental geological maps have their limitations, which stem from a number of factors, such as the scale of presentation and the availability and reliability of data. Like all geological maps, EGMs are transitory and inevitably need amendment as new or previously unavailable data become accessible. Furthermore, as McMillan and Browne (1987) pointed out, erroneous conclusions can be drawn by those who are not aware of these shortcomings. This is particularly the case with regard to ground stability in urban areas. In such areas, properties which have been constructed adequately to take into account ground conditions when they were built can be blighted if it is later revealed that they are located on ground which may be suspect, for example, may have been undermined by workings long since abandoned. In other words, it may become more difficult to obtain insurance cover or mortgages for such properties, and their values may decline. In fact, in the USA, there can be legal implications involved in the production of such maps, so that they must carry disclaimers. Hence, in the preparation of thematic maps, every attempt must be made to ensure that the

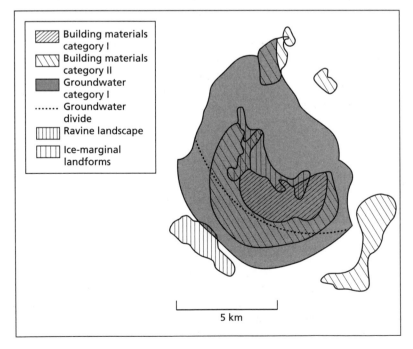

Fig. 1.10 Geoplan map. Building materials category I indicates that the sand and gravel deposits are of good quality and well documented. Building materials category II indicates that the deposits can be exploited, but the area is not as well documented. Groundwater category I is an important exploitable resource with a capacity of $200 \, l \, s^{-1}$ or more. The ravine landscape and ice-marginal landforms represent areas recommended for conservation. (After Wolden & Ericksen, 1990.)

Legend:
- Building materials category I
- Building materials category II
- Groundwater category I
- Groundwater divide
- Ravine landscape
- Ice-marginal landforms

5 km

finished product cannot be misconstrued. The limitations on their accuracy, and the interpretation and use of data in their production, must be made clear.

1.9 Engineering geological maps

An engineering geological map is produced from the information collected from various sources (literature survey, aerial photographs and imagery and fieldwork). The preparation of engineering geological maps of urban areas frequently involves systematic searches of archives (Dearman, 1991). Information from site investigation reports, records of past and present mining activity and successive editions of topographical maps may prove to be extremely useful. Once the data have been gathered, their representation on the map and the scale of the map must be decided. The latter is very much influenced by the requirement. The major differences between maps of different scales are the amount of data they show (the more detailed a map needs to be, the larger its scale) and the manner in which the data are presented. Engineering geological maps usually are produced on the scale of 1 : 10 000 or smaller, whereas engineering geological plans, being produced for particular engineering purposes, have a larger scale (Anon., 1972). As far as presentation is concerned, this may involve not only the choice of colours and symbols, but also the use of overprinting. Overprinting frequently takes the form of striped or stippled shading, both of which can be varied, for instance, according to frequency, pattern, dimension or colour.

A map represents a simplified model of the facts, and the complexity of various geological factors can never be portrayed in its entirety. The amount of simplification required is governed principally by the purpose and scale of the map, the relative importance of particular engineering geological factors or relationships, the accuracy of the data and the techniques of representation employed (Anon., 1976). Engineering geological maps should be accompanied by cross-sections, and an explanatory text and legend. More than one map of an area may be required to record all the information which has been collected during a survey. In such instances, a series of overlays or an atlas of maps can be produced. For example, De Beer *et al.* (1980) described the production of a geotechnical 'atlas' of certain urban areas in Belgium. This 'atlas' comprised a documentation map (a topographical map which shows the location of boreholes and the data derived therefrom), an individual map showing the isopachs of the upper surface of the formation in question, a hydrogeological map, a map of engineering geological zones, a number of engineering geological cross-sections and an explanatory key. The preparation of a series of engineering geological maps can reduce the amount of effort involved in the preliminary stages of a site investigation, and indeed may allow site investigations to be designed for the most economical confirmation of the ground conditions.

Engineering geological maps may serve a special purpose or a multipurpose (Varnes, 1974; Dearman & Matula, 1977). Special purpose maps provide information on one specific aspect of engineering geology, such as the grade of weathering, joint patterns, mass permeability or foundation conditions (Fig. 1.11). Alternatively, special purpose maps may serve one particular purpose, for example, the engineering geological conditions at a dam site or along a routeway, or land-use zoning in urban development. Multipurpose maps cover various aspects of engineering geology, and provide general information for planning or engineering purposes.

In addition, engineering geological maps may be analytical or comprehensive. Analytical maps provide details, or evaluate individual components, of the geological environment. Examples of such maps include those showing the degree of weathering or seismic hazard. Comprehensive maps either depict all the principal components of the engineering geological environment, or are maps of engineering geological zoning, delineating individual units on the basis of the uniformity of the most significant attributes of their engineering geological character.

An engineering geological map provides an impression of the geological environment, surveying the range and type of engineering geological conditions, their individual components and their interrelationships. These maps also provide planners and engineers with information which will assist them in land-use planning and the location, construction and maintenance of engineering structures of all types. Engineering geological maps may be simply geological maps to which engineering geological data have been added. If the engineering information is extensive, it can be represented in tabular form (Fig. 1.12; Table 1.6a,b), perhaps accompanying the map on the

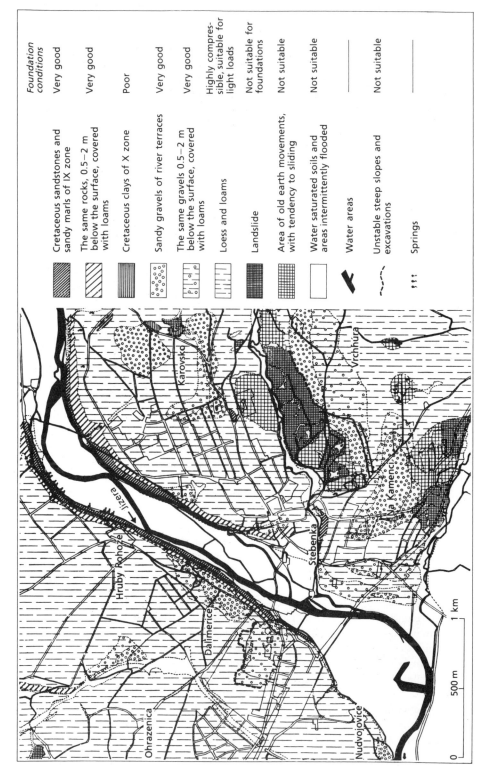

Fig. 1.11 Engineering geological map of Turnov, Czech Republic. (After Dearman, 1991.)

	Foundation conditions
Cretaceous sandstones and sandy marls of IX zone	Very good
The same rocks, 0.5–2 m below the surface, covered with loams	Very good
Cretaceous clays of X zone	Poor
Sandy gravels of river terraces	Very good
The same gravels 0.5–2 m below the surface, covered with loams	Very good
Loess and loams	Highly compressible, suitable for light loads
Landslide	Not suitable for foundations
Area of old earth movements, with tendency to sliding	Not suitable
Water saturated soils and areas intermittently flooded	Not suitable
Water areas	
Unstable steep slopes and excavations	Not suitable
Springs	

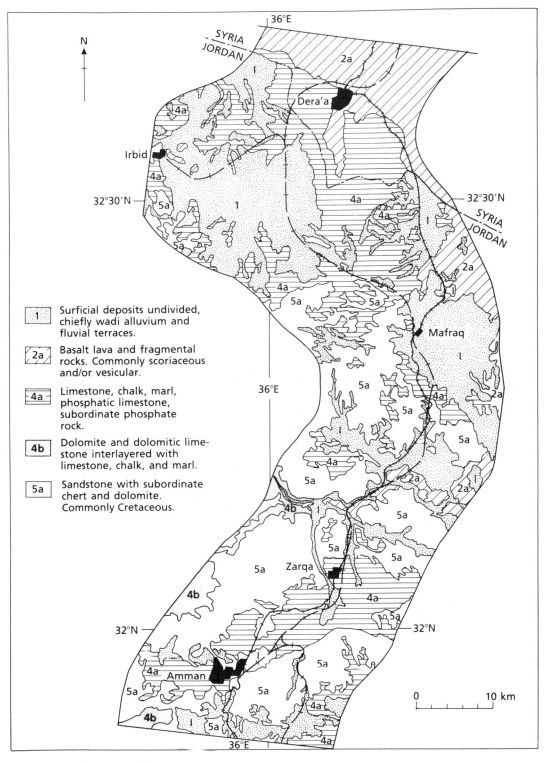

N

SYRIA
JORDAN

36°E

2a

Dera'a

4a

Irbid

4a

32°30'N   5a   4a   32°30'N

4a   SYRIA
JORDAN

1

5a

2a

5a   5a

4a

5a   Mafraq

	Surficial deposits undivided, chiefly wadi alluvium and fluvial terraces.
1	

36°E

5a

1

	Basalt lava and fragmental rocks. Commonly scoriaceous and/or vesicular.
2a	

4a   2a

5a

	Limestone, chalk, marl, phosphatic limestone, subordinate phosphate rock.
4a	

5a

	Dolomite and dolomitic limestone interlayered with limestone, chalk, and marl.
4b	

4a   2a   2a

5a

	Sandstone with subordinate chert and dolomite. Commonly Cretaceous.
5a	

4b   1   5a

5a   5a

5a

4b   5a   Zarqa

4a

5a

32°N   32°N

4a   Amman

5a   5a   5a

4b   1   5a   4a

0   10 km

36°E   4a

Fig. 1.12 Segment of the engineering geological map for the Hijaz railway in Jordan. Rock unit numbers are explained in the figure. (After Briggs, 1987.)

Table 1.6(a) Excerpts from the engineering geology table illustrating the variety of materials in the study area for the Hijaz railway. Symbols 1, 2a, 4a, 4b and 5a occur in the area shown in Fig. 1.12. (After Briggs, 1987.)

Map symbol	Geological description	Distribution	Map segments	Engineering characteristics	Suitability as source of material for	Moderate water supply favourability in shallow aquifers	Topographic expression
1	Surficial deposits, undivided, chiefly wadi alluvium and fluvial and marine terraces	Most common in Saudi Arabia and southern Jordan	Present on most map segments	Excavation: easy Stability: poor Strength: fair Tunnel support: maximum	Ballast: 0 Coarse aggregate: + Sand: +++ Embankments: 0 Rip-rap: 0	Fair to good with seasonal fluctuations. Coastal areas poor	Generally flat, locally steeply dissected
2a	Basalt lava and fragmental rocks. Commonly scoriaceous and/or vesicular	Widespread in southern Syria. Locally elsewhere	01, 02, 13, 19, 20 and 28	Excavation: difficult Stability: good Strength: good Tunnel support: moderate	Ballast: + Coarse aggregate: + Sand: + Embankments: ++ Rip-rap: +	Generally poor. Locally fair to good, depending on interlayering	Flat to mountainous. Surfaces commonly bouldery
3c	Sandstone and conglomerate with limestone and marl. Loosely cemented. Locally hard	Along coastal plain between Al Wajh and Yanbu, Saudi Arabia	26, 27, 28 and 30	Excavation: intermediate Stability: fair Strength: fair Tunnel support: moderate to maximum	Ballast: 0 Coarse aggregate: 0 Sand: + Embankments: ++ Rip-rap: 0	Poor	Flat to rolling, locally hilly and dissected
4a	Limestone, chalk, marl, phosphatic limestone, subordinate phosphate rock	Widespread in Jordan	02–05 and 29	Excavation: difficult Stability: fair to good Strength: fair to good Tunnel support: moderate to minimum	Ballast: 0 Coarse aggregate: + Sand: 0 Embankments: + Rip-rap: +	Generally poor	Hilly, locally rolling or mountainous
4b	Dolomite and dolomitic limestone interlayered with limestone, chalk and marl	Central Jordan	02 and 03	Excavation: moderately difficult Stability: fair to good Strength: fair to good Tunnel support: moderate to minimum	Ballast: + Coarse aggregate: + Sand: 0 Embankments: ++ Rip-rap: +	Generally poor	Hilly, locally mountainous
5a	Sandstone with subordinate chert and dolomite. Commonly calcareous	Widespread in Jordan	02–05 and 29	Excavation: moderately difficult Stability: fair to good Strength: good Tunnel support: moderate to minimum	Ballast: 0 Coarse aggregate: 0 Sand: + Embankments: ++ Rip-rap: 0	Poor to fair	Hilly to mountainous

Continued.

Table 1.6(a) *Continued.*

6b	Chiefly andesite lava and fragmental rocks. Common medium-grade metamorphism, greenstone	Widespread in Hijaz Mountains	10–13, 18–25, 27, 30 and 31	Excavation: difficult Stability: good Strength: good Tunnel support: minimum	Ballast: +++ Coarse aggregate: +++ Sand: 0 Embankments: ++ Rip-rap: +++	Poor	Core of Hijaz Mountains. Relief locally greater than 2000 m
7b	Early and altered granites, granodiorite, quartz monzonite. Includes some gneiss	Common in the Hijaz Mountains and southern Jordan	10–13 and 19–31	Excavation: difficult Stability: good Strength: good Tunnel support: minimum to moderate	Ballast: + Coarse aggregate: + Sand: + Embankments: ++ Rip-rap: ++	Poor	Chiefly mountainous. Mostly more resistant than other intrusive rocks

Table 1.6(b) Key to the engineering characteristics column of the engineering geology table (Table 1.6a).

Excavation facility	Stability of cut slopes	Foundation strength	Tunnel support requirements
Easy. Can be excavated by hand tools or light power equipment. Some large boulders may require drilling and blasting for their removal. Dewatering and bracing of deep excavation walls may be required	*Good*. These rocks have been observed to stand on essentially vertical cuts where jointing and fracturing are at a minimum. However, moderately close jointing or fracturing is common, so slopes not steeper than 4 : 1 (vertical : horizontal) are recommended. In deep cuts, debris-catching benches are recommended	*Good*. Bearing capacity is sufficient for the heaviest classes of construction, except where located on intensely fractured or jointed zones striking parallel to and near moderate to steep slopes	*Minimum*. Support probably required for less than 10% of length of bore, except where extensively fractured
Moderately easy. Probably rippable by heavy power equipment at least to weathered rock–fresh rock interface and locally to greater depth		*Fair*. Choice of foundation styles is largely dependent on packing of fragments, clay content and relation to the water table. If content of saturated clay is high, appreciable lateral movement of clay may be expected under heavy loads. If packing is poor, settling may occur	*Moderate*. Support may be required for as much as 50% of length of bore, more where extensively fractured
Intermediate. Probably rippable by heavy power equipment to depths chiefly limited by the manoeuvrability of the equipment. Hard rock layers or zones of hard rock may require drilling and blasting	*Fair*. Cut slopes ranging from 2 : 1 to 1 : 1 are recommended; flatter where rocks are intensely jointed or fractured. Rockfall may be frequent if steeper cuts are made. Locally, lenses of harder rock may permit steeper cuts		*Maximum*. Support probably required for entire length of bore
Moderately difficult. Probably require drilling and blasting for most deep excavations, but locally may be ripped to depths of several metres	*Poor*. Flatter slopes are recommended. Some deposits commonly exhibit a deceptive temporary stability, sometimes standing on vertical or near-vertical cuts for periods ranging from hours to more than a year	*Poor*. Foundations set in underlying bedrock are recommended for heavy construction, with precautions taken to guard against failure due to lateral stress	
Difficult. Probably require drilling and blasting in most excavations, except where extensively fractured or altered			

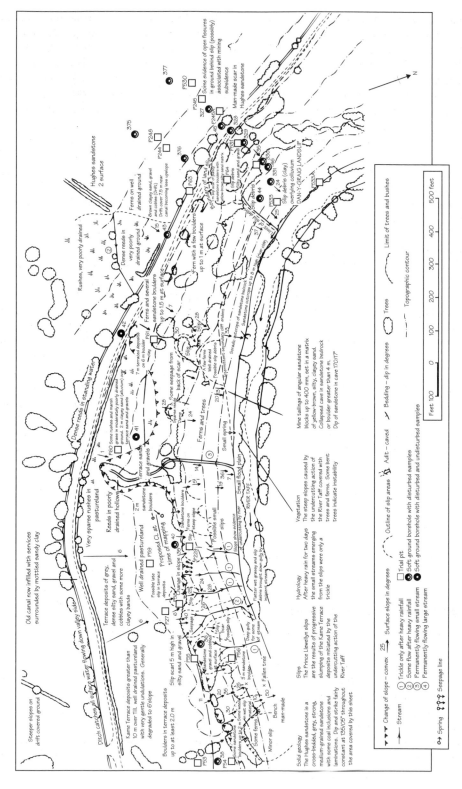

Fig. 1.13 Engineering geological plan produced at the site investigation stage, Prince Llewellyn area, Stage IV, Taff Valley trunk road, South Wales, UK. The contours are in feet above crdnance datum (AOD). (After Dearman & Fookes, 1974.)

reverse side. The superficial deposits and bedrock should be described in detail. Where possible, rocks and soils should be classified according to their engineering behaviour. Details of geological structures should be recorded, especially fault and shear zones, as should the nature of the discontinuities and grade of weathering where appropriate. Alternatively, the rocks and soils in the area concerned may be presented as mapped units, defined in terms of engineering properties. The unit boundaries are then drawn for changes in the particular property. Frequently, the boundaries of such units coincide with stratigraphical boundaries. In other instances, e.g. where rocks are deeply weathered, they may bear no relation to geological boundaries. Unfortunately, one of the fundamental difficulties in preparing such maps arises from the fact that changes in the physical properties of rocks and soils frequently are gradational. As a consequence, regular checking of visual observations by *in situ* testing or sampling is essential to produce a map based on engineering properties. This type of map often is referred to as a geotechnical map (Dearman, 1991).

There are two basic types of engineering geological or geotechnical plans, namely, the site investigation plan and the construction or foundation stage plan (Dearman & Fookes, 1974). In the case of the former type, the scale varies from 1 : 5000 to as large as 1 : 500 or even 1 : 100, depending on the size and nature of the site and the engineering requirement (Fig. 1.13). The foundation plan records the ground conditions exposed during construction operations. It may be drawn to the same scale as the site investigation plan or the construction drawings. Plans may be based on large-scale topographical maps or large-scale base maps produced by surveying or photogrammetric methods.

1.10 Geographical information systems

One means by which the power, potential and flexibility of mapping may be increased is by developing a GIS. Geographical information systems represent a form of technology which is capable of capturing, storing, retrieving, editing, analysing, comparing and displaying spatial environmental information. For instance, Star and Estes (1990) indicated that a GIS consists of four fundamental components, namely, data acquisition and verification, data storage and manipulation, data transformation and analysis, and data output and presentation. The GIS software is designed to manipulate spatial data in order to produce maps, tabular reports or data files for interfacing with numerical models. An important feature of a GIS is the ability to generate new information by the integration of existing diverse data sets sharing a compatible referencing system (Goodchild, 1993). Data can be obtained from remote sensing imagery, aerial photographs, aero-magnetometry, gravimetry and various types of map. These data are recorded in a systematic manner in a computer database. Each type of data input refers to the characteristics of recognizable point, linear or areal geographical features. Details of the features usually are stored in either vector (points, lines and polygons) or raster (grid cell) formats. The manipulation and analysis of data allows them to be combined in various ways to evaluate what will happen in certain situations.

Currently, there are many different GISs available, ranging from public domain software for personal computers (PCs) to very expensive systems for mainframe computers. Because most data sets required in environmental geology data processing still are relatively small, they can be readily accommodated by inexpensive PC-based GIS applications. The advantages of using a GIS compared with conventional spatial analysis techniques have been reviewed by Burrough (1986), and are summarized in Table 1.7.

An ideal GIS for many environmental geological situations combines conventional GIS procedures with image processing capabilities and a relational database. Because frequent map overlaying, modelling and integration with scanned aerial photographs and satellite images are required, a raster system is preferred. The system should be able to perform spatial analysis on multiple input maps and connected attribute data tables. Necessary GIS functions include map overlay, reclassification and a variety of other spatial functions incorporating logical, arithmetic, conditional and neighbourhood operations. In many cases, modelling requires the iterative application of similar analyses using different parameters. Consequently, the GIS should allow for the use of batch files and macros to assist in performing these iterations.

Mejía-Navarro and Garcia (1996) referred to several attempts to use a GIS for geological hazard

Table 1.7 Advantages and disadvantages of a GIS.

Advantages	Disadvantages
1 A much larger variety of analysis techniques are available. Because of the speed of calculation, complex techniques requiring a large number of map overlays and table calculations become feasible	**1** A large amount of time is needed for data entry. Digitizing is especially time consuming
2 It is possible to improve models by evaluating their results and adjusting the input variables. Users can achieve the optimum results by a process of trial and error, running the models several times, whereas it is difficult to use these models even once in the conventional manner. Therefore, more accurate results can be expected	**2** There is a danger in placing too much emphasis on data analysis as such at the expense of data collection and manipulation based on professional experience. A large number of different techniques of analysis are theoretically possible, but often the necessary data are missing. In other words, the tools are available, but cannot be used because of the lack of input data
3 During the course of a hazard assessment project, the input maps derived from field observations can be updated rapidly when new data are collected. Also, after completion of the project, the data can be used by others in an effective manner	

and vulnerability assessment. These were especially in relation to the assessment of landslide, seismic and fluvial hazards. They then went on to describe a decision support system for planning purposes, which evaluates a number of variables using a GIS. This integrated computer support system, termed Integrated Planning Decision Support System (IPDSS), was designed to assist urban planning by organizing, analysing and evaluating existing or needed spatial data for land-use planning. The system incorporates GIS software, which allows comprehensive modelling capabilities for geological hazards, vulnerability and risk assessment. The IPDSS uses data on topography, aspect, solid and superficial geology, structural geology, geomorphology, soil, land cover and use, hydrology and floods, and historical information on hazards. As a consequence, it has been able to delineate areas of high risk from those where future urban development could take place safely, and is capable of producing hazard susceptibility maps.

Another use of GIS for vulnerability evaluation is given by Hiscock *et al.* (1995). They used a GIS to map groundwater vulnerability. They suggested that examples of the application of a GIS in the assessment of groundwater vulnerability can be grouped into two types. In the first category, raw data or results from groundwater modelling are combined with data characterizing the hazardous activity, so that the risk

of human exposure and/or impact on the environment can be assessed. Data are handled within the GIS in the second category, and are then interfaced with a groundwater model to help input and display large amounts of groundwater data at both preprocessor and postprocessor stages of modelling. Although the resulting maps are of value to planners, they must be interpreted with caution, because the concept of general vulnerability to some universal pollutant does not have much meaning. It therefore is better to assess vulnerability to pollution in terms of a contaminant. Nonetheless, the advantages of a GIS are the ability to integrate multiple layers of data, to derive additional information, to visualize spatial data and to model results. Hiscock *et al.* carried out groundwater vulnerability surveys of the Midlands and southeast England, and recognized four classes of vulnerability, namely, extreme, high, moderate and low (Fig. 1.14).

An appreciation of the physical (e.g. landslip, subsidence) and chemical (e.g. heavy metals, toxic components, gases and organic compounds) hazards affecting an urban area can assist planning, and help to identify those areas of land requiring further investigation or possible remedial action. The British Geological Survey (Bridge *et al.*, 1996) recently completed a pilot urban environmental survey of the Wolverhampton area covering some $70\,km^2$. This multidisci-

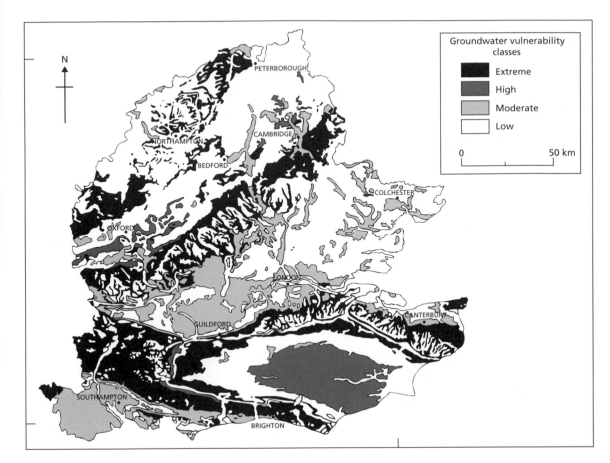

Fig. 1.14 Relative groundwater vulnerability map for southeast England. Large areas of extreme and high vulnerability are associated, respectively, with outcrops of the important Cretaceous Chalk and Oolitic Limestone, and outcrops of the Wealden Beds (Hastings Sands). Areas of moderate and low vulnerability include, respectively, the occasionally important Quaternary sands and gravels and Tertiary beds, and the London Clay and Jurassic clays. Extensive areas of low vulnerability also occur where glacial till or clay-with-flints covers underlying hydrogeological units. (After Hiscock et al., 1995.)

plinary geoscientific survey involved the collection, collation, digitization and integration of several geological, hydrogeological, geochemical and land-use data sets in a PC-based GIS. The system can be manipulated easily and updated to produce customized thematic maps. The data gathered are now being used to rank geohazards in relation to foundation conditions, and for the epidemiological evaluation of possible public health risks from soil contamination.

References

Alexander, D. (1993) *Natural Disasters*. University College London Press Limited, London.

Anon. (1972) The preparation of maps and plans in terms of engineering geology—Working Party Report. *Quarterly Journal of Engineering Geology* 5, 293–381.

Anon. (1976) *Engineering Geological Maps. A Guide to their Preparation*. UNESCO, Paris.

Bell, F.G. (1978) *Foundation Engineering in Difficult Ground*. Butterworths, London.

Bell, F.G. (1980) *Engineering Geology and Geotechnics*. Butterworth, London.

Bell, F.G. (1992) *Engineering Properties of Soils and Rocks*, 3rd edn. Butterworth–Heinemann, Oxford.

Bell, F.G. (1993) *Engineering Treatment of Soils*. Spon, London.

Bell, F.G. (1996) Dereliction: colliery spoil heaps and their restoration. *Environmental and Engineering Geosciences* 2, 85–96.

Bell, F.G., Cripps, J.C., Culshaw, M.G. & O'Hara, M. (1987) Aspects of geology in planning. In *Planning and*

Engineering Geology, Engineering Geology Special Publication No. 4, Culshaw, M.G., Bell, F.G., Cripps, J.C. & O'Hara, M. (eds). Geological Society, London, pp. 1–38.

Blikra, L.H. (1990) Geological mapping of rapid mass movement deposits as an aid to land-use planning. *Engineering Geology* **29**, 365–376.

Bridge, D.M., Brown, M.J. & Hooker, P.J. (1996) Wolverhampton Urban Environmental Survey: an integrated geoscientific study. *British Geological Survey, Technical Report* **WE/95/49**. British Geological Survey, Keyworth.

Briggs, R.P. (1987) Engineering geology and seismic and volcanic hazards in the Hijaz railway regions—Syria, Jordan and Saudi Arabia. *Bulletin of the Association of Engineering Geologists* **24**, 403–423.

Brunsden, D., Doornkamp, J.C., Fookes, P.G., Jones, D.K.C. & Kelly, J.H.M. (1975a) The use of geomorphological mapping techniques in highway engineering. *Journal of the Institution of Highway Engineers* **22**, 35–41.

Brunsden, D., Doornkamp, J.C., Fookes, P.G., Jones, D.K.C. & Kelly, J.H.M. (1975b) Large-scale geomorphological mapping and highway engineering design. *Quarterly Journal of Engineering Geology* **8**, 227–253.

Burrough, P.A. (1986) *Principles of Geographical Information Systems for Land Resource Analysis*. Oxford University Press, Oxford.

CEPAL (1990) Efectos económicos y sociales de los desatres naturales en América Latina. *Taller Regional de Capacitación para Desatres*. PNUD/UNDRO, New York.

Cooke, R.U. & Doornkamp, J.C. (1990) *Geomorphology in Environmental Management*, 2nd edn. Oxford University Press, Oxford.

Culshaw, M.G., Forster, A., Cripps, J.C. & Bell, F.G. (1990) Applied geology maps for land-use planning in Great Britain. In *Proceedings of the 6th Congress of the International Association of Engineering Geology, Amsterdam* **2**, Price, D.G. (ed). Balkema, Rotterdam, pp. 925–935.

Dearman, W.R. (1991) *Engineering Geological Maps*. Butterworth–Heinemann, Oxford.

Dearman, W.R. & Fookes, P.G. (1974) Engineering geological mapping for civil engineering practice in the United Kingdom. *Quarterly Journal of Engineering Geology* **7**, 223–256.

Dearman, W.R. & Matula, M. (1977) Environmental aspects of engineering geological mapping. *Bulletin of the International Association of Engineering Geology* **14**, 141–146.

De Beer, E., Fangnoul, A., Lousberg, E., Nuyens, J. & Maetens, J. (1980) A review on engineering geological mapping in Belgium. *Bulletin of the International Association of Engineering Geology* **21**, 91–98.

Demek, J. (1972) *Manual of Detailed Geomorphological Mapping*. Academia, Prague.

De Mulder, E.F.J. (1996) Urban geoscience. In *Urban Geoscience*, McCall, G.J.H., De Mulder, E.F.J. & Marker, B.R. (eds). Balkema, Rotterdam, pp. 1–11.

De Mulder, E.F.J. & Hillen, R. (1990) Preparation and application of engineering and environmental geological maps in the Netherlands. *Engineering Geology* **29**, 279–290.

Doornkamp, J.C., Brunsden, D., Jones, D.K.C., Cooke, R.U. & Bush, P.R. (1979) Rapid geomorphological assessments for engineering. *Quarterly Journal of Engineering Geology* **12**, 189–204.

Fobe, B. & Goosens, M. (1990) The groundwater vulnerability map for the Flemish region: its principles and uses. *Engineering Geology* **29**, 355–363.

Forster, A. & Culshaw, M.G. (1990) The use of site investigation data for the preparation of engineering geological maps and reports for use by planners and civil engineers. *Engineering Geology* **29**, 347–354.

Foster, H.D. (1980) *Disaster Planning: the Preservation of Life and Property*. Springer, New York.

Goodchild, M.P. (1993) The state of GIS for environmental problem solving. In *Environmental Modelling with GIS*, Goodchild, M.H., Parks, B.O. & Steyaert, L.T. (eds). Oxford University Press, Oxford, pp. 8–15.

Hays, W.W. & Shearer, C.F. (1981) Suggestions for improving decision making to face geological and hydrologic hazards. In *Facing Geological and Hydrologic Hazards — Earth Science Considerations, United States Geological Survey, Professional Paper* **1240-B**, Hays, W.W. (ed). United States Geological Survey, Washington, DC, pp. B103–B108.

Health and Safety Executive (1988) *The Tolerability of Risk from Nuclear Power Stations*. Her Majesty's Stationery Office, London.

Hiscock, K.M., Lovett, A.A., Brainard, J.S. & Parfitt, J.P. (1995) Groundwater vulnerability assessment: two case studies using GIS methodology. *Quarterly Journal of Engineering Geology* **28**, 179–194.

Jefferies, M., Hall, D., Hinchliff, J. & Aiken, M. (1996) Risk assessment: where are we, and where are we going? In *Engineering Geology of Waste Disposal, Engineering Geology Special Publication No.* 11, Bentley, S.P. (ed). Geological Society, London, pp. 341–359.

Kertesz, A. (1979) Representing the morphology of slopes on engineering geomorphological maps with special reference to slope morphometry. *Quarterly Journal of Engineering Geology* **12**, 235–241.

Legget, R.F. (1987) The value of geology in planning. In *Planning in Engineering Geology, Engineering Geology Special Publication No.* 14, Culshaw, M.G., Bell, F.G., Cripps, J.C. & O'Hara, M. (eds). Geological Society, London, pp. 53–58.

Leopold, L.B., Clarke, F.E., Hanshaw, B.B. & Baisley, J.R. (1971) *A Procedure for Evaluating Environmental Impact. United States Geological Survey, Circular* **745**. Department of the Interior, Washington, DC.

Marker, B.R. (1996) Urban development: identifying opportunities and dealing with problems. In *Urban Geoscience*, McCall, G.J.H., De Mulder, E.F.J. & Marker, B.R. (eds). Balkema, Rotterdam, pp. 181–213.

McMillan, A.A. & Browne, M.A.E. (1987) The use and abuse of thematic mining information maps. In *Planning and Engineering Geology, Engineering Geology Special Publication No. 4*, Culshaw, M.G., Bell, F.G., Cripps, J.C. & O'Hara, M. (eds). Geological Society, London, pp. 237–246.

Mejía-Navarro, M. & Garcia, L.A. (1996) Natural hazard and risk assessment using decision support systems, application: Glenwood Springs, Colorado. *Environmental and Engineering Geoscience* 2, 299–324.

Mora, S. (1995) The impact of natural hazards on the socioeconomic development in Costa Rica. *Environmental and Engineering Geoscience* 1, 291–298.

Nickless, E.F.P. (1982) *Environmental Geology of the Glenrothes District, Fife Region, Description of 1:25 000 Sheet No. 20. Report of the British Geological Survey* 82. Her Majesty's Stationery Office, London.

Pitts, J. (1979) Morphological mapping in the Axmouth–Lyme Regis undercliffs, Devon. *Quarterly Journal of Engineering Geology* 8, 159–176.

Robinson, G.D. & Spieker, A.M. (eds) (1978) *Nature to be Commanded—Earth Science Maps Applied to Land and Water Management. United States Geological Survey, Professional Paper* 966. United States Geological Survey, Washington, DC.

Savigear, R.A.G. (1965) A technique of morphological mapping. *Annals of the Association of American Geographers* 55, 514–538.

Seeley, M.W. & West, D.O. (1990) Approach to geological hazard zoning for regional planning, Inyo National Forest, California and Nevada. *Bulletin of the Association of Engineering Geologists* 27, 23–36.

Soule, J.M. (1980) Engineering geology mapping and potential geological hazards in Colorado. *Bulletin of the International Association of Engineering Geology* 21, 121–131.

Star, J. & Estes, J. (1990) *Geographical Information Systems*. Prentice-Hall, Englewood Cliffs, New Jersey.

UNDRO (1991) *Mitigating Natural Disasters: Phenomena and Options. A Manual for Policy Makers and Planners*. United Nations Organization, New York.

Varnes, D.J. (1974) The logic of geological maps with reference to their interpretation and use for engineering purposes. *United States Geological Survey, Professional Paper* 873. United States Geological Survey, Washington, DC.

Varnes, D.J. (1984) *Landslide Hazard Zonation: A Review of Principles and Practice. Natural Hazards* 3. UNESCO, Paris.

Villacis, C., Trucker, B., Yepes, H., Kaneko, F. & Chatelain, J.L. (1997) Use of seismic microzoning for risk management in Quito, Ecuador. *Engineering Geology* 46, 63–70.

Waters, R.S. (1958) Morphological mapping. *Geography* 43, 10–17.

Wolden, K. & Ericksen, E. (1990) Compilation of geological data for use in local planning and administration. *Engineering Geology* 29, 333–338.

Worth, D.H. (1987) Planning for engineering geologists. In *Planning and Engineering Geology, Engineering Geology Special Publication No. 4*, Culshaw, M.G., Bell, F.G., Cripps, J.C. & O'Hara, M. (eds). Geological Society, London, pp. 39–46.

Wu, T.H., Tang, W.H. & Einstein, H.H. (1996) Landslide hazard and risk assessment. In *Landslides, Investigation and Mitigation, Special Report 247*, Turner, A.K. & Schuster, R.L. (eds). Transport Research Board, National Research Council, Washington, DC, pp. 106–117.

2   Volcanic activity

2.1 Introduction

Volcanic zones are associated with the boundaries of the crustal plates (Fig. 2.1). The type of plate boundary offers some indication of the type of volcano which is likely to develop. Plates can be continental, oceanic or both oceanic and continental. Oceanic crust is composed of basaltic material, whereas continental crust varies from granitic to basaltic in composition. At destructive plate margins, oceanic plates are overridden by continental plates. The descent of the oceanic plate, together with any associated sediments, into zones of higher temperature leads to melting and the formation of magmas. Such magmas vary in composition, but some may be comparatively rich in silica. The latter form andesitic magmas and are often responsible for violent eruptions. By contrast, at constructive plate margins, where plates are diverging, the associated volcanic activity is a consequence of magma formation in the upper mantle. The magma is of basaltic composition, which is less viscous than andesitic magma. Hence, there is relatively little explosive activity and associated lava flows are more mobile. However, certain volcanoes, for example, those of the Hawaiian Islands, are located in the centres of plates. They owe their origins to hot spots in the Earth's mantle, which have 'burned' holes through the overlying plates as the plates moved over them.

Eruptions from volcanoes are spasmodic rather than continuous. Between eruptions, activity may still be witnessed in the form of steam and vapours issuing from small vents named fumaroles or solfataras. However, in some volcanoes, even this form of surface manifestation ceases, and such a dormant state may continue for centuries. To all intents and purposes, these volcanoes appear to be extinct. In old age, the activity of a volcano becomes limited to emissions of gases from fumaroles and hot water from geysers and hot springs.

According to Rittmann (1962), basaltic magmas generated in the mantle, presumably in the low-velocity layer, cannot reach the Earth's surface under their own power, but do so via open fissures which act as channels of escape. These fissures are formed by the convection currents in the mantle. The hydrostatic pressure of the magma which penetrates a fissure helps to widen it, and escaping gases from the magma help to clear the channel by blast action. Furthermore, the reduction of pressure on the magma as it ascends causes a lowering of its viscosity, which means that it can penetrate more easily along the fissure. The potential eruptive energy of the magma is governed by the quantity of volatile constituents it contains. Indeed, basaltic magmas formed in the mantle could not reach the continental surface if they lacked volatiles, because their density would be higher than that of the upper regions of the crust. They would consequently obtain a state of hydrostatic equilibrium some kilometres below the surface. The release of original gaseous constituents in the magma lowers its density and thereby aids its ascent.

2.2 Volcanic form and structure

The form and structure adopted by a volcano depend upon the type of magma feeder channel, the character of the material emitted and the number of eruptions which occur. As far as the feeder channel is concerned, this may be either a central vent or a fissure, which give rise to radically different forms. The composition and viscosity of a magma influence its eruption. For instance, acidic magmas are more viscous than basic magmas and therefore gas cannot escape as readily from them. As a consequence, the more acidic, viscous magmas are generally associated with explosive activity and the volcanoes they give rise to may be built mainly of pyroclasts. Alternatively, fluid,

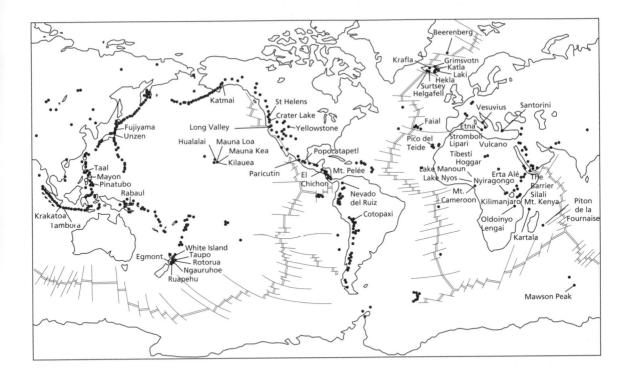

Fig. 2.1 Map of active volcanoes of the world. Tectonic plate boundaries are also indicated.

basaltic magmas construct volcanoes which consist of piles of lava with very little pyroclastic material. Volcanoes built largely of pyroclasts grow in height much more rapidly than those formed from lavas. The number of eruptions which take place from the same vent allows the recognition of monogenetic (single eruption) and polygenetic (multiple eruption) volcanoes. Monogenetic central vent volcanoes are always small and have a simple structure, the eruptive centre moving on after the eruption. Paricutin in Mexico provides an example. Monogenetic volcanoes do not occur alone, but clustered in fields. Polygenetic central vent volcanoes are much larger and more complicated. The influence of the original topography upon their form is obscured. Displacement of the vent frequently occurs in polygenetic types. Fissure volcanoes are always monogenetic.

Some initial volcanic perforations only emit gas, however, the explosive force of the escaping gas may

be sufficient to produce explosion vents (Fig. 2.2). These are usually small in size and are surrounded by angular pyroclastic material formed from the country rock. If the explosive activity is weaker, the country rocks are broken and pulverized in place rather than thrown from the vent. Accordingly, breccia-filled vents are formed.

Pyroclastic volcanoes are formed when viscous magma is erupted explosively. They are often monogenetic and are generally found in groups, their deposits interdigitating with one another. They are small when compared with some shield and composite volcanoes. Some of the earlier formed cones in this category are often destroyed or buried by later outbreaks. In pumice cones, banks of ash often alternate with layers of pumice. Cinder cones are not as common as those of pumice and ash (Fig. 2.3). They are formed by explosive eruptions of the Strombolian or Vulcanian type. Some of these cones are symmetrical, with an almost circular ground plan, the diameter of which may measure several kilometres. They may reach several hundred metres in height. Parasitic cones may arise from the sides of these volcanoes. For example, in 1943, steam began to rise from a

Fig. 2.2 Ubehebe explosion vent, Death Valley, California.

Fig. 2.3 Cinder cone in Lassen Volcanic National Park, California.

field in Mexico and, within 8 months, a cinder cone, 410m in height, that is, Paricutin, was formed. It emitted some 80Mm³ of pyroclastic material, at a maximum rate of 6Mm³ per day, with activity continuing until 1952.

Fissure pyroclastic volcanoes arising from basic and intermediate magmas are not common. By contrast, sheets of ignimbrite of rhyolitic composition have often been erupted from fissures. When enormous quantities of material are ejected from fissures associated with large volcanoes, resulting in an emp-

tying of their magma chambers, they may collapse to form huge volcano-tectonic sinks.

Mixed or composite volcanoes consist of accumulations of lava and pyroclastic material. They have an explosive index in excess of 10 (according to Rittmann (1962), the explosive index (E) is the percentage of fragmentary material in the total material erupted). They are the most common type of volcano.

Strato-volcanoes are polygenetic and consist of alternating layers of lava flows and pyroclasts. The

Fig. 2.4 Crater Lake, Oregon: a caldera.

simplest form of strato-volcano is cone shaped, with concave slopes and a crater at the summit from which eruptions take place. However, as it grows in height, the pressure exerted by the magma against the conduit walls increases, and eventually the sides are ruptured by radial fissures from which new eruptions take place. Cinder cones form around the uppermost centres, whilst lava wells from the lower. The shape of strato-volcanoes may be changed by a number of factors, for example, migration of their vents is not uncommon so that they may exhibit two or more summit craters. Significant changes in form are also brought about by violent explosions which may blow part of the volcano or even the uppermost portion of the magma chamber away. When the latter occurs, the central part of the volcanic structure collapses because of a loss of support. In this way, huge summit craters are formed, in which new volcanoes may subsequently develop.

Most calderas measure several kilometres across and are thought to be formed by the collapse of the superstructure of a volcano into the magma chamber below (Fig. 2.4), as this accounts for the small proportion of pyroclastic deposits which surround the crater. If they had been formed as a result of tremendous explosions, fragmentary material would be commonplace.

Rittmann (1962) used the term lava volcano to include those volcanic structures with an explosive index of less than 10 (usually between two and three).

Monogenetic central lava volcanoes rarely occur as independent structures and are generally developed as parasitic types on the flanks of larger lava volcanoes. The former type of volcano invariably consists of a small cone surrounding the vent from which a lava stream has issued. Polygenetic lava volcanoes are represented by shield volcanoes, such as those of the Hawaiian Islands. These volcanoes are built by successive outpourings of basaltic lava from their lava lakes. The latter occur at the summit of the volcanoes in steep-sided craters. When emitted, the lavas spread in all directions and, because of their fluidity, cover large areas. As a consequence, the slopes of these volcanoes are very shallow, usually between 4° and 6°.

A volcanic dome is a mass of rock formed when viscous lava is extruded from a volcanic vent. The lava is too viscous to flow more than a few tens or hundreds of metres. Domes are usually steep sided, for example, the Puy de Sarcoui in Auvergne reaches a height of 150 m with a base only 400 m wide. Domes may develop at the summit of a volcano, on its flanks or at vents along a fissure at or beyond its base. They may grow rapidly, reaching their maximum size in less than a year. In the centre of a dome, movement is primarily upwards, which frequently causes the sides to become unstable. As a result, the height of domes is generally limited to a few hundred metres. At times, viscous lava may block the vent of a volcano, so that subsequently rising lava accumulates beneath the obstruction and exerts an increasing pressure on the

Fig. 2.5 The basalt plateau in northern Lesotho.

sides of the volcano until they are eventually fissured. The lava is then intruded forcibly through the cracks and emerges to form streams which flow down the flanks of the dome. Cryptodomes (i.e. hidden domes) occur as a result of magmatic intrusion at shallow depth, the hydrostatic pressure of the magma causing the roof material to swell upwards. Eruption of pyroclasts and/or lava occurs if the roof fissures.

The term flood basalt was introduced by Tyrrell (1937) to describe large areas, usually at least 130 000 km², which are covered by basaltic lava flows (Fig. 2.5). These vast outpourings have tended to build up plateaux, for example, the Deccan plateau extends over some 640 000 km² and, at Bombay, reaches a thickness of approximately 3000 m. The individual lava flows which comprise these plateau basalt areas are relatively thin, varying between 5 and 13 m in thickness. In fact, some are less than 1 m thick. They form the vast majority of the sequence, pyroclastic material being of very minor importance. That the lavas were erupted intermittently is shown by their upper parts which are stained red by weathering. Where weathering has proceeded further, red earth or bole has been developed. It is likely that flood basalts were erupted by both fissures and central vent volcanoes. Flood basalts are believed to have been built up by flows from several fissures operating at different times and in different areas, the lavas from which met and overlapped to form a thick succession.

2.3 Types of central eruption

The Hawaiian type of central eruption is characterized in the earliest stages by the effusion of mobile lava flows from places where rifts intersect, the volcano growing with each emission (Fig. 2.6). When eruptive action takes place under the sea, steam-blast activity causes a much higher proportion of pyroclasts to be formed. Above sea level, Hawaiian volcanoes are typified by quiet emissions of lava. This is due to their low viscosity which permits the ready escape of gas. The low viscosity also accounts for the fact that certain flows at outbreak have been observed to travel at speeds of up to 55 km h⁻¹. Lava flows are emitted both from a summit crater and from rift zones on the flanks of the volcano.

If the lava emitted by a volcano is somewhat more viscous than the basaltic type which flows from Hawaiian volcanoes, then, because gas cannot escape as readily, moderate explosions ensue. Clots of lava are thrown into the air to form cinders and bombs, and so pyroclasts accompany the extrusion of lavas. Stromboli typifies this kind of central eruption, hence the designation Strombolian type (Fig. 2.6). After the ejection of lavas, some of the molten material congeals in the vent. At the next eruption, gas breaks through this thin crust and throws blocks of it, together with fragments of lava, into the air. Most of this material falls back into the crater.

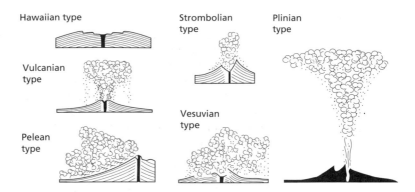

Fig. 2.6 Types of central eruption.

Vulcano gives its name to the Vulcanian type of central eruption (Fig. 2.6). In this type, the lava is somewhat more viscous than that of the Strombolian type and therefore, after an eruption, the lava remaining in the conduit quickly solidifies. This means that gases accumulate beneath the obstruction until they have gathered enough strength to blow it into the air. The fragmented crust, together with exploded lava and gas, gives rise to large clouds. The lavas which are emitted tend to fill the crater and spill over the sides. After a time, they congeal and the process begins again.

The Vesuvian type of central eruption is yet more violent and may be preceded by eruptions of the Strombolian and Vulcanian type (Fig. 2.6). Once again the conduit of the volcano is sealed with a plug of solidified lava, so that great quantities of gas become pent up in the magma below. This may cause lava to escape through fissures or parasitic vents on the slopes of the volcano, and therefore the conduit may be emptied down to a considerable depth. The release of the pressure so created allows gas, held in solution in the remaining magma, to escape. It does so with great force, thereby clearing the conduit of its obstruction, throwing its fragmented remains and exploded lava far into the air, leading to the formation of great clouds of ash. Vesuvian eruptions are followed by periods of long quiescence.

The Plinian type of central eruption is the most violent form of the Vesuvian type (Fig. 2.6). Such an eruption from Vesuvius was responsible for the destruction of Pompeii and Herculaneum in AD 79. This tremendous eruption was recorded by the Roman historian, Pliny, hence the name. The Plinian type of eruption is usually preceded by the Vesuvian and is characterized by an extremely violent upsurge of gas shooting from the neck of the volcano to heights of 12 km or more, where it levels out to form a canopy (Carey & Sigurdsson, 1989). The amount of ash involved in such eruptions is relatively small and consists chiefly of material eroded from the conduit wall by the tremendous gas activity.

In the Pelean type of central eruption, the magma concerned is extremely viscous, which leads to catastrophic explosions (Fig. 2.6). Prior to the eruption of Mt Pelée in 1902, this type of volcanic activity had been unknown. Once again, the neck of the volcano is blocked with congealed lava, so that the magma beneath becomes very highly charged with gas. The pressures developed thereby reach a point at which they have sufficient strength to fissure the weakest zone in the flanks of the volcano. The lava extruded through the fissure is violently exploded by the escape of gas which is held in solution. In this way, a cloud of very hot gas and ash is formed. This rushes with tremendous speed down the side of the volcano, destroying everything in its path. This type of glowing cloud is referred to as a nuée ardente (Fig. 2.7). The dense lower part of a nuée ardente tends to follow the valleys, filling them with pyroclastic debris. This deposits a hot, sandy and/or gravelly mass several metres thick, which may remain hot for months or years. A cloud of steam and fine ash, escaping from the debris, forms the upper part of a nuée ardente. The heavy base surge nuées ardentes, which occurred at Mt Pelée, may have moved at up to 100 km h^{-1} for distances of up to 10 km. A larger form of nuée ardente, referred to as an ash flow or pyroclastic flow, can travel at 200 km h^{-1} for distances of up to 25 km. This is also caused by the expansion of trapped gases

Fig. 2.7 A nuée ardente erupting from Mount St Helens in May 1980, Washington State.

which are released from hot pyroclastic debris. The temperature of a nuée ardente may be around 900°C where eruption occurs, falling to 400°C at the furthest extent of flow. Ash flows may extend up to 10km³ in volume.

2.4 Volcanic products: volatiles

When a magma is erupted, it separates at low pressures into incandescent lava and a gaseous phase. If the magma is viscous (the viscosity is, to a large extent, governed by the silica content), separation is accompanied by explosive activity. On the other hand, volatiles escape quietly from very fluid magmas. The composition of the gases emitted varies from one volcano to another, and from one eruption of the same volcano to another.

Steam accounts for over 90% of the gases emitted during a volcanic eruption. Other gases present include carbon dioxide, carbon monoxide, sulphur dioxide, sulphur trioxide, hydrogen sulphide, hydrogen chloride and hydrogen fluoride. Small quantities of methane, ammonia, nitrogen, hydrogen thiocyanate, carbonyl sulphide, silicon tetrafluoride, ferric chloride, aluminium chloride, ammonium chloride and argon have also been noted in volcanic gases. It has often been found that, next to steam, hydrogen chloride is the major gas emitted during an eruption but, in the later stages, sulphurous gases take over this role. These gases, depending on their concentra-

tions in the air, may be toxic to many animals and plants.

Water tends to move towards the top of a body of magma, where the temperatures and pressures are lower. For example, in a chamber extending from 7 to 14km in depth, the saturation water content would occur at the top, whereas at the bottom it would be 2%. This relationship offers an explanation of a common sequence in volcanic eruptions, that is, a highly explosive opening phase, with a resultant vigorous gas column carrying pumice to a great height, followed by the formation of ash flows and, finally, by the comparatively quiet effusion of lava flows.

Lavas frequently contain bubble-like holes or vesicles which have been left behind by gas action. Even the escape of small quantities of gas is sufficient to cause frothing of the surface of a lava, because pumice can be produced by the exsolution of less than 0.1% of dissolved water.

Solfataras and fumaroles occur in groups and are referred to as solfatara or fumarole fields. They are commonly found in the craters of dormant volcanoes or are associated with volcanoes which have reached the age of senility in their life cycle. Superheated steam may issue continuously from the fissures of larger solfataras and at irregular intervals from the smaller ones. The steam commonly contains carbon dioxide and hydrogen sulphide. Atmospheric oxygen reacts with the hydrogen sulphide to form water and

Fig. 2.8 Travertine terraces, Mammoth Hot Springs, Yellowstone National Park, Wyoming.

free sulphur, the latter being deposited around the steam holes. Sulphuric acid is also formed by the oxidation of hydrogen sulphide and this, together with superheated steam, frequently causes complete decomposition of the rocks in the immediate neighbourhood, leaching out their bases and replacing them with sulphates. Boiling mud pits may be formed on the floor of a crater where steam bubbles through fine dust and ash.

The composition of fumarolic gases depends not only upon their initial composition in the magma, but also upon their temperature, the length of time since they began to form (the more insoluble gases are more abundant in the early emanations), the place at which the gases are emitted (whether from the eruptive vent or from a lava flow), the extent of mixing with air or meteoric water and reactions with air, water and country rock. Cool fumaroles, the gas temperatures of which only exceed the boiling point of water by a few degrees Celsius, are more frequent in occurrence than solfataras. The water vapour emitted generally contains some carbon dioxide, but no hydrogen sulphide. In the coolest fumaroles, the gases arc often only a few degrees Celsius warmer than air temperature, and have either been cooled during their ascent or arise from evaporating thermal water. By contrast, the temperature of steam may reach 900 °C in very hot fumaroles which occur on active volcanoes. These emissions always contain hydrogen chloride and volatile chlorides, particularly sodium chloride and ferric chloride, together with the usual constituents of solfataric gases.

Hot springs are found in all volcanic districts, even some of those in which the volcanoes are extinct. They originate from hot steam and gases given off by masses of intruded magma which are in the last stages of crystallization. On their passage to the surface, the steam and gases often encounter groundwater which is heated and forms part of the hot springs. Many hot springs contain carbon dioxide and hydrogen sulphide, together with dissolved salts. Indeed, some very hot springs found in active volcanic districts may contain dissolved silica, which on cooling is deposited to form sinter terraces. If the hot springs contain dissolved calcium carbonate, this is precipitated to form travertine terraces (Fig. 2.8). Geysers are hot springs from which a column of hot water is explosively discharged at intervals, the water spout in some cases rising over 100 m (Fig. 2.9). The periodicity of their ejections varies from a matter of minutes to many days and changes with time. Geysers are generally short lived. Again, boiling mud pits occur in association with hot springs (Fig. 2.10).

2.5 Volcanic products: pyroclasts

The term *pyroclast* is collectively applied to material which has been fragmented by explosive volcanic action. Tephra is a synonym for the phrase 'pyroclastic material'. Pyroclasts may consist of fragments of

Fig. 2.9 Pohutu geyser, near Rotorua, North Island, New Zealand.

Fig. 2.10 Bubbling mud pits, Lassen Volcanic National Park, California.

lava exploded on eruption, fragments of pre-existing solidified lava or fragments of country rock which, in the last two instances, have been blown from the neck of the volcano. These three types have been distinguished as essential, accessory and accidental ejectamenta, respectively.

The size of pyroclasts varies enormously. It is dependent upon the viscosity of the magma, the violence of the explosive activity, the amount of gas coming out of solution during the flight of the pyroclast and the height to which it is thrown. The largest blocks thrown into the air may weigh over

100 tonnes, whereas the smallest consist of very fine ash which may take years to fall back to the Earth's surface. The largest pyroclasts are referred to as volcanic bombs. These consist of clots of lava or fragments of wall rock. They may fall in significant quantities within a 5 km radius of the eruption. The bombs may destroy structures on which they fall and incandescent bombs may ignite homes, crops or woodlands.

The term lapilli is applied to pyroclastic material which has a diameter varying from approximately 10 to 50 mm (Fig. 2.11). Cinder or scoria is irregularly shaped material of lapilli size. It is usually glassy and fairly to highly vesicular and represents the ejected froth of a magma. Lapilli can be ejected during an explosive eruption over a radius of several kilometres from a volcano.

The finest pyroclastic material is called ash. Beds of ash commonly show lateral variation as well as verti-cal. In other words, with increasing distance from the vent, the ash becomes finer and, in the latter case, because the heavier material falls first, ash frequently exhibits graded bedding, coarser material occurring at the base of a bed, becoming finer towards the top. Reverse grading may occur as a consequence of an increase in the violence of eruption or changes in wind velocity. The spatial distribution of ash is very much influenced by the wind direction, and deposits on the leeward side of a volcano may be much more extensive than those on the windward side; indeed, they may be virtually absent from the latter side. Because the fall of ash can be widespread, it can cover or bury houses and farmland, cause roofs to collapse, ruin crops and block streams and sewers (Fig. 2.12). Deposits of ash may suffer very rapid erosion by streams, which eventually may be blocked with debris. Ash is frequently metastable and, on wetting, is subject to hydrocompaction.

Rocks which consist of fragments of volcanic ejectamenta set in a fine grained groundmass are referred to as agglomerate or volcanic breccia, depending on whether the fragments are rounded or angular, respectively.

After pyroclastic material has fallen back to the surface, it eventually becomes indurated. It is then described as tuff. According to the material of which tuff is composed, distinction can be drawn between ash tuff, pumiceous tuff and tuff breccia. Tuffs are usually well bedded, and the deposits of individual eruptions may be separated by thin bands of fossil soil or old erosion surfaces. Chaotic tuffs are formed from the deposits of glowing clouds and mud streams. Glowing clouds give rise to chaotic tuffs in which blocks of all dimensions are present, together with very fine ash. Lenses of breccia, pumice and volcanic sand are found in chaotic tuffs formed by mudflows and indicate that a certain amount of incomplete sorting has occurred during flow. Pyroclastic deposits which accumulate beneath the sea are often mixed with a varying amount of sediment and are referred to as tuffites. They are generally well sorted and well bedded.

When clouds or showers of intensely heated, incandescent lava spray fall to the ground, they weld together. Because the particles become intimately fused with each other, they attain a largely pseudoviscous state, especially in the deeper parts of the deposit. The resultant massive rock frequently

Fig. 2.11 Lapilli, near Crater Lake, Oregon.

Fig. 2.12 Excavation of an old hotel which collapsed beneath the weight of ash erupted by Tarawera in 1886, North Island, New Zealand.

exhibits columnar jointing. The term ignimbrite is used to describe these rocks. If ignimbrites are deposited on a steep slope, they begin to flow, and hence resemble lava flows. The considerable mobility of some pyroclastic flows, which allows them to move over distances measured in tens of kilometres, has been explained by the process of fluidization. Fluidization involves the rapid escape of gas in which pyroclastic material becomes suspended. Ignimbrites are associated with nuées ardentes.

2.6 Volcanic products: lava flows

Lavas are emitted from volcanoes at temperatures only slightly above their freezing point (MacDonald, 1972). During the course of their flow, the temperature falls from within outwards until solidification takes place somewhere between 600 and 900°C, depending upon the chemical composition and gas content. Basic lavas solidify at a higher temperature than acidic lavas.

Generally, flow within a lava stream is laminar. The rate of flow of a lava is determined by the gradient of the slope down which it moves and by its viscosity, which, in turn, is governed by its composition, temperature and volatile content. In addition, lava flows fastest near its source and becomes progressively slower at increasing distance from the source, because it cools on contact with the ground and atmosphere. It has long been realized that the greater

the silica content of a lava, the greater is its viscosity. Thus basic lavas flow much faster and further than acidic lavas. Indeed, the former have been known to travel at speeds of over 80 km h⁻¹. Basaltic lavas may extend more than 50 km from their source, whereas andesitic flows rarely move over more than 20 km. Dacite and rhyolite lavas, because of their normally low discharge rate, rarely form long flows. They typically produce domes and short, thick flows.

The upper surface of a recently solidified lava flow develops a hummocky and ropy (*pahoehoe*), rough, fragmental, clinkery and spiny (*aa*) or blocky structure. The reasons for the formation of these different structures are not fully understood, but certainly the physical properties of the lava and the amount of disturbance it has to undergo must play an important part. *Pahoehoe* is the most fundamental type; however, a certain distance downslope from the vent, it may give way to *aa* or block lava. In other cases, *aa* or block lava may be traceable into the vent. It would appear that the change from *pahoehoe* to *aa* takes place as a result of increasing viscosity or stirring of the lava. Increasing viscosity occurs due to the loss of volatiles, cooling and progressive crystallization, whilst a lava flow may be stirred by an increase in the gradient of the slope down which it is travelling. Moreover, if strong fountaining occurs in a lava whilst it is in the vent, this increases stirring and either it may issue as an *aa* flow or the likelihood of it changing to *aa* is accordingly increased. Melts that give rise to block lavas are more viscous than those which form *aa*, for example, they are typically andesitic, although many are basaltic.

Pahoehoe is typified by a smooth, billowy or rolling and locally ropy surface (Fig. 2.13). Such surfaces are developed by dragging, folding and twisting of the still plastic crust of the flow due to the movement of the liquid lava beneath. The ridges of ropy lava are commonly curved, the convex sides pointing in the downstream direction of the flow.

With the exception of flows of flood eruptions, all large *pahoehoe* flows consist of several units. Large flows are fed by a complex of internal streams beneath the crust, each stream being surrounded by less mobile lava. When the supply of lava is exhausted, the stream of liquid may drain out of the tunnel through which it has been flowing (Fig. 2.14).

Aa lava flows are characterized by very rough fragmental surfaces (Fig. 2.15). The fragments are commonly referred to as clinker and have numerous sharp, jagged spines. Clinker is not readily compacted. *Aa* flows are fed by streams that lie approximately in the centre of each flow. These streams are usually a few metres in width. Levées, a metre or so in height, may be constructed along the sides of the streams as a result of numerous overflows congealing. These streams are only rarely roofed over by solidified lava and, if they are, it is only for a short distance. The stream retains a close connection with the interior of the flow beneath the clinker. After an eruption,

Fig. 2.13 Ropey lava flow (pahoehoe), Craters of the Moon, Idaho.

Fig. 2.14 Collapsed tunnel in lava, Craters of the Moon, Idaho.

Fig. 2.15 Aa lava flows, with a pressure ridge in the centre of the photograph, Craters of the Moon, Idaho.

much of the lava drains out of the stream and leaves behind a channel which may be several metres deep. A steep bank of clinker several metres high develops at the margin of a slowly moving *aa* flow. Movement takes place by a part of the front slowly bulging forward until it becomes unstable and separates from the parent body. This process is repeated over and over again, forming a talus of clinker deposits along the foot of the lava front. These deposits are eventually buried under the flow.

Although the term block lava refers to lavas with fragmented surfaces, it is usually restricted to those flows in which the fragments are much more regular in shape than those of *aa* flows. Most of the flows of orogenic regions are of block lava. The individual fragments of block lava are polyhedral in shape and have smooth surfaces, although some may develop spines. Block lava flows have very uneven surfaces. Moreover, they generally exhibit a series of fairly regularly arranged ridges at right angles to the direction of flow. The ridges may be several metres high and are covered with blocky rubble. Although a central massive layer is usually present, the fragmental material at times may form the whole of the flow.

Pillow lavas are formed when lavas are erupted beneath the sea. The sea rapidly chills the lava, forming an outer skin. This skin takes the form of a pillow filled with lava. Thus the outside edge of the pillow is fine grained, whilst the centre which cools more slowly is coarser grained and possesses radial cooling cracks (Fig. 2.16).

2.7 Mudflows or lahars

A lahar is a flowing slurry of volcanic debris and water which originates on a volcano. The volcanic debris accounts for 40–80% by weight. Those in which 50% of the volcanic debris is of sand size or smaller are termed mudflows. Lahars may move by turbulent flow, however, as the proportion of debris increases, laminar flow replaces turbulent flow. Mudflows or lahars (Fig. 2.17) are regarded as primary when they result directly from eruption and secondary if there are other causes. Volcanic lahars are generally formed by the spontaneous release of a large amount of water (Verstappen, 1992). For example, deposits of loose or fragmented volcanic ejectamenta, frequently in the form of ash or lapilli, but possibly including large boulders, may become saturated by torrential rain and/or melted snow associated with an eruption. Lahars also may be generated by rainfall several months or years after the last eruption. Alternatively, crater lake lahars, which occur periodically in Indonesia, result from the discharge of huge volumes of water from craters during the course of eruption, giving rise to highly destructive hot lahars. Streams may be impounded by pyroclastic material or lava, the dam eventually being breached and releasing copious quantities of water

Fig. 2.16 Pillow lava overlying mudstone at Muriwai Beach, near Auckland, North Island, New Zealand.

Fig. 2.17 Mudflow, Mount St Helens, Washington State.

for mudflow formation, or, more simply, pyroclastic flows may mix with river water to form a lahar. Lahars can occur at any time before, during or after an eruption and may consist of hot or cold material. The properties of lahars vary depending on whether the material is hot or cold. Both are very poorly sorted, but tend to fine upwards. Cold lahars may be regarded as similar to debris flows. Such debris flows are usually triggered by torrential rain storms, earthquakes or slope failures. Hot lahars are accompanied by the release of large quantities of steam and lesser amounts of volcanic gases. They are usually more destructive than cold lahars. Hot pyroclastic material may accumulate in the ravines around the summit of a volcano after nuées ardentes. Subsequent heavy rain results in lahars forming from the pyroclastic material. The latter may still be very hot (several hundred degrees Celsius), so that steam is formed, resulting in a boiling mixture of mud and water. The release of steam lessens the frictional drag, thereby allowing the lahar to travel further. Lahars can be long (over 100 km) and one mudflow can overlie another, having followed the same route as the previous one. They may be several kilometres in width. The rate of flow may reach $50 \, km \, h^{-1}$ on relatively shallow slopes with viscous material, or twice that down steep slopes with low-viscosity debris.

A lahar often moves in a series of waves or pulses. At the break of slope, lahars tend to spread out to form fans which may interfere with natural drainage channels. Further downslope, beyond the usual extent of coarse lahar flow, the streams frequently carry large quantities of sediment. This may cause the rivers they flow into to aggrade their beds, which can cause flooding. Estuaries may silt up.

2.8 Volcanic hazard and prediction

The obvious first step in any attempt at predicting volcanic eruption is to determine whether the volcano concerned is active or extinct. If a volcano has erupted during historic times, it should be regarded as active. However, many volcanoes which have long periods of repose will not have erupted in historic times. Nonetheless, they are likely to erupt at some future date. It has been estimated that the active lifespan of most volcanoes is probably between one and two million years (Baker, 1979).

The term extinct is commonly applied to volcanoes that have not erupted during historic times. However, catastrophic eruptions of volcanoes believed to have been extinct have occurred. Indeed, Crandell *et al.* (1984) maintained that one so-called extinct volcano comes to life every decade. The probability that a volcano will or will not erupt can be assessed from the prehistoric record, which involves a geological investigation of the volcanic materials associated with a volcano, as well as the consideration of any historical evidence. The probability can be expressed in terms of the ratio between the time which has elapsed since the last eruption and the length of the longest repose period of the volcano. The smaller this ratio, the greater the probability of a future eruption. Active volcanoes should be kept under surveillance, especially those which present short-term hazards, that is,

volcanic events which are likely to occur within the lifetime of a person.

Monitoring the activity of a volcano and mapping and interpreting previous eruptions aid in the delineation of volcanic hazard zones. This includes the recording of the direction and lateral distance reached by lethal and non-lethal ejectamenta, the build-up of potential lahar material, the path taken by lava flows and their maximum extension, and the distances and direction of nuées ardentes. Aerial photographs and satellite imagery can be used to record lahar, lava flow and ash plume development.

The determination of the expected recurrence interval of particular types of eruption, the distribution of deposits, the magnitude of events and the recognition of short-term cycles or patterns of volcanic activity are all of value as far as prediction is concerned. It must be admitted, however, that no volcano has revealed a recognizable cycle of activity which could be used to predict the time of an eruption within a decade. Nonetheless, individual volcanoes require mapping so that their evolution can be reconstructed. The geological data so gathered are used in conjunction with historical records, where available, to postulate future events. If it is assumed that future volcanic activity will be similar to previously observed activity, it is possible to make certain general predictions about hazard.

Tazieff (1988) pointed out that no volcanic catastrophes have occurred at the very start of an eruption. Consequently, this affords a certain length of time to take protective measures. Even so, because less than one out of several hundred eruptions proves dangerous for a neighbouring population, evacuation, presuming that an accurate prediction could be made, would not take place before an eruption became alarming. Nevertheless, it is still important to predict whether or not a developing eruption will culminate in a dangerous climax and, if it does, when and how. According to Tazieff, it is more difficult to predict the evolution of a developing eruption than to predict the initial outbreak.

2.8.1 Methods of prediction

Prediction studies are based on the detection of reliable premonitory symptoms of eruptions (Gorshkov, 1971). Such forerunners may be geophysical or geochemical in nature (Mori *et al.*, 1989). Geophysical observations, principally tiltmetry, seismography and thermometry, provide the basis for forecasting volcanic eruptions (Anon., 1971). Unfortunately, however, because of the cost involved and the problem of accessibility, intensive monitoring only takes place at a small number of volcanoes (about 12) so that these are the ones for which full-scale forecasts can be made. Even in these cases, it is not possible to determine the size and time of an eruption. If the prediction of volcanic eruptions is to be successful, it must provide sufficient time for safety measures to be put into effect.

Data from satellites can be used to monitor volcanic activity, especially in developing countries (Rothery, 1992). Three aspects of volcanism, namely the movement of eruption plumes, changes in the thermal radiation of the ground and the output of gases, can be measured using remote sensing data, so that, in some instances, prior warning of eruptions may be obtained.

A volcanic eruption involves the transfer towards the surface of millions of tonnes of magma. This leads to the volcano concerned undergoing changes in elevation, notably uplift. For instance, the summit of Kilauea, Hawaii, has been known to rise by almost a metre in the months preceding an eruption. Changes in elevation significantly larger than this have been associated with other volcanoes. For example, the side of Mount St Helens bulged at up to 2 m per day, being finally elevated through 100 m. The uplift is usually measured by a network of geodimeters and tiltmeters set up around a volcano. Gravity meters can be used to detect any vertical swelling. Tide gauges are used to measure changes in the level of the lake water in Rabaul caldera in Papua New Guinea.

Such uplift also means that rocks are fractured, so that volcanic eruptions are generally preceded by seismic activity. However, this is not always the case. For instance, this did not happen at Heimaey, Iceland, in 1973. Conversely, earthquake swarms need not be followed by eruptions; the tremors which were felt on Guadeloupe in 1976 were not followed by an eruption. Where earthquake swarms do occur, the number of tremors increases as the time of eruption approaches. For example, tremors average six per day on Kilauea but, at the beginning of 1955, these increased markedly, 600 being detected on February 26, 1955. Two days later an eruption occurred.

However, tremors may continue for a few days to a year or more. A network of seismic stations is set up to monitor the tremors and, from the data obtained, the position and depth of origin of the tremors can be ascertained. The harmonic tremor is a characteristic form of volcano-seismicity. It is a narrow band of nearly continuous seismic vibrations dominated by a single frequency, and is associated with the rise of magma or volcanically heated fluids.

Infrared techniques have been used in the prediction of volcanic eruptions because, due to the rising magma, the volcano area usually becomes hotter than its surroundings. Thermal maps of volcanoes can be produced quickly by ground-based surveys using infrared telescopes (Francis, 1979). However, consistent monitoring is necessary in order to distinguish between real and apparent thermal anomalies. Aerial surveys provide better data, but are too expensive to be used for routine monitoring.

According to Baker (1979), basaltic magmas of low viscosity prove to be an exception if their ascent is more rapid than the rate at which heat is conducted from them. Indeed, he suggested that thermal anomalies are more likely to be produced by rising andesitic or rhyolitic magmas. A more or less contrary view has been advanced by Tazieff (1979). He maintained that thermal techniques have not proved to be satisfactory when monitoring explosive andesitic and dacitic volcanoes, because the ascent of highly viscous acidic magmas is presumably too slow to give rise to easily detectable temperature changes. Perhaps both workers are correct in that the rise of magma can be too rapid or too slow to produce thermal anomalies which can be rapidly monitored.

Be that as it may, detectable anomalies are more likely to develop when heat is transferred by circulating groundwater rather than by conduction from a magma. The temperatures of hot springs may rise, as may the water temperatures in crater lakes, with the approach of a volcanic eruption. Hot groundwater also gives rise to the appearance of new fumaroles or to an increase in the temperatures at the existing fumaroles. Again, caution must be exercised, because fumarole temperatures vary as a result of other factors, such as the amount of rain which has fallen. Steam gauges are used to monitor changes in gas temperatures, pH values and amounts of suspended mineral matter. The eruption of Taal in the Philippines in 1965 was predicted because of the rise in temperature of the water in the crater lake. This allowed evacuation to take place.

An increase in temperature also leads to the demagnetization of rock, as the magnetic minerals are heated above their Curie points. This can be monitored by magnetic surveying. Changes also occur in the gravitational and electrical properties of rocks.

Geophysical methods, however, cannot detect the climax of an eruption. This is because the magma is already very close to the surface, and therefore rock fracturing, volcano inflation and increased heat transfer are not significant enough to record. As pointed out above, it is more important to forecast the climax and its character, rather than the outbreak of an eruption.

The evolution of an eruption involves changes in matter and energy. The most significant variations take place in the gas phase, and therefore it seems appropriate to gather as much information relating to this phase as possible, especially when this phase is the active agent of the eruptive phenomenon. Gas sensors can be stationed on the ground or carried in aircraft, and record the changes in the composition of gases. Unfortunately, reliable data are extremely scarce and the interpretation of what is available is in its infancy. Gas emitted from fumaroles on the flanks of a volcano may contain increases in HCl, HF and SO_2, or the ratios of, for example, S to Cl may change. These may be related to an impending eruption as the proximity of magma tends to emit more sulphur, chlorine and fluorine. Moreover, an increase in the dissolution of acidic volcanic gases in the hydrothermal system often results in small decreases in the pH value. In addition, lake water chemistry may alter, for example, at Ruapehu volcano, New Zealand, the ratio of magnesium to chlorine increased by 25–33% before an eruption.

Monitoring rainfall and storm build-up is necessary for lahar prediction. Telemetered rain gauges may be used, which are activated when a certain amount of rain falls in a given time.

2.8.2 Assessment of volcanic hazard and risk

When volcanic activity occurs in areas of high population density, it poses various kinds of hazard to the people living in the vicinity. It is impossible to restrict people from all hazardous areas around volcanoes, especially those which are active only intermittently. Hence, it is important to recognize the various types of hazard which may occur, in order to prevent or mitigate disasters, but to allow people to live on the fertile lower slopes of volcanoes.

Ten per cent of the population of the world live on or near potentially active volcanoes, at least 91 of which are in high-risk areas (42 in southeast Asia and the western Pacific, 42 in the Americas and seven in Europe and Africa). An eruption or the precarious conditions which it creates may continue for months, and therefore volcanic emergencies are often long lasting in comparison with other sudden impact natural disasters. Nonetheless, most dangerous volcanic phenomena happen very quickly. For instance, the time interval between the beginning of an eruption and the appearance of the first nuées ardentes may be only a matter of hours. Fortunately, such events are usually preceded by visible signs of eruption.

In any assessment of risk due to volcanic activity, the number of lives at stake, the capital value of the property and the productive capacity of the area concerned have to be taken into account. Evacuation from danger areas is possible if enough time is available. However, the vulnerability of property is frequently close to 100% in the case of most violent volcanic eruptions. Hazard must also be taken into account in such an assessment (Tilling, 1989). It is a complex function of the probability of eruptions of various intensities at a given volcano and of the location of the site in question with respect to the volcano (Bondal & Robin, 1989; Forgione *et al.*, 1989). Hazard is one of the most difficult factors to estimate, mainly because violent eruptions are rare events about which there are insufficient observational data for effective analysis. For example, in the case of many volcanoes, large eruptions occur at intervals of hundreds or thousands of years. Thorough stratigraphic study and dating of the deposits will help to provide the evidence needed to calculate the risk factors of such volcanoes. Because, in the foreseeable future, man is unlikely to influence the degree of hazard, the reduction of risk can only be achieved by reducing the exposure of life and property to volcanic hazards. This can be assessed by balancing the loss of income resulting from the non-exploitation of a particular area against the risk of loss in the event of an eruption.

Booth (1979) divided volcanic hazards into six categories, namely premonitory earthquakes, pyroclast falls, pyroclast flows and surges, lava flows, structural collapse and associated hazards. Each type represents a specific phase of activity during a major eruptive cycle of a polygenetic volcano and may occur singly or in combination with other types. Damage resulting from volcano-seismic activity is rare. However, intensities on the Mercalli scale in the range 6–9 have been recorded over limited areas.

Pyroclastic fall deposits may consist of bombs, scoria, lapilli, pumice, dense lithic material, crystals and/or any combination of these. There are, on average, about 60 pyroclast or tephra falls per century which are of social importance. In violent eruptions, intense falls of ash interrupt human activities and cause serious damage. The size of the area affected by pyroclastic falls depends on the amount of material ejected and the height to which it is thrown, as well as the wind speed and direction. They can affect areas up to several tens of kilometres from a volcano within a few hours from the commencement of an eruption. The principal hazards to property resulting from pyroclastic falls are burial, impact damage and fire if the material has a high temperature. The latter hazard is most dangerous within a few kilometres of the vent. The weight of the material collected on roofs may cause them to collapse.

Laterally directed blasts are among the most destructive of volcanic phenomena (Fig. 2.18). They travel at high speeds, for example, the lateral blast associated with the Mount St Helens eruption in 1980 initially travelled at 600 km h^{-1}, slowing to 100 km h^{-1} some 25 km from the volcano. They can occur with little or no warning in a period of a few minutes and can affect hundreds of square kilometres. The material carried by lateral blasts can vary from cold to temperatures high enough to scorch vegetation and start fires. Such blasts kill virtually all life by impact, abrasion, burial and heat. A high concentration of ash contaminates the air.

Pyroclastic flows are hot, dry masses of clastic volcanic material which move over the ground surface. Most pyroclastic flows consist of a dense basal flow, the pyroclastic flow proper, one or more pyroclastic surges and clouds of ash. Two major types of pyroclastic flow may be recognized. Pumiceous pyroclastic flows are concentrated mixtures of hot to incandescent pumice, mainly of ash and lapilli size. Ash flow tuffs and ignimbrites are associated with these flows. Individual flows vary in length from less than 1 km up to some 200 km, covering areas up to 20 000 km^2, with volumes from less than 0.001 km^3 to over 1000 km^3. Pyroclastic flows formed primarily of scoriaceous or lithic volcanic debris are known as hot avalanches, glowing avalanches, nuées ardentes or block and ash flows. They generally affect a narrow sector of a volcano, perhaps only a single valley. The maximum temperature of pyroclastic flow material soon after it has been deposited may range from 350 to 550 °C. Hence, pyroclastic flows are hot enough to kill anything in their path. Because of their high

Fig. 2.18 Abandoned vehicles and blown over trees near the fringe of the devastation caused by a lateral blast, Mount St Helens, Washington State.

mobility (up to $160\,km\,h^{-1}$ on the steeper slopes of volcanoes), they constitute a great potential danger to many populated areas. Other hazards associated with pyroclastic flows, apart from incineration, include burial, impact damage and asphyxiation.

A pyroclastic surge is a turbulent, low-density cloud of gases and rock debris which hugs the ground over which it moves. Hot pyroclastic surges can originate by explosive disruption of volcanic domes caused by rapidly escaping gases under high pressure or by collapse of the flank of a dome. They can also be caused by lateral explosive blast. Hot pyroclastic surges can occur together with pyroclastic flows. Generally, surges are confined to a narrow valley of a volcano, but they may reach speeds of up to $300\,km\,h^{-1}$, so that escape from them is virtually impossible. They give rise to similar hazards to pyroclastic flows. Cold pyroclastic surges are produced by phreatic and phreatomagmatic explosions. Vertical explosions can give rise to a primary surge which moves away from the volcano in all directions. Subsequently, secondary surges may be formed when volcanic material falls to the ground. The high speed of primary surges is attributable to the explosive force, whilst that of secondary surges is due to the kinetic energy gained during fall and the speed gathered as they descend the slopes of the volcano. Surges rapidly decelerate and tend not to travel more than $10\,km$ from their source. Fortunately, pyroclastic flows and surges tend to affect limited areas. Approximately 20 pyroclastic flows and surges occur every 100 years.

The distance which a lahar may travel depends on its volume and water content on the one hand, and the gradient of the slope of the volcano down which it moves on the other. Some may travel for more than $100\,km$. The speed of a lahar is also influenced by the water to sediment ratio and volume, as well as the gradient and the shape of the channel along which it moves. Because of their high bulk density, lahars can destroy structures in their path and block highways. They can reduce the channel capacity of a river and so cause flooding, as well as adding to the sediment load of a river. Hence, lahars may prove to be as destructive as pyroclastic flows over limited areas. Destructive lahars average 50 per century.

Lava effusions of social consequence average 60 per century. Fortunately, because their rate of flow is usually sufficiently slow and along courses which are predetermined by topography, they rarely pose a serious threat to life. The arrival of a lava flow along its course can be predicted if the rate of lava emission and movement can be determined. Damage to property, however, may be complete, destruction occurring by burning, crushing or burial of structures along the path of the flow. Burial of land by lava flows commonly means that its previous use is terminated. Lava flood eruptions are the most serious and may cover large areas with immense volumes of lava, for example, the Laki eruption in Iceland in 1783 produced $12.3\,km^3$ of lava which spread over $560\,km^2$, the flows reaching a maximum length of $65\,km$ from the fissure (Thorarinsson, 1970).

The likelihood of a given location being inundated with lava at a given time can be estimated from information relating to the periodicity of eruptions in time and space, the distribution of rift zones on the flanks of a volcano, the topographic constraints on the directions of flow of lava and the rate of covering of the volcano by lava. The length of a lava flow is dependent upon the rate of eruption, the viscosity of the lava and the topography of the area involved. Given the rate of eruption, it may be possible to estimate the length of flow. Each new eruption of lava alters the topography of the slopes of a volcano to a certain extent and therefore flow paths may change. Moreover, prolonged eruptions of lava may eventually surmount obstacles which lie in the path of the flow and which act as temporary dams. This may then mean that the lava invades areas which were formerly considered to be safe.

The formation of calderas and landslip scars due to the structural collapse of large volcanoes is rare (0.5–1 per century). They are frequently caused by magma reservoirs being evacuated during violent Plinian eruptions. Because calderas develop near the summits of volcanoes and subside progressively as evacuation of their magma chambers takes place, caldera collapse does not offer such a threat to life and property as sector collapse. Sector collapse involves subsidence of a large area of a volcano. It takes place over a comparatively short period of time and may involve volumes of up to tens of cubic kilometres. Collapses which give rise to landslides may be triggered by volcanic explosions or associated earthquakes. For instance, a collapse on the north of Mount St Helens in 1980 was caused by an earthquake ($M_L \approx 5$). According to Voight et al. (1981), the volume of material involved in the landslide was around 2.8 km³ and covered an area of some 60 km². It moved about 25 km from the volcano.

Hazards associated with volcanic activity include destructive floods, caused by the sudden melting of snow and ice which cap high volcanoes or by heavy downfalls of rain (vast quantities of steam may be given off during an eruption), or the rapid collapse of a crater lake. Far more dangerous are the tsunamis (see Chapter 6) generated by violent explosive eruptions and sector collapse. Tsunamis may devastate coastal areas. Dense poisonous gases normally offer a greater threat to livestock than to humans. Nonethe-

less, the large amounts of carbon dioxide released at Lake Manoun in 1984 and neighbouring Lake Nyos in 1986, in Cameroon, led to the deaths of 37 and 1887 people, respectively (Kling *et al.*, 1987). According to Le Guern *et al.* (1982), 149 people were killed by the emission of gases, notably carbon dioxide and hydrogen sulphide, on the Dieng Plateau, Indonesia, in 1979. The gases moved downslope from vents and accumulated in low-lying areas. A hazard zone map was prepared for the area after this catastrophe and was based on known sites of gas emission and topography (Fig. 2.19). In addition,

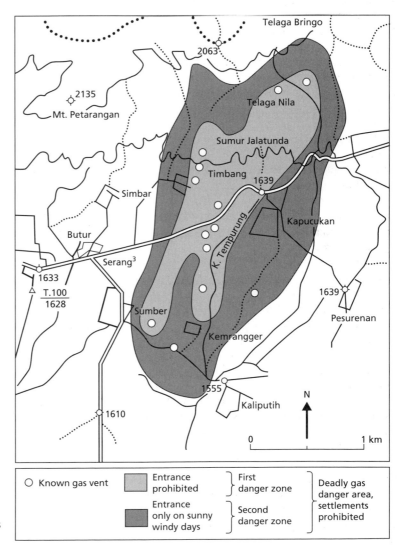

Fig. 2.19 Hazard zone map for volcanic gases for an area in the Dieng Mountains of Java. (From Crandell *et al.*, 1984.)

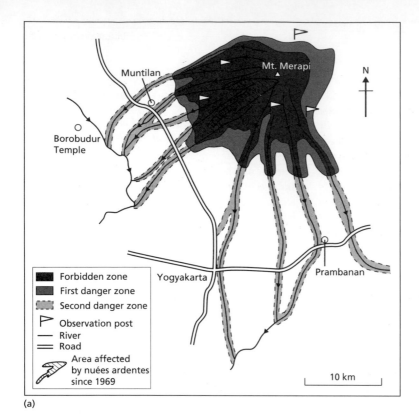

(a)

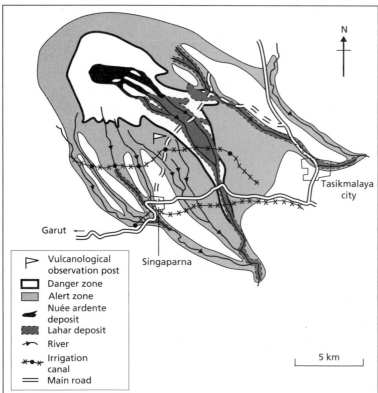

(b)

Fig. 2.20 (a) Volcanic hazard map, Merapi volcano, Java. (b) Preliminary hazard map of Galunggung volcano, Java showing extent of the nuée ardente and lahar in 1982. (After Suryo & Clarke, 1985.)

gases can be injurious to persons mainly because of the effects of acidic compounds on the eyes, skin and respiratory system. They can kill crops. Acid rain can form as a result of rain mixing with aerosols and gases adhering to tephra. Such rain can cause severe damage to natural vegetation and crops, and skin irritation to people. Air blasts, shock waves and counterblasts are relatively minor hazards, although they can break windows several tens of kilometres away from major eruptions.

Hazard zoning involves the mapping of deposits which have formed during particular phases of volcanic activity and their extrapolation to identify the areas which would be likely to suffer a similar fate at some future time. The zone limits on such maps normally assume that future volcanic activity will be similar to that recorded in the past. Unfortunately, this usually is not the case.

The volcanic hazard maps produced by the Volcanological Survey of Indonesia define three zones, namely, the forbidden zone, the first danger zone and the second danger zone (Suryo & Clarke, 1985). The forbidden zone is meant to be abandoned permanently because it is affected by nuées ardentes (Fig. 2.20a). The first danger zone is not affected by nuées ardentes, but may be affected by bombs. Finally, the second danger zone is that likely to be affected by lahars. This zone is subdivided into an abandoned zone, from which there is no escape from lahars, and an alert zone, where people are warned and from where evacuation may be necessary. Preliminary volcanic hazard maps are prepared where fewer data are available. These distinguish danger zones, which must be evacuated immediately increased activity occurs, and alert zones (Fig. 2.20b).

Volcanic risk maps indicate the specified maximum extents of particular hazards, such as lava and pyroclastic flow paths, expected ash-fall depths and the areal extent of lithic missile fallout (Fig. 2.21). They

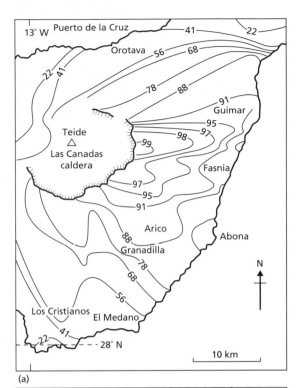

(a)

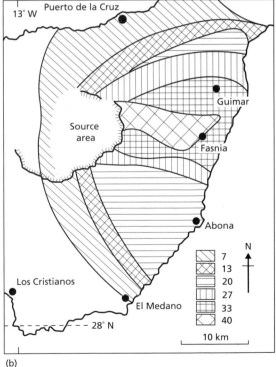

(b)

Fig. 2.21 (a) Southeastern Tenerife, showing the extent and percentage probability of burial by more than 1 m of airfall ash if a Plinian eruption occurs. Figures are based on deposits formed during 27 eruptions. The shaded area of Las Canadas shows the distribution of welded airfall deposits on the southern wall of the caldera and their probable former extent to the north. (After Booth, 1979.) (b) Frequency with which specific areas on Tenerife have been significantly affected by airfall deposits during the last 50 000 years.

are needed by local and national governments so that appropriate land uses, building codes and civil defence responses can be incorporated into planning procedures.

It has been suggested by Fournier d'Albe (1979) that events with mean recurrence intervals (MRIs) of less than 5000 years should be taken into account in the production of maps of volcanic hazard zoning. He further suggested that data on any events which have taken place in the last 50 000 years are probably significant. He proposed that two types of map would be useful for economic and social planning. One type would indicate areas liable to suffer total destruction by lava flows, nuées ardentes and/or lahars. The other type would show areas likely to be affected temporarily by damaging but not destructive phenomena, such as heavy falls of ash, toxic emissions, pollution of surface or underground waters, etc.

Losses caused by volcanic eruptions can be reduced by a combination of prediction, preparedness and land-use control (Anon., 1976). Emergency measures include alerting the public to the hazard, followed by evacuation. Unfortunately, false predictions and unnecessary evacuations do occur. Appropriate structural measures for hazard reduction are not numerous, but include building steeply pitched reinforced roofs that are unlikely to be damaged by ash fall, and constructing walls and channels to deflect lava flows. Hazard zoning, insurance, local taxation and evacuation plans are appropriate non-structural measures to put into effect. Risk management depends on the identification of hazard zones and the forecasting of eruptions. The levels of risk must be defined and linked to appropriate social responses. Volcanic zoning is constrained by the fact that the authorities and public are reluctant to make costly adjustments to a hazard that may have an MRI of 1000 years or more. Data on recurrence intervals are thus critical to zonation. In the long term, appropriate controls should be placed on land use and the location of settlements.

2.8.3 Dealing with volcanic activity

Obviously, it is impossible to control violent volcanic phenomena, such as nuées ardentes and pyroclastic fallout. Their only mitigation is via recognition of their potential extent, therefore making possible either temporary or permanent evacuation of areas likely to be badly affected. The threat of lava flows can be dealt with, with varying degrees of success, by diverting, disrupting or stopping them. For example, during the eruption from Kirkefell on the island of Heimaey, Iceland, in 1973, ash from the volcano was bulldozed to form a wall in order to divert lava flows away from the town of Vestmannaeyjar. Masonry and earth walls have been used in Japan and Hawaii for the same purpose. They should be constructed at an angle to the direction of flow so that it can be diverted rather than dammed. In the latter case, a lava flow would eventually spill over the dam. Topographic expression is therefore important; it should permit diversion so that no serious damage results. Lavas frequently flow along pre-existing channels across which such diversionary barriers can be built. Another diversionary technique is to dam the summit crater of a volcano at the usual exit, and to breach it somewhere else, so that the lavas will then flow in a different direction where they will do little or no harm.

Consolidation, check and diversion check dams have also been constructed to deal with lahars. Dams can be designed to retain the downslope movement of larger particulate material, but to allow water and suspended sediment to pass on. Such dams may be constructed of permeable gabions or of concrete with weep-holes. Suryo and Clarke (1985) suggested that the afforestation of the upper slopes of intermittently active volcanoes may inhibit the development of areas of potentially unstable soil, as well as preventing excessive run-off, and thereby reduce the likelihood of lahar formation in the event of an eruption.

Bombing was used successfully in 1935 and 1942 to disrupt lava flowing from Mauna Loa, Hawaii, away from the town of Hilo. The technique might also be used to breach a summit wall in order to release lava in a harmless direction.

Water has been sprayed onto advancing lavas to cool them and thereby cause them to solidify and stop. This technique may eventually have proved to be successful on Heimaey in 1973. However, it obviously is not known whether the lavas would have advanced much further without such treatment.

Crater lakes with a known history of volcanic eruptions have been drained, the classic example being provided by Mt Kelut, Indonesia. Following the eruption of 1919, a series of siphon tunnels was constructed to lower the surface of the lake and thereby reduce the potential lahar hazard.

References

Anon. (1971) *The Surveillance and Prediction of Volcanic Activity*. UNESCO, Paris.

Anon. (1976) *Disaster Prevention and Mitigation*, Vol. 1, *Volcanological Aspects*. United Nations, New York.

Baker, P.E. (1979) Geological aspects of volcano prediction. *Journal of the Geological Society, London* **136**, 341–346.

Bondal, C. & Robin, C. (1989) Volcano Popocatepetl: recent eruptive history and potential hazards and risks in future eruptions. In *Volcanic Hazards: Assessment and Monitoring*, Latter, J.H. (ed). Springer-Verlag, Berlin, pp. 110–128.

Booth, B. (1979) Assessing volcanic risk. *Journal of the Geological Society, London* **136**, 331–340.

Carey, S. & Sigurdsson, H. (1989) The intensity of Plinian eruptions. *Bulletin of Volcanology* **51**, 28–40.

Crandell, D.R., Booth, B., Kusumadinata, K., Shimozuru, D., Walker, G.P.L. & Westercamp, D. (1984) *Source Book for Volcanic Hazards Zonation*. UNESCO, Paris.

Forgione, G., Luongo, G. & Romano, R. (1989) Mt Etna (Sicily): volcanic hazard assessment. In *Volcanic Hazards: Assessment and Monitoring*, Latter, J.H. (ed). Springer-Verlag, Berlin, pp. 137–150.

Fournier D'Albe, E.M. (1979) Objectives of volcanic monitoring and prediction. *Journal of the Geological Society, London* **136**, 321–326.

Francis, P.W. (1979) Infrared techniques for volcano monitoring and prediction—a review. *Journal of the Geological Society, London* **136**, 355–360.

Gorshkov, G.S. (1971) General introduction. In *The Surveillance and Prediction of Volcanic Activity*. UNESCO, Paris, pp. 9–13.

Kling, G.W., Clark, M.A., Compton, H.R. *et al.* (1987) The Lake Nyos gas disaster in Cameroon, West Africa. *Science* **236**, 169–174.

Le Guern, F., Tazieff, H. & Faivre Pierret, R. (1982) An example of health hazard: people killed by gas during a phreatic eruption. *Bulletin Volcanologique* **45**, 153–156.

MacDonald, G.A. (1972) *Volcanoes*. Prentice-Hall, Englewood Cliffs, New Jersey.

Mori, J., McKee, C.O., Talai, B. & Itikarai, I. (1989) A summary of precursors to volcanic eruptions in Papua New Guinea. In *Volcanic Hazards: Assessment and Monitoring*, Latter, J.H (ed). Springer-Verlag, Berlin, pp. 260–291.

Rittmann, A. (1962) *Volcanoes and their Activity* (translated by Vincent, E.A.). Interscience, London.

Rothery, D.A. (1992) Monitoring and warning of volcanic eruptions by remote sensing. In *Geohazards: Natural and Man-Made*, McCall, G.J.H., Laming, D.J.C. & Scott, S.C. (eds). Chapman and Hall, London, pp. 25–32.

Suryo, I. & Clarke, M.G.C. (1985) The occurrence and mitigation of volcanic hazards in Indonesia as exemplified in the Mount Merapi, Mount Kelat and Mount Galunggung Volcanoes. *Quarterly Journal of Engineering Geology* **18**, 79–98.

Tazieff, H. (1979) What is to be forecast: outbreak of eruption or possible paroxysm? The example of the Guadeloupe Soufrière. *Journal of the Geological Society, London* **136**, 327–330.

Tazieff, H. (1988) Forecasting volcanic eruption disasters. In *Natural and Man-Made Hazards*, El-Sabh, M.I. & Murty, T.S. (eds). Reidel, Dordrecht, pp. 751–772.

Thorarinsson, S. (1970) The Lakigigar eruptions of 1783. *Bulletin Volcanologique* **33**, 910–927.

Tilling, R.I. (1989) Volcanic hazards and their mitigation: progress and problems. *Reviews in Geophysics* **27**, 237–269.

Tyrrell, G.W. (1937) Flood basalts and fissure eruptions. *Bulletin Volcanologique* **1**, 89–111.

Verstappan, H.Th. (1992) Volcanic hazards in Colombia and Indonesia: lahars and related phenomena. In *Geohazards: Natural and Man-Made*, McCall, G.J.H., Laming, D.J.C. & Scott, S.C. (eds). Chapman and Hall, London, pp. 33–42.

Voight, B., Glicken, H., Janda, R.J. & Douglass, P.M. (1981) Catastrophic rockslide avalanche of May 18th. In *The 1980 Eruption of Mount St Helens, Washington*, Lipman, P.W. & Mullineaux, D.R. (eds). *United States Geological Survey, Professional Paper* **1250**, pp. 347–377.

3   Earthquake activity

3.1 Introduction

Although earthquakes have been reported from all parts of the world, they are primarily associated with the edges of the plates which form the Earth's crust (Fig. 3.1). The Earth's crust is being slowly displaced at the margins of the plates. Differential displacements give rise to elastic strains which eventually exceed the strength of the rocks involved, and fault movements then occur. The strained rocks rebound along a fault under the elastic stresses until the strain is partly or wholly dissipated.

Earthquake foci are confined within a limited zone of the Earth, the lower boundary of which is located at approximately 700 km, and they rarely occur at the Earth's surface. In fact, most earthquakes originate within the upper 25 km of the Earth. Because of its significance, the depth of foci has been used as the basis of a threefold classification of earthquakes: those occurring within the upper 70 km are referred to as shallow, those located between 70 and 300 km are termed intermediate and those occurring between 300 and 700 km are considered as deep.

An earthquake propagates three types of shock wave. The first pulses that are recorded are termed primary or P waves. Sometimes they are referred to as push and pull waves, because oscillation occurs to and fro in the path of the wave. P waves are also called compression waves or longitudinal waves. The next pulses recorded are the S waves, sometimes referred to as secondary or shake waves. These waves oscillate at right angles to the path of propagation. S waves usually have a larger amplitude than P waves. The third type of vibration is known as the L wave. These waves travel from the focus of the earthquake to the epicentre immediately above at the surface, and from there they radiate over the Earth's surface. Two types of L or surface wave can occur in a solid medium, namely, Rayleigh waves and Love waves. In the former, surface displacement occurs partly in the direction of propagation and partly in the vertical plane. They can only be generated in a uniform solid. Love waves occur in non-uniform solids and oscillate in a horizontal plane, normal to the path of propagation. L waves are recorded after S waves.

As P waves travel faster than S waves (e.g. 5.95–6.75 km s⁻¹ compared with 2.9–4.0 km s⁻¹ in the crust), the further they travel from the focus of an earthquake, the greater the time interval between them. Thus the distance of a recording station from the epicentre of an earthquake can be calculated from this time interval.

P waves are not as destructive as S or L waves. This is because they have a smaller amplitude, and the force their customary vertical motion creates rarely exceeds the force due to gravity. On the other hand, S waves may develop violent, tangential vibrations strong enough to cause great destruction. The severity of an earthquake depends upon the amplitude and frequency of wave motion. S waves commonly have a higher frequency than L waves; nevertheless, the latter may be more powerful because of their larger amplitude.

Large earthquakes cause shock waves which travel throughout the world. Near the epicentre, the shocks are only felt for a matter of seconds, but, with increasing distance from the epicentre, the disturbance lasts for a progressively longer time. The vibrations of a shallow earthquake decrease rapidly in intensity away from the epicentre. Although the shocks of deep-seated earthquakes are usually weakly felt when they reach the surface, they extend over a much wider area.

It is assumed that the cause of most earthquakes is faulting. More than one rupture may occur along a fault, the second being triggered by the first. Furthermore, because it is energetically easier to make use of a pre-existing fault than to initiate a new one, faults

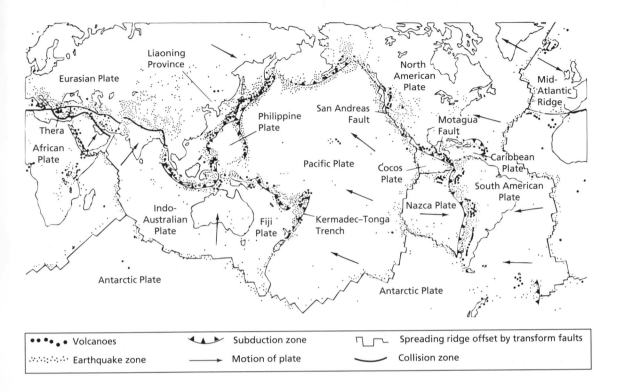

Fig. 3.1 World map showing the relation between the major tectonic plates and recent earthquakes and volcanoes. Earthquake epicentres are denoted by small dots and volcanoes by large dots.

may be repeatedly moved by successive earthquakes. There seems reason to believe that great earthquakes may consist of a rapid succession of such breaks, giving the overall impression of continuous rupturing. Some areas are affected by earthquake swarms, that is, by a series of nearly equal large shocks accompanied by small ones. Surface faulting in a particular earthquake usually only extends over a part of an existing fault.

Initial movement may occur over a small area of a fault plane, subsequently being followed by slippage over a much larger surface. Such initial movements, which represent preliminary shattering of small obstructions along a fault zone, account for the foreshocks which precede an earthquake. When these have been overcome, the main movement occurs, but complete stability is not restored immediately. The

shift of strata involved in faulting relieves the main stress, but develops new stresses in adjacent areas. Because stress is not relieved evenly everywhere, minor adjustments may arise along the fault plane and thereby generate aftershocks. The decrease in the strength of the aftershocks is irregular and, occasionally, they may continue for a year or more after the principal shocks.

Although the spatial association of larger earthquakes with faults suggests that strain accumulates along the faults, the spatial association of smaller earthquakes does not necessarily lead to the same conclusion. For example, smaller earthquakes occur almost randomly throughout southern California. Moreover, many shallow-focus earthquakes in Japan, Italy and Yugoslavia have not been accompanied by observable faulting.

Displacement along fault zones may occur by slow, differential slippage, termed fault creep, as well as by sudden rupture. Fault creep may amount to several millimetres per year. It is uncertain whether creep indicates that strain is being relieved along a major segment of a fault, thereby reducing the probability of occurrence of a strong earthquake, or whether it represents an accumulation of excess strain in spite of slight relief. In fact, both situations might apply. Be that as it may, locations where creep is taking place should be identified and monitored for abnormal movements. Obviously, creep indicates that a fault is active and, as such, is a factor to be considered in land-use planning.

Seismic waves are recorded by a seismometer which traces the oscillation caused by ground shaking on a drum, paper or magnetic tape to give a seismograph. There are various types of seismometer, and a seismic station may require an array of instruments, each of which is sensitive to a particular frequency range. Unfortunately, however, few useful data are obtained by seismic stations which are very close to the epicentre of major strong motions, as the instruments tend to be overwhelmed by the magnitude of the tremors. On the other hand, accelerometers are used to measure strong motions; indeed, they are only set in motion by strong tremors (see Section 3.5.4). Consequently, although accelerographs do not provide data on continuous, small-scale seismic activity, they do provide useful records of major tremors. Earthquake activity is now monitored in a coordinated way by the World-Wide Network of Seismic Stations. There are over 120 stations at which recording is made continuously.

3.2 Intensity and magnitude of earthquakes

In earthquake regions, it is necessary to establish the nature of the risk to structures. Thus the probability of the occurrence of earthquakes, their intensity and magnitude and the likelihood of earthquake damage to a structure must be evaluated. As soon as the intensity detrimental to a planned structure is established, the probability of an earthquake of this intensity in the given region should be estimated. The severest earthquakes wreak destruction over areas of 2500 km² or more, however, most only affect tens of square kilometres.

Earthquake intensity scales depend on human perceptibility and the destructivity of earthquakes. Whereas the degree of damage may be estimated correctly and objectively, the perceptibility of an earthquake depends on the location of the observer and his sensibility. Several earthquake intensity scales have

been proposed. That given in Table 3.1 is the Mercalli scale slightly modified by Richter (1956). However, quantification of earthquake size by the amount of damage caused to structures at the epicentre is not really satisfactory. Not only are such surface effects dependent on local conditions, but a particularly high level of damage (or intensity) can be achieved in either a small earthquake nearby or a large one further away.

The magnitude of an earthquake is an instrumentally measured quantity. In 1935, Richter devised a logarithmic scale for comparing the magnitudes of Californian earthquakes. He related the magnitude of a tectonic earthquake to the total amount of elastic energy released when overstrained rocks suddenly rebound and so generate shock waves.

The concept of magnitude has been widely extended and developed, so that now a number of magnitudes are recognized, such as the Richter local magnitude M_R, local magnitude M_L, body wave magnitude m_b and surface wave magnitude (long period) M_s. Because each of these magnitudes relates to certain wave frequencies, the character of which changes with the energy released by an earthquake, none proves to be a satisfactory index of the largest seismic events. The moment magnitude M is used for these. This is based on the surface area of a fault displaced during an earthquake, its average length of movement and the rigidity of the rocks involved. The relationship between the energy released E and magnitude M is given by the following expression:

$$\log E = a + bM \tag{3.1}$$

The values of the constants a and b have been modified several times as data have accumulated. In the case of surface and body waves, the corresponding equations are

$$\log E = 12.24 + 1.44 M_s \tag{3.2}$$

and

$$\log E = 4.78 + 2.57 m_b \tag{3.3}$$

where E is given in ergs. The seismic moment M_o is related to the area of rupture A, the rigidity modulus μ of the faulted rock and the average dislocation u caused by an earthquake as follows:

$$M_o = Au\mu \tag{3.4}$$

Table 3.1 Modified Mercalli scale, 1956 version, with Cancani's equivalent acceleration. (These are not peak accelerations as instrumentally recorded.)

Intensity	Description	Acceleration (mm s^{-2})
I	*Not felt.* Only detected by seismographs	Less than 2.5
II	*Feeble.* Felt by persons at rest, on upper floors or favourably placed	2.5–5.0
III	*Slightly felt indoors.* Hanging objects swing. Vibration similar to passing of light trucks. Duration estimated. May not be recognized as earthquake	5.0–10
IV	*Moderate.* Hanging objects swing. Vibration similar to passing of heavy trucks, or sensation of a jolt such as a heavy ball striking the walls. Standing motor cars rock. Windows, dishes, doors rattle. Glasses clink. Crockery clashes. In the upper range of IV, wooden walls and frames creak	10–25
V	*Rather strong.* Felt outdoors, direction estimated. Sleepers wakened. Liquids disturbed, some spilled. Small unstable objects displaced or upset. Doors swing, close, open. Shutters and pictures move. Pendulum clocks stop, start, change rate	25–50
VI	*Strong.* Felt by all. Many frightened and run outdoors. Persons walk unsteadily. Windows, dishes, glassware broken. Ornaments, books, etc. fall off shelves. Pictures fall off walls. Furniture moved or overturned. Weak plaster and masonry cracked. Small bells ring (church, school). Trees, bushes shaken visibly or heard to rustle	50–100
VII	*Very strong.* Difficult to stand. Noticed by drivers of motor cars. Hanging objects quiver. Furniture broken. Damage to masonry D, including cracks. Weak chimneys broken at roof line. Fall of plaster, loose bricks, stones, tiles, cornices, also unbraced parapets and architectural ornaments. Some cracks in masonry C. Waves on ponds, water turbid with mud. Small slides and caving in along sand or gravel banks. Large bells ring. Concrete irrigation ditches damaged	100–250
VIII	*Destructive.* Steering of motor cars affected. Damage to masonry C, partial collapse. Some damage to masonry B, none to masonry A. Fall of stucco and some masonry walls. Twisting, fall of chimneys, factory stacks, monuments, towers, elevated tanks. Frame houses moved on foundations if not bolted down, loose panel walls thrown out. Decayed piling broken off. Branches broken from trees. Changes in flow or temperature of springs and wells. Cracks in wet ground and on steep slopes	250–500
IX	*Ruinous.* General panic. Masonry D destroyed, masonry C heavily damaged, sometimes with complete collapse, masonry B seriously damaged. General damage to foundations. Frame structures, if not bolted, shifted off foundations. Frames cracked, serious damage to reservoirs. Underground pipes broken. Conspicuous cracks in ground. In alluviated areas, sand and mud ejected, earthquake fountains, sand craters	500–1000
X	*Disastrous.* Most masonry and frame structures destroyed with their foundations. Some well-built wooden structures and bridges destroyed. Serious damage to dams, dykes, embankments. Large landslides. Water thrown on banks of canals, rivers, lakes, etc. Sand and mud shifted horizontally on beaches and flat land. Railtracks bent slightly	1000–2500
XI	*Very disastrous.* Railtracks bent greatly. Underground pipelines completely out of service	2500–5000
XII	*Catastrophic.* Damage nearly total. Large rock masses displaced. Lines of sight and level distorted. Objects thrown into the air	Over 5000

In addition, the seismic moment is related to the total energy of a fault rupture by the expression

$$E = \sigma M_o / \mu \qquad (3.5)$$

where σ is the average stress acting on a fault during an earthquake. The stress drop is the difference between the shear stress on the fault surface before and after rupture.

The largest reported earthquakes have had a magnitude of 8.9, and these release about 700 000 times as much energy as the earthquake at the threshold of damage. Earthquakes of magnitude 5.0 or greater generate sufficiently severe ground motions to be potentially damaging to structures. It has been estimated that, in a typical year, the Earth experiences two earthquakes over magnitude 7.8, 17 with magnitudes between 7 and 7.8 and about 100 between 6 and 7.

There is little information available on the frequency of breaking along active faults. All that can be said is that some master faults have suffered repeated movement—in some cases, it has recurred in less than 100 years. By contrast, much longer intervals, totalling many thousands of years, have occurred between successive breaks. Therefore, because movement has not been recorded in association with a particular fault in an active area, it cannot be concluded that the fault is inactive.

Earthquakes resulting from displacement and energy release on one fault can sometimes trigger small displacements on other unrelated faults, many kilometres distant. Breaks on subsidiary faults have occurred at distances as great as 25 km from the main fault, but, with increasing distance from the main fault, the amount of displacement decreases. For example, displacements on branch and subsidiary faults located more than 3 km from the main fault break are generally less than 20% of the main fault displacement.

Many seismologists believe that the duration of an earthquake is the most important factor as far as the damage or failure of structures, soils and slopes is concerned. Buildings may remain standing in earthquakes which last for 30 s, whereas strong motions continuing for 100 s would bring about their collapse.

For example, the duration of the main rupture along the fault responsible for the Kobe earthquake (January 1995; $M=7.2$) was 11 s. However, the duration of strong motions was increased in some parts of the city, located on soft soils, to as long as 100 s (Esper & Tochibana, 1988). What is important in hazard assessment is the prediction of the duration of seismic shaking above a critical ground acceleration threshold. The magnitude of an earthquake affects the duration much more than the maximum acceleration, because the larger the magnitude, the greater the length of ruptured fault and, hence, the more extended the area from which the seismic waves are emitted. The idealized maximum ground accelerations in the vicinity of causative faults for earthquakes of various magnitudes are given in Table 3.2. The peak ground accelerations recorded some 35 km from the causative fault responsible for the Kobe earthquake were $0.28g$ in the north–south direction, $2.7g$ in the east–west direction and $2.4g$ in the vertical direction. In other words, the peak vertical acceleration was approximately 0.9 times the peak horizontal acceleration, which exceeds typical values recorded in past earthquakes.

With increasing distance from a fault, the duration of shaking is longer and the intensity of shaking is less, the higher frequencies being attenuated more than the lower ones. The attenuation or reduction of the acceleration amplitude, which occurs as waves travel from a causative fault, is the result of decreasing seismic energy, and is attributable to the dispersion that takes place as the outward propagating waves occupy an increasing space. Seismic energy is also reduced due to the absorption associated with internal damping in the rock material. Dispersion effects tend to increase the duration of strong shaking as the distance from the causative fault increases.

Magnitude	Rupture length (km)	Maximum acceleration (%g)	Duration (s)
5.0	—	9	2
5.5	5–10	15	6
6.0	10–15	22	12
6.5	15–30	29	18
7.0	30–60	37	24
7.5	60–100	45	30
8.0	100–200	50	34
8.5	200–400	50	37

Table 3.2 Maximum ground accelerations and durations of strong phase shaking. (After Housner, 1970.)

The physical properties of the soils and rocks through which seismic waves travel, as well as their geological structure, also influence surface ground motion. For example, if a plane wave traverses vertically through granite overlain by a thick, uniform deposit of alluvium, theoretically, the amplitude of the wave at the surface should be double that at the alluvium–granite contact. According to Ambraseys (1974), the maximum acceleration within an earthquake source area may exceed 200%g for competent bedrock. On the other hand, normally consolidated clays with low plasticity are incapable of transmitting accelerations greater than 10–15%g to the surface, whilst clays with high plasticity allow accelerations of 25–35%g to pass through. Saturated sandy clays and medium dense sands may transmit 50–60%g and, in clean gravel and dry dense sand, accelerations may reach much higher values. However, Ambraseys maintained that the amplitudes of the maximum ground velocities and, to some extent, the duration of shaking or the rate at which the energy flux is supplied to a structure are more significant than ground accelerations as far as seismic problems are concerned. Moreover, he pointed out that, for all practical purposes, there is no significant correlation between acceleration, magnitude and distance from an earthquake source in the epicentral area. The geological structure may enhance local shaking because of energy focusing effects.

3.3 Ground conditions and seismicity

The response of structures on different foundation materials has proven to be surprisingly varied. In general, structures not specifically designed for earthquake loadings have fared far worse on soft, saturated alluvium than on hard rock (Fig. 3.2). In other words, the amplitude and acceleration are much

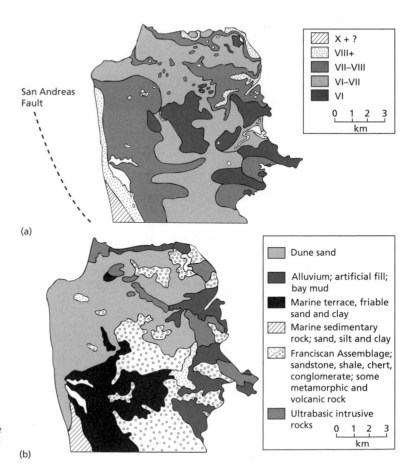

San Andreas
Fault

(a)

▨	X + ?
⬚	VIII+
▨	VII–VIII
▨	VI–VII
■	VI

0 1 2 3
km

(b)

	Dune sand
	Alluvium; artificial fill; bay mud
	Marine terrace, friable sand and clay
	Marine sedimentary rock; sand, silt and clay
	Franciscan Assemblage; sandstone, shale, chert, conglomerate; some metamorphic and volcanic rock
	Ultrabasic intrusive rocks

0 1 2 3
km

Fig. 3.2 (a) Isoseismal lines on the San Francisco Peninsula, California (based on the modified Mercalli scale) drawn by H.O. Wood after the 1906 San Francisco earthquake. (b) A generalized geological map of the San Francisco Peninsula. Note the correlation between the geology and the intensity. (From Bolt, 1978.)

greater on deep alluvium than on rock, although, by contrast, rigid buildings may suffer less on alluvium than on rock. This is because the alluvium seems to have a cushioning effect, and the motion may be changed to a gentle rocking, which is easier on such buildings than the direct effect of earthquake motions experienced on harder ground. Conversely, with any kind of poor construction, alluvial ground beneath the structure facilitates destruction, as happened in Kobe, where wooden houses built on alluvium sustained the most damage. Intensity attenuation on rock is very rapid, whereas it is extremely slow on soft

formations, and speeds up only in the fringe area of the shock. Hence, the character of intensity attenuation in any shock will depend largely on the surface geology of the shaken area. It is therefore important to try and relate the dynamic characteristics of a building to those of the subsoil on which it is founded. The vulnerability of a structure to damage is considerably enhanced if the natural frequency of vibration of the structure and the subsoil are the same.

Ground vibrations caused by earthquakes often lead to the compaction of cohesionless soil and asso-

(a)

(b)

Fig. 3.3 (a) Collapsed building in the Marina District, San Francisco, California, after liquefaction of soil due to the Loma Prieta earthquake in October 1989. (b) Slumping of the entire Turnagain Heights and subdivision in Anchorage, Alaska, which occurred when soil was liquefied during the earthquake of March 27, 1964.

ciated settlement of the ground surface (Ishihara & Okada, 1978). Loosely packed, saturated sands and silts tend to lose all strength and behave like fluids during strong earthquakes. When such materials are subjected to shock, densification occurs. During the relatively short time of an earthquake, drainage cannot be achieved, and this densification therefore leads to the development of excessive pore water pressures, which cause the soil mass to act as a heavy fluid with practically no shear strength (Peck, 1979). In other words, a quick condition develops (Fig. 3.3). Water moves upwards from the voids to the ground surface, where it emerges to form sand boils (Fig. 11.1). An approximately linear relationship exists between the relative density of sands and the stress required to cause initial liquefaction in a given number of stress cycles. Ground liquefaction has been noted in a number of earthquakes. For example, large-scale liquefaction of fills used in reclaimed areas of Osaka Bay occurred during the recent Kobe earthquake. Settlements of up to 0.75 m were recorded, and much damage was suffered by the port facilities and the adjoining industrial area.

If liquefaction occurs in a sloping soil mass, the entire mass will begin to move as a flow slide (Seed, 1970). Such slides develop in loose, saturated, cohesionless materials during earthquakes. In addition, loose, saturated silts and sands often occur as thin layers underlying firmer materials. In such instances, liquefaction of the silt or sand during an earthquake may cause the overlying material to slide over the liquefied layer. Structures on the main slide are fre-

quently moved without suffering significant damage. However, a graben-like feature (Fig. 3.4) often forms at the head of the slide, and buildings located in this area are subjected to large differential settlements and are often destroyed (Sokolenko, 1977). Buildings near the toe of the slide are commonly heaved upwards or even pushed over by the lateral thrust.

Clay soils, with the exception of quick clays, do not undergo liquefaction when subjected to earthquake activity, but, under repeated cycles of loading, large deformations can develop, although the peak strength remains about the same. Nonetheless, these deformations can reach the point where, for all practical purposes, the soil has failed. The damage caused by the 1985 earthquake which affected Mexico City was restricted almost exclusively to saturated deposits of clay (the Tacubaya Clay), which are part of an old lake bed. These deposits are characterized by low natural vibrational frequencies and were affected by the low-frequency ground motions experienced at Mexico City (the epicentre was 370 km away in the Central American Trench and high-frequency shocks tend to be attenuated rapidly with increasing distance from the epicentre). The clays amplified the shock waves between 8 and 50 times compared with the motions on solid rock in adjacent areas (Singh *et al.*, 1988). Those buildings which sustained the most severe damage were located mainly on that part of the old lake bed where the clay was greater than 37 m thick. In fact, the severity of ground motion increased in relation to the thickness of the clay.

Fig. 3.4 Graben behind L street slide area, Anchorage, Alaska due to the 1964 earthquake.

Table 3.3(a) Seismic intensity increments.

Ground conditions	Reid (1908)	Medvedev (1965)
Granite	0	0
Limestone, sandstone	0–1.2	1–1.5
Gravel	1.2–2.1	1–1.6
Sand	1.2–2.1	1.2–1.8
Clay	1.5–2.0	1.2–2.1
Fill	2.1–3.4	2.3–3.0
Wet gravel, sand	2.3	1.7–2.8
Clay	—	—
Wet fill	3.5	3.3–3.9

Table 3.3(b) Average changes in intensity associated with different types of surface geology. (From Degg, 1992.)

Subsoil	Average change in intensity
Rock (e.g. granite, gneiss, basalt)	−1
Firm sediments	0
Loose sediments (e.g. sand, alluvial deposits)	+1
Wet sediments, artificially filled ground	+1.5

Fills may fail completely or slump severely, with associated longitudinal cracking, when subjected to vibration due to earthquakes. Canal banks have a long history of slope failures during earthquakes. The tendency for earthfills to slide downhill during earthquakes results in increased pressure on retaining walls, which may be displaced.

It can be concluded from the foregoing paragraphs that microseismic observations can be used to establish seismic intensity increments for the basic categories of ground. Such an idea was first advanced by Reid (1908), who, after the San Francisco earthquake of 1906, introduced the concept of foundation coefficients for several major soil and rock types. The coefficient for the type of foundation that produced the least vibrational force, as revealed by observed earthquake damage, was designated unity. Estimates of the probable accelerations associated with these coefficients were also noted. Although it is now recognized, from the complex nature of strong motion accelerograph records, that acceleration itself has little meaning, unless the frequency is also given, there is

reason to believe from later investigations that these coefficients provide a true picture of the relative earthquake intensities experienced on the types of foundation cited.

Such seismic intensity increments can be obtained for different ground conditions by comparing the specific intensity changes for these different types of ground. The individual intensity increments for each type of ground have been related to a single standard (i.e. granite) (Table 3.3a). A more recent series of intensity increments (Table 3.3b), in which the standard ground conditions are firm unconsolidated sediments, has been proposed by the Munich Reinsurance Company and was quoted by Degg (1992).

3.4 Effects of earthquakes

The most serious direct effect of an earthquake in terms of buildings and structures is ground shaking. As pointed out in Section 3.3, the types of ground conditions are important in this regard, buildings on firm bedrock suffering less than those on saturated alluvium. Nonetheless, buildings standing on firm rock can still be affected, so that susceptible buildings should not be located near a fault trace. The type of construction also influences the amount of damage which occurs (Fig. 3.5). Poorly constructed buildings and those which are not reinforced undergo the worst damage. Indeed, strong ground shaking can reduce cities to rubble if buildings are weakly constructed. For instance, on February 29, 1960, hundreds of old unreinforced masonry buildings and many younger but poorly constructed reinforced concrete structures were destroyed in Agadir, Morocco, by an earthquake of magnitude 5.9, intensity VII. Approximately 14000 persons met their death out of a population of 33000. A number of factors can influence the death toll in a major earthquake. The time of day determines whether large numbers of people will be in offices, factories or schools. Property losses can be enormous in metropolitan areas, and this tends to rise dramatically with urban development. Fortunately, the ratio of loss of life to property damage tends to decline as more earthquake-resistant structures are built. In the Tokyo earthquake of September 1, 1923, 128000 houses were destroyed by the earthquake shock. However, a further 447000 were burnt out.

(a)

Fig. 3.5 (a) Totally destroyed office building which was poorly constructed, in the foreground, contrasts with the Tower Latino Americano building which remained intact after the Mexico City earthquake of September 19, 1985. (b) Collapse of a multi-storey carpark in Mexico City due to the earthquake of September 1985. Buildings between six and eight storeys were the worst affected, because they tend to be more sensitive to low-frequency ground motions that predominate at long distances from the earthquake epicentre (the epicentre was more than 300 km away). Note the pancaking of some floors.

(b)

Ground displacement during an earthquake may be horizontal, vertical or oblique. Displacement is associated with rupture along faults due to the accumulation of strain. Once the elastic limit of the rocks involved is exceeded, sudden movements occur. The displacement may be large, for example, the horizontal ground displacement due to the earthquake of April 18, 1906, which affected San Francisco, amounted to 6.5 m along that particular segment of the San Andreas fault. The maximum vertical displacement was less than 1 m. However, ground rupture as a cause of damage to buildings can be avoided by not building on or near major active

Fig. 3.6 Map showing the distribution of tectonic uplift and subsidence in south central Alaska caused by the Alaskan earthquake of 1964 (2–30 ft = 0.6–9.2 m). (From Eckel, 1970.)

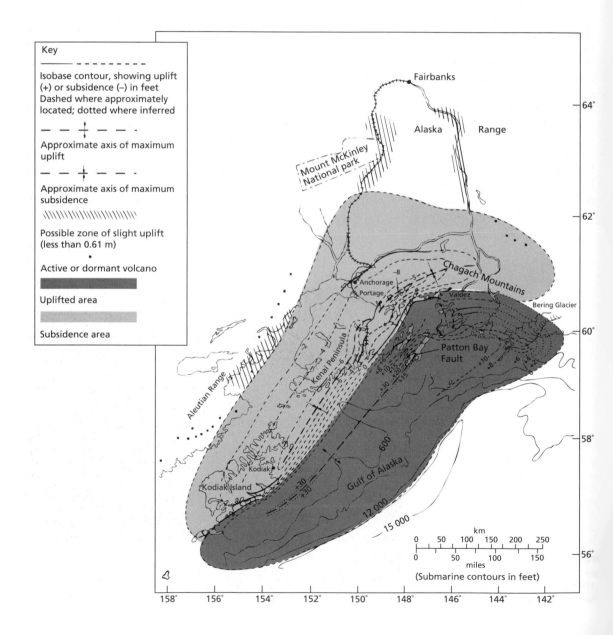

faults. On the other hand, it may prove to be impossible for roads, railways, canals or pipelines to avoid such faults. Regional crustal deformation manifested in uplift or subsidence may be associated with earthquakes. The movements which occurred at Niigata, Japan, prior to the earthquake of June 16, 1960, are referred to in Section 3.5.1. In the Alaskan earthquake of March 27, 1964, crustal deformation was due to faulting. Shorelines were uplifted by as much as 10 m in some places and sunk by 2 m elsewhere (Fig. 3.6). Some of the localities which underwent subsidence were subjected to flooding. Approximately 260 000 km^2 of coastal plain and adjoining sea floor were deformed. Furthermore, earthquakes may fissure or fracture and displace the ground, this being most prevalent next to the fault trace (Fig. 3.7). Fissures may gape up to 1 m wide at times.

Although fault creep does not lead to personal injury, it can weaken and ultimately cause structures to fail. For instance, creep movement on the Hayward fault in California is responsible for the continuous fracturing of a water tunnel which supplies Berkeley. It also gives rise to the warping of railway tracks. Furthermore, it has been suggested that fault creep

beneath the Balwin Hill Reservoir, Los Angeles, was a contributory cause of its failure.

Landslides and other types of mass movement frequently accompany earthquakes, most occurring within 40 km of the earthquake concerned (Fig. 3.8). Topography and preceding weather influence the likelihood of associated landslide occurrence. Landslides can cause great loss of life and be extremely destructive. For instance, the earthquake of May 31, 1970, which affected Ancash, Peru, triggered a slide of mud, rock and ice on the north peak of Mt Huascaran. The slide moved 3550 m downwards over a horizontal distance of 11 km at an average speed of 320 km h^{-1}. It buried the towns of Ranrahirca and Yunga, accounting for over one-third of the 67 000 deaths associated with the earthquake.

Loosely packed, fine grained sands and silt, which are saturated, may be subjected to liquefaction by earthquake shock. The resulting quick condition means that buildings can sink into the ground (see Section 3.3). In addition, lateral movement of such soils may be delineated by ruptured ground to form graben-like features, as happened at Turnagain Heights in Anchorage during the Alaskan earthquake

Fig. 3.7 Damage to a road near Irangahua after the earthquake of 1968, New Zealand. (Courtesy of the New Zealand Geological Survey.)

Fig. 3.8 Landslide in a cutting caused by the San Fernando earthquake of February 9, 1971, in California. (Courtesy of the State of California Department of Public Works.)

of March 27, 1964 (Fig. 3.4). Quick clays can also be liquefied by dynamic loading due to earthquakes.

Tsunamis are large seismic sea waves which are created by earthquakes. They can be extremely destructive along low-lying, highly developed coasts (see Section 6.9). Seiches are also produced by earthquakes if the ground is tilted or if the period of the long vertical surface waves in the ground is in resonance with the period of free oscillation of the body of water. They are vertically oscillating standing waves.

3.5 Methods of seismic investigation

Unfortunately, there is no method at present to forecast the exact location, size or time of an earthquake. However, it can be assumed that a probable prediction is reasonable and that past patterns of seismic activity will continue. Hence, earthquake risk reports should take into account an appraisal of known faults, the distance from major faults, the number of recorded earthquakes, the history of damage and an estimate of the magnitude and intensity of the strongest shock expected. The latter must take into consideration the ground conditions at the site.

3.5.1 Earthquake prediction

The purpose of earthquake forecasting is the reduction of the loss of life and damage to property. Earthquake prediction involves the use of studies of historical seismicity and the results of intensive monitoring of seismic and geological phenomena to establish the probability that a given magnitude of earthquake or intensity of damage will recur during a given period of time. Such predictions will help in the development of safety measures to be taken in relation to the degree of risk. In this way, earthquake alerts can be formed in terms of a staged gradation of prediction, from the long term to the immediate.

Earthquake precursors are phenomena which precede major ground tremors; more specifically, they are anomalous values of such phenomena which may indicate the onset of an earthquake (Rikitake, 1979). Most important is the variation in seismic wave velocity. The P waves slow down (e.g. by up to 10% in the focal region) due to the minute fracturing in rocks during dilation, but then increase as water occupies microfractures prior to an earthquake. As water is incompressible, it is an effective transmitter of compressional waves, the elasticity of the rock mass being regained. Hence, a continuous record of such changes in shock wave velocity may be of value in earthquake prediction, as a quake may occur a day or so after the wave velocities return to normal. The period of time covered by these changes indicates the size of the earthquake that can be expected. For example, an earthquake of magnitude 5.4 may be preceded by a period of 4 months of lowered velocities, and one of magnitude 4 by only 2 months. Thus, if the period of changes lasts for 14 years, it suggests a potentially violent earthquake of magnitude 7. Furthermore, the ratio of the P and S wave velocities decreases by up to 20% a few hours or days before a major quake. This is primarily due to the reduction in the compressional wave velocity. The duration, but not the amount, of the decrease in velocity ratio appears to be proportional to the earthquake magnitude. The velocity ratio returns to normal at the beginning of the 'critical' period immediately before an earthquake.

The dilatancy of rocks, due to crustal deformation prior to an earthquake, leads to an increase in their volume and results in ground shortening or lengthening, minor tilting and uplift or subsidence of the ground surface near an active fault. This ground movement can be measured by very accurate surveying or by high-precision tiltmeters and extensometers. Unfortunately, however, the point at which movements become critical is not always easy to detect. For example, Bolt (1993) referred to the uplift and subsidence which, for about 60 years, occurred along the coast of Japan, near Niigata, prior to the earthquake of June 16, 1960. The rate of movement slowed in the late 1950s and, at the time of the earthquake, a sudden subsidence was detected to the north alongside the epicentre. The amount of precursory uplift decreased gradually away from the epicentre and was not observed at distances exceeding 100 km. However, Bolt also mentioned the uplift at Palmdale in southern California which, since 1960, has amounted to over 350 mm, but no significant earthquakes have occurred in this period (although earthquakes have occurred prior to this period). Hence, he concluded that the best response was to intensify the various types of measurement in the area.

Other features brought about by rock dilation include a decrease in electrical resistivity and a change in magnetic susceptibility of the rock concerned, that is, the local magnetic field becomes stronger shortly before an earthquake. The decrease in resistivity is assumed to be due to the influx of water into cracks in the dilatant zone, flowing from its surroundings. Subsequently, the stress reduction attributable to the earthquake will allow cracks to close, thereby forcing water from the source region, so that the resistivity will rise. Changes in density are recorded by sensitive gravity meters which can detect small alterations. Changes in both magnetic and gravity measurements may help to locate hidden faults by indicating differences in the properties across sharp contacts.

The release of small quantities of the inert gases, notably radon, but also argon, helium, neon and xenon, takes place prior to an earthquake. The increase in radon has been noted in well waters, and has been attributed to increasing water flow which carries radon with it into wells after dilation and cracking in neighbouring rocks. Hence, the movement and chemistry of groundwater may change before earthquakes. Stresses in saturated strata may cause spring discharge to increase or the water in wells to alter its level or become turbid. During microfracturing due to an earthquake, migrating groundwater comes into contact with an increased surface area of rocks, dissolving radon in trace amounts. Stress in the rocks may speed the movement of water in the phreatic zone and, hence, the migration of radon and halogens to the surface. Moreover, many active faults are zones of geothermal energy release, where it may be possible to relate the temperature of hot springs to incipient seismic activity. Obviously, it is important to implement a regular monitoring programme of stream and spring discharge, water quality and water temperature.

According to Raleigh et al. (1982), severe plate boundary earthquakes recur at intervals which may vary by as much as 50% from an average value. However, it would appear that there is some evidence that this interval may be proportional to the displacement caused by faulting when the previous earthquake occurred at the same location. If this is correct, this will allow the recurrence interval to the next earthquake to be estimated more precisely, provided that the long-term displacement is constant and well known.

Because damaging earthquakes in particular seismic areas cannot be predicted exactly in time, Bolt (1993) pointed out that the best strategy is to attempt to determine the probability that such an event will occur. He went on to note that a method of determining probability can be based on the elastic rebound theory of the cause of earthquakes. In other words, as strain increases, the more probable it is that sudden slip will take place along faults and an earthquake will be generated. Recording geological and geodetic measurements should facilitate the determination of which segments of a fault are likely to slip in future. Hence, fault segments should be mapped accurately. It can be assumed that the earthquake which generates the largest magnitude will be that in which movement occurs over the complete segment. Accordingly, it is necessary to determine which fault segments along an active fault have slipped in the past, and to calculate the rate at which strain is accumulating in the area. Each time slip has taken place can be related to the magnitude of the resultant earthquake, so that the recurrence intervals between earthquakes of a given magnitude can be determined. From this, the range of probable occurrence of an earthquake of a particular size in a specific time period can be estimated.

3.5.2 Assessment of movements along faults

Most major fault zones do not consist of a single break, but are made up of a large number of parallel and interfingering breaks which may range over 1 km apart. This indicates that displacements associated with intermittent, great earthquakes have tended to migrate back and forth throughout the fault zone concerned. Likewise, fault displacements in future will not all take place along a single fracture. For example, in many areas of southern California, geologically very recent displacements have been concentrated in narrow bands within much wider fault zones, and it seems probable that, in such areas, the next displacements will follow approximately the same traces as their immediate predecessors. Furthermore, in many areas, the more abundant recurrent displacements have apparently taken place along the principal fault breaks, which suggests that these breaks extend to master faults located at depth, and thus would have a higher probability of future movements.

Rock bending at faults means that strain is accumulating in rocks, possibly foreshadowing an earthquake and sudden fault displacement. Such tectonic movement can be measured with electrical distance measurement (EDM) equipment, by triangulation, by fault movement quadrilaterals and by tiltmeters. It would appear that sudden strain release may be more likely along a segment of a fault where the movement rate differs from that of adjacent areas or from its past rate, so that this may aid earthquake prediction. Triangulation in critical fault zone areas provides precise survey control, and a basis for the determination of ground movements when observed in the future. Geodimeter measurement does not distinguish between movement due to fault slippage and that due to strain, primarily because the lines measured are long, from 13 to 32 km. Measurements over much shorter distances, 250–500 m, provide an indication of slippage. These lines are arranged in quadrilaterals across a fault. Gradual vertical movements result in some tilting at the surface. Tiltmeters can detect tilting of 0.0001 mm along a 30.5 m horizontal line. Arrays of seismographs, spread over large areas, are buried at various depths and arranged in two lines at right angles to each other. The arrays can also include electrical resistivity meters, magnetometers, gravity meters and strain gauges.

3.5.3 Aseismic investigation

Aseismic investigation prior to the design of a structure provides the designer with some idea of the frequency and size of expected earthquakes, as well as the spectrum of anticipated earthquake motion. This spectrum is an envelope of maximum motions at each specific period of frequency at a particular location.

Aseismic investigation may vary according to the site and proposed structure. A geological map of the site should be prepared and the geological structure should be illustrated by sections. Note must be taken of the regional tectonics and particular attention must be paid to the investigation of recent faulting. The smallest earthquakes cause no visible damage, but the record of their epicentres may indicate the existence of an active fault. Precautions are obviously necessary in any region where conspicuously active faults have been mapped. If a region is very seismic, most places that are going to have earthquakes will already have had one.

It may be possible to assess the relative activity of a fault using geological, seismological and historical data, and so classify it as active, potentially active, of uncertain activity or inactive. Seismological data used to recognize active faults include historical or recent surface faulting with associated strong earthquakes and tectonic fault creep or geodetic indications of fault movement. Ambraseys (1992) showed that, in regions such as the Middle East, where low to moderate rates of seismic activity are experienced, historical (i.e. pre-twentieth century) data need to be included in hazard evaluations in order to enhance their meaning. A lack of known earthquakes, however, does not necessarily mean that a fault is not active. An active fault is indicated by geological features, such as young deposits displaced or cut by faulting, fault scarps, fault rifts, pressure ridges, offset streams, enclosed depressions and fault valleys, and ground features, such as open fissures, rejuvenated streams, folding or warping of young deposits, groundwater barriers in recent alluvium, en echelon faults and fault paths on recent surfaces. Erosion features are not necessarily indicative of active faults, but may be associated with some active zones. Historical sources of displacement of man-made structures, such as roads, railways, etc., may provide further evidence of active faults.

Faults can be regarded as potentially active if there is no reliable report of historical surface faulting, faults are not known to cut or displace the most recent alluvial deposits, but may be found in older alluvial deposits, geological features characteristic of active fault zones are subdued, eroded and discontinuous, water barriers are present in older materials, and the geological setting is an area in which the geometrical relationship between active and potentially active faults suggests similar levels of activity. As far as seismic evidence is concerned, there may be alignment of some earthquake foci along the fault trace, but locations are assigned with a low degree of confidence.

Faults are of uncertain activity if data are insufficient to provide criteria definitive enough to establish fault activity. Additional studies are necessary if the fault is considered to be critical to a project.

In the case of inactive faults, there is no historical activity based on a thorough study of local sources. Geologically, features characteristic of active fault zones are not present in areas where faults are inactive, and geological evidence is available to indicate that the fault has not moved in the recent past. Although not a sufficient condition for inactivity, seis-mologically, the fault concerned should not have been recognized as a source of earthquakes.

Areas which might be exposed to the hazards of surface faulting, fissuring, liquefaction, landslide or other mass movement should be recognized. Drilling and sampling will aid in the interpretation of the geological setting, and the samples will allow a determination to be made of the physical properties of the materials concerned. The values of the static and dynamic properties of the soils may be obtained by *in situ* testing. Geophysical surveys and aerial photographs may prove to be useful.

The ground can be force vibrated by using a shaking machine and the response can be determined. By comparing this with the response of other known sites, a measure of the comparative sensitivity can be estimated. A refraction seismograph can be used to determine interval velocities, as well as accumulative velocities, from the surface to any point in a drill-hole. These up-hole seismic surveys can be made with one explosion, using pick-ups distributed along a cable placed in the drill-hole.

3.5.4 Accelerographs

Data relating to ground motion are essential for an understanding of the behaviour of structures during earthquakes. From the engineering point of view, the strong motion earthquakes are the most important, because they damage or even destroy man-made structures. Records of shocks produced by such earthquakes are obtained by using ruggedly constructed seismometers called accelerometers. These are designed to operate only where earthquake vibrations are strong enough to actuate them. The accelerograph simultaneously records the two orthogonal horizontal and vertical components of ground acceleration as a function of time. The period and damping of the pick-ups are selected so that the recorded motions are proportional to the ground acceleration over the frequency range of about 0.06–25 cycles per second (cps), which encompasses the range of periods exhibited by typical engineering structures. The instrument has a resolution of the order of $0.001\,g$ and is operative to about $1.0\,g$.

Figure 3.9 illustrates the accelerographs obtained from an area subjected to strong shaking due to a destructive earthquake. There is a rapid initial rise to

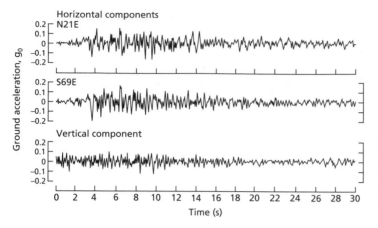

Fig. 3.9 Ground acceleration recorded at Taft, California, approximately 40 km from the causative fault of magnitude 7.7, Arvin–Tehachapi earthquake of July 21, 1952. Recorded on approximately 6 m of alluvium overlying sedimentary rock. (From Cherry, 1974.)

a strong, relatively uniform central phase of shaking for a certain duration, followed by a gradually decaying tail, during which some very strong pulses of acceleration may still occur (Bender, 1984). The records exhibit about equal intensities of motion in the two horizontal directions, which is to be expected, because of the almost random nature of the ground movement. The vertical component is normally somewhat less intense than the horizontal, and is characterized by an accentuation of the higher frequency components when compared with the horizontal motions.

Accelerometers are placed on the surface and below ground level. They have also been installed in dams and at different levels in multi-storey buildings to compare structural motion at different heights with ground motion. Strong motion records have provided information concerning the acceleration, displacement periods and duration of earthquakes (Fig. 3.10).

3.6 Seismic hazard and risk

In earthquake prone areas, any decision making relating to urban and regional planning, or for earthquake resistant design, must be based on information concerning the characteristics of probable future earthquakes. However, the lack of reliable data to be used for design purposes has been a cause for concern in engineering seismology. Hence, the engineer may, on occasions, accept an element of risk above that which would otherwise be considered to be normal. This

risk may, under certain conditions, prove to be economically acceptable for structures of relatively short life. However, for certain structures, whose failure or damage during an earthquake may lead to disaster, this is not acceptable (Lomnitz & Rosenblueth, 1976). A considerable amount of informed judgement and technical evaluation is needed to determine what is an acceptable risk for a given project, and caution is required in the selection of earthquake design parameters. There are a number of methods for selecting appropriate earthquake design parameters, which may involve estimates of strong ground motions, but all depend on the quality of the input data. The estimation of the maximum probable earthquake ground motion, and when and where it will occur, is extremely difficult to assess.

Earthquake hazard is the probability of occurrence in a given area, during a given period of time, of ground motion capable of causing significant loss of property or life (Campbell, 1984). It may be expressed by indicating the probability of occurrence of certain ground accelerations, velocities and displacements, ground movements of various duration or any other physical parameter which adversely affects a structure. Earthquake risk may be defined in terms of the probability of the loss of life or property, or the loss of function of structures or utilities. The assessment of risk must take into consideration the probability of occurrence of ground motions due to an earthquake, the value of the property and lives exposed to the hazard and their vulnerability to

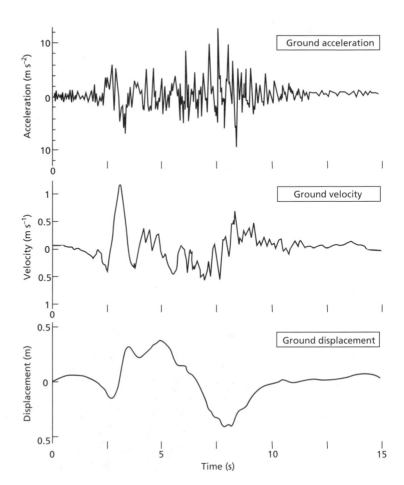

Fig. 3.10 Ground motion during the San Fernando earthquake, September 2, 1971, 06.00 h, Pacoima Dam, S 16°E component. (After Hazzard, 1978.)

damage or destruction, injury or death by ground motions associated with the hazard (Fournier d'Albe, 1982). In other words, the assessment of risk involves the assessment of value, vulnerability and hazard.

A hazard level specific to a site, using data acquired on the occurrence of earthquakes, is justified for projects in areas of significant seismicity. The preliminary source of information is the national earthquake code, if there is one. Most of these codes contain a map of the seismicity of different regions of the country. There is a diversity of earthquake maps of varied status and type (Table 3.4). If the preliminary survey shows that a significant seismic hazard exists, the hazard evaluation procedures should be commensurate both with the type of project and the quantity and quality of information available.

National agencies, such as the International Seis-

mological Centre in Britain and the National Earthquake Information Service in the USA, retain catalogues of earthquake events. Critical examination of the earthquake information is necessary. Well reported events with instrumental data may be well located. However, prior to the inception of the World-Wide Seismograph Network in the mid-1960s, the location, magnitude and, especially, focal depth of earthquakes should be viewed with caution.

Ideally, complete data sets should be used to characterize the distribution of seismicity, and only well located epicentres should be employed to make even broad association with fault zones. Rarely is an epicentre located, by historical studies, to within a 5 km radius circle and, unless an earthquake has its epicentre within a well conditioned array of seismographs

Table 3.4 Classification of earthquake-related maps. (After Skipp & Ambraseys, 1987.)

Tectonic	Show tectonic units with indication of orogenic association, main structural units, major faulting and folding
Seismo-tectonic	Show tectonic features relevant to earthquake generation: faults with indication of style and mobility, focal mechanisms, current crustal stress, neotectonic (Quaternary) features, vulcanicity
Hazard expectancy	
Magnitude based	Probabilistically based; show values of largest earthquake expected over given period within unit area
Intensity based	Probabilistically based (with or without tectonic control); show values of largest intensity expected over given period, usually expressed as contoured zoning
Effective acceleration based	Probabilistically based (with or without tectonic control); show effective peak acceleration within specified period at given confidence level
Intensity zoning	Show maximum historical intensity, usually contoured
Intensity potential	Show maximum potential intensity, contoured
Seismicity index	Show zones from which coefficients can be selected in formulation of design earthquake input

with intervening distances of less than around 50 km, instrumental location need be no better.

Once a catalogue of earthquakes has been drawn up, which may have affected the area in question, and threshold levels and uncertainties of location have been established, it is possible to apply statistical techniques. These techniques can determine whether or not spatial and temporal distributions are random or whether there have been significant periods of quiescence and activity. They can also establish whether or not there is a valid zonation within a given distance from the site where departures from background seismicity are significant.

Seismo-tectonic maps may be used to identify potentially active faults. However, such maps are rare and, even where they exist, may be so interpretative as to be highly contentious. Nonetheless, some tectonic appraisal of a site, making use of existing maps and published work, is necessary. Remote sensing surveys should be examined. Geophysical maps (e.g. seismic, gravity and magnetic) may help.

3.7 Seismic zoning

Seismic zoning and microzoning provide a means of regional and local planning in relation to the reduction of seismic hazard, and are used in earthquake-resistant design. While seismic zoning takes into account the distribution of earthquake hazard within

a region, seismic microzoning defines the distribution of earthquake risk in each seismic zone. A seismic zoning map shows the zones of different seismic hazard in a particular area. Some seismic zoning maps summarize observations of past earthquake effects, the assumption being that the same pattern of seismic activity will be valid in the future, whereas others extrapolate from areas of past earthquakes to potential earthquake source areas (Karnik & Algermissen, 1978). Unfortunately, the first type of map does not take into account any earthquake sources in areas which have been quiet during the period of observation, and the second category is more difficult to compile.

Seismic evidence obtained instrumentally and from the historical record can be used to produce maps of seismic zoning. Maximum hazard levels can be based on the assumption that future earthquakes will occur with the same maximum magnitudes and intensities as recorded at a given location in the past (Sokolenko, 1979). Hence, seismic zoning provides a broad picture of the earthquake hazard that can be involved in seismic regions, and so has led to a reduction of earthquake risk. Detailed seismic zoning maps should take into account the local engineering and geological characteristics, as well as the differences in the spectrum of seismic vibrations and, most importantly, the probability of the occurrence of earthquakes of various intensities. Expected intensity maps define the

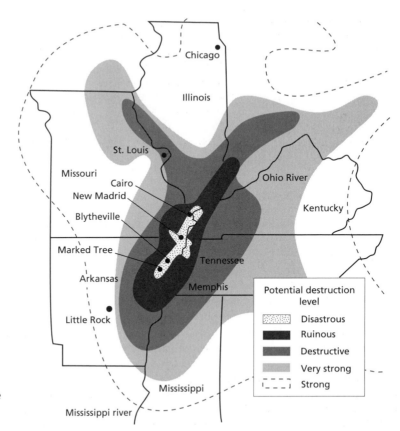

Fig. 3.11 Zones forecasting likely destruction levels (to non-resistant structures) in the New Madrid earthquake region. (Courtesy of the Federal Emergency Management Agency.)

source areas of earthquakes and either the maximum intensity which can be expected at each location or the maximum intensity to be expected in a given time interval (Donovan *et al.*, 1978).

One of the problems involved in producing seismic zoning maps is the choice of parameters to be mapped. Usually, seismic zoning maps are linked to building codes, and are commonly zoned in terms of macroseismic intensity increments or related to seismic coefficients incorporated in a particular code. The building code may specify the variation of the coefficient in terms of ground conditions and type of structure. Other maps may distinguish zones of destruction (Fig. 3.11). Karnik and Algermissen (1978), however, argued that there is a growing need for quantities related to earthquake resistant design to be taken into account, such as the maximum acceleration or peak particle velocity, predominant period of shaking or probability of occurrence. Algermissen *et al.* (1975) developed a technique for the probabilistic estimation

of ground motion, that is, for the derivation of the maximum ground shaking at a particular point in a given number of years at a given level of probability. In this way, they were able to produce maps of horizontal acceleration for a particular time period (Fig. 3.12). Such a map can be used for general site evaluation and initial design purposes.

There is no standard way of using the information for zoning in all seismic regions. Moreover, in the compilation of these maps, all engineering data concerning the surface manifestations of earthquakes must refer to identical ground conditions, as they can have a strong influence on intensity. Geological investigations may be able to throw light upon the rate of strain within the crust, and can delimit areas where faulting is widespread. However, geological data can only be used to give a qualitative assessment of seismic risk within an area.

Most studies of the distribution of damage attributable to earthquakes indicate that areas of severe

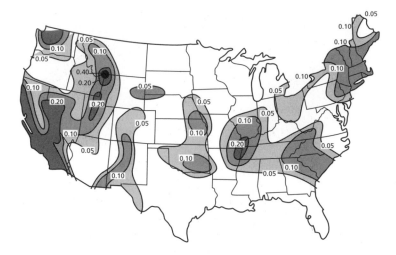

Fig. 3.12 Seismic risk map for the USA. The contours indicate the effective peak, or maximum, acceleration levels (values are in decimal fractions of gravity) that might be expected (with a probability of 1 in 10) to be exceeded during a 50-year period.

damage are highly localized and that the degree of damage can change abruptly over short distances. These differences are frequently due to changes in soil conditions or local geology. Such behaviour has an important bearing on seismic microzoning. Hence, seismic microzoning maps can be used for detailed land-use planning and for insurance risk evaluation. These maps are obviously most detailed and more accurate where earthquakes have occurred quite frequently and the local variations in intensity have been recorded. A map of microseismic zoning for Los Angeles County is shown in Fig. 3.13. It distinguishes three zones of ground response, active faults likely to give rise to ground rupture and areas of potential landslide and liquefaction risk.

3.8 Induced seismicity

Induced seismicity occurs where changes in the local stress conditions give rise to changes in strain in a rock mass. The sudden release of strain energy due to deformation and failure within a rock mass results in detectable earth movements. It is the action of man which causes the activity by bringing about these changes in the local stress conditions (McCann, 1988).

A number of events have occurred over the last 30 years or so which seem to support the idea that earthquakes might be triggered by artificially induced changes in the pore water pressures in rock masses. For example, Evans (1966) suggested that the earth-

quakes which affected Denver between April 1962 and November 1965, over 700 in number, but not exceeding $M = 4.3$, were consequent upon the injection of waste fluids into a 3660 m deep disposal well. The Precambrian rocks which received this waste from the Rocky Mountain Arsenal consisted of highly fractured gneisses. It appears that the movements took place as the pore water pressures were raised by the injection of the waste. More specifically, rising pore water pressure reduces the resistance to movement along fracture planes, causing the release of elastic wave energy and the generation of an earthquake.

Similarly, the use of water injection to maximize the yield of oil can give rise to induced seismicity, presumably due to the injection of water leading to a reduction in the resistance to faulting consequent on the increase in the pore water pressure. However, this would suggest that the rock masses involved were stressed to near failure stress prior to the injection of water. Gibbs *et al.* (1973) referred to seismic activity in the Rangely area, Colorado, associated with water injection in a nearby oilfield. They recorded 976 seismic events between 1962 and 1972; of these, 320 exceeded magnitude 1 on the Richter scale. They described an apparent correlation between the number of earth tremors recorded and the quantity of water injected. Ten years later, Rothe and Lui (1983) recorded 31 earth tremors between March 1979 and March 1980 at the Sleepy Hollow oilfield in Nebraska. Again, water had been injected to maxi-

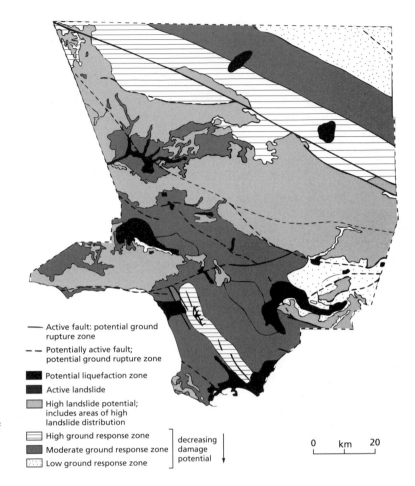

Fig. 3.13 Micro-seismic zoning map of Los Angeles County. (Courtesy of the Los Angeles County Department of Planning.)

--- Active fault: potential ground rupture zone

- - Potentially active fault; potential ground rupture zone

■ Potential liquefaction zone

■ Active landslide

▨ High landslide potential; includes areas of high landslide distribution

▤ High ground response zone

▤ Moderate ground response zone

▦ Low ground response zone

} decreasing damage potential

0   km   20

mize the yield of oil. The range of magnitude varied between 0.6 and 2.9 on the Richter scale. The source of the tremors coincided remarkably with the area of the oilfield.

More notably, a number of earthquakes have been recorded with their epicentres below or near large reservoirs. The evidence, however, is not conclusive as far as induced seismicity is concerned. In the case of reservoirs, this is not surprising, because so very few contain the instruments to record local seismic events. However, as dams are built higher and reservoirs impound larger volumes of water, this cause and effect relationship needs to be resolved, because induced earthquakes may cause serious damage, which needs to be avoided.

The longest range of data has been provided by Lake Mead, Colorado, where over 10 000 small earthquakes have been recorded. A weak correlation between seismic activity and level of impounding has been reported at Lake Mead, and the epicentres are thought to have been located along faults. The most notable earthquake occurred at Koyna Reservoir, which is located not far from Bombay. Impounding began in 1963 and small shocks were recorded a few months later. These continued, increasing in intensity, until, in December 1967, a shock with a magnitude of about 6.5 occurred. This caused significant damage and loss of life in the nearby village of Coynanagar. The dam was fissured in several places, because the epicentre of the earthquake was near the dam. The focus was located at a depth somewhere between 10 and 20 km. Like Lake Mead, there was a correlation between fluctuating water level and seismicity. Figure 3.14 shows the number of earthquakes greater than

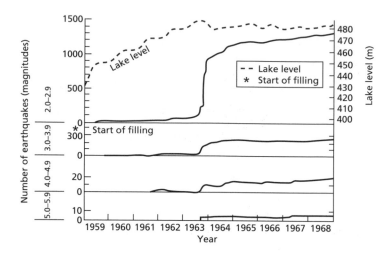

Fig. 3.14 Kariba Reservoir: distribution of earthquakes in time and in relation to lake level. (After Lane, 1972.)

magnitude 2 which were recorded in the first 10 years of life of Kariba Reservoir between Zimbabwe and Zambia. Most recently, tremors have been associated with the filling of Katse Reservoir, Lesotho (Bell & Haskins, 1997).

Seismic activity at reservoir sites may be attributable to water permeating the underlying strata, thereby increasing the pore water pressure and decreasing the effective normal stresses, so that the shear strength along local faults is reduced. In addition, increased saturation may reduce the strength of rock masses sufficiently to facilitate the release of crustal strains. Such activity would not appear to be related to the weight of water stored, as some large impounded reservoirs have not been associated with noticeable seismic activity. Seismic activity may be initiated almost as soon as impounding begins, but the time of maximum activity sometimes shows an appreciable delay compared with the time at which the reservoir reaches its maximum level.

Induced seismicity has been associated with many mining operations, where changes in local stress conditions have given rise to corresponding changes in strain and deformation in the rock mass concerned. These changes have been responsible for movements along discontinuities, which may be on a macroscopic or microscopic scale. In the former case, they generate earth movements which are detectable at the ground surface. Such movements can cause damage to buildings, but generally this is minor, although very occasionally high local intensities are generated. Obviously, the extent of damage is related to the mag-

nitude of the earth tremor and the distance from the source, as well as the nature of the surface rock and the strength of the structure. Nonetheless, such seismic events are often a cause of concern to the general public.

When mining, particularly hard rock at depth, violent rock failures, known as rockbursts, can occur in the workings. However, although all rockbursts generate seismic events, all seismic events associated with mining activity are not rockbursts. Cook (1976) suggested that the changes in stress consequent on mining were responsible for a sudden loss of stored strain energy, which resulted in brittle fracture in the rock mass in the excavation. These changes in strain energy may be related to a decrease in potential energy of the rock mass as the rock is mined. On the other hand, it may be that mining triggers latent seismic events in rock masses which are in a near-unstable condition.

Some damage has occurred to surface structures. For example, in 1976, an earth tremor of magnitude 5.1 damaged several buildings in Welkom in the gold mining area of the Free State, South Africa. Neff *et al.* (1992) referred to severe cracking of some houses in a gold mine township in the western Transvaal; in fact, some houses had to be evacuated. The seismic tremors had magnitudes ranging up to 3.4.

Early studies of seismic records of tremors in the Johannesburg area, South Africa, clearly showed that a peak occurred in the daily distribution of seismic events on weekdays around the time of blasting (De Bruyn & Bell, 1997). The focal depths from which

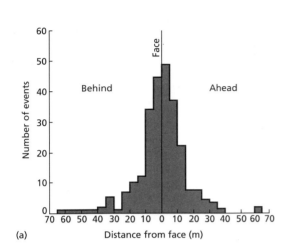

(a)

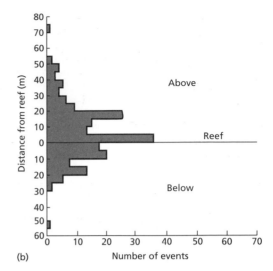

(b)

Fig. 3.15 Earth tremors in the Johannesburg area. (a) Histogram showing the position of the working face relative to seismic events. (b) Histogram showing the relation of seismic events to the position of the reef. (After Cook, 1976.)

the tremors emanated were shown to be in close proximity to the position where mining was taking place (Fig. 3.15). The magnitude of these events ranged from 2 to 4 on the Richter scale.

Seismic events have also been associated with coal mining in the UK. Redmayne (1988) suggested that it seems likely that all deep coal mining is accompanied by such activity of varying magnitude, depending on the geological conditions, local faulting, local stress field, rate and type of extraction, seam thickness, past mining in the area and nature of the overburden.

References

Algermissen, S.T., Perkins, D., Isherwood, W., Gordon, D., Rengor, G. & Howard, C. (1975) Seismic risk evaluation of the Balkan region. *Survey of the Seismicity of the Balkan Region. UNDP/UNESCO Report.* United States Geological Survey, Denver.

Ambraseys, N.N. (1974) Notes on engineering seismology. In *Engineering Seismology and Earthquake Engineering*, Solnes, J. (ed). *NATO Advanced Study Institute Series, Applied Sciences* 3, 33–54.

Ambraseys, N.N. (1992) Long-term seismic hazard in the eastern Mediterranean region. In *Geohazards: Natural and Man-Made*, McCall, G.J.H., Laming, D.J.C. & Scott, S.C. (eds). Chapman and Hall, London, pp. 88–92.

Bell, F.G. & Haskins, D.R. (1997) A geotechnical overview of the Katse Dam and Transfer Tunnel, Lesotho, with a note on basalt durability. *Engineering Geology* 41, 175–198.

Bender, B. (1984) Incorporating acceleration variability into seismic hazard analysis. *Bulletin of the Seismological Society of America* 74, 1451–1462.

Bolt, B.A. (1978) *Earthquakes: A Primer.* W.H. Freeman, San Francisco.

Bolt, B.A. (1993) *Earthquakes.* W.H. Freeman, New York.

Campbell, K.W. (1984) Probabilistic evaluation of seismic hazard for sites located near active faults. In *Proceedings of the 8th World Conference on Earthquake Engineering* 1, 231–238.

Cherry, S. (1974) Design input for seismic analysis. In *Engineering Seismology and Earthquake Engineering*, Solnes, J. (ed). *NATO Advanced Study Institute Series, Applied Sciences* 3, 151–162.

Cook, N.G.W. (1976) Seismicity associated with mining. *Engineering Geology* 10, 99–122.

De Bruyn, I.A. & Bell, F.G. (1997) Mining induced seismicity in South Africa: a survey. In *Proceedings of the International Symposium on Engineering Geology and the Environment, Athens* 3, Marinos, P.G., Koukis, G.C., Tsiamboas, G.C. & Stournaras, G.C. (eds). Balkema, Rotterdam, pp. 2321–2326.

Degg, M.R. (1992) The ROA Earthquake Hazard Atlas project: recent work from the Middle East. In *Geohazards: Natural and Man-Made*, McCall, G.J.H., Laming, D.J.C. & Scott, S.C. (eds). Chapman and Hall, London, pp. 93–104.

Donovan, N.C., Bolt, B.A. & Whitman, R.V. (1978) Development of expectancy maps and risk analysis. In *Proceedings of the American Society of Civil Engineers, Journal Structural Division* **1**, 1170–1192.

Eckel, E.B. (1970) The Alaskan earthquake, March 27, 1964: lessons and conclusions. *US Geological Survey Professional Paper* **546**, Washington, DC.

Esper, P. & Tochibana, E. (1988) The lesson of Kobe earthquake. In *Geohazards and Engineering Geology, Engineering Geology Special Publication* **14**, Eddleston, M. & Maund, J. (eds). Geological Society, London.

Evans, D.M. (1966) Man-made earthquakes in Denver. *Geotimes* **10**, 11–18.

Fournier d'Albe, E. (1982) An approach to earthquake risk management. *Engineering Structures* **4**, 147–152.

Gibbs, J.F., Healy, J.H., Raleigh, G.B. & Cookley, J. (1973) Seismicity in Rangely, Colorado, area: 1962–70. *Bulletin of the Seismological Society of America* **63**, 1557–1570.

Hazzard, A.O. (1978) Earthquake and engineering: structural design. *The Consulting Engineer* **May**, 14–23.

Housner, G.W. (1970) Strong ground motion. In *Earthquake Engineering*, Weigel, R.L. (ed). Prentice-Hall, Englewood Cliffs, New Jersey, pp. 75–92.

Ishihara, R. & Okada, S. (1978) The effects of stress history on cyclic behaviour of sand. *Soils and Foundation* **14**, 31–45.

Karnik, V. & Algermissen, S.T. (1979) Seismic zoning. In *The Assessment and Mitigation of Earthquake Risk*. UNESCO, Paris, pp. 11–47.

Lane, R.G.T. (1972) Seismic activity at man-made reservoirs. *Proceedings of the Institution of Civil Engineers, Paper* **7416**, 15–24.

Lomnitz, C. & Rosenblueth, E. (1976) *Seismic Risk and Engineering Decisions*. Elsevier, Amsterdam.

McCann, D.M. (1988) Induced seismicity in engineering. In *Engineering Geology of Underground Movements, Engineering Geology Special Publication* **5**, Bell, F.G., Culshaw, M.G., Cripps, J.C. & Lovell, M.A. (eds). Geological Society, London, pp. 405–413.

Medvedev, S.V. (1965) *Engineering Seismology*. Israel Program for Scientific Translations, Jerusalem.

Neff, P.A., von Wagener, R.M. & Green, R.W.E. (1992) Houses damaged by mine tremors—distress repair and design. In *Proceedings of the Symposium on Construction over Mined Areas, Pretoria*. South African Institution of Civil Engineers, Yeoville, pp. 133–137.

Peck, R.B. (1979) Liquefaction potential: science vs. practice. In *Proceedings of the American Society of Civil Engineers, Journal Geotechnical Engineering Division* **105**, 393–398.

Raleigh, G.B., Sieh, K.E., Sykes, L.R. & Anderson, D.L. (1982) Forecasting southern Californian earthquakes. *Science* **217**, 1097–1104.

Redmayne, D.W. (1988) Mining induced seismicity in U.K. coalfields identified on the BGS National Seismograph Network. In *Engineering Geology of Underground Movements, Engineering Geology Special Publication* **5**, Bell, F.G., Culshaw, M.G., Cripps, J.C. & Lovell, M.A. (eds). Geological Society, London, pp. 405–413.

Reid, H.F. (1908) *The Californian Earthquake of April 18, 1906. Report of the State Earthquake Investigation Commission*, Vol. 2, *The Mechanics of the Earthquake*. Carnegie Institution, Washington, DC.

Richter, C.F. (1935) An instrumental earthquake scale. *Bulletin of the Seismological Society of America* **25**, 1–32.

Richter, C.F. (1956) *Elementary Seismology*. W.H. Freeman, San Francisco.

Rikitake, T. (1979) Classification of earthquake precursors. *Tectonophysics* **54**, 293–309.

Rothe, G.H. & Lui, C. (1983) Possibility of induced seismicity in the vicinity of Sleepy Hollow oilfield, south western Nebraska. *Bulletin of the Seismological Society of America* **73**, 1357–1367.

Seed, H.B. (1970) Soil problems and soil behaviour. In *Earthquake Engineering*, Wiegel, R.L. (ed). Prentice-Hall, Englewood Cliffs, New Jersey, pp. 227–252.

Singh, S.K., Mena, E. & Castro, R. (1988) Some aspects of source characteristics of the 19 September 1985 Michoacan earthquake and ground motion amplification in and near Mexico City from strong motion data. *Bulletin of the Seismological Society of America* **78**, 451–477.

Skipp, B.O. & Ambraseys, N.N. (1987) Engineering seismology. In *Ground Engineer's Reference Book*, Bell, F.G. (ed). Butterworth, London, pp. 18/1–18/26.

Sokolenko, V.P. (1977) Landslides and collapses in seismic zones and their prediction. *Bulletin of the Association of Engineering Geologists* **15**, 4–8.

Sokolenko, V.P. (1979) Mapping the after effects of disastrous earthquakes and estimation of hazard for engineering construction. *Bulletin of the Association of Engineering Geologists* **19**, 138–142.

4   Mass movements

4.1 Soil creep and valley bulging

Mass movements on slopes can range in magnitude from soil creep, on the one hand, to instantaneous and colossal landslides on the other. Sharpe (1938) defined creep as the slow downslope movement of superficial rock or soil debris, which is usually imperceptible except by observations of long duration. Walker *et al.* (1987) suggested that creep could be regarded as mass movement which occurs at less than 0.06 m per year. Creep is a more or less continuous process, and is a distinct surface phenomenon. It occurs on slopes with gradients somewhat in excess of the angle of repose of the material involved. Like landslip, its principal cause is gravity, although it may be influenced by seasonal changes in temperature, and by swelling and shrinkage in surface rocks. Other factors which contribute towards creep include interstitial rainwashing, ice crystals heaving stones and particles during frost and the wedging action of rootlets. The liberation of stored strain energy in the weathered zone, particularly of overconsolidated clays with strong diagenetic bonds, is another contributory cause of creep. Although creep movement is exceedingly slow, there are occasions on record when it has carried structures with it.

Evidence of soil creep may be found on almost every soil covered slope. For example, it occurs in the form of small terracettes, the downslope tilting of poles, the curving downslope of trees and soil accumulation on the uphill sides of walls (Fig. 4.1). Indeed, walls may be displaced or broken and, sometimes, roads may be moved out of alignment. The rate of movement depends not only on the climatic conditions and the angle of slope, but also on the soil type and parent material.

Talus (scree) creep occurs wherever a steep talus exists. Its movement is quickest and slowest in cold and arid regions, respectively.

Solifluction is a form of creep which occurs in cold climates or high altitudes, where masses of saturated rock waste move downslope. Generally, the bulk of the moving mass consists of fine debris, but blocks of appreciable size may also be moved. Saturation may be due to either water from rain or melting snow. Moreover, in periglacial regions associated with permafrost, water cannot drain into the ground because it is frozen permanently. Solifluction differs from mudflow in that it moves much more slowly, the movement is continuous and it occurs over the whole slope. Solifluction processes vary at different altitudes owing to the progressive comminution of materials in their downward migration and to differences in the growth of vegetation. At higher elevations, surfaces are irregular and terraces are common; at lower elevations, the most common solifluction phenomena are continuous aprons of detritus which skirt the bases of all the more prominent relief features.

Valley bulges consist of folds (Fig. 4.2) formed by the mass movement of argillaceous material in valley bottoms, the argillaceous material being overlain by thick competent strata. The amplitude of the fold can reach 30 m in those instances where a single anticline occurs along the line of the valley. Alternatively, the valley floor may be bordered by a pair of reverse faults or a belt of small-scale folding.

The valleyward movement of argillaceous material results in cambering of the overlying competent strata, blocks of which become detached and move down the hillside. Fracturing of cambered strata produces deep, debris-filled cracks or 'gulls' which run parallel to the trend of the valley. Some gulls may be several metres wide. Small gulls are sometimes found in relatively flat areas away from the slopes with which they are associated.

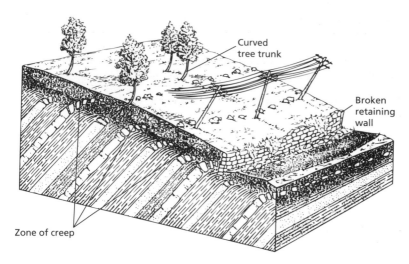

Fig. 4.1 Diagram of creep showing associated features, tops of beds being turned downslope, trees and telegraph poles being tilted downslope, stones orientated downslope and soil piling against retaining wall.

Fig. 4.2 Valley bulging in interbedded shales and thin sandstones of Namurian age revealed during the excavation for the dam for Howden reservoir in 1933, South Yorkshire, England. (Courtesy of the Severn–Trent Water Authority.)

4.2 Causes of landslides

Varnes (1978) defined landslides as the downward and outward movement of slope-forming materials composed of rocks, soils or artificial fills. The dis-

placed material has well defined boundaries. Movement may take place by falling, sliding or flowing, or some combination of these factors. This movement generally involves the development of a slip surface between the separating and remaining masses. However, rockfalls, topples and debris flows involve little or no true sliding on a slide surface. The majority of stresses found in most slopes are the sum of the gravitational stress from the self-weight of the material plus the residual stress.

In most landslides, a number of causes contribute towards movement, and any attempt to decide which factor finally produces the failure is not only difficult but pointless. Often, the final factor is nothing more than a trigger mechanism that sets in motion a mass which is already on the verge of failure. Hence, elements which influence slope stability are numerous and varied, and interact in complex and often subtle ways. Basically, however, landslides occur because the forces creating movement, the disturbing forces (M_D), exceed those resisting it, the resisting forces (M_R), that is, the shear strength of the material concerned. In general terms, therefore, the stability of a slope may be defined by a factor of safety (F) where

$$F = M_R / M_D \qquad (4.1)$$

If the factor of safety exceeds unity, the slope is stable, whereas, if it is less than unity, the slope is unstable.

The causes of landslides were grouped into two categories by Terzaghi (1950), namely, internal causes and external causes (Table 4.1). The former included

Table 4.1 Processes leading to landslides. (After Terzaghi, 1950.)

Name of agent	Event or process which brings agent into action	Mode of action of agent	Slope materials most sensitive to action	Physical nature of significant actions of agent	Effects on equilibrium conditions of slope
Transporting agent	Construction operations or erosion	(1) Increase in height or rise of slope	Every material	Changes state of stress in slope-forming material	Increases shearing stresses
			Stiff, fissured clay, shale	Changes state of stress and causes opening of joints	Increases shearing stresses and initiates process (8)
Tectonic stresses	Tectonic movements	(2) Large scale deformations of Earth's crust	Every material	Increases slope angle	Increases shearing stresses
Tectonic stresses or explosives	Earthquakes or blasting	(3) High frequency vibrations	Every material	Produces transitory change in stress	Increases shearing stresses
			Loess, slightly cemented sands and gravel	Damages intergranular bonds	Decrease in cohesion and increase in shearing stresses
			Medium or fine, loose sand in saturated state	Initiates rearrangement of grains	Spontaneous liquefaction
Weight of slope-forming material	Process which created the slope	(4) Creep on slope	Stiff, fissured clay, shale remnants of old slides	Opens up closed joints, produces new ones	Reduces cohesion, accelerates process (8)
		(5) Creep in weak stratum below foot of slope	Rigid materials resting on plastic ones	Opens up closed joints, produces new ones	Reduces cohesion, accelerates process (8)
Water	Rains or melting snow	(6) Displacement of air in voids	Moist sand	Increases pore water pressure	Decrease in frictional resistance
		(7) Displacement of air in open joints	Jointed rock, shale	Increases pore water pressure	Decrease in frictional resistance
		(8) Reduction of capillary pressure associated with swelling	Stiff, fissured clay and some shales	Causes swelling	Decrease in cohesion
		(9) Chemical weathering	Rock of any kind	Weakens intergranular bonds (chemical weathering)	Decrease in cohesion
	Frost	(10) Expansion of water due to freezing	Jointed rock	Widens existing joints, produces new ones	Decrease in cohesion

Continued on p.86

Table 4.1 *Continued.*

Name of agent	Event or process which brings agent into action	Mode of action of agent	Slope materials most sensitive to action	Physical nature of significant actions of agent	Effects on equilibrium conditions of slope
		(11) Formation and subsequent melting of ice layers	Silt and silty sand	Increases water content of soil in frozen top layer	Decrease in frictional resistance
	Dry spell	(12) Shrinkage	Clay	Produces shrinkage cracks	Decrease in cohesion
	Rapid drawdown	(13) Produces seepage towards foot of slope	Fine sand, silt, previously drained	Produces excess pore water pressure	Decrease in frictional resistance
	Rapid change of elevation of water table	(14) Initiates rearrangement of grains	Medium or fine, loose sand in saturated state	Spontaneous increase in pore water pressure	Spontaneous liquefaction
	Rise of water table in distant aquifer	(15) Causes a rise of piezometric surface in slope-forming material	Silt or sand layers between or below clay layers	Increases pore water pressure	Decrease in frictional resistance
	Seepage from artificial source of water (reservoir or canal)	(16) Seepage towards slope	Saturated silt	Increases pore water pressure	Decrease in frictional resistance
		(17) Displaces air in the voids	Moist, fine sand	Eliminates surface tension	Decrease in cohesion
		(18) Removes soluble binder	Loess	Destroys intergranular bonds	Decrease in cohesion
		(19) Subsurface erosion	Fine sand or silt	Undermines the slope	Increase in shearing stresses

those mechanisms within the mass which brought about a reduction of its shear strength to a point below the external forces imposed on the mass by its environment, thus inducing failure. External mechanisms were those outside the mass involved, which were responsible for overcoming its internal shear strength, thereby causing it to fail.

The common force tending to generate movements on slopes is, of course, gravity. Generally, the steeper the slope, the greater the likelihood that landslides will occur. Obviously, there is no universal threshold value at which slides take place, because this must be related to the ground conditions. Many steep slopes on competent rock are more stable than comparatively gentle slopes on weak material. Nonetheless, numerous authors have reported critical angles below

which sliding does not occur in particular soils or weak rock types (Walker *et al.*, 1987).

Climatic conditions act in a number of ways to promote the occurrence of landslides. Rainfall is the most important climatic factor. Landslides can be triggered by rainfall if some threshold intensity is exceeded, so that pore water pressures are increased by a required amount (Olivier *et al.*, 1994). The process is also affected by the duration of the rainfall, the slope angle, the distribution of shear strength and permeability, particularly within the regolith, the effects of soil pipes, variations in regolith thickness and antecedent weather and the pore water pressure conditions. Jahns (1978) suggested that larger, deeper slides were more likely to occur during long, continuously wet periods, whereas shallow slides were

Table 4.2 Effects of vegetation on slope stability (Greenway, 1987).

Hydrological mechanisms	Influence
1 Foliage intercepts rainfall, causing absorptive and evaporative losses that reduce rainfall available for infiltration	B
2 Roots and stems increase the roughness of the ground surface and the permeability of the soil, leading to increased infiltration capacity	A
3 Roots extract moisture from the soil, which is lost to the atmosphere via transpiration, leading to lower pore water pressures	B
4 Depletion of soil moisture may accentuate desiccation cracking in the soil, resulting in higher infiltration capacity	A

Mechanical mechanisms	
5 Roots reinforce the soil, increasing soil shear strength	B
6 Tree roots may anchor into firm strata, providing support to the upslope soil mantle through buttressing and arching	B
7 Weight of trees surcharges the slope, increasing normal and downhill force components	A/B
8 Vegetation exposed to the wind transmits dynamic forces into the slope	A
9 Roots bind soil particles at the ground surface, reducing their susceptibility to erosion	B

A, adverse to stability; B, beneficial to stability.

more characteristic of short-duration, high-intensity rainfall.

The influence of vegetation on slope stability has been examined by several authors and is summarized in Table 4.2. Greenway (1987) provided details of how to assess the relative importance of the various factors outlined in Table 4.2 in relation to slope stability. According to Varnes (1984), distressed vegetation may help to demonstrate slope movement, either in the field or on aerial photographs. Removal of vegetative cover alters the hydrological and hydrogeological conditions of a slope, and frequently leads to accelerated run-off and, consequently, increased erosion and increased probability of slides and debris flow.

An increase in the weight of slope material means that shearing stresses are increased, leading to a decrease in the stability of a slope, which may ultimately give rise to a slide. This can be brought about by natural or artificial (man-made) activity. For instance, removal of support from the toe of a slope, either by erosion or excavation, is a frequent cause of slides, as is overloading the top of a slope. Such slides are external slides in that an external factor causes failure.

In many parts of the world, marine erosion on near-present coastlines was halted by the glacio-ustatic lowering of the sea level during Pleistocene times and recommenced on subsequent recovery. For example, landslides around the English coast generally were reactivated by rising sea levels in Flandrian times, that is, some 4000–8000 years BP. Hutchinson (1992) stated that, once the sea level became reasonably constant, erosion continued at a steady pace, giving rise to coastal landslides. A cyclic situation then develops in which landslide material is removed by the sea, steepening the cliffs and leading to further landsliding. Hence, extended periods of slow movement are succeeded by sudden first-time failures.

Other external mechanisms include earthquakes (see Chapter 3) or other shocks and vibrations. Earthquake shocks and vibrations in granular soils not only increase the external stresses on slope material, but can cause a reduction in the pore space, which effectively increases the pore water pressure. The area affected by landslides caused by an earthquake is influenced by the magnitude of the earthquake. For example, Keefer (1984) suggested that an earthquake with a Richter magnitude of 4 probably would not generate landslides, whereas an earthquake with a magnitude of 9.2 would cause landslides to take place over an area as large as $500000\,km^2$. He further suggested that rockfalls, rock slides, soil falls and soil slides are triggered by the weaker seismic tremors, whereas deep-seated slides and earthflows generally are the result of stronger earthquakes. Materials which are particularly susceptible to earthquake motions include loess, volcanic ash on steep slopes, saturated sands of low density, quickclays and loose boulders on slopes. The most severe losses of life generally have been caused by earthquake-induced landslides, for example, that which occurred in Khansu Province, China, in 1920 killed around 200 000 people.

Internal slides generally are caused by an increase in pore water pressure within the slope material, which causes a reduction in the effective shear strength. Indeed, it generally is agreed that, in most landslides, groundwater constitutes the most important single contributory cause. Therefore, the identification of the source and amount of water, water movement and development of excess pore water pressure is important. An increase in water content also means an increase in the weight of the slope material or its bulk density, which can induce slope failure. Significant volume changes may occur in some materials, notably clays, on wetting and drying out. Not only does this weaken the clay by developing desiccation cracks within it, but the enclosing strata may also be affected adversely. Rises in the levels of water tables because of short-duration, intense rainfall or prolonged rainfall of lower intensity are a major cause of landslides (Bell, 1994). At times, the water table may perch on the failure surface of a landslide. Seepage forces within a granular soil can produce a reduction in strength by reducing the number of contacts between grains. Water can also weaken slope material by causing minerals to alter or by bringing about their solution.

Precipitation is a time-dependent factor which influences the pore water pressure. Shallow slides can develop after high rainfall. For example, proximate rainfall, according to Brand *et al.* (1984), tends to trigger shallow slides in Hong Kong, whereas antecedent rainfall is an important factor for deeper slides. Another factor which influences saturated, low permeability soils is the long duration of equalization of pore water pressures after unloading. For example, for the London Clay, it has been suggested by Vaughan and Walbanke (1973) that the equilibration of the pore water pressure could take thousands of years. Hence, this can exercise significant control on the delayed failure of certain slopes, such as those developed in heavily overconsolidated clay.

Weathering can cause a reduction in the strength of slope material, leading to sliding. The necessary breakdown of equilibrium to initiate sliding may take decades. For example, Chandler (1974) quoted a case of slope failure in Lias Clay in Northamptonshire, which was primarily due to swelling in the clay. It took 43 years to reduce the strength of the clay below the critical level at which sliding occurred. Indeed, in relatively impermeable, cohesive soils, the swelling process is probably the most important factor leading to a loss of strength and therefore to delayed failure. On the other hand, progressive softening of fissured clays involved in a constructed slope can occur whilst construction is taking place.

4.3 Classification of landslides

There are many classifications of landslides which, to some extent, is due to the complexity of slope movements (Hansen, 1984). However, by far the most widely used classification is that of Varnes (1978). Varnes classified landslides according to the type of movement that occurred, on the one hand, and the type of materials involved, on the other (Fig. 4.3). Types of movement were grouped into falls, slides and flows. The materials concerned were simply grouped as rocks and soils. Obviously, one type of slope failure may grade into another, for example, slides often turn into flows. Complex slope movements are those in which there is a combination of two or more principal types of movement. Multiple movements are those in which repeated failures of the same type occur in succession, and compound movements are those in which the failure surface is formed from a combination of curved and planar sections.

4.3.1 Falls

Falls are very common (Fig. 4.4). The moving mass travels mostly through the air by free fall, saltation or rolling, with little or no interaction between the moving fragments. The movements are very rapid and may not be preceded by minor movements. A rockfall event involves a single block or group of blocks which become detached from a rock face; each block may be a falling block behaving more or less independently of other blocks (Culshaw & Bell, 1992). Blocks may be broken during the fall. There is a temporary loss of ground contact and high acceleration during the descent, with blocks attaining signifi-

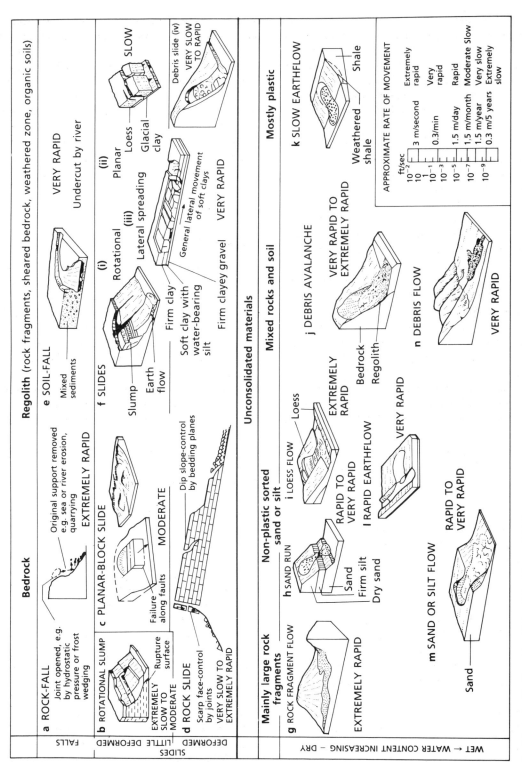

Fig. 4.3 A classification of landslides. (After Varnes, 1978.)

Fig. 4.4 Rockfall on Table Mountain, Cape Town, South Africa.

cant kinetic energy. Blocks accumulate at the bottom of a slope as a scree deposit. If a rockfall is active or very recent, the slope from which it was derived is scarped. Freeze–thaw action is one of the major causes of rockfall.

Toppling failure is a special type of rockfall which can involve considerable volumes of rock. The danger of slope toppling increases with increasing discontinuity angle, and steep slopes in vertically jointed rocks frequently exhibit signs of toppling failure.

4.3.2 Slides

In true slides, the movement results from shear failure along one or several surfaces, such surfaces offering the least resistance to movement. The mass involved may or may not experience considerable deformation. One of the most common types of slide occurs in clay soils, where the slip surface is approximately spoon shaped. Such slides are referred to as rotational slides (Fig. 4.5). They are commonly deep seated (0.15 × depth/length < 0.33). Backward rotation of the failed mass is the dominant characteristic, and the failed material remains intact to the extent that only one or a few discrete blocks are likely to form.

Rotational slides usually develop from tension scars in the upper part of a slope. The tension cracks at the head of a rotational slide are generally concentric and parallel to the main scar. Undrained depressions and perimeter lakes, bounded upward by the

Fig. 4.5 Rotational slide in Edale Shales, Mam Tor, Derbyshire, England. No attempt is to be made to reconstruct the road.

main scar, characterize the head regions of many rotational slides.

When the scar at the head of a rotational slide is almost vertical and unsupported, further failure usually is just a matter of time. As a consequence, successive rotational slides occur until the slope is stabilized. These are retrogressive slides and they develop in a headward direction. All multiple retrogressive slides have a common basal shear surface in which the individual planes of failure are combined.

Non-circular slips occur in overconsolidated clays in which weathering has led to the development of quasi-planar slide surfaces, or in unweathered, structurally anisotropic clays. Circular and noncircular, shallow rotational slips tend to form on moderately inclined slopes in weathered or colluvial clays.

Translational slides occur in inclined, stratified deposits, the movement taking place along a planar surface, frequently a bedding plane. The mass involved in the movement becomes dislodged because the force of gravity overcomes the frictional resistance along the potential slip surface, the mass having been detached from the parent rock by a prominent discontinuity, such as a major joint. Slab slides, in which the slip surface is roughly parallel to the ground surface, are a common type of translational slide. Such a slide may progress almost indefinitely if the slip surface is sufficiently inclined and the resistance along it is less than the driving force. Slab slides can occur on gentler surfaces than rotational slides and may be more extensive.

Failures which involve lateral spreading may develop in clays, quickclays and varved clays. This type of failure is due to high pore water pressure in a more permeable zone at a relatively shallow depth, the dissipation of pore water pressure leading to the mobilization of the clay above. The movement usually is complex, being dominantly translational, although rotation and liquefaction, and consequent flow, may also be involved. Such masses, however, generally move over a planar surface and may split into a number of semi-independent units. Like other landslides, these generally are sudden failures, although sometimes movement takes place slowly.

Rock slides and debris slides usually are the result of a gradual weakening of the bonds within a rock mass, and generally are translational in character. Most rock slides are controlled by the discontinuity patterns within the parent rock. Water is seldom an important direct factor in causing rock slides, although it may weaken bonding along joints and bedding planes. Freeze–thaw action, however, is an important cause. Rock slides commonly occur on steep slopes, and most are of single rather than multiple occurrence. They are composed of rock boulders. Individual fragments may be very large and may move great distances from their source. Debris slides usually are restricted to the weathered zone or to surficial talus (Fig. 4.6). With increasing water content, debris slides grade into mudflows. These slides often are limited by the contact between the loose material and the underlying firm bedrock.

4.3.3 Flows

In a flow, the movement resembles that of a viscous fluid. In other words, as movement downslope

Fig. 4.6 Debris slide, Arthur's Pass, South Island, New Zealand.

continues, intergranular movements become more important than shear surface movements. Slip surfaces usually are not visible or are short lived, and the boundary between the flow and the material over which it moves may be sharp or may be represented by a zone of plastic flow. Some content of water is necessary for most types of flow movement, but dry flows can and do occur. Consequently, the water content in flows must be regarded as ranging from dry at one extreme to saturated at the other. Dry flows, which consist predominantly of rock fragments, are simply referred to as rock fragment flows or rock avalanches, and generally result from a rock slide or rockfall turning into a flow. They usually are very rapid and short lived, and frequently are composed mainly of silt or sand. As would be expected, they are of frequent occurrence in rugged mountainous regions, where they usually involve the movement of many millions of tonnes of material. Wet flows occur when fine grained soils, with or without coarse debris, become mobilized by an excess of water. They may be of great length.

Progressive failure is rapid in debris avalanches, and the whole mass, either because it is quite wet or is on a steep slope, moves downwards, often along a stream channel, and advances well beyond the foot of a slope. Lumb (1975) reported speeds of $30\,\mathrm{m\,s^{-1}}$ for debris avalanches in Hong Kong. Debris avalanches generally are long and narrow, and frequently leave V-shaped scars tapering headwards. These gullies often become the sites of further movement.

Debris flows are distinguished from mudflows on the basis of particle size, the former containing a high percentage of coarse fragments and the latter consisting of at least 50% sand size or less. Almost invariably, debris flows follow unusually heavy rainfall or the sudden thaw of frozen ground. These flows are of high density, perhaps 60–70% solids by weight, and are capable of carrying large boulders. Like debris avalanches, they commonly cut V-shaped channels, at the sides of which coarser material may accumulate as the more fluid central area moves downchannel. Debris may move over many kilometres.

Mudflows may develop when a rapidly moving stream of stormwater mixes with a sufficient quantity of debris to form a pasty mass (Fig. 4.7). Because such mudflows frequently occur along the same courses, they should be kept under observation when significant damage is likely to result. Mudflows frequently move at rates ranging between 10 and $100\,\mathrm{m\,min^{-1}}$, and can travel over slopes inclined at 1° or less. Indeed, they usually develop on slopes with shallow inclinations, that is, between 5° and 15°. Skempton and Hutchinson (1969) observed that mudflows also develop along discretely sheared boundaries in fissured clays and varved or laminated fluvio-glacial deposits, where the ingress of water has led to softening at the shear zone. Movement involves the development of forward thrusts, due to undrained loading of the rear part of the mudflow, where the basal shear surface is inclined steeply downwards. A mudflow continues to move down shallow slopes due to this

Fig. 4.7 Mudflow in colluvial ground, Durban, South Africa.

undrained loading, which is implemented by frequent small falls or slips of material from a steep rear scarp onto the head of the moving mass. This not only aids instability by loading, but also raises the pore water pressures along the back part of the slip surface.

An earthflow involves mostly cohesive or fine-grained material, which may move slowly or rapidly. The speed of movement is, to some extent, dependent on the water content, in that the higher the content, the faster the movement. Slowly moving earthflows may continue to move for several years. These flows generally develop as a result of a build-up of pore water pressure, so that part of the weight of the material is supported by interstitial water with a consequent decrease in the shearing resistance. If the material is saturated, a bulging frontal lobe is formed, and this may split into a number of tongues which advance with a steady rolling motion. Earthflows frequently form the spreading toes of rotational slides due to the material being softened by the ingress of water. Skempton and Hutchinson (1969) restricted the term earthflow to slow movements of softened weathered debris, as forms at the toe of a slide. They maintained that movement was transitional between a slide and a flow, and that earthflows accommodated less breakdown than mudflows.

4.4 Landslides in soils

Displacement in soil, usually along a well defined plane of failure, occurs when the shear stress rises to the value of the shear strength. The shear strength of the material along the slip surface is reduced to its residual value, so that subsequent movement can take place at a lower level of stress. The residual strength of a soil is of fundamental importance in the behaviour of landslides (Skempton, 1964) and in progressive failure (Bishop, 1971).

A slope in dry, frictional soil should be stable provided that its inclination is less than the angle of repose. Slope failure tends to be caused by the influence of water. For instance, seepage of groundwater through a deposit of sand in which slopes exist can cause them to fail. Failure on a slope composed of granular soil involves the translational movement of a shallow surface layer. The slip often is appreciably longer than its depth. This is because the strength of granular soils increases rapidly with depth. If, as generally is the case, there is a reduction in the density of

the granular soil along the slip surface, the peak strength is reduced ultimately to the residual strength. The soil will continue shearing without a further change in volume once it has reached its residual strength. Although shallow slips are common, deep-seated shear slides can occur in granular soils. They usually are due either to rapid drawdown or to the placement of heavy loads at the top of the slope.

In cohesive soils, the slope and height are interdependent, and can be determined when the shear characteristics of the material are known. Because of their water retaining capacity, due to their low permeability, pore water pressures are developed in cohesive soils. These pore water pressures reduce the strength of an element of the failure surface within a slope in cohesive soil, and the pore water pressure at that point needs to be determined to obtain the total and effective pressure. This effective pressure is then used as the normal stress in a shear box or triaxial test to assess the shear strength of the soil concerned. Skempton (1964) showed that, on a stable slope in clay, the resistance offered along a slip surface, that is, its shear strength (s), is given by

$$s = c' + (\sigma - u)\tan\phi' \tag{4.2}$$

where c' is the cohesion intercept, ϕ' is the angle of shearing resistance (these are average values around the slip surface and are expressed in terms of the effective stress), σ is the total overburden pressure and u is the pore water pressure. In a stable slope, only part of the total available shear resistance along a potential slip surface will be mobilized to balance the total shear force (τ); hence

$$\Sigma\tau = \Sigma c'/F + \Sigma(\sigma - u)\tan\phi'/F \tag{4.3}$$

If the total shear force equals the total shear strength, a slip can occur (i.e. $F = 1.0$).

Cohesive soils, especially in short-term conditions, may exhibit relatively uniform strength with increasing depth. As a result, slope failures, particularly short-term failures, may be comparatively deep seated, with roughly circular slip surfaces. This type of failure is typical of relatively small slopes. Landslides on larger slopes often have non-circular failure surfaces following bedding planes or other weak horizons.

When shear failure occurs for the first time in an unfissured clay, the undisturbed material lies very nearly at its peak strength, and it is the effective peak friction angle (ϕ'_p) and the effective peak cohesion (c'_p) that are used in the analysis. In slopes excavated in fissured, overconsolidated clay, although stable initially, there is a steady decrease in the strength of the clay towards a long-term residual condition. During the intermediate stages, swelling and softening, due to the dissipation of residual stress and the ingress of water along fissures which open on exposure, take place. Large strains can occur locally due to the presence of the fissures, and considerable non-uniformity of shear stress along the potential failure surface and local overstressing lead to progressive slope failure. In 1964, Skempton showed that, if a clay is fissured, initial sliding occurs at a value below the peak strength. However, the residual strength is reached only after considerable slip movement has taken place, so that the strength relevant to first-time slip lies between the peak and residual values. Skempton accordingly introduced the term residual factor (R), which he defined as

$$R = \frac{\text{Peak shear strength} - \text{mean shear stress at failure}}{\text{Peak shear strength} - \text{residual shear strength}} \qquad (4.4)$$

The residual factor represents the proportion of the slip surface over which the strength has deteriorated to the residual value. Therefore, the residual factor is used in analysis when the residual strength has not developed over the entire slip surface. It allows the mean shear stress at failure to be used in the calculation of the factor of safety.

Chandler (1977) showed that, where drainage is impeded in cohesive soils, the threshold slope for landsliding approximates to half the angle of the residual shearing resistance (ϕ'_r). However, he went on to point out that, as the normal stresses increase, the value of ϕ'_r frequently falls. This means that, where there is a potential for deep-seated landsliding on clay slopes, the range of normal stress will be large and there will not be a unique value of ϕ'_r. As a consequence, larger landslides will move on flatter slopes than smaller landslides, other factors being equal.

4.5 Landslides in rock masses

Landsliding of steep slopes in hard, unweathered rock (defined as rock with an unconfined compressive strength of 35 MPa and over) is largely dependent on the incidence, orientation and nature of the discontinuities present. It is only in very high slopes and/or weak rocks that failure in intact material becomes significant. Data relating to the spatial relationships between discontinuities afford some indication of the modes of failure which may occur, and information relating to the shear strength along discontinuities is required for use in stability analysis. As in soil, the shearing resistance of rock with a random pattern of jointing can be obtained from the Coulomb equation. The value of the angle of shearing resistance (ϕ) depends on the type and degree of interlocking between the blocks on either side of the surface of sliding but, in such rock masses, interlocking is independent of the orientation of the surface of sliding.

In a bedded and jointed rock mass, if the bedding planes are inclined, the critical slope angle depends upon their orientation in relation to the slope and the orientation of the joints. The relation between the angle of shearing resistance (ϕ) along a discontinuity at which sliding would occur under gravity and the inclination of the discontinuity (α) is important. If $\alpha < \phi$, the slope would be stable at any angle, whereas, if $\phi < \alpha$, gravity would induce movement along the dis-

Table 4.3 Sensitivity of factor of safety to various parameters. (After Richards et al., 1978.)

Functions affecting the factor of safety (for a given value of slope height and angle).

Function	Probable range of magnitude
Unit weight	$0-300\,\text{kN m}^{-3}$
Cohesion	$15-30\,\text{kPa}$
Water pressure	$0-H\,\text{m}$
Friction angle	$0-60°$
Joint inclination	$10-50°$

Order of importance of functions

	Slope height		
Rank	10 m	100 m	1000 m
1	Joint inclination	Joint inclination	Joint inclination
2	Cohesion	Friction angle	Friction angle
3	Unit weight	Cohesion	Water pressure
4	Friction angle	Water pressure	Cohesion
5	Water pressure	Unit weight	Unit weight

continuity surface, and the slope would not exceed the critical angle, which would have a maximum value equal to the inclination of the discontinuities. It must be borne in mind, however, that rock masses generally are interrupted by more than one set of discontinuities.

Hard rock masses are liable to sudden and violent failure if their peak strength is exceeded in an excessively steep or high slope. On the other hand, soft materials, which exhibit small differences between peak and residual strengths, tend to fail by gradual sliding. The relative sensitivity of the factor of safety to the variation in importance of each parameter

which influences the stability of slopes depends initially on the height of the slope. For example, Richards *et al.* (1978) graded each parameter concerned in order of importance with respect to its effect on the factor of safety in relation to slopes with heights of 10, 100 and 1000 m. The results are shown in Table 4.3 and Fig. 4.8. The heights and angles of slopes in hard rocks can be estimated roughly from Fig. 4.8. The joint inclination is always the most important parameter as far as the slope stability is concerned. Friction is the next most important parameter for slopes of medium and large height, whereas the unit weight is more important than friction for

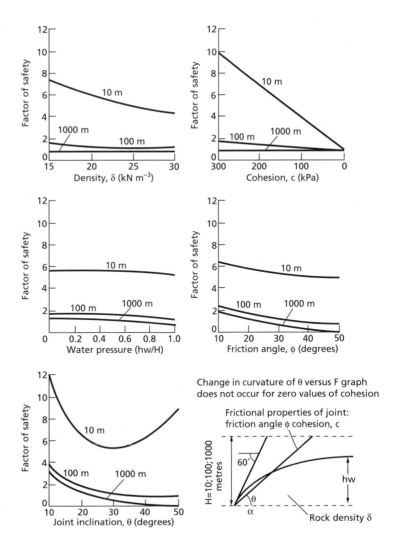

Fig. 4.8 Sensitivity analysis for slope stability calculations. (After Richards *et al.*, 1978.)

small slopes. Cohesion becomes less significant with increasing slope height, whilst the converse is true as far as the effects of water pressure are concerned.

The shear strength along a joint is mainly attributable to the basic frictional resistance which can be mobilized on opposing joint surfaces. Normally, the basic friction angle (ϕ_b) approximates the residual strength of the discontinuity. An additional resistance is consequent upon the roughness of the joint surface. Shearing at low normal stresses occurs when the asperities are overridden; at higher confining conditions and stresses, they are sheared through. The shear strength along a discontinuity is also influenced by the presence and type of any fill material, and by the degree of weathering along the discontinuity.

According to Hoek and Bray (1981), in most hard rock masses, neither the angle of friction nor the cohesion is dependent upon the moisture content to a significant degree. Consequently, any reduction in shear strength is almost solely attributable to a reduc-

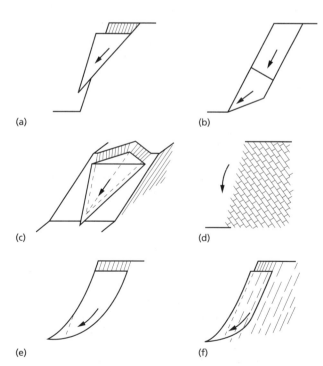

Fig. 4.9 Idealized failure mechanisms in rock slopes: (a) plane; (b) active and passive blocks; (c) wedge; (d) toppling; (e) circular; (f) non-circular.

tion in normal stress across the failure plane, and it is the water pressure, rather than the moisture content, which influences the strength characteristics of a rock mass.

The principal types of failure which are generated in rock slopes are rotational, translational and toppling modes (Fig. 4.9). Rotational failures normally only occur in structureless overburden, highly weathered material or in very high slopes in closely jointed rock. They may develop either circular or non-circular failure surfaces. Circular failures take place where rock masses are intensely fractured, or where the stresses involved override the influence of the discontinuities in the rock mass. Relict jointing may persist in highly weathered materials, along which sliding may take place. These failure surfaces are often intermediate in geometry between planar and circular slides.

There are three kinds of translational failure: plane failure, active and passive block failure and wedge failure. Plane failure is a common type of translational failure, and occurs by sliding along a single plane which daylights into the slope face (i.e. the dip of the failure plane is less than that of the slope). When considered in isolation, a single block may be stable. Forces imposed by unstable adjacent blocks may give rise to active and passive block failures (Fig. 4.9). In wedge failure, two planar discontinuities intersect, the wedge so formed daylighting into the face (Fig. 4.10). In other words, failure may occur if the line of intersection of both planes dips into the slope at an angle less than that of the slope.

Toppling failure generally is associated with steep slopes in which the jointing is near vertical. It involves the overturning of individual blocks, and is therefore governed by the discontinuity spacing as well as the orientation. The likelihood of toppling increases with increasing inclination of the discontinuities. The water pressure within the discontinuities helps to promote the development of toppling.

Rockfalls have been referred to above. Failures involving relatively small volumes of rock often pose great problems, particularly in regions where steep terrain is in close proximity to developed areas or transportation corridors. For instance, Martin (1988) reported that rockfalls, small rock slides and ravelling are the most chronic problems on transportation routes in mountainous areas of North America; millions of dollars are spent annually on maintenance

Fig. 4.10 Wedge failure in Victoria Range, South Island, New Zealand.

and remedial measures to provide protection against such hazards.

4.6 Monitoring slopes

Small movements usually precede slope failure, particularly a catastrophic failure, and accelerating displacement frequently precedes collapse. If these initial small movements are detected in sufficient time, remedial action can be taken to prevent or control further movement. A slope monitoring system provides a means of early warning, and involves the use of sensitive instruments. Other adverse conditions which give rise to instability, notably excess pore water pressures, also require recording.

When there is a lack of adequate data, uncertainties are likely to arise in design. Under such circumstances, if the stability of a slope is in doubt, the expense of a monitoring programme may be justified provided that remedial measures following the detection of incipient failure are feasible and that the cost of monitoring and remedial action is less than the cost of slope failure. Even if a complete picture of the ground conditions is available, the analytical methods may not be able to deal with the complexity of a real situation. Consequently, data must be simplified into an idealized model with a resulting loss of accuracy. Such uncertainties are normally taken into account in the selection of the factor of safety. Monitoring can justify the use of a lower factor of safety

than would otherwise be permissible provided that it is accompanied by contingency plans for remedial action should the slope in question prove to be unstable. Accordingly, the cost of a monitoring system has to be measured against the cost of operating at an uneconomically high safety factor, necessitating either flatter slopes or expensive remedial work. The total value of the project concerned and the cost and effect on the project if slopes fail must also be considered.

Two of the initial steps in the planning and design of a monitoring system are to assess the extent and depth of potentially unstable rock material and to determine the factors of safety against sliding for various modes of failure. This indicates whether there is a problem of slope stability, and aids the choice of instrumentation and its location within the slope.

4.6.1 Monitoring movement

Monitoring of movement provides a direct check on the stability of a slope. Instruments indicate the location, direction and maximum depth of movement, and these results help to determine the extent and depth of treatment which are necessary. Moreover, the same instruments can then be used to determine the effect of this treatment. Monitoring of surface movement can be performed using conventional surveying techniques, electronic distance measurement or laser equipment, providing accurate results.

Surveys should be designed to suit the topography and the anticipated directions of movement. Surveying should extend beyond the limits of possible movement into the surrounding stable area. In this way, any development of surface strain in advance of the appearance of tension cracks can be detected, as can any toe heave. Automated slope monitoring procedures, using total station surveying instruments, can be programmed to take various measurements across a slope (Tran-Duc *et al.*, 1992).

Precise results can be obtained when close-up photographs are taken from ground stations and are measured in a stereocomparator. Movements may be revealed by an examination of a sequence of photographs taken at suitable time intervals. Photographs can be used to evaluate pre-existing ground topography, for the back-analysis of previous landslips and as a basis for mapping.

The appearance of tension cracks at the crest of a slope may provide the first indication of instability. Crack measurements, that is, their width and vertical offsets, should be taken, because they may provide an indication of slope behaviour. Dunnicliff (1988) described a number of ways in which cracks can be measured. These included using a survey tape, a tensional wire crack gauge or a surface extensometer.

Single point tiltmeters can also be used to monitor surface movements, when the surface of the landslide has a rotational component. Portable tiltmeters can be used with a series of reference places, or tiltmeters can be left in place and connected to a datalogger. Dunnicliff (1992) mentioned multipoint, liquid level gauges which have been employed to detect slope movements.

Measurements of subsurface horizontal movements are more important than measurements of subsurface vertical movements. Subsurface movements can be recorded by using extensometers, inclinometers and deflectometers. Borehole extensometers are used to measure the vertical displacement of the ground at different depths. Fixed borehole extensometers include the single rod and multirod types. A single rod extensometer is anchored in a borehole, and movement between the rod and the reference sleeve is monitored. Multirod installations monitor displacements at various depths using rods of varying lengths. Each rod is isolated by a close-fitting sleeve, and the complete assembly is grouted into place, fixing the anchors to the ground, while allowing free movement of each rod within its sleeve. A precise borehole extensometer essentially consists of circular magnets embedded in the ground, which act as markers, and reed switch sensors move in a central access tube to locate the positions of the magnets. A slope extensometer is a multipoint borehole extensometer which uses tensioned wires instead of rods to monitor the deformation perpendicular to the axis of the borehole.

An inclinometer is used to measure horizontal movements below ground (Mikkelsen, 1996). High-accuracy inclinometer measurements frequently represent the initial data relating to subsurface movement. Inclinometers can detect differential movements of 0.17–3.4 mm per 10 m run of hole. Inclinometers designed for permanent installation in a hole usually comprise a chain of pivoted rods. Angular movements between rods may be measured at the pivot points. Another type of permanently installed inclinometer uses a flexible metal strip onto which resistance strain gauges are bonded, which record any bending in the strip induced by ground movements. Fixed position inclinometers monitor the differential lateral movement between the borehole collar and a deep datum, and are used most frequently in slope stability work. Probe inclinometers are inserted into a special casing in a borehole each time a set of readings is required. They incorporate a pendulum, the deflection of which indicates movement.

4.6.2 Monitoring groundwater

Groundwater is one of the most influential factors governing the stability of slopes. Instability problems may be associated with either excessive discharge or excessive pore water pressure. Pore water pressures are recorded by a piezometer, the simplest type comprising a standpipe installed in a borehole. If the minimum head recorded is less than 8 m below ground level, 'closed system' piezometers connected to mercury manometers normally are used. Pressure transducers are necessary where greater heads have to be measured, especially in low permeability ground.

There are important differences between the monitoring of water pressures in rock and in soil. Usually, in rock, the majority of flow takes place via discontinuities rather than through intergranular pore space.

The predominance of fissure flow means that piezometer heads in rock slopes often vary considerably from point to point, and therefore a sufficient number of piezometers must be installed to define the overall conditions. They should be located with reference to the geology, especially with regard to the intersection of major discontinuities in rock masses. This can be facilitated by examining the fracture index, by inspecting the drill-hole with a television camera, by packer testing or by logging the velocity of flow in the drill-hole using micropropeller or dilution methods. The piezometer test section in rock, that is, the permeable filter material between sections of grouted hole, may need to be as long as 4 m in order to incorporate a representative number of water-bearing fissures.

4.6.3 Monitoring acoustic emissions or noise

Movements in rock or soil masses are accompanied by the generation of acoustic emissions or noise. The detection of acoustic emissions is most effective when the amplitude of the signals is high. Hence, detection is more likely in rock masses or cohesionless soils than in cohesive soils. Obviously, when a slope collapses, noise is audible, but subaudible noises are produced at earlier stages in the development of instability. Normally, the rate of these microseismic occurrences increases rapidly with the development of instability. Such noises can be picked up by an array of geophones located in the vicinity of the slope or in shallow boreholes. Most movements generating noise originate near or along the plane of failure, so that seismic detection helps to locate the depth and extent of the surface of sliding.

4.7 Landslide hazard, investigation and mapping

Landslide hazard refers to the probability of a landslide of a given size occurring within a specified period of time within a particular area. The associated risk is the associated loss of lives or damage to property. Consequently, landslide hazard must be assessed before landslide risk can be estimated.

Schuster and Fleming (1986) estimated that landslides in the USA cause between 25 and 50 deaths per year and 1–2 billion dollars in annual economic losses. The costs of landslides are both direct and indirect, and range from the expense of clean-up operations and the repair or replacement of structures to lost tax revenues and reduced productivity and property values. Nonetheless, landslides are considered to be one of the most potentially predictable of geological hazards, and slope failures tend to affect discrete areas of land. Furthermore, landslides, of all the geological hazards, are perhaps the most amenable to avoidance, prevention and corrective measures. Be that as it may, and proceeding with the USA as an example, Wold and Jochim (1989) pointed out that, despite the availability of successful techniques for landslide management and control, landslide losses are increasing. This primarily is because of the increasing pressure of development in areas of hazardous terrain, as well as the failure of state

Table 4.4 Slope stability classification: frequency and potential criteria. (After Crozier, 1984.)

Class I	Slopes with active landslides. Material is continually moving, and landslide forms are fresh and well defined. Movement may be continuous or seasonal
Class II	Slopes frequently subject to new or renewed landslide activity. Movement is not a regular, seasonal phenomenon. Triggering of landslides results from events with recurrence intervals of up to 5 years
Class III	Slopes infrequently subject to new or renewed landslide activity. Triggering of landslides results from events with recurrence intervals greater than 5 years
Class IV	Slopes with evidence of previous landslide activity, but which have not undergone movement in the preceding 100 years. Subclass IVa: erosional forms still evident Subclass IVb: erosional forms no longer present —previous activity indicated by landslide deposits
Class V	Slopes which show no evidence of previous landslide activity, but which are considered likely to develop landslides in the future. Landslide potential indicated by stress analysis or analogy with other slopes
Class VI	Slopes which show no evidence of previous landslide activity and which, by stress analysis or analogy with other slopes, are considered to be stable

Table 4.5 Features indicating active and inactive landslides. (After Crozier, 1984.)

Active	Inactive
Scarps, terraces and crevices with sharp edges	Scarps, terraces and crevices with rounded edges
Crevices and depressions without secondary infilling	Crevices and depressions infilled with secondary deposits
Secondary mass movement on scarp faces	No secondary mass movement on scarp faces
Surface of rupture and marginal shear planes show fresh slickensides and striations	Surface of rupture and marginal shear planes show old or no slickensides and striations
Fresh fractured surfaces on blocks	Weathering on fractured surfaces of blocks
Disarranged drainage system; many ponds and undrained depressions	Integrated drainage system
Pressure ridges in contact with slide margin	Marginal fissures and abandoned levées
No soil development on exposed surface of rupture	Soil development on exposed surface of rupture
Presence of fast-growing vegetation	Presence of slow-growing vegetation
Distinct vegetation differences 'on' and 'off' slide	No distinction between vegetation 'on' and 'off' slide
Tilted trees with no new vertical growth	Tilted trees with new vertical growth above inclined trunk
No new supportive, secondary tissue on trunks	New supportive, secondary tissue on trunks

authorities and private developers to recognize landslide hazards and so apply appropriate measures for their mitigation. This is in spite of the fact that there is overwhelming evidence that landslide hazard mitigation programmes serve both public and private interests by saving the cost of implementation many times over.

Investigations of slope stability involve the selection of specific criteria upon which stability assessment is to be based, the recognition and measurement of field evidence for instability, the assessment and classification of the degree of stability and the mapping of stability conditions. If a large area is involved, field investigation and mapping are divided into land unit areas. In terms of the assessment of stability, the frequency with which a slope fails and the potential for failure are fundamental criteria, as are the magnitude, rate and type of movement. Crozier (1984) recognized six stability classes based on the frequency or potential landslide criteria (Table 4.4). Different land-use activities can tolerate different degrees of landslide activity.

Although particular slopes which lack a history of previous landslides can be designated as unstable, most stability assessment is based on the recognition of features which indicate former or active mass movement. Crozier (1984) listed the features which indicate whether landsliding areas can be regarded as active or inactive (Table 4.5). However, different zones within a landslide may become stabilized at different times. As slopes which have moved previously are likely to move again, the identification of landslide features forms an integral part in the prediction of potential instability. As pointed out above, the occurrence of a landslide rarely can be attributed to one factor alone, and so it is important to establish the critical combination of factors.

Forecasting the time of occurrence of a failure generally is considerably more difficult than assessing the degree of stability of a slope, because of the precision with which the instability process needs to be known. Unfortunately, the history of landslide activity is rarely sufficiently well documented for the direct assessment of landslide recurrence intervals. The recurrence interval is the average time between landslides of the same magnitude. If the variation in recurrence interval is plotted against the magnitude of the event, a recurrence curve is produced. Methods of determining recurrence intervals include the use of radio-isotopes, infrared photography, dendrochronology, debris mapping and tree damage assessments. Recurrence intervals of landslides often are difficult to evaluate unless they can be correlated with some other factor, such as rainfall, for which recurrence data may exist. Nonetheless, catastrophic phenomena with a recurrence interval in excess of 100 years are extremely difficult to evaluate. In the case of rainstorm-triggered events, it may be possible, by examination of the record, to establish the minimum rainfall intensity required to generate landslides. In this way, the recurrence interval may be derived from the meteorological record. Even so, this method suffers drawbacks in terms of the length of climatic records available and the assumption of climatic constancy. Furthermore, other factors influence slope stability.

The magnitude of movement can be derived from aerial photographs of a landslide or from field investigations. However, in many unstable areas, the form and limits to the erosional and depositional parts of a landslide are difficult to define. As a result, accurate measurements of volume cannot be determined.

The rate and depth of movement, together with the amount of deformation, are important criteria in determining the stability of an area for land use. Rates of movement which exceed 1.5 m annually probably cannot be tolerated by buildings or routeways.

Most landslide investigations are local and site specific in character, being concerned with establishing the nature and degree of stability of a certain slope or slope failure or group of these. Such investigations involve desk studies, geomorphological mapping from aerial photographs or satellite imagery and mapping in the field, subsurface investigation by trenches, shafts or boreholes, sampling and testing, especially of slip surface material, monitoring of surface movements, monitoring of pore water pressures with piezometers and analysis of the data obtained. There are a number of data sources from which background information may be obtained. These include reports, records, papers, topographical and geological maps and aerial photographs and remote sensing imagery. Field investigation, monitoring, sampling and laboratory testing may provide more accurate, and therefore more valuable data.

Existing landslides can be mapped directly, primarily by the use of aerial photographs or imagery, with field checks made when necessary. The simplest form of landslide mapping is a record of landslide locations. Landslides should be classified as active or inactive, historical data being of importance, as well as recent evidence of movement and freshness of form. Problems of recognition of ancient landslides may arise, as they may have little surface expression or be hidden by a subsequently developed slope mantle.

The chief advantage of aerial photography over field investigation is that terrain characteristics can be seen in relation to each other over a large area at the same time. This permits the boundaries of large landslides to be determined, together with an appreciation of the environmental setting in which the landslide occurred. Photographic coverage made at intervals over a number of years facilitates the determination of changes in topography, and allows an approximate establishment of the times when landslide events occurred. The data derived from aerial photographs should be checked in the field.

Digitization of data from aerial photographs or remote sensing imagery enables them to be handled in a Geographical Information System (GIS) through which the landslide-controlling factors can be evaluated (Carrara *et al.*, 1990). Expert systems and artificial intelligence are also being employed (Wislocki & Bentley, 1991).

Quantitative results appear to be obtainable in forested areas by using airborne vertical laser sensing. The airborne laser topography profiler (ALTP) system is operated from a helicopter with a laser pulse frequency of 2000 Hz. This provides a reading about every 20 mm along the flight path, the aim being to penetrate the forest canopy with a proportion of readings.

Hutchinson (1992) suggested that the terrain evaluation approach was the best way of extrapolating available subsurface data and ensuring that physical insights were fully considered in arriving at a landslide hazard assessment. In Hong Kong, terrain evaluation constitutes an important part of the Geotechnical Areas Studies Programme (GASP), in which Geotechnical Land Use Map (GLUM) classes define areas of varying geotechnical limitations on development (Brand, 1988).

A useful starting point in landslide investigation, as well as an efficient method of data collection, is provided by the use of checklists (Fig. 4.11) or inventory forms. The principal advantage of using checklists or

RELIEF

Factor				
Valley depth	Small	Moderate	Large	Very large
Slope steepness	Low	Moderate	Steep	Very steep
Cliffs	Absent			Present
Height difference between different valleys	Small	Moderate	Large	Very large
Valley-side shape	Spur	Straight	Shallow cove	Deep cove

DRAINAGE

Factor				
Drainage density	Low	Moderate	High	Very high
River gradient	Gentle	Moderate	Steep	Very steep
Slope undercutting	None	Moderate	Severe	Very severe
Concentrated seepage flow	Absent	Present at local base level	Present slowly draining	Present rapidly draining
Standing water	Absent	Small	Moderate	Large
Recent incision	Absent	Small	Moderate	Large
Pore-water pressure	Low	Moderate	High	Very high

BEDROCK

Factor				
Jointing density	Low	Moderate	High	Very high
Direction of major joints (faults or bedding planes) with respect to steepest slopes	Away	Normal		Towards
Amount of dip for steepness of joint and/or fault planes	Horizontal	Small	Moderate	Large
Strong beds over weak beds	Absent			Present
Degree of weathering	None	Small	Moderate	Large
Compressive strength	High	Moderate	Low	Very low
Coherence (particularly of lower beds)	High	Moderate	Low	Very low

SOILS (incl. drift materials)

Factor				
Site	Valley floor	Gentle slopes	Moderate slopes	Steep slope
Coherent over incoherent beds	Absent			Present
Angle of rest	Low	Moderate	Steep	Very steep
Depth	Small	Moderate	Large	Very large
Shear strength	High	Moderate	Low	Very low
Liquidity index	Low	Moderate	High	Very high

EARTHQUAKE ZONE

Factor				
Tremors felt	Never	Seldom	Some	Many

LEGACIES FROM THE PAST

Factor				
Fossil solifluction lobes and sheets	Absent	Rare	Some	Many
Previous landslides	Absent	Rare	Some	Many
Deep weathering	None	Slight	Moderate	Much

MAN-MADE FEATURES

Factor				
Excavations—depth	None	Small	Moderate	Large
Excavations—position	Hillcrest	High valley	Low valley	Bottom valley
Reservoir	Absent	Small	Moderately deep	Very deep
Drainage diversion across hillside	Absent			Present
Lowering of reservoir level	None	Small	Moderate	Large
Loading of upper valley side	None	Some	Moderate	Large

Fig. 4.11 A checklist for sites liable to large scale instability. (After Cooke & Doornkamp, 1990.)

inventories is that each of the main categories of influencing and controlling parameters can be examined systematically. They also enable each separate slope unit to be classified according to a stability rating. The checklist can be used either during an investigation of aerial photographs or during a field survey, and provides a systematic examination of the main factors influencing mass movement. The more boxes which are ticked on the right-hand side of the checklist, the more the slope concerned is approaching an unstable state.

Because most landslides occur in areas previously affected by instability, and because few take place without prior warning, Cotecchia (1978) emphasized the importance of carrying out careful surveys of areas which appear to be potentially unstable and of making systematic records of the relevant phenomena. He provided a review of the techniques involved in mapping mass movements, as well as itemizing which data should be included on such maps. He maintained that the ultimate aim should be the production of maps of landslide hazard zoning. Landslide hazard maps delineate areas which probably will be affected by slope instability within a given period of time. Landslide risk maps attempt to quantify the vulnerability of an area, in relation either to the probability of occurrence of a landslide or to the likelihood of damage to property or injury or death to persons.

Generally, the purpose of mapping landslide hazards is to locate problem areas and to help understand why, when and where landslides are likely to occur. According to Varnes (1984), three basic principles have guided the production of landslide hazard zonation maps. The first assumption involves the concept that slope failures in the future will take place for reasons similar to those which gave rise to the failures in the past. In fact, the evidence of past slope instability frequently is the best guide to the future behaviour of a slope. However, the absence of landslides in the past or present does not mean that they will not occur in the future. Second, the basic causes of slope instability are fairly well known, so that most can be recognized and mapped. It often is possible to estimate the relative contributions to slope instability of the conditions and processes responsible once they have been identified. Third, in this way, the degree of potential hazard can be assessed, depending on the correct recognition of the number of failure-inducing

factors present, their severity and interaction. Data processing can range from subjective evaluation to sophisticated processing by computers.

Walker *et al.* (1987) itemized a number of questions which require answers when designing a landslide hazard mapping scheme. These included, should the map be wholly concerned with actual or potential landslides; should the emphasis be placed on the nature of the slope failure, the extent of instability, the frequency of occurrence or the consequences of slope failure; and should the map be concerned with hazard or risk? The answers help to determine how extensive the investigation needs to be and the scale at which the mapping will be undertaken. Regional maps tend to be general purpose, whereas community maps or site maps often are concerned with more specific problems. In the latter case, it is more cost effective to represent only the necessary data for the particular purpose.

Many landslide maps only show the hazards known at a particular time, whereas others provide an indication of the possibility of landslide occurrence (Fig. 4.12). The latter are landslide susceptibility maps, and involve some estimation of relative risk. An assessment of the level of risk attributable to landslide occurrence for a particular area involves the classification of the data obtained into risk groups. Most risk classifications recognize low, medium and high levels of risk, but the categories usually are poorly defined. Walker *et al.* (1987) provided a summary of a number of risk classifications or stability ratings of slopes which have been used in Australia. However, these examples tend to be specific to the areas for which they were developed, and so cannot be applied readily to other areas.

Multivariate landslide susceptibility maps, obtained by the statistical evaluation of the physical factors which influence slope instability, generally are based on grid cells and produced by computer. Cell sizes usually are square, with sides between 60 m and several hundred metres in length. The end result of the assessment of the various data normally is a numerical rating for each cell, which forms the basis of landslide hazard assessment (Fig. 4.13). However, grid cells of fixed size have the disadvantage of often relating poorly to geomorphological slope units. Con-sequently, later work, facilitated by the use of GISs, has associated land characteristics with geomorphological units (Carrara *et al.*, 1990). Discrimi-

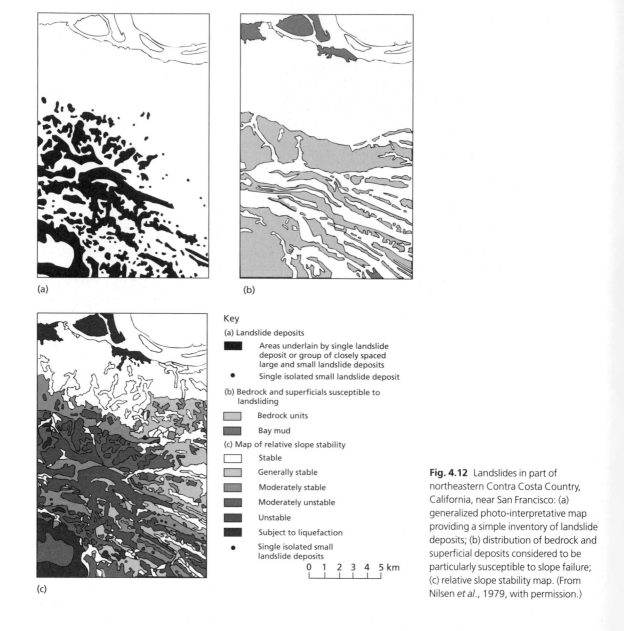

Key

(a) Landslide deposits

▮ Areas underlain by single landslide deposit or group of closely spaced large and small landslide deposits

• Single isolated small landslide deposit

(b) Bedrock and superficials susceptible to landsliding

▭ Bedrock units

▨ Bay mud

(c) Map of relative slope stability

□ Stable

▫ Generally stable

▨ Moderately stable

▨ Moderately unstable

▨ Unstable

▮ Subject to liquefaction

• Single isolated small landslide deposits

0 1 2 3 4 5 km

Fig. 4.12 Landslides in part of northeastern Contra Costa Country, California, near San Francisco: (a) generalized photo-interpretative map providing a simple inventory of landslide deposits; (b) distribution of bedrock and superficial deposits considered to be particularly susceptible to slope failure; (c) relative slope stability map. (From Nilsen *et al.*, 1979, with permission.)

nant analysis has been used, employing combinations of measured parameters, to distinguish stable from unstable slopes (Payne, 1985).

Effective landslide hazard management has done much to reduce economic and social losses due to slope failure, by avoiding the hazards or by reducing the damage potential (Schuster, 1992). This has been accomplished by restrictions placed on development in landslide prone areas, the application of excavation grading, landscaping and construction codes, the use of remedial measures to prevent or control slope failure and landslide warning systems.

Future policy should develop and promote techniques and initiate hazard recognition and reduction schemes as a preventative measure. Risk reduction can be achieved with reference to either the process

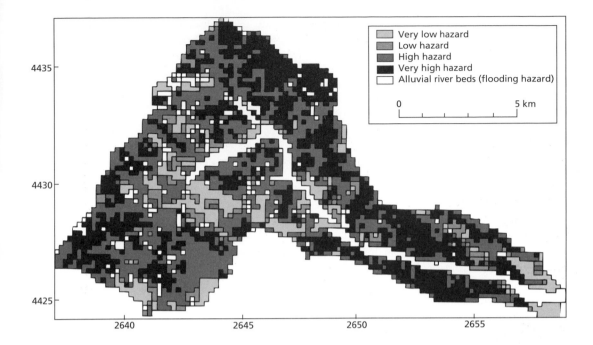

Fig. 4.13 Landslide hazard and erosion zones, Ferro River basin, Calabria, southern Italy. (After Carrara *et al.*, 1977.)

system or the degree of vulnerability of the land use. The final choice of risk reduction measure will depend on the type of development, either in existence or proposed, and the type, magnitude and time-scale of the hazardous process.

Kockelman (1986) dealt with some of the techniques used to reduce landslide hazards. These may be used in a variety of combinations to help solve both existing and potential landslide problems. The techniques generally are applicable to all types of surface ground failure, including flows, slides and falls. The effectiveness of each hazard reduction technique varies with the time, place and persons involved in the planning and implementation of the programme for reducing the hazard. The control of the landslide hazard system is easier for new developments as the vulnerability can be restricted.

There are several methods to discourage development in hazardous areas. These include disclosing hazards to real estate buyers, posting warning signs, adopting appropriate utility and public facility service area policies, informing and educating the public and recording the hazard in public documents.

Recurrent damage from landslides can be avoided by permanently evacuating areas that continue to have slope failures. If there are obvious indications of instability, evacuations must be considered. This may be easier if the choice is on an individual scale, but social and political forces, including hazard perception, may restrict whole settlements from relocating, although it should be forced if necessary. Structures may be removed or converted to a use which is less vulnerable to landslide damage. Techniques for removal or conversion include acquiring or exchanging hazardous areas and relocating their occupants, incorporating non-conforming use provisions into zoning ordinances, clearing and redeveloping damaged areas, removing unsafe structures or public nuisances and urban redevelopment.

Financial techniques of encouraging or discouraging development in landslide areas include amending government assistance programmes, increasing the awareness of legal liability, adopting appropriate lending policies, requiring insurance and providing tax credits or lower assessments.

Various types of land-use and land development regulations can be used to reduce landslide hazards.

They often are the most economical and most effective means available to a local government. It is unrealistic to assume that development can be indirectly discouraged for a long period, and other techniques, such as protecting existing development or purchasing hazardous areas, can be very costly. Development can be prohibited, restricted or regulated in landslide areas. These areas can be used as open spaces, or the density of development can be kept to a minimum to reduce landslide effects and the potential for damage. Zoning and subdivision regulations, as well as moratoria on rebuilding, can be used to meet these objectives.

Property damage from landslides often leads to a demand for costly public works to provide protection for existing development. This demand usually is limited to smaller landslide areas because of the costs involved, and the necessity for careful and accurate engineering design, construction and maintenance.

Potentially unstable land can be monitored so that residents can be warned and, if necessary, evacuated. Immediate relay of information is vital in areas in which landslides occur rapidly.

4.8 Methods of slope control and stabilization

As the same preventative or corrective work cannot always be applied to different types of slide, it is important to identify the type of slide which is likely to take place, or which has taken place. In this context, however, it is important to bear in mind that landslides may change in character, and that they usually are complex, frequently changing their physical characteristics as time proceeds. When it comes to the correction of a landslide, as opposed to its prevention, because the limits and extent of the slide generally are well defined, the seriousness of the problem can be assessed. Nevertheless, in such instances, consideration must be given to the stability of the area immediately adjoining the slide. Obviously, any corrective treatment must not adversely affect the stability of the area about the slide.

Slide prevention may be brought about by reducing the activating forces, by increasing the forces resisting movement or by avoiding or eliminating the slide. In the first case, reduction of the activating forces can be accomplished by removing material from that part of the slide which provides the force giving rise to movement, and by drainage, which reduces the pore water pressure and bulk density. Drainage also brings about an increase in shearing resistance.

The most frequently used methods of slope stabilization include retention systems, buttresses, slope modification and drainage. Properly designed retention systems can be used to stabilize most types of slope, where large volumes of earth materials are not involved and where lack of space excludes slope modification. The control of subsurface water frequently represents a major component in slope stabilization works.

4.8.1 Rockfall treatment

Single-mesh fencing supported by rigid posts will contain small rockfalls. Larger, heavy duty catch fences or nets are required for larger rockfalls. Rolling rocks up to 0.6 m in diameter can be restrained by a chain-link fence, but this can suffer severe damage when hit by rocks of this size and is not able to stop larger rocks. Fences have been developed so that, when rocks collide with them, the nets engage energy-absorbing friction brakes which extend the time of collision and, in this way, increase the capacity of the nets to restrain the falling rocks.

Wire meshing of rock slopes is one of the most effective methods of preventing rockfalls from steep slopes. The wire panels are laced together with binding wire. If a more robust method of linking panels is required, horizontal and vertical steel cables can be shackled to the mesh and fixed to hooks and dowels. The latter provide a strong cable grid at 2–3 m centres, offering a greater resistance to large-scale block movement.

The use of cable lashing and cable nets to restrain loose rock blocks was referred to by Piteau and Peckover (1978). Areas of potential instability can be covered with mesh fixed with light cables. High-capacity horizontal cables are then strung across the block using anchoring and tensioning methods.

Rock traps in the form of a ditch and/or barrier can be installed at the foot of a slope. Benches may also act as traps to retain rockfalls, especially if a barrier is placed at their edge. Wire mesh suspended from the top of the face provides another method for controlling rockfalls. Where a road or railway passes along the foot of a steep slope, protection from rockfall is

Fig. 4.14 Canopy to protect traffic from rockfall, north of Masjöen, Norway.

afforded by the construction of a rigid canopy from the face (Fig. 4.14).

4.8.2 Alteration of slope geometry

Often, altering the geometry of a slope is the most efficient way of increasing the factor of safety, especially in deep-seated slides (Leventhal & Mostyn, 1987). However, such an approach may not be easy to adopt for long translational slides where there is no obvious toe or crest, where the geometry is determined by engineering constraints or where the unstable area is complex, so that a change in topography which improves the stability of one area adversely affects the stability of another. Unstable material can be removed from near the crest of the slope and material can be added to the toe. In fact, it usually is more practical to load the toe.

Although partial removal is suitable for dealing with most types of mass movement, for some types it is inappropriate. For example, removal of material from the head has little influence on flows or slab slides. On the other hand, this treatment is eminently suitable for rotational slips (Bell & Maud, 1996). Slope flattening, however, is rarely applicable to rotational or slab slides.

Benching brings about stability by dividing a slope into segments. Benches ideally should be over 5 m wide to allow access for inspection, and should be kept clear. If rock faces are to be scaled efficiently, benches should not be higher than 12 m. Drainage systems can be installed on benches.

4.8.3 Reinforcement of slopes

Dentition refers to masonry or concrete infill placed in fissures or cavities in a rock slope (Fig. 4.15). The use of the same rock material for the masonry as forms the slope provides a more attractive finish than otherwise. It is often necessary to remove soft material from fissures and to pack the void with permeable material prior to constructing the dentition. Drainage should be provided through the latter.

Thin to medium bedded rocks dipping parallel to the slope can be held in place by steel dowels, which are up to 2 m in length. Holes are drilled beneath the slip surface and are normal to the bedding. The dowels are grouted into place and are not stressed. They are used where low loads are needed to increase stability and where the joint surfaces are at least moderately rough.

Rock bolts may be used as reinforcement to enhance the stability of slopes in jointed rock masses (Fig. 4.15). They provide additional strength on critical planes of weakness within the rock mass. Rock bolts inclined to the potential plane of failure provide greater resistance than those installed normal to the plane. Rock bolts may be up to 8 m in length with a tensile working load of up to 100 kN. They are put in tension, so that the compression induced in the rock

Fig. 4.15 Rock bolts and dentition used to stabilize a slope in limestone, North Wales.

mass improves the shearing resistance on potential failure planes. Bearing plates, light steel sections or steel mesh may be used between bolts to support the rock face.

Rock anchors are used for major stabilization works, especially in conjunction with retaining structures. They may exceed 30m in length. Because the stress levels are far greater than those involved in rock bolting, anchor loads are more dependent upon the rock type and structure. Anchors can be installed at different inclinations in order to dissipate the load within a rock mass.

Gunite or shotcrete frequently is used to preserve the integrity of a rock face by sealing the surface and inhibiting the action of weathering (Fig. 4.16). The former is pneumatically applied mortar and the latter is pneumatically applied concrete. Gunite/shotcrete adapts to the surface configuration and can be coloured to match the surrounding rocks. Coatings may be reinforced with wire mesh and/or used in combination with rock bolts. Groundwater must be allowed to drain through the protective cover, otherwise it may be affected by frost action and groundwater pressures within the rock mass.

4.8.4 Restraining structures

Restraining structures control sliding by increasing the resistance to movement. They include retaining walls, cribs and gabions.

Fig. 4.16 Use of shotcrete and rock bolts to protect a road cutting in Natal Group Sandstone, near Mont-aux-Sources, Natal, South Africa.

Retaining walls often are used where there is a lack of space for the full development of a slope, such as along many roads and railways. As retaining walls are subjected to unfavourable loading, a large wall width is necessary to increase slope stability, which means that they are expensive. Retaining structures should be designed for a predetermined load, which they are to transmit to the foundation of known bearing capacity. Retaining walls are located at the foot of a slope, and should include adequate provision for drainage, for example, weep-holes through the wall and pipe drainage in any backfill. This will not only prevent the build-up of pore water pressures, but will also reduce the effects of frost. Nonetheless, there are certain limitations which must be considered before retaining walls are used as a method of landslide control. These involve the ability of the structure to resist shearing action, overturning and sliding on or below its base.

Reinforced earth can be used for retaining earth slopes (Fig. 4.17). Reinforced soil structures are flexible, and so can tolerate large deformations and are resistant to seismic loadings. Thus, reinforced earth can be used on poor ground, where conventional alternatives would require expensive foundations. Reinforced earth structures are constructed by erecting facing panels at the face of the wall at the same time as the earth is placed. Strips of galvanized steel or geogrids are fixed to the panels. The system relies on the transfer of shear forces to mobilize the tensile capacity of closely spaced reinforcing strips.

Soil nailing has been used to retain slopes, the nails consisting of steel bars, metal rods or metal tubes which are driven into *in situ* soil or soft rock or grouted into bored holes. The nails are passive elements that are not post-tensioned. Normally, one nail is used for each 1–6 m² of ground surface. The ground surface between the nails is covered with a layer of shotcrete reinforced with wire mesh.

Cribs may be constructed of precast reinforced concrete or steel units set up in cells which are filled with gravel or stone (Fig. 4.18). The systems are reasonably flexible due to the segmental nature of the

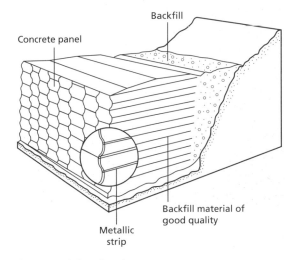

Fig. 4.17 Reinforced earth system.

Fig. 4.18 Crib wall, San Antonio, Texas.

elements which comprise the walls and, therefore, are not particularly sensitive to differential settlements. Plant growth can occur on the faces of crib walls, which masks their presence. They only serve to support shallow translational slides. Gabions consist of strong wire mesh surrounding placed stones. Like crib walls, gabions can also be constructed readily, especially in difficult terrain. Gabions are flexible. The gabion filling provides for good subsurface drainage conditions in the vicinity of the wall, and filtration protection between the gabions and the wall backfill or soil can be afforded by geotextiles.

4.8.5 Drainage

Drainage is the most generally applicable method for improving the stability of slopes or for the corrective treatment of slides, regardless of type, because it reduces the effectiveness of one of the principal causes of instability, namely, excess pore water pressure. In rock masses, groundwater also tends to reduce the shear strength along discontinuities. Moreover, drainage is the only economic way of dealing with slides involving the movement of several million cubic metres.

Surface run-off should not be allowed to flow unrestrained over a slope. This usually is prevented by the installation of a drainage ditch at the top of an excavated slope to collect water drainage from above. The ditch, especially in soils, should be lined to prevent erosion, otherwise it will act as a tension crack. It may be filled with cobble aggregate. Herringbone ditch drainage is usually employed to convey water from the surfaces of slopes. These drainage ditches lead into an interceptor drain at the foot of the slope (Fig. 4.19). Trench drains are filled with free-draining materials and may be lined with geotextiles. They are used for shallow subsurface drainage. Infiltration can be lowered by sealing the cracks in a slope by regrading or filling with cement, bitumen or clay. A surface covering has a similar purpose and function. For example, the slope may be covered with granular material resting upon filter fabric.

Water may be prevented from reaching a zone of potential instability by a cutoff. Cutoffs may take the form of a trench backfilled with asphalt or concrete, sheet piling, a grout curtain or a well curtain, whereby water is pumped from a row of vertical wells.

Support and drainage may be afforded by counterfort-drains, where an excavation is made in sidelong ground, likely to undergo shallow, parallel slides. Deep trenches are cut into the slope, lined with filter fabrics and filled with granular filter material. The granular fill in each trench acts as a supporting buttress or counterfort, as well as providing drainage. However, counterfort-drains must extend beneath the potential failure zone, otherwise they merely add unwelcome weight to the slipping mass.

Fig. 4.19 Surface drainage of a slope in fill deposits, near Loch Lomond, Scotland. The aggregate-filled ditch drainage leads to an interceptor drain.

Fig. 4.20 Internal drainage gallery in restored slope, near Aberfan, South Wales.

Deep wells are used to drain slopes where the depths involved are too large for the construction of trench drains (Bianco & Bruce, 1991). Usually, the collected water flows from the base of the drainage wells under gravity but occasionally pumps may be installed in the bottom of the wells to remove the water.

The successful use of subsurface drainage depends on the tapping of the source of water, locating the presence of permeable material which aids free drainage, the location of the drain on relatively unyielding material to ensure continuous operation (flexible polyvinylchloride, PVC drains are now frequently used) and the installation of a filter to minimize silting in the drainage channel. Drainage galleries are costly to construct, generally being placed by tunnelling techniques, and in slipped areas may experience caving. Most galleries are designed to drain by gravity (Fig. 4.20). In certain cases, collector galleries may be needed from which to pump water.

Subhorizontal drainage holes are much cheaper than galleries and are satisfactory over short lengths, but it is more difficult to intercept water-bearing layers with them. Subhorizontal drains can be inserted from the ground surface or by drilling from drainage galleries, large-diameter wells or caissons. The drainage hole typically is 120–150 mm in diameter (Sembenelli, 1988). The drainage holes are lined with slotted PVC pipe. The pipes are lined on the outside with a geotextile covering which acts as a filter. When individual benches are drained by horizontal holes, the latter should lead into a properly graded interceptor trench, which is lined with impermeable material.

Afforestation may help to stabilize shallow slides, but it cannot prevent further movements occurring in large landslip areas. It can, however, lower infiltration. The most satisfactory trees are those which consume most water and have high transpiration rates; thus, deciduous trees are better than conifers. The root system helps to bind the soil together, and foliage and plant residues reduce the impact of rainfall.

References

Bell, F.G. (1994) Floods and landslides in Natal and notably the greater Durban region, September 1987: a retrospective view. *Bulletin of the Association of Engineering Geologists* **31**, 59–74.

Bell, F.G. & Maud, R.R. (1996) Landslides associated with the Pietermaritzburg Formation in the greater Durban area, South Africa. *Environmental and Engineering Geoscience* **2**, 557–573.

Bianco, B. & Bruce, D.A. (1991) Large landslide stabilization by deep drainage wells. In *Proceedings of the International Conference on Slope Stability Engineering: Applications and Development, Isle of Wight*. Thomas Telford Press, London, pp. 319–326.

Bishop, A.W. (1971) The influence of progressive failure on the choice of the method of stability analysis. *Geotechnique* **21**, 169–172.

Brand, E.W. (1988) Landslide risk assessment in Hong Kong. In *Proceedings of the Fifth International Symposium on Landslides, Lausanne* 2, Bonnard, C. (ed). Balkema, Rotterdam, pp. 1059–1074.

Brand, E.W., Premchitt, J. & Philipson, H.B. (1984) Relationship between rainfall and landslides in Hong Kong. In *Proceedings of the Fourth International Symposium on Landslides, Toronto* 1, pp. 377–384.

Carrara, A., Catalano, E., Sorriso Valvo, M., Reali, C., Meremda, L. & Rizzo, V. (1977) Landslide morphology and topology in two zones, Calabria, Italy. *Bulletin of the International Association of Engineering Geology* 16, 8–13.

Carrara, A., Cardinali, M., Detti, R., Guzzetti, F., Pasqui, V. & Reichenbach, P. (1990) Geographical Information Systems and multivariate models in landslide hazard evaluation. In *Proceedings of the Sixth International Conference and Field Workshop on Landslides: Alps '90, Milan*, Cancelli, A. (ed), pp. 17–28.

Chandler, R.J. (1974) Lias Clay: the long term stability of cutting slopes. *Geotechnique* 24, 21–38.

Chandler, R.J. (1977) The application of soil mechanics methods to the study of slopes. In *Applied Geomorphology*, Hails, J.R. (ed). Elsevier, Amsterdam, pp. 157–182.

Cooke, R.M. & Doornkamp, J.C. (1990) *Geomorphology in Environmental Management*, 2nd edn. Clarendon Press, Oxford.

Cotecchia, V. (1978) Systematic reconnaissance mapping and registration of slope movements. *Bulletin of the International Association of Engineering Geology* 17, 5–37.

Crozier, M.J. (1984) Field assessment of slope instability. In *Slope Instability*, Brunsden, D. & Prior, D.B. (eds). Wiley–Interscience, Chichester, pp. 103–142.

Culshaw, M.G. & Bell, F.G. (1992) The rockfalls of James Valley, St Helena. In *Proceedings of the Sixth International Symposium on Landslides, Christchurch* 2, Bell, D.H. (ed). Balkema, Rotterdam, pp. 925–935.

Dunnicliff, J. (1988) *Geotechnical Instrumentation for Monitoring Field Performance*. Wiley, New York.

Dunnicliff, J. (1992) Monitoring and instrumentation of landslides. In *Proceedings of the Sixth International Symposium on Landslides, Christchurch* 3, Bell, D.H (ed). Balkema, Rotterdam, pp. 1881–1893.

Greenway, D.R. (1987) Vegetation and slope stability. In *Slope Stability*, Anderson, M.G. & Richards, K.S. (eds). Wiley, New York, pp. 187–230.

Hansen, A. (1984) Landslide hazard analysis. In *Slope Instability*, Brunsden, D. & Prior, D.B. (eds). Wiley–Interscience, Chichester, pp. 523–602.

Hoek, E. & Bray, J.W. (1981) *Rock Slope Engineering*. Institution of Mining and Metallurgy, London.

Hutchinson, J.N. (1992) Landslide hazard assessment. In *Proceedings of the Sixth International Symposium on Landslides, Christchurch* 3, Bell, D.H. (ed). Balkema, Rotterdam, pp. 1805–1841.

Jahns, R.H. (1978) Geophysical predictions. In *Landslides*

Analysis and Control, Schuster, R.L. & Krizek, R.J. (eds). *Transportation Research Board, Special Report* 176. National Academy of Sciences, Washington, DC, pp. 58–65.

Keefer, D.K. (1984) Landslides caused by earthquakes. *Bulletin of the American Geological Society* 95, 406–421.

Kockelman, W.J. (1986) Some techniques for reducing landslide hazards. *Bulletin of the Association of Engineering Geologists* 23, 29–52.

Leventhal, A.R. & Mostyn, G.R. (1987) Slope stabilization techniques and their application. In *Proceedings of the Extension Course on Soil Slope Stability and Stabilization, Sydney*, Walker, B.F. & Fell, R. (eds). Balkema, Rotterdam, pp. 121–181.

Lumb, P. (1975) Slope failures in Hong Kong. *Quarterly Journal of Engineering Geology* 8, 31–65.

Martin, D.C. (1988) Rockfall control: an update: technical note. *Bulletin of the Association of Engineering Geologists* 25, 137–144.

Mikkelsen, P.E. (1996) Field instrumentation. In Turner, A.K. & Schuster, R.L. (eds). *Landslides Investigation and Mitigation. Transportation Research Board, Special Report* 247. National Research Council, Washington, DC, pp. 278–318.

Nilsen, T.H., Wright, R.H., Vlasic, T.C. & Spangle, W. (1979) Relative slope stability and land-use planning in the San Francisco Bay region, California. *United States Geological Survey, Professional Paper 944*. Washington, DC, 96pp.

Olivier, M., Bell, F.G. & Jermy, C.A. (1994) The effect of rainfall on slope failure, with examples from the greater Durban area. In *Proceedings of the Seventh Congress of the International Association of Engineering Geology, Lisbon* 3. Balkema, Rotterdam, pp. 1629–1636.

Payne, H.R. (1985) Hazard assessment and rating methods. In *Landslides in South Wales Coalfields*, Morgan, C.S. (ed). Polytechnic of Wales, Pontypridd, pp. 59–71.

Piteau, D.R. & Peckover, F.L. (1978) Rock slope engineering. In *Landslides: Analysis and Control*, Schuster, R.L. & Krizek, R.J. (eds). *Transportation Research Board, Special Report* 176. National Academy of Sciences, Washington, DC, pp. 192–228.

Richards, L.R., Whittle, R.A. & Ley, G.M.M. (1978) Appraisal of stability conditions in rock slopes. In *Foundation Engineering in Difficult Ground*, Bell, F.G. (ed). Butterworth, London, pp. 449–512.

Schuster, R.L. (1992) Recent advances in slope stabilization. In *Proceedings of the Sixth International Symposium on Landslides, Christchurch* 3, Bell, D.H. (ed). Balkema, Rotterdam, pp. 1715–1745.

Schuster, R.L. & Fleming, R.W. (1986) Economic losses and fatalities due to landslides. *Bulletin of the Association of Engineering Geologists* 23, 11–28.

Sembenelli, P. (1988) Stabilization and drainage. In *Proceedings of the Fifth International Symposium on*

Landslides, Lausanne **2**, Bonnard, C. (ed). Balkema, Rotterdam, pp. 813–819.

Sharpe, C.F.S. (1938) *Landslides and Related Phenomena.* Columbia University Press, New York.

Skempton, A.W. (1964) Long-term stability of clay slopes. *Geotechnique* **14**, 77–101.

Skempton, A.W. & Hutchinson, J.N. (1969) Stability of natural slopes and embankment foundations. In *Proceedings of the Seventh International Conference on Soil Mechanics and Foundation Engineering, Mexico City*, State-of-the-Art Volume, pp. 291–340.

Terzaghi, K. (1950) Mechanisms of landslides. In *Applications of Geology to Engineering Practice*, Paige, S. (ed). Berkey Volume, American Geological Society, New York, pp. 83–124.

Tran-Duc, P.O., Ohno, M. & Mawatari, Y. (1992) An automated landslide monitoring system. In *Proceedings of the Sixth International Symposium on Landslides, Christchurch* **2**, Bell, D.H. (ed). Balkema, Rotterdam, pp. 1163–1166.

Varnes, D.J. (1978) Slope movement types and processes. In *Landslides, Analysis and Control*, Schuster, R.L. &

Krizek, R.J. (eds). *Transportation Research Board, Special Report* **176**. National Academy of Sciences, Washington, DC, pp. 11–33.

Varnes, D.J. (1984) *Landslide Hazard Zonation: A Review of Principles and Practice. Natural Hazards* **3**. UNESCO, Paris.

Vaughan, P.R. & Walbanke, H.J. (1973) Pore pressure changes and delayed failure of cutting slopes in overconsolidated clay. *Geotechnique* **23**, 531–539.

Walker, B.F., Blong, R.J. & MacGregor, J.P. (1987) Landslide classification, geomorphology and site investigations. In *Proceedings of the Extension Course on Soil Slope Stability and Stabilization, Sydney*, Walker, B.F. & Fell, R. (eds). Balkema, Rotterdam, pp. 1–52.

Wislocki, A.P. & Bentley, S.P. (1991) An expert system for landslide hazard and risk assessment. *Computers and Structures* **40**, 169–172.

Wold, R.L. & Jochim, C.L. (1989) *Landslide Loss Reduction: A Guide for State and Local Government Planning. Special Publication* **33**. Colorado Geological Survey, Department of Natural Resources, Denver.

5   River activity

All rivers form part of a drainage system, the form of which is influenced by the rock type and structure, the nature of the vegetation cover and the climate. An understanding of the processes which underlie river development forms the basis of proper river management.

Rivers also form part of the hydrological cycle in that they carry precipitation run-off. This run-off is the surface water which remains after evapotranspiration and infiltration into the ground have taken place. Some precipitation may be frozen, only to contribute to run-off at some other time, while any precipitation that has infiltrated into the ground may reappear as springs where the water table meets the ground surface. Although, as a result of heavy rainfall, or in areas with few channels, the run-off may occur as a sheet, usually it becomes concentrated into channels which are eroded by the flow of water and eventually form valleys.

5.1 The development of drainage systems

It is assumed that the initial drainage pattern which develops on a new surface consists of a series of subparallel rills flowing down the steepest slopes. The drainage pattern then becomes integrated by micropiracy (the beheading of the drainage system of a small rill by that of a larger rill) and cross-grading. Micropiracy occurs when the ridges which separate the initial rills are overtopped and broken down. When the divides are overtopped, the water tends to move towards those rills at a slightly lower elevation and, in the process, the divides are eroded. Eventually, water drains from rills of higher elevation into adjacent ones of lower elevation (Fig. 5.1). The flow towards the master rill steadily increases, and its development across the main gradient is termed cross-grading. The tributaries which flow into the master stream are subsequently subjected to cross-grading and so a dendritic system is developed.

As noted, the texture of the drainage system is influenced by the rock type and structure, the nature of the vegetation cover and the type of climate. The drainage density affords a measure of comparison between the development of one drainage system and another. It is calculated by dividing the total length of a stream by the area it drains, and is generally expressed in kilometres per square kilometre.

Horton (1945) classified streams into orders. First-order streams are unbranched and, when two such streams become confluent, they form a second-order stream. It is only when streams of the same order meet that they produce one of the higher rank, for example, a second-order stream flowing into a third-order stream does not alter its rank. The frequency with which streams of a certain order flow into those of the next order above them is referred to as the bifurcation ratio. The bifurcation ratio for any consecutive pair of orders is obtained by dividing the total number of streams of the lower order by the total number in the next highest order. Similarly, the stream length ratio is found by dividing the total length of streams of a lower order by the total length of those in the next highest order. The values of the stream length ratio depend mainly on the drainage density and stream entrance angles, and increase somewhat with increasing order. A river system is also assigned an order, which is defined numerically by the highest stream order it contains.

In the early stages of development, in particular, rivers tend to accommodate themselves to the local geology. For instance, tributaries may develop along fault zones. Moreover, the rock type has a strong influence on the drainage texture or channel spacing. In other words, a low drainage density tends to form on resistant or permeable rocks, whereas weak

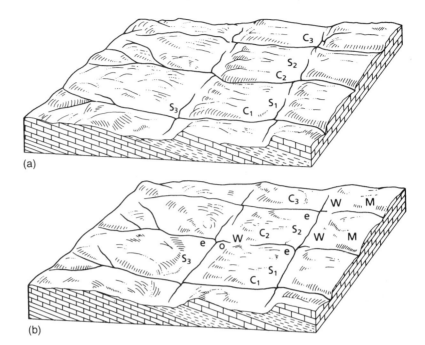

Fig. 5.1 (a) Trellised drainage pattern of consequent streams (C) and their subsequents (S), showing the erosion of a gently dipping series of hard and soft beds into escarpments. (b) Later development illustrating river capture or micropiracy by the headward growth of the more vigorous subsequent streams: e, elbow of capture; W, wind gap; M, misfit stream; o, obsequent stream.

highly erodible rocks are characterized by a high drainage density.

The initial dominant action of master streams is vertical downcutting, which is accomplished by the formation of pot-holes, which ultimately coalesce, and by the abrasive action of the load. Hence, in the early stages of river development, the cross-profile of the valley is sharply V shaped. As time proceeds, valley widening due to soil creep, slippage, rainwash and gullying becomes progressively more important, and eventually lateral corrasion replaces vertical erosion as the dominant process. A river possesses few tributaries in the early stages but, as the valley widens, their numbers increase, which affords a growing increment of rock waste to the master stream, thereby enhancing its corrasive power.

During valley widening, the stream erodes the valley sides by causing undermining and slumping to occur on the outer concave curves of meanders where steep cliffs or bluffs are formed. These are most marked on the upstream side of each spur. Deposition usually takes place on the convex side of a meander. The meanders migrate both laterally and downstream, and their amplitude is progressively increased. In this manner, spurs are continually

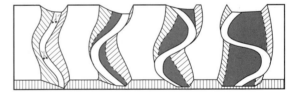

Fig. 5.2 Widening of valley floor by lateral corrasion.

eroded, first becoming more asymmetrical until they are eventually truncated (Fig. 5.2). The slow deposition which occurs on the convex side of a meander will, as lateral migration proceeds, produce a gently sloping area of alluvial ground called the floodplain. The floodplain gradually grows wider as the river bluffs recede, until it is as broad as the amplitude of the meanders. It was at this period in river development that Davis (1909) regarded maturity as having been reached (Fig. 5.3). Thereafter, the continual migration of meanders slowly reduces the valley floor to an almost flat plain, which slopes gently downstream and is bounded by shallow valley sides.

Throughout its length, a river channel has to adjust to several factors which change independently of the

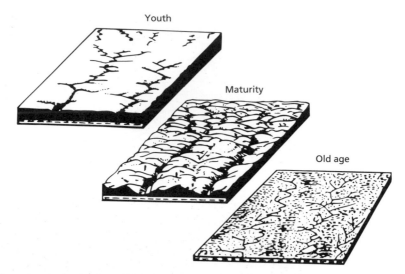

Fig. 5.3 Diagram illustrating the three main periods in the denudation of an uplifted land surface according to the Davisian interpretation of the 'normal' cycle of erosion: in youth, parts of the initial surface survive; in maturity, most or all of the initial surface has vanished and the landscape is mainly composed of slopes, apart from the valley floors; in old age or senility, the landscape becomes subdued and gently undulating, rising only to residual hills representing the divides between adjoining drainage basins. Eventually, such hills are worn down and the region becomes a peneplain.

channel itself. These include the different rock types and structures across which the river flows. The tributaries and inflow of water from underground sources affect the long profile, but are independent of the channel. Other factors which bring about the adjustment of a river channel are the flow resistance, which is a function of the particle size and the shape of transitory deposits such as bars, the method of load transport and the channel pattern, including meanders and islands. Lastly, the river channel must also adjust itself to the river slope, width, depth and velocity of flow.

As the longitudinal profile or thalweg of a river is developed, the differences between the upper, middle and lower sections of its course become more clearly defined until three distinctive tracts are observed. These are the upper or torrent, the middle or valley and the lower or plain stages. The torrent stage includes the headstreams of a river, where small, fast-flowing streams are engaged principally in active downwards and headwards erosion. They possess steep-sided cross-profiles and irregular thalwegs. The initial longitudinal profile of a river reflects the irregularities which occur in its path. For instance, it may exhibit waterfalls or rapids where it flows across resistant rocks. However, such features are transient in the life of a river. In the valley tract, the predomi-

nant activity is lateral corrasion. The shape of the valley sides depends upon the nature of the rocks being excavated, the type of climate, the rate of rock wastage and meander development. Some reaches in the valley tract may approximate to grade, and there the meanders may have developed alluvial flats, whilst other stretches may be steep sided with irregular longitudinal profiles. The plain tract is formed by the migration of meanders and deposition is the principal river activity.

Meanders, although not confined to floodplains, are characteristic of them. The consolidated veneer of alluvium, spread over a floodplain, offers little resistance to continual meander development, and so the loops become more and more accentuated. As time proceeds, the swelling loops approach one another. During flood, the river may cut through the neck separating two adjacent loops and thereby straighten its course. As it is much easier for the river to flow through this new course, the meander loop is silted off and abandoned as an oxbow lake (Fig. 5.4).

Deformed and compressed meanders commonly develop when a river migrates freely back and forth across a valley floor. Such features reflect the influence of more resistant alluvium or bedrock. In other words, a meander deforms when its downstream limb is fixed in place by resistant materials. The continuing

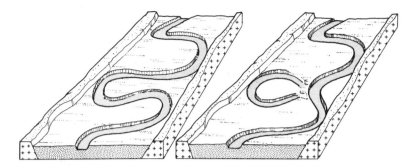

Fig. 5.4 Formation of an oxbow lake.

Fig. 5.5 The Rakaia River flowing through the Canterbury Plains, South Island, New Zealand. This braided river flows over a thick sequence of gravel deposits. (Courtesy of the New Zealand Geological Survey.)

downstream movement of the upstream limb leads to the formation of a compressed meander.

Meander lengths vary from seven to ten times the width of the channel, whilst cross-overs occur at about every five to seven channel widths. The ampli-

tude of a meander bears little relation to its length, but is largely determined by the erosion characteristics of the river bed and local factors. For instance, in uniform material, the amplitudes of meanders do not increase progressively, nor do meanders form oxbow

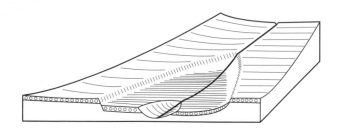

(a)

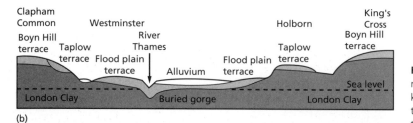

(b)

Fig. 5.6 (a) Paired river terraces due to rejuvenation: note valley in valley and knick points. (b) Section across London to show the paired terraces and one of the buried 'gorges' of the Thames Valley.

lakes during the downstream migration of bends. Higher sinuosity is associated with small width relative to depth and a larger percentage of silt and clay, which afford greater cohesiveness, in the river banks. Relatively sinuous channels with a low width to depth ratio are developed by rivers transporting large quantities of suspended sediment. By contrast, the channel tends to be wide and shallow, and less sinuous, when the amount of bedload discharge is high.

A river is described as being braided if it splits into a number of separate channels or anabranches to adjust to a broad valley (Fig. 5.5). The areas between the anabranches are occupied by islands built of gravel and sand. For the islands to remain stable, the river banks must be more erodible, so that they give way rather than the islands. Braided channels occur on steeper slopes than meanders.

Climatic changes and earth movements alter the base level to which a river grades. When a land surface is elevated, the downcutting activity of rivers flowing over it is accelerated. The rivers begin to regrade their courses from their base level and, as time proceeds, their newly graded profiles are extended upstream until they are fully adjusted to the new conditions. Until this time, the old longitudinal profile intersects with the new to form a knick point. The upstream migration of knick points tends to be retarded by outcrops of resistant rock; consequently, after an interval of time, they are usually located at

hard rock exposures. The acceleration of downcutting consequent upon uplift frequently produces a new valley within the old, the new valley extending upstream to the knick point.

River terraces are also developed by rejuvenation. In the lower course of a river uplift leads to the river cutting into its alluvial plain. The lateral and downstream migration of meanders means that a new floodplain is formed, but very often paired alluvial terraces, representing the remnants of the former floodplain, are left at the sides of the river (Fig. 5.6).

Incised meanders are also associated with rejuvenation and are often found together with river terraces. When uplift occurs, the downcutting action of meanders is accelerated and they carve themselves into the terrain over which they flow. The landforms which are then produced depend upon the character of the terrain and the relative rates of downcutting and meander migration. If vertical erosion is rapid meander shift has little opportunity to develop and consequently the loops are not greatly enlarged. The resulting incised meanders are described as entrenched (Fig. 5.7a,c). However, when time is afforded for meander migration, they incise themselves by oblique erosion and the loops are enlarged they are then referred to as ingrown meanders (Fig. 5.7b,d).

When incision occurs in the alluvium of a river plain, the meanders migrate back and forth across the

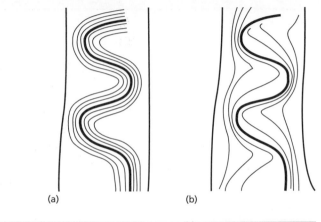

(a)   (b)

(c)

Fig. 5.7 Incised meanders. (a)
Entrenched. (b) Ingrown. (c) The Great
Gooseneck, an entrenched meander on
the San Juan River, Utah. (d) Incised
meanders of the Rheidol Valley, Cardigan,
Wales.

(d)

Fig. 5.8 Alluvial terraces cut into thick deposits of gravel, marking successive stages of erosion. South Island, New Zealand.

floor. On each successive occasion that a meander swings back to the same side, it does so at a lower level. As a consequence, small remnant terraces may be left above the newly formed plain. These terraces are not paired across the valley and their position and preservation depend on the swing of meanders over the valley (Fig. 5.8). If downcutting is very slow, erosion terraces are unlikely to be preserved.

Conversely, when sea level rises, rivers again have to regrade their courses to the new base level. For instance, during glacial times, the sea level was much lower and rivers carved out valleys accordingly. As the last glaciation retreated, so the sea level rose, and the rivers had to adjust to these changing conditions. This frequently meant that their valleys were filled with sediments. Hence, buried channels are associated with the lowland sections of many rivers.

5.2 Fluvial processes

5.2.1 River flow

Generally, it is considered that, except at times of flood, a stream has a steady, uniform flow, that is, one in which, at any given point, the depth does not vary with time. If the flow is uniform, the depth is constant over the length of the stream concerned. Although these two assumptions are not strictly true, they allow simple and satisfactory solutions for river engineering problems. Assuming a steady, uniform flow, the rate at which water passes through successive cross-sections of a stream is constant. River flow is measured as the discharge of the volume of water passing a given point per unit time. The height or stage of water in a channel depends on the discharge, as well as the shape and capacity of the river channel. Bankfull discharge refers to the maximum volume of water at a certain velocity of flow which a river can sustain without overbank spillage. The hydraulic geometry and bankfull discharge alter if the channel is changed by erosion.

If water flows along a smooth, straight channel at very low velocities, it moves in laminar flow, with parallel layers of water shearing one over the other. In laminar flow, the layer of maximum velocity lies below the water surface, whilst around the wetted perimeter it is least. Laminar flow cannot support particles in suspension and, in fact, is not found in natural streams except near the bed and banks. When the velocity of flow exceeds a critical value it becomes turbulent. Fluid components in turbulent flow follow a complex pattern of movement: components mix with each other and secondary eddies are superimposed on the main forward flow. The Reynolds number (N_R) is the starting point for the calculation of erosion and transportation within a channel system and is commonly used to distinguish between laminar and turbulent flow; it is expressed as

$$N_R = \rho \frac{(vR)}{(\mu)} \tag{5.1}$$

where ρ is the fluid density, v is the mean velocity, R is the hydraulic radius (the ratio between the cross-sectional area of a river channel and the length of its wetted perimeter) and μ is the viscosity. The flow is laminar for small values of the Reynolds number and turbulent for higher ones. The Reynolds number for flow in streams is generally over 500, varying from 300 to 600.

There are two kinds of turbulent flow, namely, streaming and shooting flow. Streaming flow refers to the ordinary turbulence found in most streams, whereas shooting flow occurs at higher velocities such as those found in rapids. Whether turbulent flow is shooting or streaming, it is determined by the Froude number (F_n)

$$F_n = \frac{v}{\sqrt{(gZ)}} \qquad (5.2)$$

where v is the mean velocity, g is the force of gravity and Z is the depth of water. If the Froude number is less than unity, the stream is in the streaming flow regime, whereas, if it is greater than unity, it is in the shooting flow regime.

Generally, the highest velocity of a stream is at its centre, below or extending below the surface. The exact location of maximum velocity depends upon the channel shape, roughness and sinuosity, but usually lies between 0.05 and 0.25 of the depth. In a symmetrical river channel, the maximum water velocity is below the surface and centred. Regions of moderate velocity, but high turbulence, occur outwards from the centre, being greatest near the bottom. Near the wetted perimeter, the velocities and turbulence are low. On the other hand, in an asymmetrical channel, the zone of maximum velocity shifts away from the centre towards the deeper side. In such instances, the zone of maximum turbulence is raised on the shallow side and lowered on the deeper side. Consequently, the channel morphology has a significant influence on erosion.

The turbulence and velocity are very closely related to the erosion, transportation and deposition. The work done by a stream is a function of the energy it possesses. Potential energy is converted by downflow to kinetic energy, which is mostly dissipated in friction. Stream energy is therefore lost owing to friction from turbulent mixing and such frictional losses are dependent upon the channel roughness and shape. The total energy is influenced mostly by velocity,

which, in turn, is a function of the stream gradient, volume and viscosity of water in the flow, and the characteristics of the channel cross-section and bed. This relationship has been embodied in the Chezy formula, which expresses the velocity as a function of the hydraulic radius (R) and slope (S)

$$v = C\sqrt{(RS)} \qquad (5.3)$$

where v is the mean velocity and C is a constant which depends upon gravity and other factors contributing to the friction force. The minimum bed erosion takes place when the gradient of the channel is low and the wetted perimeter is large compared with the cross-sectional area of the channel.

The Manning formula represents an attempt to refine the Chezy equation in terms of the constant C

$$v = \frac{1.49}{n} R^{2/3} S^{1/2} \qquad (5.4)$$

where the terms are the same as in the Chezy equation and n is a roughness factor. The velocity of flow increases as the roughness decreases for a channel of particular gradient and dimensions. The roughness factor has to be determined empirically and varies not only for different streams, but also for the same stream under different conditions and at different times. In natural channels, the value of n is 0.01 for smooth beds, about 0.02 for sand and 0.03 for gravel. Anything which affects the roughness of the channel changes n, including the size and shape of grains on the bed, the sinuosity and obstructions in the channel section. A variation in discharge also affects the roughness factor, because the depth of water and volume influence the roughness.

The quantity of flow can be estimated from measurements of the cross-sectional area and current speed of a river. Generally, channels become wider relative to their depth and adjusted to larger flows with increasing distance downstream. Bankfull discharges also increase downstream in proportion to the square of the width of the channel or of the length of individual meanders, and to the 0.75 power of the total drainage area focused at the point in question.

Statistical methods are used to predict river flow and assume that recurrence intervals of extreme events bear a consistent relationship to their magnitude. A recurrence interval, generally of 50 or 100

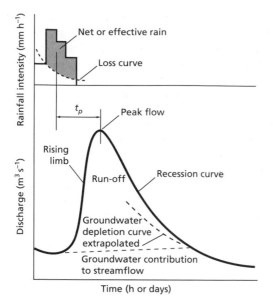

Fig. 5.9 Component parts of a hydrograph. When rainfall commences, there is an initial period of interception and infiltration before any measurable run-off reaches the stream channels. During the period of rainfall, these losses continue in a reduced form so that the rainfall graph has to be adjusted to show effective rain. When the initial losses are met, surface run-off begins and continues to a peak value which occurs at time t_p, measured from the centre of gravity of the effective rain on the graph. Thereafter, surface run-off declines along the recession limb until it disappears. Base-flow represents the groundwater contribution along the banks of the river.

years, is chosen in accordance with given hydrological requirements. The unit hydrograph concept of Sherman (1932) postulates that the most important hydrological characteristics of any basin can be seen from the direct run-off hydrograph resulting from 25 mm of rainfall evenly distributed over 24 h. This is produced by drawing a graph of the total stream flow at a chosen point as it changes with time after such a storm, from which the normal base-flow caused by groundwater is subtracted (Fig. 5.9).

There is a highly significant relationship between the mean annual flood discharge per unit area and the drainage density. The peak discharge and lag time of discharge (the time which elapses between maximum precipitation and maximum run-off) are also influenced by the drainage density, as well as by the shape and slope of the drainage basin. Generally, stream flow is most variable and flood discharge is at

a maximum per unit area in small basins. This is because storms tend to be local in occurrence. An inverse relationship exists between the drainage density and the base-flow or groundwater discharge. This is related to the permeability of the rocks present in a drainage basin. In other words, the greater the quantity of water which moves on the surface, the higher the drainage density which, in turn, means that the base-flow is lower. In areas of high drainage density, the soils and rocks are relatively impermeable and water runs off rapidly. The amount of infiltration is reduced accordingly.

5.2.2 River erosion

The work undertaken by a river is threefold, it erodes soils and rocks, transports the products thereof, which it eventually deposits. Erosion occurs when the force provided by the river flow exceeds the resistance of the material over which it flows. Thus the erosion velocity is appreciably higher than that required to maintain stream movement.

The amount of erosion accomplished by a river in a given time depends upon the quantity of energy it possesses which, in turn, is influenced by its volume and velocity of flow, and the character and size of its load. The soil, rock type and geological structure over which it flows, the infiltration capacity of the area it drains and the vegetation cover, which directly affects the stability and permeability of the soil, also influence the rate of erosion.

Once bank erosion starts, and the channel is widened locally, the process is self-sustaining. The recession of the bank permits the current to wash more directly against the downstream side of the eroded area, and so erosion and bank recession continue. Sediment accumulates on the opposite side of the channel, or on the inside of a bend, so that the channel gradually shifts in the direction of bank attack. In such a way, meandering is initiated, and this process is responsible for most of the channel instability.

Stream meanders usually occur in series and there is normally a downstream progression of meander loops. Meander growth is often stopped by the development of shorter chute channels across the bars formed on the inside of the bends. Chutes may develop because the resistance to flow around the lengthening bend becomes greater than that across a bar, or because changes in alignment caused by channel shifting upstream tend to direct flow across a bar inside the bend. Meander loops may be abandoned because cutoffs (Fig. 5.4) develop from adjacent bends. The loops either migrate into each other or channel avulsions form across the necks between adjacent bends during periods of overbank flooding. Meander cutoffs and shortening by chute developments reduce channel lengths and increase slopes, and hence are generally beneficial for reducing flood heights and improving drainage. However, they may cause much local damage by channel shifting and bank erosion, and the resulting unstable bed conditions may interfere with navigation.

During flood, the volume of a river is greatly increased, which leads to an increase in its velocity. The principal effect of flooding in the upper reaches of a river is to accelerate the rate of erosion; much of the material so produced is then transported downstream and deposited over the floodplain (Lewin, 1989). The vast increase in erosive strength during maximum flood is well illustrated by the devastating floods which occurred on Exmoor, England in August 1952. It was estimated that these moved $153\,000\,m^3$ of rock debris into Lynmouth, some of the boulders weighing up to 10 tonnes. Scour and fill are characteristic of flooding. Often a river channel is filled during the early stages of flooding but, as discharge increases, scour takes over. For example, streams flooding on alluvial beds normally develop an alternating series of deep and relatively narrow pools, typically formed along the concave sides of bends, together with shallower, wider reaches between bends where the main current crosses the channel diagonally from the lower end of one pool to the upper end of the next. During high flows, the pools or bends tend to scour more deeply, while the crossing bars are built higher by sediment deposition, although deposition does not equal the rise in stage and hence the water depth increases on the bars. When the stage falls, erosion takes place from the top of crossing bars leading to some infilling of the pools. However, as low stage activity is less effective, the general shape of the bed usually reflects the influence of the flood stage.

5.2.3 River transport

The load which a river carries is transported by traction, saltation, suspension or solution. The compe-

tence of a river to transport its load is demonstrated by the largest boulder it is capable of moving. This varies according to its velocity and volume, being at a maximum during flood. Generally, the competence of a river varies as the sixth power of its velocity. The capacity of a river refers to the total amount of sediment which it carries, and varies according to the size of the particles which form the load on the one hand and its velocity on the other. When the load consists of fine particles, the capacity is greater than when it is made up of coarse material. Usually, the capacity of a river varies as the third power of its velocity. Both the competence and capacity of a river are influenced by changes in the weather, and by the lithology and structure of the rocks over which it flows.

The sediment discharge of a river is defined as the mass rate of transport through a given cross-section, measured as the mass per second per metre width, and can be divided into the bedload and suspended load. The force necessary to entrain a given particle is referred to as the critical tractive force, and the velocity at which this force operates on a given slope is the erosion velocity. The critical erosion velocity is the lowest velocity at which loose grains of a given size on the bed of a channel will move. The value of the erosion velocity varies according to the characteristics and depth of the water, the size, shape and density of the particles being moved, and the slope and roughness of the floor. The graph of Hjulstrom (1935) shows the threshold boundaries for erosion, transportation and deposition (Fig. 5.10a). Modified versions of these curves have appeared subsequently (Fig. 5.10b).

It can be seen from Fig. 5.10(a) that fine to medium grained sands are more easily eroded than clays, silts or gravels. The fine particles are resistant because of the strong cohesive forces which bind them and because fine particles on the channel floor give the bed a smoother surface. There are accordingly few protruding grains to aid entrainment by giving rise to local eddies or turbulence. However, once silts and clays are entrained, they can be transported at much lower velocities. For example, particles of 0.01 mm in diameter are entrained at a critical velocity of about $600\,mm\,s^{-1}$, but remain in motion until the velocity drops below $1\,mm\,s^{-1}$. Gravel is hard to entrain simply because of the size and weight of the particles involved. Particles in the bedload move slowly and intermittently. They generally

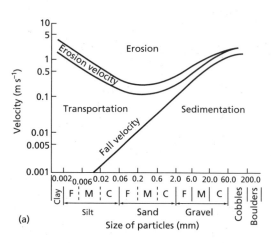

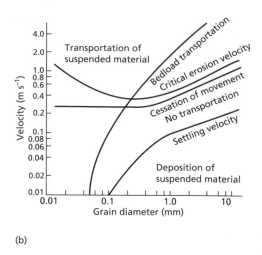

Fig. 5.10 (a) Curves for erosion, transportation and deposition of uniform sediment: F, fine; M, medium; C, coarse. Note that fine sand is the most easily eroded material. (After Hjulstrom, 1935.) (b) Relation between flow velocity, grain size, entrainment and deposition for uniform grains with a specific gravity of 2.65. The velocities are those 1.0 m above the bottom of the body of water. The curves are not valid for high sediment concentrations, which increase the fluid viscosity. (After Sundberg, 1956.)

move by rolling or sliding, or by saltation if the instantaneous hydrodynamic lift is greater than the weight of the particle. Deposition takes place wherever the local flow conditions do not re-entrain the particles.

Bedload transport in sand bed channels depends upon the regime of flow, that is, on streaming or shooting flow. When the Froude number is much smaller than unity, the flow is tranquil, the velocity is low, the water surface is placid and the channel bottom is rippled. In this streaming regime, resistance to flow is great and sediment transport is small, with only single grains moving along the bottom. As the Froude number increases, but remains within the streaming flow regime, the form of the bed changes to dunes or large-scale ripples (Fig. 5.11). Turbulence is now generated at the water surface and eddies form in the lee of the dunes. The movement of grains takes place up the leeside of the dunes to cascade down the steep front, causing the dunes to move downstream. When the Froude number exceeds unity, the flow is rapid, the velocity is high, the resistance to flow is small and the bedload transport is great. At the transition to the upper flow regime, planar beds are formed. As the Froude number increases further, standing waves form and antidunes are developed. The parti-

cles in the suspended load have settling velocities which are less than the buoyant velocity of the turbulence and vortices. Once particles are entrained and are part of the suspended sediment load, little energy is required to transport them. Indeed, as mentioned above, they can be carried by a current with a velocity less than the critical erosion velocity needed for their entrainment. Moreover, the suspended load decreases turbulence which, in turn, reduces the frictional losses of energy and makes the stream more efficient.

The distribution of suspended load increases rapidly with depth below the surface of a stream, the highest concentration generally occurring near the bed. However, there is a variation in suspended sediment concentrations at various depths of a stream for grains of different sizes. Most of the sand grains are carried in suspension near the bottom, whereas there is little change in silt concentration with depth.

The suspended load concentration usually increases with an increase in stage of a stream. In large streams, the peak of the sediment concentration is generally close to the peak of the discharge. In fact, during flood, the amount of suspended sediment load generally increases more quickly than the discharge and reaches a peak concentration some several hours before the floodwater peak. In such cases, the sus-

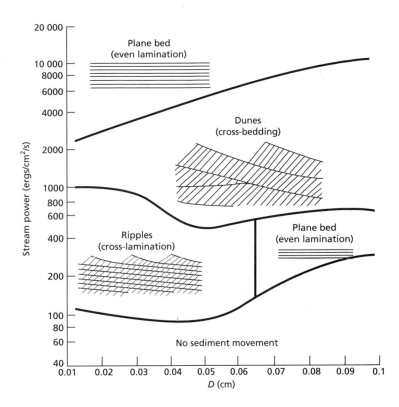

Fig. 5.11 Bed forms in relation to stream power and calibre of bedload material. (After Simons & Richardson, 1961.)

pended load carried during the highest water flow is considerably less than capacity.

Because the discharge of a river varies, sediments are not transported continuously, for example, boulders may be moved only a few metres during a single flood. In other words, there is a threshold discharge below which no movement of bedload occurs; thus there is a direct relationship between bedload movement and flood discharge. Once the rate of flow falls below the threshold value, the bedload remains stationary until the next flood of equal or higher magnitude occurs. Nonetheless, the contribution of the bedload to the total streamload is frequently high. Often, the amount of bedload moved downstream during times of flood exceeds that of the suspended load by several fold.

As described above, the total amount of sediment carried by a river increases significantly during flood. Part of the increase in load is derived from within the channel and part from outside the channel. Consequently, the channel form changes during flood, the depth and width of the channel adapting to scour and

deposition (Gupta, 1988). Scour at one point is accompanied by deposition at another. Initially, during a flood, the bedload may be moved into small depressions in the floor of a river and cannot be removed easily. Major floods can bring about notable changes in channel form by bank scour and slumping. However, after a flood, deposition may make good some of the areas which were removed.

5.2.4 Deposition of sediments

Deposition occurs where turbulence is at a minimum or where the region of turbulence is near the surface of a river. For example, lateral accretion occurs with the deposition of a point bar on the inside of a meander bend. A point bar grows as the meander moves downstream or new ones are built as the river changes course during or after floods. Old meander scars can often be seen on floodplains. The combination of point bar and filled slough results in what is called ridge and swale topography. The ridges are composed of sand bars and the swales are the depressions which were subsequently filled with silt and clay.

Some alluvial deposits, such as channel bars, are transitory, existing for a matter of days or even minutes. Hence, the channels of most streams are excavated mainly in their own sedimentary deposits, which streams continually rework by eroding the banks in some places and redepositing the sediment further downstream. Indeed, sediments which are deposited over a floodplain may be regarded as being stored there temporarily.

An alluvial floodplain is the most common depositional feature of a river. Ward (1978) suggested that a floodplain can be regarded as a store of sediment across which channel flow takes place and which, in the long term, is comparatively unchanging in amount. However, dramatic changes may occur in the short term (Rahn, 1994). In fact, the morphology of a floodplain often displays an apparent adjustment to flood discharge in that different floodplain or terrace levels may be related to different frequencies of flood discharge. Obviously, the higher levels are inundated by the largest floods.

As noted above, the peak concentration of suspended load usually occurs prior to the discharge peak. Hence, much of the load of a stream is contained within its channel. The water which overflows onto the floodplain accordingly possesses less suspended sediment (Macklin et al., 1992). As a consequence, the processes within the channel responsible for lateral accretion are usually the most important in the formation of a floodplain, compared with overbank flow, which results in vertical accretion. In fact, lateral accretion may account for between 60% and 80% of the sediment which is deposited (Leopold et al., 1964).

The alluvium of floodplains is made up of many kinds of deposit, laid down both in the channel and outside it (Marsland, 1986). Vertical accretion on a floodplain is accomplished by in-channel filling and the growth of overbank deposits during and immediately after floods (Fig. 5.12a). Gravels and coarse sands are moved chiefly at flood stages and deposited in the deeper parts of a river. As the river overtops its banks, its ability to transport material is lessened, so that coarser particles are deposited near the banks to form levées. Levées therefore slope away from the channels into the flood basins, which are the lowest part of a floodplain. At high stages, low sections and

breaks in levées may mean that there is a concentrated outflow of water from the channel into the floodplain. This outflow rapidly erodes a crevasse, leading to the deposition in the flood basin of a crevasse splay. Finer material is carried further and laid down as backswamp deposits (Fig. 5.12b). At this stage, a river sometimes aggrades its bed, eventually raising it above the level of the surrounding plain. In such a situation, when the levées are breached by floodwater, hundreds of square kilometres may be inundated.

5.3 Floods

Floods represent the commonest type of geological hazard (Fig. 5.13a). They probably affect more individuals and their property than all the other hazards put together. However, the likelihood of flooding is more predictable than some other types of hazard, such as earthquakes, volcanic eruptions and landslides.

Most disastrous floods are the result of excessive rainfall or snow melt, that is, they are due to excessive surface run-off. In most regions, floods occur more frequently in certain seasons than others. Meltwater floods are seasonal and are characterized by a substantial increase in river discharge causing a single flood wave (Church, 1988). The latter may have several peaks. The two most important factors governing the severity of floods due to snow melt are the depth of snow and the rapidity with which it melts.

Identical flood-generating mechanisms, especially those associated with climate, can produce different floods within different catchments, or within the same catchment at different times. Such differences, according to Newson (1994), are attributable to the effects of flood intensifying factors, such as certain basin, network and channel characteristics which may increase the rate of flow through the system. For instance, flood peaks may be low and attenuated in a long, narrow basin with a high bifurcation ratio, whereas, in a basin with a rounded shape and low bifurcation ratio, the flood peaks may be higher. Generally, drainage patterns which give rise to the merging of flood flows from major tributaries in the lower part of a river basin are associated with high-magnitude flood peaks in that part of the system. In addition, the capacity of the ground to allow water to

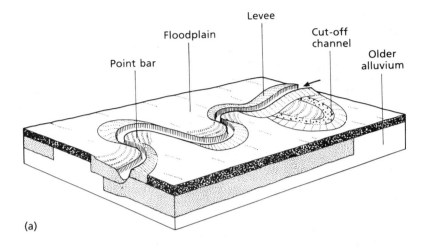

(a)

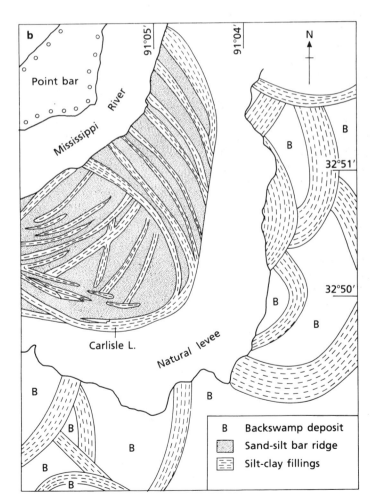

Fig. 5.12 (a) The main depositional features of a meandering channel. (b) Map of a portion of the Mississippi River floodplain, showing various kinds of deposits. (After Fisk, 1944.)

(a)

(b)

Fig. 5.13 (a) Floods in Ladysmith, Natal, South Africa in February 1994. (b) Flash flood at Klamath in northern California. (Courtesy of Eureka Newspapers.)

infiltrate and the amount which can be stored influence the size and timing of a flood. Frozen ground, by inhibiting infiltration, is an important flood intensifying factor and may extend the source area over the whole of a catchment. A catchment area with highly permeable ground conditions may have such a high infiltration capacity that it is rarely subjected to floods. Flood discharges are at a maximum per unit area in small drainage basins. This is because storms tend to be local in occurrence. Watersheds with a high drainage density produce simultaneous flooding in tributaries, which then overload junction capacity to transmit flow. In basins with a low drainage density, the procession of flood waves have longer delay periods, allowing the main channel to absorb incoming flows one at a time. Hence, it usually takes some time to accumulate enough run-off to cause a major disaster.

Flash floods prove the exception (Fig. 5.13b). Flash floods are short-lived extreme events. They usually occur under slowly moving or stationary thunderstorms which last for less than 24 h. The resulting rainfall intensity exceeds the infiltration capacity so that run-off takes place very rapidly. Flash floods are frequently very destructive as the high-energy flow can carry much sedimentary material (Clarke, 1991).

On the other hand, long-rain floods are associated with several days or even weeks of rainfall, which may be of low intensity. They are the most common cause of major flooding (Bell, 1994). Single event floods have a single main peak, but are of notably longer duration than flash floods. They represent the commonest type of flooding. Nevertheless, some of the most troublesome floods are associated with a series of flood peaks which follow closely upon each other. Such multiple event floods are caused by more complex weather conditions than those associated with single event floods. The effects of multiple event floods are usually severe because of the duration over which they extend. This could be several weeks or even months in the case of certain monsoon or equatorial regions, where seasonal floods are frequently of extended duration.

As the volume of water in a river is greatly increased during times of flood, so its erosive power increases accordingly. Thus, the river carries a much higher sediment load. Deposition of the latter where it is not wanted also represents a serious problem.

The influence of human activity can bring about changes in drainage basin characteristics, for example, the removal of forest from parts of a river basin can lead to higher peak discharges which generate increased flood hazard. A most notable increase in flood hazard can arise as a result of urbanization; the impervious surfaces created mean that infiltration is reduced and, together with stormwater drains, give rise to increased run-off. Not only does this produce higher discharges, but lag times are reduced. The problem of flooding is particularly acute where rapid expansion has led to the development of urban sprawl without proper planning or, worse, where informal settlements have sprung up. Heavy rainfall can prove to be disastrous where informal settlements are concerned.

A flood can be defined in terms of the height or stage of water above a certain given point, such as the banks of a river channel. However, rivers are generally considered to be in flood when the level has risen to such an extent that damage occurs. The discharge rate provides the basis for most methods of predicting the magnitude of flooding, the most important factor being the peak discharge which is responsible for maximum inundation. Not only does the size of a flood determine the depth and area of inundation, but it also primarily determines the duration of a flood. These three parameters, in turn, influence the velocity of flow of floodwaters, all four being responsible for the damage potential of a flood. As noted above, the physical characteristics of a river basin, together with those of the stream channel, affect the rate at which discharge downstream occurs. The average time between a rainstorm event and the consequent increase in river flow is referred to as the lag time. The lag time can be measured from the commencement of rainfall to the peak discharge or from the time when actual flood conditions have been attained (e.g. bank-full discharge) to the peak discharge. This lag time is an important parameter in flood forecasting. However, the calculation of the lag time is a complicated matter. Nonetheless, once enough data on rainfall and run-off versus time have been obtained and analysed, an estimate of where and when flooding will occur along a river system can be made.

An estimate of future flood conditions is required for either forecasting or design purposes. In terms of flood forecasting, more immediate information is needed regarding the magnitude and timing of a flood so that appropriate evasive action can be taken. In terms of design, planners and engineers require data on the magnitude and frequency of floods. Hence, there is a difference between flood forecasting for warning purposes and flood prediction for design purposes. A detailed understanding of the run-off processes involved in a catchment and stream channels is required for the development of flood forecasting, but is less necessary for long-term prediction. The ability to provide enough advance warning of a flood means that it may be possible to reduce the resulting damage as people will be able to evacuate the area likely to be affected, together with some of their possessions. The most reliable forecasts are based on data from rainfall or melt events which have just taken place or are still occurring. Hence, advance warning is generally measured only in hours or sometimes several days. The longer-term forecasts are commonly associated with snow melt. Basically, flood forecasting involves the determination of the amount of precipitation and resultant run-off within a given catchment area. The volume of run-off is then converted into a time-distributed hydrograph, and flood routeing procedures are used, where appropriate, to estimate the changes in the shape of the hydrograph as the flood moves downstream. In other words, flood routeing involves the determination of the height and time of arrival of the flood wave at successive locations along a river. As far as prediction for design purposes is concerned, the design flood, namely, the maximum flood against which protection is being designed, is the most important factor to determine. Methods of flood forecasting and prediction have been reviewed by Ward (1978).

There are two specific problems related to run-off predictions. Firstly, there is the need to forecast peak flows associated with sudden increases in surface run-off and, secondly, there is the prediction of the minimum flow which very much involves the decreasing volume of base-flow. The accuracy of run-off predictions tends to improve as the time interval is increased. Estimates of annual or even seasonal run-off totals for given catchment areas may, in some cases, be made from annual or seasonal rainfall totals, using a simple straight-line regression between the two variables.

However, correlations between rainfall and run-off may generally be expected to yield forecasts of only token accuracy for they do not take into account the contributions made by interflow and base-flow. Increases in surface run-off after rainfall or snow melt tend to be rapid, leading to a short-lived peak, which is followed by a rather longer period of declining run-off. Because it is the peak flow which brings havoc, the principal object of prediction concerns the magnitude and timing of this peak, and the frequency with which it is likely to occur.

In order to carry out a flood frequency analysis, either the maximum discharge for each year or all discharges greater than a given discharge are recorded from gauging stations according to magnitude (Dalrymple, 1960; Benson, 1968; Bobee & Ashkar, 1988). Then, the recurrence interval, that is, the period of years within which a flood of a given magnitude or greater occurs, is determined from

$$T = \frac{n+1}{m} \tag{5.5}$$

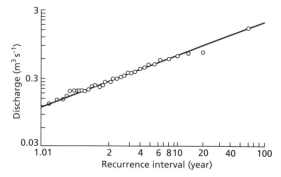

Fig. 5.14 Flood frequency curve of the Licking River, Tobaso, Ohio. (After Dalrymple, 1960.)

where T is the recurrence interval, n is the number of years of record and m is the rank of the magnitude of the flood, with $m = 1$ as the highest discharge on record. Each flood discharge is plotted against its recurrence interval and the points are joined to form the frequency curve (Fig. 5.14). The flood frequency curve can be used to determine the probability of the size of the discharge that could be expected during any given time interval. The larger the recurrence interval, the longer the return period and the greater the magnitude of flood flow. The probability that a given magnitude of flow will occur or be exceeded in a given time is the reciprocal of the recurrence interval. The probability P of a flood with a recurrence interval of T years being equalled or exceeded in x years is given by

$$P = 1 - (1 - 1/T)^{x} \tag{5.6}$$

Estimates of the recurrence intervals of floods of different sizes can be improved by using alluvial stratigraphy to date the sediments they deposited, the more widespread, thicker deposits being formed by larger floods (Jarret, 1990). The technique can be used to reconstruct past changes in climate and so to determine whether, in the past, floods were more or less frequent at a particular location (Costa, 1978). Clarke (1996) described how to use boulder size to estimate the probable maximum flood which has occurred in the past.

The duration and magnitude of precipitation tend to be inversely related to the storm intensity. If a river basin has been saturated by precipitation, a low intensity can be compensated for by a high magni-

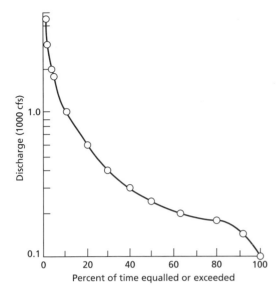

Fig. 5.15 Flow duration curve for Bowie Creek near Hattiesburg, Missouri, for the period 1939–1948. (After Searcy, 1959.)

tude. Consequently, the quantity of antecedent moisture in the ground is important in relation to flooding. In other words, if the ground is saturated, no moisture can infiltrate and so the quantity of run-off is enhanced.

A flow duration curve shows the percentage of time a specified discharge is equalled or exceeded. In order to prepare such a curve, all flows during a given period are listed according to their magnitude. The percentage of time each one equalled or exceeded a given discharge is calculated and plotted (Fig. 5.15). The shape of the curve affords some insight into the characteristics of the drainage basin concerned. For instance, if the curve has an overall steep slope, this means that there is a large amount of direct run-off. On the other hand, if the curve is relatively flat, there is substantial storage, either on the surface or as groundwater. This tends to stabilize stream flow.

5.4 Hazard zoning, warning systems and adjustments

The lower stage of a river, in particular, can be divided into a series of hazard zones based on flood stages and risk (Fig. 5.16). A number of factors have to be taken into account when evaluating flood hazard, such as the loss of life and property, erosion and structural damage, disruption of socio-economic activity, including transport and communications, contamination of water, food and other materials, and damage to agricultural land and loss of livestock. A floodplain management plan involves the determination of such hazard zones. These are based on historical evidence related to flooding, including the magnitude of each flood and the elevation it reached, the recurrence intervals, the amount of damage involved, the effects of urbanization and any further development, and an engineering assessment of flood potential. Maps are then produced from such investigations which, for example, show the zones of most frequent flooding and the elevation of the floodwaters. Flood hazard maps provide a basis for flood management schemes. Kenny (1990) recognized four geomorphological flood hazard map units in central Arizona, which formed the basis of a flood management plan (Table 5.1).

Floodplain zones can be designated for specific types of land use; therefore in the channel zone, water should be allowed to flow freely without obstruction, for example, bridges should allow sufficient waterway capacity. Another zone could consist of the area with a recurrence interval of 1–20 years, which could be used for parks, and agricultural and recreational purposes. Such areas act as washlands during times of flood. Buildings would be allowed in the zone encompassing the 20–100-year recurrence interval, but they would need to have some form of protection against flooding. However, a line which is drawn on a map to demarcate a floodplain zone may encourage a false sense of security and, as a consequence, development in the upslope area may be greater than otherwise expected. At some point in time, this line is likely to be transgressed, which will cause more damage than would have been the case without a floodplain boundary being so defined. In addition, urbanization, as described above, is a flood intensifying land use. The replacement of permeable with impermeable surfaces and artificial drainage systems are responsible for more rapid run-off.

Land-use regulation seeks to obtain the beneficial use of floodplains with minimum flood damage and minimum expenditure on flood protection. In other words, the purpose of land-use regulation is to maintain an adequate floodway (i.e. the channel and those

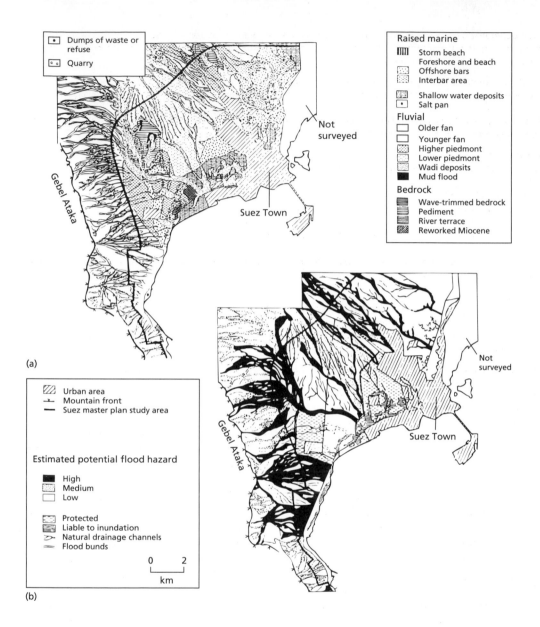

Fig. 5.16 Flood hazard map of Suez, Egypt, derived from geomorphological field mapping: (a) geomorphological map; (b) flood hazard map. (From Cooke *et al.*, 1982.)

adjacent areas of the floodplain which are necessary for the passage of a given flood discharge) and to regulate land-use development alongside it (i.e. in the floodway fringe which, in the USA, is the land between the floodway and the maximum elevation subject to flooding by the 100-year flood). Land use in the floodway in the USA is now severely restricted to, for example, agriculture and recreation. The floodway fringe is also a restricted zone in which land use is limited by zoning criteria based upon the degree of flood risk. However, land-use control involves the cooperation of the local population and authorities, and often central government. Any relocation of

Table 5.1 Generalized flood hazard zones and management strategies. (After Kenny, 1990.)

Flood hazard zone I (active floodplain area)

Prohibit development (business and residential) within floodplain
Maintain area in a natural state as an open area or for recreational uses only

Flood hazard zone II (alluvial fans and plains with channels less than 1 m deep; bifurcating and intricately interconnected systems subject to inundation from overbank flooding)

Flood proofing to reduce or prevent loss to structures is highly recommended
Residential development densities should be relatively low; development in obvious drainage channels should be prohibited
Dry stream channels should be maintained in a natural state and/or the density of native vegetation should be increased to facilitate superior water drainage retention and infiltration capabilities
Installation of upstream stormwater retention basins to reduce peak water discharges
Construction should be at the highest local elevation site where possible

Flood hazard zone III (dissected upland and lowland slopes; drainage channels where both erosional and depositional processes are operative along gradients of generally less than 5%)

Similar to flood hazard zone II
Roadways which traverse channels should be reinforced to withstand the erosive power of a channelled stream flow

Flood hazard zone IV (steep gradient drainage consisting of incised channels adjacent to outcrops and mountain fronts characterized by relatively coarse bedload material)

Bridges, roads and culverts should be designed to allow unrestricted flow of boulders and debris up to 1 m or more in diameter
Abandon roadways which at present occupy the wash floodplains
Restrict residential dwellings to relatively level building sites
Provisions for subsurface and surface drainage on residential sites should be required
Stormwater retention basins in relatively confined upstream channels to mitigate against high peak discharges

settled areas which are at high risk involves costly subsidization. Such cost must be balanced against the cost of alternative measures and the reluctance of some to move. Bell and Mason (1998) outlined such a problem at Ladysmith, South Africa. In fact, a change in land use in intensely developed areas is usually so difficult and so costly that the inconvenience of occasional flooding is preferred. The purchase of land by government agencies to reduce flood damage is rare.

In some situations, properties of high value which are likely to be threatened by floods can be flood proofed. For instance, an industrial plant can be protected by a flood wall, or buildings may have no windows below high water level and possess watertight doors, with valves and cutoffs on drains and sewers to prevent backing-up. Other structures may be raised above the most frequent flood levels. Most flood-proofing measures can be more readily and eco-

nomically incorporated into buildings at the time of their construction rather than subsequently. Hence, the adoption and implementation of suitable design standards should be incorporated into building codes and planning regulations to ensure the integrity of buildings during flood events.

Reafforestation of slopes denuded of woodland tends to reduce run-off and thereby lowers the intensity of flooding. As a consequence, forests are commonly used as a watershed management technique. They are most effective in relation to small floods, where the possibility exists of reducing flood volumes and delaying flood response. Nonetheless, if the soil is saturated, differences in the interception and soil moisture storage capacity due to forest cover will be ineffective in terms of flood response. Agricultural practices, such as contour ploughing and strip cropping, are designed to reduce soil erosion by reducing the rate of run-off. Accordingly, they can influence flood response. These practices are referred to in Chapter 10.

Emergency action involves the erection of temporary flood defences and possible evacuation. The success of such measures depends on the ability to predict floods and the effectiveness of the warning systems. A flood warning system, broadcast by radio or television, or from helicopters or vehicles, can be used to alert a community to the danger of flooding. Warning systems may use rain gauges and stream sensors equipped with self-activating radio transmitters to convey data to a central computer for analysis (Gruntfest & Huber, 1989). Alternatively, a radar rainfall scan can be combined with a computer model of a flood hydrograph to produce real time forecasts as a flood develops (Collinge & Kirby, 1987). However, widespread use of flood warning is usually only available in highly sensitive areas. The success of a flood warning system depends largely on the hydrological characteristics of the river. Warning systems often work well in large catchment areas which allow enough time between the rainfall or snow melt event and the resultant flood peak to enable evacuation and any other measures to be put into effect. By contrast, in small tributary areas, especially those with steep slopes or appreciable urban development, the lag time may be so short that, although prompt action may save lives, it is seldom possible to remove or protect property.

Financial assistance, in the form of government relief or insurance payouts, does nothing to reduce flood hazard. Indeed, by attempting to reduce the economic and social impact of a flood, it encourages the repair and rebuilding of damaged property, which may lead to the next flood of similar size giving rise to more damage. Moreover, the expectation that financial aid will be made available in such an emergency may result in further development of flood-prone areas. Hence, it would be more realistic to adjust insurance premiums for flooding in relation to the degree of risk. Be that as it may, the basic principle of insurance needs to be modified in the case of floods, so that premiums paid over many years cover the large losses which will be encountered in a few years. In the USA, the Federal Flood Insurance Program is administered by the Federal Emergency Management Agency (FEMA) from which interested parties can purchase subsidized flood insurance. The Program provides incentives to local governments to plan and regulate land use in flood hazard areas. The boundary of the 100-year flood defines the area that is subject to flood damage compensation and, if a community wishes to qualify for federal aid, it must join the Federal Flood Insurance Program. All property owners within the 100-year flood boundary must then purchase flood insurance. The hope was to achieve flood damage abatement and the efficient use of the floodplain.

5.5 The design flood and flood control

No structure of any importance, either in or adjacent to a river, should ever be planned or built without due consideration being given to the damage it may cause by its influence on floodwaters or the damage to which it may be subjected by those same waters. To avoid disaster, bridges must have the required waterway opening, flood walls and embankments must be high enough for overtopping not to occur, reservoirs must have sufficient capacity and dams must have sufficient spillway capacity as well as adequate protection against scour at the toe.

The maximum flood that any such structure can safely pass is called the design flood. If a flood of a given magnitude occurs on average once in 100 years, there is a 1% chance that such a flood will occur during any one year. The important factor to be determined for any design flood is not simply its magnitude, but the probability of its occurrence. In other words, is the structure safe against the 2%, 1% or the 0.1% chance flood or against the maximum flood that may ever be anticipated? Once this has been answered, the magnitude of the flood that may be expected to occur with that particular average frequency has to be determined. When assessing the future effectiveness of a specific flood mitigation project, its likely reduction of damage and safety afforded to the community over the entire spectrum of possible floods should be evaluated. The design flood should therefore be chosen in relation to all relevant economic, social and environmental factors.

A relatively simple solution to the problem of flooding is to build flood defences consisting of either earth embankments or masonry or concrete walls around the area to be protected (Thampapillai & Musgrave, 1985). Levées have been used extensively in the USA to protect floodplains from overflow (Fig. 5.17). Levées are earth embankments, the slopes of

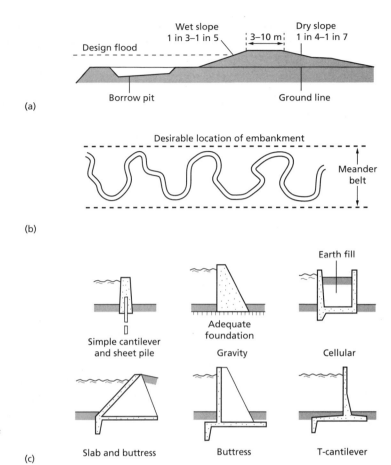

Fig. 5.17 (a) Cross-section of a typical flood embankment (levée). (b) Preferred location of embankments along a meandering channel. (c) Cross-sections of typical flood walls.

which should be protected against erosion by planting trees and shrubs, by paving or with rip-rap (Smith & Winkley, 1996). Without protection against bank erosion, a river will probably begin to meander again. The rate at which river channels revert to their former condition depends upon many factors. Nevertheless, this means that there is a need for maintenance. Levées reduce the storage of floodwater by eliminating the natural overflow basins of a river on the floodplain. Furthermore, they contract the channel and so increase flood stages within, above and below the levéed reach. Accordingly, wherever possible, levées should be located away from river channels and ideally outside the meander belt. In rural areas, this normally poses no great problems, but it is impractical in urban areas. On the other hand, because levées confine a river to its channel, this means that its efficiency is increased, hence they expedite run-off. Therefore, consideration must be given to the fact that more rapid movement of water through a section of river which has been hydraulically improved can enhance flood peaks further downstream and be responsible for accelerated erosion. Levées often encourage new development at lower levels where previously no one was inclined to build. Consequently, when the exceptional flood occurs and overtops the levées, the hazard may be reduced by building fuseplugs into the levées, that is, making certain sections deliberately weaker than the standard levée section, thereby determining that, if breaks occur, they do so at locations where they cause minimum damage.

Flood walls may be constructed in urban areas where there is not enough space for embankments.

They should be designed to withstand the hydrostatic pressure (including uplift pressure) exerted by the water when at design flood level. If the wall is backed by an earthfill, it must also act as a retaining wall.

Diversion is another method used to control flooding. This involves opening a new exit for part of the river water. Diversion schemes may be temporary or permanent. In the former case, the river channel is supplemented or duplicated by a flood relief channel, a bypass channel or a floodway. These operate during times of flood. Floodwater is diverted into a diversion channel via a sluice or fixed crest spillway. A permanent diversion acts as an intercepting or cutoff channel, which replaces the existing river channel, diverting either all or a substantial part of the flow away from a flood prone reach or river. Such schemes are normally used to protect intensively developed areas. Old river channels can be used as diversions to relieve the main channel of part of its floodwater (Fig. 5.18). Usually, it is necessary to return the diverted water into the river at a point downstream. Therefore, the diversion channel should be long enough to minimize backwater effects in the stretch of river being protected.

The flood hazard can often be lessened by stage reduction by improving the hydraulic capacity of the channel without affecting the rate of discharge. This may be accomplished by straightening, widening and deepening a river channel. Inasmuch as the quantity of discharge through any given cross-section of a river during a given time depends upon its velocity, the stage can be reduced by increasing the velocity (Brookes, 1988). River channels may be enlarged to carry the maximum flood discharges within their banks without overspill. However, overwidened channels eventually revert to their natural sizes unless continuously dredged.

Flood routeing is a procedure by which the variation of discharge with time at a point on a stream channel may be determined by the consideration of similar data for a point upstream (Wilson, 1983). In other words, it is a process which shows how a flood wave may be reduced in magnitude and lengthened in time by the use of storage in a reach of the river between the two points. Flood routeing therefore depends on a knowledge of storage in the reach. This can be evaluated by either making a detailed topographical and hydrographical survey of the river

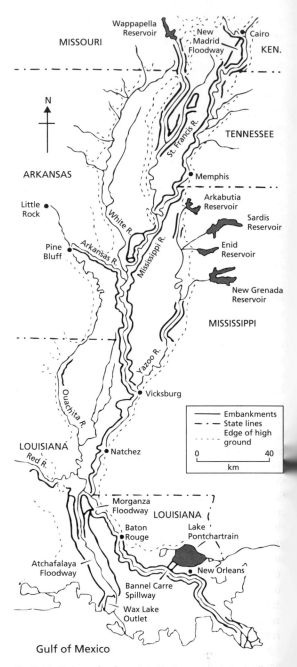

Fig. 5.18 Major embankments in the Lower Mississippi Valley and Atchafalaya floodway.

reach and the riparian land, thereby determining the storage capacity of the channel at different levels, or by using records of past levels of flood waves at the limits of the reach and hence deducing the storage capacity.

Peak discharges can be reduced by temporarily storing a part of the surface run-off until the crest of the flood has passed. This is done by inundating areas where flood damage is not important, such as water-meadows or wasteland. If, however, storage areas are located near or in towns and cities, they sterilize large areas of land which, if usable for other purposes, could be extremely valuable. On the other hand, it may be feasible to develop these storage areas as recreational centres. This method is seldom sufficient in itself and should be used to complement other measures.

Reservoirs aid in the regulation of run-off, so helping to control floods and improve the utility of a river (Hager & Sinniger, 1985). There are two types of storage, regardless of the size of the reservoir, controlled and uncontrolled. In controlled storage, gates in the impounding structure may regulate the outflow. Only in unusual cases does such a reservoir have sufficient capacity to eliminate completely the peak of a major flood. As a result, the regulation of the outflow must be planned carefully. This necessitates an estimation of how much of the early portion of a flood may be safely impounded which, in turn, requires an assessment of the danger which can arise if the reservoir is filled before the peak of a flood is reached. Where reservoirs exist on several tributaries, the additional problem of the timing of the release of the stored waters becomes a matter of great importance, because to release these waters in such a way that the peak flows combine at a downstream point can bring disaster. In uncontrolled storage, there is no regulation of the outflow capacity, and the only flood benefits result from the modifying and delaying effects of storage above the spillway crest.

Reservoirs for flood control should be operated so that the capacity required for storing floodwater is available when needed (Bell & Mason, 1998). This can generally be accomplished by lowering the water level of the reservoir as soon as practicable after the flood passes. On the other hand, the greatest effectiveness of reservoirs for increasing the value of a river for utilization is realized by keeping the reservoirs as near to full as possible. Hence, there must be some compromise in operation between these two purposes.

Retarding basins are much less common than reservoirs. They are provided with fixed, ungated outlets which regulate outflow in relation to the volume of water in storage. An ungated sluiceway acting as an orifice tends to be preferable to a spillway functioning as a weir. The discharge from the outlet at full reservoir capacity should ideally be equal to the maximum flow which the downstream channel can accept without causing serious flood damage. One of the advantages of retarding basins is that only a small area of land is permanently removed from use after their construction. Although land will be inundated at times of flood, this will occur infrequently, and so the land can be used for farming, but obviously not for permanent habitation.

References

Bell, F.G. (1994) Floods and landslides in Natal and notably the greater Durban area, September 1987, a retrospective view. *Bulletin of the Association of Engineering Geologists* **31**, 59–74.

Bell, F.G. & Mason, T.R. (1998) The problems of flooding in Ladysmith, Natal, South Africa. In *Geohazards and Engineering Geology, Engineering Geology Special Publication No.* **14**, Eddleston, M. & Maund, J.G. (eds). Geological Society, London.

Benson, M.A. (1968) Uniform flood frequency estimating methods for federal agencies. *Water Resources Research* **4**, 891–908.

Bobee, B. & Ashkar, F. (1988) Review of statistical methods for estimating flood risk with special emphasis on the log Pearson type 3 distribution. In *Natural and Man-Made Hazards*, El-Sabh, M.I. & Murty, T.S. (eds). Reidel, Dordrecht, pp. 357–368.

Brookes, A. (1988) *Channelized Rivers; Perspectives for Environmental Management*. Wiley, Chichester.

Church, M. (1988) Floods in cold climates. In *Flood Geomorphology*, Baker, V.R., Kochel, R.C. & Patton, P.C. (eds). Wiley, New York, pp. 205–229.

Clarke, A.O. (1991) A boulder approach to estimating flash flood peaks. *Bulletin of the Association of Engineering Geologists* **28**, 45–54.

Clarke, A.O. (1996) Estimating probable maximum floods in the upper Santa Ana Basin, southern California, from stream boulder size. *Environmental and Engineering Geoscience* **2**, 165–182.

Collinge, V. & Kirby, C. (eds) (1987) *Weather Radar and Flood Forecasting*. Wiley, Chichester.

Cooke, R.U., Brumsden, D., Doornkamp, J.C. & Jones,

D.K.C. (1982) *Urban Geomorphology in Drylands*. Oxford University Press, Oxford.

Costa, J.E. (1978) Holocene stratigraphy and flood frequency analysis. *Water Resources Research* **14**, 626–632.

Dalrymple, T. (1960) Flood frequency analysis. *United States Geological Survey Water Supply Paper* **1543A**, 1–80.

Davis, W.M. (1909) *Geographical Essays*. Dover, New York.

Fisk, H.N. (1944) *Geological Investigation of the Alluvial Valley of the Lower Mississippi River*. United States Army Corps of Engineers, Mississippi River Commission, Vicksburg, Mississippi.

Gruntfest, E.C. & Huber, C. (1989) Status report on flood warning systems in the United States. *Environmental Management* **13**, 357–368.

Gupta, A. (1988) Large floods as geomorphic events in the humid tropics. In *Flood Geomorphology*, Baker, V.R., Kochel, R.C. & Patton, P.C. (eds). Wiley, New York, pp. 301–315.

Hager, W.H. & Sinniger, R. (1985) Flood storage in reservoirs. *Proceedings of the American Society of Civil Engineers, Journal of Irrigation and Drainage Engineering* **111**, 76–85.

Hjulstrom, F. (1935) Studies of the morphological activity of rivers, as illustrated by the river Fynis. *Uppsala University Geological Institute Bulletin* **25**.

Horton, R.E. (1945) Erosion development of streams and their drainage basins: hydrological approach to quantitative morphology. *Bulletin of the Geological Society of America* **56**, 275–370.

Jarret, R.D. (1990) Palaeohydrologic techniques used to define the special occurrence of floods. *Geomorphology* **3**, 181–195.

Kenny, R. (1990) Hydrogeomorphic flood hazard evaluation for semi-arid environments. *Quarterly Journal of Engineering Geology* **23**, 333–336.

Leopold, L.B., Wolman, M.G. & Miller, J.P. (1964) *Fluvial Processes in Geomorphology*. Freeman, San Francisco.

Lewin, J. (1989) Floods in fluvial geomorphology. In *Floods: Hydrological, Sedimentological and Geomorphological Implications*, Beven, K. & Carling, F. (eds). Wiley, Chichester, pp. 265–284.

Macklin, M.G., Rumsby, M.T. & Newson, M.D. (1992) Historic overbank floods and floodplain sedimentation in the lower Tyne valley, north east England. In *Gravel Bed Rivers*, Hey, R.D. (ed). Wiley, Chichester.

Marsland, A. (1986) The flood plain deposits of the Lower Thames. *Quarterly Journal of Engineering Geology* **19**, 223–247.

Newson, M.D. (1994) *Hydrology and the River Environment*. Clarendon Press, Oxford.

Rahn, P.H. (1994) Flood plains. *Bulletin of the Association of Engineering Geologists* **31**, 171–183.

Searcy, J.M. (1959) Flow duration curves. *United States Geological Survey Water Supply Paper* **1542A**.

Sherman, L.K. (1932) Streamflow from rainfall by unit graph method. *Engineering News Record* **108**, 501–505.

Simons, D.B. & Richardson, E.V. (1961) Forms of bed roughness in alluvial channels. *Proceedings of the American Society of Civil Engineers, Journal of Hydraulics Division* **87**, 87–105.

Smith, L.M. & Winkley, B.R. (1996) The response of the Lower Mississippi River to river engineering. *Engineering Geology* **45**, 433–455.

Sundberg, A. (1956) The river Klarelren, a study of fluvial processes. *Geografiska Annaler* **38**, 127–316.

Thampapillai, D.J. & Musgrave, W.F. (1985) Flood damage and mitigation: a review of structural and non-structural measures and alternative decision frameworks. *Water Resources Research* **21**, 411–424.

Ward, R.C. (1978) *Floods: a Geographical Perspective*. Macmillan, London.

Wilson, E.M. (1983) *Engineering Hydrology*. Macmillan London.

6   **Marine activity**

Johnson (1919) distinguished three elements in a shoreline, the coast, the shore and the offshore. The coast was defined as the land immediately behind the cliffs, whilst the shore was regarded as that area between the base of the cliffs and low-water mark; the area which extended seawards from the low-water mark was termed the offshore. The shore was further divided into the foreshore and backshore, the former embracing the intertidal zone, and the latter extending from the foreshore to the cliffs. Those deposits which cover the shore usually are regarded as constituting the beach.

Marine activity in coastal environments acts within a narrow and often varying vertical zone. Significant changes in sea level, associated with the Pleistocene glaciation, have taken place in the recent geological past, which have influenced the character of present-day coastlines. When the ice sheets expanded, the sea level fell, rising when the ice melted. In addition, as the ice sheets retreated, the land beneath began to rise in an attempt to regain isostatic equilibrium, so complicating the situation. Hence, coasts can be characterized by features developed in response to submergence on the one hand or emergence on the other. Such changes in the relative level of the sea are not only a feature of the past, but still are occurring at present. Tides also have an influence upon the coast, the tidal range governing the vertical interval over which the sea can act. In addition, tidal currents can influence the distribution of sediments on the sea floor.

Waves acting on beach material are a varying force. They vary with time and place due to changes in the wind force and direction over a wide area of the sea, and with changes in coastal configuration and offshore relief. This variability means that the beach is rarely in equilibrium with the waves, in spite of the fact that it may take only a few hours for equilibrium to be attained under new conditions. Such a more or less constant state of disequilibrium occurs most frequently where the tidal range is considerable, as waves are continually acting at a different level on the beach.

6.1 Waves

When wind blows across the surface of deep water, it causes an orbital motion in those water particles in the plane normal to the wind direction (Fig. 6.1). The motion decreases in significance with increasing depth, dying out at a depth equal to that of the wave length. Because adjacent particles are at different stages in their circular course, a wave is produced. However, there is no progressive forward motion of the water particles in such a wave, although the form of the wave profile moves rapidly in the direction in which the wind is blowing. Such waves are described as oscillatory waves.

Forced waves are those formed by the wind in the generating area. They usually are irregular. On moving out of the area of generation, these waves become long and regular. They are then referred to as free waves. As these waves approach a shoreline, they feel bottom, which disrupts their pattern of motion, changing them from oscillatory to translation waves. Where the depth is approximately one-half of the wave length, the water particle orbits become ellipses with their major axes horizontal.

The forward movement of the water particles, as a whole, is not compensated for entirely by the backward movement. As a result, there is a general movement of the water in the direction in which the waves are travelling. This is known as mass transport. The time required for any one particle to complete its orbital revolution is the same as the period of the wave form. The orbital velocity u is equal to the length of orbital travel divided by the wave period

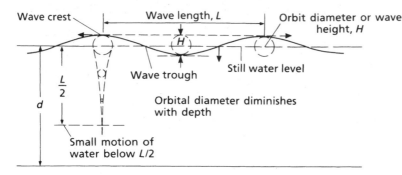

Direction of travel of wave profile →

Wave crest

Wave length, L

Orbit diameter or wave height, H

Wave trough

Still water level

$\frac{L}{2}$

d

Orbital diameter diminishes with depth

Small motion of water below L/2

Fig. 6.1 Orbital motion of water particles during the passage of an idealized sinusoidal wave in deep water. The orbital diameter decreases with depth and disappears at a depth of approximately one-half the wave length.

T. Hence

$$u = \frac{2\pi r}{T} \qquad (6.1)$$

Similarly, the wave velocity or celerity c is

$$c = \frac{2\pi R}{T} \qquad (6.2)$$

where R is the rolling circle required to generate a trochoidal wave form and r is the amplitude of particle orbit. The celerity of waves may also be derived from the following expression:

$$c = \left[\frac{gL}{2\pi} \tanh\left(\frac{2\pi Z}{L}\right)\right]^{\frac{1}{2}} \qquad (6.3)$$

where g is the acceleration due to gravity, Z is the still water depth and L is the wave length. If Z exceeds L, then $\tanh(2Z\pi/L)$ becomes equal to unity, so that in deep water

$$c = \sqrt{\frac{gL}{2\pi}} = 2.26\sqrt{L} \qquad (6.4)$$

When the depth is less than one-tenth of the wave length, $\tanh(2\pi Z/L)$ approaches $2\pi Z/L$ and then $c = \sqrt{(gZ)}$. As waves move into shallowing water, their velocity is reduced and so their wave length decreases.

6.1.1 Force and height of waves

The effectiveness of wave impact on the shoreline or marine structure depends on the depth of water and the size of the wave, dropping sharply with increasing depth. If deep water occurs alongside cliffs or sea walls, waves may be reflected without breaking and, in so doing, they may interfere with incoming waves.

In this way, standing waves, which do not migrate, are formed, in which the water surges back and forth between the obstruction and a distance equal to half the wave length away. The crests are much higher than in the original wave. This form of standing wave is known as clapotis. The oscillation of standing waves causes an alternating increase and decrease in pressure along any discontinuities in rocks or cracks in marine structures which occur below the water line. Such action gradually dislodges blocks of material. It has been estimated that translation waves, reflected from a vertical face, exert six times as much pressure on the wall as oscillatory waves of equal dimensions. When waves break, jets of water are thrown at approximately twice the wave velocity, which also causes increases in the pressure in discontinuities and cracks, thereby producing damage.

The fetch is the distance blown by a wind over a body of water, and is the most important factor determining the wave size and efficiency of transport. For instance, winds of moderate force, which blow over a wide stretch of water, generate larger waves than do strong winds which blow across a short reach. For a given fetch, the stronger the wind, the higher are the waves (Fig. 6.2a). Their period also increases with increasing fetch. The wind duration also influences waves, the longer it blows, the faster the waves move, whilst their lengths and periods increase (Fig. 6.2b). Usually, wave lengths in the open sea are less than 100 m and the speed of propagation is approximately 48 km h^{-1}. Normally, they do not exceed 8.5 m in height. Those waves which are developed in storm centres in the middle of an ocean may journey outwards to the surrounding landmasses. This explains why large waves may occur along a coast during fine

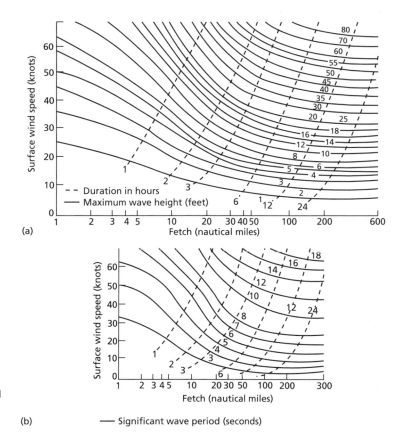

Fig. 6.2 Graphs relating wave height (a) and wave period (b) to wind speed, wind duration and fetch in oceanic waters. (After Darbyshire & Draper, 1963.)

weather. Waves frequently approach a coastline from different areas of generation; if they are in opposition, their height is decreased, whereas their height is increased if they are in phase. Moreover, when the wind shifts in direction or intensity, the new generation of waves differs from the older waves, which still persist in their original course. The more recent waves may overtake or, if the storm centre has migrated meanwhile, may intercept the earlier waves, so complicating the wave pattern.

6.1.2 Wave refraction

Wave refraction is the process whereby the direction of wave travel changes because of changes in the topography of the nearshore sea floor (Fig. 6.3). When waves approach a straight beach at an angle, they tend to swing parallel to the shore, due to the retarding effect of the shallowing water. At the break point, such waves seldom approach the coast at an

angle exceeding 20°, irrespective of the offshore angle to the beach. As waves approach an irregular shoreline, refraction causes them to turn into shallower water, so that the wave crests run roughly parallel to the depth contours. Along an indented coast, shallower water is first met with off headlands. This results in wave convergence and an increase in wave height, with wave crests becoming concave towards headlands. Conversely, where waves move towards a depression in the sea floor, they diverge, are decreased in height and become convex towards the shoreline. In both cases, the wave period remains constant. The concentration of erosion on headlands leads to the coast being gradually smoothed.

Refraction diagrams often form part of a shoreline study, graphic or computer methods being used to prepare them from hydrographic charts or aerial photographs. These diagrams indicate the direction in which waves flow, and the spacing between the orthogonal lines is inversely proportional to the

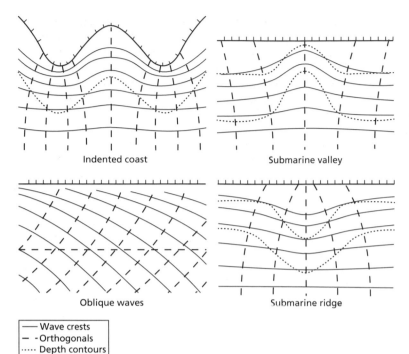

Indented coast

Submarine valley

Oblique waves

Submarine ridge

—— Wave crests
— - Orthogonals
····· Depth contours

Fig. 6.3 Diagrammatic wave refraction patterns.

energy delivered per unit length of shore. Because of refraction, it is possible, by constructing one or more raised areas on the sea bed, to deviate waves passing over them, so that calm water occurs at harbour entrances.

Waves are smaller in the lee of a promontory or off a partial breakwater. The nature of the wave pattern within this sheltered area is determined by diffraction as well as refraction phenomena. That portion of the advancing wave crest which is not intercepted by the barrier immediately spreads out into the sheltered area, and the wave height shrinks correspondingly. This lateral dissipation of wave energy is termed diffraction. Diffraction results from interference with the horizontal components of wave motion. The depth of water is not a relevant factor.

6.2 Tides

Tides are the regular periodic rise and fall of the surface of the sea, observable along shorelines. They are formed under the combined attraction of the Moon and the Sun, chiefly the Moon. Normal tides have a dual cycle of about 24 h and 50 min. The tide height is at a maximum or minimum when the Sun, Moon and Earth are in line, so that the attractive force of both the Sun and Moon upon the ocean waters is combined, thus creating spring tides. Neap tides are generated when the attractive forces of the Sun and Moon are aligned in opposing directions. In this instance, the range between high and low tide is lowest. The tidal range is also affected by the configuration of a coastline. For example, in the funnel-shaped inlet of the Bay of Fundy, a range of 16 m is not uncommon. By contrast, along relatively straight coastlines or in enclosed seas, such as the Mediterranean, the range is small.

The tidal current which flows into bays and estuaries along the coast is called the flood current. The current which returns to the sea is named the ebb current. These are of unequal duration, as the higher flood tide travels more quickly than the lower ebb tide.

The inshore current all along the coast need not necessarily be in the same direction as the offshore tidal current, for example, a headland may cause an eddy, giving an inshore current locally in the opposite direction to the tidal current. Information with regard to the directions and velocities of tidal currents at definite points and at tabulated periods in hours before and after high water at a nearby port may be obtained in Britain from the relevant Admiralty charts. Moreover, data on local tidal conditions are available in the Admiralty tide tables. Nonetheless, local conditions affecting tidal estuaries are often so variable that a detailed study of each case usually is required. Scale models frequently are used in such studies.

6.3 Beach zones

Four dynamic zones have been recognized within the nearshore current system of the beach environment; they are the breaker zone, the surf zone, the transition zone and the swash zone (Fig. 6.4). The breaker zone is that in which waves break. The surf zone refers to that region between the breaker zone and the effective seaward limit of backwash. The presence and width of a surf zone is primarily a function of the beach slope and tidal phase. Those beaches which have gentle foreshore slopes often are characterized by

Fig. 6.4 Summary diagram schematically illustrating the effect of the four major dynamic zones in the beach environment. Shaded areas represent zones of high concentrations of suspended grains. The surf zone is bounded by two high-energy zones—the breaker and the transition zones. MWLW, mean water low water. (After Ingle, 1966.)

wide surf zones during all tidal phases, whereas this zone may not be represented on steep beaches. The transition zone includes that region where backwash interferes with the water at the leading edge of the surf zone, and is characterized by high turbulence. The region in which water moves up and down the beach is termed the swash zone.

The breaking of a wave is influenced by its steepness, the slope of the sea floor and the presence of an opposing or supplementary wind. When waves enter water equal in depth to approximately one-half of their wave length, they begin to feel bottom, their velocity and length decrease and their height, after an initial decrease, increases. The wave period remains constant. As a result, the wave grows steeper until it eventually breaks.

Three types of breaking wave can be distinguished, namely, plunging breakers, spilling breakers and surging breakers (Fig. 6.5). The beach slope influences the type of breaking wave which develops, for example, spilling breakers are most common on gentle beaches, whereas surging breakers are associated with steeper beaches.

Plunging breakers collapse suddenly when their wave height is approximately equal to the depth of the water. At the plunge point, the wave energy is transformed into energy of turbulence and kinetic energy, and a tongue of water rushes up the beach. The greater part of the energy released by the waves is used in overcoming frictional forces or in generating

Water motion	Oscillatory waves	Wave collapse	Waves of translation (bores); longshore currents, seaward return flow; rip currents		Collision	Swash, backwash	Wind
Dynamic zone	Offshore	Breaker	Surf		Transition	Swash	Berm crest
Profile							MWLW
Sediment size trends	Coarser ⟶	Coarsest grains	⟵ Coarser	∿	Bi-modal lag deposit	⟵ Coarser ⟶	Wind-winnowed lag deposit
Predominant action	Accretion	Erosion	Transportation		Erosion	Accretion and erosion	
Sorting	⟵ Better ⟶	Poor	Mixed		Poor	Better ⟶	
Energy	⟶ Increase ⟶	High	⟵ Gradient ⟶		High	⟵ ⟶	

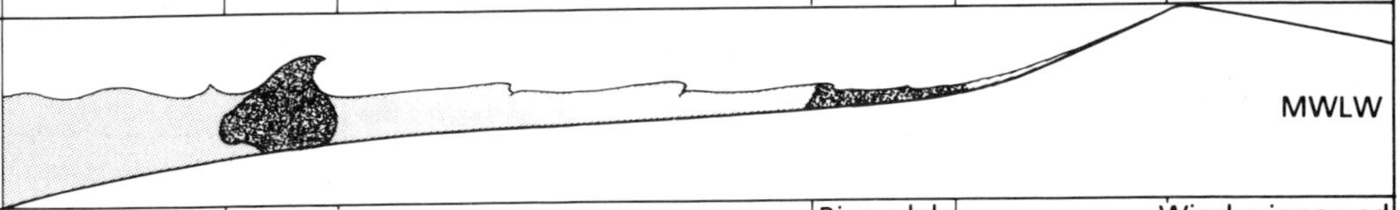

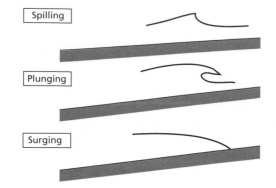

Fig. 6.5 Cross-sections of three breaker types.

turbulence, and relatively smaller amounts are utilized in shifting bottom materials or in developing longshore currents. The plunge point is defined as the final breaking point of a wave just before water rushes up the beach. For any given wave, there is a plunge line along which it breaks, but the variation in the position at which the waves strike the shore usually results in a plunge zone of limited width. At the plunge line, the distribution of sand is almost uniform from the bottom to the water surface during the breaking of a wave, but seaward of the plunge line, the sand content of the surface water rapidly decreases. Accordingly, the maximum disturbance of the sand due to turbulence occurs in the vicinity of the plunge line, it tending to migrate from the plunge line towards less turbulent water on both sides. This may result in a trough along the plunge line. Seaward of the plunge line, shingle generally is not moved by waves in depths much greater than the wave height, but sand is moved for a considerable distance offshore. Spilling breakers begin to break when the wave height is just over one-half of the water depth. They do so gradually over some distance. Surging breakers or swash rushes up the beach and is usually encountered on beaches with a steep profile. The height to which the swash rises determines the height of shingle crests and whether or not sea walls or embankments are overtopped. The more impermeable and steep the slope, the higher is the swash height. The term backwash is used to describe the water which subsequently descends the beach slope.

Swash tends to pile water against the shore and, thereby, gives rise to currents which move along it, termed longshore currents. After flowing parallel to the beach, the water runs back to the sea in narrow flows called rip currents (Fig. 6.6). In the neighbourhood of the breaker zone, rip currents extend from the surface to the floor, whereas in the deeper reaches, they override the bottom water which still maintains an overall onshore motion. The positions of rip currents are governed by the submarine topography, coastal configuration, and the height and period of the waves. They frequently occur on the upcurrent sides of points and on either side of convergences, where the water moves away from the centre of the convergence and turns seawards.

Material is sorted according to its size, shape and density, and the variations in the energy of the transporting medium. During the swash, material of various sizes may be swept along in traction or suspension, whereas during the backwash, the lower degree of turbulence results in a smaller lifting effect, so that most of the movement of grains is by rolling along the bottom. This can mean that the maximum size of particles thrown on a beach is larger than the maximum size washed back to the surf zone. However, grains of larger diameter may roll downslope further than smaller particles. The continued operation of waves on a beach, accompanied by the winnowing action of wind on dry sand, tends to develop patterns of variation in average particle size, sorting, firmness, porosity, permeability, moisture content, mineral composition and other attributes of the beach.

6.4 Coastal erosion

Those waves with a period of approximately 4 s are usually destructive, whereas those with a lower frequency, that is, a period of about 7 s or more, are constructive. When high-frequency waves collapse, they form plunging breakers, and the mass of water accordingly is directed downwards at the beach. In such instances, swash action is weak and, because of the high frequency of the waves, is impeded by backwash. As a consequence, material is removed from the top of the beach. The motion within waves of a lower frequency is more elliptical, and produces a strong swash, which drives material up the beach. In this case, the backwash is reduced in strength because water percolates into the beach

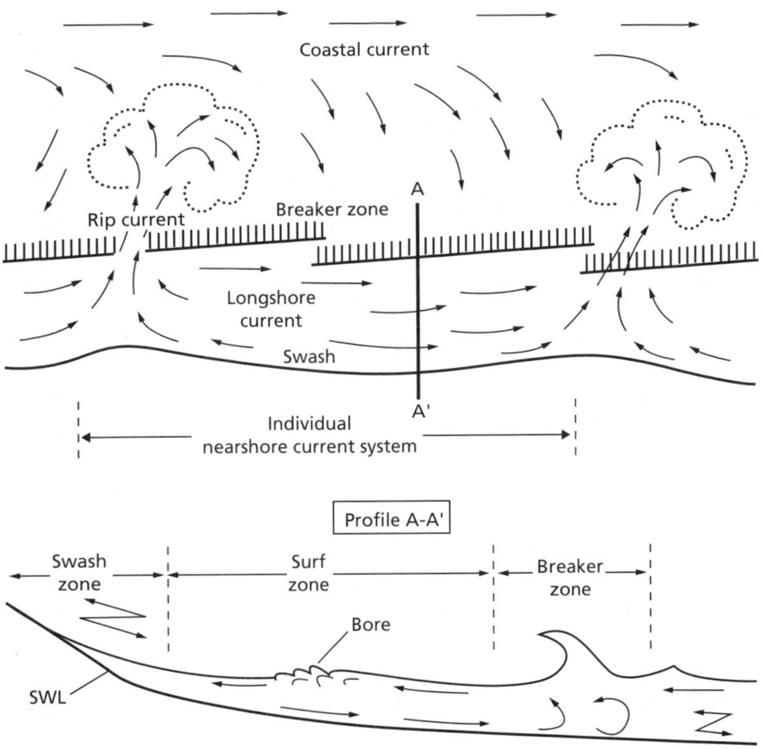

Fig. 6.6 Terminology of nearshore current systems. Each individual system begins and terminates with a seaward-flowing rip current. Arrows indicate the direction of water movement in plan and profile. The existence of the controversial seaward return flow along the foreshore–inshore bottom (profile A–A) has recently been confirmed by electromechanical measurements in the surf zone. These measurements indicated that a seaward bottom flow exceeding 1.2 ft s⁻¹ (0.366 m s⁻¹) often occurs at the same time as the surface flow is shoreward. The surf zone is defined here as the area between the seaward edge of the swash zone and the breaker zone. SWL, still water line. (From Ingle, 1966.)

deposits, and therefore little material is returned down the beach. Although large waves may throw material above the high water level and thus act as constructive agents, they nevertheless have an overall tendency to erode the beach, whereas small waves are constructive. For example, when steep storm waves attack a sand beach, they usually are entirely destructive, and the coarser the sand, the greater the quantity which is removed to form a submarine bar offshore. In some instances, a vertical scarp is left on the beach at the high tide limit. It is by no means a rarity for the whole beach to be removed by storm waves. Storm waves breaking on shingle throw some shingle to the backshore to form a storm beach ridge, which may extend far above high tide level. Storm waves may also bring shingle down the foreshore, so that a step is developed at their break point.

The rate at which coastal erosion proceeds is influenced by the nature of the coast itself. Marine erosion is most rapid where the sea attacks soft,

unconsolidated sediments (Fig. 6.7). Some such stretches of the coast may disappear at a rate of a metre or more annually. In addition, beaches which are starved of sediment supply due to, for example, natural or artificial barriers inhibiting the movement of longshore drift are exposed to erosion.

Steep cliffs made of unconsolidated materials are prone to landsliding (Hutchinson *et al.*, 1991). For example, clay may be weakened by wetting due to waves or spray, or the toe of the cliffs may be removed by wave action. Thus rotational slips and mudflows may develop (Fig. 6.8) and may carry away protective structures. For erosion to continue, the debris produced must be removed by the sea. This usually is accomplished by longshore currents. On the other hand, if material is deposited to form extensive beaches and the detritus is reduced in size, the submarine slope, because of the small angle of rest, is very wide. Therefore, wave energy is dissipated as the water moves over the beach and cliff erosion ceases.

Fig. 6.7 Marine erosion of cliffs formed of till, leading to the destruction and disappearance of holiday homes, Holderness coast, England.

Fig. 6.8 Landsliding on the south coast of the Isle of Wight, England.

Marine erosion obviously is concentrated in areas along a coast where the rocks offer least resistance. In most cases, a retreating coast is characterized by steep cliffs, at the base of which a beach is excavated by wave action. Erosive forms of local relief include features such as wave-cut notches, caves, blow-holes, marine arches and stacks (Fig. 6.9). Debris from the cliff is washed seawards to form a terrace, which marks the outer limit of the beach. The degree to which rocks are traversed by discontinuities affects the rate at which they are removed. In particular, the attitude of joints and bedding planes is important. Where the bedding planes are vertical or dip inland, the cliff recedes vertically under marine attack. However, if beds dip seawards, blocks of rock are more readily dislodged, because the removal of material from the cliff means that the rock above lacks support and tends to slide into the sea. Marine erosion also occurs along fault planes. The height of a cliff also influences the rate at which erosion takes place. The higher the cliff, the more material that falls when its base is undermined by wave attack. This, in turn, means that a greater amount of debris has to be broken down and removed before the cliff is once more attacked with the same vigour.

6.5 Beaches and longshore drift

Beaches are supplied with sand which usually is derived entirely from the adjacent sea floor although, in some areas, a larger proportion is produced by cliff erosion. During periods of low waves, the differential velocity between onshore and offshore motion is sufficient to move sand onshore, except where rip currents are operational. Onshore movement is notable particularly when long-period waves approach a coast, whereas sand is removed from the foreshore during high waves of short period. Of course, all beaches continually are changing. Some develop during periods of small waves and disappear during periods of high waves, whereas others change in height and width during stormy seasons. If seasonal

(a)

Fig. 6.9 (a) Wave-cut platform, coast of California, near Cambrai. (b) Blow-hole, Housel Bay, Cornwall, England. (c) (*Opposite*) Marine arch, north of Santa Barbara, California. (d) The Apostles (stacks), south coast of Victoria, Australia.

(b)

(c)

(d)

Fig. 6.9 *Continued*.

changes bring about changes in the wave approach, then sand is shifted along the beach, beaches tending to form at right angles to the direction of the wave approach. If sand forms dunes and these migrate landwards, then sand is lost to the beach.

The beach slope is produced by the interaction of swash and backwash. Beaches undergoing erosion tend to have steeper slopes than prograding beaches. The beach slope is also related to the grain size, in general, the finer the material, the gentler the slope, and the lower the permeability of the beach. For example, the loss of swash due to percolation into beaches composed of grains of 4 mm in median diameter is 10 times greater than that into beaches where the grains average 1 mm. As a result, there is almost as much water in the backwash on fine beaches as there is in the swash, so that the beach profile is gentle and the sand is hard packed. The grain size of beach material continually is changing, however, because of the breakdown and removal or addition of material.

Storm waves produce the most conspicuous constructional features on a shingle beach, but they remove material from a sandy beach. A small foreshore ridge develops on a shingle beach at the limit of the swash when constructional waves are operative. Similar ridges or berms may form on a beach composed of coarse sand, these being built by long, low swells. Berms represent a marked change in slope, and usually occur at a small distance above high water mark. They are not conspicuous features on beaches composed of fine sand. A fill on the upper part of a beach may be balanced by a cut lower down the foreshore.

Dunes are formed by onshore winds carrying sand-sized material landwards from the beach along low-lying stretches of coast, where there is an abundance of sand on the foreshore. Dunes act as barriers, energy dissipators and sand reservoirs during storm conditions, for example, the broad, sandy beaches and high dunes along the coast of the Netherlands present a natural defence against inundation during storm surges. Because dunes provide a natural defence against erosion, once they are breached, the ensuing coastal changes may be long lasting (Gares, 1990). On the other hand, along parts of the coast of North Carolina, where dune protection is limited, washovers associated with storm tides have been responsible for high rates of erosion (Cleary & Hosier, 1987).

In spite of the fact that dunes inhibit erosion, Leathermann (1979) stressed that, without beach nourishment, they cannot be relied upon to provide protection, in the long term, along rapidly eroding shorelines. Beach nourishment widens the beach and maintains the proper functioning of the beach–dune system during normal and storm conditions.

When waves approach a coast at an angle, material is moved up the beach by the swash, in a direction normal to that of wave approach, and is then rolled down the beach slope by the backwash. In this manner, material is moved in a zig-zag path along the beach, the phenomenon being referred to as long-shore or littoral drift. Currents, as opposed to waves, seem incapable of moving material coarser than sand grade. Because the angle of the waves to the shore affects the rate of drift, there is a tendency for erosion or accretion to occur where the shoreline changes direction. The determining factors are the rate of arrival and the rate of departure of beach material over the length of the foreshore concerned. Moreover, the projection of any solid structure below mean tide level results in the build-up of drift material on the updrift side of the structure and, perhaps, in erosion of material from the other side. If the drift is of any magnitude, the effects can be serious, especially in the case of works protecting harbour entrances. On the other hand, sand can be trapped by deliberately stopping longshore drift in order to build up a good beach section. In normal longshore drift, the bulk of the sand is moved in a relatively shallow zone, with perhaps 80% of shore drift being moved within depths of 2 m or less.

The amount of longshore drift along a coast is influenced by the coastal outline and the wave length. Short waves can approach the shore at a considerable angle and generate consistent downdrift currents. This is particularly the case on straight or gently curving shores, and can result in serious erosion where the supply of beach material reaching the coast from updrift is inadequate. Conversely, long waves suffer appreciable refraction before they reach the coast.

An indication of the direction of longshore drift is provided by the orientation of spits along a coast. Spits are deposits which grow from the coast and chiefly are supplied by longshore drift. Their growth is spasmodic and alternates with episodes of retreat. The distal end of a spit frequently is curved (Fig. 6.10a). A spit which extends from the mainland to link up with an island is referred to as a tombolo. A bay-bar is constructed across the entrance to a bay by the growth of a spit being continued from one headland to the other (Fig. 6.10b). A cuspate bar arises when a change in the direction of spit growth takes place, so that it eventually joins the mainland again, or where two spits coalesce. If progradation occurs, bars give rise to cuspate forelands (Fig. 6.10c). Bay-head beaches are one of the most common types of coastal deposit and tend to

(a)

(b)

Fig. 6.10 (a) Hurst Castle Spit, showing recurved laterals, Hampshire, England. (b) Bay-bar, south of Monterey, California. (c) (*Opposite*) Cuspate foreland, Dungeness, Kent, England.

Fig. 6.10 *Continued.*   (c)

straighten a coastline. Wave refraction causes long-shore drift from headlands to bays, sweeping sediments with it. Because the waves are not so steep in the innermost reaches of bays, they tend to be constructive. Marine deposition also helps to straighten coastlines by building beach plains.

Offshore bars consist of ridges of shingle or sand which may extend for several kilometres and may be located up to a few kilometres offshore. They usually project above sea level and may cut off a lagoon on their landward side. The distance from the shore at which a bar forms is a function of the deep water wave height, the water depth and the deep water wave steepness. If the water depth and wave steepness remain constant, and the wave height is reduced, the bar moves shorewards. If the wave height and wave length stay the same, and the depth of water increases, again the bar moves shorewards. If the water depth and wave height are unchanged, but the steepness is reduced, the bar moves seawards. During storms, sand is shifted outwards to form an offshore bar and, during the ensuing quieter conditions, this sand is wholly or in large part moved back to the beach.

6.6 Storm surges and marine inundation

Except where caused by failure of protection works, marine inundation is almost always attributable to severe meteorological conditions giving rise to abnormally high sea levels, referred to as storm surges. For example, Murty (1987) described storm surges as oscillations of coastal waters in the period range from a few minutes to a few days which were brought about by weather systems, such as hurricanes, cyclones and deep depressions. Low pressure and driving winds during a storm may lead to marine inundation of low-lying coastal areas, particularly if this coincides with high spring tides. This is especially the case when the coast is unprotected. Floods may be frequent, as well as extensive, where floodplains are wide and the coastal area is flat (Fig. 6.11). Coastal areas which have been reclaimed from the sea and are below high tide level are particularly suspect if coastal defences are breached. A storm surge can be regarded as the magnitude of sea level along the shoreline which is above the normal seasonally adjusted high tide level. Storm surge risk often is associated with a particular season. The height and location of storm damage along a coast over a period of time, when analysed, provides some idea of the maximum likely elevation of surge effects. The seriousness of the damage caused by storm surge tends to be related to its height and the velocity of water movement.

Factors which influence storm surges include the intensity of the fall in atmospheric pressure, the length of water over which the wind blows, the storm motion and the offshore topography. Obviously, the principal factor influencing storm surge is the intensity of the causative storm, that is, the speed of the wind piling

Fig. 6.11 Flooded polder at St Philipsland, the Netherlands, in January 1953. (Courtesy of the Netherlands Embassy.)

up the sea against the coastline. For instance, threshold wind speeds of approximately 120 km h⁻¹ tend to be associated with central pressure drops of around 34 mbar. Normally, the level of the sea rises with reductions in atmospheric pressure associated with intense, low-pressure systems. In addition, the severity of a surge is influenced by the size and track of a storm and, especially in the case of open coastline surges, by the nearness of the storm track to the coastline. The wind direction and the length of fetch are also important, both determining the size and energy of the waves. Because of the influence of the topography of the sea floor, wide, shallow areas on the continental shelf are more susceptible to damaging surges than those where the shelf slopes steeply. Surges are intensified by converging coastlines, which exert a funnel effect as the sea moves into such inlets.

Ward (1978) distinguished two types of major storm surge. First, there are storm surges on open coastlines which travel as running waves over large areas of the sea. Such surges are associated with tropical cyclones, typhoons and hurricanes. Second, surges occur in enclosed or partially enclosed seas, where the area of sea is small compared with the size

of the atmospheric disturbance. Thus the surge more or less affects the whole sea at any one time. This type of surge may be frequent, as well as damaging.

Most deaths have occurred as a result of storm surges generated by hurricanes. The height of a storm surge associated with a hurricane is influenced by the distance from the centre of the storm to the point at which maximum wind speeds occur, the barometric pressure in the eye of a hurricane, the rate of forward movement of the storm, the angle at which the centre of the storm crosses the coastline and the depth of sea water offshore. In addition to the storm surge, enclosed bays may experience seiching. Seiches involve the oscillatory movement of waves generated by a hurricane. Seiching may also occur in open bays if the storm moves forwards very quickly and, in low-lying areas, this can be highly destructive.

Obviously, the prediction of the magnitude of a storm surge in advance of its arrival can help to save lives. Storm tide warning services have been developed in various parts of the world. Warnings usually are based upon comparisons between predicted and observed levels at a network of tidal gauges. To a large extent, the adequacy of a coastal flood forecasting and warning system depends on the accuracy with which the path of the storm responsible can be determined. Storms can be tracked by satellite, and satellite and ground data can be used for forecasting. Numerous attempts have been made to produce models which predict storm surges (Jelesmanski, 1978; Johns & Ali, 1980).

Protective measures are not likely to be totally effective against every storm surge. Nevertheless, flood protection structures are used extensively, and often prove to be substantially effective. Sand dunes and high beaches offer natural protection against floods, and can be maintained by artificial nourishment with sand. Dunes can also be constructed by tipping and bulldozing sand or by aiding the deposition of sand by the use of screens and fences. Coastal embankments usually are constructed of local material. Earth embankments are susceptible to erosion by wave action and, if overtopped, can be scoured on the landward side. Slip failures may also occur. Sea walls may be constructed of concrete or sometimes of steel sheet piling. Permeable sea walls may be formed of rip-rap or concrete tripods. The roughness of their surfaces reduces the swash height and the scouring effect of backwash on the foreshore.

Barrages may be used to shorten the coastline along highly indented coasts and at major estuaries. Although expensive, barrages can serve more than one function, not just flood protection. They may produce hydro-electricity, carry communications, and fresh water lakes can be created behind them. The Delta Scheme in the Netherlands provides one of the best examples. This was a consequence of the disastrous flooding which resulted from the storm surge of February 1953. This inundated some 150 000 ha of polders and damaged over 400 000 buildings. The death toll amounted to 1835 and 72 000 people were evacuated. Some 47 000 cattle were killed. The scheme involved the construction of four massive dams to close off sea inlets, and a series of secondary dams, with a flood barrier, northwest of Rotterdam (Fig. 6.12). The length of the coastline was reduced by some 700 km and the coastal defences were improved substantially.

Estuaries which have major ports require protection against sea flooding. The use of embankments may be out of the question in urbanized areas, as space is not available without expensive compulsory purchase of waterfront land, and barrages with locks

Fig. 6.12 Main Delta Works, Oosterschelde flood gates, the Netherlands.

would impede sea-going traffic. In such situations, movable flood barriers may be used, which can be placed into position when storm surges threaten. The Thames Barrier in London offers an example.

6.7 Protective barriers

Training walls are designed to protect inlets and harbours, as well as to prevent serious wave action in the area they protect. They also impound any longshore drift material upbeach of an inlet and thereby prevent sanding of the channel. However, this may cause or accelerate erosion downdrift of the walls. Training walls usually are built at right angles to the shore, although their outer segments may be set at an angle. Two parallel training walls may extend from each side of a river mouth for some distance out to sea and, because of this confinement, the velocity of river flow

Fig. 6.13 Sea wall at Newbiggin-on-Sea, Northumberland, England.

is increased, which lessens the amount of deposition taking place between the walls. Training walls may be built on rubble mounds.

Sea walls may rise vertically or may be curved or stepped in cross-section (Fig. 6.13). They are designed to prevent wave erosion and overtopping by high seas by dissipating or absorbing destructive wave energy. Sea walls are expensive, so that they tend to be used where property requires protection. They must be stable, and this is usually synonymous with weight. They should also be impermeable and resist marine abrasion. Sea walls can be divided into two main classes, those from which waves are reflected and those on which waves break (Thorn & Simmons, 1971). A wall which combines reflection and breaking tends to set up severe erosive action immediately in front of it. Consequently, in the layout of a sea wall, particular attention should be given to wave reflection and the possibility of the crests of the waves increasing in height as they travel along the wall due to their high angle of obliquity (Pilarczyk, 1990). The toe of the wall should not be allowed to become exposed.

Breakwaters disperse the waves of heavy seas and provide shelter for harbours. As such, they commonly run parallel to the shore or at slight angles to it, and are attached to the coast at one end, their orientation being chosen with respect to the direction from which storm waves approach the coast. The detached offshore breakwater is constructed as a barrier parallel

to the shoreline and is not connected to the coast. Attached breakwaters cause sand or shingle of longshore drift, on shorelines so affected, to accumulate against them and so rob downdrift beaches of sediment. Consequently, severe erosion may take place in the downdrift area after the construction of a breakwater. Furthermore, the accumulation of sand eventually can move around the breakwater to be deposited in the protected area. Komar (1976) described this as happening at Santa Barbara, California, after a breakwater had been built to protect the harbour. As a result, the harbour entrance has to be dredged, and the sand is deposited on the downdrift shoreline. Although long offshore breakwaters shelter their leeside, they cause wave refraction, and may generate currents in opposite directions along the shore towards the sheltered area, with the resultant impounding of sand.

The character of the sea bed at the site of a proposed breakwater must be investigated carefully. Not only must the sea bed be investigated fully with respect to its stability, bearing capacity and ease of removal if dredging is involved, but the adjacent stretches of coast must be studied in order to determine the local currents, longshore drift and any features that may in any way be affected by the proposed structure.

The oldest form of breakwater is a simple pile of dumped rock, that is, a rubble mound (Fig. 6.14a). Rubble mound breakwaters dissipate storm waves by

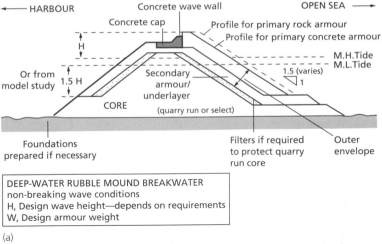

Fig. 6.14 (a) Cross-section of a deep water, rubble mound breakwater. (b) Tripods used for coastal protection, Seaford, England.

(b)

turbulence, as sea water penetrates the voids between the blocks of rock (Silvester, 1974). Most of the attenuation occurs near the surface of the water, with oscillatory motions of water at depth being reduced somewhat. Rubble mound breakwaters are adaptable to any depth of water, are suitable for nearly all types of foundation from solid rock to soft mud and can be repaired readily. The quality and durability of the stone is also important (Latham, 1991). Tripods are three-legged concrete blocks which weigh up to 40 tonnes (Fig. 6.14b). They have been used mainly in harbour works.

Embankments have been used for centuries as a means of coastal protection (Fig. 6.15). The weight of an embankment must be sufficient to withstand the pressure of the water. The material composing the embankment should be impervious or, at least, should be provided with a clay core to ensure that seepage does not occur. An embankment requires a protective stone apron to withstand the destructive

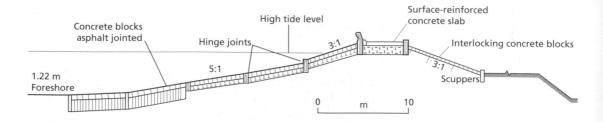

Concrete blocks
asphalt jointed

Hinge joints

High tide level

Surface-reinforced
concrete slab

Interlocking concrete blocks

3:1

5:1

3:1

1.22 m
Foreshore

Scuppers

0   m   10

Fig. 6.15 A concrete crest on an earth embankment, covered with concrete blocks, used for coastal protection.

power of the waves. Clay is the usual material for embankment construction.

A revetment affords an embankment protection against wave erosion. Stone is used most commonly for revetment work, although concrete is also employed. The chief factor involved in the design of a stone revetment is the selection of stone size, it being important to guard against erosion between stones (Fig. 6.16). Consequently, coarse rip-rap must be isolated from the earth embankment by one or more courses of filter stone. Stone pitching is an ancient form of embankment protection, consisting of stone properly positioned on the clay face of the wall and keyed firmly into place. Revetments of this type are flexible and have proved to be very satisfactory in the past. Flexibility is an important requirement of a revetment, because slow settlement is likely to occur in a clay embankment. As an alternative to keying, stones may be grouted or asphalt jointed. Individual blockwork units may be used so that, in the event of a washout, the spread of damage is limited. Blockwork formed of igneous rock is one of the finest forms of revetment for resisting severe abrasion. On walls where wave action is not severe, interlocking concrete blocks provide a very satisfactory form of revetment. The effectiveness of a revetment lies more in its water tightness than in the weight of the stones.

Bulkheads are vertical walls, either of timber or of steel sheet piling, which support an earth bank. A bulkhead may be cantilevered out from the foundation soil or it may be additionally restrained by tie rods. Except for installations of moderate height and stable foundation, a cantilever bulkhead will be rejected in favour of a design having one or two anchorages. The tie rod on an anchored bulkhead

Fig. 6.16 Rip-rap used as stone revetment for embankment protection, Freemantle, Western Australia.

must be carried to a secure anchorage; generally, this is provided by a deadman, the placement of which must be such that the potential slip surface behind it does not invade the zone influencing the bulkhead pressures. Foundation conditions for bulkheads must be given careful attention, and due consideration must be paid to the likelihood of scour occurring at

Fig. 6.17 Groynes at Eastbourne, England.

the foot of the wall and to changes in beach conditions. Cutoff walls of steel sheet piling below reinforced concrete superstructures provide an effective method of construction ensuring protection against scouring.

6.8 Stabilization of longshore drift

Before any scheme for beach stabilization is put into operation, it is necessary to determine the prevailing direction of longshore drift, its magnitude and whether there is any seasonal variation. Also, it is necessary to know whether the foreshore is undergoing a net gain or loss of beach material and what is the annual rate of change.

The groyne is the most frequent structure used to stabilize or increase the width of the beach by arresting longshore drift (Fig. 6.17). Groynes are used to limit the movement of beach material, and to stabilize the foreshore by encouraging accretion. However, groynes do not usually halt all drift. Groynes should be constructed transverse to the mean direction of the breaking crest of storm waves, which usually means that they should be approximately at right angles to the coastline. Standard types usually slope at about the same angle as the beach. With abundant longshore drift and relatively mild storm conditions, almost any type of groyne appears to be satisfactory, whereas, when the longshore drift is lean, the choice is much more difficult.

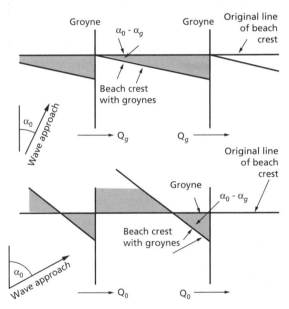

Fig. 6.18 Diagram of groyned shore. (After Muir-Wood & Flemming, 1981.) Q_g, longshore drift with groynes; Q_0, longshore drift without groynes; α_0 and α_g, wave approach without and with groynes respectively.

By arresting longshore drift, groynes cause the beach line to be re-orientated between successive groynes, so that the prevailing waves arrive more nearly parallel to the beach. If the angle of wave incidence is greater than that for the maximum rate of drift, groynes serve no useful purpose, because the accumulation of beach material against groynes orientates the beach immediately updrift of them in such a way as to lead initially to increased longshore drift (Fig. 6.18). The beach crest rapidly swings around until the original rate of drift is restored, resulting in the loss of beach material (Muir Wood & Flemming, 1983).

The height of a groyne determines the maximum beach profile updrift of it. In general, there is little advantage in building groynes which are higher than the highest level normally reached by the sea. Low groynes are more simple to construct. The stability must take into account the storm profile and local scour. On a sandy foreshore, the height of a groyne should not exceed 0.5–1.0 m above the beach if it is to minimize scour from wave and tidal currents, particularly at the seaward end. Closely spaced groynes on a sandy beach have the additional advantage that they tend to control rip currents, so that these occur at correspondingly more frequent intervals and consequently the longshore drift in the surf zone is reduced. Where long groynes are too close, a proportion of the beach material will not be captured by each groyne bay. The effect is more marked for sand than shingle, because of the beach gradient and the mode of transport.

The length of groynes often is determined by the tidal range and beach slope. Limiting groynes to the beach above the level of low spring tides is utterly defenceless in terms of efficiency, because the total rate of sand movement in the coastwise direction is greater below the level of low spring tides than above it. Groynes should extend landwards to the cliff or protective structure or, on shingle shores, to somewhat higher than the highest swash height at high water.

The maximum groyne spacing frequently is governed by the resulting variation in beach level on either side (Flemming, 1990). For instance, in bays in which the direction of wave attack is confined, groynes may be more widely spaced than on exposed promontories. Indeed, it may be necessary to provide a more substantial groyne where a change in the direction of the coastline occurs, to counter the accentuated wave attack and change in the rate of drift. The common spacing rule for groynes is to arrange them at intervals of one to three groyne lengths.

Permeable groynes have openings which increase in size seawards, and thereby allow some drift material to pass through them. The use of permeable groynes for a sand foreshore may help to reduce local scour, but may only be considered where Q_g/Q_0 is to remain fairly high, because they cause no more than a slight reduction in the rate of longshore drift. On a shingle foreshore, groyne-induced scour need not be a serious problem and the permeable groyne is not an economic expedient.

Groynes reduce the amount of material passing downdrift and, therefore, can prove to be detrimental to those areas of the coastline. Thus, their effect on the coastal system should be considered before installation.

Artificial replenishment of the shore by building beach fills is used when it is economically preferable or when artificial barriers fail to defend the shore adequately from erosion (Whitcombe, 1996). In fact, beach nourishment represents the only form of coastal protection which does not affect adversely other sectors of the coast. Unfortunately, it often is difficult to predict how frequently a beach should be renourished. Ideally, the beach fill used for renourishment should have a similar particle size distribution to the natural beach material.

The material may be obtained from a borrow pit away from the coastline, in which case it is necessary to consider the effects of using a material of different grain size and sorting on the stability of the beach, or it may be obtained from points of accretion on the foreshore. A common method of beach replenishment is to take the material from the downdrift end of a beach and return it to the updrift end. This is usually the case when beach feeding is combined with groynes. However, investigation of the foreshore may indicate that drift increases in the downdrift direction and therefore that recharge is required at intermediate points to make good such a deficiency. Special attention should be given to areas in which there is a change in the direction of the coastline. The immediate advantage of beach feeding is experienced updrift rather than downdrift to the supply point.

Because of the large cost of engineering works and the expense of rehabilitation, coastal management represents an alternative way of tackling the problem of coastal erosion, and involves prohibiting or even removing developments which are likely to destabilize the coastal regime (Gares & Sherman, 1985; Ricketts, 1986). The location, type and intensity of development must be planned and controlled. Public safety is the primary consideration. Generally, the level at which risk becomes unacceptable is governed by the socio-economic cost. Conservation areas and open space zones may need to be established. However, in areas which have been developed, the existing properties along a coastline may be expensive, so that public purchase may be out of the question. Moreover, owners of private property may resist acquisition attempts.

6.9 Tsunamis

One of the most terrifying phenomena which occurs along coastal regions is inundation by large masses of water called tsunamis (Fig. 6.19). Most tsunamis originate as a result of fault movement on the sea floor, although they can also be developed by submarine landslides or volcanic activity. However, even the effects of large earthquakes are relatively localized compared with the impact of tsunamis. Seismic tsunamis are most common in the Pacific Ocean, and usually are formed when submarine faults have a significant vertical movement. Faults of this type occur along the coasts of South America, the Aleutian Islands and Japan. The resulting displacement of the water surface generates a series of tsunami waves, which travel in all directions across the ocean. As with other forms of wave, it is the energy of

tsunamis which is transported not the mass. Oscillatory waves are developed with periods of 10–60 min which affect the whole column of water from the bottom of the ocean to the surface. Together with the magnitude of an earthquake and its depth of focus, the amount of vertical crustal displacement determines the size, orientation and destructiveness of tsunamis. Horizontal fault movements, such as those which occur along the Californian coast, do not result in tsunamis.

In the open ocean, tsunamis normally have a very long wave length and their amplitude is hardly noticeable. Successive waves may be from 5 min to 1 h apart. They travel rapidly across the ocean at speeds of around 650 km h^{-1}, so that it is possible to predict arrival times along coasts. Unfortunately, however, the vertical distance between the maximum height reached by the water along a shoreline and mean sea level, that is, the tsunami run-up, is not possible to predict. Because their speed is proportional to the depth of water, the wave fronts always move towards shallower water. In addition, in coastal areas, the waves slow down and increase in height, rushing onshore as highly destructive breakers. Waves have been recorded up to nearly 20 m in height above normal sea level. Tsunamis, like other waves, are refracted by offshore topography and by the differences in the configuration of the coastline.

Fig. 6.19 Wreckage of the clubhouse, Kanehameha Avenue, Hilo, Hawaii. Every house on the main street was smashed against the buildings on the other side by tsunamis in 1946. (Courtesy of the US Corps of Engineers.)

Table 6.1 Scale of tsunami intensity (Soloviev, 1978).

Intensity	Run-up height (m)	Description of tsunami	Frequency in Pacific Ocean
I	0.5	*Very slight*. Waves so weak as to be perceptible only on tide gauge records	
II	1	*Slight*. Waves noticed by people living along the shore and familiar with the sea. On very flat shores, waves generally noticed	One per hour, months
III	2	*Rather large*. Generally noticed. Flooding of gently sloping coasts. Light sailing vessels carried away or on shore. Slight damage to light structures situated near the coast. In estuaries, reversal of river flow for some distance upstream	One per 8 months
IV	4	*Large*. Flooding of the shore to some depth. Light scouring on made ground. Embankments and dykes damaged. Light structures near the coast damaged. Solid structures on the coast lightly damaged. Large sailing vessels and small ships swept inland or carried out to sea. Coasts littered with floating debris	One per year
V	8	*Very large*. General flooding of the shore to some depth. Quays and other heavy structures near the sea damaged. Light structures destroyed. Severe scouring of cultivated land and littering of the coast with floating objects, fish and other sea animals. With the exception of large ships, all vessels carried inland or out to sea. Large bores in estuaries. Harbour works damaged. People drowned; waves accompanied by a strong roar	Once in 3 years
≥VI	16	*Disastrous*. Partial or complete destruction of man-made structures for some distance from the shore. Flooding of coasts to great depths. Large ships severely damaged. Trees uprooted or broken by the waves. Many casualties	Once in 10 years

Due to the long period of tsunamis, the waves are of great length (e.g. 200–700 km in the open ocean; 50–150 km on the continental shelf). Therefore, it is almost impossible to detect tsunamis in the open ocean because their amplitudes (0.1–1.0 m) are extremely small in relation to their length. They can only be detected near the shore.

The size of a wave as it arrives at the shore depends upon the magnitude of the original displacement of water at the source and the distance it has travelled, as well as the underwater topography and coastal configuration. Soloviev (1978) devised a classification of tsunami intensity which is given in Table 6.1.

Free oscillations develop when tsunamis reach the continental shelf, which modify their form. Usually, the largest oscillation level is not the first, but one of the subsequent oscillations. However, to the observer on the coast, a tsunami appears not as a sequence of waves, but as a quick succession of floods and ebbs

(i.e. a rise and fall of the ocean as a whole) because of the great wave length involved. Shallow water permits tsunamis to increase in amplitude without a significant reduction in velocity and energy. On the other hand, where the water is relatively deep off a shoreline, the growth in the size of the wave is restricted. Large waves, several metres in height, are most likely when tsunamis move into narrowing inlets. Such waves cause terrible devastation and can wreck settlements. For example, if the wave is breaking as it crosses the shore, it can destroy houses merely by the weight of water. In fact, the water flow exerts a force on obstacles in its path which is proportional to the product of the depth of water and the square of its velocity. The subsequent backwash may carry many buildings out to sea and may remove several metres depth of sand from dune coasts. Damage is also caused when the resultant debris smashes into buildings and structures.

Usually, the first wave is like a very rapid change in tide, for example, the sea level may change by 7 or 8 m in 10 min. A bore occurs where there is a concentration of wave energy by funnelling, as in bays, or by convergence, as on points. A steep front rolls over relatively quiet water. Behind the front, the crest of such a wave is broad and flat, and the wave velocity is about 30 km h^{-1}. Along rocky coasts, large blocks of material may be dislodged and moved shorewards.

Because of development in coastal areas within the last 30 years or more, the damaging effects of future tsunamis will probably be much more severe than in the past. Hence, it is increasingly important that the hazard is evaluated accurately and that the potential threat is estimated correctly. This involves an analysis of the risk for the purpose of planning and putting into place mitigation measures.

Nonetheless, the tsunami hazard is not frequent and, when it does affect a coastal area, its destructiveness varies both with location and time. Accordingly, an analysis of the historical record of tsunamis is required in any risk assessment. This involves a study of the seismicity of a region to establish the potential threat from earthquakes of local origin. In addition, tsunamis generated by distant earthquakes must be evaluated. The data gathered may highlight spatial differences in the distribution of the destructiveness of tsunamis, which may form the basis for zonation of the hazard. If the historical record provides sufficient data, it may be possible to establish the frequency of recurrence of tsunami events, together with the area which would be inundated by a 50-, 100- or even 500-year tsunami event. On the other hand, if insufficient information is available, tsunami modeling may be resorted to using either physical or computer models. The latter are now much more commonly employed and provide reasonably accurate predictions of potential tsunami inundation which can be used in the management of tsunami hazard. Such models permit the extent of damage to be estimated and the limits for evacuation to be established. The ultimate aim is to produce maps indicating the degree of tsunami risk which, in turn, aids the planning process, thereby allowing high risk areas to be avoided or used for low-intensity development (Pararas-Carayannis, 1987). Models also facilitate the design of defence works.

Various instruments are used to detect and monitor the passage of tsunamis. These include sensitive seismographs which can record waves with long-period oscillations, pressure recorders placed on the sea floor in shallow water and buoys anchored to the sea floor and used to measure changes in the level of the sea surface. The locations of places along a coast affected by tsunamis can be hazard mapped which, for example, shows the predicted heights of tsunamis at a certain location for given return intervals (e.g. 25, 50 or 100 years). Homes and other buildings can be removed to higher ground, and new construction can be prohibited in the areas of highest risk. However, the resettlement of all coastal populations away from possible danger zones is not a feasible economic proposition. Hence, there are occasions when evacuation is necessary. This depends on the estimation of how destructive any tsunami will be when it arrives on a particular coast. Furthermore, evacuation requires that the warning system is effective and that there is an adequate transport system to convey the public to safe areas.

Breakwaters, coastal embankments and groves of trees tend to weaken a tsunami wave, reducing its height and the width of the inundation zone. Sea walls may offer protection against some tsunamis. Buildings which need to be located at the coast can be constructed with reinforced concrete frames and elevated on reinforced concrete piles with open spaces at ground level (e.g. for carparks). Consequently, the tsunami may flow through the ground floor without adversely affecting the building. Buildings usually are orientated at right angles to the direction of approach of the waves, that is, perpendicular to the shore. It is, however, more or less impossible to protect a coastline fully from the destructive effects of tsunamis.

Ninety per cent of destructive tsunamis occur within the Pacific Ocean, averaging more than two each year. For example, in the past 200 years, the Hawaiian Islands have been subjected to over 150 tsunamis. Hence, a long record of historical data provides the basis for the prediction of tsunamis in Hawaii, and allows an estimation to be made of the events following a tsunamigenic earthquake.

The Pacific Tsunami Warning System (PTWS) is a communications network covering the countries bordering the Pacific Ocean, and is designed to give advance warning of dangerous tsunamis (Dohler, 1988). The system uses 69 seismic stations and 65 tide stations in major harbours about the Pacific

Ocean. Earthquakes with a magnitude of 6.5 or over cause alarms to sound, and those over 7.5 give rise to an around-the-clock tsunami watch. Nevertheless, it is difficult to predict the size of waves which will be generated and to avoid false alarms. Clearly, the PTWS cannot provide a warning of an impending tsunami to those areas which are very close to the earthquake epicentre responsible for the generation of the tsunami. In fact, 99% of the deaths and much of the damage due to a tsunami occur within 400 km of the area where it was generated. Considering the speed of travel of a tsunami wave, the warning afforded in such instances is less than 30 min. Recently, however, a system has been developed whereby an accelerometer transmits a signal, via a satellite, over the eastern Pacific Ocean, to computers when an earthquake of magnitude 7 or more occurs within 100 km of the coast. The computers decode the signal, cause water level sensors to start monitoring and transmit messages to those responsible for carrying out evacuation plans. Waves generated off the coast of Japan take 10 h to reach Hawaii. In such instances, the PTWS can provide a few hours for evacuation to take place if it appears to be necessary (Bernard, 1991).

References

Bernard, E.N. (1991) Assessment of Project THRUST: past, present and future. *Natural Hazards* **4**, 285–292.

Cleary, W.J. & Hosier, P.E. (1987) North Carolina coastal geologic hazards, an overview. *Bulletin of the Association of Engineering Geologists* **24**, 469–488.

Darbyshire, J. & Draper, L. (1963) Forecasting wind-generated sea waves. *Engineering* **195**, 482–484.

Dohler, G.C. (1988) A general outline of the ITSU Master Plan for the tsunami warning system in the Pacific. *Natural Hazards* **1**, 295–302.

Flemming, C.A. (1990) Principles and effectiveness of groynes. In *Coastal Protection*, Pilarczyk, K.W. (ed). Balkema, Rotterdam, pp. 121–156.

Gares, P.A. (1990) Predicting flooding probability for beach–dunes systems. *Environmental Management* **14**, 115–123.

Gares, P.A. & Sherman, D.J. (1985) Protecting an eroding shoreline: the evolution of management response. *Applied Geography* **5**, 55–69.

Hutchinson, J.N., Bromhead, E.N. & Chandler, M.P. (1991) Investigation of the coastal landslides at St Catherine's Point, Isle of Wight. In *Proceedings of the Conference on Slope Stability Engineering—Applications and Developments*. Thomas Telford Press, London, pp. 151–161.

Ingle, J.G. (1966) *The Movement of Beach Sand*. Elsevier, Amsterdam.

Jelesmanski, C.P. (1978) Storm surges. In *Geophysical Predictions*. National Academy of Sciences, Washington, DC, pp. 185–192.

Johns, B. & Ali, A. (1980) The numerical modelling of storm surges in the Bay of Bengal. *Quarterly Journal of the Royal Meteorological Society* **106**, 1–18.

Johnson, D.W. (1919) *Shoreline Processes and Shoreline Development*. Wiley, New York.

Komar, P.D. (1976) *Beach Processes and Sedimentation*. Prentice Hall, Englewood Cliffs, New Jersey.

Latham, J.-P. (1991) Degradation model for rock armour in coastal engineering. *Quarterly Journal of Engineering Geology* **24**, 101–118.

Leathermann, S.P. (1979) Beach and dune interactions during storm conditions. *Quarterly Journal of Engineering Geology* **12**, 281–290.

Muir Wood, A.M. & Flemming, C.A. (1983) *Coastal Hydraulics*, 2nd edn. Macmillan, London.

Murty, T.S. (1987) Mathematical modelling of global storm surge problems. In *Natural and Man-Made Hazards*, El-Sabh, M.I. & Murty, T.S. (eds). Reidel, Dordrecht, pp 183–192.

Pararas-Carayannis, G. (1987) Risk assessment of the tsunami hazard. In *Natural and Man-Made Hazards*, El-Sabh, M.I. & Murty, T.S. (eds). Reidel, Dordrecht, pp 183–192.

Pilarczyk, K.W. (1990) Design of sea wall and dikes-including an overview of revetments. In *Coastal Protection*, Pilarczyk, K.W. (ed). Balkema, Rotterdam, pp 197–288.

Ricketts, P.J. (1986) National policy and management responses to the hazard of coastal erosion in Britain and the United States. *Applied Geography* **6**, 197–221.

Silvester, R. (1974) *Coastal Engineering*, Vols. 1 and 2. Elsevier, Amsterdam.

Soloviev, S.L. (1978) Tsunamis. In *The Assessment and Mitigation of Earthquake Risk*. UNESCO, Paris, pp. 91–143.

Thorn, R.B. & Simmons, J.C.F. (1971) *Sea Defence Works*. Butterworths, London.

Ward, R.C. (1978) *Floods: a Geographical Perspective*. Macmillan, London.

Whitcombe, L.J. (1996) Behaviour of an artificially replenished shingle beach at Hayling Island, U.K. *Quarterly Journal of Engineering Geology* **29**, 265–272.

7   Arid and semi-arid lands

7.1 Introduction

Arid and semi-arid regions account for one-third of the land surface of the Earth (Fig. 7.1). These are regions which experience a deficit of precipitation, and where the range of climatic conditions, particularly the magnitude, frequency and duration of precipitation and wind activity, vary appreciably. Consequently, run-off usually is ephemeral, varying between rainfall events and type of flow, and aeolian processes commonly are dominant. Nonetheless, rainfall can be torrential at times and can generate flash floods. Because of the lack of vegetation, especially in very arid regions, the landscape is vulnerable to erosion.

Although the overall distribution of arid or desert regions is governed by the global pattern of atmospheric circulation, in that they tend to coincide with high pressure belts, the position is much more complicated. Additional influences affecting their position include the distribution of continents and oceans, the size and shape of water and landmasses, the influence of warm and cold ocean currents, the topography of the landmasses, the seasonal migration of the climatic belts and the seasonal presence of high and low pressure cells. Hence, there are interior deserts far from ocean supplies of moisture and mountain barrier deserts on the leeward side of high mountains which rob winds of their moisture, as well as low latitude deserts which occur within the high pressure belts. Some of the most arid regions occur in the low latitudes on the western sides of continents, where the winds are offshore and the coasts are washed by cold currents.

The semi-arid transition lands on the borders of deserts are hazardous because of the vagaries of their climate. Periods of drought are common in these regions and can last for weeks or months. A number of dry years often alternate with a wetter cycle of years. Precipitation is unreliable, and the only consistent factor about the climate is its inconsistency.

7.2 Wind action

In arid regions, because there is little vegetation and the ground surface may be dry, wind action is much more significant and sediment yield may be high. The most serious problem attributable to wind action is soil erosion, which is dealt with in Chapter 10. This makes the greatest impact in relation to agriculture. Dust or sand storms and migrating sand dunes also give rise to problems. As mentioned in Chapter 10, hazards such as dust storms and migrating dunes frequently are made more acute because of the intervention of humans. Commonly, the problems are most severe on desert margins, where rainfall is uncertain and where there may be significant human pressure on land.

By itself, wind can only remove uncemented debris up to a certain size, which it can perform more effectively if the debris is dry rather than wet. However, once armed with rock particles, the wind becomes a noteworthy agent of abrasion. The size of individual particles which the wind can transport depends on the strength of the wind, and on the particle shape and density. The distance which the wind, given that its velocity remains constant, can carry particles depends principally upon their size. Fine grained particles can be transported for hundreds or even thousands of kilometres.

As the shear strength of the wind increases, loose particles on the ground surface are subjected to increasing stress, so that they ultimately begin to move. According to Bagnold (1941), the critical velocity at which particle movement is initiated is given by

$$V_{*_t} = A \left(\frac{\rho_g - \rho_a}{\rho_a} \right) gD \tag{7.1}$$

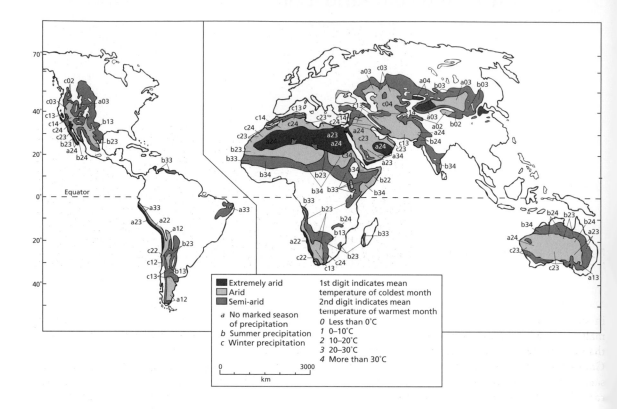

Fig. 7.1 World distribution of arid lands. (After Meigs, 1953.)

where V_{*t} is the threshold velocity, ρ_g is the relative density of the grains, ρ_a is the relative density of air, D is the diameter of the grain (cm) and A is a coefficient, which is 0.1 for particles larger than 0.1 mm in diameter. Particles smaller than this do not conform to this law, and the velocities increase with decreasing grain size. This probably is due to the increased cohesion between fine particles, their greater water retention and the lower values of the surface roughness. Bagnold proposed a value of 16 km h⁻¹ as the threshold velocity for most desert sands. Chepil (1945) subsequently showed that few particles with diameters exceeding 0.84 mm are eroded. Once particle movement is initiated, the ground is impacted by moving grains, thereby causing further movement. In this situation, sediment movement can be maintained at lower velocities than required to start movement.

At first, particles are moved by saltation, that is, in a series of jumps. The impact of saltating particles on others may cause them to move by creep, saltation or suspension. Saltation accounts for up to three-quarters of the grains transported by wind, most of the remainder being carried in suspension; the rest are moved by creep or traction. Saltating grains may rise to heights of up to 2 m, their trajectory then being flattened by faster moving air and tailing off as the grains fall to the ground. The saltation height generally is inversely related to the particle size and directly related to the roughness. The length of the trajectory is roughly 10 times the height. Particles are moved in suspension when the terminal velocity of fall is less than the mean upward eddy currents in the air flow.

One of the most important factors in wind erosion is its velocity. Its turbulence, frequency, duration and direction are also important. As far as the mobility of particles is concerned, the important factors are the size, shape and density. It would appear that particles less than 0.1 mm in diameter usually are transported

Fig. 7.2 Desert pavement, Damaraland, Namibia.

in suspension, those between 0.1 and 0.5 mm normally are transported by saltation and those larger than 0.5 mm tend to be moved by traction or creep. Grains with a relative density of 2.65, such as quartz sand, are most susceptible to wind erosion in the size range 0.1–0.15 mm. A wind blowing at 12 km h^{-1} will move grains of 0.2 mm in diameter—a lesser velocity will keep the grains moving.

Because wind can only remove particles of limited size range, if erosion is to proceed beyond the removal of existing loose particles, the remaining material must be sufficiently reduced in size by other agents of erosion or weathering, if it is not to inhibit seriously further wind erosion. Removal of fine material leads to a proportionate increase in that of larger size which cannot be removed. The latter affords increasing protection against continuing erosion and eventually a wind-stable surface is created. Desert pavements therefore consist of stones left behind after the finer sediment has been winnowed out (Fig. 7.2). The stones may be flattened and polished by sand blasting to provide a smooth surface.

Binding agents, such as silt, clay and organic matter, hold particles together, and so make wind erosion more difficult. Soil moisture also contributes to cohesion between particles. The particle aggregates in the soil may represent non-erodible or less erodible obstacles to wind erosion. These aggregates may be reduced in size by weathering or erosive processes. Their resistance to wind erosion varies inversely according to their mechanical stability, which depends on the cohesion between the particles, and therefore varies with the type of soil.

Generally, a rough surface tends to reduce the velocity of the wind immediately above it. Consequently, particles of a certain size are not as likely to be blown away as they would on a smooth surface. Even so, Bagnold (1941) found that grains of sand less than 0.03 mm in diameter were not lifted by the wind if the surface on which they lay was smooth. On the other hand, particles of this size can easily remain suspended by the wind. Vegetation affords surface cover and increases the surface roughness. The taller and denser the vegetation, the more effective its protection against wind erosion. The characteristics of the wind, the organic content in the soil, the soil moisture content and the vegetation cover can change over short periods of time, particularly according to the seasonal variation. The longer the surface distance over which a wind can blow without being interrupted, the more likely it is to attain optimum efficiency.

Particles removed by the wind primarily are of sand and dust size. Bagnold (1941) vividly described the difference between the two in terms of their transportation. He indicated that, in dust storms, dust rises in dense clouds to a great height, and dust particles, because of their very low terminal velocity of fall, are kept in suspension by the turbulence of the wind. By contrast, sand is moved principally by saltation, and travels across the surface as a thick, low-lying cloud, with the bulk of the sand movement taking place near the ground. This can produce sand blast effects on obstacles in the path of movement.

Bagnold (1941) noted that sand flow can increase downwind until it becomes saturated, that is, the amount of material in motion is the maximum sustainable by the wind. Beyond this point, no further removal downwind takes place. The downwind increase in the quantity of material transported is referred to as avalanching, and leads to increased surface abrasion. As noted above, as wind erosion continues, the proportion of non-erodible particles increases until, eventually, they protect the erodible particles from wind erosion. At this point, a wind-stable surface is produced. The area of removal accordingly migrates downwind. Hence, particles may move across a surface which may not suffer a net loss of material. In fact, the transport of particles by wind is related to supply sources and sediment stores, which may be permanent or temporary.

There are three types of wind erosion, namely deflation, attrition and abrasion (Chepil & Woodruff, 1963). Deflation results in the lowering of land surfaces by loose, unconsolidated rock waste being blown away by the wind. The effects of deflation are seen most acutely in arid and semi-arid regions. Small deflation hollows may be formed as a result of strong winds blowing over bare, dry surfaces of unconsolidated sediment, giving rise to differential erosion. The latter may be aided by the sediment being less well bound in the areas where hollows form. Basin-like depressions are formed by deflation in the Sahara and Kalahari deserts. However, downward lowering almost invariably is arrested when the water table is reached, because the wind cannot readily remove moist rock particles. Moreover, deflation of sedimentary material, particularly alluvium, creates a protective covering if the material contains pebbles. The fine particles are removed by the wind, leaving a surface formed of pebbles, which are too large to be blown away.

The suspended load carried by the wind is further comminuted by attrition. Turbulence causes the particles to collide vigorously with one another, leading to breakdown.

When the wind is armed with grains of sand, it possesses great erosive force, the effects of which are best displayed in rock deserts. Wind abrasion occurs as a result of windblown grains impacting against a surface, and involves the transfer of kinetic energy. The collision may damage either or both the particle and the surface. Basically, the greater the kinetic energy involved (depending on the grain size and/or wind velocity), the larger the amount of abrasion. Accordingly, any surface subjected to prolonged attack by windblown sand is polished, etched or fluted. Abrasion has a selective action, picking out the weaknesses in rocks. For example, discontinuities are opened and rock pinnacles developed. Because the heaviest rock particles are transported near the ground, abrasion is at its maximum there, and rock pedestals are formed.

The differential effects of wind erosion are well illustrated in areas in which alternating beds of hard and soft rock are exposed. If strata are steeply tilted, then, because soft rocks are more readily worn away than hard rocks, a ridge and furrow relief develops. Such ridges are called yardangs. Yardangs are elongated in the direction of the wind, with rounded upwind faces and long, pointed downwind projections. Conversely, when an alternating series of hard and soft rocks is more or less horizontally bedded, features known as zeugens are formed. In such cases, the beds of hard rock act as resistant caps, affording protection to the soft rocks beneath. Nevertheless, any weaknesses in the hard caps are picked out by weathering, and the caps are eventually breached, exposing the underlying soft rocks. Wind erosion rapidly eats into the latter and, in the process, the hard cap is undermined. As the action continues, tabular masses, known as mesas and buttes, are left in isolation (Fig. 7.3).

7.3 Desert dunes

About one-fifth of the land surface of the Earth is desert (Fig. 7.1). Approximately four-fifths of this desert area consists of exposed bedrock or weathered rock waste. The rest mainly is covered with deposits of sand (Glennie, 1970). Desert regions may have very little sand, for example, about one-ninth of the Sahara desert is covered by sand. Most of the sand which occurs in deserts does so in large masses referred to as sand seas or ergs.

Bagnold (1941) recognized five main types of sand accumulation, namely, sand drifts and sand shadow whalebacks, low-scale undulations, sand sheets and

Fig. 7.3 Mesas and buttes, Monument Valley, Utah.

true dunes. He further distinguished two kinds of true dune: the barkhan and the seif. Several factors control the form which an accumulation of sand adopts. First, there is the rate at which sand is supplied, second, there is the wind speed, frequency and constancy of direction, third, there is the size and shape of the sand grains, and, fourth, there is the nature of the surface across which the sand is moved. Sand drifts accumulate at the exits of the gaps through which wind is channelled and are extended downwind. However, such drifts, unlike true dunes, are dispersed if they are moved downwind. Whalebacks are large mounds of comparatively coarse sand, which are thought to represent the relics of seif dunes. Presumably, the coarse sand is derived from the lower parts of seifs, where accumulations of coarse sand are known to exist. These features develop in regions devoid of vegetation. By contrast, undulating mounds are found in the peripheral areas of deserts, where the patchy cover of vegetation slows the wind and creates sand traps. Large undulating mounds are composed of fine sand. Sand sheets are also developed in the marginal areas of deserts. These sheets consist of fine sand which is well sorted. Indeed, they often present a smooth surface which is capable of resisting wind erosion. A barkhan is crescentic in outline (Fig. 7.4) and is orientated at right angles to the prevailing wind direction. A seif is a long, ridge-shaped dune running parallel to the direction of the wind. Seif

dunes are much larger than barkhans; they may extend lengthwise for up to 90 km and reach heights up to 100 m (Fig. 7.5). Barkhans are rarely more than 30 m in height and their width is usually about 12 times their height. Generally, seifs occur in great numbers, running approximately equidistant from each other, the individual crests being separated from one another by anything from 30 to 500 m or more.

It commonly is believed that sand dunes come into being when an obstacle prevents the free flow of sand, sand piling up on the windward side of the obstacle to form a dune. However, in areas where there is an exceptionally low rainfall and therefore little vegetation to impede the movement of sand, observation has revealed that dunes develop most readily on flat surfaces, devoid of large obstacles. It seems that, where the size of the sand grains varies, or where a rocky surface is covered with pebbles, dunes grow over areas of greater width than 5 m. Such patches exert a frictional drag on the wind, causing eddies to blow sand towards them. Sand is trapped between the larger grains or pebbles and an accumulation results. If a surface is strewn with patches of sand and pebbles, deposition normally takes place over the pebbles. However, patches of sand exert a greater frictional drag on strong winds than do patches of pebbles, and so deposition under such conditions takes place over the former.

Fig. 7.4 Barkhan dune, near Luderitz, Namibia.

Fig. 7.5 Seif dunes, near Sossusvlei, Namibia.

When strong winds sweep over a rough surface, they become transversely unstable and barkhans may develop. It appears that barkhans do not form unless a pile of sand exceeds 0.3 m in height. At this critical height, the heap of sand develops a shallow windward and a steep leeward slope. Sand is driven up the former slope and emptied down the latter, however, eddying of the wind occurs along the leeward slope, impeding the fall of sand and imparting a concave outline to it. At a height smaller than 0.3 m, the sand deposit cannot maintain its form, because individual grains, when moved by the wind, fall beyond its boundaries. It is then dispersed to form sand ripples. Windblown sand commonly is moved by saltation, that is, in a series of jumps. The distance which an individual particle can leap depends on its size, shape and weight, the wind velocity and the angle of lift, which, in turn, is influenced by the nature of the surface. If a deposit of sand reaches a height of 0.3 m, the grains which leap from its windward face usually land on the leeward slope. As the infant barkhan grows, an increasing amount of sand is piled at the top of the leeward slope, because the trajectories of sand grains cannot carry them further. The leeward slope is increasingly steepened until the angle of rest is reached, which is approximately 35°. Subsequent deposition leads to sand slumping down the face. The dune advances in this manner.

As there is more sand to move in the centre of a barkhan than at its tails, the latter advance at a faster rate and the deposit gradually assumes an arcuate shape. The tails are drawn out until they reach a length at which their obstructive power is the same as that of the centre of the dune. At this point, the dune adopts a stable form, which is maintained as long as the factors involved in dune development do not alter radically. Barkhans tend to migrate downwind in waves, their advance varying from about 6 to 16 m annually, depending on their size (the rate of advance decreases rapidly with increasing size). If there is a steady supply of material, the dune tends to advance at a constant speed and maintains the same shape, whereas, when the supply is increased, the dune grows and its advance decelerates. The converse happens if the supply of sand decreases.

Longitudinal dunes may develop from barkhans. Suppose that the tails of a barkhan, for some reason, become fixed, for example, by vegetation or by the water table rising to the surface; the wind will continue to move the central part until the barkhan eventually loses its convex shape, becoming concave towards the prevailing wind. As the central area becomes further extended, the barkhan may split. The two separated halves are gradually rotated by the eddying action of the wind until they run parallel to one another, in line with the prevailing wind direction. Dunes that develop in this manner often are referred to as blow-outs.

Seif dunes appear to form where winds blow in two directions, that is, where the prevailing winds are gentle and carry sand into an area, the sand then being driven by occasional strong winds into seif-like forms. Seifs may also develop along the bisectrix between two diagonally opposed winds of roughly equal strength.

Because of their size, seif dunes can trap coarse sand much more easily than can barkhans. This material collects along the lower flanks of the dune. Indeed, barkhans sometimes occur in the troughs of seif dunes. On the other hand, the trough may be floored by bare rock. The wind is strongest along the centre of the troughs, as the flanks of the seifs slow the wind by frictional drag, so creating eddies. The

wind blowing along the flanks of seifs accordingly is diverted up their slopes.

7.4 Stream action in arid and semi-arid regions

The processes of weathering, erosion and sedimentation in arid regions are similar to those in humid regions, but differ in their relative importance, intensity and superficial effects. Weathering, although widespread, dominantly is physical, because of the deficiency of moisture and vegetation. Although intermittent, streams are the most important agents of erosion on barren slopes, because of the intense run-off.

In fact, stream activity plays a significant role in the evolution of the landscape in arid and semi-arid regions. Admittedly, the amount of rainfall occurring in arid regions is small and falls irregularly, whereas that of semi-arid regions is markedly seasonal. Consequently, many drainage courses in arid lands may be dry for a good part of the year, and run-off events do not necessarily activate whole drainage systems. The extent to which river courses are occupied after rainfall events depends on the magnitude, duration and frequency of the events. Moreover, river channels on alluvial ground frequently migrate during flow events, so that they flow along different courses during successive events. However, large rivers may enter arid lands from outside, such as the Fish River in Namibia and, more notably, the Nile in Egypt.

Rain often falls as heavy and sometimes violent showers in these regions. The result is that the river channels cannot cope with the amount of rain water, and extensive flooding takes place. These floods develop with remarkable suddenness, and form either raging torrents, which tear their way downslope excavating gullies as they go, or sheet floods. Dry wadis are filled rapidly with swirling water and therefore are enlarged. However, such floods are short lived, because the water soon becomes choked with sediment, and the consistency of the resultant mudflow eventually reaches a point when further movement is checked. Much water is lost by percolation, and mudflows are also checked where there is an appreciable slackening in gradient. Gully development frequently is much in evidence.

Some of the most notable features produced by stream action in arid and semi-arid regions are found

Fig. 7.6 Death Valley from Dantes View. Note the alluvial fan at the foot of the slope left-centre, and other fans merged to form a bahada on the opposite side of the playa.

in intermontane basins, that is, where mountains circumscribe a basin of inland drainage (Fig. 7.6). The mountain catchment area has an appreciable influence upon the quantity and duration of run-off, and so on the floods which develop and associated erosion. Mechanical weathering also plays a significant role in the mountain zone. Channels in mountain catchments may be largely devoid of debris, as it has been swept away by floods, or bedrock may be overlain by thin deposits of sediment of mixed origin. They may contain braided stream courses or debris flow deposits. The types of channel which occur affect the nature of a flood, as the character of the deposits influences the sediment transport, base-flow, recharge and the initiation of debris flows. Short-term changes in stream channels may be brought about by changes in the catchment area, such as overgrazing or construction operations, as well as by flooding.

Run-off generally is very rapid in arid regions, and so the nature of slopes can have a significant effect on flood and sediment problems. For instance, run-off is especially rapid from the bare rock slopes found in the upper reaches of mountain catchments. Accumulations of debris occur further downslope. Fluvial processes only are important on the lower parts of these slopes, where the materials are finer and may have been cemented to form duricrusts. Rills and gullies develop on the lowermost slopes, which are formed of fine grained sediment or stone pavements. They may also be washed over by sheet flows. Rockfalls and rock slides occur from the upper slopes. After rainfall or snow melt, debris flows can originate in small mountain catchments which have a dense network of gullies carved into relatively loose sediment. The flows build levées along their channels. They consist mainly of sandy material, and tend to spread out on the lower slopes to form lobate deposits or fans. Even on gently sloping fans, the run-out can be up to 30 km, but if the fan deposits are highly permeable, rapid drainage from the flow restricts its movement. In addition, the rapid infiltration of water into highly permeable fan deposits can give rise to undulating, irregular topography formed of sieved deposits, that is, the fine sediments are washed out of the coarser sediments. This frequently means that such deposits have variable load-carrying capacity. The particles composing the fans are almost all angular in shape, boulders and cobbles being more frequent upslope, grading downslope into fine gravels, sands and silts. A decrease occurs in channel depth downslope, with a tendency for the flow to be directed into unstable, shifting channels.

Debris flows represent a hazard on the debris slopes and, as well as descending from the adjoining mountains, develop especially on the finer material which can become saturated by rainfall or snow melt. Furthermore, ancient landslides which formed when the climate was wetter, where present, can be reactivated by human interference, notably construction operations.

Alluvial fans commonly are made up of distinctive features, such as contemporary deposition areas, abandoned surfaces and an array of channels. The contemporary areas of deposition occur at the ends of active channels. Abandoned surfaces may be distinguished by desert varnish or by duricrusts. Small, localized stream channels are present, which are sup-

(a)

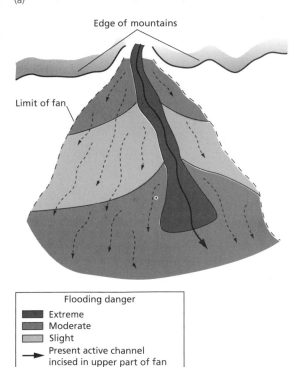

Edge of mountains

Limit of fan

Flooding danger
- ◼ Extreme
- ◼ Moderate
- ◻ Slight
- → Present active channel incised in upper part of fan
- --→ Older discontinuous channels

(b)

Fig. 7.7 (a) Alluvial fan below the mouth of Hannapah Canyon. Note that the fan is interrupted by a fault scarp formed as a result of an earthquake. Death Valley, California. (b) The pattern of flood hazard on a typical alluvial fan in western USA. (From Cooke *et al.*, 1982.)

plied with water and sediment from the catchment areas of these old surfaces. Consequently, flood risk in these areas is low. Channels include entrenched feeder channels, distributary channels and small dendritic systems, which develop on the fan itself. These have a relatively high flood risk and tend to migrate laterally. Although stream flow on alluvial fans is ephemeral, flooding nevertheless can constitute a serious problem, occurring along the margins of the main channels and in the zone of deposition beyond the ends of supply channels (Fig. 7.7). The floodwaters are problematic because of their high velocities, variable sediment content and tendency to change location with successive floods, abandoning and creating channels in a relatively short time. Hydrocompaction may occur on alluvial fans, particularly if they are irrigated. The dried surface layer of these fans may contain many voids. Percolating water frequently reduces the strength of this material which, in turn, causes collapse of the void space, giving rise to settlement. The streams which descend from the mountains rarely reach the centre of the basin, because they become choked by their own deposits and split into numerous distributaries, which spread a thin veneer of gravel over the pediment. When alluvial fans merge into one another, they form a bahada.

Pediments are graded plains, which are formed by the lateral erosion of mainly ephemeral streams. They occur at the foot of mountain slopes, and may be covered with a thin veneer of alluvium. Pediments are adjusted to dispose of water in the most efficient way and, when heavy rain falls, this often means that it takes the form of sheet wash. Although true laminar flow occurs during sheet wash, as the flowing water deepens, laminar flow yields to turbulent flow. The latter possesses much greater erosive power and occurs during and immediately after heavy rainfall. This is why these pediments carry only a thin veneer of rock debris. With a lesser amount of rainfall, there is insufficient water to form sheets, and it is confined to rills and gullies. Distributary channels may develop in the upper parts of pediments. Streams on pediments occasionally are active and liable to flood, scour and fill, and lateral migration. The rock waste transported across the pediment is relatively fine and is deposited in hollows, thereby smoothing the slope. The abrupt change in the slope at the top of the pediment is caused by a change in the principal processes

Fig. 7.8 Inselberg, north of Bulawayo, Zimbabwe.

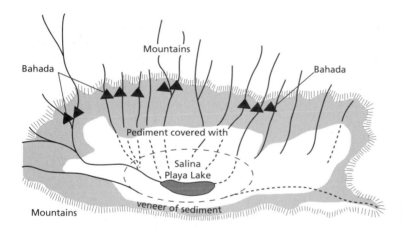

Fig. 7.9 Intermontane basin showing bahadas, pediment, salina and playa lake.

of earth sculpture, the nature of the pediment being governed by sheet erosion, whilst that of the steep hillsides being controlled by the downward movement of rock debris.

The extension of pediments on opposing sides of a mountain mass means that the mountains slowly are reduced until a pediplain is formed. The pediments are first connected through the mountain mass by way of pediment passes. The latter become progressively enlarged, forming pediment gaps. Finally, opposing pediments meet to form a pediplain on which there are residual hills. Such isolated, steep-sided residual hills have been termed inselbergs or bornhardts (Fig. 7.8). They characteristically are developed in the semi-arid regions of Africa, where

they usually are composed of granite or gneiss, that is, of more resistant rock than that which forms the surrounding pediplain. The pediplains represent extensive planation surfaces, which have been subjected to long periods of denudation, where tectonic activity frequently has been of little consequence or absent.

The central and lowest area of a basin is referred to as the playa and sometimes contains a lake (Fig. 7.9). This area may be subject to seasonal flooding, and is covered with deposits of sand, silt, clay and evaporites. The silts and clays often contain crystals of salt, whose development further comminutes their host. Silts usually exhibit ripple marks, whilst clays frequently are laminated. Desiccation structures, such as

Fig. 7.10 Grand Canyon, Arizona.

mudcracks, are developed on an extensive scale in these fine grained sediments (Lister & Secrest, 1985). If the playa lake has contracted to leave a highly saline tract, this area is termed a salina. The capillary rise generally extends to the surface, leading to the formation of a salt crust. Where the capillary rise is close to, but does not normally reach the surface, desiccation ground patterns (some cracks are extremely large) provide an indication of its closeness. Aeolian and fluvial deposits, notably sand, may also be laid down in the intermediate zone between the pediment and the playa. However, if deflation is active, this zone may be barren of sediments. Sands commonly are swept into dunes.

Yet another type of fluvial landscape developed in arid regions is found where there are thick surface formations of more or less horizontally bedded sedimentary rocks. Large rivers coming from outside the region frequently flow in canyons, such as the Colorado River in Arizona and the Fish River in Namibia (Fig. 7.10). The plateau lands which characterize such areas are interrupted by buttes and mesas, and by escarpments or cuestas. Such landscapes are typified by the Colorado Plateau, and are found in the semi-arid and arid areas of southern Africa.

Duricrusts are surface or near-surface hardened accumulations or encrusting layers, formed by the precipitation of salts or the evaporation of saline groundwater. When describing duricrusts, those terms ending in *crete* refer to hardened surfaces usually occurring on hard rock, and those ending in *crust* represent softer accumulations, which usually are found in salt playas, salinas or sabkhas. Duricrusts may be composed of calcium or magnesium carbonate, gypsum, silica, alumina or iron oxide, or even halite, in varying proportions (Goudie, 1973). However, enrichment of silica or iron to form silcrete or ferricrete respectively, occurs very occasionally in arid regions. Duricrusts may occur in a variety of forms, ranging from a few millimetres in thickness to over a metre. A leached, cavernous, porous or friable zone frequently is found beneath the duricrust.

The most commonly precipitated material is calcium carbonate. These caliche deposits are referred to as calcrete (Netterberg, 1980; Braithwaite, 1983). Calcrete occurs where soil drainage is reduced due to long and frequent periods of deficient precipitation and high evapotranspiration. Nevertheless, the development of calcrete is inhibited beyond a certain aridity, because the low precipitation is unable to dissolve and drain calcium carbonate towards the water table. Consequently, gypcrete tends to take the place of calcrete where the annual rainfall is less than 100 mm.

7.5 Flooding and sediment problems

Desert rainfall is erratic both in time and space. In any one year, the annual amount of rainfall may be four times the average or more, and may occur in a single storm. Conversely, in other years, very little rain may fall. A high proportion of rainfall in arid regions is lost by evaporation or infiltrates into the ground. That proportion of rain which remains for run-off in extremely arid regions may be as little as 10%. In fact, not all low intensity storms give rise to run-off, so that the intensity becomes a more important factor in this regard than the total rainfall. Intensities of probably at least $1\,\text{mm}\,\text{min}^{-1}$ over a minimum of 10 min may be needed to generate channelled run-off. On the mountain or fan catchments, run-off may either occur as sheet flow or be concentrated into rills and gullies. Overland flow may be concentrated quickly to give rise to floods where rills and gullies occur. Such rapid flow from hillslopes governs the major characteristics of flow generated in ephemeral stream channels. Ephemeral run-off in small channels rapidly attains peak discharge, and wave-like rapid advance of a flood is typical. Hence, rapid flash floods of relatively small size are associated with small- to medium-sized streams, and these may have catastrophic consequences.

Run-off is rapid from mountain catchments because of the steep slopes, which frequently are bare or covered with thin deposits of sediment and lack vegetation. However, extensive, deep deposits of coarse grained alluvium may occur in many ephemeral stream channels, and such material possesses a high infiltration capacity and a potential for significant storage of water. The amount of water which infiltrates into such alluvium depends on a number of factors, including the permeability, moisture content and position of the water table within the alluvium, and the size of the channel, flow duration and peak discharge. At times, all the water involved in run-off eventually may infiltrate into the bed of the stream system. Most infiltration occurs in the early stages of a flood, and the distance downstream achieved by floods obviously is very much influenced by the size of the rainfall event and the amount of run-off generated. Because of the high infiltration in the upper parts of stream systems, groundwater recharge downstream is affected adversely which, in turn, may mean that less water is available for any irrigation schemes.

The movement of sediment on slopes primarily is caused by run-off, supplemented by raindrop impact and splash erosion. High run-off associated with torrential rainfall can flush sediment from gullies. Small fans may develop at the exits from gullies spreading into stream channels. This, in turn, may lead to the formation of gravel sheets, lobes and bars, on the one hand, and/or bank erosion on the other. Major flows can transport massive quantities of sediment which, when deposited, raise valley floor levels. Channels frequently are flanked by terraces which provide evidence of previous flood events. Sediment load increases downstream as slopes and tributaries contribute more load, and as water is lost by infiltration. Hence, the lower reaches of streams are dominated by deposition. The deposits exhibit some lateral grading, in that the larger material, that is, the bedload, is deposited first, the suspended sediment being carried further downstream. Individual flood deposits also tend to show an upward-fining sequence. In fact, the surface of channel deposits may consist of a thin layer of sandy or silty material, which can affect infiltration. An important factor to be borne in mind is that the annual sediment yield in such regions is highly variable.

7.6 Movement of dust and sand

As noted previously, the movement of dust and sand by wind is different. Nonetheless, the movement of dust and sand is governed by the same factors, namely, the velocity and turbulence of the wind, the particle size distribution of the grains involved and the roughness and cohesion of the surface over which the wind blows. In other words, movement is concerned with the erosivity of the wind, on the one hand, and with the erodibility of the surface on the other. As far as the erosivity of the wind is concerned, the most important factors are its mean velocity, direction and frequency, the period and intensity of gusting and the vertical turbulence exchange. The two main factors governing the erodibility are the nature of the surface and its vegetation cover. As dust particles are bound, to varying degrees, by cohesive forces, high velocity winds may be needed to entrain them, although saltating sand particles can entrain dust as a result of impact. Once entrained, dust can be carried aloft by fairly gentle winds. Dust storms are common in arid and semi-arid regions, and dust from deserts often is transported hundreds of kilometres. These storms, at times, cover huge areas, for example, the dust storm which affected North America on November 12, 1933 covered an area greater than France, Italy and Hungary combined (Goudie, 1978). As mentioned above, sand moves principally by saltation and, as such, is concentrated near the ground. The problems of sand and dust movement are most acute in areas of active sand dunes, or where dunes have been destabilized by human interference, disturbing the surface cover and vegetation of fixed dunes.

The rate of movement of sand dunes depends on the velocity and persistence of the winds, the constancy of the wind direction, the roughness of the surface over which they move, the presence and density of vegetation, the size of the dunes and the particle size distribution of the sand grains. Large dunes may move up to 6 m per year and small ones may exceed 23 m per year. For instance, Watson (1985) recorded that barkhans, up to 25 m in height, which occur in the Eastern Province of Saudi Arabia, have an average rate of movement of 15 m per year, and drift rates reach 30 m³ per metre width annually. In fact, the largest amounts of material in motion occur in the sand deserts. This is because the grains tend to be loose and not aggregated together.

Deflation leads to the removal of finer material, that is, silt and clay, together with organic matter from soil, leaving behind coarser particles, which are less capable of retaining moisture. Once lost, soil in arid and semi-arid regions does not re-form quickly. Deflation may lead to the scour and undermining of railway lines, roads and structures, causing their collapse. However, deflation can give rise to a wind-stable surface, such as a stone pavement or surface crust. Human interference can initiate deflation on a stable surface by causing its destruction, for example, by the construction of unpaved roads and other forms of urbanization.

Sand and dust storms can reduce visibility, bringing traffic and airports to a halt. During severe sand storms, the visibility can be reduced to less than 10 m. In addition, dust storms may cause respiratory problems and even lead to the suffocation of animals, disrupt communications and spread disease by the transport of pathogens. The abrasive effect of moving sand is most notable near the ground (i.e. up to a height of 250 mm). However, over hard, man-made surfaces, the abrasion height may be higher, because the velocity of sand movement and saltation increase. Structures may be pitted, fluted or grooved depending on their orientation in relation to the prevailing wind direction.

The movement of sand in the form of dunes can bury obstacles in its path, such as roads and railways, or accumulate against large structures. Such moving sand necessitates continuous and often costly maintenance activities. Indeed, areas may be abandoned as a result of sand encroachment (Fig. 7.11). Stipho (1992) pointed out that only a few centimetres on a road surface can constitute a major driving hazard. Deep burial of pipelines by sand makes their inspection and maintenance difficult, and unsupported pipes on active dunes can be left high above the ground as dunes move on, causing the pipes to move and possibly to fracture (Cooke *et al.*, 1982).

Fig. 7.11 Houses in Gonghe County, Qinghai, China, threatened by windblown sand.

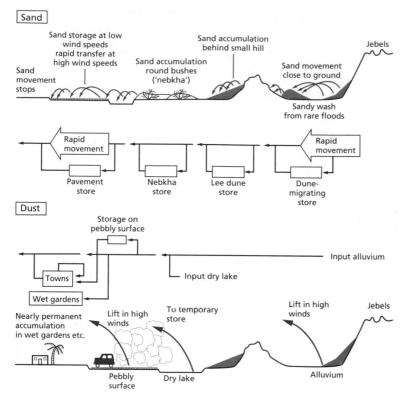

Fig. 7.12 Sources and stores for sand and dust. (After Jones *et al.*, 1986.)

Aerial photographs and remote sensing imagery of the same area taken at successive time intervals can be used to study dune movement, the rate of land degradation and the erodibility of surfaces (Jones *et al.*, 1986). The recognition of these various features of arid landscapes by the use of aerial photographs should be checked in the field. Field mapping involves the identification of erosional and depositional evidence of sediment movement (e.g. sand drifts, dunes, surface grooving, ventifacts and yardangs), which can provide useful information on the direction of sand movement, but less information on dust transport. For instance, indirect evidence of the wind regime can be derived from the trends of active dunes. However, care needs to be taken when assessing the data obtained from the field, in that the features identified should represent forms which are developing under present-day conditions. Fixed dunes, which have been developed in the past, are not necessarily related to the contemporary wind regime. As their destabilization by human activity should be avoided, it is important that dunes should be identified and monitored.

Of particular importance in an assessment of the sand and dust hazard is the recognition of sources and stores of sediment. Jones *et al.* (1986) defined sources as sediments which contributed to the contemporary aeolian transport system, and included weathered rock outcrops, fans, dunes, playa/sabkha deposits and alluvium. Stores are unconsolidated or very weakly consolidated surface deposits, which have been moved by contemporary aeolian processes, and have accumulated temporarily in surface spreads, dunes, sabkhas and stone pavements (Fig. 7.12).

Erodibility can be assessed in the field by the use of experimental plots or by soil classification. The environmental conditions can be controlled, to a large extent, when using an experimental plot. Soil classification according to erodibility takes into consideration the depth and structure of the surface horizon, as well as the depth of non-erodible subsurface material.

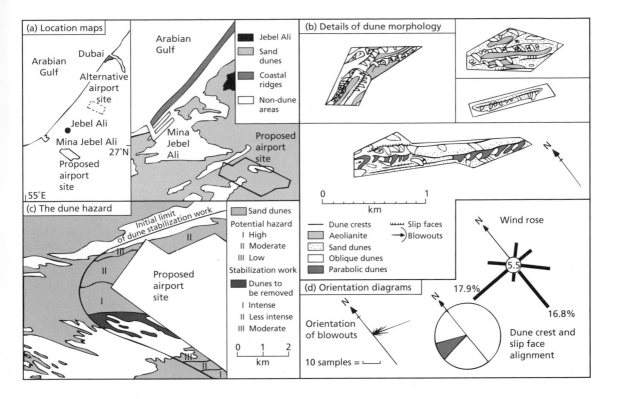

Fig. 7.13 Geomorphological analysis of a proposed airport site in Dubai with respect to the threat from mobile sand dunes. (After Doornkamp *et al.*, 1979.)

Examples of hazard maps of geomorphology, dune morphology and migrating dunes have been provided by Doornkamp *et al.* (1979). For example, they produced a number of such maps for two proposed sites for a new airport in Dubai (Fig. 7.13). The field survey revealed that the dune characteristics, notably their form and vegetation density, could be related to their mobility, and these characteristics could also be recognized on aerial photographs. High, angular, fresh dunes, lacking vegetation, were regarded as potentially or actively mobile. By contrast, broad, rounded, low, well-vegetated dunes were regarded as posing a low hazard risk. It was suggested that the small unstable dunes could be removed, whilst the larger dunes required extensive stabilization. The dunes of low hazard required moderate control measures to check their migration. Jones *et al.* (1986) assessed the relative importance of geomorphological

units as potential sources of aeolian sand and dust, recognizing high, medium and low classes of potential (Fig. 7.14).

The best way to deal with a hazard is to avoid it, this being both more effective and cheaper than resorting to control measures. This is particularly relevant in the case of aeolian problems, notably those associated with large, active dune fields, where even extensive control measures are likely to prove to be only temporary. Hence, sensible site selection, based on a thorough site investigation, is necessary prior to any development.

The removal of moving sand can only happen where the quantities of sand involved are small, and so can only apply to small dunes (a dune 6 m in height may incorporate 20 000–26 000 m³ of sand with a mass of between 30 000 and 45 000 tonnes). Even so, this is expensive, and the excavated material has to be disposed of. Often the removal of dunes is only practical when the sand can be used as fill or ballast, however, the difficulty of compacting aeolian sand frequently precludes its use. In addition, flattening dunes does not represent a solution, because the wind

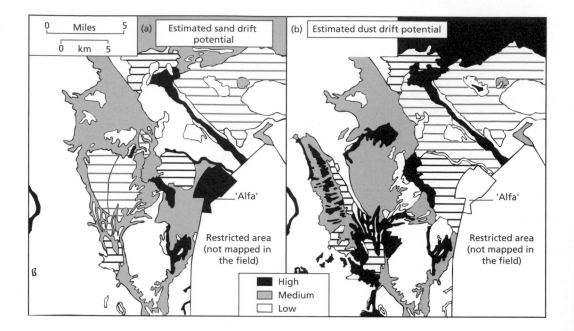

Fig. 7.14 (a) Map of estimated sand drift potential for area to north and west of 'Alfa'. (b) Map of estimated dust drift potential for area to north and west of 'Alfa'. (After Jones *et al.*, 1986.)

will develop a new system of dunes on the flattened surface within a short period of time.

Accordingly, a means of stabilizing mobile sand must be employed. One of the best ways to bring this about is by the establishment of a vegetative cover. Plants used for the stabilization of dunes must be able to exist on moving sand, and either survive temporary burial or keep pace with deposition. The plants normally used have coarse stiff stems to resist sand blasting and are unpalatable to livestock, as well as being fast growing. Cooke *et al.* (1982) noted that, if successful stabilization is to be achieved, the character of the substrate, the thickness of the sand deposit, the quantity and quality of the water in the sand and the substrate, the position of the water table and the rate of sand movement should be investigated. Obviously, vegetation is more difficult to establish in arid than semi-arid regions. In arid regions, young plants have to be protected from burial by windblown sand or from being blown out of the soil. Generally, the more degraded the vegetation, the longer it takes for natural recovery, recovery being confined to areas

where groundwater occurs at shallow depth and where the threat of deflation is small. When a planting programme is undertaken, it usually is necessary to apply fertilizers liberally. Mulches can be used to check erosion and to provide organic material. Brush matting has been used to check sand movement temporarily, especially in deflation hollows. The brush matting is laid in rows so that succeeding rows overlap. Trees may be planted in the brush matting. However, brush matting generally is impractical where wind velocities exceed 65 km h^{-1}.

Natural geotextiles can be placed over sand surfaces after seeding, to protect the seeds, to help retain moisture and to provide organic matter eventually. These geotextiles can contain seeds within them (see Chapter 10).

Gravel or coarse aggregate can be placed over a sand surface to prevent its deflation. A minimum particle diameter of around 20 mm is needed for the gravels to remain unaffected by strong winds. In addition, the gravel layer should be at least 50 mm in thickness because, if it is disrupted, the underlying sand can be subjected to the scouring action of the wind. The gravel material should not be susceptible to abrasion.

Artificial stabilization, which provides a protective coating or bonds grains together, may be necessary on loose sand. In such cases, rapid-curing cutback asphalt and rapid-setting asphalt emulsion may be used. Asphalts and oils are prone to oxidation over a period of 1 or 2 years and, once this has occurred, the treatment rapidly becomes ineffective. Cutback asphalts are less prone to oxidation, and their tendency to become brittle is reduced by additives such as liquid latex. Nonetheless, the performance of cutbacks is limited by their low degree of penetration. Chemical sprays have been used to stabilize loose sand surfaces, and the thin protective layer which develops at the surface helps to reduce water loss. Many chemical stabilizers are only temporary, breaking down in a year or so, and therefore they tend to be used together with other methods of stabilization, such as vegetation. Seeds sometimes are sprayed together with the chemical spray and fertilizers in a type of hydroseeding. For example, wood cellulose fibres can be used to make a suspension with water, seed and fertilizer, together with an asphalt or resin emulsion, and sprayed onto sand. Resin-in-water emulsion, latex-in-water emulsion and gelatinized starch solutions have been used as stabilizers. Large aggregates are formed when emulsions are mixed with sand. This helps to increase the rate of infiltration, thereby decreasing run-off and enhancing the moisture content in the soil. The large aggregates are also more resistant to wind erosion. Polyacrylamide solutions, when applied to soil, also give rise to high rates of infiltration, irrespective of whether large or small aggregates are formed (Gabriels *et al.*, 1979). A dune will not migrate if its windward side is stabilized. Polyurea polymers have been used to stabilize dune sands successfully. These may be injected to appreciable depth and may extend the life expectancy of the treatment to more than 5 years. Nonetheless, sand will be blown over a stabilized surface. For example, barkhans form oval-shaped deposits of sand, which are several times larger than the original dune (Fig. 7.15).

Gravels, cobbles and crushed rock provide a stable cover for dunes, and can be employed where environmental considerations preclude the use of oils or chemicals. However, spreading such material over unstable sand may present a problem.

Vegetation windbreaks frequently are used to control wind erosion, and obviously are best developed where groundwater is close enough to the surface for trees and shrubs to have access to it. It is essential to select species of trees which have growth rates exceeding the rate of sand accumulation and which have a bushy shape. *Tamarix* and *Eucalyptus* species have proved to be most successful (Watson, 1985). Windbreaks can act as effective dust and sand traps. Their shape, width, height and permeability influence their trapping efficiency and the amount of turbulence generated as the wind blows over them. For example, a low permeability enhances the trapping efficiency, but gives rise to eddying on the leeward side of a windbreak.

Fences can be used to impound or divert moving sand. Most sand accumulation occurs on the leeside of the fence, chiefly in a zone four times as wide as the fence is high. Fences can be constructed of various materials, for example, brushwood or palm fronds, wood or metal panels, stone walls or earthworks. The surfaces around fences should be stabilized in order to avoid erosion and undermining with consequent collapse. Individual fence designs have different sand trapping abilities at different wind velocities, so that

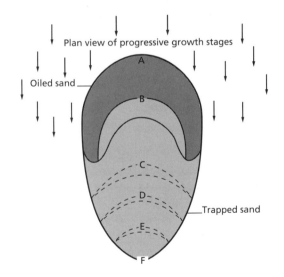

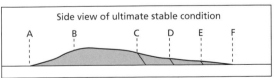

Fig. 7.15 Dune stabilization by oiling or polymers. (After Kerr & Nigra, 1952.)

the fence design must be tailored to the local environmental conditions. The alignment of fences relative to the direction of sand drift, and the number and spacing of rows of fences, are critical (Watson, 1985). The optimum spacing of rows of fences is about four times their height, so that 1 m high fences have a maximum capacity to trap approximately 16 m³ per metre width, assuming that there are four rows of fences 4 m apart. Multiple rows of fencing can trap more than 80% of wind-borne sand, even under variable wind conditions, if optimal alignment and porosity for the specific area are employed. The practical number of rows of fences is about four.

The most effective fences are semi-permeable. Although such fences reduce the wind velocity to a lesser extent than impermeable fences, they restrict diffusion and eddying effects and their influence extends further downwind. In areas which experience a wide range of wind speeds, the optimum fence porosity is around 40%. If the wind speed exceeds 18 m s⁻¹, no sand is trapped by a single fence with a porosity of 40%. Under such conditions, a double row will trap about 30% of the sand being moved. Sand accumulates against an impounding fence, until it reaches the top, when a further fence needs to be installed or the fence must be heightened. The next fence installed on top of the sand mound has a greater sand trapping capacity. Hence, the use of fences over a long period of time represents a commitment to a sand control policy based on dune building.

Panels may be used to direct air flow over a road or around a building. They only are used where the rate of deposition of sand is low. However, although panels can be arranged to increase wind velocity and so keep surfaces sand free, when the wind leaves the panelling, it decreases in speed and some deposition of sand occurs.

Stipho (1992) mentioned the use of trenches and pits to reduce the amount of blown sand. These must be wider than the horizontal leap of saltating particles (up to 3–4 m) and sufficiently deep to prevent scouring. As the regular removal of accumulated sand is necessary, pits and trenches are expensive to maintain. In effect, trenches are useful only for short-term protection. Stipho also referred to the use of sand traps to collect sand to prevent its accumulation on highways. He suggested that the sand could be removed subsequently. Again in relation to highways, Stipho suggested that, in some situations, it may be necessary to elevate a carriageway well above the surrounding dunes in order to allow the wind to accelerate over the road, while reducing rutting in the sand to a minimum.

The deposition of sand can be reduced either by removing areas of low wind energy or by increasing the velocity of the wind over the surface by shaping the land surface (Watson, 1985). For example, flat slopes in the range of 1:5–1:6, with rounded shoulders, are necessary for small to medium embankments, such slopes helping to streamline wind flow. Cuttings require flatter slopes of perhaps 1:10 to allow for free sand transport, and usually are accompanied by means to collect blown sand, such as a wide ditch at the base of a slope.

7.7 Sabkha soil conditions

Sabkhas are salt-encrusted, fairly level surfaces which are common in arid regions (Fig. 7.16a; Table 7.1). They occur in both coastal (Fig. 7.16b) and inland areas, but the mode of development and the properties of the sabkhas in these areas are different. Inland sabkhas have also been referred to as playas and salinas. Coastal sabkhas may consist of cemented and uncemented layers of reworked aeolian sand and muddy sand of varying thickness. These sabkhas are highly variable in both horizontal and vertical extent. The horizontal variation may be related to the changing position of the shoreline, whilst the vertical variation is connected with the depositional sequence and subsequent diagenesis. Groundwater is saline, containing calcium, sodium, chloride and sulphate ions. Evaporative pumping, whereby brine moves upwards from the water table under capillary action, appears to be the most effective mechanism for the concentration of brine in groundwater and for the precipitation of minerals in sabkhas. The aragonite and calcite in a sabkha, according to Akili and Torrance (1981), originated primarily during a previous subaqueous state, when lagoon waters covered part of the coastal areas and were concentrated by evaporation. However, some diagenetic aragonite forms during the early stages of brine concentration in the sabkha and may give rise to a thin, lithified crust. Nevertheless, gypsum is the commonest of the diagenetic minerals formed during this stage. Anhydrite is the dominant sulphate mineral above the standing level of the brines in the sabkha. Halite forms as a crust at or near

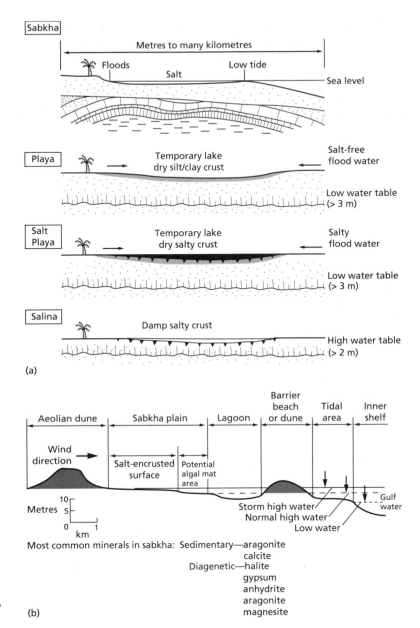

Fig. 7.16 (a) Idealized cross-sections of sabkha, playa, salt playa and salina terrain. (After Fookes & Collis, 1975.) (b) Generalized cross-section across a typical coastal sabkha with typical surface features. (After Akili & Torrance, 1981.)

the surface, or sometimes as massive layers. However, it is removed by rain, floods, inundating high tides or even strong winds. In fact, wind may move surface salt crystals to form drifts or even dunes. In sabkhas which are only intermittently invaded by the sea, halite forms a surface crust which may be in the shape of polygonal ridges. The salt may be deposited only

decimetres below the surface, decreasing downwards in amount (Fookes *et al.*, 1985). Dolomite is another diagenetic mineral and, where the concentration of magnesium in the brine is very high, magnesite may make its appearance. The formation of these diagenetic minerals and their positions within a sabkha depend mainly on its hydrogeology and the way in

Table 7.1 Ground conditions associated with sabkhas, playas and salinas. (After Fookes & Collis, 1975.)

Name	Terrain	Groundwater table	Salts	Special significance	Construction notes
Sabkha	Coastal flat, inundated by sea water during exceptional floods	Very near the surface	Thick, surface salt crusts from evaporating brines. Salts usually include carbonates, sulphates, chlorides and others	Generally aggressive to all types of foundations by salt weathering of stone and concrete and/or sulphate attack on cement-bound materials. Evaluate bearing capability carefully	Carefully investigate; consider tanking concrete foundations; use dense concrete. For surfaced roads, consider using inert aggregate, raised embankment or positive cutoff below sub-bases. Use as fill may be suspect. May not be deleterious to unsurfaced roads
Playa	Inland, shallow, centrally draining basin of any size	Too deep for the capillary moisture zone to reach the ground surface, but area will be a temporary lake during floods	None if temporary lake is of salt-free water	Non-special. Ground surface may be silt/clay or covered by windblown sands. Evaluate bearing capability	Non-special
Salt playa	As playa, but sometimes smaller	As above, but lake of salty water	Surface salt deposits from evaporating temporary salty lake water. Salts usually include chlorides and sometimes sulphates and carbonates	Can be slightly to moderately aggressive to all types of foundations by salt weathering and sulphate attack. More severe near water table	As sabkha
Salina	Inland basin of any size	Near surface; capillary moisture zone from salty groundwater can reach the surface	Surface crusts from evaporating salty groundwater. Salts include sulphates and others	Can be slightly to exceptionally aggressive to all types of foundations by salt weathering and sulphate attack	As sabkha

which groundwater at a given location is replenished: from fresh water from flash floods or infrequent rains, or from marine inundation.

One of the main problems with sabkhas is the decrease in strength which occurs, particularly in the uppermost layers, after rainfall, flash floods or marine inundation, due to the dissolution of soluble salts which act as cementing materials. This decrease in strength can render normally stable surface crusts impassable to traffic. Hence, when wetted, sabkhas may not be capable of supporting heavy equipment. There is a possibility of differential settlement occurring on loading due to the different compressibility characteristics resulting from the differential cemen-

tation of sediments. Excessive settlement can also occur due to the removal of soluble salts by flowing groundwater. This can cause severe disruption to structures within months or a few years. Movement of groundwater can also lead to the dissolution of minerals to the extent that small caverns, channels and surface holes can be formed. In addition, gypsum can undergo dehydration in sabkha environments, leading to volume changes in the soil. Heave, resulting from the precipitation and growth of crystals, according to Bathurst (1971), can elevate the surface of a sabkha in places by as much as 1 m.

Continental sabkhas, playas or salinas are saline deposits found in inland areas, which do not have any hydrological connection allowing groundwater replenishment by the sea. Akili and Torrance (1981) indicated that, in Arabia, they tend to occur in areas which are dominated by dune sand. Groundwater brines near the surface are concentrated by evaporation, from which minerals, such as gypsum and anhydrite, are precipitated. Gypsum tends to occur below the water table, whereas anhydrite tends to occur above it. Some of these sabkhas, however, contain little carbonate. Salts are precipitated at the ground surface when the capillary fringe extends from the water table to the surface. Although halite frequently forms at the ground surface, it commonly is dispersed by wind. The sediments may consist of layers of fine to medium grained quartz sand, alternating with muddy sand, in which the grains are partially cemented by evaporative minerals. Salt-bearing soils tend to be strongly hygroscopic, depending on the type of salt.

Coarse grained soils formed in arid regions differ from their counterparts developed in temperate and humid climates. Such soils frequently are characterized by easy transportation by wind and a lack of vegetation and, where the water table occurs at depth, often are unsaturated and highly cemented. On the other hand, where the water table is at a shallow depth, the soils may possess a salty crust and are chemically aggressive due to the precipitation of salts from saline groundwater. Windblown carbonate silt- to clay-sized particles generally occur between coarser grains. Fookes and Gahir (1995) indicated that these coarse grained soils exhibit large variations in grain size distribution, are free draining and achieve high densities upon compaction. Occasional wetting and subsequent evaporation have been responsible for the patchy development of weak, mainly carbonate and occasionally gypsum cement, often with clay deposited between and around the coarser particles. This has given rise to a metastable structure. These soils therefore may undergo collapse, especially where localized changes in the soil water regime are brought about by construction activity. Collapse is attributed to a loss of strength in the binding agent, and the amount of collapse which occurs depends upon the initial void ratio.

Stipho (1985) found that an increased salt content increased the shear strength of sandy soils as a result of increased cohesion, but led to a reduction in the angle of friction. Remoulding gave rise to a significant decrease in soil strength. He also found that the residual strength of salt-cemented sand was close to that of uncemented sand.

An excess of evaporation over infiltration gives rise to a negative pore water pressure (soil suction) as water is removed from the soil by evaporation. If the soil suction is large enough, air enters the soil. Under these circumstances, the soil behaviour has to be described in terms of the total applied stress and soil water suction. Changes in the total applied stress and soil water suction separately bring about changes in the volume and strength. Blight (1994) maintained that the soil water suction has physical (matrix) and osmotic components, and that the former governs the mechanical behaviour of the soil. If the soil possesses an unstable structure, collapse may take place once the soil water suction falls below a critical value.

The occurrence of calcareous expansive clays in eastern Saudi Arabia has been described by Abduljauwad (1994). These clays have a moderate to very high swelling potential, depending on their smectite content. They have been formed by the weathering of limestone and marl in an alkaline environment with an absence of leaching. In some cases, the natural moisture content of the clays is higher than that of the plasticity index.

7.8 Salt weathering

Because of the high rate of evaporation in hot, arid regions, the capillary rise of near-surface groundwater normally is very pronounced (Fookes *et al.*, 1985). The type of soil governs the height of capillary rise, but, in clay soils, it may be a few metres (Table 7.2). Normally, however, the capillary rise in desert conditions does not exceed 2–3 m. Although the capillary fringe can be located at depth within the ground, obviously depending on the position of the water table, it can also be very near or at the ground surface. For instance, Al Sanad *et al.* (1990) referred to the occurrence of saline groundwater (chloride content, 1000–3000 mg l[-1]; sulphate content, 2000–5500 mg l[-1]) within less than 1 m of the ground surface at the coast in Kuwait. In the latter case, evidence is provided in the form of efflorescences on the ground surface due to the precipitation of salts. Salts are also precipitated in the upper layers of the soil. Among the most frequently occurring salts are calcium carbonate, gypsum, anhydrite and halite. Nonetheless, the

Table 7.2 Capillary rise, capillary pressure and suction pressure in soil.

(a) Capillary rises and pressures.

Soil	Capillary rise (mm)	Capillary pressure (kPa)
Fine gravel	Up to 100	Up to 1.0
Coarse sand	100–150	1.0–1.5
Medium sand	150–300	1.5–3.0
Fine sand	300–1000	3.0–10.0
Silt	1000–10 000	10.0–100.0
Clay	Over 10 000	Over 100.0

(b) Suction pressure and pF values.

pF value	Equivalent suction	
	(mm water)	(kN m^{-2})
0	10	0.1
1	100	1.0
2	1000	10.0
3	10 000	100.0
4	100 000	1000.0
5	1 000 000	10 000.0

Table 7.3 Typical compositions of sea water and sabkha water in mg l^{-1}. (After Fookes *et al.*, 1985.)

Ion (mg l^{-1})	Open sea	Coastal sea water, Arabian Gulf	Sabkha water
Ca	420	420	1 250
Mg	1 320	1 550	4 000
Na	10 700	20 650	30 000
K	380	650	1 300
SO_4	2 700	3 300	9 950
Cl	19 300	35 000	56 600
HCO_3	75	170	150

occurrence of salts is extremely variable from place to place.

Groundwater flow causes the dissolution of soluble phases in older sediments beneath the water table. The rate of dissolution or precipitation varies with the individual properties of the rock–soil water system, and usually is rapid enough to be of engineering significance. New layers of minerals can be formed within months, and thin layers can be dissolved just as quickly (Fookes *et al.*, 1985). The groundwater regime can be changed by construction operations, and so may lead to changes in the positions at which mineral solution or precipitation occurs. Groundwaters are high in sodium, chloride, potassium, sulphate and magnesium, and generally low in calcium, but normally are not saturated in these ions (Table 7.3).

Salt weathering leading to rock disintegration is brought about as a result of the stresses set up in the pores, joints and fissures in rock masses due to the growth of salts, the hydration of particular salts and the volumetric expansion which occurs due to the high diurnal range of temperature. The aggressive-

ness of the groundwater depends on the position of the water table and the capillary fringe above in relation to the ground surface, the chemical composition of the groundwater and the concentration of salts within it, the type of soil and the soil temperature. The pressures produced by the crystallization of salts in small pores are appreciable, for instance, gypsum ($CaSO_4.H_2O$) exerts a pressure of 100 MPa; anhydrite ($CaSO_4$), 120 MPa; kierserite ($MgSO_4.H_2O$), 100 MPa; and halite (NaCl), 200 MPa. Some salts are more effective in the breakdown process than others. For example, Goudie *et al.* (1970) showed that, from a series of experiments, Na_2SO_4, followed closely by $MgSO_4$, were much more effective in bringing about the disintegration of cubes of sandstone than other salts, and that NaCl brought about little change. Limestone also proved to be susceptible to this form of breakdown, but igneous rocks were unaffected. In addition, it appears that, when some salts, (such as $CaSO_4$ and NaCl) are combined in solution, their effectiveness as agents of breakdown is increased.

Ground surface temperatures in hot deserts may have a diurnal range exceeding 50°C. This, together with the fact that the coefficients of thermal expansion of the common salts found in desert regimes are greater than those of most common rock-forming minerals, suggests that the potential for disruptive disintegration is significant. The hydration, dehydration and rehydration of hydrous salts may occur several times throughout a year, and depend upon the temperature and relative humidity conditions, on the one hand, and the dissociation vapour pressures of the salts, on the other. In fact, Obika *et al.* (1989) indicated that the crystallization and hydration-dehydration thresholds of the more soluble salts, such as sodium chloride, sodium carbonate, sodium sulphate and magnesium sulphate, may be crossed at least once daily.

Salt weathering also attacks structures and buildings, leading to cracking, spalling and disintegration of concrete, brick and stone. The extent to which damage due to salt weathering occurs depends upon the climate, soil type and groundwater conditions, because they determine the type of salt, its concentration and mobility in the ground and the type of building materials used. In addition, faulty design or poor workmanship can increase the susceptibility of structures to salt attack by leaving cracks or hollows in which salts can accumulate (Fookes & Collis, 1975). One of the most notable forms of damage to buildings and structures is that brought about by salt weathering which is attributable to sulphate attack. For example, Robinson (1995) described damage done to concrete house slabs due to corrosion and heaving. The deterioration of the concrete arose because the capillary fringe, and so the aggressive groundwater, extended to the surface, the Portland cement reacting with the calcium sulphate in solution. Heaving was due to the precipitation of salts beneath the slabs. The doming of concrete slabs as a result of heaving leads to cracking, and hence to the concrete becoming more susceptible to corrosion. Where slabs have suffered moderate to severe damage, they may be removed, and some 0.65 m of underlying soil excavated. This is replaced with gravel over which a vapour barrier (e.g. 6 mm visqueen or polythene) is placed. The latter should be contained within a layer of sand. A new slab of reinforced concrete is then laid.

The most serious damage caused to brickwork and limestone and sandstone building stone takes place in low-lying salinas, playas and sabkhas where saline groundwater occurs at a shallow depth, giving rise to aggressive ground conditions. In such instances, the capillary rise may continue into the lower parts of buildings, giving rise to dampness, subflorescence, efflorescence, surface disintegration, spalling and the development of cavities in the building materials. Salt attack on brick foundations has sometimes led to their disintegration to a powdery material which, in turn, results in the building undergoing settlement. If this is to be avoided, the position of the upper surface of the capillary fringe in relation to the ground surface should be determined, because this will indicate whether or not the foundations of buildings and structures will be subjected to saline attack. If the potential capillary rise is above the ground surface, the buildings and structures themselves will be affected. Obviously, any such damage can be avoided by not undertaking construction in low-lying areas with aggressive ground conditions. Alternatively, foundations can be provided with a protective covering, together with damp courses to prevent capillary rise in buildings. In addition, it is advisable to use building materials which offer a high degree of resistance to salt weathering.

Salt weathering of bituminous paved roads, built over areas where saline groundwater is at or near the surface, is likely to result in notable signs of damage, such as heaving, cracking, blistering, stripping, pot-holing, doming and disintegration. For example, Blight (1976) mentioned that the surface layer can be damaged in damp soils as a result of upward capillary migration and the concentration of calcium, magnesium and sulphate salts due to evaporation from the surface. Such movements probably are encouraged in the immediate vicinity of the road by the fact that the road has a higher temperature than the surrounding ground. The physical and chemical consequences of the attack on roads mainly depend on the salinity of the groundwater, the type of aggregates used and the design of the road. The assessment of local groundwater regimes and salt profiles in relation to pavement location is necessary before construction commences in areas where the water table is high. Aggregates used for base and sub-base courses can be attacked, and this can mean that roads may undergo settlement and cracking (Fookes & French, 1977). Once cracks appear, the intensity and extent of salt damage tend to increase with time. As a result of the upward migration of capillary moisture, salts may be precipitated beneath the bituminous surface, leading to its degradation, and gradually to it being heaved to form ripples in the road, with the ripples containing tension cracks.

Wherever possible, roads should avoid areas in which saline groundwater occurs at or near the surface, and aggregate sources should be investigated for salt content prior to use, those with high salt content being rejected. As it is not always possible to avoid the construction of roads on saline ground or the use of materials containing salts, a number of precautions should be taken. Usually, pavement damage attributable to soluble salts is confined to thin, bituminous surfaces. Consequently, Blight (1976) suggested that this could be avoided by laying a surface layer of at least 30 mm of dense asphalt concrete to prevent evaporation, and thereby the migration and crystallization of salts at or near the surface. However, salts may accumulate beneath such surfaces, leading to the degradation of base and sub-base material. It appears that cutback bitumen and emulsion bitumen primers perform better than tar primers in relation to the reduction of surface degradation. This may be due to the fact that emulsion primers rest on the surface rather than penetrating the pavement layer, and thereby provide a lower permeability. Januszke and Booth (1984) suggested that evaporation, and therefore the accumulation of salts at the surface, could be avoided by placing a bituminous surface immediately after compaction. Alternatively, French et al. (1982) have suggested the placement of an impermeable membrane in the base course. A granular layer in the base course may also help to reduce the capillary rise.

If the problems associated with aggressive, salty ground are to be avoided, the first object must be to identify the limits of the hazard zone and the spatial variability of the hazard within it. Fookes and French (1977) recognized five moisture zones (Fig. 7.17a; Table 7.4), of which zones C, D and E generally can be identified and mapped (Fig. 7.17b), thereby enabling a thematic hazard map of aggressive, salty ground conditions to be produced. Such a map indicates those areas which should be avoided as far as construction is concerned or where precautionary measures must be adopted. The concentration of salts in the groundwater can allow hazard zones to be subdivided into those of high, moderate or low salinity thereby expressing the relative hazard between different areas. Cooke et al. (1982) described the production of groundwater level, electrical conductivity and saline hazard intensity maps for northern Bahrain. The position where the capillary fringe intersected the ground surface was also marked on the groundwater level map, as was the position of the 10 m contour. The latter marked the inland limit of the capillary fringe hazard. The groundwater salinity was assessed

Table 7.4 Summary of the influence of moisture zones A to E on the behaviour of roads. (After Fookes & French, 1977.)

Zone	Moisture conditions	Salt conditions	Possible damage
Moisture zone E	Transient water from rain, dew, etc.	Salts may be removed in solution and may accumulate by subsequent evaporation or by vapour transfer, etc.	Damage not serious, unless aggregate is rich in salt or road is of thin construction with unsound aggregates, in the long term
Moisture zone D	Water present by capillary moisture movement	Salts may be precipitated at all levels of road construction and in large quantities	Aggregate and bitumen may decompose, blisters may develop, small holes and cracks likely. Serious damage only in thin construction
Moisture zone C	Water present by capillary movement or ground may be saturated at times of high water table	Salts precipitated and may be redissolved	Large pot-holes develop, aggregate and bitumen decompose rapidly. Irregular surface develops. Maximum damage in thin construction
Moisture zone B	Permanently saturated zone below capillary fringe	Soil and rock properties may be changed in the long term	Damage by long-term deformation possible
Moisture zone A	Saturated zone below water	May create sabkha conditions in reclaimed ground or embankments	Damage as moisture zones E, D and C depending on elevation of construction

Note: for explanation of moisture zones, see Fig. 7.17(a).

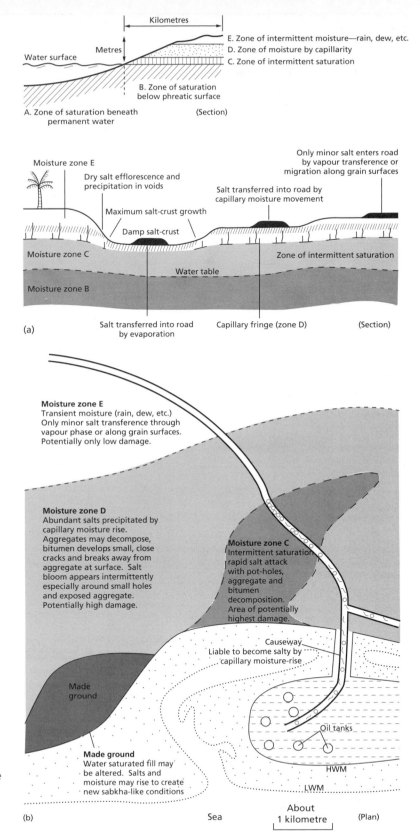

Fig. 7.17 (a) Soil moisture zones in lowlands. (After Fookes & French, 1977.) (b) Hypothetical map showing how the identification of soil moisture zones can assist in road construction and in the determination of maintenance priority areas. (After Fookes & French, 1977.)

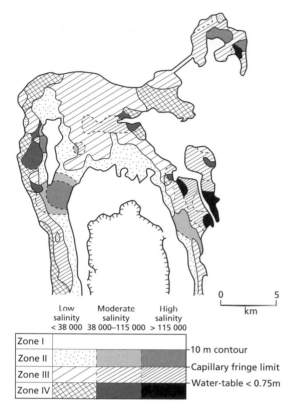

	Low salinity < 38 000	Moderate salinity 38 000–115 000	High salinity > 115 000
Zone I			
Zone II			
Zone III			
Zone IV			

— 10 m contour

— Capillary fringe limit

— Water-table < 0.75m

Fig. 7.18 Aggressive ground conditions in northern Bahrain: predicted hazard intensity. (From Cooke *et al.*, 1982.)

at the same sampling points as where the water level was determined, by means of a conductivity meter, the specific conductance being directly related to the concentration of salts present. In addition, the ionic concentrations of chloride, sulphate, sodium, potassium, calcium and magnesium in groundwater samples were determined. The areas of highest ionic concentrations corresponded to those of highest electrical conductivity. The map of hazard intensity was based on the assumption that the hazard is related to the depth of the water table, on the one hand, and the conductivity (salinity) of the groundwater, on the other (Fig. 7.18). Similar hazard maps have been produced for aggressive ground conditions encountered in other areas of the Middle East, and although, as with any other type of hazard map, they represent broad generalizations of the ground conditions, they nevertheless can be used for preliminary planning purposes.

7.9 Desertification

Desertification is a process of environmental degradation which occurs mainly in arid and semi-arid regions, and causes a reduction in the productive capacity of the soil. The prime cause of such degradation is excessive human activity and demand in these regions with fragile ecosystems. The net result is that the productivity of agricultural land declines significantly, grasslands no longer produce sufficient pasture, dry farming fails and irrigated areas may be abandoned. Deserts are encroaching into semi-arid regions largely as a consequence of poor farming practices, which include overstocking. The improper and inefficient use of water resources, and the drying up of streams, aggravates the problem still further. Excessive abstraction of water from wells lowers the water table, which adversely affects plant growth. Desertification can occur within a short time, that is in 5–10 years. Whereas droughts come and go desertification can be permanent if, in order to reverse the situation, substantial capital and resources are not available. Nonetheless, when prolonged period of drought are coupled with environmental mismanagement, they may result in the permanent degradation of the land.

Desertification brings with it associated problems such as the removal of soil, as well as a reduction in it fertility, the deposition of windblown sand and silt which can bury young plants and block irrigatio canals and rivers, and moving sand dunes. In addition, when rain does fall, a greater proportion of contributes towards run-off, so that erosion become more aggressive. This, in turn, means that the amour of sediment carried by streams and rivers increases. also means that less water infiltrates into the groun for plant growth.

Initially, desertification may go unrecognized, ar it may not be until significant changes have occurre in the fertility of the soil that it is identified. Hence, may be looked upon as a creeping disaster (Biswas Biswas, 1978). Some indicators of desertification a given in Table 7.5. It has been suggested that t world loses some 20–60 million hectares of land desertification each year, and surveys have classifi 18% of desertification as slight, 4% as modera and 28% as severe. However, as noted abov desertification is not always readily identified w accuracy. There are a number of reasons for th

Table 7.5 Forms and severities of desertification. (After Warren & Maizels, 1977.)

Form	Severity			
	Slight	Moderate	Severe	Very severe
Water erosion	Rills, shallow runnels	Soil hummocks, silt accumulations	Piping, coarse washout deposits, gullying	Rapid reservoir siltation, landslides, extensive gullying
Wind erosion	Rippled surfaces, fluting and small-scale erosion	Wind mounts, wind sheeting	Pavements	Extensive active dunes
Water and wind erosion			Scalding	Extensive scalding
Irrigated land	Crop yield reduced less than 10%	Minor white saline patches, crop yield reduced 10–50%	Extensive white saline patches, crop yield reduced more than 50%	Land unusable through excessive salinization, soils nearly impermeable, encrusted with salt
Plant cover	Excellent–good range condition	Fair range condition	Poor range condition	Virtually no vegetation

First, the deterioration of land can take numerous forms, and it may be irregular in pattern, rather than advancing as a recognizable front. In addition, the deterioration of soil affects areas of greater size than those which strictly can be regarded as desert.

The expansion of desertification cannot be estimated easily, because several years may need to elapse before it can be distinguished from an area which has been subjected to a prolonged drought. Nevertheless, it has been estimated that 14% of the world's population live in drylands which are under threat, and that over 60 million people are affected by desertification. More or less one-third of the land surface (Fig. 7.19) and one-seventh of the population of the world are affected directly, but the rest of the population has to face the indirect effects of lower agricultural production in the form of international aid. Furthermore, in many desert margin regions, there have been dramatic increases in population during the twentieth century, and many formerly nomadic peoples in Africa have become settled. Hence, the increased population pressure on soil resources has led to environmental degradation and a decline in productivity.

The loss of vegetation at the margins of deserts leads to diminishing rainfall, increasing dust content in the air and accelerates the rate of desertification.

An increased amount of dust in the atmosphere may adversely affect the radiation balance and deplete food production even further. It may also affect people with respiratory problems adversely. In the marginal and degraded areas, there will be a tendency for the number of plant and animal species to decline. For the people involved, desertification can mean a loss of income and eventual starvation which, in turn, may lead to population migrations, with severe social and economic repercussions. In other words, desertification may be viewed as the interaction between socio-economic pressures and ecosystem fragility (Nnoli, 1990).

Deserts may expand or contract naturally as precipitation varies over time. This process can be charted by various means. Dendrochronology can be used to identify seasons when moisture availability was restricted and growth reduced. Limnology can indicate the expansion and contraction of lakes over time, while archaeology can indicate changes in early settlement patterns in relation to water availability. Nevertheless, most desertification would appear to be attributable to the unwise use of agricultural land by humans (Cloudsley-Thompson, 1978).

One of the common causes of desertification is the overgrazing of animals on a limited supply of forage

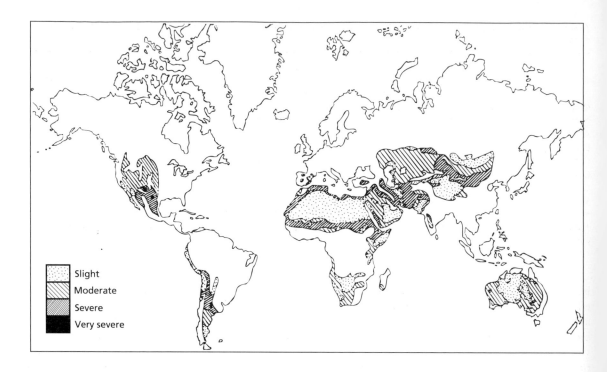

Fig. 7.19 World distribution of desertification. (After Dregne, 1983.)

and, in many cases, especially in the developing world, the complete absence of range management. Even in the western USA, overgrazing in 70% of the rangelands has led to a decline in the original forage potential by about 50%. Drought puts a severe strain on such regions. Recovery from drought may be a slow process and, in some cases, irreversible damage may be done to an ecosystem that already is undergoing serious degradation. It therefore is necessary to attempt to determine the carrying capacity of such grasslands, that is, the number of animals which they can support without the vegetation being adversely affected.

As desertification is a process of gradual degradation, it is important that it is monitored effectively, so that the changes it brings about are detected as they occur. In this context, remote sensing imagery or aerial photographs, taken at successive time intervals, can prove to be useful, in that they can record changes in the vegetation cover. Agricultural production may also be of some value in assessing degradation.

The cost of reversing desertification is very high, so that it obviously is better to prevent it in the first place. In very degraded drylands, retrenchment may be the only solution, and recovery may be out of the question in the short term. In areas which have suffered less deterioration, rotational cropping, a reduction in the numbers of livestock, the use of special equipment for ploughing and sowing, the use of specially adapted crops or the establishment of irrigation may be appropriate strategies. Over and above this, the pressure on the soil resources must be reduced which, in turn, must inevitably mean a reduction in population in the region.

7.10 Irrigation

Irrigation has been used in semi-arid and arid regions for centuries in order to increase agricultural production. At present, large-scale irrigation schemes depend upon large sources of water, which can be supplied by large reservoirs or abstracted by deep wells from groundwater sources. However, increasing demands for water, limited availability and concerns about water quality mean that water must be used effectively. Therefore, an irrigation system must

Table 7.6 Seasonal evapotranspiration and irrigation requirements for crops near Deming, New Mexico*. (After Jensen, 1973.)

Crop	Length of growing season (days)	Evapotranspiration depth (mm)	Effective rainfall depth (mm)	Evapotranspiration less rainfall (mm)	Water application efficiency (%)	Irrigation requirement depth (mm)
Alfalfa	197	915	152	763	70	1090
Beans (dry)	92	335	102	233	65	358
Corn	137	587	135	452	65	695
Cotton	197	668	152	516	65	794
Grain (spring)	112	396	33	363	65	558
Sorghum	137	549	135	414	65	637

* Average frost-free period is from April 15 to October 29. Irrigation prior to the frost-free period may be necessary for some crops.

be properly planned and designed, and operated efficiently. In order to make the best use of the water available, the seasonal water requirements of the crops grown must be known, as plant requirements vary with the season. Furthermore, weather conditions (i.e. temperature, wind, humidity) affect the water demand. In particular, estimates of evapotranspiration must be determined when planning an irrigation system. The duration of periods of insufficient rainfall in humid and subhumid regions influences the economic feasibility of irrigation. Table 7.6 illustrates the water requirements and evapotranspiration of some crops grown in New Mexico. It should be noted that the expected effective rainfall is used in determining the field irrigation requirement. Not all rainfall is effective. It is that proportion which contributes towards evapotranspiration.

Easily the most common method of applying irrigation water, particularly in arid regions, is by flooding the ground surface (Hansen et al., 1980). For example, wild flooding is where water is allowed to flow uncontrolled over the soil. Water may also be conveyed to the soil in a more regular manner via canals, ditches and furrows or basins. In the latter cases, the soil should be prepared prior to the irrigation water being applied. Efficient surface irrigation requires that the ground surface is graded in order to control the flow of water. Obviously, the extent of grading is governed by the nature of the topography. Sprinkler irrigation systems provide a fairly uniform method of applying water, and are more efficient than surface irrigation. For example, conveying water via

canals, ditches and furrows is regarded, at most, as being only 60% efficient, whereas sprinkler systems are about 75% efficient. Sprinkler irrigation does not entail the land surface being graded, and it allows the rate of water application to be controlled easily. Microirrigation or drip systems deliver water in small amounts (1–10 l h[-1]) via special porous tubes to the plant roots, and are 90% efficient. However, because of the high cost of drip irrigation, it tends to be restricted to high-value crops. Alternatively, perforated or porous tubes often are installed 0.1–0.3 m below the surface of the soil, supplying 1–5 l min[-1] per 100 m of tube. A comparison of the different types of irrigation system in relation to site and situation factors is given in Table 7.7.

Irrigation can raise the soil water content to its field capacity. In either ditch or sprinkler irrigation, the infiltration capacity and the permeability of the soil determine how fast water can be applied. The field capacity is the water content after a soil is wetted and allowed to drain for 1 or 2 days. It represents the upper limit of water available to plants and, in terms of soil suction, can be defined as a pF value of about 2. At each point where moisture menisci are in contact with soil particles, the forces of surface tension are responsible for the development of capillary or suction pressure (Table 7.2). Soil suction is a negative pressure, and indicates the height to which a column of water could rise due to such suction. Because this height or pressure may be very large, a logarithmic scale has been developed to express the relationship between soil suction and moisture

Table 7.7 Comparison of irrigation systems in relation to site and situation factors. (From Schwab *et al.*, 1993.)

Site and situation factors	Improved surface systems		Sprinkler systems			Microirrigation systems
	Redesigned surface systems	Level basins	Intermittent mechanical move	Continuous mechanical move	Solid-set and permanent	Emitters and porous tubes
Infiltration rate	Moderate to low	Moderate	All	Medium to high	All	All
Topography	Moderate slopes	Small slopes	Level to rolling	Level to rolling	Level to rolling	All
Crops	All	All	Generally shorter crops	All but trees and vineyards	All	High value required
Water supply	Large streams	Very large streams	Small streams, nearly continuous	Small streams, nearly continuous	Small streams	Small streams, continuous and clean
Water quality	All but very high salts	All	Salty water may harm plants	Salty water may harm plants	Salty water may harm plants	All—can potentially use high-salt waters
Efficiency	Average 60–70%	Average 80%	Average 70–80%	Average 80%	Average 70–80%	Average 80–90%
Labour requirement	High, training required	Low, some training	Moderate, some training	Low, some training	Low to seasonal high, little training	Low to high, some training
Capital requirement	Low to moderate	Moderate	Moderate	Moderate	High	High
Energy requirement	Low	Low	Moderate to high	Moderate to high	Moderate	Low to moderate
Management skill	Moderate	Moderate	Moderate	Moderate to high	Moderate	High
Machinery operations	Medium to long fields	Short fields	Medium field length, small interference	Some interference, circular fields	Some interference	May have considerable interference
Duration of use	Short to long	Long	Short to medium	Short to medium	Long term	Long term, but durability unknown
Weather	All	All	Poor in windy conditions	Better in windy conditions than other sprinklers	Windy conditions reduce performance, good for cooling	All
Chemical application	Fair	Good	Good	Good	Good	Very good

content. The latter is referred to as the pF value. The level at which soil moisture is no longer available to plants is termed the permanent wilting point, and this corresponds to a pF value of about 4.2. The moisture characteristic of a soil provides valuable data concerning the moisture content corresponding to the field capacity and the permanent wilting point, as well as the rate at which changes in soil suction take place with variations in moisture content. This enables an assessment to be made of the ranges of soil suction and moisture content which are likely to occur in the zone of the soil affected by seasonal changes in climate. The difference between the field capacity and the permanent wilting point is termed the available water. Plants can only remove a proportion of available water before growth and yield are affected.

The use of irrigation in semi-arid and arid regions to increase crop production can lead to the deterioration of water quality and salinity problems (Hoffman *et al.*, 1990). As far as the quality of water for irrigation is concerned, the most important factors are the total concentration of salts, the concentration of potentially toxic elements, the bicarbonate concentration as related to the concentration of calcium and magnesium, and the proportion of sodium to other ions. Most water used for irrigation contains dissolved salts, some of which remain in the soil moisture as a result of evapotranspiration. If crop production levels are to be maintained, drainage systems are required to remove excess water and associated salts from the plant root zone. Capillary action also brings salts to the surface. The capillary rise for different soil types is given in Table 7.2.

Inefficient irrigation leads to salinization of the soil, that is, the accumulation of salts near the ground surface. Dissolved nitrates from the application of fertilizers are another source of contamination. In addition, the presence of trace elements derived from the rocks in the area, in amounts greater than their threshold values, can prove to be toxic to plants. Such trace elements include arsenic, boron, cadmium, chromium, lead, molybdenum and selenium. High concentrations of bicarbonate ions can lead to the precipitation of calcium and magnesium bicarbonate from the soil water content, thereby increasing the relative proportions of sodium.

The suitability of water for irrigation depends on the effect that the salt concentration contained therein has on the plants (Table 7.8) and soil. One of the principal effects of salinity is to reduce the availability of water to plants. In semi-arid and arid regions, the presence of soluble salts in the root zone can pose a serious problem, unlike in subhumid and

Table 7.8 Relative tolerance of crops to salt concentrations expressed in terms of the specific electrical conductance. (After Ayers, 1975.)

Crop	Low salt tolerance	Medium salt tolerance	High salt tolerance
Fruit crops	Avocado Lemon Strawberry Peach Apricot Almond Plum Prune Grapefruit Orange Apple Pear	Cantaloupe Date Olive Fig Pomegranate	Date palm
Vegetable crops*	$300\,\mu S\,cm^{-1}$ Green beans Celery Radish $4000\,\mu S\,cm^{-1}$	$4000\,\mu S\,cm^{-1}$ Cucumber Squash Peas Onion Carrot Potatoes Sweetcorn Lettuce Cauliflower Bell pepper Cabbage Broccoli Tomato $10\,000\,\mu S\,cm^{-1}$	$10\,000\,\mu S\,cm^{-1}$ Spinach Asparagus Kale Garden beet $12\,000\,\mu S\,cm^{-1}$

Continued on p. 194

Table 7.8 *Continued.*

Crop	Low salt tolerance	Medium salt tolerance	High salt tolerance
Forage crops*	2000 µS cm⁻¹	4000 µS cm⁻¹	12 000 µS cm⁻¹
	Burnet	Sickle milkvetch	Bird's-foot trefoil
	Ladino clover	Sour clover	Barley (hay)
	Red clover	Cicer milkvetch	Western wheat
	Alsike clover	Tall meadow oatgrass	grass
	Meadow foxtail	Smooth brome	Canada wild rye
	White Dutch clover	Big trefoil	Rescue grass
	4000 µS cm⁻¹	Reed canary	Rhodes grass
		Meadow fescue	Bermuda grass
		Blue grama	Nuttall alkali grass
		Orchard grass	Salt grass
		Oats (hay)	Alkali sacaton
		Wheat (hay)	18 000 µS cm⁻¹
		Rye (hay)	
		Tall fescue	
		Alfalfa	
		Hubam clover	
		Sudan grass	
		Dallis grass	
		Strawberry clover	
		Mountain brome	
		Perennial rye grass	
		Yellow sweet clover	
		White sweet clover	
		12 000 µS cm⁻¹	
Field crops*	4000 µS cm⁻¹	6000 µS cm⁻¹	10 000 µS cm⁻¹
	Field beans	Castor beans	Cotton
		Sunflower	Rape
		Flax	Sugar beet
		Corn (field)	Barley (grain)
		Sorghum (grain)	16 000 µS cm⁻¹
		Rice	
		Oats (grain)	
		Wheat (grain)	
		Rye (grain)	
		10 000 µS cm⁻¹	

* Specific electrical conductance values represent salinity levels at which a 50% decrease in yield may be expected as compared with the yield on non-saline soil under comparable growing conditions. Concentrations refer to soil water. Specific electrical conductance is measured in microsiemens per centimetre (µS cm⁻¹) which is equivalent to micromhos per centimetre.

humid regions, where irrigation acts as a supplement to rainfall, and salinity usually is of little concern because rainfall tends to leach salts out of the soil. Salts may harm plant growth, in that they reduce the uptake of water either by modifying osmotic processes or by metabolic reactions, such as those caused by toxic constituents. The effects produced by salts on some soils, notably the changes brought about in the soil fabric which, in turn, affect the permeability and aeration, also influence plant growth. In other words, cations can cause deflocculation of clay minerals in a soil, which damages its crumb structure and reduces its infiltration capacity. The quality of water used for irrigation varies according to the climate, the type and drainage characteristics of the soil and the type of crop grown. Plants growing in adverse climatic conditions are susceptible to injury if poor quality water is used for irrigation. In addition, in hot, dry climates, plants abstract more moisture from the soil, and so tend to concentrate dissolved solids in the soil moisture quickly. Clayey soils may cause problems when poor quality water is used for irrigation because they are poorly drained, which means that the capacity for removing excess salts by leaching is reduced. On the other hand, if a soil is well drained, crops may be grown on it with the application of generous amounts of saline water.

In addition to the potential dangers due to high salinity, a sodium hazard sometimes exists, in that sodium in irrigation water can bring about a reduction in soil permeability and cause the soil to harden. Both effects are attributable to cation exchange of calcium and magnesium ions by sodium ions on clay minerals and colloids. The sodium content can be expressed in terms of the percentage of sodium as follows:

$$\%Na = \frac{(Na)100}{Ca + Mg + Na + K} \qquad (7.2)$$

where all ionic concentrations are expressed in milli-equivalents per litre. The extent of replacement of calcium and magnesium ions by sodium ions, that is, the amount of sodium adsorbed by a soil, can be estimated from the sodium adsorption ratio (*SAR*), which is defined as

$$SAR = \frac{Na}{\sqrt{(Ca + Mg)/2}} \qquad (7.3)$$

where, again, the concentrations are expressed in milli-equivalents per litre. The sodium hazard as defined in terms of the *SAR* is shown in Fig. 7.20.

Generally, soil with more than 0.1% of soluble salts within 0.2 m of the surface is regarded as salinized. Heavily salinized land has been abandoned to agriculture in many parts of the world (e.g. in parts of China, Pakistan and the USA). To avoid salinization, an adequate system of drainage must be installed prior to the commencement of irrigation. This lowers the water table, as well as conveying water away more quickly. Subsurface drainage can be used, together with controlled surface drainage. Excess water can be applied to fields during the non-growing season to flush salts from the soil. Usually, the quantity of salt removed by crops is so small that it does not make a significant contribution to salt removal. Water use must be managed, and planting trees as windbreaks and rotating crops, where possible, helps. The other problem which can result from inadequate drainage is waterlogging, which generally is brought about by a rising water table caused by irrigation.

Salt-affected soils have been classified as saline, sodic and saline–sodic (Schwab *et al.*, 1993). Saline soils contain enough soluble salt to interfere with the growth of most plants. They can be recognized by the presence of white crusts on the soil, and by stunted and irregular plant growth (Fig. 7.21). Sodium salts occur in relatively low concentration as compared with calcium and magnesium salts. Sodic soils are relatively low in soluble salts, but contain enough exchangeable sodium to interfere with plant growth. Unlike saline soils, which generally have a flocculated structure, sodic soils develop a dispersed structure as the amount of exchangeable sodium increases. Saline–sodic soils contain sufficient quantities of soluble salts and adsorbed sodium to reduce the yields of most plants. Both sodic and sodic–saline soils can be improved by the replacement of excessive adsorbed sodium by calcium.

The only way in which salts, which have accumulated in the soil due to irrigation, can be removed satisfactorily is by leaching. Hence, sufficient water must be applied to the soil so as to dissolve and flush out excess salts. Not only is adequate drainage required to convey away excess water, but it is also needed for water moving through the root zone. The traditional concept of leaching involves the ponding of water, so that uniform salt removal is achieved from the root zone. However, high frequency watering should be effective as far as salinity control is concerned. The salt content of the soil should be monitored to ensure that the correct amount of water is being used for irrigation.

Irrigation should be scheduled according to water availability and crop need. Provided that an adequate supply of water is available, sufficient water is applied to give optimum or maximum yield. If too much water is applied, this reduces soil aeration, heightens the water table and may cause waterlogging. It can also flush away fertilizers, thereby decreasing crop yields. On the other hand, when water supply is

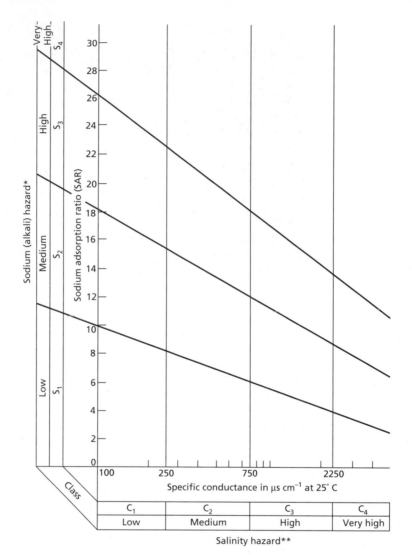

Fig. 7.20 *Description of sodium hazard. S_1: low-sodium water can be used for irrigation on almost all soils with little danger of the development of harmful levels of exchangeable sodium, however, sodium-sensitive crops, such as stonefruit trees and avocado, may accumulate injurious concentrations of sodium. S_2: medium-sodium water will present an appreciable sodium hazard in fine textured soils having a high cation exchange capacity, especially under low leaching conditions, unless gypsum is present in the soil; this water may be used on coarse textured or organic soils with good permeability. S_3: high-sodium water may produce harmful levels of exchangeable sodium in most soils and will require special soil management—good drainage, high leaching and organic matter additions; gypsiferous soils may not develop harmful levels of exchangeable sodium, but amendments may not be feasible with waters of very high salinity. S_4: very high-sodium water is generally unsatisfactory for irrigation purposes, except at low and perhaps medium salinity, where the dissolving of calcium from the soil, or the use of gypsum or other additives, may make the use of these waters feasible.
**Description of salinity hazard. C_1: low-salinity water—can be used for irrigation with most crops on most soils with little likelihood that a salinity problem will develop; some leaching is required, but this occurs under normal irrigation practices, except in soils of extremely low permeability. C_2: medium-salinity water—can be used if a moderate amount of leaching occurs; plants with moderate salt tolerance can be grown in most instances without special practices for salinity control. C_3: high-salinity water—cannot be used on soils with restricted drainage; special management for salinity control may be required and plants with good salt tolerance should be selected. C_4: very high-salinity water—is not suitable for irrigation under ordinary conditions, but may be used occasionally under very special circumstances; the soil must be permeable, drainage must be adequate, irrigation water must be applied in excess to provide considerable leaching and very salt-tolerant crops should be selected.

Fig. 7.21 Salinized land on the Huanghe River Plain, China.

limited, the irrigation scheduling strategy is one of maximizing economic return.

As the purpose of irrigation is to provide water for crops, their growth and appearance offer an indication of water need. Various methods can be used to obtain an assessment of the soil moisture content. For example, the simplest way to obtain the moisture content is to heat the soil at 105°C, after weighing, until a constant weight is achieved. The moisture content is then the loss in weight, expressed as a percentage of the dry mass. A tension meter can be used to measure the soil water tension which, in turn, can indicate the need for irrigation. A neutron probe provides a direct measurement of the soil moisture content. Water balance techniques can also be used to obtain an estimate of water in the root zone (Schwab *et al.*, 1993).

References

Abduljanwad, S.N. (1994) Swelling behaviour of calcareous clays from the Eastern Province of Saudi Arabia. *Quarterly Journal of Engineering Geology* **27**, 333–351.

Akili, W. & Torrance, J.K. (1981) The development and geotechnical problems of sabkha, with preliminary experiments on the static penetration resistance of cemented sands. *Quarterly Journal of Engineering Geology* **14**, 59–73.

Al-Sanad, H.A., Shagour, F.M., Hencher, S.R. & Lumsden, A.C. (1990) The influence of changing groundwater levels on the geotechnical behaviour of desert sands. *Quarterly Journal of Engineering Geology* **23**, 357–364.

Ayers, R.S. (1975) Quality of water for irrigation. In *Proceedings Speciality Conference*. American Society of Civil Engineers, Irrigation Drainage Division, Logan, UT, pp. 24–56.

Bagnold, R.A. (1941) *The Physics of Wind Blown Sand and Desert Dunes*. Methuen, London.

Bathurst, R.G.C. (1971) *Carbonate Sediments and their Diagenesis*. Elsevier, Amsterdam.

Biswas, M.R. & Biswas, A.R. (eds) (1978) Loss of productive soil. *International Journal of Development Studies* **12**, 189–197.

Blight, G.E. (1976) Migration of subgrade salts damages thin pavements. *Proceedings of the American Society of Civil Engineers, Transportation Engineering Journal* **102**, 779–791.

Blight, G.E. (1994) The geotechnical behaviour of arid and semi-arid zone soils—South African experience. In *Proceedings of the First International Symposium on Engineering Characteristics of Arid Soils, London*, Fookes P.G. & Parry, R.H.G. (eds). Balkema, Rotterdam, pp. 221–235.

Braithwaite, C.J.R. (1983) Calcrete and other soils in Quaternary limestones: structures, processes and applications. *Journal of the Geological Society* **140**, 351–364.

Chepil, W.S. (1945) Dynamics of wind erosion: initiation of soil movement. *Soil Science* **60**, 397–411.

Chepil, W.S. & Woodruff, N.P. (1963) The physics of sand erosion and its control. *Advances in Agronomy* **15**, 211–302.

Cloudsley-Thompson, J.L. (1978) Human activities and desert expansion. *Geographical Journal* **144**, 416–423.

Cooke, R.U., Brunsden, D., Doornkamp, J.C. & Jones, D.K.C. (1982) *Urban Geomorphology in Drylands*. Oxford University Press, Oxford.

Doornkamp, J.C., Brunsden, D., Jones, D.K.C., Cooke, R.U. & Bush, P.R. (1979) Rapid geomorphological assessments for engineering. *Quarterly Journal of Engineering Geology* 12, 189–204.

Dregne, H.E. (1983) *Desertification of Arid Lands*. Harwood, New York.

Fookes, P.G. & Collis, L. (1975) Problems in the Middle East. *Concrete* 9 (7), 12–17.

Fookes, P.G. & French, W.J. (1977) Soluble salt damage to surfaced roads in the Middle East. *Journal of the Institution of Highway Engineers* 24 (12), 10–20.

Fookes, P.G. & Gahir, J.S. (1995) Engineering performance of some coarse grained arid soils in the Libyan Fezzan. *Quarterly Journal of Engineering Geology* 28, 105–130.

Fookes, P.G., French, W.J. & Rice, S.M.M. (1985) The influence of ground and groundwater chemistry on construction in the Middle East. *Quarterly Journal of Engineering Geology* 18, 101–128.

French, W.J., Poole, A.B., Ravenscroft, P. & Khiabani, M. (1982) Results from preliminary experiments on the influence of fabrics on the migration of groundwater and water soluble minerals in the capillary fringe. *Quarterly Journal of Engineering Geology* 15, 187–199.

Gabriels, D., Maene, L.J., Leavain, J. & De Boodt, M. (1979) Possibilities of using soil conditioners for soil erosion control. In *Soil Conservation and Management in the Humid Tropics*, Greenland, D.J. & Lal, R. (eds). Wiley, Chichester, pp. 99–108.

Glennie, K.N. (1970) *Desert Sedimentary Environments*. Elsevier, Amsterdam.

Goudie, A.S. (1973) *Duricrusts in Tropical and Subtropical Latitudes*. Oxford University Press, Oxford.

Goudie, A.S. (1978) Dust storms and their geomorphological implications. *Journal of Arid Environments* 1, 291–310.

Goudie, A.S., Cooke, R.U. & Evans, I.S. (1970) Experimental investigation of rock weathering by salts. *Area* 4, 42–48.

Hansen, V.E., Israelsen, O.W. & Stringham, G.E. (1980) *Irrigation Principles and Practice*, 4th edn. Wiley, New York.

Hoffman, G.T., Howell, T.A. & Solomon, K.H. (1990) *Management of Farm Irrigation Systems. American Society of Agricultural Engineers Monograph*. American Society of Agricultural Engineers, St Joseph, MI.

Januszke, R.M. & Booth, E.H.S. (1984) Soluble salt damage to sprayed seals on the Stuart highway. In *Proceedings of the Twelfth Australian Road Research Board Conference, Hobart* 3, 18–30.

Jensen, M.E. (1973) *Consumptive Use of Water and Irrigation Water Requirements*. American Society of Civil Engineers, New York.

Jones, D.K., Cooke, R.U. & Warren, A. (1986) Geomorphological investigation, for engineering purposes, of blowing sand and dust hazard. *Quarterly Journal of Engineering Geology* 19, 251–270.

Kerr, R.C. & Nigra, J.O. (1952) Eolian sand control. *Bulletin of the American Association of Petroleum Geologists* 36, 1541–1573.

Lister, L.A. & Secrest, C.D. (1985) Giant desiccation cracks and differential surface subsidence, Red Lake playa, Mohave County, Arizona. *Bulletin of the Association of Engineering Geologists* 22, 299–314.

Meigs, P. (1953) World distribution of arid and semi-arid homoclimates. In *Arid Zone Hydrology*. UNESCO, Paris, pp. 203–209.

Netterberg, F. (1980) Geology of South African calcretes: terminology, description and classification. *Transactions of the Geological Society of South Africa* 83, 255–283.

Nnoli, O. (1990) Desertification, refugees and regional conflict in west Africa. *Disasters* 14, 132–139.

Obika, B., Freer-Hewish, R.J. & Fookes, P.G. (1989) Soluble salt damage to bituminous road and runway surfaces. *Quarterly Journal of Engineering Geology* 22, 59–73.

Robinson, D.M. (1995) Concrete corrosion and slab heaving in a sabkha environment: Long Beach–Newport Beach, California. *Environmental and Engineering Geoscience* 1, 35–40.

Schwab, G.O., Fangmeier, D.D., Elliot, W.J. & Frevert, R.K. (1993) *Soil and Water Conservation Engineering*, 4th edn. Wiley, New York.

Stipho, A.S. (1985) On the engineering properties of salina soil. *Quarterly Journal of Engineering Geology* 18, 129–137.

Stipho, A.S. (1992) Aeolian sand hazards and engineering design for desert regions. *Quarterly Journal of Engineering Geology* 25, 83–92.

Warren, A. & Maizels, J. (1977) Ecological change and desertification. In *Desertification: Its Causes and Consequences*. United Nations, Pergamon, Oxford, pp. 171–260.

Watson, A. (1985) The control of wind blown sand and moving dunes: a review of methods of sand control in deserts with observations from Saudi Arabia. *Quarterly Journal of Engineering Geology* 18, 237–252.

8 Glacial and periglacial terrains

8.1 Introduction

At the present day, glaciation is less important in shaping landscapes than it was formerly. During Pleistocene times, ice masses were much more extensive, and they have left their imprint on over 12 million square kilometres of the Earth's surface. Recent work suggests that there probably were about 20 major events during Pleistocene times, which lasted for some 2 million years. Each glacial episode produced its own effects and its own suite of deposits. Deposits of earlier glacial episodes were reshaped by later glacial advances. In this way, complex glacial sequences developed due to glaciers advancing and retreating over the same areas.

A glacier may be defined as a mass of ice which is formed from recrystallized snow and refrozen meltwater, and which moves under the influence of gravity. Glaciers develop above the snow-line, that is, in regions of the world which are cold enough to allow snow to remain on the surface throughout the year. The snow-line varies in altitude from sea level in polar regions to above 5000 m in equatorial regions.

Glaciers can be grouped into three types, namely, valley glaciers, piedmont glaciers, and ice sheets and ice caps (Flint, 1967). Valley glaciers flow down pre-existing valleys from mountains where snow has collected and formed ice. As the area of a valley glacier exposed to wastage is small compared with its volume, this accounts for the fact that glaciers penetrate into the warmer zones below the snow-line. They disappear where the rate of melting exceeds the rate of supply of ice. When a number of valley glaciers emerge from a mountain region onto a plain, where they coalesce, they form a piedmont glacier. At the present day, piedmont glaciers are found in Alaska and Antarctica. Ice sheets are huge masses of ice that extend over areas which may be of continental size;

ice caps are of smaller dimensions. At the present day, there are two ice sheets in the world, one extends over the Antarctic continent, whilst the other covers most of Greenland. Ice caps are found in mountainous areas.

Permafrost is an important characteristic, although it is not essential to the definition of periglacial conditions, the latter referring to conditions under which frost action is the predominant weathering agent. Permafrost covers 20% of the Earth's land surface and, during Pleistocene times, it was developed over an even larger area (Fig. 8.1). Ground cover, surface water, topography and surface materials all influence the distribution of permafrost. The temperature of perennially frozen ground below the depth of seasonal change ranges from slightly less than 0 °C to −12 °C. Generally, the depth of thaw is less, the higher the latitude. It is at a minimum in peat or highly organic sediments, and increases in clay, silt and sand, to a maximum in gravel, where it may extend to 2 m in depth.

8.2 Glacial erosion

Although pure ice is a comparatively ineffective agent as far as eroding massive rocks is concerned, it does acquire rock debris, which enhances its abrasive power. The larger fragments of rock embedded in the sole of a glacier tend to carve grooves in the path over which it travels, whilst the finer material smooths and polishes rock surfaces. Ice also erodes by a quarrying process, whereby fragments are plucked from rock surfaces. Generally, 'quarrying' is a more effective form of glacial erosion than abrasion.

The rate of glacial erosion is extremely variable and depends upon the velocity of the glacier, the weight of the ice, the abundance and physical character of the rock debris carried at the bottom of the glacier and the resistance offered by the rocks of the glacier

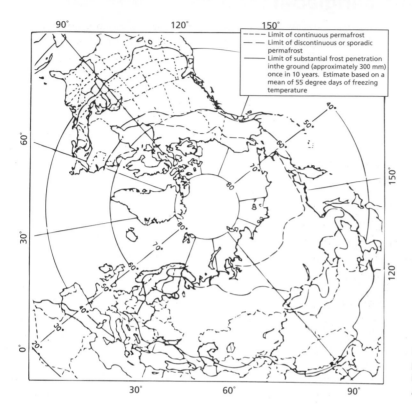

Fig. 8.1 Distribution of seasonally and perennially frozen ground (permafrost) in the northern hemisphere. (From Andersland, 1987.)

channel. The erodibility of the surface over which a glacier travels varies with depth and hence with time. Once the weathered overburden and open jointed bedrock have been removed, the rate of glacial erosion slackens. This is because 'quarrying' becomes less effective and hence the quantity of rock fragments contributed to abrasive action gradually is reduced.

In the case of continental ice sheets, these move very slowly, and may be effective agents of erosion only temporarily, removing the weathered mantle from, and smoothing off the irregularities of, a landscape. The preglacial relief features consequently are afforded some measure of protection by the overlying ice against denudation, although the surface is modified somewhat by the formation of hollows and hummocks.

The commonest features produced by glacial abrasion are striations on rock surfaces, which are formed by rock fragments embedded in the base of a glacier. These striations provide some indication of the direction in which the ice moved (Fig. 8.2).

Many glaciated slopes, formed of resistant rocks which are well jointed, display evidence of erosion in the form of ice-moulded hummocks. These hummocks, which are known as roches moutonnées, vary in size, but are usually asymmetrical in outline (Fig. 8.3). Again, these features help to enable the direction of former ice movement to be determined.

Large, highly resistant obstructions, such as volcanic plugs, which lie in the path of advancing ice, give rise to features individually called crag and tail (Fig. 8.4). The resistant obstruction forms the crag, and offers protection to the softer rocks which occur on its leeside, hence indicating the direction in which the ice moved. The latter therefore form a tail which slopes gently away from the crag. The tail may or may not possess a covering of till.

Drumlins are mounds which are rather similar in shape to the inverted bowl of a spoon (Fig. 8.5). Their long axis is orientated in the direction of ice movement, and their stoss ends (these face the direction from which the ice came) frequently are blunted. Drumlins vary in composition, ranging from 100%

bedrock to 100% glacial deposit. It has been suggested that they were developed under thick ice some distance from the snout of a glacier (Gravenor, 1953). Obviously, those types formed of bedrock originated

Fig. 8.2 Striations formed by a glacier, Crater Lake, Oregon.

as a consequence of glacial erosion, however, even those composed of glacial debris were, at least in part, moulded by glacial action. Drumlins range up to a kilometre in length and some may be over 70m in height. Usually, they do not occur singly, but in scores or even hundreds in drumlin fields. Some drumlins may be indistinctly separated from each other; they may form double or treble ridges, which are united at their stoss ends, but which possess distinct tails. The tail of one may rise from the flank of another, or small drumlins may arise from the flanks of larger ones. A hummocky drumlin landscape, with its irregular drainage, is commonly referred to as 'basket of eggs' topography.

Corries are located at the head of glaciated valleys, being the features in which ice accumulated (Fig. 8.6). Hence, they formed at or close to the snow-line. Corries frequently are arranged in tiers and, in such instances, give rise to a corrie stairway up a mountain side. Because of their shape, corries are often likened to amphitheatres, in that they are characterized by steep backwalls and steep sides. Their floors are generally rock basins. Corries vary in size, some of the largest being about 1km across. The dominant factor influencing their size is the nature of the rock in which they were excavated. Corries begin life as small nivation hollows. As erosion proceeds, they first adopt a circular outline but, when mature, their outline becomes rectangular. As corries about a mountain mass grow, the area between them is progressively

Fig. 8.3 Roches moutonnées in a valley downstream of Brinsdalsbreen Glacier, Norway.

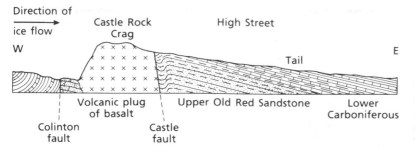

Direction of
ice flow

Castle Rock
Crag

High Street

W

E

Tail

Volcanic plug
of basalt

Upper Old Red Sandstone

Lower
Carboniferous

Colinton
fault

Castle
fault

Fig. 8.4 Crag and tail, Edinburgh, Scotland. The Castle Rock probably represents an early phase of volcanic activity associated with the ancient volcano, Arthur's Seat.

Fig. 8.5 Drumlins, Tweed Valley, Berwickshire, Scotland.

Fig. 8.6 Corrie near Athabasca Glacier, Alberta, Canada.

Fig. 8.7 Flinsch Peak, a pyramidal peak in Glacier National Park, Montana.

reduced, until they are separated from each other by sharp ridges termed arêtes. Furthermore, the headward extension of the backwalls of each corrie into the mountain side eventually produces a pyramidal peak (Fig. 8.7).

The cross-profile of a glaciated valley is typically steep sided, with a comparatively broad, flat bottom, and is commonly referred to as U-shaped (Fig. 8.8). Most glaciated valleys are straighter than those of rivers, because their spurs have been truncated by ice. In some glaciated valleys, a pronounced bench or shoulder occurs above the steep walls of the trough. Tributary streams of ice flow across the shoulders to the main glacier. When the ice disappears, the tributary valleys are left hanging above the level of the trough floor. The valleys are then occupied by streams. Those in the hanging valleys cascade down the slopes of the main trough as waterfalls. An alluvial cone may be deposited at the base of the waterfall.

Generally, glaciated valleys have a scalloped or stepped long profile, and sometimes the head of the valley is terminated by a major rock-step, known as a trough's end. Such rock-steps develop where a number of tributary glaciers, descending from corries at the head of the valley, converge and thereby effectively increase the erosive power. A simple explanation of a scalloped valley floor can be found in the character of the rock type. Not only is a glaciated valley stepped, but reversed gradients are also encountered within its path. The reversed gradients are located in rock floored basins which occur along the valley. Rock basins appear to be formed by localized ice action.

Fiords are found along the coasts of glaciated highland regions which have suffered recent submergence; they represent the drowned part of glaciated valleys (Fig. 8.9). Frequently, a terminal rock barrier, the threshold, occurs near the entrance of a fiord. Some thresholds rise very close to sea level. Indeed, some may be uncovered at low tide. However, water landwards of the threshold very often is deeper than the known postglacial rise in sea level. For example, depths in excess of 1200 m have been recorded in some Norwegian fiords. Many fiords occur along belts of structural weakness, such as areas shattered by faults or heavily dissected by joints.

8.3 Glacial deposits: tills and moraines

Glacial deposits form a more significant element of the landscape in lowland areas than in highlands. Two kinds of glacial deposit are distinguished, namely, unstratified drift or till and stratified drift (Flint, 1967). However, one type commonly grades into the other. Till usually is regarded as being synonymous with boulder clay, and is deposited directly by ice, whilst stratified drift or tillite is deposited by meltwaters issuing from the ice (Boulton, 1975).

The nature of a till deposit depends on the lithology of the material from which it was derived, on the position in which it was transported in the glacier and on the mode of deposition (Boulton & Paul, 1976). The underlying bedrock usually constitutes up to about 80% of basal tills, depending on its resistance to abrasion. Argillaceous rocks, such as shales and mudstones, are more easily abraded, and produce fine grained tills, which tend to be richer in clay minerals and therefore more plastic than other tills. The mineral composition also influences the natural moisture content, which is slightly higher in tills contain-

(a)

(b)

Fig. 8.8 (a) 'U'-shaped valley just west of Mount Edith Cavell, Alberta, Canada. (b) Stepped glaciated valley, Yosemite National Park, California.

Fig. 8.9 Geiranger Fiord, Norway.

ing appreciable quantities of clay minerals. The uppermost tills in a sequence contain a high proportion of far-travelled material, and may not contain any of the local bedrock. Till sheets can comprise one or more layers of different material, not all of which are likely to be found at any one locality. Shrinking and reconstituting of an ice sheet can complicate the sequence.

Deposits of till consist of a variable assortment of rock debris, ranging from fine rock flour to boulders. They characteristically are unsorted. The shape of the rock fragments found in till varies, but is conditioned largely by the initial shape of the fragment at the moment of incorporation into the ice. Angular boulders are common, their irregular sharp edges resulting from crushing. Crushing or grinding of a rock fragment occurs when it comes into contact with another fragment or the rock floor. The random occurrence in a till of boulders has been referred to as a raisin cake structure, however, more frequently, fragments are aligned crudely in rows. The larger elongated fragments generally possess a preferred orientation in the path of ice movement. On the one hand, tills may consist essentially of sand and gravel with very little binder; alternatively, they may have an excess of clay. Lenses and pockets of sand, gravel and highly plastic slickensided clay frequently are encountered in some tills. Some of the masses of sand and gravel are interconnected, due to the action of meltwater, but many are isolated. Small distorted pockets

of sand and silt have been termed flame structures. Most tills contain a significant amount of quartz in their silt–clay fractions.

The compactness of a till varies according to the degree of consolidation which has occurred, the amount of cementation and the size of the grains. Tills which contain less than 10% clay usually are friable, whilst those with over 10% clay tend to be massive and compact. Frequently, there is a concentration of boulders near the surface of a till. Such concentrations may have been brought about by postglacial erosion, either by sheet water action or by deflation winnowing out the finer particles, prior to the establishment of vegetation.

Distinction has been made between tills derived from rock debris carried along at the base of a glacier and those deposits which were transported within or at the terminus of the ice (Fookes *et al.*, 1975). The former sometimes is referred to as lodgement till, whilst the latter is termed ablation till. Lodgement till commonly is compact and fissile, and the fragments of rock it contains frequently are orientated in the path of ice movement. Lodgement till is thought to be plastered onto the ground beneath the moving glacier in small increments as the basal ice melts. Because of the overlying weight of the ice, such deposits are overconsolidated. Ablation till accumulates on the surface of the ice when englacial debris melts out and, as the glacier decays, the ablation till is slowly lowered to the ground. Therefore, it is normally consolidated and non-fissile, and the boulders present usually display no particular orientation. Lodgement till contains fewer, smaller stones (they generally possess a preferred orientation) than ablation till, and they are rounded and striated. Clast orientation in ablation till varies from almost random to broadly parallel to the ice flow direction.

Due to abrasion and grinding, the proportion of silt and clay sized material is relatively high in lodgement till (e.g. the clay fraction varies from 15 to 40%). Lodgement till is commonly stiff, dense and relatively incompressible. Hence it is practically impermeable. Oxidation of lodgement till takes place very slowly, so that it usually is grey.

Fissures frequently are present in lodgement till, especially if it is clay matrix dominated. Subhorizontal fissures have been developed as a result of incremental loading and periodic unloading, whilst subvertical fissures owe their formation to the overriding effects of ice and stress relief. McGown et al. (1974) noted a very definite preferred orientation of fissures in till, and that their intensity increased as the surface of the till was approached. This was attributed to the greater stresses caused by ice movement and to the effects of weathering.

Because it has not been subjected to much abrasion, ablation till is characterized by abundant large stones that are angular and not striated, the proportion of sand and gravel is high and clay is present only in small amounts (usually less than 10%). Because the texture is loose, ablation till oxidizes rapidly, and commonly is brown or yellowish brown. It may also have an extremely low *in situ* density. Because ablation till consists of the load carried at the time of ablation, it usually forms a thinner deposit than lodgement till.

McGown and Derbyshire (1977) devised an elaborate system for the classification of tills. Their classification is based upon the mode of formation, transportation and deposition of glacial material, and provides a general basis for the prediction of the engineering behaviour of tills (Fig. 8.10; Table 8.1). However, a glacial deposit can undergo considerable changes after deposition due to the influence mainly of water and wind.

A moraine is an accumulation of drift deposited directly from a glacier, which gives rise to till (Embleton & King, 1975). There are six types of moraine deposited by valley glaciers. Rock debris which a glacier wears from its valley sides, and which is sup-

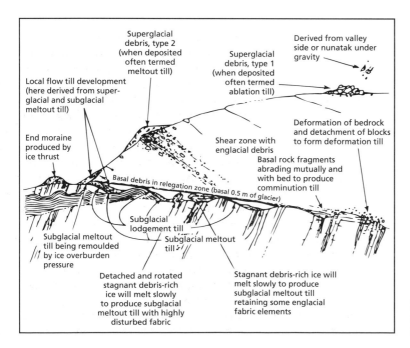

Fig. 8.10 Acquisition, transportation and deposition of tills by a glacier. (After McGown & Derbyshire, 1977.)

Table 8.1 Till types and processes. (After McGown & Derbyshire, 1977.)

Formative processes	Transportation processes	Depositional processes
Comminution till. Produced by abrasion and interaction between particles in the basal zone of a glacier. It is a common element in most tills. Pebbles show a preferred orientation and are surrounded by a compact matrix in which there is a high concentration of fines	*Superglacial till*. Derived from frost riving of adjacent rocks or by differential melting out of glacier dirt beds. It may or may not become incorporated in the glacier, and may suffer frost shattering and washing by meltwaters as it is transported on the top of the glacier	**Ablation till*. Accumulated by melting out on the surface of a glacier or as a coating on inert ice
Deformation till. Produced by plucking, thrusting, folding and brecciation of the glacier bed. Varies from mixtures of local material and erratic debris to disrupted, but little transported, masses of material derived from local bedrock. The materials are porous, and some initially contain more water than required for compaction to maximum density. Hence, they are soft tills which tend to be deformed rather than crushed	*Englacial till*. Derived from superglacial till subsequently buried by accumulating snow or entrained in shear zones. It is transported within the ice mass and is more abundant in polar regions than in temperate zones. Englacial debris occurs mainly in the lower 30–60 m of a glacier where rock detritus may comprise as much as 10 to 20% of its volume. Consequently, an appreciable thickness of englacial till can be melted out from the base of a glacier, although this suffers a reduction in volume of anything up to 90%. Such till may be precompressed by overlying ice and may be sliced by thrust planes. Elongate stones may possess preferred orientation. The average grain size is much smaller than that of superglacial till	**Meltwater till*. Accumulated as the ice of an ice–debris mixture melts out. Meltout tills exhibit a relatively low bulk density, but show some variation depending on the particle size distribution. Generally, the material is poorly sorted, although occasionally thin layers or lenses of washed sediment occur. The microfabric is normally rather open. The clast fabrics of meltout tills show a wide variation depending on the disposition and debris concentration of the melting out mass
	Basal till. Derived from comminution products in the ice–rock contact zones, particularly the lowermost regions of a glacier. It generally is transported in concentrated bands in the bottom metre or so of a glacier	*Lodgement till*. Accumulated subglacially by accretion from debris-rich basal ice. Lodgement tills generally possess a wide range of particle sizes, and frequently are anisotropic, fissured or jointed, especially when well graded or clay matrix dominated. They usually are stiff, dense, relatively incompressible soils. Macro, meso and microfabric patterns show high consistency. Subhorizontal fissuring is due to incremental lodgement and periodic unloading, whilst subvertical fissures are evidence of ice overriding and stress relief. Low-angle shear failure planes also occur
		Flow till. Consists essentially of melted out superglacial comminution debris, but also occurs due to flow of subglacial meltout tills in subglacial cavities and at the ice margin. It usually contains a wide range of particle sizes. Flow till frequently is interbedded with fluvio-glacial material. Orientation of clasts is broadly parallel to the deposition plane and imbricated upflow. The fines also reflect the flow mechanism in the parallelism of clay–silt fractions, which produces a locally dense microfabric
		Waterlain till. Accumulates on the subaqueous surface under a variety of depositional processes, and may thus show a wide variety of characters. These tills vary from rather soft lodgement tills to subaqueous mudflows and to crudely stratified lacustrine clay silts. Stratification, deformation and very diffuse to random clast fabrics are common

* This distinction between superglacial meltout and ablation tills is based on the degree of disturbance of their ice-inherited fabric, including loss of fines. The term ablation till is thus best avoided.

(a)

(b)

Fig. 8.11 (a) Athabasca Glacier, Alberta, showing lateral and terminal moraines. (b) Terminal moraine, Fox Glacier, South Island, New Zealand.

plemented by material that falls from the valley slopes above the ice, forms the lateral moraine (Fig. 8.11a). When two glaciers become confluent, a medial moraine develops from the merger of the two inner lateral moraines. Material which falls onto the surface of a glacier and then makes its way via crevasses into the centre, where it becomes entombed, is termed englacial moraine. Some of this debris, however, eventually reaches the base of the glacier and there enhances the material eroded from the valley floor. This constitutes the subglacial moraine. The ground moraine often is distributed irregularly,

because it is formed when basal ice becomes overloaded with rock debris and is forced to deposit some of it. That material which is deposited at the snout of a glacier when the rate of wastage is balanced by the rate of outward flow of ice is known as a terminal moraine (Fig. 8.11b). Terminal moraines possess a curved outline, impressed upon them by the lobate nature of the snout of the ice. They usually are discontinuous, being interrupted where streams of melt water issue from the glacier. Frequently, a series of terminal moraines may be found traversing a valley, the furthest downvalley marking the point o

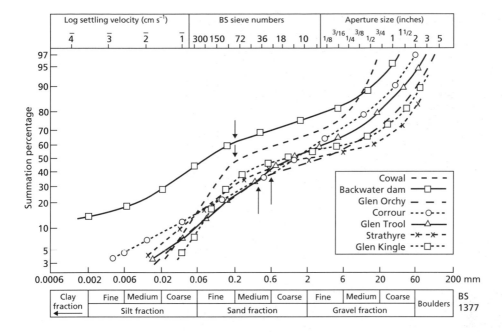

g. 8.12 Typical gradings of some Scottish morainic soils. (After cGown, 1971.)

aximum extension of the ice, and the others indicat-
g pauses in glacial retreat. Sometimes, the latter
es are called recessional moraines.

Ground moraines and terminal moraines are the
o principal types of moraine deposited by ice sheets
ich spread over lowland areas (Charlesworth,
57). In lowland areas, the terminal moraines of ice
ets may rise to a height of some 60 m. In plan, they
mmonly form a series of crescents, each crescent
responding to a lobe at the snout of the ice. If
ious amounts of meltwater drained from the ice
nt, morainic material was washed away. Hence, a
minal moraine either did not develop or, if it did,
s of inconspicuous dimensions.

Basic properties of tills

s frequently are gap graded, the gap generally
urring in the sand fraction (Fig. 8.12). Large, often
local, variations can occur in the gradings of
which reflect local variations in the formation
cesses, particularly the comminution processes.

Table 8.2 Sand, silt and clay fractions of till from north Norfolk, England. (After Bell, 1991.)

	Hunstanton Till	Marly Drift	Contorted Drift	Cromer Till
Sand (%)	34–58	15–45	26–40	38–64
Silt (%)	36–27	30–23	54–48	32–18
Clay (%)	30–15	55–32	20–12	30–18

The clast size consists principally of rock fragments
and composite grains, and was formed by frost action
and crushing by ice. Single grains predominate in the
matrix. The range of the proportions of coarse and
fine fractions in tills dictates the degree to which the
properties of the fine fraction influence the properties
of the composite soil. The variation in the engineering
properties of the fine soil fraction is greater than that
of the coarse fraction, and this often tends to dominate the behaviour of the till.

In a survey of tills in north Norfolk, England, Bell
(1991) noted that their clast fraction accounts for
less than 40% of the deposits and usually less than
20%. Accordingly, they are matrix-dominated tills,
the approximate proportions of sand, silt and clay
varying as shown in Table 8.2.

A similar situation was found in the tills of Holderness in Humberside, England, by Bell and Forster (1991). There, the fine fraction of the tills generally accounts for over 60% and frequently over 80% of the individual deposits. Hence, they similarly are matrix-dominated tills. The Basement Till is the finest, containing the largest amount of clay sized material, usually between 22 and 40%. The propor-

A North Norfolk.

		m	PL (%)	LL (%)	PI (%)	LI	CI	A
1	**Hunstanton Till (Holkham)**							
	Max	18.6	23	40 (I)	23	0.07	0.97 (S)	1.00 (N)
	Min	16.8	15	34 (L)	15	−0.19	0.89 (S)	0.75 (N)
	Mean	17.6	18	37 (I)	20	−0.02	0.92 (S)	0.85 (N)
2	**Marly Drift (Weybourne)**							
	Max	25.2	21	45 (I)	26	0.48	0.85 (S)	0.50 (I)
	Min	22.4	18	32 (L)	14	0.15	0.50 (F)	0.40 (I)
	Mean	23.6	20	37 (I)	18	0.32	0.68 (F)	0.45 (I)
3	**Contorted Drift (Trimingham)**							
	Max	18.9	18	29 (L)	13	0.33	0.86 (S)	0.80 (N)
	Min	13.2	9	19 (L)	8	0.07	0.72 (F)	0.65 (N)
	Mean	15.6	14	25 (L)	11	0.16	0.78 (S)	0.75 (N)
4	**Cromer Till (Happisburgh)**							
	Max	15.8	20	40 (I)	24	−0.16	1.16 (VS)	0.95 (N)
	Min	11.9	14	27 (L)	13	−0.18	0.98 (S)	0.65 (N)
	Mean	13.2	17	35 (I)	19	−0.17	1.09 (VS)	0.80 (N)

B Holderness.

		m	PL (%)	LL (%)	PI (%)	LI	CI	A
1	**Hessle Till (Dimlington, Hornsea)**							
	Max	26.6	26	53 (H)	32	0.072	1.147 (VS)	2.10 (A)
	Min	18.5	20	38 (I)	17	−0.019	0.794 (S)	0.06 (N)
	Mean	22.6	22	47 (I)	25	0.044	0.972 (S)	1.24 (N)
2	**Withernsea Till (Dimlington)**							
	Max	19.3	21	39 (L)	20	−0.276	1.016 (VS)	1.21 (N)
	Min	12.3	15	22 (L)	12	−0.095	0.828 (S)	0.72 (I)
	Mean	16.9	18	34 (L)	17	0.164	0.986 (S)	0.93 (N)
3	**Skipsea Till (Dimlington)**							
	Max	18.2	19	36 (I)	18	−0.294	1.288 (VS)	0.67 (I)
	Min	13.5	14	20 (L)	9	−0.044	0.978 (S)	0.51 (I)
	Mean	15.5	16	30 (L)	14	−0.188	1.108 (VS)	0.56 (I)
4	**Basement Till (Dimlington)**							
	Max	20.4	23	42 (I)	22	−0.158	1.081 (VS)	0.59 (I)
	Min	15.6	16	28 (L)	12	−0.032	0.984 (S)	0.53 (I)
	Mean	17.0	20	36 (I)	19	−0.127	1.009 (VS)	0.55 (I)

Table 8.3 Natural moisture content (*m*), plastic limit (*PL*), liquid limit (*LL*), plasticity index (*PI*), liquidity index (*LI*), consistency index (*CI*) and activity (*A*) of tills from north Norfolk and Holderness. (After Bell, 1991; Bell & Forster, 1991.)

Liquid limit: L, low plasticity, less than 35%; I, intermediate plasticity, 35–50%; H, high plasticity, 50–70%.
Consistency index: VS, very stiff, above 1; S, stiff, 0.75–1; F, firm, 0.5–0.75.
Activity: I, inactive, less than 0.75; N, normal, 0.75–1.25; A, active, over 1.25.

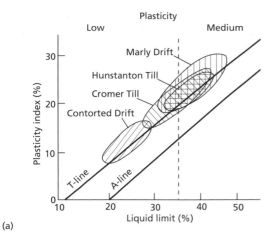

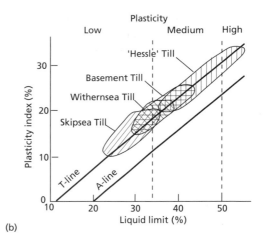

(a)

(b)

Fig. 8.13 Plasticity data for tills of (a) north Norfolk (after Bell, 1991) and (b) Holderness (after Bell & Forster, 1991).

...ion of silt tends to vary from 27 to 35% and fine ...and from 15 to 20%. The remaining proportion of ...and is always less than 15%. In the Skipsea Till, the ...ne sand fraction can range between 20 and 30%, ...hilst medium and coarse sands together do not ...ccount for more than 15% and frequently for less ...nan 10%. The silt fraction varies between 30 and ...0%, whilst that of clay constitutes between 15 and ...5%. The Withernsea Till contains between 20 and ...5% silt and between 10 and 27% clay. The fine sand ...action is between 15 and 20%, and the remaining ...nd amounts to between 12 and 18%. A wider range ...f particle size distribution occurs in the Hessle Till ...an in the other three tills. The proportion of clay ...ries from 6 to 37%, silt from 28 to 32% and fine ...nd from 17 to 21%. Medium and coarse sand may ...mprise from 9 to 18%. This wide spread of particle ...ze distribution can be attributed to the fact that the ...essle Till is the weathered product of the Skipsea ...d Withernsea Tills. Like all tills, these characteristi...lly are unsorted.

The consistency limits of tills are dependent upon ...e water content, grain size distribution and the ...operties of the fine grained fraction. Generally, ...wever, their plasticity index is small, and the liquid ...it of tills decreases with increasing grain size. The ...stic and liquid limits of the tills of north Norfolk ...d Holderness are set out in Table 8.3, from which it

Table 8.4 Typical geotechnical properties for Northumberland lodgement tills. (After Eyles & Sladen, 1981.)

Property	Weathering zones	
	I	III and IV
Bulk density (Mg m^{-3})	2.15–2.30	1.90–2.20
Natural moisture content (%)	10–15	12–25
Liquid limit (%)	25–40	35–60
Plastic limit (%)	12–20	15–25
Plasticity index	0–20	15–40
Liquidity index	−0.20 to −0.05	III −0.15 to +0.05
		IV 0 to +30
Grading of fine (<2 mm) fraction		
% clay	20–35	30–50
% silt	30–40	30–50
% sand	30–50	10–25
Average activity	0.64	0.68
c' (kPa)	0–15	0–25
ϕ' (degrees)	32–37	27–35
ϕ' (degrees)	30–32	15–32

can be seen that in the Cromer, Hunstanton, Basement, Skipsea and Withernsea Tills, the natural moisture content generally is slightly below that of the plastic limit, and hence they commonly have negative liquidity indices. The frequent positive liquidity index and the higher liquid and plastic limits of the Hessle Till, compared with the other tills of Holderness, help to confirm the view that this till is a weathered derivative of the two tills below it.

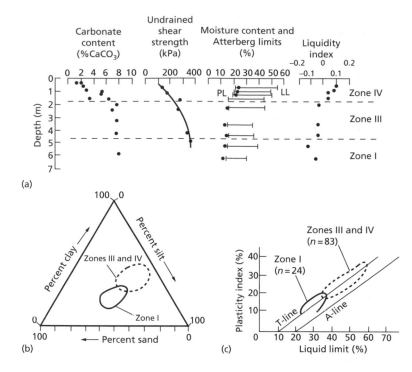

(a)

(b)

(c)

Fig. 8.14 (a) Northumberland lodgement tills: carbonate content, undrained strength, moisture content, Atterberg indices and liquidity index vs. the depth at a single representative site (Sandy Bay). (b) Particle size distribution envelopes for weathered and unweathered Northumberland lodgement tills. (c) A plasticity chart showing envelopes for weathered and unweathered Northumberland lodgement tills (n = number of determinations). (After Eyles & Sladen, 1981.)

In heavily overconsolidated, unweathered lodgement tills, the natural moisture content generally is rather low and slightly below that of the plastic limit. Furthermore, the liquidity indices of such tills typically lie within the range –0.1 to –0.35.

When the plasticity indices of these tills are plotted against their liquid limits, they all fall well above the A-line on the plasticity chart (Fig. 8.13). In fact, they tend to fall along the T-line of Boulton (1975), indicating the unsorted nature of the tills, their somewhat different locations reflecting the composition and particle size distribution of their matrix material. The position of the Hessle Till along the T-line again suggests that this is a weathered till (Sladen & Wrigley, 1983). Generally, these tills are inactive, the Hessle Till once more proving to be the exception.

In fact, Eyles and Sladen (1981) noted that, as the degree of weathering of a till increases, so does the clay fraction and moisture content. This, in turn, leads to changes in the liquid and plastic limits and in the shear strength (Table 8.4; Fig. 8.14). They recognized four zones of weathering within the soil profile of lodgement till in the coastal area of Northumberland, England (Table 8.5).

8.5 Compressibility and strength of tills

The compressibility and consolidation characteristic of tills principally are determined by the clay content. For example, the value of the compressibility index tends to increase linearly with increasing clay content whilst, for tills of very low clay content, less than 2%, this index remains about constant ($C_c = 0.01$).

Klohn (1965) noted that dense, heavily overconsolidated till is relatively incompressible and that, when loaded, it undergoes very little settlement, most of which is elastic. For the average structure, such elastic compressions are too small to consider and can therefore be ignored. However, for certain structures, these are critical, and their magnitude must be estimated prior to construction. In another survey of dense till, Radhakrishna and Klym (1974) found that the undrained shear strength ranged between 0.75 and 1.3 MPa. The average value of the initial modulus of deformation, as determined in the field from plate load tests, was 215 MPa, which was approximately twice the laboratory value. These differences between field and laboratory results were attributed to the stress relief of material on sampling and sampling disturbance.

Table 8.5 A weathering scheme for Northumberland lodgement tills. (After Eyles & Sladen, 1981.)

Weathering zone	Zone	Description	Maximum depth (m)
Highly weathered	IV	Oxidized till and surficial material Strong oxidation colours High rotten boulder content Leaching of most primary carbonate Prismatic gleyed jointing Pedological profile usually leached brown earth	3
Moderately weathered	III	Oxidized till Increased clay content Low rotten boulder content Little leaching of primary carbonate Usually dark brown or dark red brown Base commonly defined by fluvio-glacial sediments	8
Slightly weathered	II	Selective oxidation along fissure surfaces where present, otherwise as Zone I	10
Unweathered	I	Unweathered till No post-depositionally rotted boulders No oxidation No leaching of primary carbonate Usually dark grey	

Significantly lower values of shear strength were obtained by Bell (1991) and Bell and Forster (1991) for tills from north Norfolk and Holderness when tested in unconfined compression. When remoulded and tested again, all tills experienced only small losses in strength on remoulding, and therefore were of low sensitivity. These authors also subjected these tills to a series of shear box and triaxial tests, a summary of the results being provided in Table 8.6.

The presence of fissures influences the shear strength of till and therefore its stability. For example, opening of fissures sympathetically orientated to cut slopes, and softening of till along fissures as a result of weathering, frequently are responsible for small slip failures. In other words, these two factors give rise to rapid reduction of the undrained shear strength along the fissures. In fact, the undrained shear strength of fissures in till may be as little as one-sixth of that of the intact soil (McGown et al., 1977).

8.6 Fluvio-glacial deposits; stratified drift

Stratified deposits of drift are often subdivided into two categories, namely, those which develop in

contact with the ice and those which accumulate beyond the limits of the ice, forming in streams, lakes or seas. The former are referred to as ice-contact deposits, whilst the latter are termed proglacial deposits.

Most meltwater streams which deposit outwash fans do not originate at the snout of a glacier, but from within or upon the ice. Many of the streams which flow through a glacier have steep gradients, and therefore are efficient transporting agents, but when they emerge at the snout, they do so onto a shallower incline and deposition results. Outwash deposits typically are cross-bedded and range in size from coarse sand to boulders. When first deposited, the porosity of these sediments varies from 25 to 50%. They therefore are very permeable and so can resist erosion by local run-off. The finer silt–clay fraction is transported further downstream. In addition, in this direction, an increasing amount of stream alluvium is contributed by tributaries so that, eventually, the fluvio-glacial deposits cannot be distinguished from the alluvium. Most outwash masses are terraced.

Valley trains are outwash deposits which are confined within long, narrow valleys. Because deposi-

Table 8.6 Strength of tills of north Norfolk and Holderness. (After Bell, 1991; Bell & Forster, 1991.)

A North Norfolk.

	Unconfined compressive strength (kPa)			Cohesion (kPa)			Angle of friction (degrees)		
	Intact	Remoulded	Sensitivity	c_u	c'	c_r	ϕ_u	ϕ'	ϕ_r
1 Hunstanton Till (Holkham)									
Max	184	164	1.22 (L)	43	18	4	9	34	28
Min	152	128	1.18 (L)	22	8	0	3	26	21
Mean	158	134	1.19 (L)	29	12	2	5	29	23
2 Marly Drift (Weybourne)									
Max	120	94	1.49 (L)	49	16	0	3	28	25
Min	104	70	1.28 (L)	16	7	0	0	21	16
Mean	110	81	1.34 (L)	27	11	0	1	24	21
3 Contorted Drift (Trimingham)									
Max	180	168	1.67 (L)	46	20	3	10	33	25
Min	124	76	1.08 (L)	20	6	0	3	27	20
Mean	160	136	1.23 (L)	26	11	1	6	30	22
4 Cromer Till (Happisburgh)									
Max	224	188	1.19 (L)	48	19	5	6	32	29
Min	154	140	1.10 (L)	26	12	0	2	26	18
Mean	176	156	1.13 (L)	35	14	3	4	29	23

B Holderness.

	Unconfined compressive strength (kPa)			Direct shear				Triaxial			
	Intact	Remoulded	Sensitivity	c	ϕ	c_r	ϕ_r	c_u	ϕ_u	c'	ϕ'
1 Hessle Till (Dimlington, Hornsea)											
Max	138	116	1.31 (L)	30	25	3	23	98	8	80	24
Min	96	74	1.10 (L)	16	16	0	13	22	5	10	13
Mean	106	96	1.19 (L)	20	24	1	20	35	7	26	25
2 Withernsea Till (Dimlington)											
Max	172	148	1.18 (L)	38	30	2	27	62	19	42	34
Min	140	122	1.15 (L)	21	20	0	18	17	5	17	16
Mean	160	136	1.16 (L)	26	24	1	21	30	9	23	25
3 Skipsea Till (Dimlington)											
Max	194	168	1.15 (L)	45	38	5	35	50	21	25	36
Min	182	154	1.08 (L)	25	20	0	19	17	10	22	2
Mean	186	164	1.13 (L)	27	26	1	25	29	12	28	3
4 Basement Till (Dimlington)											
Max	212	168	1.27 (L)	47	34	2	30	59	17	42	3
Min	163	140	1.19 (L)	23	20	0	18	22	6	19	2
Mean	186	156	1.21 (L)	29	24	1	23	38	9	34	2

c, cohesion in kPa; ϕ, angle of friction; L, low sensitivity.

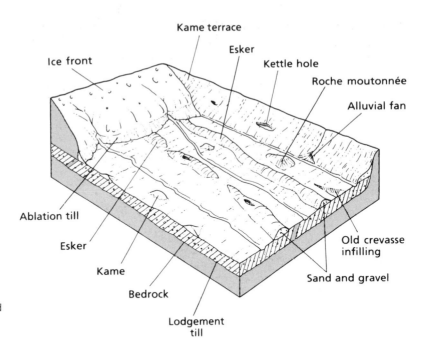

Fig. 8.15 Block diagram of a glaciated valley showing typical glacial and fluvio-glacial deposits.

tion occurs more rapidly at the centres of the valleys than at the sides, the deposits are thickest there. If outwash quickly accumulates in a valley, it may eventually dam tributary streams, so that small lakes form along the sides of the main valley. Such deposition may bury small watersheds and divert proglacial streams.

Deposition which takes place at the contact of a body of ice frequently is sporadic and irregular. Locally, the sediments possess a wide range of grain size, shape and sorting. Most are granular, and variations in their engineering properties reflect differences in particle size distribution and shape. Deposits often display abrupt changes in lithology and consequently in the density index (relative density). They characteristically are deformed, as they sag, slump or collapse as the ice supporting them melts.

Kame terraces are deposited by meltwater streams which flow along the contact between the ice and the valley side (Fig. 8.15). The drift is derived principally from the glacier, although some is supplied by tributary streams. They occur in pairs, one each side of the valley. If a series of kame terraces occurs on the valley slopes, then each pair represents a pause in the process of glacier thinning. The surfaces of these terraces often are pitted with kettle holes (these are

depressions where large blocks of ice remained unmelted whilst material accumulated around them). Narrow kame terraces usually are discontinuous, spurs having impeded deposition.

Kames are mounds of stratified drift which originate as small deltas or fans built against the snout of a glacier, where a tunnel in the ice, along which meltwater travels, emerges (Fig. 8.16). Other small ridge-like kames accumulate in crevasses in stagnant or near-stagnant ice. Many kames do not survive deglaciation for any appreciable period of time.

Eskers are long, narrow, sinuous, ridge-like masses of stratified drift which are unrelated to surface topography (Fig. 8.17). For example, eskers may climb up valley sides and cross low watersheds. They represent sediments deposited by streams which flowed within channels in a glacier. Although eskers may be interrupted, their general continuity is easily discernible and indeed some may extend lengthwise for several hundred kilometres. Eskers may reach up to 50 m in height, and they range up to 200 m in width. Their sides often are steep. Eskers are composed principally of sands and gravels, although silts and boulders are found within them. These deposits generally are cross-bedded.

Fig. 8.16 Working a kame for sand, north of Lillehammer, Norway.

Fig. 8.17 Esker in centre ground, with trees, east of Kamloops, British Columbia, Canada.

Lacustrine clays and silts often occur as layers interbedded with fluvio-glacial sand and gravel deposits, but may also occur as pockets and lenses. These interbedded deposits may reduce the vertical permeability significantly. Inclusions of till may have similar effects. Lenses of openwork gravel may mean that the permeability of a fluvio-glacial deposit is higher than expected.

Five different types of stratified drift deposited in glacial lakes have been recognized, namely terminal moraines, deltas, ice-rafted erratics, beach deposits

and bottom deposits. Terminal moraines formed in glacial lakes differ from those which arose on land in that lacustrine deposits are interstratified with drift. Glacial lake deltas usually are composed of sands and gravels which typically are cross-bedded. Large boulders on the floors of glacial lakes were transported on rafts of ice and were deposited when the ice melted. Usually, the larger the glacial lake, the larger the beach deposits developed about it. If changes in the lake level took place, these may be represented by a terraced series of beach deposits. Those sediments which accumulated on the floors of glacial lakes are fine grained, consisting of silts and clays. These fine grained sediments sometimes are composed of alternating laminae of finer and coarser grain size. Each couplet has been termed a varve, and sediments so stratified consequently are described as varved.

The most familiar proglacial deposits are varved clays. The thickness of the individual varve frequently is less than 2 mm, although much thicker layers have been noted. Generally, the coarser layer is of silt size and the finer layer is of clay size. In a survey of the varved clays of the Elk Valley, British Columbia, George (1986) noted that the clay layers invariably contained very fine silty to sandy laminations and/or lenses. According to Bell and Coulthard (1991), clay is the dominant particle size in the Tees Laminated Clay of northeast England, comprising between 44 and 76%, with an average of 61%. The average silt content is 37%, varying from 27 to 43%, and the fine sand fraction ranges up to 10%.

Taylor *et al.* (1976) showed that, in varved clays from Gale Common in Yorkshire, England, the clay minerals were well orientated around silt grains such that, at boundaries between silty partings and matrix, the clay minerals tended to show a high degree of orientation parallel to the laminae. A similar situation occurs in the Tees Laminated Clay and in the varved clay of the Elk Valley, British Columbia. In the latter, the fabric of the clay material typically possesses a preferred orientation parallel to bedding and consists of elongated clay platelets and microaggregates. Voids are present due to the edge-to-face contacts between particles.

Usually, very finely comminuted quartz, feldspar and mica form the major part of varved clays rather than clay minerals. For example, the clay mineral content may be as low as 10%, although instances where it has been as high as 70% have been recorded. Moreover, montmorillonitic clay has also been found in varved clays. Bell and Coulthard (1991) showed that the bulk mineralogical composition of the Tees Laminated Clay was dominated by illite and kaolinite, with lesser amounts of chlorite. The quartz content of the bulk specimens ranged from 4 to 26%, reflecting the differing thicknesses of the silt–fine sand layers, and the carbonate content was always less than 15%. Traces of potash feldspar, muscovite and expandable mixed-layer clays were also recorded. Similarly, George (1986) found that kaolinite and

illite were the dominant minerals in the clay material of the varved clay of the Elk Valley, British Columbia, with minor amounts of mixed-layer clays. Calcite, dolomite and quartz were also present in the clay fraction.

Varved clays tend to be normally consolidated or lightly overconsolidated, although it usually is difficult to make the distinction (Bell & Coulthard, 1997). In many cases, the precompression may have been due to ice loading. However, Saxena *et al.* (1978) reported that the upper part of the varved clay of Hackensack Valley, New Jersey, was highly overconsolidated. This they attributed to the effects of fluctuating water levels or to desiccation.

The two normally discrete layers formed during the deposition of the varve present an unusual problem in that they may invalidate the normal soil mechanics analyses based on homogeneous soils. As far as the Atterberg limits are concerned, the assessment of the liquid and plastic limits of a bulk sample may not yield a representative result. However, Metcalf and Townsend (1961) suggested that the maximum possible liquid limit obtained for any particular varved deposit must be that of the clayey portion, whereas the minimum value must be that of the silty portion. Hence, they assumed that the maximum and minimum values recorded for any one deposit approximate to the properties of the individual layers. The range of the liquid limit for varved clays tends to vary between 30 and 80%, whilst that of the plastic limit often varies between 15 and 30%. These limits, obtained from varved clays in Ontario, allow the material to be classified as inorganic, silty clay of medium to high plasticity. In some varved clays in Ontario, the natural moisture content would appear to be near the liquid limit. They consequently are soft, and have sensitivities generally of the order of four. Because triaxial and unconfined compression tests tend to give very low strains at failure, around 3%, Metcalf and Townsend presumed that this indicated a structural effect in the varved clays. The average strength reported was about 40 kPa, with a range of 24–49 kPa. The effective stress parameters of apparent cohesion and angle of shearing resistance ranged from 5 to 19 kPa and 22° to 25°, respectively.

Some of the geotechnical properties of the varved clays of the Elk Valley, British Columbia, are given in Table 8.7. The failure envelopes produced by consolidated, undrained, triaxial tests possessed both over-

Table 8.7 Some properties of the varved clays of the Elk Valley, British Columbia. (After George, 1986.)

(a) Basic soil properties.

Varved clay description (typical)	Yellowish brown Typical thicknesses clay: 33–37 mm silt: 2–8 mm Usually fine, silty to sandy laminations and lenses (<0.5 mm thick) in clay layers, 2–24 mm apart	
Soil properties	Clay	Bulk
USCS classification	CL	CL
Natural moisture content (%)	41.0	35.4
Liquid limit (%)	39.7	34.3
Plastic limit (%)	24.2	22.4
Plasticity index (%)	15.5	11.9
Liquidity index	1.08	1.09
% Clay (<0.002 mm)	43	—
Activity	0.36	—

(b) Results of oedometer tests.

Parameter	Value
Compression index	0.405–0.587* (0.496)†
Recompression index	0.030–0.060 (0.045)
Effective preconsolidation pressure (kPa)	
Schmertmann	250–330 (290)
Casagrande	350–400 (375)
Initial void ratio	1.09–1.14 (1.02)
Unit weight (kPa)	17.46–18.23 (17.85)
Natural water content (%)	37.9–41.0 (39.5)

CL, clay of low plasticity; USCS, Unified Soil Classification System.
* Observed range for two tests.
† Average value.

consolidated and normally consolidated segments, which gave values of the cohesion and angle of friction as $c' = 53.2$ kPa, $\phi' = 4.7°$ and $c' = 0$ kPa, $\phi' = 14.8°$, respectively. The overconsolidated effect was attributed to cementation bonding. This is supported by the values of the preconsolidation pressure obtained (Table 8.7). These values, according to George (1986), are significantly greater than any overburden stress the site has undergone throughout its geological history.

Turning to the Tees Laminated Clay, Bell and Coulthard (1991, 1997) showed that it displays a wide range of plastic and liquid limit values (Table 8.8). The plastic limits usually are less than the natural moisture content, and so the liquidity indices normally are positive. From the plasticity chart (Fig. 8.18), it can be seen that these clays range from low to very high plasticity. The wide spread reflects the different proportions of clay and silt in the specimens tested; obviously, the higher the content of the latter, the lower the plasticity. However, most of the results show that the Tees Laminated Clay is of high plasticity. Indeed, the average value of the liquid limit is 56%. All the specimens tested fell above the A-line and straddled the T-line (Boulton, 1975). In fact, the distribution of the results for this clay bears a close similarity to that of other laminated clays in northern England (Taylor et al., 1976) and Norfolk (Kazi & Knill, 1969).

Quick, undrained, triaxial tests carried out on the Tees Laminated Clay indicated that it tended to fail at less than 10% strain (Bell & Coulthard, 1991). In some of those specimens in which the test was continued beyond failure, the stress, after attaining a peak, tended to fall away quickly. In others, however, this was not the case (Fig. 8.19). The average shear strength of the Tees Laminated Clay tends to be about

Table 8.8 Index properties of Tees Laminated Clay. (After Bell & Coulthard, 1991.)

	Moisture content (%)	Plastic limit (%)	Liquid* limit (%)	Plasticity index	Liquidity index	Linear shrinkage (%)	Consistency index†
Max	35	31	78 (VH)	49	0.35	14	0.89 (S)
Min	22	18	29 (L)	16	−0.12	9	0.42 (SO)
Mean	30	26	56 (H)	33	0.15	11	0.70 (F)

* L, low plasticity (<35%); I, intermediate plasticity (35–50%); H, high plasticity (50–75%); VH, very high plasticity (70–90%).
† SO, soft (<0.5); F, firm (0.5–0.75); S, stiff (0.75–1.0).

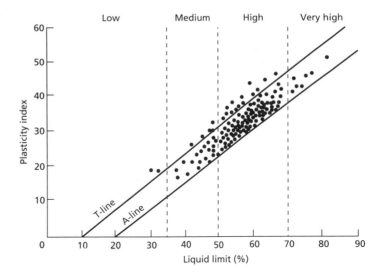

Fig. 8.18 Plasticity chart showing the distribution of Tees Laminated Clay. (After Bell & Coulthard, 1991.)

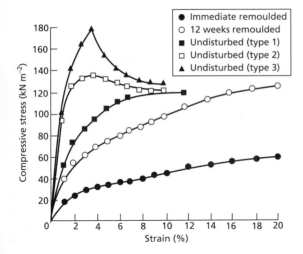

Fig. 8.19 Examples of stress–strain curves of undisturbed and remoulded Tees Laminated Clay tested in quick, undrained, triaxial conditions. (After Bell & Coulthard, 1991.)

60 kPa, although there is a wide variation—from the upper 20s to 102 kPa. The residual strength is between 20 and 30 kPa lower than the peak strength. Some specimens of laminated clay were remoulded and retested to determine their sensitivity. This ranged between 3.5 and 4.2, with an average value of 4.0. Hence, these clays are medium sensitive to sensitive and undergo a notable reduction in strength when remoulded. However, the remoulded strength increases with time (Fig. 8.19).

8.7 Other glacial effects

Ice sheets have caused diversions of drainage in areas of low relief. In some areas, which were completely covered with glacial deposits, the postglacial drainage pattern may bear no relationship to the surface beneath the drift; indeed, moraines and eskers may form minor water divides (Charlesworth, 1957). As would be expected, notable changes occurred at or near the margin of the ice. There, lakes were formed, which were drained by streams whose paths disregarded the preglacial relief. Evidence of the existence of proglacial lakes is to be found in the lacustrine deposits, strandlines and overflow channels which they left behind.

Where valley glaciers extend below the snow-line, they frequently pond back streams that flow down the valley sides, and thereby give rise to lakes. If any col between two valleys is lower than the surface of the glacier occupying one of them, the water from an adjacent lake dammed by this glacier eventually spills into the adjoining valley and, in so doing, erodes an overflow channel. Marginal spillways may develop along the side of a valley at the contact with the ice. Lakes of glacial origin form significant features of the postglacial landscape. They were formed by the scouring action of ice, by being dammed by morainic debris or in depressions in the surface of glacial drift.

The enormous weight of an overlying ice sheet causes the Earth's crust beneath it to sag. Once the ice sheet disappears, the land slowly rises to recover its

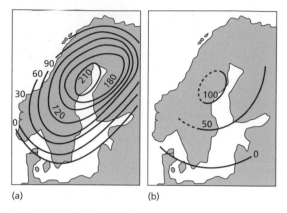

Fig. 8.20 (a) Total elevation of Scandinavia, in metres, during the last 9700 years. (b) Present rate of rise in millimetres per 10 years.

former position, and thereby restores isostatic equilibrium. Consequently, the areas of northern Europe and North America presently affected by isostatic uplift more or less correspond to those areas which were formerly covered with ice. At the present day, the rate of isostatic recovery in, for example, the centre of Scandinavia is approximately 1 m per century (Fig. 8.20). Evidence concerning crustal recovery is demonstrated by river and lake terraces, and raised beaches. Isostatic uplift is neither regular nor continuous. Therefore, the rise in the land surface has at times been overtaken by a rise in sea level. The latter was caused by meltwater from the retreating ice sheets. With the advance and retreat of ice sheets in Pleistocene times, the level of the sea fluctuated. Marine terraces (strandlines) were produced during interglacial periods when the sea was at a much higher level. The postglacial rise in sea level has produced drowned coastlines, such as rias and fiords, young developing cliff lines, aggraded lower stretches of river valleys, buried channels, submerged forests, marshlands, shelf seas, straits and the re-formation of numerous islands. Buried channels represent abandoned erosional features occupied by coarse stream bedload deposits or, in the case of channels formed by glacial erosion, by coarse granular deposits derived from subglacial drainage or a preglacial river.

8.8 Glacial hazards

Although the potential hazards of glaciers may be appreciable, their impact on humans is not significant, because less than 0.1% of the world's glaciers occur in inhabited areas. Reynolds (1992) divided glacial hazards into two groups, that is, those which are a direct action of ice or snow, such as avalanches, and those which give rise to indirect hazards. The latter include glacier outbursts and flooding. Reynolds also mentioned that the cumulative annual cost in some areas affected by glacial hazards can be substantial, and quoted as an example the flooding of the Reuss Valley in Switzerland in August 1987. This was the consequence of the deforestation of the hillslopes, which led to an increase in the rate of avalanching. The resulting damage cost some 400 million dollars. The time involved in the different types of glacial hazard varies significantly, as can be seen from Table 8.9.

The rapid movement of masses of snow or ice down slopes as avalanches can pose a serious hazard in many mountain areas. For example, avalanches, particularly when they contain notable amounts of debris, can damage buildings and routeways, and may lead to the loss of life. The path of an avalanche consists of three parts, namely, the starting area, the track and the run-out–deposition zone. Reynolds (1992) recognized two types of avalanche based on the type of initiation mechanism and the pattern of failure in the starting zone. The first type, the loose snow avalanche, generally occurs in the cohesionless surface layers of dry snow or wet snow containing water. In many instances, a rotational failure occurs when the angle of repose is exceeded, that is, the weight of the snow mass exceeds the frictional resistance of the surface on which it lies. Wet or slush avalanches do not need steep slopes in the areas of initiation as do dry snow avalanches; the critical angle of repose can be as low as 15°. Slab avalanches represent the second type, and these take place when a slab of cohesive snow fails. These tend to be the more dangerous type. Dry avalanches commonly travel along a straight path, as compared with wet snow avalanches, which tend to hug the contours and become channelled in small valleys. Ice avalanches are released by frontal block failure, ice slab failure and ice bedrock failure from a mass of ice.

Avalanche location often can be predicted from historical evidence relating to previous avalanches combined with topographical data. As a consequence, hazard maps of avalanche-prone areas can be produced. Avalanche forecasting is related to weather conditions and the monitoring of snow.

Table 8.9 Types of snow and ice hazard. (After Reynolds, 1992.)

Time-scale	Hazard	Description
Minutes	Avalanche	Slide or fall of large mass of snow, ice and/or rock
Hours	Glacier outburst	Catastrophic discharge of water under pressure from a glacier
	Jökulhlaup	Outburst which may be associated with volcanic activity (Icelandic)
	Débâcle	Outburst, but from a proglacial lake (French)
	Aluvión	Catastrophic flood of liquid mud, irrespective of its cause, generally transporting large boulders (Spanish; plural; aluviones)
Days to weeks	Flood	Areal coverage mostly by water
Months to years	Glacier surge	Rapid increase in rate of glacier flow
Years to decades	Glacier fluctuations	Variations in ice front positions due to climatic changes, etc.

Glacier floods result from the sudden release of water which is impounded in, on, under or adjacent to a glacier. Glacier outburst and jökulhlaup both refer to a rapid discharge of water, which is under pressure, from within a glacier. Water pressure may build up within a glacier to a point where it exceeds the strength of the ice, with the result that the ice is ruptured. In this way, water drains from water-pockets, so that the subglacial drainage is increased and may be released as a frontal wave many metres in height. Progressive enlargement of internal drainage channels also leads to an increase in discharge. In both cases, the quantity of discharge declines after the initial surge. Water which is dammed by ice eventually may cause the ice to become detached from the rock mass on which it rests. The ice is then buoyed up, allowing the water to drain from the ice-dammed lake. Discharges of several thousand cubic metres per second have occurred from such lakes, and can cause severe damage to any settlements located downstream. Most of these outbursts occur during the summer months. Nevertheless, it can be difficult to predict the location and timing of the release of water, so that the floods are unexpected. The resulting floods from these outbursts often carry huge quantities of debris. Débâcles are rapid discharges of water from proglacial moraine-dammed lakes, whilst aluvións involve the rapid discharge of liquid mud which may contain large boulders. Such releases of large quanti-ties of water on occasions have led to thousands of people being killed.

Glacier fluctuations, in which glaciers either advance or retreat, occur in response to climatic changes (Tufnell, 1984). The advance of a glacier into a valley can lead to the river which occupies the valley being dammed, with land being inundated as a lake forms. Hence, land use and routeways may be affected. Rapid changes in the position of the snout of a glacier are referred to as surges. Mayo (1989) described the blocking of Russell Fiord in Alaska by a lobe of the Hubbard Glacier in May 1986. In October of that year, the ice dam failed, releasing 380 000 000 m^3 in an hour.

8.9 Frozen ground phenomena in periglacial environments

Frozen ground phenomena are found in regions which experience a tundra climate, that is, in those regions where the winter temperatures rarely rise above freezing point and the summer temperatures are only warm enough to cause thawing in the upper metre or so of the soil (Derbyshire, 1977). Beneath the upper or active zone, the subsoil is permanently frozen and so is known as the permafrost. The upper surface of perennially frozen ground is termed the permafrost level or table. Continuous permafrost generally has a mean annual temperature of less than

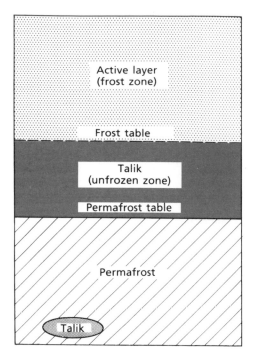

Fig. 8.21 Terminology of some features associated with permafrost.

−5 °C at a depth of 10–15 m, with the permafrost table being some 0.6 m below the ground surface. In granular material, this may be up to 1.8 m in depth. Discontinuous permafrost is interrupted by thawed areas and is thinner, with a lower permafrost table. Because of the permafrost layer, summer meltwater cannot seep into the ground; the active zone then becomes waterlogged and the soils on gentle slopes are liable to flow. Layers or lenses of unfrozen ground, termed taliks, may occur, often temporarily, in the permafrost (Fig. 8.21). Occasionally taliks may be formed of viscous liquid and as such can flow.

Cold, long winters with little snow, short, dry and relatively cool summers and low precipitation in all seasons favour the development of permafrost. Other factors which influence the distribution of permafrost include the topography, vegetation and ground cover, hydrology and snowfall, and the type of ground conditions (Harris, 1986). Permafrost is thinner or may be absent beneath rivers and lakes, which are not frozen completely in winter. This is especially the case in areas of discontinuous permafrost.

Frost action in a soil is, of course, not restricted to tundra regions. Its occurrence is influenced by the initial temperature of the soil, as well as the air temperature, the intensity and duration of the freezing period, the depth of frost penetration, the depth of the water table, and the type of ground and ground cover. If frost penetrates down to the capillary fringe in fine grained soils, especially silts, then, under certain conditions, lenses of ice may be developed. The formation of such ice lenses may, in turn, cause frost heave and frost boil, which may lead to the break-up of roads, the failure of slopes, etc.

Ice may occur in frozen soil as small disseminated crystals whose total mass exceeds that of the mineral grains, as large tabular masses which range up to several metres thick or as ice wedges. The last may be several metres wide and may extend to 10 m or so in depth. As a consequence, frozen soils need to be described and classified for engineering purposes. A recent method of classifying frozen soils involves the identification of the soil type and the character of the ice (Andersland & Anderson, 1978). First, the character of the actual soil is classified according to the Unified Soil Classification System. Second, the soil characteristics consequent upon freezing are added to the description. The frozen soil characteristics are divided into two basic groups on the basis of whether or not segregated ice can be seen with the naked eye (Table 8.10). Third, the ice present in the frozen soil is classified; this refers to inclusions of ice which exceed 25 mm in thickness.

The amount of segregated ice in a frozen mass of soil depends largely upon the intensity and rate of freezing. When freezing takes place quickly, no layers of ice are visible, whereas slow freezing produces visible layers of ice of various thicknesses. Ice segregation in soil also takes place under cyclic freezing and thawing conditions.

The presence of masses of ice in a soil means that, as far as its behaviour is concerned, the properties of both must be taken into account. Ice has no long-term strength, that is, it flows under very small loads. If a constant load is applied to a specimen of ice, instantaneous elastic deformation occurs. This is followed by creep, which eventually develops a steady state. Instantaneous elastic recovery takes place on removal of the load, followed by recovery of the transient creep.

The mechanical properties of frozen soil are very

Table 8.10 Description and classification of frozen soils. (From Andersland & Anderson, 1978.)

I Description of soil phase (independent of frozen state)	Classify soil phase by the Unified Soil Classification System			
	Major group		Subgroup	
	Description	Designation	Description	Designation
			Poorly bonded or friable	NF
	Segregated ice not visible by eye	N	Well bonded — No excess ice	Nb n
			Well bonded — Excess ice	Nb e
II Description of frozen soil			Individual ice crystals or inclusions	Vx
			Ice coatings on particles	Ve
	Segregated ice visible by eye (ice 25 mm or less thick)	V	Random or irregularly oriented ice formations	Vr
			Stratified or distinctly oriented ice formations	Vs
III Description of substantial ice strata	Ice greater than 25 mm thick	ICE	Ice with soil inclusions	ICE + soil type
			Ice without soil inclusions	ICE

much influenced by the grain size distribution, mineral content, density, frozen and unfrozen water contents and presence of ice lenses and layers. The strength of frozen ground develops from cohesion, interparticle friction and particle interlocking, much the same as in unfrozen soils. However, cohesive forces include the adhesion between soil particles and ice in the voids, as well as the surface forces between particles. More particularly, the strength of frozen soils is sensitive to the particle size distribution, particle orientation and packing, impurities (air bubbles, salts or organic matter) in the water–ice matrix, temperature, confining pressure and rate of strain. Obviously, the difference in the strength between frozen and unfrozen soils is derived from the ice component.

The relative density influences the behaviour of frozen granular soils, especially their shearing resistance, in a manner similar to that when they are unfrozen. The cohesive effects of the ice matrix are superimposed on the latter behaviour, and the initial deformation of frozen sand is dominated by the ice matrix. Sand in which all the water is more or less frozen exhibits a brittle type of failure at low strains, for example, at around 2% strain. However, the presence of unfrozen films of water around particles of soil not only means that the ice content is reduced, but leads to a more plastic behaviour of the soil during deformation. For instance, frozen clay, as well as often containing a lower content of ice than sand, has layers of unfrozen water (of molecular proportions) around the clay particles. These molecular layers of water contribute towards a plastic type of failure.

Lenses of ice frequently are formed in fine grained soils frozen under a directional temperature gradient. The lenses impart a laminated appearance to the soil. In such situations, the strength of the bond between the soil particles and ice matrix is greater than that between the particles and adjacent ice lenses. Under very rapid loading, the ice behaves as a brittle material, with strengths in excess of those of fine grained frozen soils. By contrast, the ice matrix deforms continuously when subjected to long-term loading, with no limiting long-term strength. The laminated texture of the soil in rapid shear possesses the greatest strength when the shear zone runs along the contact between the ice lenses and the frozen soil.

When loaded, stresses at the point of contact between soil particles and ice bring about pressure melting of the ice. Because of differences in the surface tension of the meltwater, it tends to move into regions of lower stress, where it refreezes. The processes of ice melting and the movement of unfrozen water are accompanied by a breakdown of the ice and the bonding with the grains of soil. This leads to plastic deformation of the ice in the voids and to a rearrangement of particle fabric. The net result is a time-dependent deformation of the frozen soil, namely, creep. Frozen soil undergoes appreciable deformation under sustained loading, the magnitude and rate of creep being governed by the composition of the soil, especially the amount of ice present, the temperature, the stress and the stress history.

The creep strength of frozen soils is defined as the stress level, after a given time, at which rupture, instability leading to rupture or extremely large defor-

mations without rupture occur. Frozen, fine grained soils can suffer extremely large deformations without rupturing at temperatures close to freezing point. Hence, the strength of these soils must be defined in terms of the maximum deformation which a particular structure can tolerate. As far as laboratory testing is concerned, axial strains of 20%, under compressive loading, frequently are arbitrarily considered as amounting to failure. The creep strength is then defined as the level of stress producing this strain after a given interval of time.

In fine grained sediments, the intimate bond between the water and the clay particles results in a significant proportion of soil moisture remaining unfrozen at temperatures as low as $-25\,°C$. The more clay material that is in the soil, the greater the quantity of unfrozen moisture. Nonetheless, there is a dramatic increase in structural strength with decreasing temperature. Indeed, it appears to increase exponentially with the relative proportion of moisture frozen. Taking silty clay as an example, the amount of moisture frozen at $-18\,°C$ is only 1.25 times that frozen at $-5\,°C$, but the increase in compressive strength is more than fourfold.

By contrast, the water content of granular soils is almost wholly converted into ice at a few degrees Celsius below freezing point. Hence, frozen granular soils exhibit a reasonably high compressive strength only a few degrees Celsius below freezing. The order of increase in the compressive strength with decreasing temperature is shown in Fig. 8.22.

Because frozen ground is more or less impermeable, this increases the problems due to thaw, by impeding the removal of surface water. Moreover, when thaw occurs, the amount of water liberated may greatly exceed that originally present in the melted out layer of the soil (see below). As the soil thaws downwards, the upper layers become saturated and, because water cannot drain through the frozen soil beneath, the soil may suffer a complete loss of strength. Indeed, under some circumstances, excess water may act as a transporting agent, thereby giving rise to soil flows which can move on gentle slopes, that is, slopes with gradients as low as 2°. This movement downslope as a viscous flow of saturated debris is referred to as solifluction. The movement generally is extremely slow, most measurements indicating rates between 10 and 300 mm per year. Nonetheless, solifluction is probably the most significant process of

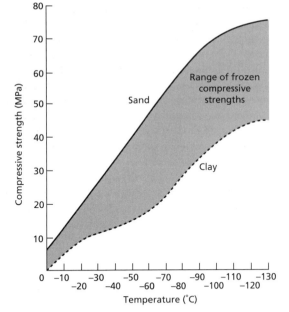

Fig. 8.22 Increase in compressive strength with decreasing temperature.

mass wastage in tundra regions. Solifluction deposits commonly consist of gravels, which are characteristically poorly sorted, sometimes gap graded and poorly bedded. These gravels consist of fresh, poorly worn, locally derived material. Individual deposits rarely are more than 3 m thick and frequently display flow structures. Sheets and lobes of solifluction debris, transported by mudflow activity, commonly are found at the foot of slopes. These materials may be reactivated by changes in drainage, stream erosion, sediment overloading or during construction operations. Solifluction sheets may be underlain by slip surfaces, the residual strength of which controls their stability. Such material in England commonly is referred to as head. Head deposits are usually of late Pleistocene age and have not been consolidated by ice loading. They tend to be relatively permeable, weak and compressible materials. However, their strength is affected by the proportion of rock fragments they contain. Those deposits derived from calcareous rocks may be weakly cemented by calcium carbonate.

Settlement is associated with the thawing of frozen ground. As ice melts, settlement occurs, water being squeezed from the ground by overburden pressure or by any applied loads. Excess pore water pressures

develop when the rate of ice melt is greater than the discharge capacity of the soil. Because excess pore water pressures can lead to the failure of slopes and foundations, both the rate and amount of thaw settlement should be determined. Pore water pressures should also be monitored.

Further consolidation, due to drainage, may occur on thawing. If the soil was previously in a relatively dense state, the amount of consolidation is small. This situation only occurs in coarse grained frozen soils containing very little segregated ice. On the other hand, some degree of segregation of ice is always present in fine grained frozen soils. For example, lenses and veins of ice may be formed when silts have access to capillary water. Under such conditions, the moisture content of the frozen silts significantly exceeds the moisture content present in their unfrozen state. As a result, when such ice-rich soils thaw under drained conditions, they undergo large settlements under their own weight.

Thermokarst features are formed by the degradation of ground ice in permafrost areas. They include pits, ground-ice slumps and lakes, that is, subsidence features which are formed when supersaturated icy soils or ice lenses located near the top of the permafrost thaw. Seasonal thaw can lead to thermokarst development. The degradation of permafrost may take place laterally, for example, by the horizontal erosion by water as lakes are extended.

Prolonged freezing gives rise to shattering in the frozen layer, fracturing taking place along joints and cracks. Frost shattering, due to ice action in Pleistocene times, has been found to extend to significant depths. For instance, Higginbottom and Fookes (1970) mentioned depths of up to 30 m in the Chalk of southern England. In this way, the rock concerned suffers a reduction in bulk density and an increase in deformability and permeability. Fretting and spalling are particularly rapid where the rock is closely fractured. Frost shattering may be concentrated along certain preferred planes, if joint patterns are suitably orientated. Preferential opening takes place most frequently in those joints which run more or less parallel with the ground surface. Silt and clay frequently occupy the cracks in frost-shattered ground, down to appreciable depths, having been deposited by meltwater. Their presence may cause stability problems. If the material possesses a certain range of grain size and permeability, the freezing of intergranular water causes expansion and disruption of previously intact material, and frost shattering may be very pronounced.

Knill (1968) described flat-lying discontinuities within a zone of more general shattering. These discontinuities ran more or less subparallel to the bedrock surface and possessed an anastomosing curvi-planar form. They enclosed lenticular units of frost-shattered rock. Knill noted that such discontinuities commonly are associated with slightly cleaved rocks of Lower Palaeozoic age, such as those of the Borrowdale Volcanic Series in Cumbria, England, and that they have only been reported from glaciated areas. They are best developed where the strike of the cleavage runs at an appreciable angle to the ground surface, and in flat terrain. Silt and clay infill usually is present in the discontinuities. Knill maintained that these discontinuities are shear surfaces developed initially by glacial drag. Such shear surfaces can penetrate to depths of 30 m. Frost heave under subsequent periglacial conditions is thought to have caused the general frost shattering and fracture enlargement.

Stress relief following the disappearance of ice on melting may cause the enlargement of joints. This may aid failure on those slopes which were oversteepened by glaciation.

Stone polygons are common frozen ground phenomena and fossil forms are found in Pleistocene strata (Fig. 8.23). They consist of marginal rings of stone which embrace mounds of finer material. Their diameters range up to 12 m. The stones are raised gradually by frost heaving through the active zone to the surface. Once lifted to the surface, the stones are moved slowly to the peripheries of individual mounds where they help to construct polygons. The polygons

are regular in pattern on slopes up to 2°, but become elongated on slopes of 3–7°, and form alternating stripes of stones and finer debris when the gradient exceeds 7°. Ice wedges frequently surround polygons. Indeed, their expansion may be responsible for producing the mound-like appearance of polygons. In the process of being moved upwards through the ground, stones can be rotated so that their long axes are vertical (Harris, 1986). This upward movement can damage pipelines, culverts and foundations.

Shrinkage, which gives rise to polygonal cracking in the ground, presents another problem when soil is subjected to freezing. The formation of these cracks is attributable to thermal contraction and desiccation. Water which accumulates in the cracks is frozen and consequently helps to increase their size. This water may also aid the development of lenses of ice. Individual cracks may be 1.2 m wide at their top, may penetrate to depths of 10 m and may be up to 12 m apart. They form when, because of exceptionally low temperatures, shrinkage of the ground occurs. Ice wedges occupy these cracks and cause them to expand. When the ice in an ice wedge disappears, an ice wedge pseudomorph is formed by sediment, frequently sand, filling the crack. In addition, frozen soils may undergo notable disturbance as a result of mutual interference of growing bodies of ice or from excess pore water pressures developed in confined water-bearing lenses. Involutions are plugs, pockets or tongues of highly disturbed material, generally possessing inferior geotechnical properties, which have been intruded into overlying layers. They are formed as a result of hydrostatic uplift in water trapped under a refreezing surface layer. They usually are confined to the active layer. Ice wedges, pseudomorphs and involutions usually mean that one material suddenly replaces another. This can cause problems in shallow excavations (Morgan, 1971).

The segregation of lenticular masses of ground ice produces uplift and fissuring in the surface layers, and leads to the formation of ice mounds or hummocks. These measure several metres across and a few metres thick. Poor drainage facilitates the development of ice mounds, and they frequently are associated with peaty areas. On the other hand, solifluction leads to wastage of ice mounds. Consequently, water-filled depressions are formed where the ice melts. These depressions eventually are occupied by peat, silt and clay.

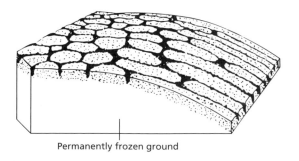

Permanently frozen ground

Fig. 8.23 Diagram illustrating the merging of stone polygons on a flat surface into stone stripes on a slope.

Pingos or hydrolaccoliths are found in Canada, Greenland and Siberia. They are large mounds, being anything up to 300 m in diameter and 50 m in height, which have been upheaved by intrusions of ground ice. Such cone-like features usually are deeply fissured and a crater may occur at their summit. Boreholes sunk into these craters have penetrated several metres of soil and then drilled into ice.

Periglacial action accelerates hill creep, the latter being particularly well developed on thinly bedded or cleaved rocks. Creep material may give way to solifluction deposits on approaching the surface. These deposits consist mainly of flat rock fragments orientated parallel with the hillside and are interrupted by numerous shallow slips.

Oversteepening of glaciated valleys and meltwater channels occurs when ground is stabilized by deep permafrost or supported by ice masses. Frost sapping at the bottom of scarp features also causes oversteepening. When the support disappears, the oversteepened slopes become potentially unstable. Meltwater in a shattered rock mass gives rise to an increase in pore water pressure which, in turn, leads to movement or instability along bedding planes and joints. An increase in the moisture content of cohesive material brings about a reduction in its strength, and may cause it to swell, thereby aggravating the instability, due to oversteepening, in the near-surface zone. As a result, landsliding on a large scale is associated with such oversteepened slopes.

The solubility of carbon dioxide in water varies inversely with temperature; for example, it is 1.7 times greater at 0 °C than at 15 °C. Accordingly, cold meltwaters frequently have had a strong leaching effect on calcareous rocks. Some pipes and swallow holes in chalk in England may have been produced by such meltwaters. The problem of pipes and swallow holes in chalk is aggravated by their localized character and the frequent absence of surface evidence. They often are undetected by a conventional site investigation.

The development of ice means that expansive forces are set up, which not only shatter frozen material, but cause ground heave. In England, chalk proved especially susceptible to frost heave during Pleistocene times. For example, chalk frequently possesses a surface zone of very closely jointed material in which the fractures tend to be crazed, with curviplanar surfaces typical of frost shattering. Continued

Fig. 8.24 Frost-shattered chalk, near Weybourne, Norfolk, England.

frost churning frequently led to the obliteration of the macrostructure in the upper metre or so of chalk, and to the formation of a mass of pasty remoulded chalk (putty chalk), enclosing angular fragments of unaltered material, which increase in size with depth as the undisturbed parent chalk is approached (Fig. 8.24).

8.10 Construction in permafrost regions

There are two methods of construction in permafrost, namely, passive and active. In the former, the frozen ground is not disturbed, and heat from a structure is prevented from thawing the ground below, thereby reducing its stability. Certainly, settlement of notable proportions can occur if the permafrost is melted (Thomson, 1980). The prevention of heat flow to permafrost can be accomplished by providing an air space beneath the structure. Placing an insulating layer, such as a gravel blanket, between the structure and the frozen soil delays, but does not stop, thawing. By contrast, the ground is thawed prior to construction in the active method. It is either kept thawed or removed and replaced by materials not affected by frost action. The latter method is used where permafrost is thin, sporadic or discontinuous, and where thawed ground has an acceptable bearing capacity. On the other hand, if permafrost is well developed, the removal of frozen ground probably will prove to be impracticable, and hence the passive method is employed.

Certain conditions are preferable for the location of buildings, such as a thin active layer with bedrock near the surface, good drainage, non-frost-susceptible soil, soil which, when it thaws, has an adequate bearing capacity and areas not liable to solifluction (Harris, 1986). Foundations frequently are taken through the active layer into the permafrost beneath. Hence, piles often are used as foundations (Johnson, 1981). The refreezing time for piles in permafrost depends upon the time of emplacement, the ground temperature and the soil moisture. For example, it may take 2–3 months for piles placed in early spring to refreeze and therefore to be loaded, whereas those sunk in autumn may take 6 months. The adfreezing of the ground in the active layer against the piles must be prevented. This can be achieved by insulating, collaring or lubricating the piles in the active layer. In this way, the upward-acting force during freezing does not affect the piles. The piles usually extend to a depth of at least twice the active layer into the permafrost. Floors commonly are raised above the ground to allow the circulation of air.

As far as the construction of roads and runways is concerned, the object is to provide a stable surface across frequently unstable ground conditions. This entails either the use of insulating layers or the replacement of surface material with non-frost-susceptible material. Water must be prevented from accumulating beneath a road or runway by the provision of drainage ditches.

8.11 Frost heave

The following factors are necessary for the occurrence of frost heave, namely, capillary saturation at the beginning and during the freezing of the soil, a plentiful supply of subsoil water and a soil possessing fairly high capillarity, together with moderate permeability. According to Kinosita (1979), the ground surface experiences an increasingly larger amount of heave, the higher the initial water table. Indeed, it has been suggested that frost heave could be prevented by lowering the water table (Andersland & Anderson, 1978).

Grain size is another important factor influencing frost heave. For example, gravels, sands and clays are not particularly susceptible to heave, whilst silts definitely are. The reason for this is that silty soils are associated with high capillary rises but, at the same time, their voids are sufficiently large to allow moisture to move quickly enough for them to become saturated rapidly. If ice lenses are present in clean gravels or sands, they simply represent small pockets of moisture which have been frozen. Indeed, Taber (1930) gave an upper size limit of 0.007 mm, above which, he maintained, layers of ice do not develop. However, Casagrande (1932) suggested that the particle size critical to heave formation was 0.02 mm. If the quantity of such particles in a soil is less than 1%, no heave is expected, but considerable heaving may take place if this amount is over 3% in non-uniform soils and over 10% in very uniform soils. This 0.02 mm criterion has been used by the US Army Corps of Engineers (Anon., 1965), together with data from frost heave tests, to develop a frost susceptibility system which is outlined in Fig. 8.25. Figure 8.25 shows that soil groups exhibit a range of susceptibilities, reflecting variations in particle size distribution, density and mineralogy.

Croney and Jacobs (1967) suggested that, under the climatic conditions experienced in Britain, well-drained cohesive soils with a plasticity index exceeding 15% could be looked upon as being non-frost susceptible. They suggested that, where the drainage is poor and the water table is within 0.6 m of the formation level of a road, the limiting value of the plasticity index should be increased to 20%. In addition, in experiments with sand, they noted that, as the amount of silt added was increased up to 55% or the clay fraction up to 33%, the decrease in permeability at the freezing front was the overriding factor, and heave tended to increase. Beyond these values, the decreasing permeability below the freezing zone became dominant and progressively reduced the heave. This indicates that the permeability below the frozen zone principally is responsible for controlling heave.

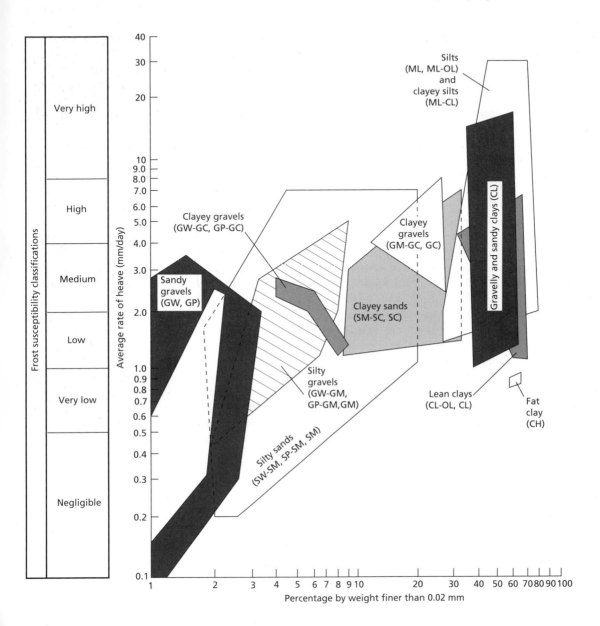

Fig. 8.25 Range of the degree of frost susceptibility of soils according to the US Army Corps of Engineers. (From Anon., 1965.)

Horiguchi (1979) demonstrated, from experimental evidence, that the rate of frost heave increased as the rate of heat removal from the freezing front was increased. However, removal of heat does not increase the rate of heave indefinitely; it reaches a maximum, after which it declines. The maximum rate of heave was shown to be influenced by the particle size distribution, in that it increased in soils with a finer grain size. Horiguchi also found that, when the particles were the same size in different test specimens, the maximum rate of heave depended upon the types of exchangeable cations present in the soil. The rate of heave is also influenced by the thickness of the overburden. For instance, Penner and Walton (1979) indicated that the maximum rate of ice accumulation at lower overburden pressures occurred at temperatures closer to 0 °C than those for ice accumulation at higher overburden pressures. However, it appears that the rate of heave for various overburden pressures tends to converge as the temperature below freezing is lowered. As the overburden pressure increases, the zone over which heaving takes place becomes greater, and it extends over an increasingly larger range of temperature.

Maximum heaving does not necessarily occur at the time of maximum depth of penetration of the 0 °C line, there being a lag between the minimum air temperature prevailing and the maximum penetration of the freezing front. In fact, soil freezes at temperatures slightly lower than 0 °C.

As heaves amounting to 30% of the thickness of the frozen layer have frequently been recorded, moisture, other than that initially present in the frozen layer, must be drawn from below, as water increases in volume by only 9% when frozen. In fact, when a soil freezes, there is an upward transfer of heat from the groundwater towards the area in which freezing is occurring. The thermal energy, in turn, initiates an upward migration of moisture within the soil. The moisture in the soil can be translocated upwards either in the vapour or liquid phase or by a combination of both.

Before freezing, soil particles develop films of moisture about them due to capillary action. This moisture is drawn from the water table. As the ice lens grows, the suction pressure it develops exceeds that of the capillary attraction of moisture by the soil particles. Hence, moisture moves from the soil to the ice lens. However, the capillary force continues to draw moisture from the water table and so the process continues.

Jones (1980) suggested that, if heaving is unrestrained, the heave (H) can be estimated as follows:

$$H = 1.09\, kit \qquad (8.1)$$

where k is the permeability, i is the suction gradient (this is difficult to derive) and t is the time.

In a discussion of frost heaving, Reed et al. (1979) noted that predictions failed to take account of the fact that soils can exist at different states of density and therefore porosity, yet have the same grain size distribution. Moreover, the pore size distribution controls the migration of water in the soil and hence, to a large degree, the mechanism of frost heave. They accordingly derived expressions, based upon pore space, to predict the amount of frost heave.

Where there is a likelihood of frost heave occurring, it is necessary to estimate the depth of frost penetration. Once this has been done, provision can be made for the installation of adequate insulation or drainage within the soil, and the amount by which the water table may need to be lowered can be determined so that it is not affected by frost penetration. The base of footings should be placed below the estimated depth of frost penetration, as should water supply lines and other services. Frost-susceptible soils may be replaced by gravels. The addition of certain chemicals to soil can reduce its capacity for water absorption and so can influence its frost susceptibility. For example, Croney and Jacobs (1967) noted that the addition of calcium lignosulphate and sodium tripolyphosphate to silty soils was effective in reducing frost heave. The freezing point of the soil may be lowered by mixing in solutions of calcium chloride or sodium chloride, at concentrations of 0.5–3.0% by weight of the soil mixture. The heave of non-cohesive soils containing appreciable quantities of fines can be reduced or prevented by the addition of cement or bituminous binders. The addition of cement both reduces the permeability of a soil mass and gives it sufficient tensile strength to prevent the formation of small ice lenses as the freezing isotherm passes through.

References

Andersland, O.B. (1987) Frozen ground engineering. In Ground Engineer's Reference Book, Bell, F.G. (ed). Butterworths, London, 8/1–8/24.

Andersland, O.B. & Anderson, C.M. (eds) (1978) Geotechnical Engineering for Cold Regions. McGraw–Hill, New York.

Anon. (1965) Soils and Geology — Pavement Design for Frost Conditions. Technical Manual TM 5-818-2. US Corps of Engineers, Department of Army, Washington DC.

Bell, F.G. (1991) A survey of the geotechnical properties of some till deposits occurring on the north coast of Norfolk. In Quaternary Engineering Geology, Special Publication No. 7, Forster, A., Culshaw, M.G., Cripps, J.C., Little J.I. & Moon, C.F. (eds). Geological Society, London, pp 103–110.

Bell, F.G. & Coulthard, J.M. (1991) The Tees Laminated

Clay. In *Quaternary Engineering Geology, Engineering Geology, Special Publication No. 7*, Forster, A., Culshaw, M.G., Cripps, J.C., Little, J.L. & Moon, C.F. (eds). Geological Society, London, pp. 111–180.

Bell, F.G. & Coulthard, J.M. (1997) A survey of some geotechnical properties of the Tees Laminated Clay of the central Middlesbrough area, north east England. *Engineering Geology* 48, 117–133.

Bell, F.G. & Forster, A. (1991) The till deposits of Holderness. In *Quaternary Engineering Geology, Engineering Geology, Special Publication No. 7*, Forster, A., Culshaw, M.G., Cripps, J.C., Little, J.L. & Moon, C.F. (eds). Geological Society, London, pp. 339–348.

Boulton, G.S. (1975) The development of geotechnical properties in glacial tills. In *Glacial Till—An Interdisciplinary Study*, Legget, R.F. (ed). *Special Publication No. 12*. Royal Society of Canada, Ottawa, pp. 292–303.

Boulton, G.S. & Paul, M.A. (1976) The influence of genetic processes on some geotechnical properties of glacial tills. *Quarterly Journal of Engineering Geology* 9, 159–194.

Casagrande, A. (1932) Discussion on frost heaving. In *Proceedings Highway Research Board, Bulletin No. 12*, 169.

Charlesworth, J.K. (1957) *The Quaternary Era*. Arnold, London.

Croney, D. & Jacobs, J.C. (1967) *The Frost Susceptibility of Soils and Road Materials. Transport and Road Research Laboratory, Report LR 90*. Transport and Road Research Laboratory, Crowthorne, Berkshire.

Derbyshire, E. (1977) Periglacial environments. In *Applied Geomorphology*, Hails, J.R. (ed). Elsevier, Amsterdam, pp. 227–276.

Embleton, C. & King, C.A.M. (1975) *Periglacial Geomorphology*. Arnold, London.

Eyles, N. & Sladen, J.A. (1981) Stratigraphy and geotechnical properties of weathered lodgement till in Northumberland, England. *Quarterly Journal of Engineering Geology* 14, 129–142.

Flint, R.F. (1967) *Glacial and Pleistocene Geology*. Wiley, New York.

Fookes, P.G., Gordon, D.L. & Higginbottom, I.E. (1975) Glacial landforms, their deposits and engineering characteristics. In *The Engineering Behaviour of Glacial Materials, Proceedings Symposium*. Midland Soil Mechanics and Foundation Engineering Society, Birmingham University, pp. 18–51.

George, H. (1986) Characteristics of varved clays of the Elk Valley, British Columbia, Canada. *Engineering Geology* 23, 59–74.

Gravenor, C. (1953) The origin of drumlins. *American Journal of Science* 251, 624–681.

Harris, S.A. (1986) *The Periglacial Environment*. Croom Helm, London.

Higginbottom, I. & Fookes, P.G. (1970) Engineering aspects of periglacial features in Britain. *Quarterly Journal of Engineering Geology* 3, 85–117.

Horiguchi, K. (1979) Effect of rate of heat removal on the rate of frost heaving. *Engineering Geology, Special Issue on Ground Freezing* 13, 63–72.

Johnson, G.H. (ed). (1981) *Permafrost, Engineering Design and Construction*. Wiley, Toronto.

Jones, R.H. (1980) Frost heave on roads. *Quarterly Journal of Engineering Geology*. 12, 77–86.

Kazi, A. & Knill, J.L. (1969) The sedimentation and geotechnical properties of the Cromer Till between Happisburgh and Cromer, Norfolk. *Quarterly Journal of Engineering Geology* 2, 63–86.

Kinosita, S. (1979) Effects of initial soil water conditions on frost heaving characteristics. *Engineering Geology, Special Issue on Ground Freezing* 13, 53–62.

Klohn, E.J. (1965) The elastic properties of a dense glacial till deposit. *Canadian Geotechnical Journal* 2, 115–128.

Knill, J.L. (1968) Geotechnical significance of some glacially induced rock discontinuities. *Bulletin of the Association of Engineering Geologists* 5, 49–61.

Mayo, L.R. (1989) Advance of Hubbard Glacier and 1986 outburst of Russell Fiord, Alaska, USA. *Annals of Glaciology* 13, 189–194.

McGown, A. (1971) The classification for engineering purposes of tills from moraines and associated landforms. *Quarterly Journal of Engineering Geology* 4, 115–130.

McGown, A. & Derbyshire, E. (1977) Genetic influences on the properties of tills. *Quarterly Journal of Engineering Geology* 10, 389–410.

McGown, A., Saldivar-Sali, A. & Radwan, A.M. (1974) Fissure patterns and slope failures in till at Hurlford, Ayrshire. *Quarterly Journal of Engineering Geology* 7, 1–26.

McGown, A., Radwan, A.M. & Gabr, A.W.A. (1977) Laboratory testing of fissured and laminated soils. In *Proceedings of the Ninth International Conference on Soil Mechanics Foundation Engineering, Tokyo* 1, 205–210.

Metcalf, J.B. & Townsend, D.L. (1961) A preliminary study of the geotechnical properties of varved clays as reported in Canadian case records. In *Proceedings of the Fourteenth Canadian Conference on Soil Mechanics, Section 13*, 203–225.

Morgan, A.V. (1971) Engineering problems caused by fossil permafrost features in the English Midlands. *Quarterly Journal of Engineering Geology* 4, 111–114.

Penner, E. & Walton, T. (1979) Effects of temperature and pressure on frost heaving. *Engineering Geology, Special Issue on Ground Freezing* 13, 29–40.

Radhakrishna, H.S. & Klym, T.W. (1974) Geotechnical properties of a very dense glacial till. *Canadian Geotechnical Journal* 11, 396–408.

Reed, M.A., Lovell, C.W., Altschaeffe, A.G. & Wood, L.E. (1979) Frost heaving rate predicted from pore size distribution. *Canadian Geotechnical Journal* 16, 463–472.

Reynolds, J.M. (1992) The identification and mitigation of

glacier related hazards: examples from the Cordillera Blanca, Peru. In *Geohazards, Natural and Man-made*, McCall, G.J.H., Laming, D.J.C. & Scott, S.C. (eds). Chapman and Hall, London, pp. 143–157.

Saxena, S.K., Helberg, J. & Ladd, C.C. (1978) Geotechnical properties of Hackensack Valley varved clays of New Jersey. *Geotechnical Testing* **1** (3), 148–161.

Sladen, J.A. & Wrigley, W. (1983) Geotechnical properties of lodgement till—a review. In *Glacial Geology: An Introduction for Engineers and Earth Scientists*, Eyles, N. (ed). Pergamon Press, Oxford, pp. 184–212.

Taber, S. (1930) The mechanics of frost heaving. *Journal of Geology* **38**, 303–317.

Taylor, R.K., Barton, R., Mitchell, J.E. & Cobb, A.E. (1976) The engineering geology of Devensian deposits underlying P.F.A. lagoons at Gale Common, Yorkshire. *Quarterly Journal of Engineering Geology* **9**, 195–218.

Thomson, S. (1980) A brief review of foundation construction in the western Canadian Arctic. *Quarterly Journal of Engineering Geology* **13**, 67–76.

Tufnell, L. (1984) *Glacial Hazards*. Longman, Harlow, Essex.

9 Water resources

9.1 The hydrological cycle

The hydrological cycle involves the movement of water in all its forms over, on and through the Earth (Fig. 9.1). The cycle can be visualized as starting with the evaporation of water from the oceans, and the subsequent transport of the resultant water vapour by winds and moving air masses. Some water vapour condenses over land and falls back to the surface of the Earth as precipitation. To complete the cycle, this precipitation must then make its way back to the oceans via streams, rivers or underground flow, although some precipitation may be evapotranspired and may describe several subcycles before completing its journey. Groundwater forms an integral part of the hydrological cycle. Most groundwater recharge is the result of precipitation. Less significant, although important locally, is the direct contribution from inland waters (including rivers) and from adjacent, commonly less permeable, water-bearing strata by leakage. The cycle can be regarded as a series of storage components, with water moving slowly from one to another until one circuit has been completed. Table 9.1 shows the estimated amount of water available within the various storage components, and the total quantity that would be available if it could all be released from storage. Only 0.5% of the total water resources of the world are in the form of groundwater. Not all this is available for exploitation, because about half is below 800 m, and therefore is too deep for economic utilization. However, the capacity of the underground resource should not be underestimated; about 98% of the usable fresh water of the Earth is stored underground.

Run-off is made up of two basic components: surface water run-off and groundwater discharge. The former usually is the more important and is responsible for the major variations in river flow. Run-off generally increases in magnitude as the time

from the beginning of precipitation increases. This can be represented on a hydrograph as shown in Fig. 5.9.

Infiltration refers to the process whereby water penetrates the ground surface and starts to move down through the zone of aeration. The subsequent gravitational movement of the water down to the zone of saturation is termed percolation, although there is no clearly defined point at which infiltration becomes percolation. Whether infiltration or run-off is the dominant process at a particular time depends upon several factors, such as the intensity of the rainfall and the porosity and permeability of the surface. For example, if the rainfall intensity is much greater than the infiltration capacity of the soil, run-off is high. On the other hand, if the rainfall is fairly gentle, infiltration may be increased at the expense of run-off.

If lower strata are less permeable than the surface layer, the infiltration capacity is reduced, so that some of the water that has penetrated the surface moves parallel to the water table and is called interflow. The water that becomes interflow will probably be discharged to a river channel at some point, and will form part of the base-flow of the river. The remaining water may continue down through the zone of aeration until it reaches the water table and becomes groundwater recharge. This can be a slow process (typically about 1 m per year), because the percolating water may temporarily become suspended in the zone of aeration as a result of the various dynamic forces that operate in this region.

Although infiltration may be high in a dry soil, the fact that the soil is dry means that the water is more likely to be held in the surface layers of the soil, and either evaporated, or transpired by plants, and therefore is less likely to reach the water table. Consequently, most groundwater recharge takes place when the ground is comparatively wet and evapotranspiration is relatively insignificant. When evapo-

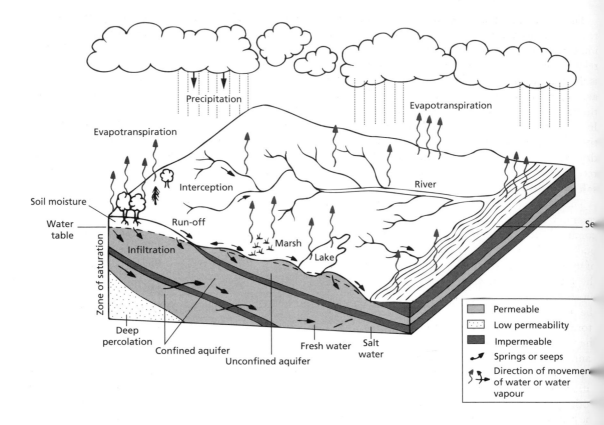

Fig. 9.1 The hydrological cycle.

Table 9.1 The water inventory of the Earth. (From Nace, 1969.)

Storage component	Volume of water (10^{12} m³)	Total water (%)	
Oceans	1 350 400	97.6	
Saline lakes and inland seas	105	0.008	
Ice caps and glaciers	26 000	1.9	
Soil moisture	150	0.01	
Groundwater	7 000	0.5	
Fresh water lakes	125	0.009	Usable fresh
Rivers	2	0.0001	water = 51%
Atmosphere	13	0.009	
Total	1 384 000	100	

All figures are approximate estimates and rounded.

transpiration exceeds precipitation, and vegetation has to draw on reserves of water in the soil to satisfy transpiration requirements, soil moisture deficits (*SMD*s) occur. The soil moisture deficit at any time is the difference between the moisture remaining and the field capacity of the soil, which is the amount of water retained in the soil by capillary forces after all excess water has been drained from it. Before this can happen, any soil moisture deficits must be made up so that the soil is returned to its field capacity. Once this has been achieved, any additional or excess water may become groundwater recharge.

9.2 Reservoirs

Although most reservoirs today are multipurpose, their principal function, no matter what their size, is to stabilize the flow of water, in order to satisfy a varying demand from consumers or to regulate water supplied to a river course. In other words, water is stored at times of excess flow to conserve it for later release at times of low flow.

The most important physical characteristic of a reservoir is its storage capacity. Probably the most important aspect of storage in reservoir design is the relationship between the capacity and yield. The yield is the quantity of water which a reservoir can supply at any given time. The maximum possible yield equals the mean inflow less evaporation and seepage loss. In any consideration of the yield, the maximum quantity of water which can be supplied during a critical dry period (i.e. during the lowest natural flow on record) is of prime importance and is defined as the safe yield.

The maximum elevation to which the water in a reservoir basin will rise during ordinary operating conditions is referred to as the top water or normal pool level. For most reservoirs, this is fixed by the top of the spillway. Conversely, the minimum pool level is the lowest elevation to which the water is drawn under normal conditions, this being determined by the lowest outlet. Between these two levels, the storage volume is termed the useful storage, whilst the water below the minimum pool level, because it cannot be drawn upon, is the dead storage.

In any adjustment of a river regime to the new conditions imposed by a reservoir, problems may emerge both upstream and downstream. Deposition around the head of a reservoir may cause serious aggradation upstream, resulting in a reduced capacity of the stream channels to contain flow. Hence, flooding becomes more frequent and the water table rises. The removal of sediment from the outflow of a reservoir can lead to erosion in the river regime downstream of the dam, with consequent acceleration of headward erosion in tributaries and a lowering of the water table.

In an investigation of a potential reservoir site, consideration must be given to the amount of rainfall, run-off, infiltration and evapotranspiration which occur in the catchment area. The climatic, topographical and geological conditions therefore are important, as is the type of vegetative cover. Accordingly, the two essential types of basic data needed for reservoir design studies are adequate topographical maps and hydrological records. Indeed, the location of a large, impounding direct supply reservoir is very much influenced by the topography, because this governs its storage capacity. Initial estimates of storage capacity can be made from topographical maps or aerial photographs, more accurate informa-

tion being obtained, where necessary, from subsequent surveying. Catchment areas and drainage densities can also be determined from maps and aerial photographs.

The reservoir volume can be estimated by planimetering areas on a topographic map upstream of the dam site for successive contours up to the proposed top water level. Then, the mean of two successive contour areas is multiplied by the contour interval to give the interval volume, the summation of the interval volumes providing the total volume of the reservoir site.

Records of stream flow are required to determine the amount of water available for conservation purposes. Such records contain flood peaks and volumes, which are used to determine the amount of storage needed to control floods, and to design spillways and other outlets. Records of rainfall are used to supplement stream flow records or as a basis for computing the stream flow where no flow records are obtainable. Losses due to seepage and evaporation must also be taken into account.

The field reconnaissance provides indications of the areas in which detailed geological mapping may be required and of where to locate drill-holes, such as in low, narrow saddles or other seemingly critical areas in the reservoir rim. Drill-holes on the flanks of reservoirs should be drilled at least to floor level. Permeability and pore water pressure tests can be carried out in these drill-holes.

The most attractive site for a large impounding reservoir is a valley constricted by a gorge at its outfall, with steep banks upstream, so that a small dam can impound a large volume of water with a minimum extent of water spread. However, two other factors must be taken into consideration, namely, the watertightness of the basin and the bank stability. The question of whether or not significant water loss will take place is chiefly determined by the groundwater conditions, more specifically by the hydraulic gradient. Consequently, once the groundwater conditions have been investigated, an assessment can be made of the watertightness and possible groundwater control measures. Seepage is a more discrete flow than leakage, which spreads out over a larger area, but may be no less in total amount.

Apart from the conditions in the immediate vicinity of a dam, the two factors which determine the retention of water in reservoir basins are the piezometric

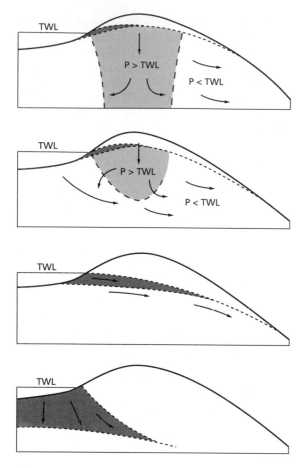

Fig. 9.2 Reservoir conditions in relation to differing water table and piezometric situations. (After Knill, 1971.) TWL, top water level; P, groundwater pressure; heavily shaded area, area flooded by reservoir water; lightly shaded area, groundwater pressures in excess of top water level. Note that diagrams apply to two-dimensional, uniform conditions, and are not rigorous.

conditions in, and the natural permeability of, the floor and flanks of the basin. Knill (1971) pointed out that the following four groundwater conditions exist on the flanks of a reservoir (Fig. 9.2):

1 The groundwater divide and piezometric level are at a higher elevation than that of the proposed top water level. In this situation, no significant water loss takes place.

2 The groundwater divide, but not the piezometric level, is above the top water level of the reservoir. In these circumstances, seepage can take place through the separating ridge into the adjoining valley. Deep

seepage can take place, but the rate of flow is determined by the *in situ* permeability.

3 Both the groundwater divide and piezometric conditions are at a lower elevation than the top water level, but higher than that of the reservoir floor. In this case, the increase in the groundwater head is low, and the flow from the reservoir may be initiated under conditions of low piezometric pressure in the reservoir flanks.

4 The water table is depressed below the base of the reservoir floor. This indicates deep drainage of the rock mass or very limited recharge. A depressed water table does not necessarily mean that reservoir construction is out of the question, but groundwater recharge will take place on filling, which will give rise to a changed hydrogeological environment as the water table rises. In such instances, the impermeability of the reservoir floor is important.

When impermeable beds are more or less saturated, particularly when they have no outlet, seepage is appreciably decreased. At the same time, the accumulation of silt on the floor of the reservoir tends to reduce seepage. If, however, any permeable beds present contain large pore spaces or discontinuities, and they drain from the reservoir, seepage continues.

The economics of reservoir leakage vary. Although a highly leaky reservoir may be acceptable in an area where run-off is evenly distributed throughout the year, a reservoir basin with the same rate of water loss may be of little value in an area where run-off is seasonally deficient. Serious water loss has, in some instances, led to the abandonment of a reservoir scheme.

Leakage from a reservoir downstream of the dam site can take the form of sudden increases in stream flow, with boils in the river and the appearance of springs on the valley sides. It may be associated with major defects in the geological structure, such as solution channels, fault zones or buried channels through which large and essentially localized flows take place.

Serious leakage has taken place at reservoirs as a result of cavernous conditions in limestone, which were not fully revealed or appreciated at the site investigation stage. Sites are best abandoned where large, numerous solution cavities extend to considerable depths. Where the problem is not so severe, solution cavities can be cleaned and grouted. Sinkholes and caverns can develop in thick beds of gypsum more rapidly than they can in limestone.

Buried channels may be filled with coarse, granular stream deposits or deposits of glacial origin and, if they occur near the perimeter of a reservoir, they almost invariably pose leakage problems. Indeed, leakage through buried channels, via the perimeter of a reservoir, usually is more significant than through the main valley. Hence, the bedrock profile, the types of deposit and the groundwater conditions should be determined.

A thin layer of relatively impermeable, superficial deposits does not necessarily provide an adequate seal against seepage. A controlling factor in such a situation is the groundwater pressure immediately below the deposits. Where artesian conditions exist, springs may break the thinner parts of the superficial cover. If the water table below the deposits is depressed, there is a risk that the weight of water in the reservoir may puncture them. Moreover, on filling a reservoir, there is a possibility that the superficial material may be ruptured or partially removed to expose the underlying rocks.

Because of the occurrence of permeable contacts, close jointing, pipes and vesicles, and the possible presence of tunnels and cavities, recent accumulations of basaltic lava flows can prove to be highly leaky and treacherous with respect to watertightness.

Lava flows frequently are interbedded, often in an irregular fashion, with pyroclastic deposits. Deposits of ash and cinders tend to be highly permeable.

Leakage along faults generally is not a serious problem as far as reservoirs are concerned, because the length of the flow path usually is too long. However, fault zones occupied by permeable fault breccia running beneath the dam must be given special consideration. When the reservoir basin is filled, the hydrostatic pressure may cause the removal of loose material from such fault zones, and thereby accentuate leakage. Permeable fault zones can be grouted, or if a metre or so wide, excavated and filled with rolled clay or concrete.

Troubles from seepage can usually be controlled by exclusion or drainage techniques. Cutoff trenches, carried into bedrock, may be constructed across cols occupied by permeable deposits. Grouting may be effective where localized fissuring is the cause of leakage. Impervious linings (Fig. 9.3) consume large amounts of head near the source of water, thereby reducing the hydraulic gradients and saturation at the points of exit, and increasing the resistance to seepage loss. Clay blankets or layers of silt have been used to seal exits from reservoirs.

The formation of a reservoir upsets the groundwater regime and represents an obstruction to water flowing downhill. The greatest change involves the

Fig. 9.3 An asphalt membrane at the side of Mornos Reservoir, near Lidoniki, Greece. This is to prevent leakage into folded and broken strata which include limestone.

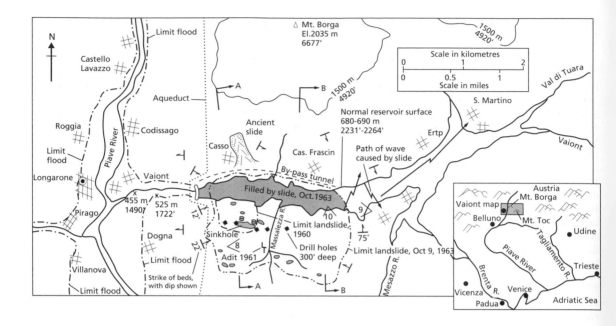

Fig. 9.4 Map of Vaiont Dam area and Piave River valley showing geographical features, and limits of slide and of destructive floodwaters.

raising of the water table. Some soils or rocks, which are brought within the zone of saturation, may then become unstable and fail, as saturated material is weaker than unsaturated. This can lead to slumping and sliding on the flanks of a reservoir (Fig. 9.4). In glaciated valleys, morainic material generally rests on a rock slope smoothed by glacial erosion, which accentuates the problem of slip. Landslides which occur after a reservoir is filled reduce its capacity. In addition, ancient landslipped areas which occur on the rims of a reservoir could be reactivated, as well as presenting a potential leakage problem.

Although it is seldom a decisive factor in determining location, sedimentation in reservoirs is an important problem in certain countries (Table 9.2). Sedimentation in a reservoir may lead to one or more of its major functions being seriously curtailed, or even to it becoming inoperative. In a small reservoir, sedimentation may seriously affect the available carry-over water supply and ultimately necessitate abandonment.

In those areas in which streams carry heavy sediment loads, the rates of sedimentation must be esti-

mated accurately in order that the useful life of any proposed reservoir may be determined. The volume of sediment carried varies with the stream flow, but usually the peak sediment load occurs prior to the peak stream flow discharge. Frequent sampling accordingly must be made to ascertain changes in sediment transport. Volumetric measurements of sediment in reservoirs are made by soundings taken to develop the configuration of the reservoir sides and floor below the water surface.

The size of a drainage basin is the most important consideration as far as the sediment yield is concerned, the rock types, drainage density and gradient of slope also being significant. The sediment yield is also influenced by the amount and seasonal distribution of precipitation and the vegetative cover. Poor cultivation practices, overgrazing, improper disposal of mine waste and other human activities may accelerate erosion or contribute directly to streamloads.

The ability of a reservoir to trap and retain sediment is known as the trap efficiency, and is expressed as the percentage of incoming sediment which is retained. The trap efficiency depends on the total inflow, rate of flow, sediment characteristics and the size of the reservoir.

Pumped storage reservoirs for direct water supply commonly are sited adjacent to the lower reaches of

Table 9.2 Rates of sediment accumulation in some reservoirs in the USA. (After Gottschalk, 1964.)

Reservoir (state)	Net drainage area (km²)	Year storage began	Original capacity (×1000 m³)	Annual sediment production rate (Mg km⁻²)	Loss of storage (annual)	Per cent of total
Schoharie (NY)	808.1	1926	78 489	570.7	0.07	1.75
Byllesby (VA)	3 392.9	1912	10 937	625.9	2.54	60.2
Norris (TN)	7 311.6	1936	2 515 719	1 183.5	0.05	0.54
Abilene (TX)	252.5	1921	12 700	720.6	0.19	5.22
Dallas (TX)	2 996.6	1928	222 334	3 429.5	0.72	7.57
Morena (CA)	282.3	1910	82 123	6 427.7	0.31	11.7
Roosevelt (AZ)	14 918.4	1909	1 872 306	2 924.6	0.25	9.2
Mead (NV)	434 084	1935	38 437 500	2 306.5	0.33	4.6
Black Canyon (ID)	6 578.6	1924	46 321	454.9	0.89	10.7
Seminoc (WY)	18 951	1939	1 254 600	402.4	0.08	0.9
Bennington (KS)	3.6	1929	92	13 967.9	5.00	56.0
Springfield (IL)	668.2	1934	75 078	1 735.8	0.30	4.36
Altus (OK)	5 480.4	1948	192 702	2 046.1	0.70	8.81
Elephant Butte (NM)	66 992.9	1915	3 240 804	2 098.7	0.51	16.6
Cold Springs (OR)	481.7	1908	61 142	28 141.1	0.24	10.1
Franklinton (NC)	2.9	1925	43	1 954.1	1.60	21.3
Upper Pine (IA)	35.7	1934	812	3 918.7	2.38	31.5

rivers from which the water is abstracted, hence taking advantage of the greatest catchment area and maximum available run-off. The yield, however, is influenced by the capacity of the pumping plant to abstract water from the river.

9.3 Dam sites

The type and size of dam constructed depend upon the need for and the amount of water available, the topography and geology of the site, and the construction materials which are readily obtainable. Dams can be divided into two major categories according to the type of material with which they are constructed, namely, concrete dams and earth dams. Concrete dams can be subdivided into gravity, arch and buttress dams (Fig. 9.5). Earth dams (Fig. 9.6) comprise rolled-fill and rockfill embankments. In dam construction, the prime concern is safety (this coming before cost), that is, the foundations and abutments must be adequate for the type of dam selected.

Some sites which are geologically unsuitable for a specific type of dam design may support one of composite design. For instance, a broad valley which has strong rocks on one side and weaker rocks on the other possibly can be spanned by a combined gravity and embankment dam, that is, a composite dam (Fig. 9.7).

The construction of a dam and the filling of a reservoir behind it impose a load on the sides and floor of a valley, creating new stress conditions. These stresses must be analysed, so that there is ample assurance that there will be no possibility of failure. A concrete dam behaves as a rigid, monolithic structure; the stress acting on the foundation is a function of the weight of the dam distributed over the total area of the foundation. By contrast, earthfill dams exhibit semi-plastic behaviour, and the pressure on the foundation at any point depends on the thickness of the dam above that point. Vertical static forces act downwards, and include the weight of both the structure and the reservoir water, although a large part of the dam is submerged and therefore the buoyancy effect reduces the influence of these two forces. The most important dynamic forces acting on a dam are wave action, overflow of water and seismic shocks.

Horizontal forces are exerted on a dam by the lateral pressure of water behind it. These, if excessive, may cause concrete dams to slide. The tendency towards sliding at the base of such dams is of particu-

(a)

(b)

Fig. 9.5 (a) Nagle Dam, Valley of One Thousand Hills, Natal, South Africa. This is a concrete gravity dam. Note the second dam on the far left-hand side. This helps to control the flow into the reservoir, as well as trapping silt. Excess flow and silt can be removed through a sluice to the right of this dam, the water and silt flowing through a cutting into the river below Nagle Dam. (b) Hoover Dam, on the Colorado River, USA. When completed in the mid-1930s, this was the largest arch dam in the world. Over 60 years later, it is still a most impressive structure. (c) (*Opposite*) Evrochty Dam, an example of a buttress dam, Scotland. (Courtesy of the North of Scotland Hydro-Electric Board.)

Fig. 9.5 *Continued.* (c)

Fig. 9.6 An earth dam nearing the final stages of completion in southern Spain.

lar significance in fissile rocks, such as shales, slates and phyllites. Weak zones, such as interbedded ashes in a sequence of basalt lava flows, can prove troublesome. The presence of flat-lying joints may destroy much of the inherent shear strength of a rock mass, and reduce the problem of resistance of a foundation to horizontal forces to one of sliding friction, so that the roughness of joint surfaces becomes a critical factor. The rock surface should be roughened to prevent sliding, and keying the dam some distance into the foundation is advisable. Another method of reducing sliding is to give a downward slope to the

base of the dam in the upstream direction of the valley.

Variations in pore water pressure cause changes in the state of stress in rock masses. Increasing pore water pressure may lift beds and the dam itself, and so decrease the shear strength and resistance to sliding within the rock mass. Pore water reduces the compressive strength of rocks and causes an increase in the amount of deformation they undergo. It may also be responsible for swelling in certain rocks and for an acceleration in their rate of alteration. Pore water in the stratified rocks of a dam foundation reduces the

Fig. 9.7 Inanda Dam, a composite dam in the final stages of completion, near Durban, South Africa.

coefficient of friction between the individual beds, and between the foundation and the dam.

Percolation of water through the foundations of a concrete dam, even when the rock masses concerned are of good quality and of minimum permeability, is always a decisive factor in the safety and performance of a dam. Such percolation can remove filler material which may be occupying joints which, in turn, can lead to differential settlement of the foundations. It may also open joints, which decreases the strength of the rock masses.

In highly permeable rock masses, excessive seepage beneath a dam may damage the foundations. Seepage rates can be lowered by reducing the hydraulic gradient beneath the dam by incorporating a cutoff into the design. A cutoff lengthens the flow path, so reducing the hydraulic gradient. It extends to an impermeable horizon or to a certain specified depth, and usually is located below the upstream face of the dam. The rate of seepage can also be effectively reduced by placing an impervious earthfill against the lower part of the upstream face of a dam.

Uplift pressure acts against the base of a dam, and is caused by water seeping beneath it, which is under hydrostatic head from the reservoir. The uplift pressure should be distinguished from the pore water pressure in the material beneath the dam. The uplift pressure on the heel of a dam is approximately equal to the depth of the foundation below water level multiplied by the unit weight of the water. In the simplest

case, it is assumed that the difference in hydraulic heads between the heel and the toe of the dam is dissipated uniformly between them. The uplift pressure can be reduced by allowing water to be conducted downstream by drains incorporated into the foundation and base of the dam.

When load is removed from a rock mass on excavation, it is subject to rebound. The amount of rebound depends on the modulus of elasticity of the rocks concerned, the larger the modulus of elasticity, the smaller the rebound. The rebound process in rocks generally takes a considerable time to achieve completion, and will continue after a dam has been constructed, if the rebound pressure or heave developed by the excavation of foundation material exceeds the effective weight of the dam. Hence, if heave is to be counteracted, a dam should impose a load on the foundation equal to or slightly in excess of the load removed.

All foundation and abutment rocks yield elastically to some degree. In particular, the modulus of elasticity of the rock is of primary importance as far as the distribution of stresses at the base of a concrete dam is concerned. In addition, tensile stresses may develop in concrete dams when the foundations undergo significant deformation. The modulus of elasticity is used in the design of gravity dams for comparing the different types of foundation rocks with each other and with the concrete of the dam. In the design of arch dams, if the Young's modulus of the foundation has a lower value than that of the concrete, or varies widely in the rocks against which the dam abuts, dangerous stress conditions may develop in the dam. The elastic properties of a rock and the existing strain conditions assume importance in proportion to the height of a dam, because this influences the magnitude of the stresses imparted to the foundations and abutments. The influence of geological structures in lowering Young's modulus must be accounted for by the provision of adequate safety factors. It should also be borne in mind that blasting during the excavation of the foundations can open up fissures and joints, which leads to a greater deformability of the rock mass.

9.4 Geology and dam sites

Of the various natural factors which directly influence the design of dams, none is more important than geology. Not only does this control the character of the foundations, but it also governs the materials available for construction. The major questions which need to be answered include the depth at which adequate foundations exist, the strength of the rocks involved, the likelihood of water loss and any special features which have a bearing on excavation. The character of the foundations upon which dams are built and their reaction to the new conditions of stress and strain, hydrostatic pressure and exposure to weathering must be ascertained, so that the correct factors of safety may be adopted to ensure against subsequent failure. Excluding the weaker types of compaction shales, mudstones, marls, pyroclasts and certain very friable types of sandstone, there are few foundation materials, deserving of the name rock, that are incapable of resisting the bearing loads even of high dams.

In their unaltered state, plutonic rocks are essentially sound and durable, with adequate strength for any engineering requirement. In some instances, however, intrusives may be highly altered by weathering or hydrothermal attack. Generally, the weathered product of plutonic rocks has a large clay content, although that of granitic rocks is sometimes porous, with a permeability comparable with that of medium-grained sand, so that it requires some type of cutoff or special treatment of the upstream surface.

Thick, massive basalts make satisfactory dam sites (Bell & Haskins, 1997), but many basalts of compar-

atively young geological age are highly permeable, transmitting water via their open joints, pipes, cavities, tunnels and contact zones. Foundation problems in young, volcanic sequences are twofold. Weak beds of ash and tuff may occur between the basalt flows, which give rise to problems of differential settlement or sliding, and weathering during periods of volcanic inactivity may have produced fossil soils, these being of much lower strength.

Pyroclastics usually give rise to extremely variable foundation conditions due to wide variations in strength, durability and permeability. Their behaviour very much depends upon their degree of induration. Ashes invariably are weak and often highly permeable, and they may also undergo hydrocompaction on wetting.

Fresh, metamorphosed rocks, such as quartzite and hornfels, are very strong and afford excellent dam sites. Marble has the same advantages and disadvantages as other carbonate rocks. Generally speaking, gneiss has proved to be a good foundation rock for dams. Cleavage and schistosity in regional metamorphic rocks may adversely affect their strength and make them more susceptible to decay. Moreover, areas of regional metamorphism have usually suffered extensive folding, so that rocks may be fractured and deformed. Certain schists, slates and phyllites are variable in quality, some being excellent for dam site purposes; others, regardless of the degree of deformation or weathering, are so poor as to be wholly undesirable in foundations and abutments. Some schists become slippery upon weathering, and therefore fail under a moderately light load. Particular care is required in blasting slates, phyllites and schists, otherwise considerable overbreak or shattering may result. It may be advantageous to use smooth blasting for final trimming purposes.

Joints and shear zones are responsible for the unsound rock encountered at dam sites on plutonic and metamorphic rocks. Unless they are sealed, they may permit leakage through foundations and abutments. The slight opening of joints on excavation leads to the imperceptible rotation and sliding of rock blocks, large enough to reduce appreciably the strength and stiffness of the rock mass. Sheet or flat-lying joints tend to be approximately parallel to the topographical surface, and introduce a dangerous element of weakness into valley slopes. Their width varies and, if they remain untreated, large quantities of water may escape through them from the reservoir.

Sandstones have a wide range of strength depending largely upon the amount and type of cement-matrix material occupying the voids. With the exception of shaly sandstone, sandstone is not subject to rapid surface deterioration on exposure. As a foundation rock, even poorly cemented sandstone is not susceptible to plastic deformation. However, friable sandstones introduce problems of scour within the foundation. Moreover, sandstones are highly vulnerable to the scouring and plucking action of the overflow from dams, and have to be adequately protected by suitable hydraulic structures. A major problem of dam sites located in sandstones results from the fact that they are generally transected by joints, which reduce the resistance to sliding. Generally, however, sandstones have high coefficients of internal friction, which give them high shear strengths when restrained under load. Sandstones frequently are interbedded with shale. These layers of shale may constitute potential sliding surfaces. The permeability of sandstone depends upon the amount of cement in the voids and, especially, on the incidence of discontinuities.

Limestone dam sites vary widely in their suitability. Thick-bedded, horizontally lying limestones, relatively free from solution cavities, afford excellent dam sites. On the other hand, thin-bedded, highly folded or cavernous limestones are likely to present serious foundation or abutment problems involving the bearing capacity, watertightness or both (Soderburg, 1979). If the rock mass is thin bedded, a possibility of sliding may exist. This should be guarded against by suitably keying the structure into the foundation rock. Beds separated by layers of clay or shale, especially those inclined downstream, may under certain conditions, serve as sliding planes and give rise to failure. Some solution features will always be present in limestone. The size, form, abundance and downward extent of these features depend upon the geological structure and the presence of interbedded impervious layers. Individual cavities may be open, may be partially or completely filled with clay, silt, sand or gravel mixtures, or may be water-filled conduits. Solution cavities present numerous problems in the construction of large dams, of which the bearing capacity and watertightness are paramount. Few dam sites are so bad that it is impossible to construct safe

and successful structures upon them, but the cost of the necessary remedial treatment may be prohibitive (Vick & Bromwell, 1989). Dam sites should be abandoned where the cavities are large and numerous, and extend to considerable depths.

Well cemented shales, under structurally sound conditions, present few dam site problems, although their strength limitations and elastic properties may be factors of importance in the design of concrete dams of appreciable height. However, they have lower moduli of elasticity and lower shear strength values than concrete, and therefore are unsatisfactory foundation materials for arch dams. Moreover, if the lamination is horizontal and well developed, the foundations may offer little shear resistance to the horizontal forces exerted by a dam. A structure keying the dam into such a foundation is then required. Severe settlements may take place in low-grade compaction shales. Thus, such sites generally are developed with earth dams, but associated concrete structures, such as spillways, will involve these problems. Rebound in deep spillway cuts may cause buckling of spillway linings, and differential rebound movements in the foundations may require special design provisions. The stability of slopes in cuts is one of the major problems in shale both during and after construction.

The opening of joints and the development of shear planes in shales for considerable distances behind the normal zones of creep on valley sides result from a combination of elastic rebound and the oversteepening of slopes. These deep-seated disturbances may give rise to dangerous hydrostatic pressures on the abutment rocks downstream from the dam, leakage around the ends of the dam and reduced resistance of the rock to horizontal forces.

Earth dams usually are constructed on clay soils, as such soils lack the load-bearing properties necessary to support concrete dams. Beneath valley floors, clays frequently are contorted, fractured and softened due to valley creep, so that the load of an earth dam may have to spread over wider areas than is the case with shales and mudstones. Rigid ancillary structures necessitate spread footings or raft foundations. Deep cuts involve problems of rebound if the weight of removed material exceeds that of the structure. Slope stability problems also arise, with rotational slides a hazard.

The many manifestations of glaciation include the

presence of buried channels, disrupted drainage systems, deeply filled valleys, sand–gravel terraces, narrow overflow channels connecting open valleys and extensive deposits of lacustrine silts and clays, till, and outwash sands and gravels. Deposits of peat and head (solifluction debris) may be interbedded with these glacial deposits. Consequently, glacial deposits may be notoriously variable in composition, both laterally and vertically. As a result, dam sites in glaciated areas are among the most difficult to appraise on the basis of surface evidence. A knowledge of the preglacial, glacial and postglacial history of a locality is of vital importance in the search for the most practicable sites. A primary consideration in glacial terrains is the discovery of sites where rock foundations are available for spillway, outlet and powerhouse structures. Generally, earth dams are constructed in areas of glacial deposits. Concrete dams, however, are feasible in postglacial, rock-cut valleys, and composite dams are practicable in valleys containing rock benches.

The major problems associated with foundations on alluvial deposits generally result from the fact that the deposits are poorly consolidated. Silts and clays are subject to plastic deformation or shear failure under relatively light loads, and undergo consolidation for long periods of time when subjected to appreciable loads. Many large earth dams have been built upon such materials, but this demands a thorough exploration and testing programme in order to design safe structures. Soft, alluvial clays at ground level generally have been removed, if economically feasible. The slopes of an embankment dam may be flattened in order to mobilize greater foundation shear strength, or berms may be introduced. Where soft, alluvial clays are no more than 2–3 m thick, they should consolidate during construction if covered with a drainage blanket, especially if resting on sand and gravel. With thicker deposits, it may be necessary to incorporate vertical sand drains within the clays. However, coarser sands and gravels undergo comparatively little consolidation under load, and therefore afford excellent foundations for earth dams. Their primary problems result from their permeability. Alluvial sands and gravels form natural drainage blankets under the higher parts of an earth or rockfill dam, so that seepage through them beneath the dam must be cut off. Problems relating to underseepage through pervious strata may be met by a grout curtain. Alternatively, underseepage may be checked by the construction of an impervious upstream blanket to lengthen the path of percolation, and the installation on the downstream side of suitable drainage facilities to collect the seepage.

Landslides are a common feature of valleys in mountainous areas, and large slips often cause the narrowing of a valley which thus appears to be topographically suitable for a dam. Unless landslides are shallow seated and can be removed or effectively drained, it is prudent to avoid landslipped areas in dam location, because their unstable nature may result in movement during construction or subsequently on inundation by reservoir water.

Fault zones may be occupied by shattered or crushed material, and so represent areas of weakness, which may give rise to landsliding upon excavation for a dam. The occurrence of faults in a river is not unusual, and this generally means that the material along the fault zone is highly altered, thus necessitating a deep cutoff.

In most known instances of historic fault breaks, the fracturing has occurred along a pre-existing fault. Movement along faults occurs not only in association with large and infrequent earthquakes, but also in association with small shocks and continuous slippage, known as fault creep. Earthquakes resulting from displacement and energy release on one fault can sometimes trigger small displacements on other unrelated faults many kilometres distant. The longer fault breaks have greater displacements and generate larger earthquakes. The maximum displacement is less than 6 m for the great majority of fault breaks, and the average displacement along the length of a fault is less than half the maximum. These values suggest that zoned embankment dams can be built with safety at sites with active faults (see Chapter 3).

All major faults located in regions in which strong earthquakes have occurred should be regarded as potentially active, unless convincing evidence exists to the contrary. In stable areas of the world, little evidence exists of fault displacements in the recent past. Nevertheless, an investigation should be carried out to confirm the absence of active faults at or near any proposed major dam in any part of the world.

Wherever possible, construction materials for an earth dam should be obtained from within the future reservoir basin (Anderson & McNicol, 1989).

Accordingly, the investigation of a dam site and the surrounding area should determine the availability of impervious and pervious materials for the embankment, sand and gravels for drains and filter blankets, and stone for rip-rap.

In some cases, only one type of soil is easily obtainable for an earth dam. If this is impervious, the design will consist of a homogeneous embankment, which incorporates a small amount of permeable material, in the form of filter drains, to control internal seepage. On the other hand, where sand and gravel are in plentiful supply, a very thin earth core may be built into a dam if enough impervious soil is available; otherwise, an impervious membrane may be constructed of concrete or interlocking steel sheet piles. However, because concrete can withstand very little settlement, such core walls must be located on sound foundations.

Sites which provide a variety of soils lend themselves to the construction of zoned dams. The finer, more impervious materials are used to construct the core, whereas the coarser materials provide strength and drainage in the upstream and downstream zones.

Embankment soils need to develop high shear strength, low permeability and low water absorption, and undergo minimal settlement. This is achieved by compaction.

9.5 Groundwater

The principal source of groundwater is meteoric water, that is, precipitation (rain, sleet, snow and hail). However, two other sources are very occasionally of some consequence: juvenile water and connate water. The former is derived from magmatic sources, whilst the latter was trapped in the pore spaces of sedimentary rocks as they were formed.

The amount of water that infiltrates into the ground depends upon how precipitation is dispersed, that is, on what proportions are assigned to immediate run-off and to evapotranspiration, the remainder constituting the proportion allotted to infiltration/percolation. The infiltration capacity is influenced by the rate at which rainfall occurs, the vegetation cover, the porosity of the soils and rocks, their initial moisture content and the position of the zone of saturation. Accordingly, the rate at which groundwater is replenished basically is dependent

upon the quantity of precipitation falling on the recharge area of an aquifer, although the rainfall intensity is also very important. Frequent rainfall of moderate intensity is more effective in recharging groundwater resources than short, concentrated periods of high intensity. This is because the rate at which the ground can absorb water is limited, any surplus water tending to become run-off. Some precipitation will be lost through evaporation and transpiration. The rate at which water can be lost from a surface through evapotranspiration is dependent, to some extent, upon the amount of water that is present in the soil.

The retention of water in a soil depends upon the capillary force and the molecular attraction of the particles. As the pores in a soil become thoroughly wetted, the capillary force declines, so that gravity becomes more effective. In this way, downward percolation can continue after infiltration has ceased but, as the soil dries, so capillarity increases in importance. No further percolation occurs after the capillary and gravity forces are balanced. Thus water percolates into the zone of saturation when the retention capacity is satisfied. This means that the rains which occur after the deficiency of soil moisture has been catered for are those which count as far as the supplementation of groundwater is concerned.

The pores within the zone of saturation are filled with water, generally referred to as phreatic water. The upper surface of this zone therefore is known as the phreatic surface, but more commonly is termed the water table. Above the zone of saturation is the zone of aeration, in which both air and water occupy the pores. The water in the zone of aeration commonly is referred to as vadose water.

The geological factors which influence percolation not only vary from one soil or rock outcrop to another, but may do so within the same one. This, together with the fact that rain does not fall evenly over a given area, means that the contribution to the zone of saturation is variable, which influences the position of the water table, as do the points of discharge. A rise in the water table as a response to percolation partly is controlled by the rate at which water can drain from the area of recharge. Mounds and ridges form in the water table under the areas of greatest recharge. When the influence of water, which may drain from lakes and streams, is superimposed on this, it can be appreciated that a water table continually is adjusting.

The water table fluctuates in position, particularly in those climates in which there are marked seasonal changes in rainfall. Thus, permanent and intermittent water tables can be distinguished, the former marking the level beneath which the water table does not sink, and the latter expressing the fluctuation. Usually, water tables fluctuate within the lower and upper limits, rather than between them, especially in humid regions, because the periods between successive recharges are small. The position at which the water table intersects the surface is termed the spring line. Intermittent and permanent springs similarly can be distinguished.

A perched water table is one which forms above a discontinuous impermeable layer, such as a lens of clay in a formation of sand, the clay impounding a water mound.

An aquifer is the term given to a rock or soil mass which not only contains water, but from which water can be abstracted readily in significant quantities. The ability of an aquifer to transmit water is governed by its permeability. Indeed, the permeability of an aquifer usually is in excess of $10^{-5}\,\mathrm{m\,s^{-1}}$. By contrast, a formation with a permeability of less than $10^{-9}\,\mathrm{m\,s^{-1}}$ is one which, in engineering terms, is regarded as impermeable and is referred to as an aquiclude. An aquitard is a formation which transmits water at a very slow rate but which, over a large area of contact, may permit the passage of large amounts of water between adjacent aquifers which it separates.

An aquifer is described as unconfined when the water table is open to the atmosphere, that is, the aquifer is not overlain by material of lower permeability (Fig. 9.8). Conversely, a confined aquifer is one which is overlain by impermeable rocks. Confined aquifers may have relatively small recharge areas

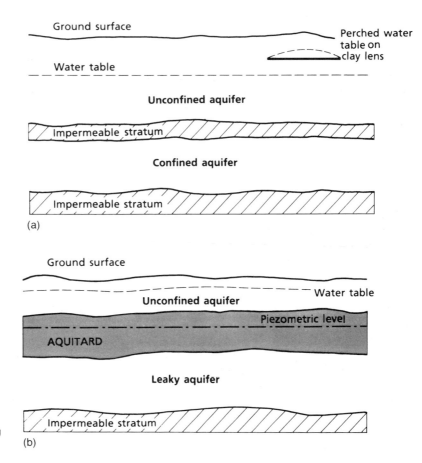

Fig. 9.8 (a) Diagram illustrating unconfined and confined aquifers, with a perched water table in the vadose zone. (b) Diagram illustrating a leaky aquifer.

compared with unconfined aquifers, and therefore may yield less water. An aquifer which is overlain and/or underlain by aquitards is described as a leaky aquifer (Fig. 9.8b).

Very often, the water in a confined aquifer is under piezometric pressure, that is, there is an excess of pressure sufficient to raise the water above the base of the overlying bed when the aquifer is penetrated by a well. Piezometric pressures are developed when the buried upper surface of a confined aquifer is lower than the water table in the aquifer at its recharge area. Where the piezometric surface is above ground level, water overflows from a well. Such wells are described as artesian. A synclinal structure is the commonest cause of artesian conditions (Fig. 9.9).

9.6 A note on basement aquifers

Basement aquifers are developed within the weathered overburden and fractured bedrock of rocks of intrusive igneous or metamorphic origin. Because basement aquifers generally provide a low yield, development usually is from point sources, utilizing handpumps or bucket and windlass systems. Viable aquifers which occur completely within fractured bedrock, according to Clark (1985), are rare due to the low storativity of fracture systems. Hence, to be effective, the bedrock component should be in continuity with storage in the weathered regolith, or in other suitable formations such as alluvium. Furthermore, because of their low storativity, basement aquifers may be depleted significantly during periods of drought. The fact that the groundwater in such aquifers commonly is contained within fissures in rock at shallow depth means that it is susceptible to surface pollution. There is a close relationship between groundwater occurrence, on the one hand, and relief, soil, surface water hydrology and vegetation cover on the other. Because of the localized nature of many basement aquifers, there frequently is a high failure rate associated with boreholes, that is, in the range of 10 to 40%, and there is a wide range of yield (Wright, 1992).

The thickness of the weathered horizon or regolith depends on the nature of the bedrock, notably the rock type and structure, the age of the land surface, its relief and the climate. The degree of weathering increases towards the surface, with an increasing proportion of minerals being broken down to form clay minerals. As weathering is most effective in the vadose zone and the zone within which the water table fluctuates, the upper part of the saprolite usually contains more clay material and so has a low permeability. The boundary with the underlying saprock, the weathered bedrock, may be sharp (e.g. against coarse grained massive rocks) or transitional (e.g. against finer grained rocks). Regolith permeability values, quoted by Wright (1992) for certain areas in Malawi and Zimbabwe, generally did not exceed $0.6^{-5}\,m\,s^{-1}$.

The regolith may contain throughflow channels, such as basal breccias, conglomerates and gravel beds, old termite tunnels, tree roots and residual quartz veins. Residual fractures can be orientated more or less in any direction from vertical to horizontal, and frequently are open below the water table.

Fracture or discontinuity systems are related to the dissipation of residual stress, as overburden is removed by erosion, or to tectonic forces. In the former type of system, the discontinuities tend to be subhorizontal and decline in number with depth. Discontinuities developed by tectonic forces normally are subvertical and often are in zonal concentrations. The fissure permeability depends upon the nature of the discontinuities, in particular on their frequency

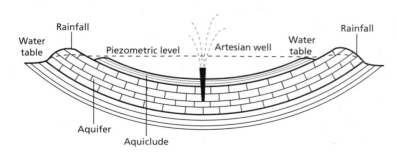

Fig. 9.9 An artesian basin.

and whether they are open or filled (if open, the amount of gape is important). The sealing of discontinuities, especially in the weathered zone, often is attributable to clay illuviation. Whether tectonic discontinuities close with depth is debatable, however, Black (1987) observed no relationship between depth and decreasing permeability to a depth of 700 m, and groundwater flow via discontinuities at times is a problem in deep mines.

The development of basement aquifers, particularly in much of Africa, is principally for rural water supply, with little usage for irrigation or urban supply. Recharge usually appreciably exceeds the maximum requirement of rural water supply. Hence, some of the excess could be made available for other uses. Boreholes, dug wells and, to a much lesser extent, collector wells are used to abstract water from basement aquifers. Substantially larger yields can be obtained from collector wells than from boreholes, with the added advantage of low drawdowns. Radial collectors can be drilled at any angle from large-diameter (2–3 m) wells.

9.7 Springs

Springs develop at the points at which underground conduits discharge water at the surface. Water rarely moves uniformly throughout an entire rock mass, and most springs therefore issue as concentrated flows. Springs which percolate from many small openings have been termed seepage springs, and they may discharge so little water that they barely are noticeable.

The location of a spring is dependent upon a number of factors, climate and geology being two of the most important. Intermittent springs are very much influenced by climate. After heavy rainfall, the water table may rise to intersect the ground surface, and so produce a spring. In dry weather, the water table sinks and the stream disappears.

The commonest geological setting for a spring is at the contact between two beds of differing permeability. This setting often gives rise to a spring line, and these springs are referred to as stratum springs. There are two types of stratum spring (Fig. 9.10a). If the downward percolation of water in a permeable horizon is impeded by an underlying impermeable layer, the spring which issues is termed a contact spring. Conversely, an overflow spring may be

formed where a permeable bed dips beneath an impermeable one. Unconformities may give rise to springs along their outcrop. When a permeable formation is thrown against an impermeable formation by faulting, water may issue at the surface along the fault plane (Fig. 9.10b). Water table or valley springs emerge where a valley is carved beneath the water table in thick, permeable formations (Fig. 9.10c). The discharge from such springs usually is small.

The most common type of thermal spring is the hot spring, however, the temperatures of thermal springs may range from lukewarm to near boiling point. Hot springs are found in all volcanic districts, even some

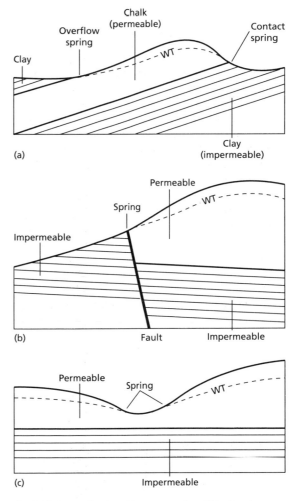

Fig. 9.10 Types of spring: (a) stratum spring; (b) fault spring; (c) valley spring. WT, water table.

of those where the volcanoes are extinct. They presumably originate from steam given off by a magmatic source. On its passage to the surface, the steam commonly encounters groundwater, which consequently is heated and contributes to the hot springs. A geyser is a hot spring which periodically discharges a column of hot water (Fig. 2.9). In several places in the world, hot springs are utilized for the generation of electric power, for example, at Larderello in Italy, at Wairakei in New Zealand and in Somona County, California. In Iceland, hot springs are used for domestic heating. However, hot springs generally contain dissolved gases and minerals, such as carbon dioxide, hydrogen sulphide, calcium carbonate and silica. These may cause serious problems of corrosion and precipitation in plant which harnesses such springs.

9.8 Water budget studies

Water budget studies attempt to quantify the average annual flow through an aquifer by considering the disposition of the rain that falls on the recharge area. Basically, the rain must either run off the surface and become stream flow, be evapotranspired or infiltrate the surface and thereby add to groundwater storage. The potential infiltration rate and, subsequently, the change in groundwater storage can be estimated from the following expression:

$$dS_g = (P - E_{ta} - R_o) \qquad (9.1)$$

where dS_g is the change in groundwater storage (metres per year), P is the precipitation (metres per year), E_{ta} is the actual evapotranspiration (metres per year) and R_o is the total run-off, including groundwater discharge and interflow (cubic metres per metre of catchment per year).

If the area of the recharge zone of an aquifer is known, the annual volume of recharge to the aquifer is given by

$$Q = (P - E_{ta} - R_o)A \qquad (9.2)$$

where Q is the annual volume of groundwater recharge (cubic metres per year) and A is the area of the recharge zone (square metres). If the potential evapotranspiration (E_{tp}) is used in eqn (9.2) instead of E_{ta}, a more conservative estimate of groundwater recharge is obtained, because E_{tp} always exceeds E_{ta}. Alternatively, if the infiltration rate over the recharge

area has been determined using a percolation gauge or by analysing groundwater discharge, then

$$Q = f_e A \qquad (9.3)$$

where f_e is the effective infiltration rate (metres per year).

The reliability of water budget methods depends, to some extent, upon how accurately the values of the variables are known or can be estimated. For good results, quite intensive instrumentation of the recharge area is necessary. Another limitation of this technique is the assumption that water infiltrating the surface of the recharge area eventually will reach any well located down gradient. Some potential groundwater recharge may become interflow and be discharged to a surface water course without even reaching the water table. In addition, some proportion of the water that has succeeded in percolating down to the zone of saturation may be lost as groundwater discharge to rivers and springs. The perennial yield to wells will only be a fraction of the total recharge volume calculated. Conversely, recharge may take place as a result of seepage from rivers or other bodies of surface water. In such circumstances, the use of river gauging or tracer techniques may be necessary to identify and quantify the leakage.

The water budget technique attempts to quantify groundwater recharge by considering the balance between precipitation, actual evapotranspiration and surface run-off in a given area. However, in areas which have highly permeable surface strata, such as limestone areas, there may be little or no surface run-off. Consequently, any increase in groundwater storage is equal to the difference between the precipitation (P) and the actual evapotranspiration (E_{ta}) over the recharge area (A), so that the annual volume of recharge (Q) is given by

$$Q = (P - E_{ta})A \qquad (9.4)$$

Unfortunately, some water is likely to be lost as interflow or groundwater discharge to neighbouring catchments, and this adds some element of uncertainty to the use of eqn (9.4).

If a region has very permeable surface deposits, but also has some sizeable streams and rivers, it is quite possible (or even probable) that these water courses are maintained by groundwater discharge. In an area in which a very high proportion of the available groundwater resource discharges naturally to surface streams, it may be possible to quantify the surplus capacity of an aquifer from an analysis of river discharge hydrographs. The assumption inherent in this method is that the aquifer is overflowing into the surface water courses, and that this water could be diverted to wells instead. Thus the amount of water that is available, assuming that natural overflow is stopped, is equal to the total groundwater component of river discharge.

Most aquifers discharge either directly or indirectly to rivers and seas by way of seepage and springs. The most common form of groundwater discharge is that to a river. The groundwater discharge, which becomes the base-flow of a river, is the outflow from unconfined or artesian aquifers bordering the river, which go on discharging more and more slowly with time as the differential head falls. Bank storage is the water temporarily held in store in the ground adjacent to the river between the low and high water levels. This water is released as the river level falls.

The groundwater contribution to stream flow is most important during the dry season, when surface run-off is reduced as a result of soil moisture deficits. During the dry months, a very high proportion of stream flow may be derived from groundwater sources. In fact, it may be the groundwater contribution to flow during this period that prevents streams from drying up.

9.9 Hydrogeological properties

The porosity and permeability are the two most important factors governing the accumulation, migration and distribution of groundwater. Both may change within a rock or soil mass during the course of its geological evolution.

The porosity can be defined as the percentage pore space within a given volume of rock. The total or absolute porosity is a measure of the total void volume, and is the excess of bulk volume over grain volume per unit of bulk volume. The effective or net porosity is a measure of the effective void volume of a porous medium, and is determined as the excess of bulk volume over grain volume and occluded pore (i.e. a pore with no connection to others) volume. It may be regarded as the pore space from which water can be removed. The porosity of a deposit does not

necessarily provide an indication of the amount of water that can be obtained therefrom.

The capacity of a material to yield water is of greater importance than its capacity to hold water, as far as supply is concerned. Even though a rock or soil may be saturated, only a certain proportion of water can be removed by drainage under gravity or by pumping, the remainder being held in place by capillary or molecular forces. The ratio of the volume of water retained to the total volume of rock or soil, expressed as a percentage, is referred to as the specific retention. The amount of water retained varies directly in accordance with the surface area of the pores and indirectly with regard to the pore space.

The specific yield of a rock or soil refers to its water-yielding capacity, attributable to gravity drainage, that occurs when the water table declines. It is the ratio of the volume of water, after saturation, that can be drained by gravity to the total volume of the aquifer, expressed as a percentage. The specific yield plus the specific retention is equal to the porosity of the material when all the pores are interconnected. The relationship between the specific yield and particle size distribution is shown in Fig. 9.11. Examples of the specific yields of certain common types of soil and rock are given in Table 9.3 (it must be appreciated that individual values of the specific yield can vary considerably from those quoted).

The storage coefficient or storativity of an aquifer is defined as the volume of water released from or taken into storage per unit surface area of the aquifer per unit change in head normal to that surface. It is a dimensionless quantity. Changes in storage in an unconfined aquifer represent the product of the

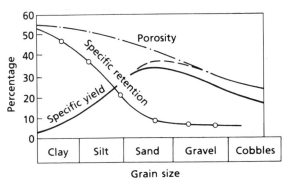

Fig. 9.11 Relationship between grain size, porosity, specific retention and specific yield. (From Bear, 1979.)

Table 9.3 Some examples of specific yield.

Materials	Specific yield (%)
Gravel	15–30
Sand	10–30
Dune sand	25–35
Sand and gravel	15–25
Loess	15–20
Silt	5–10
Clay	1–5
Till (silty)	4–7
Till (sandy)	12–18
Sandstone	5–25
Limestone	0.5–10
Shale	0.5–5

volume of the aquifer between the water table before and after a given period of time and the specific yield. The storage coefficient of an unconfined aquifer virtually corresponds to the specific yield, as more or less all the water is released from storage by gravity drainage, and only an extremely small part results from compression of the aquifer and the expansion of water. In confined aquifers, water is not yielded simply by gravity drainage from the pore space, because there is no falling water table, and the material remains saturated. Hence, other factors are involved with regard to yield, such as the consolidation of the aquifer and the expansion of water consequent upon the lowering of the piezometric surface. Therefore, much less water is yielded by confined than unconfined aquifers.

In ordinary hydraulic usage, a material is termed permeable when it permits the passage of a measurable quantity of fluid in a finite period of time, and impermeable when the rate at which it transmits that fluid is slow enough to be negligible under existing temperature–pressure conditions. The permeability of a particular soil or rock type is defined by its coefficient of permeability or hydraulic conductivity, this being the flow in a given time through a unit cross-sectional area of a material (Table 9.4). The transmissivity, or flow in cubic metres per day, through a section of aquifer 1 m wide, under a hydraulic gradient of unity, sometimes is used as a convenient quantity in the calculation of groundwater flow instead of the hydraulic conductivity.

The permeability of intact rock (primary permeability) is usually several orders less than the *in situ*

permeability (secondary permeability). In other words, permeability in the field generally is governed by fissure flow.

Other things being equal, the velocity of flow of water between two points is directly proportional to the difference in head between them, water flowing from areas of high head to areas of low head. The hydraulic gradient refers to the loss of head or energy of water flowing through the ground. This loss of energy by the water is due to the friction resistance of the ground material. If it is assumed that the resistance to flow is constant, then for a given difference in head, the flow velocity is directly proportional to the length of the flow path.

The basic law concerned with flow is that enunciated by Darcy (1856), which states that the rate of flow v per unit area is proportional to the gradient of the potential head i measured in the direction of flow

$$v = ki \qquad (9.5)$$

and, for a particular rock or soil, or part of it, of area A

$$Q = vA = Aki \qquad (9.6)$$

where Q is the quantity in a given time and k is the coefficient of permeability. Darcy's law is valid as long as laminar flow exists. Departures from this law therefore occur when the flow is turbulent, such as when the velocity of flow is high. Such conditions exist in very permeable media, normally when the Reynolds number exceeds four. Accordingly, it usually is accepted that this law can be applied to those soils which have finer textures than gravels. Furthermore, Darcy's law probably does not accurately represent the flow of water through a porous medium of extremely low permeability, because of the influence of surface and ionic phenomena, and the presence of gases.

When considering the general case of flow in a porous medium, it is assumed that the medium is isotropic and homogeneous as far as the permeability is concerned. If an element of saturated material is taken, with the dimensions dx, dy and dz (Fig. 9.12), and flow is taking place in the x–y plane, then the generalized form of Darcy's law is

$$v_x = k_x i_x \qquad (9.7)$$

$$v_x = k_x \frac{\delta h}{\delta x} \qquad (9.8)$$

Table 9.4 Relative values of permeability.

Rock type	Porosity Primary (grain) (%)	Porosity Secondary (fracture)*	Permeability range (m s⁻¹) / Well yield	Type of water-bearing unit
Sediments, unconsolidated				
Gravel	30–40		Very high / High	Aquifer
Coarse sand	30–40		High / Medium	Aquifer
Medium to fine sand	25–35		Medium / Medium	Aquifer
Silt	40–50	Occasional	Low / Low	Aquiclude
Clay, till	45–55	Often fissured	Very low / Low	Aquiclude
Sediments, consolidated				
Limestone, dolostone	1–50	Solution joints, bedding planes	Very high–Low / High	Aquifer or aquiclude
Coarse, medium sandstone	<20	Joints and bedding planes	High / Medium	Aquifer or aquiclude
Fine sandstone	<10	Joints and bedding planes	Medium /	Aquifer or aquiclude
Shale, siltstone	—	Joints and bedding planes	Low / Low	Aquifer or aquiclude
Volcanic rocks, e.g. basalt	—	Joints and bedding planes	Very high–Medium / High	Aquifer or aquiclude
Plutonic and metamorphic rocks	—	Weathering and joints decreasing as depth increases	Medium–Very low /	Aquiclude or aquifer

Permeability range (m s⁻¹): 10^0 (Very high), 10^{-2} (High), 10^{-4} (Medium), 10^{-6} (Low), 10^{-8} (Very low), 10^{-10} (Impermeable). Well yield: High, Medium, Low.

* Rarely exceeds 10%.

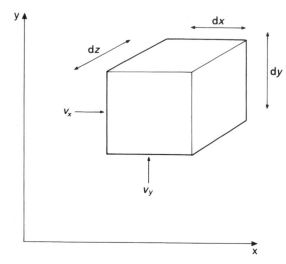

Fig. 9.12 Seepage through a soil element.

and

$$v_y = k_y i_y \tag{9.9}$$

$$v_y = k_y \frac{\delta h}{\delta y} \tag{9.10}$$

where h is the total head under steady-state conditions and k_x, i_x and k_y, i_y are the coefficients of permeability and the hydraulic gradients in the x and y directions, respectively. Assuming that the fabric of the medium does not change and that the water is incompressible, the volume of water entering the element is the same as that leaving in any given time; hence

$$v_x \mathrm{d}y\mathrm{d}x + v_y \mathrm{d}x\mathrm{d}z = \left(v_x + \frac{\delta v_x}{\delta x}\mathrm{d}x \right)\mathrm{d}y\mathrm{d}z$$
$$+ \left(v_y + \frac{\delta v_y}{\delta y}\mathrm{d}y \right)\mathrm{d}x\mathrm{d}z \tag{9.11}$$

In such a situation, the difference in volume between the water entering and leaving the element is zero, therefore

$$\frac{\delta v_x}{\delta x} + \frac{\delta v_y}{\delta y} = 0 \tag{9.12}$$

Generally, it is the interconnected system of discontinuities which determines the permeability of a particular rock mass. The permeability of a jointed rock

mass is usually several orders higher than that of intact rock. According to Serafim (1968), the following expression can be used to derive the filtration through a rock mass intersected by a system of parallel-sided joints with a given opening e separated by a given distance d:

$$k = \frac{e^3 \gamma_w}{12 d \mu} \tag{9.13}$$

where γ_w is the unit weight of water and μ is its viscosity.

9.10 Groundwater exploration

A groundwater investigation requires a thorough appreciation of the hydrology and geology of the area concerned, and a groundwater inventory needs to determine the possible gains and losses affecting the subsurface reservoir. Of particular interest is information concerning the lithology, stratigraphical sequence and geological structure, as well as the hydrogeological characteristics of the subsurface materials. Also of importance are the positions of the water table and piezometric level, and their fluctuations.

In major groundwater investigations, records of precipitation, temperatures, wind movement, evaporation and humidity may provide essential or useful supplementary information. Similarly, data relating to stream flow may be of value in helping to solve the groundwater equation, because seepage into or from streams constitutes a major factor in the discharge or recharge of groundwater. The chemical and bacterial qualities of groundwater obviously require investigation.

Essentially, an assessment of groundwater resources involves the location of potential aquifers within economic drilling depths. Whether or not an aquifer will be able to supply the required amount of water depends on its thickness and spatial distribution, porosity and permeability, whether it is fully or partially saturated and whether or not the quality of the water is acceptable. The pumping lift and effect of drawdown upon an aquifer must also be considered.

The desk study involves a consideration of the hydrological, geological, hydrogeological and geophysical data available concerning the area in question. Particular attention should be given to assessing the lateral and vertical extent of any potential aquifers, their continuity and structure, any possible variations in formation characteristics and possible areas of recharge and discharge. Additional information relating to groundwater chemistry, the outflow of springs and surface run-off, data from pumping tests, mine workings and waterworks and meteorological data should be considered. Information on vegetative cover, land utilization, topography and drainage pattern can prove to be of value at times.

Aerial photographs may aid the recognition of broad rock and soil types, and thereby help to locate potential aquifers. The combination of topographical and geological data may help to identify areas of probable groundwater recharge and discharge. In particular, the nature and extent of superficial deposits may provide some indication of the distribution of areas of recharge and discharge. Aerial photographs allow the occurrence of springs to be recorded.

Variations in water content in soils and rocks, which may not be readily apparent on black and white photographs, often are clearly depicted by false colour. The specific heat of water usually is 2–10 times greater than that of most rocks, which therefore facilitates its detection in the ground. Indeed, the specific heat of water can cause an aquifer to act as a heat sink which, in turn, influences the near-surface temperatures.

Furthermore, because the occurrence of groundwater is influenced strongly by the nature of the ground surface, aerial photographs can yield useful information. In addition, the vegetative cover may be identifiable from aerial photographs, and as such may provide some clue as to the occurrence of groundwater. In arid and semi-arid regions, the presence of phreatophytes, that is, plants which have a high transpiration capacity and derive water directly from the water table, indicates that the water table is near the surface. By contrast, xerophytes can exist at low moisture contents in soil, and their presence suggests that the water table is at an appreciable depth. Thus, groundwater prediction maps sometimes can be made from aerial photographs. These can be used to help to locate the sites of test wells.

Geological mapping frequently forms the initial phase of exploration, and should identify potential aquifers, such as sandstones and limestones, and dis-

tinguish them from aquicludes. Superficial deposits may perform a confining function in relation to the major aquifers that they overlie or, because of their lithology, they may play an important role in controlling the recharge to major aquifers. Moreover, geological mapping should locate igneous intrusions and major faults, and it is important during the mapping programme to establish the geological structure.

Isopachyte maps can be drawn to show the thickness of a particular aquifer and the depth below the surface of a particular bed. They can be used to estimate the positions and depths of drill-holes. They also provide an indication of the distribution of potential aquifers.

Maps showing groundwater contours are compiled when there are a sufficient number of observation wells to determine the configuration of the water table (Fig. 9.13). Data on the surface water levels in reservoirs and streams that have free connection with the water table should also be used in the production of such maps. These maps are usually compiled for the periods of maximum, minimum and mean annual positions of the water table. A water table contour map is most useful for studies of unconfined groundwater.

As groundwater moves from areas of higher potential towards areas of lower potential, and as the contours on groundwater contour maps represent lines of equal potential, the direction of groundwater flow moves from highs to lows at right angles to the contours or equipotential lines. The analysis of the conditions revealed by groundwater contours is made in accordance with Darcy's law. Accordingly, the spacing of contours is dependent on the flow rate, aquifer thickness and aquifer permeability. If continuity of the flow rate is assumed, the spacing depends upon the aquifer thickness and permeability. Hence, areal changes in contour spacing may be indicative of changes in aquifer conditions. However, because of the heterogeneity of most aquifers, changes in gradient must be carefully interpreted in relation to all factors. The shape of the contours portraying the position of the water table helps to indicate where areas of recharge and discharge of groundwater occur. Groundwater mounds can result from the downward seepage of surface water. In an ideal situation, the gradient from the centre of such a recharge area will decrease radially and at a declining rate. An

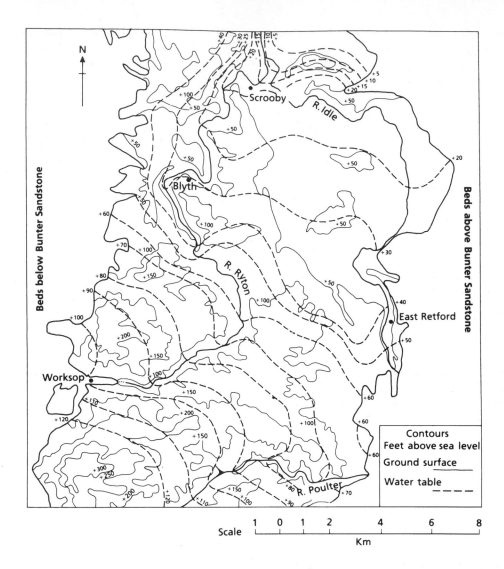

Fig. 9.13 Sketch map of part of Nottinghamshire showing the water table in the Bunter Sandstone (now Sherwood Sandstone).

impermeable boundary or change in transmissivity will affect this pattern.

Depth-to-water table maps show the depth to water from the ground surface. They are prepared by overlaying a water table contour map on a topographical map of the same area and scale, and recording the differences in values at the points at which the two types of contour intersect. Depth-to-water contours are then interpolated in relation to these point. A map indicating the depth to the water table can al provide an indication of areas of recharge and discharge. Both are most likely to occur where the wat table approaches the surface.

Water level change maps are constructed by plotting the change in the position of the water tab recorded at wells during a given interval of time (Fi 9.14). The effect of local recharge or discharge oft shows as distinct anomalies on water level chan maps, for example, the maps may indicate that t

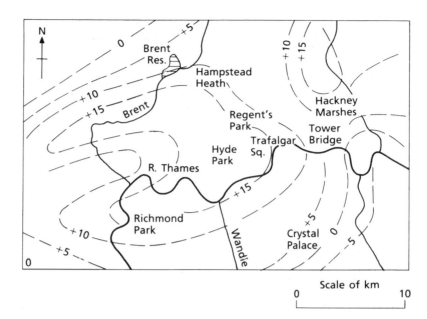

9.14 Changes in groundwater els in the Chalk below London, 55–1980 (all contours in metres). er Marsh & Davies, 1983.)

undwater levels beneath a river have remained stant while falling everywhere else. This would gest an influent relationship between the river and aquifer. Hence, such maps can help to identify the ations at which there are interconnections between ace water and groundwater. These maps also nit an estimation to be made of the change in undwater storage which has occurred during the e of time involved.

eophysical methods of exploration, together with ing, are dealt with in Chapter 16.

Assessment of permeability and flow he field

nitial assessment of the magnitude and variability e *in situ* hydraulic conductivity can be obtained tests carried out in boreholes as the hole is nced. By artificially raising the level of water in orehole (falling head test) above that in the sur- ding ground, the flow rate from the borehole e measured. However, in very permeable soils, it not be possible to raise the level of water in the nole. Conversely, the water level in the borehole be artificially depressed (rising head test), so ing the rate of water flow into the borehole to sessed. Wherever possible, a rising and a falling

head test should be carried out at each required level, and the results averaged.

In a rising or falling head test in which the piezo- metric head varies with time, the permeability is determined from the expression

$$k = \frac{A}{F(t_2 - t_1)} \ln\left(\frac{h_1}{h_2}\right) \tag{9.14}$$

where h_1 and h_2 are the piezometric heads at times t_1 and t_2, respectively, A is the inner cross-sectional area of the casing in the borehole and F is an intake or shape factor. Where a borehole of diameter D is open at the base and cased throughout its depth, $F = 2.75D$. If the casing extends throughout the perme- able bed to an impermeable contact, then $F = 2D$.

The constant head method of *in situ* permeability testing is used when the rise or fall in the level of water is too rapid for accurate recording (i.e. occurs in less than 5 min). This test normally is conducted as an inflow test, in which the flow of water into the ground is kept under a sensibly constant head (e.g. by adjusting the rate of flow into the borehole so as to maintain the water level at a mark on the inside of the casing near the top) (Sutcliffe & Mostyn, 1983). The method is only applicable to permeable ground, such as gravels, sands and broken rock, when there is a negligible or zero time for equalization. The rate of flow Q is measured once a steady flow into and out of the borehole has been attained over a period of some 10 min. The permeability k is derived from the following expression

$$k = Q/Fh_c \qquad (9.15)$$

where F is the intake factor and h_c is the applied constant head.

The permeability of an individual bed of rock can be determined by a water-injection or packer test carried out in a drill-hole (Bliss & Rushton, 1984; Brassington & Walthall, 1985). This is performed by sealing off a length of uncased hole with packers, and injecting water under pressure into the test section (Fig. 9.15). Usually, because it is more convenient, packer tests are carried out after the entire length of a hole has been drilled. Two packers are used to seal off selected test lengths, and the tests are performed from the base of the hole upwards. The hole must be flushed to remove sediment prior to a test being performed. With double packer testing, the variation in hydraulic conductivity throughout the test hole can be determined. The rate of flow of water over the test length is measured under a range of constant pressures, and recorded. The permeability is calculated from a flow–pressure curve. Water generally is pumped into the test section at steady pressures for periods of 15 min. The test usually consists of five cycles at successive pressures of 6, 12, 18, 12 and 6 kPa for every metre depth of the packer below the surface. The evaluation of the permeability from packer tests normally is based upon methods using a relationship of the form

$$k = \frac{Q}{C_s r h} \qquad (9.16)$$

where Q is the steady flow rate under an effective applied head h (corrected for friction losses), r is the radius of the drill-hole and C_s is a constant depending upon the length and diameter of the test section.

Field pumping tests allow the determination of the coefficients of permeability and storage, as well as the transmissivity, of a larger mass of ground than the aforementioned tests. A pumping test involves the abstraction of water from a well at known discharge rate(s) and the observation of the resulting water levels as drawdown occurs (Lovelock et al., 1975; Anon., 1983). At the same time, the behaviour of the

water table in the aquifer can be recorded in observation wells radially arranged about the abstraction well. There are two types of pumping test, the constant-pumping-rate aquifer test and the step-performance test. In the former test, the rate of discharge is constant, whereas in a step-performance

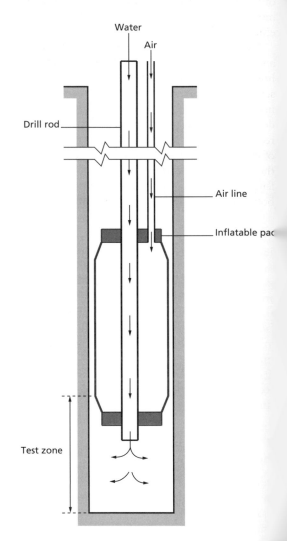

Fig. 9.15 Packer test equipment. The zone of rock to be tested in a drill-hole is isolated by using two packers which seal off the drill-hole, the water being pumped into the space between the packers. An alternative method, as shown here, which can be carried out only as drilling proceeds, is to use a single packer, testing the zone between the bottom of the packer and the base of the drill-hole. The average flow under equilibrium conditions is obtained from a metered water supply acting under a known pressure and gravity head.

test, there are a number of stages, each of equal length of time, but at different rates of discharge (Clark, 1977). The step-performance test usually is carried out before the constant-pumping-rate aquifer test. Yield drawdown graphs are plotted from the information obtained (Fig. 9.16). The hydraulic efficiency of the well is indicated by the nature of the curve(s), more vertical and straighter curves indicate a more efficient well.

A flowmeter log provides a record of the direction and velocity of groundwater movement in a drill-hole. Flowmeter logging requires the use of a velocity-sensitive instrument, a system for lowering the instrument into the hole, a depth measuring device to determine the position of the flowmeter and a recorder located at the surface. The direction of flow of water is determined by slowly lowering and raising the flowmeter through a section of hole, 6–9 m in length, and recording velocity measurements during both traverses. If the velocity measured is greater

during the downward than the upward traverse, the direction of flow is upwards, and vice versa. A flowmeter log made while a drill-hole is being pumped at a moderate rate, or by allowing water to flow if there is sufficient artesian head, permits the identification of the zones contributing to the discharge. It also provides information on the thickness of these zones and the relative yield at that discharge rate. Because the yield varies approximately directly with the drawdown of the water level in a well, flowmeter logs made by pumping should be pumped at least at three different rates. The drawdown of the water level should be recorded for each rate.

A number of different types of tracer have been used to investigate the movement of groundwater and the interconnection between surface and groundwater resources. The ideal tracer should be easy to detect quantitatively in minute concentrations, should not change the hydraulic characteristics or be adsorbed by the medium through which it is flowing,

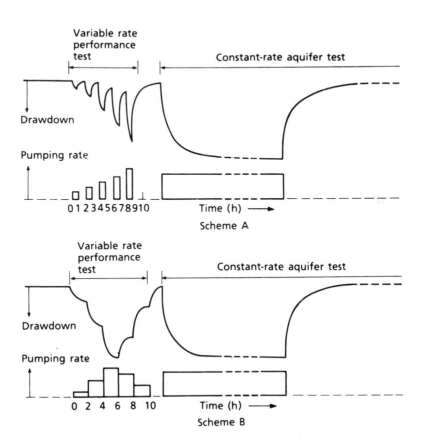

Fig. 9.16 Pumping test procedure.

should be more or less absent from, and should not react with, the groundwater concerned and should have a low toxicity. The types of tracer in use include water-soluble dyes which can be detected by colorimetry, sodium chloride or sulphate salts which can be detected chemically, strong electrolytes which can be detected by electrical conductivity and radioactive tracers. One of the advantages of the latter is that they can be detected in minute quantities in water. Radioactive tracers should have a useful half-life and should present a minimum hazard, for example, tritium is not one of the best tracers because of its relatively long half-life. In addition, because it is introduced as tritiated water, it is preferentially adsorbed by any montmorillonite present.

When a tracer is injected via a drill-hole into groundwater, it is subject to diffusion, dispersion, dilution and adsorption. Dispersion is a result of very small variations in the velocity of laminar flow through porous media. Molecular diffusion probably is negligible unless the velocity of flow is unusually low. Even if these processes are not significant, flow through an aquifer may be stratified or concentrated along discontinuities. Therefore, a tracer may remain undetected unless the observation drill-holes intersect these discontinuities.

The determination of the hydraulic conductivity in the field can be performed by measuring the time it takes for a tracer to move between two test holes. Like pumping tests, this tracer technique is based on the assumption that an aquifer is homogeneous and that observations taken radially at the same distance from the well are comparable. This method of assessing the hydraulic conductivity requires that the injection and observation wells are close together so as to avoid excessive travel time and so that the direction of flow is known.

9.12 Water quality and uses

In any evaluation of water resources, the quality of the water is of almost equal importance to the quantity available. In other words, the physical, chemical and biological characteristics of the water are of major importance in determining whether or not it is suitable for domestic, industrial or agricultural use. However, the number of major dissolved constituents in groundwater is quite limited, and natural variations are not as large as might be expected. Finally,

the quality of water in the zone of saturation reflects the quality of the water which has percolated to the water table, and the subsequent reactions that occur between water and rock. The factors which influence the solute content include the original chemical quality of the water entering the zone of saturation, the distribution, solubility, exchange capacity and exchange selectivity of the minerals involved in the reaction, the porosity and permeability of the rocks and the flow path of the water. Of critical importance in this context is the residence time of the water, because this determines whether there is sufficient time for the dissolution of minerals to proceed to the point at which the solution is in equilibrium with the reaction. The residence time depends on the rate of groundwater movement, and this usually is very slow beneath the water table.

The quality of groundwater often compares favourably with that of surface sources, and the groundwater from deep aquifers, in particular, can be remarkably pure. However, this does not mean that the quality can be relied upon. In fact, groundwater can undergo cyclic changes in quality, and natural variations in groundwater quality can occur with depth and rock type. Thus, the purity of groundwater should not be taken for granted, and untreated water should not be used directly for supply.

The quality of water in the zone of saturation is affected by the fluctuation of the water table. In particular, if the water table occurs at shallow depth losses by transpiration and, possibly, evaporation will increase when it rises. This means that the salt content will increase. Conversely, when the water table is lowered, this may cause lateral inflow from surrounding areas, with a consequent change in salinity.

The uppermost layers of soil and rock act as a purifying agent. In the soil, organisms such as fungi and bacteria attack pathogenic bacteria, and react with certain other harmful substances. The other important factor in purifying groundwater is the filtering action of soil and rock. This depends on the size of the pores, the proportion of argillaceous and organic matter present and the distance travelled by the water involved. Unconfined water at shallow depth is highly susceptible to pollution but, at greater depth recharge is by water which has been partially or wholly purified.

There is a frequent tendency for the salt content of

groundwater to increase with depth. The reasons for this are several-fold. First, the greater the depth at which the groundwater occurs, the slower its movement, and so the less likely it is to be replaced by other water, especially that infiltrating from the surface. Second, a longer residence time of water provides more time for reaction with the host rock, and so more material goes into solution until an equilibrium condition is attained. In addition, connate or fossil water may occur at greater depth. As the character of groundwater in an aquifer frequently changes with depth, it is possible at times to recognize zones of different quality of groundwater. With increasing depth, cation exchange reactions increase in importance, and there is a gradual replacement of calcium and magnesium in the water by sodium. Any nitrates present near the surface of an aquifer invariably decrease with depth. On the other hand, sulphates tend to increase with depth. However, at appreciable depth, sulphates are reduced, which produces a low-sulphate–high-bicarbonate water. The chloride content also tends to increase with depth. In fact, with increasing depth, groundwater may become non-potable, due to its high chloride content. Most highly saline, chloride groundwater (not associated with evaporites) occurring at depth, where groundwater circulation is restricted, is connate or fossil water.

The temperature, colour, turbidity, odour and taste are the most important physical properties of water in relation to water supply. Groundwater undergoes appreciable fluctuations in temperature only at shallow depth, beneath which temperatures remain relatively constant. In fact, the depth at which temperatures are more or less uniform occurs at about 10 m in the tropics, increasing to about 20 m in polar regions, although the rock type, elevation, precipitation and wind can produce significant local deviations. Below the zone of surface influence, groundwater temperatures increase by approximately 1 °C for every 30 m of depth, that is, in accordance with the geothermal gradient. The colour of groundwater may be attributable to organic or mineral matter carried in solution. For example, light to dark brown discolorations can occur in groundwater which has been in contact with peat or other organic deposits. Brownish discoloration can also result from groundwater which contains dissolved ferrous iron being exposed to the atmosphere. This leads to the formation of insoluble ferric hydroxides.

The turbidity of groundwater is mainly caused by the presence of clay and silt particles derived from an aquifer. The oxidation of dissolved ferrous iron to form insoluble ferric hydroxides also contributes towards the turbidity. The natural filtration which occurs when groundwater flows through unconsolidated deposits largely removes such material from groundwater. Tastes and odours may be derived from the presence of mineral matter, organic matter, bacteria or dissolved gases.

Standard tests used to determine the safety of water for drinking purposes involve identifying whether or not bacteria belonging to the coliform group are present. One of the reasons for this is that this group of bacteria (*Escherichia coli* in particular) is relatively easy to recognize. Because the whole coliform group is foreign to water, a positive *E. coli* test indicates the possibility of bacteriological contamination. If *E. coli* is present, it is possible that the less numerous or harmful pathogenic bacteria, which are much more difficult to detect, are also present. On the other hand, if there are no *E. coli* in a sample of water, the chances of faecal contamination and of pathogens being present generally are regarded as negligible. The results of coliform tests are reported in terms of the most probable number (*MPN*) of coliform group organisms present in a given volume of water. Viruses in groundwater are more critical than bacteria, in that they tend to survive longer and some are more resistant to disinfectant. In addition, one virus unit (one plague-forming unit in a cell culture) may cause an infection when ingested. By contrast, the ingestion of thousands of pathogenic bacteria may be required before clinical symptoms are developed. Fortunately, groundwater, except perhaps from very shallow aquifers, generally is free from pathogenic bacteria and viruses. Microorganisms can be carried by groundwater, but tend to attach themselves by adsorption to the surfaces of clay particles. In fine grained soils, bacteria generally move less than a few metres, but they can migrate over much larger distances in coarse grained soils or discontinuous rocks. The maximum rate of travel of bacterial pollution appears to be about two-thirds of the rate of groundwater movement. Viruses which retain all their characteristics for more than 50 days may migrate 250 m or more in soils where organic matter is present to supply a food source. The recommended safe distances between domestic wells and sources of pollution in non-karstic terrains are indicated in Table 9.5. Nonetheless, the biological pollution of groundwater generally does not occur because the soil represents a fairly effective filter between a source of pollution and the water table. Pathogenic bacteria, viruses and other microorganisms not native to the subsurface environment generally do not multiply underground and eventually die.

Most cases of contamination are associated with poor well construction, overabstraction or aquifers which possess large pores such as gravel deposits, or open discontinuities such as some limestones. Both afford connection between the surface water, which may be polluted, and groundwater. In cavernous or fissured limestone, the distances travelled may be several kilometres.

The chemical elements present in groundwater are derived from precipitation which infiltrates into the ground, organic processes which take place in the soil and the breakdown of minerals in the rocks through which the groundwater flows (Table 9.6). The solution of carbonates, especially calcium and magnesium carbonate, is due principally to the formation of weak carbonic acid in the soil horizons where CO_2 is dis-

Table 9.5 Recommended safe distances between domestic water wells and sources of pollution. (From Hamill & Bell, 1986.)

Source of pollution	Distance (m)
Septic tank	15
Cess pit	45
Sewage farm	30
Infiltration ditch	30
Percolation zone	30
Pipe with watertight joints	3
Other pipes	15
Dry well	15

Table 9.6 Examples of groundwater quality from different types of rock mass (in mg l⁻¹).

Rock type	TDS	Ca	Mg	HCO₃	Na	K	Cl	SO₄	Fe	SiO₄	Location
Granite	223	27	6.2	93	9.5	1.4	5.2	32	1.6	39	Maryland
Rhyolite	148	12	2.2	80	6.8	0.6	2.0	0.1	1.1	39	North Carolina
Gabbro	359	32	16	203	25	1.1	13	10	0.06	56	North Carolina
Basalt	505	62	28	294	24	—	37	30	—	30	Hyderabad
Diorite	346	72	4.1	114	10	2.8	6.5	115	0.04	22	North Carolina
Syenite	80	9.5	2.3	38	2.8	0.6	2.1	2.8	0.14	19	New York
Andesite	70	12	0.5	38	1.8	2.6	—	6.3	—	8.9	Idaho
Quartzite	52	1.6	5.8	18	2.8	—	9.9	2.0	—	8	Transvaal
Marble	236	39	10	162	2.7	0.3	3.8	2.4	0.03	9.9	Alabama
Schist	221	27	5.7	138	16	0.7	2.5	9.6	0.11	21	Georgia
Gneiss	135	19	5.1	39	4.4	3.2	5.8	30	0.09	13	Connecticut
Sandstone	—	60	60	—	—	—	22	38	—	—	Northumberland
Sandstone	210	40	12	67	7.6	0.4	19	26	—	12	Worcestershire
Sandstone	439	65	38	326	44	—	63	79	—	14	France
Arkose	101	9.6	1.9	38	5.1	—	1.8	7.4	0.2	35	Colorado
Greywacke	553	74	20	381	34	1.2	2.7	26	0.62	12	New York
Limestone	247	48	5.8	168	4	0.7	0.1	4.8	0.05	8.9	Florida
Limestone	720	124	28	460	14	3	18	57	0.22	9.2	Tennessee
Chalk	384	115	5	152	10.2	1.2	20	39	—	1.1	Hertfordshire
Chalk	491	125	9	—	18	4	24	150	—		Lincolnshire
Dolostone	546	67	39	390	7.6	0.4	—	17	—	24	Transvaal
Gypsum	2480	636	43	143	16.1	0.9	24	1570	—	29	New Mexico
Lignite	2580	74	53	702	624	5.4	25	1080	0.9	11	North Dakota
Shale	260	29	16	126	12	1.1	12	22	0.02	16	New Jersey
Shale	1100	123	70	539	61	2.2	3.5	283	1.3	19	Ohio
Alluvium	371	45	20	207	16	2.6	17	35	0.05	25	Nevada
Glacial deposits	548	86	27	33	5.1	3	6	60	—	24	Minnesota

TDS, total dissolved solids.

solved by soil water. Calcium in sedimentary rocks is derived from calcite, aragonite, dolomite, anhydrite and gypsum. In igneous and metamorphic rocks, calcium is supplied by the feldspars, pyroxenes and amphiboles, and the less common minerals, such as apatite and wollastonite.

Because of its abundance, calcium is one of the most common ions in groundwater. When calcium carbonate is attacked by carbonic acid, bicarbonate is formed. Calcium carbonate and bicarbonate are the dominant constituents found in the zone of active circulation, and for some distance under the cover of younger strata. The normal concentration of calcium in groundwater ranges from 10 to 100 mgl^{-1}. Such concentrations have no effect on health, and it has been suggested that as much as 1000 mgl^{-1} may be harmless.

Magnesium, sodium and potassium are less common cations, and sulphate and chloride and, to some extent, nitrate are less common anions, although in some groundwater the latter may be present in significant concentrations. Dolomite is the common source of magnesium in sedimentary rocks. The rarer evaporite minerals, such as epsomite, kieserite, kainite and carnallite, are not significant contributors. Olivine, biotite, hornblende and augite are among those minerals which make significant contributions in igneous rocks, and serpentine, talc, diopside and tremolite are amongst the metamorphic contributors. Despite the higher solubilities of most of its compounds (magnesium sulphate and magnesium chloride are both very soluble), magnesium usually occurs in lesser concentrations in groundwater than calcium. In this case, common concentrations of magnesium range from about 1 to 40 mgl^{-1}; concentrations above 100 mgl^{-1} rarely are encountered.

Sodium does not occur as an essential constituent of many of the principal rock forming minerals, plagioclase feldspar being the exception. Plagioclase is the primary source of most sodium in groundwater; in areas of evaporitic deposits, halite is important. Sodium salts are highly soluble and will not precipitate unless concentrations of several thousand parts per million are reached. The only common mechanism for the removal of large amounts of sodium ions from water is ion exchange, which operates if the sodium ions are in great abundance. The conversion of calcium bicarbonate to sodium bicarbonate accounts for the removal of some sodium ions from sea water which has invaded fresh water aquifers. This process is reversible. All groundwater contains measurable amounts of sodium, usually up to 20 mgl^{-1}.

Common sources of potassium are the feldspars and micas of the igneous and metamorphic rocks. Potash minerals, such as sylvite, occur in certain evaporitic sequences, but their contribution is not important. Although the abundance of potassium in the Earth's crust is similar to that of sodium, its concentration in groundwater usually is less than one-tenth that of sodium. Most groundwater contains less than 10 mgl^{-1}. Like sodium, potassium is highly soluble, and therefore is not easily removed from water except by ion exchange.

Sedimentary rocks, such as shales and clays, may contain pyrite or marcasite from which sulphur can be derived. Most sulphate ions probably are derived from the solution of calcium and magnesium sulphate minerals found in evaporitic sequences, gypsum and anhydrite being the most common. The concentration of sulphate ions in water can be affected by sulphate-reducing bacteria, the products of which are hydrogen sulphide and carbon dioxide. Hence, a decline in sulphate ions frequently is associated with an increase in bicarbonate ions. The concentration of sulphate in groundwater usually is less than 100 mgl^{-1}, and may be less than 1 mgl^{-1} if sulphate-reducing bacteria are active.

The chloride content of groundwater may be due to the presence of soluble chlorides from rocks, saline intrusion, connate and juvenile water or contamination by industrial effluent or domestic sewage. In the zone of circulation, the chloride ion concentration normally is relatively small, and it is a minor constituent in the Earth's crust, sodalite and apatite being the only igneous and metamorphic minerals containing chlorite as an essential constituent. Halite is one of the principal mineral sources. As with sulphate ions, the atmosphere probably makes a significant contribution to the chloride content of surface waters. These, in turn, contribute to the groundwater. Usually, the concentration of chloride in ground water is less than 30 mgl^{-1}, but concentrations of 1000 mgl^{-1} or more are common in arid regions.

Nitrate ions generally are derived from the oxidation of organic matter with a high protein content. Their presence may indicate a pollution source, and their occurrence usually is associated with shallow groundwater sources. Concentrations in fresh water do not generally exceed $5 \, mg \, l^{-1}$, although in rural areas, where nitrate fertilizer is liberally applied, concentrations may exceed $600 \, mg \, l^{-1}$.

Although silicon is the second most abundant element in the Earth's crust and is present in almost all the principal rock forming minerals, its low solubility means that it is not one of the most abundant constituents of groundwater. Groundwater generally contains between 5 and $40 \, mg \, l^{-1}$, although high values may be recorded in water from volcanic rocks.

Iron forms approximately 5% of the Earth's crust, and is contained in a great many minerals in rocks, as well as occurring as ore bodies. Most iron in solution is ionized. Normally, iron occurs in groundwater in the form of Fe^{2+}, $Fe(OH)_3$ or $FeOH^+$. When iron occurs at concentrations of $1 \, mg \, l^{-1}$ or above, it does so in the ferrous state. If such groundwater is exposed to air, the iron is oxidized to the ferric condition and precipitated as ferric hydroxide. Concentrations of ferrous iron in groundwater are typically in the range $1–10 \, mg \, l^{-1}$.

Ion exchange affects the chemical nature of groundwater. The most common natural cation exchangers are clay minerals, humic acids and zeolites. The replacement of Ca^{2+} and Mg^{2+} by Na^+ may occur when groundwater moves beneath argillaceous rocks into a zone of more restricted circulation. This produces soft water. Changes in temperature–pressure conditions may result in precipitation (e.g. a decrease in pressure may liberate CO_2 causing the precipitation of calcium carbonate).

Certain dissolved gases, such as oxygen and carbon dioxide, will affect the groundwater chemistry. Others will affect the use of water, for example, hydrogen sulphide at concentrations of more than $1 \, mg \, l^{-1}$ will render water unfit for consumption because of the objectionable odour. Methane coming out of solution may accumulate and present a fire or explosion hazard. The minimum concentration of methane in water sufficient to produce an explosive methane–air mixture above the water from which it has escaped depends on the volume of air into which the gas evolves. Theoretically, water containing as little as $1–2 \, mg \, l^{-1}$ of methane can produce an explosion in poorly ventilated air space.

Several minor elements are a matter of concern because of their toxic effects. For instance, arsenic should not exceed $0.01–0.1 \, mg \, l^{-1}$ in drinking water, barium and copper $1 \, mg \, l^{-1}$, chromium and lead $0.05 \, mg \, l^{-1}$, cadmium $0.01 \, mg \, l^{-1}$ and mercury $0.001 \, mg \, l^{-1}$. Although boron is essential to healthy plant growth, it is injurious if present in groundwater in significant quantities. However, the sensitivies of plants to boron vary widely. For example, citrus trees may be damaged by as little as $0.5 \, mg \, l^{-1}$ if soil drainage is good. The normal amount contained by groundwater varies from 0.01 to $1.0 \, mg \, l^{-1}$.

The total dissolved solids (*TDS*) in a sample of water includes all solid material in solution, whether ionized or not. Water for most domestic and industrial uses should contain less than $1000 \, mg \, l^{-1}$, and the *TDS* of water for most agricultural purposes should not be above $3000 \, mg \, l^{-1}$. Groundwater has been classified by Hem (1985) according to its *TDS* as follows:

Fresh	Less than $1000 \, mg \, l^{-1}$
Slightly saline	$1000–3000 \, mg \, l^{-1}$
Moderately saline	$3000–10\,000 \, mg \, l^{-1}$
Very saline	$10\,000–35\,000 \, mg \, l^{-1}$
Briny	Over $35\,000 \, mg \, l^{-1}$

The hardness of water relates to its reaction with soap, and to the scale and encrustations which form in boilers and pipes where water is heated and transported. It is attributable to the presence of divalent metallic ions, calcium and magnesium being the most abundant in groundwater. Water derived from limestone or dolostone aquifers containing gypsum or anhydrite may contain $200–300 \, mg \, l^{-1}$ hardness or more. Water for domestic use should not contain more than $80 \, mg \, l^{-1}$ total hardness. The total hardness H_T generally is expressed in terms of the equivalent of calcium carbonate; hence

$$H_T = Ca \times (Ca \times O_3 / Ca)$$
$$+ Mg \times (CaCO_3 / Mg) \qquad (9.17)$$

where H_T, Ca and Mg are measured in milligrams per litre and the ratios in equivalent weights. This equation reduces to

$$H_T = 2.5Ca + 4.1Mg \qquad (9.18)$$

The degree of hardness in water has been described as follows

Description	Hardness (mg l^{-1} as CaCO$_3$)	
	(After Sawyer & McCarty, 1967)	(After Hem, 1985)
Soft	Below 75	Below 60
Moderately hard	75–150	61–120
Hard	150–300	121–180
Very hard	Over 300	Over 180

Water for human consumption must be free from organisms and chemical substances in concentrations large enough to affect health adversely. In addition, drinking water should be aesthetically acceptable in that it should not possess unpleasant or objectionable taste, odour, colour or turbidity. For example, the maximum concentration of chloride in drinking water is 250 mg l^{-1}, primarily for reasons of taste. Again for reasons of taste, and also to avoid staining, the recommended maximum concentration of iron in drinking water is 0.3 mg l^{-1}. Oxidation of manganese in groundwater can produce black stains, and therefore the maximum permitted concentration of manganese for domestic water is 0.05 mg l^{-1}. The pH value of drinking water should be close to pH 7, but treatment can cope with a range of pH 5–9. Beyond this range, treatment to adjust the pH to pH 7 becomes less economical.

The bacterial quality of drinking water which has been established by the Environmental Protection Agency in the USA requires that tests reveal no more than one total coliform organism per 100 ml as the arithmetic mean of all water samples examined per month, with no more than 4 per 100 ml in more than one sample if the number of samples is less than 20 per month, or no more than 4 per 100 ml in 5% of the samples if the number of samples is more than 20 per month. Samples should be taken at frequent intervals of time. These vary according to the population supplied, for example, one sample per month if fewer than 1000 are supplied to 300 per month if 1 000 000 are supplied. In addition, drinking water should contain less than 1 virus unit per 400–4000 l. Viruses can be eliminated by effective chlorination. The water should be free from suspended solids in which viruses can harbour and thereby be protected against disinfectant.

European and international standards of drinking water, the latter laid down by the World Health Organization (WHO), are given in Table 9.7. Most drinking water in the USA conforms with the standards established by the Environmental Protection Agency (Table 9.8). The permissible and mandatory limits of trace elements present in drinking water are recorded in Table 9.9. The values quoted cannot be universally applied.

The quality of water required in different industrial processes varies appreciably. Indeed, it can differ within the same industry. Nonetheless, the salinity, hardness and silica content are the three parameters that usually are most important in terms of industrial water. Water used in the textile industry should contain a low amount of iron, manganese and other heavy metals likely to cause staining. The hardness, *TDS*, colour and turbidity must also be low. However, the quality of water required by the chemical industry varies widely depending on the processes involved. Similarly, the water required in the pulp and paper industry is governed by the types of product manufactured. Groundwater generally is preferable to surface water, because it shows less variation in chemical and physical quality. The suitability of water for irrigation is discussed in Chapter 7.

9.13 Wells

The commonest way of recovering groundwater is to sink a well and lift water from it (Fig. 9.17, p. 268). The most efficient well is developed so as to yield the greatest quantity of water with the least drawdown and the lowest velocity in the vicinity of the well. The specific capacity of a well is expressed in litres of yield per metre of drawdown when the well is being pumped. It is indicative of the relative permeability of the aquifer. The location of the well obviously is important if an optimum supply is to be obtained, and a well site should always be selected after a careful study of the geological setting.

The completion of a well in an unconsolidated formation requires that it be cased so that the surrounding deposits are supported. Sections of the casing must be perforated to allow the penetration of water from the aquifer into the well, or screens can be used. The casing should be as permeable as, or more permeable than, the deposits it confines.

Wells which supply drinking water should be properly sealed. However, an important advantage of groundwater is its comparative freedom from

Table 9.7 Standards for drinking water. (After WHO, 1993; CEC, 1980; Brown et al., 1972.)

	European standards (1971)	International standards (1972)
*Biology**		
Coliform bacteria	Nil	Nil
Escherichia coli	Nil	Nil
Streptococcus faecalis	Nil	Nil
Clostridium perfringens	Nil	
Virus	Nil	
	Less than 1 plague-forming unit per litre per examination in 10 l of water	
Microscopic organisms	Nil	
Radioactivity		
Overall α radioactivity	<3 pCi l^{-1}	<3 pCi l^{-1}
Overall β radioactivity	<30 pCi l^{-1}	
Chemical elements	(mg l^{-1})	(mg l^{-1})
Pb	<0.05	<0.1
As	<0.05	<0.05
Se	<0.01	<0.01
Hexavalent Cr	<0.05	<0.05
Cd	<0.01	<0.01
Cyanides (in CN)	<0.05	<0.05
Ba	<1.00	<1.00
Cyclic aromatic hydrocarbon	<0.20	
Total Hg	<0.01	<0.01
Phenol compounds (in phenol)	<0.001	<0.001–0.002
NO_3 recommended	<50	
acceptable	50–100	
not recommended	>100	
Cu	<0.05	0.05–1.5
Total Fe	<0.1	0.10–1.0
Mn	<0.05	0.10–0.5
Zn	<5	5.00–15
Mg if SO_4 >250 mg l^{-1}	<30	<30
if SO_4 <250 mg l^{-1}	<125	<125
SO_4	<250	250–400
H_2S	0.05	
Cl recommended	<200	
acceptable	<600	
NH_4	<0.05	
Total hardness	2–10 meq l^{-1}	2–100 meq l^{-1}
Ca	75–200	75–200
F	The limits depend upon the air temperature (see below)	

Mean annual maximum daytime temperature (°C)	Lower limit (mg l^{-1})	Optimum (mg l^{-1})	Upper limit (mg l^{-1})	Unsuitable (mg l^{-1})
10–12	0.9	1.2	1.7	2.4
12.1–14.6	0.8	1.1	1.5	2.2
14.7–17.6	0.8	1.0	1.3	2.0
17.7–21.4	0.7	0.9	1.2	1.8
21.5–26.2	0.7	0.8	1.0	1.6
26.3–32.6	0.6	0.7	0.8	1.4

* No 100 ml sample to contain *E. coli* or more than 10 coliform organisms.

Table 9.8 Drinking water standards in the USA.*

Physical characteristics

Criterion	Recommended limit†	Tolerance limit‡
Colour, units	15	—
Odour, threshold number	3, inoffensive	—
Residue		
filtrable (mg l⁻¹)	500	—
Taste	Inoffensive	—
Turbidity, units	5	—

Inorganic chemicals (mg l⁻¹)

Substance	Recommended limit†	Tolerance limit
Alkyl benzene sulphonate (ABS)	0.5	—
Arsenic (As)	0.01	0.05
Barium (Ba)	—	1.0
Cadmium (Cd)	—	0.01
Carbon chloroform extract (CCE)	0.2	—
Chloride (Cl)	250	—
Chromium, hexavalent (Cr^{6+})	—	0.05
Copper (Cu)	1.0	—
Cyanide (CN)	0.01	0.2
Fluoride (F)	0.8–1.7§‖	1.4–2.4‡
Iron (Fe)	0.3	—
Lead (Pb)	—	0.05
Manganese (Mn)	0.05	—
Mercury (Hg)	—	0.002
Nitrate (as N)	10	—
Phenolic compounds (as phenol)	0.001	—
Selenium (Se)	—	0.01
Silver (Ag)	—	0.05
Sulphate (SO_4)	250	—
Zinc (Zn)	5	—

Organic chemicals (mg l⁻¹)

Substance	Tolerance limit
(A) Chlorinated hydrocarbons	
Endrin	0.0002
Lindane	0.004
Methoxychlor	0.1
Toxaphene	0.005
(B) Chlorophenoxys	
2,4-D	0.1
2,4,5-TP Silvex	0.01

Continued on p. 268

Table 9.8 *Continued.*

Biological standards

Substance examined	Maximum permissible limit
Standard 10 ml portions	Not more than 10% in one month shall show coliforms¶
Standard 100 ml portions	Not more than 60% in one month shall show coliforms¶

Radioactivity (pCi l⁻¹)

Substance	Recommended limit
Radium-226	3
Strontium-90	10
Gross β activity	1000**

* Based on maximum contaminant levels of the Environmental Protection Agency (1975).
† Concentrations that should not be exceeded where more suitable water supplies are available.
‡ Concentrations above this constitute grounds for rejection of the supply.
§ Dependent on annual maximum daily air temperature.
‖ Where fluoridation is practised, minimum recommended limits are also specified.
¶ Subject to further specified restrictions.
** In absence of strontium-90 and α emitters.

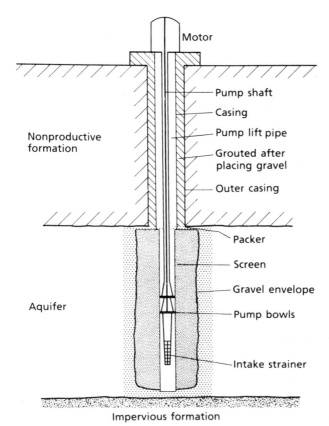

Fig. 9.17 Gravel-packed well installation.

Table 9.9 Minor and trace elements and compounds in drinking water. (From UNESCO; see Brown *et al.*, 1972.)

	Comment	Recommended limit (mg l^{-1})	Mandatory limit (mg l^{-1})	Unofficial limit (mg l^{-1})
Alkyl benzene sulphonate (ABS)		0.5		
Arsenic (As)	Serious cumulative systemic poison; 100 mg usually causes severe poisoning	0.01	0.05	
Antimony (Sb)	Similar to As, but less acute. Recommended limit 0.1 mg l^{-1}, routinely below 0.05 mg l^{-1}; over long periods, below 0.01 mg l^{-1}	0.05	0.05	
Barium (Ba)	Muscle (including heart) stimulant. Fatal dose is 550–600 mg as chloride		1	
Beryllium (Be)	Poisonous in some of its salts in occupational exposure			None
Bismuth (Bi)	A heavy mineral in the arsenic family—avoid in water supplies			None
Boron (B)	Ingestion of large amounts can affect central nervous system	1	5	1
Cadmium (Cd)	13–15 mg kg^{-1} in food has caused illness	0.003	0.01	
Carbon chloroform extract (CCE)	At limit stated, organics in water are not considered a health hazard	0.200		
Chromium (hexavalent)	Limit provides a safety factor. Carcinogenic when inhaled		0.05	
Cobalt (Co)	Beneficial in small amounts; about 7 µg per day			None
Copper (Cu)	Body needs copper at level of about 1 mg per day for adults; not a health hazard unless ingested in large amounts	1.0		
Cyanide (CN)	Rapid fatal poison, but limit set provides safety factor of about 100	0.01	0.20	
Fluoride (F$^-$)	Beneficial in small amounts; above 2250 mg dose can cause death	0.7–1.2	1.4–2.4	
Iron (Fe)		0.3		
Lead (Pb)	Serious cumulative body poison		0.01	
Manganese (Mn)		0.05		
Mercury (Hg)	Continued ingestion or large amounts can damage brain and central nervous system		0.07	0.005
Molybdenum (Mo)	Necessary for plants and ruminants. Excessive intakes may be toxic to higher animals; acute or chronic effects not well known			None
Nickel (Ni)	May cause dermatitis in sensitive people; doses of 30–73 mg of $NiSO_4 \cdot 6H_2O$ have produced toxic effects		0.02	None
Radium (radium-226)	A bone-seeking, internal α emitter that can destroy bone marrow	3 pCi l^{-1}		
Selenium (Se)	Toxic to both humans and animals in large amounts. Small amounts may be beneficial		0.01	
Silver (Ag)	Can produce irreversible, adverse cosmetic changes		0.05	
Sodium (Na)	A beneficial and needed body element, but can be harmful to people with certain diseases			200
Strontium-90	A bone-seeking internal β emitter	10 pCi l^{-1}		
Zinc (Zn)	Beneficial in that a child needs 0.3 mg kg^{-1} per day; 675–2 280 mg l^{-1} may be an emetic		5	

UNESCO, United Nations Educational, Scientific and Cultural Organization.

bacterial pollution. Abandoned wells should be sealed to prevent aquifers being contaminated.

The yield from a well in granular material can be increased by surging, which removes the finer particles from the zone about the well. The water supply from wells in rock can be increased by driving galleries or adits from the bottom of deep wells. Yields from rock formations can also be increased by fracturing the rocks with explosives, by pumping fluid into the well under high pressure, or, in the case of carbonate rocks, such as chalk, by using acid to enlarge the discontinuities. The use of explosives in sandstones has led to increases in yield of up to 40%, whilst the acidification of wells in carbonate rocks has increased yields by over 100%.

From the hydrological point of view, the long-term yield of a well depends upon the following factors:

1 The annual rate of groundwater recharge. This determines the rate of flow in the aquifer, and thus the amount of water available for abstraction.

2 The location of the well in the aquifer. There may be some advantage in siting a well near a recharge area, so that a surface water resource is diverted underground to augment the aquifer flow by induced infiltration. This could increase the well yield. Alternatively, a well could be sited near the discharge area of the aquifer, with the objective of diverting as much of the natural discharge as possible to the well. Neither of these options should be undertaken without a careful evaluation of the possible consequences. An alternative strategy may be to exploit the storage of the aquifer without interfering with the flow of groundwater through the aquifer. In this case, the wells should be sited some distance from the areas of natural recharge and discharge, so that it takes a significant time for the pumping effects to reach them. An indication of the delay to be expected is given by the response time, which is SL^2/T, where S is the storage coefficient, T is the transmissivity and L is the length of the flow path from the well to the zone of affected natural discharge.

3 The permeability of the aquifer in the area surrounding the well. The higher the permeability, the easier it is for water to flow to a well during periods of abstraction.

4 The thickness of the aquifer at the well site. The well should be located where the saturated thickness is greatest. If the aquifer material has only a small variation in permeability, the transmissivity will increase with increasing saturated thickness. This again facilitates flow to a well.

5 The location and orientation of any faults or notable discontinuities. These may act as preferred flow channels and greatly increase the flow to a well. Many wells in relatively impermeable material have been successful as a result of flow through secondary permeability features. However, a fault may also act as a barrier to flow if it is filled with impermeable gouge material, or if the throw of the fault places the aquifer against an impermeable stratum. Faults or other potential boundaries should be evaluated using pumping test analysis techniques.

6 The location of wells with respect to any features that may jeopardize the quality and quantity of the discharge, or the groundwater resource as a whole. It is important that a well should be able to operate at its design discharge and drawdown without the quantity and quality of the abstracted water being adversely affected, and without the abstraction having an adverse effect upon the ecological or environmental features or resulting in the derogation of existing groundwater sources. For example, induced infiltration can be used to augment a groundwater supply, but if the surface source is polluted, it can also mean that the pumping rates of wells in the area have to be severely limited. In coastal aquifers, saline intrusion may cause a progressive decrease in the water abstracted from wells and, if the discharge becomes unacceptably saline, the wells may have to be abandoned.

9.14 Safe yield

The yield of a surface water resource generally is related to a return interval, which is defined as the steady supply that can be maintained through a drought of specified severity. For example, the yield that can be maintained through a 1 in 50-year (2%) drought is greater than that which can be maintained through a 1 in 100-year (1%) drought. Thus, the yield is not an absolute quantity, but a variable that depends upon the specified frequency of occurrence of the limiting drought conditions. The yield may also be defined as the steady supply that can be maintained through the worst drought on record. In this case, the severity of the limiting drought conditions depends upon the rainfall recorded in the years preceding the drought period, the length of the record available for analysis and chance—a short record may contain a particularly severe drought, while a much longer record elsewhere may not. For this reason, the first definition is preferred.

The abstraction of water from the ground at a rate greater than it is being recharged leads to a lowering of the water table, and upsets the equilibrium between discharge and recharge. The concept of a safe yield has been used for many years to express the quantity of water that can be withdrawn from the ground without impairing the aquifer as a water source. A draught in excess of the safe yield is an overdraught. The estimation of the safe yield is a complex problem which must take into account the climatic, geological and hydrogeological conditions. As such, the safe yield is likely to vary appreciably with time. Nonetheless, the recharge–discharge equation, the transmissivity of the aquifer, the potential sources of contamination and the number of wells in operation must all be given consideration if an answer is to be found. The safe yield G often is expressed as follows:

$$G = P - Q_s - E_T + Q_g - \Delta S_g - \Delta S_s \qquad (9.19)$$

where P is the precipitation on the area supplying the aquifer, Q_s is the surface stream flow over the same area, E_T is the evapotranspiration, Q_g is the net groundwater inflow to the area, ΔS_g is the change in groundwater storage and ΔS_s is the change in surface storage. With the exception of precipitation, all the terms in this expression can be subjected to artificial change. The equation cannot be considered as an equilibrium equation and cannot be solved in terms of the mean annual values. It can be solved correctly only on the basis of specified assumptions for a stated period of years.

The transmissivity of an aquifer may place a limit on the safe yield even though this equation may indicate a potentially large draught. This can only be realized if the aquifer is capable of transmitting water from the source area to wells at a rate high enough to sustain the draught. Where contamination of the groundwater is possible, the location of wells, their type and the rate of abstraction must be planned in such a way that conditions permitting contamination cannot be developed.

Once an aquifer is developed as a source of water supply, effective management becomes increasingly necessary if it is not to suffer deterioration. Manage-

ment should not merely be concerned with the abstraction of water, but should also consider its utilization, because different qualities of water can be put to different uses. Pollution of the water supply is most likely to occur when the level of the water table has been so lowered that all the water that goes underground within a catchment area drains quickly and directly to the wells. Such lowering of the water table may cause reversals in drainage, so that water drains from rivers into the groundwater system rather than the other way around. This river water may be contaminated.

A progressive decline in groundwater level frequently is an indication of future management problems, because this is often the consequence of exceeding the safe or perennial yield of the aquifer. The result is likely to be an unacceptable pumping lift, a reduced yield due to the restricted drawdown available and possibly a deterioration in water quality. The latter often occurs as a result of old, highly mineralized water being drawn from deep in the aquifer into the wells, or through induced infiltration or saline intrusion. These problems can necessitate a reduction in the output of a well field, or even its abandonment. Falling groundwater levels may also result in the loss of natural marshes and wetlands, with potentially serious agricultural and ecological implications.

9.15 Artificial recharge

Artificial recharge may be defined as an augmentation of the natural replenishment of groundwater storage by artificial means. Its main purpose is water conservation, often with improved quality as a second aim, for example, soft river water may be used to reduce the hardness of groundwater. Artificial recharge therefore is used to reduce the overdraught, to conserve and improve surface run-off and to increase the available groundwater supplied.

The suitability of a particular aquifer for artificial recharge must be investigated. For instance, it must have adequate storage, and the bulk of the water recharged should not be lost rapidly by discharge into a nearby river. The hydrogeological and groundwater conditions must be amenable to artificial replenishment. An adequate and suitable source of water for recharge must be available. The source of water for artificial recharge may be storm run-off, river or lake water, water used for cooling purposes, industrial waste water or sewage water. Many of these sources require some kind of pretreatment.

Interaction between artificial recharge and groundwater may lead to precipitation, for example, of calcium carbonate and iron and magnesium salts, resulting in a lower permeability. Nitrification or denitrification, and possibly even sulphate reduction, may occur during the early stages of infiltration. Bacterial action may lead to the development of sludges, which reduce the rate of infiltration.

Artificial recharge may be accomplished by various surface-spreading methods utilizing basins, ditches or flooded areas, by spray irrigation or by pumping water into the ground via vertical shafts, horizontal collector wells, pits or trenches. The most widely practised methods involve water spreading, which allow increased infiltration to occur over a wide area when the aquifer outcrops at or near the surface. Therefore, these methods require that the ground has a high infiltration capacity. In the basin method, water is contained in a series of basins formed by a network of dykes, constructed to take maximum advantage of local topography. Care must be taken in spray irrigation not to flush salts and nutrients from the soil into the groundwater. In regions with hot, dry climates, there may be a danger of excessive evaporation of recharge water leading to salinization. Recharge wells are employed most frequently when the aquifer to be recharged is deep or confined, or where there is insufficient space for recharge basins.

9.16 Groundwater pollution

Pollution can be regarded as the impairment of water quality by chemicals, heat or bacteria to an extent that does not necessarily create an actual public health hazard, but does adversely affect such waters for domestic, agricultural or industrial use. In fact, most human activities have a direct and usually adverse effect upon water quality.

The greatest danger of groundwater pollution is from surface sources, such as excessive application of fertilizers, leaking sewers, polluted streams, mining and mineral wastes, domestic and industrial waste and so on. Areas with thin soil cover or where an aquifer is exposed, such as the recharge area, are the most critical from the point of view of pollution

potential. Any possible source of pollution in these areas should be carefully evaluated, both before and after well construction, and the viability of groundwater protection measures should be considered. However, it should be appreciated that the slow rate of travel of pollutants in underground strata means that a case of pollution may go undetected for a number of years. During this period, a large part of the aquifer may become polluted and cease to have any potential as a source of water (Selby & Skinner, 1979).

The attenuation of a pollutant occurs as it moves through the ground as a result of four major processes. First, the soil has an enormous purifying power due to the communities of bacteria and fungi which live there. These organisms are capable of attacking pathogenic bacteria and also can react with certain other harmful substances. Second, as water passes through fine grained porous media, suspended impurities are removed by filtration. Third, some substances react with minerals in the soil/rock, and some are oxidized and precipitated from solution. Adsorption may also occur in argillaceous or organic material. Fourth, the dilution and dispersion of a pollutant may lead to its concentration being reduced, until it eventually becomes negligible at some distance from the source.

The form of the pollutant is an important factor with regard to its susceptibility to the various purifying processes. For instance, pollutants which are soluble, such as fertilizers and some industrial wastes, are not affected by filtration. Metal solutions may not be susceptible to biological action. Solids, on the other hand, are amenable to filtration provided that the transmission medium is not coarse grained, fractured or cavernous. Non-soluble liquids, such as hydrocarbons, generally are transmitted through porous media, although a certain fraction may be retained in the host material. Usually, however, the most dangerous forms of groundwater pollution are those which are miscible with the water in the aquifer.

Biological pollution in the form of microorganisms, viruses and pathogens is quite common. Not all bacteria are harmful; on the contrary, many are beneficial and perform valuable functions, such as attacking and biodegrading pollutants as they migrate through the soil.

The bacteria which normally inhabit the soil thrive

at temperatures of around 20 °C. Bacteria which are of animal origin prefer temperatures of around 37 °C, and so generally die quite quickly outside the host body. Consequently, it sometimes is erroneously assumed that pathogenic bacteria cannot survive long underground. However, Brown *et al.* (1972) pointed out that some bacteria in groundwater may have a lifespan of up to 4 years. It generally is assumed that bacteria move at a maximum rate of about two-thirds of the water velocity. Because most groundwater only moves a few metres per year, the distances travelled usually are quite small and, in general, it is unusual for bacteria to spread more than 33 m from the source of pollution. Of course, in openwork gravel, cavernous limestone or fissured rock, the bacteria may spread over distances of many kilometres.

Viruses are parasites which require a host organism before they can reproduce, but some are capable of retaining all their characteristics for 50 days or more in other environments. Because of their very small size, viruses are not greatly affected by filtration, but are prone to adsorption, particularly if pH is around 7. Brown *et al.* (1972) suggested that viruses are capable of spreading over distances exceeding 250 m, although 20–30 m may be a more typical figure.

In general, the concentration of a pollutant decreases as the distance it has travelled through the ground increases. Thus, the greatest pollution potential exists for wells tapping shallow aquifers. In particular, an aquifer that is exposed or overlain by a relatively thin formation in the recharge area is at risk, particularly when the overlying material consists of sand or gravel. Conversely, deeply buried aquifers overlain by relatively impermeable beds of shale or clay generally can be considered to have a low pollution potential, and are less prone to severe pollution.

For non-karstic terrains, the minimum recommended safe distance between a domestic well and a source of pollution is shown in Table 9.5. In karst limestone areas, however, pollutants may be able to travel quickly over large distances. For instance, Hagerty and Pavoni (1973) observed the spread of polluted groundwater over a distance of 30 km through limestone in approximately 3 months. They also noted that the degradation, dilution and dispersion of harmful constituents were less effective than in surface waters.

If a pollutant is evenly distributed over a large area, its probable effect will be less than that of the same amount of pollutant concentrated at one point. Concentrated sources are most undesirable, because the self-cleansing ability of the soil/rock concerned is likely to be exceeded.

Perhaps the greatest risk of groundwater pollution arises when pollutants can be transferred from the ground surface to an aquifer. In this case, the purifying processes that take place within the soil are bypassed, and the attenuation of the pollutant is reduced. Therefore a common cause of such groundwater pollution is poor well design, construction and maintenance (Campbell & Lehr, 1973). When a drill-hole penetrates only one aquifer, the principal concern is the transfer of pollution from the ground surface. However, with multiple aquifers, there is the additional possibility of interaquifer flow. In such cases, each aquifer (or potential source of pollution) must be isolated using cement-grout seals in the intervening strata. The failure to do this, particularly when the well incorporates a gravel pack, provides a conduit through which water can be transferred from the ground surface and from one aquifer to another. This could be potentially disastrous where a shallow, polluted aquifer overlies a deeper, unpolluted aquifer.

Induced infiltration occurs where a stream is hydraulically connected to an aquifer and lies within the area of influence of a well. When water is abstracted from a well, the water table in the immediate vicinity is lowered and assumes the shape of an inverted cone, which is referred to as the cone of depression (Fig. 9.18). The steepness of the cone is governed by the soil/rock type, it being flatter in highly permeable materials. The size of the cone depends on the rate of pumping, equilibrium being achieved when the rate of abstraction balances the rate of recharge. However, if abstraction exceeds recharge, the cone of depression increases in size. As the cone of depression spreads, groundwater is withdrawn from storage over a progressively increasing area of influence, and the groundwater levels about the well continue to be lowered. Cones associated with some larger wells may have radii of up to 1.5 km. With time, the aquifer is recharged by influent seepage from the surface stream. As pumping continues, the proportion of water derived from the stream which enters the cone of depression increases pro-

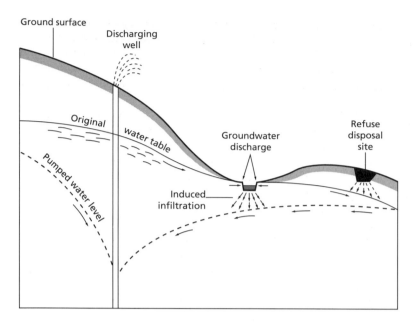

Fig. 9.18 Example of induced infiltration as a result of pumping. Over much of the area, the original hydraulic gradient has been reversed, so that pollutants can travel in the opposite direction to that expected. Additionally, the aquifer has become influent instead of effluent, as it was originally.

gressively. Whether or not induced infiltration gives rise to pollution depends upon the quality of the surface water source, the nature of the aquifer, the quantity of infiltration involved and the intended use of the abstracted groundwater. Induced infiltration at one extreme can have potentially disastrous consequences, whilst at the other can provide a valuable addition to the overall water resources of an area.

9.16.1 Landfill and groundwater pollution

The disposal of wastes in landfill sites leads to the production of leachate and gases (see Chapter 14), which may present a health hazard as a consequence of pollution of groundwater supply. Leachate often contains high concentrations of dissolved organic substances, resulting from the decomposition of organic material, such as vegetable matter and paper. Recently emplaced wastes may have a chemical oxygen demand (COD) of around $11\,600\,\mathrm{mg\,l^{-1}}$ and a biochemical oxygen demand (BOD) in the region of $7250\,\mathrm{mg\,l^{-1}}$. The concentration of chemically reduced inorganic substances, such as ammonia, iron and manganese, varies according to the hydrology and the chemical and physical conditions at the site. As an illustration, Brown *et al.* (1972) calculated that $1000\,\mathrm{m^3}$ of waste can yield 1.25 tonnes of potassium and sodium, 0.8 tonnes of calcium and magnesium, 0.7 tonnes of chloride, 0.19 tonnes of sulphate and 3.2 tonnes of bicarbonate. Furthermore, Barber (1982) estimated that a small landfill site with an area of 1 ha located in southern England could produce up to $8\,\mathrm{m^3}$ of leachate per day, mainly between November and April, from rainfall of 900 mm per year, assuming that evaporation was close to the average for the region and run-off was minimal.

Gray *et al.* (1974) considered that the major criterion for the assessment of a landfill site as a serious risk was the presence of toxic or oily liquid waste, although sites on impermeable substrata often merited a lower assessment of risk, depending on local conditions. In such cases, serious risk means that there is a serious possibility of an aquifer being polluted and not necessarily that there is a danger to life. As such, restriction of the type of materials that can be tipped at the site concerned may have to be considered. Therefore, site selection for waste disposal must take into account the character of the material which is likely to be tipped.

The selection of a landfill site for a particular waste or a mixture of wastes involves a consideration of the geological and hydrogeological conditions. Argillaceous sedimentary, massive igneous and metamorphic rocks have low permeabilities, and therefore afford most protection to the water supply. By contrast, the least protection is offered by rocks intersected by open discontinuities or in which solution features are developed. In this respect, limestones and some sequences of volcanic rocks may be suspect. Granular material may act as a filter.

There are two ways in which pollution by leachate can be tackled, first, by concentrating and containing and, second, by diluting and dispersing. Infiltration through granular ground of liquids from a landfill may lead to their decontamination and dilution. Hence, sites for the disposal of domestic refuse can be chosen where decontamination has the maximum chance of reaching completion, and where groundwater sources are located far enough away to enable dilution to be effective. Consequently, domestic waste can be tipped, according to Gray *et al.* (1974), at dry sites on granular material, which has a thickness of at least 15 m. Water supply sources should be located at least 0.8 km away from the landfill site. Landfill sites should not be located on discontinuous rocks unless overlain by 15 m of clay deposits.

As far as potentially toxic waste is concerned, it is best to contain it. Gray *et al.* (1974) recommended that such sites should be underlain and confined by at least 15 m of impermeable strata, and that any source abstracting groundwater for domestic use should be at least 2 km away. Furthermore, the topography of the site should be such that run-off can be diverted from the landfill, so that it can be disposed of without causing pollution of surface waters. Containment can be achieved by an artificial impermeable lining being placed over the base of a site. However, there is no guarantee that this will remain impermeable. Thus the migration of leachate from a landfill into the substrata will occur eventually, only the length of time before this happens being in doubt. In some instances, this will be sufficiently long for the problem of pollution to be greatly diminished.

In order to reduce the amount of leachate emanating from a landfill, it is advisable to construct cutoff drains around the site to prevent the flow of surface water into the landfill area. Leachate should be collected in a sump and pumped to a sewer, transported away by tanker or treated on site.

9.16.2 Saline intrusion

Although saline groundwater, originating as connate water or from evaporitic deposits, may be encountered, the problem of saline intrusion is specifically related to coastal aquifers. Near a coast, an interface exists between the overlying fresh groundwater and the underlying saline groundwater (Fig. 9.19). Excessive lowering of the water table along a coast leads to saline intrusion, the salt water entering the aquifer via submarine outcrops, thereby displacing fresh water. However, the fresh water still overlies the saline water, and continues to flow from the aquifer to the sea. In the past, the two groundwater bodies usually have been regarded as immiscible, from which it was assumed that a sharp interface exists between them. In fact, there is a transition zone which may vary from 0.5 m or so to over 100 m in width.

Overpumping is not the only cause of salt water encroachment, continuous pumping or the inappropriate location and design of wells also being contributory factors. In other words, the saline–fresh water interface is the result of a hydrodynamic balance; hence, if the natural flow of fresh water to the sea is interrupted or significantly reduced by abstraction, saline intrusion is almost certain to occur.

The shape of the interface is governed by the hydrodynamic relationship between the flowing fresh and saline groundwater. However, if it is assumed that hydrostatic equilibrium exists between the immiscible fresh and salt water, the depth of the interface can be approximated by the Ghyben–Herzberg expression

$$Z = \frac{\rho_w}{\rho_{sw} - \rho_w} \times h_w \qquad (9.20)$$

where ρ_{sw} is the density of sea water, ρ_w is the density of fresh water and h_w is the head of fresh water above sea level at the point on the interface (Fig. 9.19a). If ρ_{sw} is taken as 1025 kg m^{-3} and ρ_w as 1000 kg m^{-3}, $Z = 40 h_w$. Thus, at any point in an unconfined aquifer, there is approximately 40 times as much fresh water below mean sea level as there is above it. The above expression, however, implies no flow, but groundwater invariably is moving in coastal areas. Where flow is moving upwards, near the coast, the relationship gives too small a depth to salt water, but further inland, where the flow lines are nearly horizontal, the error is negligible. This relationship can also be applied to confined aquifers if the height of the water table is replaced by the elevation of the piezometric surface above mean sea level. If the aquifer overlies an

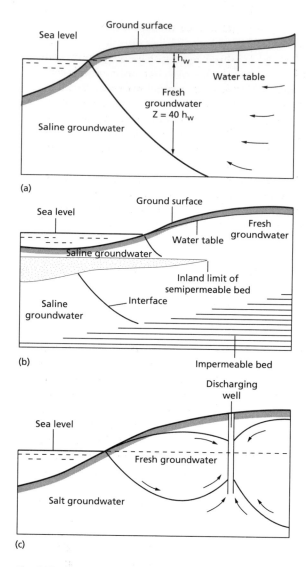

Fig. 9.19 Diagrams illustrating, in a simplified manner, the Ghyben–Herzberg hydrostatic relationship in (a) a homogeneous coastal aquifer, (b) a layered coastal aquifer and (c) a pumped coastal aquifer.

fore is referred to as 'upconing' (Fig. 9.19c). This is a dangerous condition, which can occur even if the aquifer is not overpumped and a significant proportion of the fresh water flow still reaches the sea. A well may be ruined by an increase in salt content even before the actual 'cone' reaches the bottom of the well, as a result of 'leaching' of the interface by fresh water.

The encroachment of salt water may extend for several kilometres inland, leading to the abandonment of wells. The first sign of saline intrusion is likely to be a progressively upward trend in the chloride concentration of the water obtained from the affected wells. Chloride levels may increase from a normal value of around $25\,mg\,l^{-1}$ to something approaching $19\,000\,mg\,l^{-1}$, which is the concentration in sea water. The recommended limit for chloride concentration in drinking water in Europe is $200\,mg\,l^{-1}$, above which the water has a salty taste.

Once saline intrusion develops in a coastal aquifer, it is not easy to control. The slow rates of groundwater flow, the density differences between fresh and salt waters and the flushing required usually mean that pollution, once established, may take years to remove under natural conditions. The encroachment of salt water, however, can be checked by maintaining a fresh water hydraulic gradient towards the sea. This gradient can be maintained naturally, or by artificial recharge or an extraction barrier. Artificial recharge involves the injection of water into the aquifer via wells so as to form a groundwater mound between the coast and the area where abstraction is taking place (Brown & Signor, 1973). However, the technique requires an additional supply of fresh water. An extraction barrier abstracts encroaching salt water before it reaches the inland well field. It consists of a line of wells parallel to the coast. The abstracted water is brackish, and generally is pumped back into the sea. There will probably be a progressive increase in salinity at the extraction wells. These wells must be pumped continuously if an effective barrier is to be created.

Neither of these methods of controlling saline intrusion offers an inexpensive or foolproof solution. Consequently, when there is a possibility of saline intrusion, the best policy is to locate wells as far from the coast as possible, select the design discharge with care and use an intermittent seasonal pumping regime (Ineson, 1970). Ideally, at the first sign of

impermeable stratum, this formation will intercept the interface and prevent any further saline intrusion (Fig. 9.19b).

The problem of saline intrusion usually starts with the abstraction of groundwater from a coastal aquifer, which leads to the disruption of the Ghyben–Herzberg equilibrium condition. Generally, saline water is drawn up towards a well, and there-

progressively increasing salinity, pumping should be stopped.

9.16.3 Nitrate pollution

There are at least two ways in which nitrate pollution of groundwater is known or suspected to be a threat to health. First, the build-up of stable nitrate compounds in the bloodstream reduces its oxygen-carrying capacity. Infants under 1 year of age are most at risk, and excessive amounts of nitrate can cause methaemoglobinaemia, commonly called 'blue-baby'. Consequently, if the limit of $50\,mg\,l^{-1}$ of NO_3 recommended by WHO for European countries is exceeded frequently, or if the concentration lies within the minimum acceptable range of $50-100\,mg\,l^{-1}$, bottled, low-nitrate water should be provided for infants. Second, there is the possibility that a combination of nitrates and amines, through the action of bacteria in the digestive tract, may result in the formation of nitrosamines, which are potentially carcinogenic.

Nitrate pollution basically is the result of intensive cultivation. The major source of nitrate is the large quantity of synthetic nitrogenous fertilizer that has been used since around 1959, although overmanuring with natural organic fertilizer can have the same result. Foster and Crease (1974) estimated that, in the 11 years between 1956 and 1967, the application of nitrogen fertilizer to all cropped land in certain areas of east Yorkshire, England, increased by about a factor of four (from around 20 to $80\,kgN\,ha^{-1}$ per year). Regardless of the form in which the nitrogen fertilizer is applied, within a few weeks it will have been transformed to NO_3^-. This ion is neither adsorbed nor precipitated in the soil, and therefore is easily leached by heavy rainfall and infiltrating water. However, nitrate does not have an immediate effect on the groundwater quality, possibly because most of the leachate that percolates through the unsaturated zone as intergranular seepage has a typical velocity of about 1 m per year. Thus there may be a considerable delay between the application of the fertilizer and the subsequent increase in the concentration of nitrate in the groundwater. Therefore, although the use of nitrogenous fertilizer increased sharply in east Yorkshire after 1959, a corresponding increase in groundwater nitrate was not apparent until after 1970, when the level rose from around $14\,mg\,l^{-1}$ NO_3 (about 3 mg

l^{-1} NO_3-N) to between 26 and $52\,mg\,l^{-1}$ NO_3 (6–11.5 $mg\,l^{-1}$ NO_3-N).

The effect of the time lag, which is frequently of the order of 10 years or more, is to make it very difficult to correlate fertilizer application with nitrate concentration in groundwater. In this respect, one of the greatest concerns is that if nitrate levels are unacceptably high now, they may be even worse in the future, because the quantity of nitrogenous fertilizer used has continued to increase.

Measures that can be taken to alleviate nitrate pollution include the better management of land use, mixing of water from various sources or the treatment of high-nitrate water before it is put into supply. In general, the ion exchange process has been recommended as the preferred means of treating groundwater, although this may not be considered cost effective at all sources.

9.16.4 Other causes of groundwater pollution

The list of potential groundwater pollutants is almost endless. For example, sewage sludge arises from the separation and concentration of most of the waste materials found in sewage. Because the sludge contains nitrogen and phosphorus, it has value as a fertilizer. As a result, about 50% of the 1.24 million dry tonnes per year of sludge produced in the UK is used on agricultural land. Although this does not necessarily lead to groundwater pollution, the presence in the sludge of contaminants, such as metals, nitrates, persistent organic compounds and pathogens, does mean that the practice must be carefully controlled (Davis, 1980). In particular, cadmium may give cause for concern. Too much cadmium can cause kidney damage in humans, and this metal was implicated in itai-itai disease in Japan. Food is the usual source of cadmium found in humans, although small amounts are also present in water. The European standard for drinking water recommends a cadmium concentration of less than $0.01\,mg\,l^{-1}$.

The widespread use of chemical and organic pesticides and herbicides is another possible source of groundwater pollution. For instance, Zaki *et al.* (1982) recorded a groundwater pollution incident in Suffolk County, New York, involving the pesticide aldicarb.

According to Mackay (1998), volatile organic chemicals (VOCs) are the most frequently detected organic contaminants in water supply wells in the USA. Of the VOCs, by far the most commonly found are chlorinated hydrocarbon compounds. Conversely, petroleum hydrocarbons rarely are present in supply wells. This may be due to their *in situ* biodegradation.

Many of the VOCs are liquids, and usually are referred to as non-aqueous phase liquids (NAPLs), which are sparingly soluble in water. Those which are lighter than water, such as the petroleum hydrocarbons, are termed LNAPLs, whereas those which are denser than water, that is, the chlorinated solvents, are called DNAPLs. Of the VOCs, the DNAPLs are the least amenable to remediation. If they penetrate the ground in a large enough quantity, depending on the hydrogeological conditions, DNAPLs may percolate downwards into the saturated zone. This can occur in granular soils or discontinuous rock masses. Plumes of dissolved VOCs develop from the source of pollution. Although dissolved VOCs migrate more slowly than the average velocity of groundwater, Mackay and Cherry (1989) indicated that there are many examples of chlorinated VOC plumes of several kilometres in length in the USA occurring in sand and gravel aquifers. Such plumes contain billions of litres of contaminated water. Because VOCs are sparingly soluble in water, the time taken for complete dissolution, especially of DNAPLs, by groundwater flow in granular soils is estimated to be decades or even centuries.

In some parts of the world, such as Canada and Scandinavia, 'acid rain', originating from the combination of industrial gases with atmospheric water, is acidifying not only the surface water resources, but groundwater as well. The results from several sites in Norway indicate that, in the affected areas, the groundwater has a pH lower than normal (Henriksen & Kirkhusmo, 1982).

The potential of run-off from roads to cause pollution often is overlooked. This water can contain chemicals from many sources, including those that have been dropped, spilled or deliberately spread onto the road. For instance, hydrocarbons from petroleum products and urea and chlorides from de-icing agents are all potential pollutants, and have caused groundwater contamination. There is also the possibility of accidents involving vehicles carrying large quantities of chemicals. The run-off from roads can cause bacteriological contamination.

Cemeteries and graveyards form another possible health hazard. The minimum distance between a potable water well and a cemetery required by law in England is 91.4 m (100 yards). However, a distance of around 2500 m is better, because the purifying processes in the soil can sometimes break down. Decomposing bodies produce fluids that can leak to the water table if a non-leakproof coffin is used. Typically, the leachate produced from a single grave is of the order of 0.4 m³ per year, and this may constitute a threat for about 10 years. It is recommended that the water table in cemeteries should be at least 2.5 m deep, and that an unsaturated depth of 0.7 m should exist below the bottom of the grave.

Careful consideration must be given to the selection of the pumping rate and the duration of the abstraction of a well. The danger of overpumping an aquifer lies in the natural variations of water quality with depth within the aquifer, the reversal of hydraulic gradients so that pollutants travel in the opposite direction from that normally expected, induced infiltration, saline intrusion and the exhaustion of supply. Overpumping or 'groundwater mining' during periods of water shortage, on the assumption that groundwater levels will recover at a later date, should be avoided if possible. The groundwater level may recover, but the quality may not. Essentially, this is a question of good aquifer management, part of which includes the routine analysis and monitoring of groundwater quality. This provides an early warning system for overabstraction and pollution incidents.

A mass of fluid introduced into an aquifer tends to remain intact rather than mix with, or be diluted by, the groundwater. In many instances of groundwater pollution, the contaminating fluid is discharged into the aquifer as a continuous or nearly continuous flow. Hence, the polluted groundwater often takes the form of a plume or plumes. An important part of an investigation of groundwater pollution is to locate and define the extent of the contaminated body of groundwater, in order to establish the magnitude of the problem and, if necessary, to design an efficient abatement system.

Frequently, it is not practical to initiate countermeasures to combat groundwater pollution once it has occurred, the eventual attenuation of the pollutant being the result of time, degradation, dilution and dispersion. In the case of a recent accidental spill, however, it may be possible to excavate affected soil, and use scavenger wells to intercept and recover some of the pollutant before it has dispersed significantly. Alternatively, in a few situations, artificial recharge may be undertaken in an attempt to dilute the pollutant, or to form a hydraulic pressure barrier that will divert the pollutant away from abstraction wells. Unfortunately, none of these options provides a reliable method of dealing with groundwater pollution, and they should only be considered as a last resort. Remedial methods used to deal with contaminated groundwater are referred to in Chapter 14. The problem of pollution due to acid mine drainage is covered in Chapter 13.

9.17 Groundwater monitoring

In almost all situations in which groundwater monitoring is undertaken, it is important that adequate background samples are obtained before a groundwater abstraction scheme is inaugurated. Without these background data, it will be impossible to assess the effects of a new development. It should also be remembered that groundwater can undergo cyclic changes in quality, and so any apparent changes must be interpreted with caution. Hence, routine all-year-round monitoring is essential.

Pfannkuch (1982) pointed out that the first important step in designing an efficient groundwater monitoring system is the proper understanding of the mechanics and dynamics of contaminant propagation (e.g. soluble or multiphase flow), the nature of the controlling flow mechanism (e.g. vadose or saturated flow), and the aquifer characteristics (e.g. permeability, porosity). He listed the objectives of a monitoring programme as follows:

1 The determination of the extent, nature and degree of contamination (source monitoring).

2 The determination of the propagation mechanism and hydrological parameters, so that the appropriate countermeasures can be initiated.

3 The detection and warning of movement into critical areas.

4 The assessment of the effectiveness of the immediate countermeasures undertaken to offset the effects of contamination.

5 The recording of data for long-term evaluation and compliance with standards.

6 The initiation of research monitoring to validate and verify the models and assumptions upon which the immediate countermeasures are based. Obviously, these objectives may have to be changed to suit the physical, political or other conditions prevailing at the site of a particular pollution incident.

The design of a water quality monitoring well and its method of construction must be related to the geology of the site. The depth and diameter of the well should be as small as possible so as to reduce the cost, but not so small that the well becomes difficult to use or ineffective. However, an additional requirement when water quality is concerned is that the well structure should not react with the groundwater. Clearly, if the groundwater does react with the well casing or screen, any subsequent analysis of the samples taken from the well will be affected. Thus monitoring wells frequently are constructed using plastic casing and screens, partly for economy and partly because plastic is relatively inert. Plastic does react with some types of pollutant, so that well materials must be selected to suit the anticipated conditions. The well screen should be provided with a gravel or sand pack to prevent the migration of fine material into the well. The selection of a suitable screen slot width and particle size distribution for the pack is accomplished using the same procedures as for a water supply well, although the design entrance velocity should be lower for a monitoring well than for a water well. Generally, a sample of the aquifer material is obtained, its grain size distribution is determined and the appropriate particle size for the pack is decided. The pack should extend at least 0.3 m above and below the screened zone.

When deciding the diameter of a monitoring well, some consideration must be given to the method that is to be used to obtain the water sample. If a bailer or some form of sampler is to be inserted down the casing, the well must be of a sufficiently large diameter to permit this. There are, however, several alternatives. Water samples can be obtained using a specially designed submersible pump, which is small enough to fit into a well of 50 mm in diameter.

After a well has been completed, it should be developed, by pumping or bailing, until the water becomes clear. Background samples should then be collected over a lengthy period, prior to the commencement of whatever it is that the well has been constructed to monitor.

When designing a monitoring network to detect pollution, the problem is to ensure that there are sufficient wells to allow the extent, configuration and concentration of the pollution plume to be determined, without incurring the unnecessary expense of constructing more wells than actually are required. The network of monitoring wells must be designed to suit a particular location, and modified, as necessary, as new information is obtained or in response to changing conditions.

Diefendorf and Ausburn (1977) recommended that at least three monitoring wells should be used to observe the effects of a new development, such as a landfill site. However, as a result of aquifer heterogeneity, non-uniform flow, variations in quality with depth and so on, it was suggested that three well clusters (rather than individual wells) may be required. Each well cluster should contain two or more wells located at different depths within the aquifer, or within different aquifers in the case of a multiple aquifer system (Fig. 9.20). One cluster should be located close to the source of the pollution, for early warning purposes, with another installation some suitable distance down gradient to assess the propagation of the plume. A third installation should be located up gradient of the monitored site to detect changes in background quality attributable to other causes.

Under favourable conditions, the resistivity method can be used to determine the boundaries of a plume of polluted groundwater. Vertical electrical soundings are made in areas of known pollution in an attempt to define the top and bottom of the plume. Drill-hole logs are used to establish geological control. Next, resistivity profiling is carried out to determine the lateral extent of the polluted groundwater. In this way, a quantitative assessment can be made of groundwater pollution. The method is based on the fact that the formation resistivity depends on the conductivity of the pore fluid as well as on the properties of the porous medium. Generally, the resistivity of a rock is proportional to the water it is saturated with. The groundwater resistivity decreases if its salinity increases, and hence the resistivity of the rock concerned decreases. Obviously, there must be a contrast in resistivity between the contaminant and groundwater in order to obtain useful results. Contrasts in resistivity may be attributed to mineralized groundwater with a higher than normal specific conductance due to pollution.

However, the resistivity of a saturated, porous sandstone or limestone aquifer depends not only on the salinity of the saturated groundwater, but also on the porosity and the amount of conductive minerals, notably clay, in the rock matrix. The accurate determination of the water quality only can take place if the effects of porosity and clay content are insignificant, or at least understood (Anon., 1988). Fortunately, it is often the case, especially in coastal aquifers, that extreme variations in groundwater

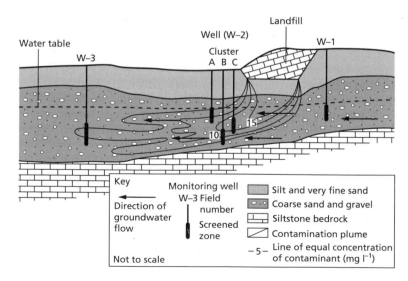

Fig. 9.20 Hypothetical example of a well field to monitor a landfill contamination plume. (After Diefendorf & Ausburn, 1977.)

salinity preclude the necessity of considering porosity and matrix conduction effects.

In simple situations, ground conductivity profiling can be used to detect plumes of polluted water. For example, a ground conductivity survey can be carried out if saline groundwater is near the surface. If the depth to water is too great, the thickness of the overlying unsaturated sediments can mask any contrasts between polluted and natural groundwater. In addition, the geology of the area must be relatively uniform so that the conductivity values and profiles can be compared with others. Reported uses of this technique include tracing polluted groundwater from landfills, septic tanks, oilfield brine disposal pits, acid mine drainage, sewage treatment effluent, industrial process waters and spent sulphur liquor.

The ionic content of the groundwater in a drill-hole can be monitored by measuring the resistivity of the fluid at a short electrode spacing (Brown, 1971). The quality of groundwater can also be estimated from the spontaneous potential deflection on an electric log, provided that the specific conductance of the drilling mud is less than the specific conductance of the pore water. This method tends to give better results as the salinity of the pore water increases.

9.18 Conjunctive use

The combined use of surface and groundwater is referred to as conjunctive use. By adopting a conjunctive use approach, the differing characteristics of surface and groundwater can be used to optimize the yield of the total water resource. For instance, surface waters are available seasonally, but usually some uncertainty surrounds the time and amount available. Additionally, surface systems are characterized by floods that cannot be captured by impounding reservoirs or used for water supply. Although surface reservoirs can be filled rapidly, they are subject to losses by evaporation and seepage. On the other hand, groundwater usually is available in large aquifers in large quantities, with relatively little variation over time. Groundwater reservoirs tend to react comparatively slowly to changes in inflow or outflow. Hence, less uncertainty is involved in predicting future groundwater availability than in predicting surface stream flow. A conjunctive use approach to water supply aims to manage the surface and groundwater resources of an area to obtain a net gain in

yield. As demand levels increase towards the upper limits of available resources, a conjunctive use strategy becomes more and more attractive.

Some of the considerations involved in the design and operation of a conjunctive use scheme include the following:

1 Groundwater can be used to augment river flow during the dry part of the year or during droughts. The quantity of groundwater required will depend upon the variability of the river flow and the level of river regulation adopted.

2 The drawdown experienced by the aquifer, and thus the time required for groundwater levels to recover, will depend not only upon the properties of the aquifer, but also upon the level of river regulation adopted.

3 Some idea regarding the rate at which groundwater levels will recover naturally can be obtained from a consideration of the aquifer response time. This parameter also gives an indication of the seasonal variation in groundwater flow to a river.

4 Groundwater levels can be increased during periods of surplus river flow using artificial recharge techniques if natural recharge is insufficient or too slow.

5 A reduction in river flow will probably accompany groundwater abstraction from wells. Pumping may lower the groundwater level, with the result that spring discharge and any other effluent discharge from the aquifer will be diminished, while losses through the river bed may be increased (possibly by induced infiltration) and some river base-flow may be intercepted. Obviously, the reduction in aquifer discharge will depend upon whether or not the aquifer is hydraulically connected to the river, the hydraulic characteristics of the aquifer (S and T), the response time of the aquifer and the distance between the wells and the river.

6 The efficiency of the conjunctive aquifer–river system is expressed as the net gain, that is, the net increase in river flow taking into account any reduction in river flow that occurs as a result of groundwater abstraction. Thus

$$\text{Net gain} = \frac{\text{Groundwater abstraction rate} - \text{reduction in river flow}}{\text{Groundwater abstraction rate}} \quad (9.21)$$

7 The best results are obtained when the aquifer has a relatively low permeability and a high storage

coefficient (i.e. a slow response time). Additionally, the wells should be concentrated in restricted areas to limit the area affected by pumping, and hence reduce the length of river over which decreased flow can be expected.

8 With unconfined aquifers, particularly when they have a fast response time, it may be desirable to site the abstraction wells some distance away from the river. If the wells are too close to the river, induced infiltration may cause a very rapid circulation of water around the aquifer–river system, with a negligible net gain. Siting wells remote from the river does, unfortunately, increase pumping and pipeline costs.

9 Confined aquifers, because of their small coefficient of storage and fast response time, are not always suitable for conjunctive use, although the apparent isolation of the surface and groundwater resources may, at first, make them appear to be attractive propositions.

Conjunctive use schemes are quite complex to operate, and require a detailed preliminary study to assess their feasibility. Consequently, before deciding to initiate a conjunctive use scheme, it should be certain that the advantages of doing so outweigh the disadvantages. The principal merits and demerits of conjunctive use are summarized below.

Advantages

1 Optimization of water use. Employing both surface and underground reservoirs provides a larger storage capacity and reduces 'wasted' run-off.
2 Smaller impounding reservoirs are needed, because groundwater storage can satisfy the additional demand during critical drought periods.
3 Greater flood control. Because water can be transferred from impounding reservoirs to underground storage, the level in surface reservoirs can be dropped to allow for increased flood storage.
4 Greater flexibility when responding to an increase in demand, because more than one source is available. This can lead to greater efficiency when the travel distance of releases is reduced.

Disadvantages

1 Higher running costs as a result of greater power consumption through increased pumping. Most surface regulating schemes operate under gravity, whereas conjunctive use schemes require pumps to recover water from underground, transport it to the river and possibly to recharge groundwater artificially as well. Monitoring costs will also be increased.

2 Decreased pumping efficiency due to fluctuations in groundwater levels. This is most significant when artificial recharge is practised.

3 Management problems are increased, because there are a greater variety of options, such as which source to use at any time, when to stop groundwater abstraction and switch to a surface source, when to initiate groundwater recharge and so on.

4 Economic assessment of the scheme is more difficult, because there are a number of surface and groundwater sources that can be used independently or simultaneously. Selecting the least cost option at any time may be difficult and may not always provide the most efficient use of water or satisfy other management constraints.

5 If water is derived from different sources at different times, the water supplied to the consumer may change from soft to hard. This may cause problems. Some blending of the water from the different sources may be required.

As conjunctive use schemes have some significant disadvantages, it may be advisable not to attempt such a scheme unless there is adequate justification, namely, a shortage of water that cannot be satisfied by any other reasonable means. If such a scheme is undertaken, it should be on a small scale initially, and models should be developed to help to predict future trends, select the most appropriate options and generally assist the management process.

References

Anderson, J.G.C. & McNicol, R. (1989) The engineering geology of the Kielder Dam. *Quarterly Journal of Engineering Geology* **22**, 111–130.

Anon. (1983) *Code of Practice for Test Pumping Water Wells* BS 6316. British Standards Institution, London.

Anon. (1988) Engineering geophysics. Engineering Group Working Party Report. *Quarterly Journal of Engineering Geology* **27**, 207–273.

Barber, C. (1982) *Domestic Waste and Leachate, Notes of Water Research No. 31*. Water Research Centre, Medmenham, UK.

Bear, J. (1979) *Hydraulics of Groundwater*. McGraw-Hill, New York.

Bell, F.G. & Haskins, D.R. (1997) A geotechnical overview of the Katse Dam and Transfer Tunnel, Lesotho, with a note on basalt durability. *Engineering Geology* **46**, 175–198.

Black, J.H. (1987) Flow and flow mechanisms in crystalline rock. In *Fluid Flow in Sedimentary Basins and Aquifers, Special Publication No. 34*, Goff, J.C. & Williams, B.P.J. (eds). Geological Society, London, pp. 185–200.

Bliss, J.C. & Rushton, K.R. (1984) The reliability of packer tests for estimating the hydraulic conductivity of aquifers. *Quarterly Journal of Engineering Geology* **17**, 81–91.

Brassington, F.C. & Walthall, S. (1985) Field techniques using borehole packers in hydrogeological investigations. *Quarterly Journal of Engineering Geology* **18**, 181–194.

Brown, D.L. (1971) Techniques for quality of water interpretation from calibrated geophysical logs, Atlantic Coastal Area. *Ground Water* **9** (4), 25–38.

Brown, R.F. & Signor, D.C. (1973) Artificial recharge—state of the art. *Ground Water* **12** (3), 152–160.

Brown, R.H., Konoplyantsev, A.A., Ineson, J. & Kovalevsky, V.S. (eds) (1972) *Groundwater Studies. An International Guide for Research and Practice, Studies and Reports in Hydrology* **7**. UNESCO, Paris.

Campbell, M.D. & Lehr, J.H. (1973) *Water Well Technology*. McGraw-Hill, New York.

CEC (1980) *Directive related to Quality of Water for Human Consumption*, 80/778/EEC. Commission of the European Community, Brussels.

Clark, L. (1977) The analysis and planning of step-drawdown tests. *Quarterly Journal of Engineering Geology* **10**, 125–143.

Clark, L. (1985) Groundwater abstraction from basement complex areas in Africa. *Quarterly Journal of Engineering Geology* **18**, 25–34.

Darcy, H. (1856) *Les Fontaines Publiques de la Ville de Dijon*. Dalmont, Paris.

Davis, R.D. (1980) *Control of Contamination Problems in the Treatment and Disposal of Sewage Sludge, Technical Report* **156**. Water Research Centre, Medmenham, UK.

Diefendorf, A.F. & Ausburn, R. (1977) Groundwater monitoring wells. *Public Works* **108** (7), 48–50.

Environmental Protection Agency (1975) *Federal Register* **40** (248), December 24. Environmental Protection Agency, Washington, DC, pp. 59566–59588.

Foster, S.S.D. & Crease, R.I. (1974) Nitrate pollution of chalk groundwater in east Yorkshire—a hydrological appraisal. *Journal of the Institution of Water Engineers and Scientists* **28**, 178–194.

Gottschalk, L.C. (1964) Reservoir sedimentation. In *Applied Hydrology*, Chow, V.T. (ed.), McGraw-Hill, New York, 17/1–17/33.

Gray, D.A., Mather, J.D. & Harrison, I.B. (1974) Review of groundwater pollution for waste sites in England and Wales with provisional guidelines for future site selection. *Quarterly Journal of Engineering Geology* **7**, 181–196.

Hagerty, D.J. & Pavoni, J.L. (1973) Geological aspects of landfill refuse disposal. *Engineering Geology* **7**, 219–230.

Hamill, L. & Bell, F.G. (1986) *Groundwater Resource Development*. Butterworths, London.

Hem, J.D. (1985) Study and interpretation of the chemical characteristics of natural water. *United States Geological Survey, Water Supply Paper* **2254**. United States Geological Survey, Washington, DC.

Henriksen, A. & Kirkhusmo, L.A. (1982) Acidification of groundwater in Norway. *Nordic Hydrology* **13**, 183–192.

Ineson, J. (1970) Development of groundwater resources in England and Wales. *Journal of the Institution of Water Engineers and Scientists* **24**, 155–177.

Knill, J.L. (1971) Assessment of reservoir feasibility. *Quarterly Journal of Engineering Geology* **4**, 355–372.

Lovelock, P.E.R., Price, M. & Tate, T.K. (1975) Groundwater conditions in the Penrith Sandstone at Cliburn, Westmoreland. *Journal of the Institution of Water Engineers and Scientists* **29**, 157–174.

Mackay, D.M. (1998) Is clean-up of VOC-contaminated groundwater feasible? In *Contaminated Land and Groundwater—Future Directions, Engineering Geology Special Publication No. 14*, Lerner, D.N. & Walton, N. (eds). Geological Society, London, 3–11.

Mackay, D.M. & Cherry, J.A. (1989) Groundwater contamination: limits of pump-and-treat remediation. *Environmental Science and Technology* **23**, 630–636.

Marsh, T.J. & Davies, P.A. (1983) The decline and partial recovery of groundwater levels below London. In *Proceedings of the Institution of Civil Engineers*, Part 1 (74), pp. 263–276.

Nace, R.L. (1969) World water inventory and control. In *Water, Earth and Man*, Chorley, R.J. (ed). Methuen, London, pp. 31–42.

Pfannkuch, H.A. (1982) Problems of monitoring network design to detect unanticipated contamination. *Ground Water Monitoring Review* **2** (1), 67–76.

Sawyer, C.N. & McCarty, P.L. (1967) *Chemistry for Sanitary Engineers*, 2nd edn. McGraw-Hill, New York.

Selby, K.H. & Skinner, A.C. (1979) Aquifer protection in the Severn–Trent region: policy and practice. *Water Pollution Control* **78**, 320–326.

Serafim, J.L. (1968) Influence of interstitial water on rock masses. In *Rock Mechanics in Engineering Practice*, Stagg, K.G. & Zienkiewicz, O.C. (eds). Wiley, London, pp. 55–77.

Soderberg, A.D. (1979) Expect the unexpected: foundations for dams in karst. *Bulletin of the Association of Engineering Geologists* **16**, 409–427.

Sutcliffe, G. & Mostyn, G. (1983) Permeability testing for the O K Tedi Project (Papua New Guinea). *Bulletin of the International Association of Engineering Geology* **26–27**, 501–508.

Vick, S.G. & Bromwell, L.G. (1989) Risk analysis for dam design in karst. *Proceedings of the American Society of*

Civil Engineers, Journal Geotechnical Engineering Division **115**, 819–835.

WHO (1993) *Guidelines for Drinking Water Quality.* World Health Organization, Geneva.

Wright, E.P. (1992) The hydrogeology of crystalline basement aquifers in Africa. In *The Hydrogeology of Crystalline Basement Aquifers in Africa, Special Publication No.* **66**, Wright, E.P. & Burgess, W.G. (eds). Geological Society, London, pp. 1–28.

Zaki, M.H., Moran, D. & Harris, D. (1982) Pesticides in groundwater, the aldicarb story in Suffolk County, New York. *American Journal of Public Health* **72**, 1391–1395.

10   Soil resources

10.1 Origin of soil

Soils are one of the most important resources available to man. Although they are of most value in terms of agriculture, they also play an important role as construction materials.

Soil is derived from the breakdown of rock material by weathering and/or erosion, and it may have suffered a certain amount of transportation prior to deposition. It may also contain organic matter. The type of breakdown process(es) and the amount of transport undergone by sediments influence the nature of the macrostructure and microstructure of the soil (Table 10.1). The same type of rock can give rise to different types of soil, depending on the climatic regime and the vegetative cover. Time is also an important factor in the development of a mature soil, especially from the agricultural point of view.

Probably the most important methods of soil formation are mechanical and chemical weathering. The agents of weathering, however, are not capable of transporting material. Transport is brought about by gravity, water, wind or moving ice. If sedimentary particles are transported, this affects their character, particularly their grain size distribution, sorting and shape. For instance, stream channel deposits commonly are well graded, although the grain size characteristics may vary erratically with location. On the other hand, windblown deposits usually are uniformly sorted with well rounded grains.

Plants affect the soil in several ways. For instance, when their roots die and decay, they leave behind a network of passages, which allows air and water to move through the soil more readily. Roots, especially those of grasses, help to bind the soil together and so reduce erosion. However, perhaps the major contribution to soil made by plants is the addition of organic matter. Microorganisms occur in greatest numbers in the surface horizons of the soil, where there are the largest concentrations of food supply. They are particularly important in terms of the decomposition of organic matter, which leads to the formation of humus. The plant and animal communities and microorganisms help to release or convert nutrients in the soil, so that they become available for other plants growing in the soil.

The total organic content of soils can vary from less than 1% in the case of some immature or desert soils to over 90% in the case of peats. Normally, the upper horizons of a soil contain less than 15% of organic matter. Humus is capable of absorbing large quantities of moisture, so that the more humus present, the greater the water retaining capacity of the soil. The cation exchange capacity of humus is greater than that of clay minerals, being some $3000\,meq\,kg^{-1}$, compared with $30–150\,meq\,kg^{-1}$ for kaolinite, $200–400\,meq\,kg^{-1}$ for illite and $600–1000\,meq\,kg^{-1}$ for montmorillonite. It therefore enhances the cation-holding capacity of soil.

Changes occur in soils after they have accumulated. In particular, seasonal changes take place in the moisture content of sediments above the water table. Volume changes associated with alternate wetting and drying occur in cohesive soils with high plasticity indices. The exposure of a soil to dry conditions means that its surface dries out, and water is drawn from deeper zones by capillary action. The capillary rise is associated with a decrease in the pore water pressure in the layer beneath the surface, and a corresponding increase in effective pressure. This supplementary pressure is known as capillary pressure, and it has the same mechanical effect as a heavy surcharge. Therefore, surface evaporation from very compressible soils produces a conspicuous decrease in the void ratio of the layer undergoing desiccation. If the moisture content in this layer reaches the shrinkage limit, air begins to invade the voids and the soil structure begins to break down. The decrease in

Table 10.1 Effects of transportation on sediments.

	Gravity	Ice	Water	Air
Size	Various	Varies from clay to boulders	Various sizes from boulder gravel to muds	Sand size and less
Sorting	Unsorted	Generally unsorted	Sorting takes place both laterally and vertically Marine deposits often uniformly sorted River deposits may be well sorted	Uniformly sorted
Shape	Angular	Angular	From angular to well rounded	Well rounded
Surface texture	Striated surfaces	Striated surfaces	Gravel: rugose surfaces Sand: smooth, polished surfaces Silt: little effect	Impact produces frosted surfaces

void ratio, consequent upon the desiccation of a cohesive sediment, leads to an increase in its bearing strength. Thus, if a dry crust is located at or near the surface above softer material, it acts as a raft. The thickness of dry crusts often varies erratically.

Chemical changes which take place in the soil due, for example, to the action of weathering may bring about an increase in its clay mineral content, the latter developing from the breakdown of less stable minerals. In such instances, the plasticity of the soil increases, whereas its permeability decreases. Leaching, whereby soluble constituents are removed from the upper horizons, to be precipitated in the lower horizons, occurs when rainfall exceeds evaporation. The porosity may be increased in the zone undergoing leaching.

With continuing exposure, the soil develops a characteristic profile from the surface downwards. This development involves the accumulation and decay of organic matter, leaching, precipitation, oxidation or reduction, and further mechanical and biological breakdown. The profile which forms is influenced by the character of the parent material, but climatic conditions, vegetative cover, groundwater level and relief also play their part, and the time factor allows a distinction to be made between immature and mature soils.

10.2 Soil horizons

A soil profile is divided into zones or horizons which differ in character from the parent material to the soil surface (Fig. 10.1). The master horizons are denoted by a capital letter as follows:

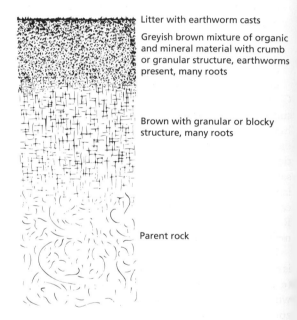

Litter with earthworm casts

Greyish brown mixture of organic and mineral material with crumb or granular structure, earthworms present, many roots

Brown with granular or blocky structure, many roots

Parent rock

Fig. 10.1 Diagram of a Cambisol profile which is the characteristic soil of temperate deciduous forests.

H Formed by the accumulation of organic material deposited on the soil surface. Generally, it contains 20% or more organic matter. It may be saturated with water for prolonged periods.

O Formed by the accumulation of organic material (plant litter) deposited on the surface. It contains 35% or more organic matter. It is not saturated with water for more than a few days in a year.

A A mineral horizon occurring at or adjacent to the surface, which has a morphology acquired by soil formation. Generally possesses an accumulation of humified organic matter intimately associated with the mineral fraction.

E A pale coloured mineral horizon with a concentration of sand and silt fractions, rich in resistant minerals, from which clay, iron or aluminium, or some combination thereof, has been leached. E horizons, if present, are eluvial horizons, which generally underlie an H, O or A horizon and overlie a B horizon.

B A mineral horizon in which parent material is absent or faintly evident. It is characterized by one or more of the following features:

   (a) an illuvial concentration of clay minerals, iron, aluminium or humus, alone or in combination;

   (b) a residual concentration of sesquioxides relative to the source materials;

   (c) an alteration of material from its original condition, to the extent that clay minerals are formed, oxides are liberated, or both; or a granular, blocky or prismatic structure is formed.

C A mineral horizon of unconsolidated material similar to that from which the solum is presumed to have formed, and which does not show properties diagnostic of any other master horizon. Accumulations of carbonates, gypsum or other more soluble salts may occur in C horizons. Carbonate often forms laminar layers which run parallel to the surface, and carbonate may fill the pores between soil particles.

R Rock which is sufficiently coherent when moist to make hand digging impracticable.

If two master horizons merge, the resultant horizon is indicated by the combination of two capital letters (e.g. AE). The first letter marks the master horizon to which the transitional horizon is most similar. Horizons that contain mixtures of two horizons, which are identifiable as different master horizons, are designated by two capital letters separated by a solidus (e.g. E/B). Not all the horizons mentioned above necessarily are present in a given soil profile.

The H, O and A horizons tend to be dark in colour, because of the presence of plentiful organic matter. The E horizon is light in colour due to leaching of iron oxides from it. Sometimes notable changes in colour may be displayed by the B horizon, varying from yellowish-brown to light reddish-brown to dark red. This depends primarily on the presence of iron oxides and clay minerals. The presence of carbonates may lighten the colour. Well drained soils are well aerated, so that iron is oxidized to give a red colour. On the other hand, wet, poorly drained soils generally mean that iron is reduced, which may give rise to a yellow colour.

As remarked above, it takes time to develop a mature soil. Hence, different grades of soil profile may be recognized from immature to mature. A poorly developed, immature soil profile is one in which the A horizon may overlie the C horizon directly, and the C horizon may show signs of oxidation. In a moderately developed soil profile, the A horizon may rest upon an argillic B horizon (i.e. one that contains illuvial clay and in which the peds may be coated with clay) which, in turn, is underlain by the C horizon. At times, a carbonate B horizon may also be present. A mature, well developed soil profile possesses a good soil structure, in which the B horizon may contain a notable amount of clay material.

10.3 Soil fertility

Soil fertility refers to the capacity of a soil to supply nutrients needed for plant growth. In this context, more or less all soils have a certain inherent fertility. For example, soils which are acidic, alkaline, waterlogged or deficient in particular elements can support specific plant communities and, as such, can be regarded as fertile in relation to those communities. However, when man attempts to cultivate such soils, the inherent fertility may be unsuitable for the particular crop(s). It then becomes necessary to alter the fertility by, for example, the addition of fertilizers or the installation of drainage measures to suit the needs of the crop(s).

Be that as it may, there are a number of factors which influence plant growth. These include soil aeration and moisture content, soil temperature, pH value and essential elements. Good aeration usually is facilitated by a granular or crumb structure and free drainage. An adequate and balanced supply of moisture is essential for plant growth. The moisture within the soil is derived from precipitation or irrigation. The particle size distribution, texture, organic matter content and structure of a soil affect its retention of moisture. Generally, clays and organic soils have the highest moisture-retaining capacity, while organic soils have the highest available moisture. The moisture retained in the soil is lost mainly by evapotranspiration, so that the rate of loss depends upon the temperature and plant cover. As they increase, moisture losses increase. On the other hand, wet or waterlogged soils need drainage for the satisfactory cultivation of most crops. Drainage is beneficial to agriculture, as it improves the soil structure, facilitates aeration, increases the rate of decomposition of organic matter, leads to an increase in soil temperature and increases the bacterial population in the soil. In addition, the thickness of soil available for root growth and penetration is very important.

Moisture is taken up by plants together with nutrients and is lost by transpiration. It is estimated that a 1 kg dry weight increase in plants requires about 500 kg of transpired water. Two important moisture characteristics of soil are the field capacity and the permanent wilting point. When soil is saturated and the excess water drains away, the soil is described as being at its field capacity. Plants extract moisture from the soil until, as the soil dries out, it becomes impossible for them to continue doing so; they then wilt and die if the soil is not rewetted. The point at which permanent wilting starts is known as the permanent wilting point (Fig. 10.2). The field capacity and permanent wilting point have been defined in terms of soil suction, the pF values being 2.0 and 4.2, respectively. However, the permanent wilting point varies from soil to soil and from plant to plant. Water held between the field capacity and the permanent wilting point is the water available to plants and the amount varies considerably from soil to soil.

The pH value for soils ranges between pH 3 and pH 9, although for cultivated soils, the range is more limited, that is, pH 5.5–7.5. Very low values sometimes can be found in swamps, where the breakdown of organic material produces humic acid and sulphides give rise to sulphuric acid. At the other extreme, very high pH values may be attributable to the presence of sodium carbonate. This may be the case in some salinas or sabkhas. Generally, pH values of about neutral are associated with large amounts of exchangeable calcium and some magnesium. High acidity may lead to increasing amounts of manganese and aluminium in the pore water, with little formation of ammonia or nitrate, a low availability of phosphorus and molybdenum deficiency.

A number of elements are essential for healthy plant growth. However, they must be present in the soil in the correct proportions. If there is an excess or deficiency in any one element, plant growth can be

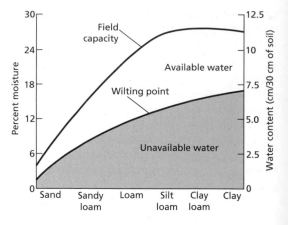

Fig. 10.2 General relationship between soil moisture characteristics and soil texture. Note that the wilting point increases as the texture becomes finer. The field capacity increases up to silt loams and then levels off.

affected seriously, plants developing symptoms of toxicity or nutrient starvation. Of the essential elements, calcium, carbon, hydrogen, magnesium, nitrogen, oxygen, phosphorus, potassium and sulphur should be present in the soil in relatively large amounts, whereas boron, chlorine, cobalt, copper, iron, manganese, molybdenum and zinc are necessary in small amounts. Most of these elements (i.e. except those obtained from the air) are derived initially from the breakdown of minerals within the parent rock, and are taken up by the plant roots. They subsequently are returned to the soil in the plant litter when it decomposes, thereby releasing the elements into the soil for the cycle to begin again.

10.4 Pedological soil types

Comprehensive pedological classifications of soils have been developed by the United States Department of Agriculture (USDA) (Anon., 1975a) and the Food and Agriculture Organization (FAO) (Anon., 1980). There are 10 orders in the USDA system of soil taxonomy, which usually are differentiated by the presence or absence of diagnostic horizons or features that show the dominant set of soil forming processes (Table 10.2). The following description of the major soil types is according to the FAO, the equivalent USDA terminology being given in parentheses.

Table 10.2 Soil orders according to the United States Department of Agriculture. (Anon., 1975a.)

Name	Description
Alfisols	Characterized by a brown–grey surface horizon and an argillic horizon, and a high base saturation, they tend to occur under forests in humid regions of mid-latitudes. Their organic content is low, and may have calcic horizon Luvisols
Aridisols	These are the soils of the dry regions, and include Xerosols, Yermosols and Solonchaks
Entisols	These are young soils with no horizon development (azonal) which form on recent sediments, such as alluvium and sand dunes
Histosols	These soils are composed predominantly of organic matter, and include peats, mucks, bogs and moors
Inceptisols	Show a modest amount of development, in which one or more horizons may have developed quickly, but horizons often difficult to differentiate. Frequently occur on young, but not recent, land surfaces. Most common in humid climates, but range from the Arctic to the tropics. Natural vegetation usually forest, and have appreciable accumulation of organic material. Include the Andosols, Cambisols and Gleysols
Mollisols	These have a well developed soil structure. They are characterized by a black, organic-rich A horizon with surface horizons rich in bases. Usually occur in subhumid or semi-arid regions (prairie soils). Include the Chernozems and Kastanozems
Oxisols	Strongly weathered soils of tropical and subtropical areas. Often deep soils, which are leached of bases, hydrated, containing oxides of iron and aluminium (plinthite, i.e. laterite), as well as kaolinite. Include Ferrasols and Nitosols
Spodosols	Characterized by a grey subsoil which has been leached, with accumulations of amorphous iron–aluminium sesquioxides and humus. Hard pan may have developed. Occur under coniferous forests in humid and subhumid areas. These include all the Podzols
Ultisols	Soils of mid-to-low humid latitudes that have an argillic (clay) horizon and low base saturation. These soils are usually strongly weathered with high amounts of kaolinite and gibbsite. Often have a reddish-yellow or reddish-brown colour. They equate with the Acrisols
Vertisols	Clayey soils that tend to swell and shrink according to the season; generally occur in regions with distinct wet and dry seasons. Often dark in colour with high amounts of montmorillonite and a high base saturation

The Gleysols (Cryosols) of tundra regions are characterized by a thin accumulation of organic matter at the surface, beneath which there is a dark, greyish-brown mixture of mineral and organic material, lying on top of a wet, mottled horizon about 500 mm thick. This is followed by a sharp change to permafrost, composed of thin, bifurcating horizontal veins of ice surrounding lenticular peds of frozen soil. The upper horizons freeze in winter and thaw in summer, with accompanying expansion and contraction, and the formation of stone polygons. Ice wedges may extend to depths of 3 m or more.

South of the tundra, there are extensive areas of Histosols, with marshy vegetation dominated by *Juncus*, *Carex* and *Sphagnum* moss. These are peat soils. Peat is an accumulation of organic matter which forms under wet, anaerobic conditions where plant detritus decomposes slowly (see Chapter 11). Nevertheless, some degree of humification does take place in peats, so that they range from very fibrous to amorphous types. Peat is, in fact, widely distributed in humid, temperate and tropical regions, where land is badly drained, boggy and swampy. Peats tend to be acidic, and peaty soils need to be drained and limed if they are to be cultivated. Where sand underlies peat, the two frequently are mixed to form a light, well-drained soil suitable for horticulture.

The Podzols (Spodosols) tend to coincide with the coniferous forest regions. There is an accumulation of plant litter at the surface, below which there is a layer of dark brown, partially decomposed plant material (Fig. 10.3a). This grades into very dark brown, amorphous organic matter. A dark grey layer of mixed organic and mineral material occurs beneath the

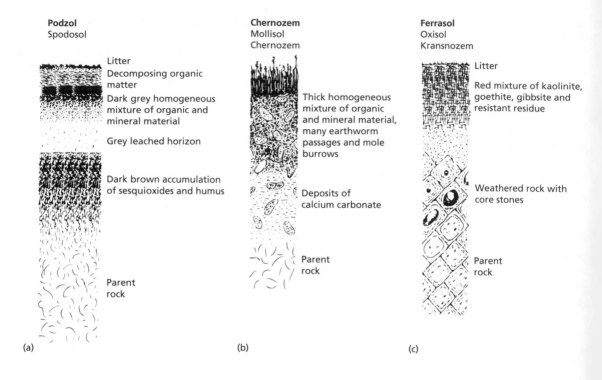

Podzol
Spodosol

Litter
Decomposing organic
matter
Dark grey homogeneous
mixture of organic and
mineral material

Grey leached horizon

Dark brown accumulation
of sesquioxides and humus

Parent
rock

Chernozem
Mollisol
Chernozem

Thick homogeneous
mixture of organic
and mineral material,
many earthworm
passages and mole
burrows

Deposits of
calcium carbonate

Parent
rock

Ferrasol
Oxisol
Kransnozem

Litter
Red mixture of kaolinite,
goethite, gibbsite and
resistant residue

Weathered rock with
core stones

Parent
rock

(a)   (b)   (c)

Fig. 10.3 Soil profiles of a Podsol, a Chernozem and a Ferrasol.

organic matter. A pale grey horizon then follows, which has been leached of iron compounds and organic matter by downward percolating water. The leached material is precipitated beneath this horizon to form sesquioxides. A layer of hard pan may be present at this level. Finally, the unaltered parent material occurs below this horizon. Podzols may be up to 2 m thick. These soils are usually acid, with pH values of less than pH 4.5 at the surface, increasing to about pH 5.5 in the lowest horizons. They are also very deficient in plant nutrients, except at the surface, where nutrients are released by the decomposition of organic matter. Hence, Podzols are inherently infertile in terms of arable farming. Constant additions of lime and fertilizers are necessary if they are to be used for growing crops.

The Cambisols (Altosols), commonly known as brown earths, occur beneath some deciduous woodlands of the cool, temperate regions. There is a continuous gradation from Podzols to Cambisols. At the surface of a Cambisol, there is a loose layer of plant litter resting on a brown or greyish-brown horizon of mixed organic and mineral material. This overlies a brown, friable, loamy B horizon, which grades into unaltered material. This parent material at times is basic or calcareous, and gives the soil a high base saturation. These soils generally are slightly acidic to mildly alkaline. Organic matter rapidly is broken down and incorporated into the soil. As a result, Cambisols have a high natural fertility. Even so, they need to be fertilized and limed to support arable farming.

Luvisols (Alfisols) are also found in cool, temperate regions beneath deciduous forests or mixed deciduous and coniferous forests. A thin layer of loose plant litter rests on a greyish-brown mixture of organic and mineral material. This changes into a grey, sandy horizon, beneath which occurs a sharp change to a brown, blocky or prismatic horizon, which has a high content of clay. There is then a gradation to the unaltered material. These soils have a middle horizon which contains more clay than those above or below (argillic B horizon), as a result of clay particles being washed from the upper horizons. Luvisols are weakly acidic at the surface, with medium base saturation which means that they have a moderate natural fertility. They similarly need lime and fertilizers for arable farming.

As the precipitation declines to less than 400mm on moving towards the equator, the deciduous forests are replaced by grasslands. Here are found the black earths or Chernozems (Mollisols). These soils have a root mat at the surface, which overlies a thick, black horizon with a high humus content, which may be up to 2m thick (Fig. 10.3b). Thin, thread-like deposits of calcium carbonate occur in the lower part of this horizon as well as below. Concretions of calcium carbonate may also be present. The black horizon grades into yellowish-brown material. One of the notable features of the Chernozems found in Russia is the presence of burrows (crotovinas) which are filled by dark soil and extend into the underlying material. Chernozems have neutral pH values, very high base saturation and a very high inherent fertility. These soils have a well-developed crumb structure and high water-holding capacity, due to their high organic content from annual increments of dead grass. They represent major regions of grain production.

As the precipitation gradually decreases, the vegetation changes from tall grass, to short grass, to bunchy grass and then to species that can withstand long, dry periods. Accordingly, there are similar changes in the soils, with the thick, black horizon becoming lighter in colour and shallower, and the calcium carbonate horizon approaching the surface. Generally, three soil types can be recognized, namely, the Kastanozems, the Xerosols and the Yermosols. The Kastanozems (Ustols) are covered with short grass. They have a dark brown horizon, which may be up to about 300mm thick, below which the calcium carbonate horizon is present, followed by the unaltered material. Xerosols (Aridisols) are characterized by bunchy grass and a thinner, brown upper horizon than the lower calcium carbonate horizon. Yermosols (Aridisols) are covered with typical, sparse, xerophytic, desert-type vegetation. They possess a thin, brownish-grey upper horizon, which rests on a carbonate or sometimes a gypsiferous layer. These three soils frequently have a high inherent fertility but, unless irrigated, crop production usually is restricted by the lack of moisture. Dry farming is undertaken on Kastanozem soils when irrigation is impossible. Dry farming is not successful on Xerosols

or Yermosols, irrigation being required for crop growth.

Soils with high salinity or alkalinity are found in semi-arid or arid regions where evapotranspiration greatly exceeds precipitation. Ions, such as calcium, chloride, magnesium, potassium, sodium and sulphate, accumulate in the soils. Solonchaks (Salorthids) are saline soils which often have salt efflorescences on the ped surfaces and on the ground surface. Normally, salt-tolerant plants (the upper limit of salt tolerance for most plants is 0.5%) are present, but do not cover the surface completely. The soil profile often contains an upper, grey, organic–mineral material mixture which rests on a mottled horizon, beneath which is a grey or olive, completely reduced horizon. Many Solonchaks are potentially useful for agriculture if the excess ions can be dissolved and removed by large amounts of water provided by irrigation. One of the problems in such regions is salinization. This occurs when irrigated soils are not well drained. The irrigation water percolates to the water table, with the result that it may rise. This may mean that capillary rise eventually extends to the surface, thereby causing salinity problems, which severely reduce crop production.

Solonetz (Naturargids) soils also occur in arid and semi-arid areas. A thin layer of litter may occur at the surface, to be quickly followed by a thin, very dark mixture of organic and mineral matter, and then a dark grey, frequently sandy horizon. Below this is a distinctive middle horizon with a marked increase in clay. The latter is characterized by a prismatic or columnar structure. These soils have pH values that often are above pH 8.5 due to high exchangeable sodium and magnesium, but generally have a low salt content. The high pH is harmful to plant growth and must be reduced if arable farming is to be practised. This usually is brought about by adding calcium sulphate, the calcium replacing sodium, which is removed by leaching by rainfall or irrigation.

Vertisols or black cotton soils occur in the savannah grassland regions of the tropics and subtropics, which are characterized by alternating wet and dry seasons. They are dark-coloured, clayey soils in which montmorillonite is an important constituent. The surface horizon usually is granular, but can be massive, and rests upon a dense horizon, with a prismatic or angular, blocky structure, that grades into similar material with a marked wedge structure. This is caused by the montmorillonite shrinking and expanding in response to seasonal drying and wetting. Many of the peds are slickensided, that is, have shiny surfaces which form as one ped slips over another during soil expansion. The pH value of Vertisols is about neutral. They have a high base saturation and high cation exchange capacity. Vertisols are fertile soils with a high potential for agriculture. Irrigation frequently proves necessary in such regions as they tend to be affected by droughts.

The Ferrasols (Oxisols) occur in the humid tropics, being highly weathered soils that are often red in colour. These soils possess a thin layer of plant litter at the surface, followed by a greyish-red mixture of organic and mineral matter, which is not more than 100 mm thick. This grades quickly into a bright red, clayey horizon which may be several metres thick, that overlies the underlying rock (Fig. 10.3c). The red horizon consists largely of kaolinite, with iron and aluminium, together with hydrated oxides of gibbsite. The change from lateritic soils to the parent rock can be gradual or sharp. There often is a red and cream, mottled horizon beneath the red topsoil, or a thick, white horizon, known as the pallid zone, may be present. Because of the effectiveness of chemical weathering, these soils are deficient in essential plant elements, even though they often carry forests. This is because the elements are constantly being recycled, and there is a small reservoir in the top few centimetres where they are released from the decomposing plant detritus. In fact, in many instances, when forest has been cleared for agriculture, this has failed within a short period of time due to the soil becoming exhausted. Nevertheless, when fertilizers are applied to these soils, a variety of tropical crops can be grown.

10.5 Basic properties of soil

Soil consists of an unconsolidated assemblage of particles between which are voids, which may contain air or water or both. Therefore, soil can contain three phases, solids, water and air. The interrelationships between the weights and volumes of these three phases are important, because they help to define the character of a soil.

One of the most fundamental properties of a soil is the void ratio, which is the ratio of the volume of the voids to the volume of the solids. The porosity is a similar property, it being the ratio of the volume of the voids to the total volume of the soil, expressed as a percentage. Both the void ratio and the porosity indicate the relative proportion of void volume in a soil sample.

Water plays a fundamental part in determining the behaviour of any soil, and the moisture content is expressed as a percentage of the weight of the solid material in the soil sample. The degree of saturation expresses the relative volume percentage of water in the voids.

The range of values of the phase relationships for cohesive soils is much larger than that for granular soils. For instance, saturated sodium montmorillonite at low confining pressure can exist at a void ratio of more than 25, its moisture content being some 900%. Peat soils have similar exceptionally high void ratios and moisture contents (see Chapter 11). On the other hand, saturated clays under high stresses that exist at great depth may have void ratios of less than 0.2 with about 7% moisture content.

The unit weight of a soil is its weight per unit volume, whereas the specific gravity of a soil is the ratio of its weight to that of an equal volume of water. The density of a soil is the ratio of its mass to its volume. A number of types of density can be distinguished. The dry density is the mass of the solid particles divided by the total volume, whereas the bulk density is simply the mass of the soil (including its natural moisture content) divided by its volume. The saturated density is the density of the soil when saturated, whereas the submerged density is the ratio of the effective mass to the volume of soil when submerged. The submerged density can be derived by subtracting the density of water from the saturated density.

The density of a soil is governed by the manner in which its solid particles are packed. For example, granular soils may be densely or loosely packed. Indeed, a maximum and minimum density can be distinguished. The smaller the range of particle sizes present and the more angular the particles, the smaller the minimum density. Conversely, if a wide range of particle sizes is present, the void space is reduced accordingly and, hence, the maximum density is higher. A useful way to characterize the density of a granular soil is by its relative density (D_r), which is given by

$$D_r = \frac{e_{max} - e}{e_{max} - e_{min}} \qquad (10.1)$$

Table 10.3 Particle size distribution of soils.

Types of material	Sizes (mm)
Boulders	Over 200
Cobbles	60–200
Gravel	
Coarse	20–60
Medium	6–20
Fine	2–6
Sand	
Coarse	0.6–2
Medium	0.2–0.6
Fine	0.06–0.2
Silt	
Coarse	0.02–0.06
Medium	0.006–0.02
Fine	0.002–0.006
Clay	Less than 0.002

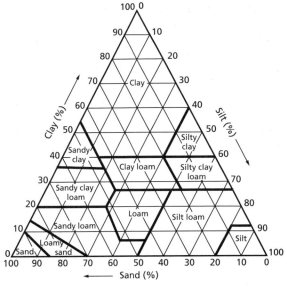

Fig. 10.4 Triangular diagram relating particle size distribution to texture. (After Anon., 1971.)

where e is the naturally occurring void ratio, e_{max} is the maximum void ratio and e_{min} is the minimum void ratio.

10.5.1 Particle size distribution

The particle size distribution expresses the size of particles in a soil in terms of the percentages by weight of boulders, cobbles, gravel, sand, silt and clay. The United Soil Classification (Wagner, 1957) and Anon. (1981a) give the limits shown in Table 10.3 for these size grades. The USDA (1971) classification of particle size distribution, as related to soil texture, is given in Fig. 10.4 for comparison.

In nature, there is a deficiency of soil particles in the fine gravel and silt ranges, and boulders and cobbles are, quantitatively speaking, not significant. Sands and clays therefore are the most important soil types.

The results of particle size analysis are given in the form of a series of fractions, by weight, of different size grades. These fractions are expressed as a percentage of the whole sample, and generally are summed to obtain a cumulative percentage. Cumulative curves are then plotted on semi-logarithmic paper to give a graphical representation of the particle size distribution. The slope of the curve provides

an indication of the degree of sorting. If, for example, the curve is steep, as in curve A in Fig. 10.5, the soil is uniformly sorted, whereas curve B represents a well sorted soil. Curve C represents a gap-graded soil, and both this type of sorting and uniform sorting are regarded as examples of poor sorting (Anon., 1981a). The sorting of a particle size distribution has been expressed in many ways, but one simple statistical measure is the coefficient of uniformity (U). This makes use of the effective size of the grains (D_{10}), that is, the size on the cumulative curve where 10% of the particles are passing, and is defined as

$$U = D_{60}/D_{10} \qquad (10.2)$$

Similarly, D_{60} is the size on the curve where 60% of the particles are passing. A soil having a coefficient of uniformity of less than two is considered to be uniform, whereas one with a value of 10 is described as well graded. In other words, the higher the coefficient of uniformity, the larger the range of particle sizes.

Soil particles may form aggregates referred to as peds. Angular, blocky peds have sharp, angular corners with flat, convex and/or concave faces. Sub-angular, blocky peds have convex and/or concave faces and rounded corners. Large peds which are

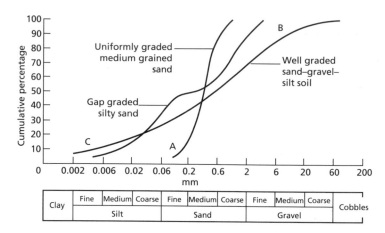

Fig. 10.5 Grading curves.

elongated vertically are described as columnar. These peds commonly have domed upper surfaces and grade into the underlying material. Irregularly shaped peds with rough surfaces characterize soils with a crumb structure. Such soils normally have a high porosity and are well drained. Soils with granular structures contain subspherical peds. They also have an open texture. Horizontally elongate peds, with flat, parallel surfaces, are described as laminar. Lenticular peds are lens shaped, with convex surfaces, and tend to overlap one another. Vertically elongate peds have three or more vertically flat faces and, where the peds have somewhat indeterminate upper and lower boundaries, they are called prismatic peds.

10.5.2 Consistency limits

The Atterberg or consistency limits of cohesive soils are founded on the concept that such soils can exist in any of four states, depending on their water content. These limits are also influenced by the amount and character of the clay mineral content. In other words, a cohesive soil is solid when dry but, as water is added, it first turns into a semi-solid, then a plastic and, finally, a liquid state. The water contents at the boundaries between these states are referred to as the shrinkage limit (SL), the plastic limit (PL) and the liquid limit (LL) respectively.

The shrinkage limit is defined as the percentage moisture content of a soil at the point at which it suffers no further decrease in volume on drying. The plastic limit is the percentage moisture content at

Table 10.4 Plasticity according to liquid limit.

Description	Plasticity	Range of liquid limit
Lean or silty	Low plasticity	Less than 35
Intermediate	Intermediate plasticity	35–50
Fat	High plasticity	50–70
Very fat	Very high plasticity	70–90
Extra fat	Extra high plasticity	Over 90

Table 10.5 Plasticity of soils. (After Anon., 1979.)

Class	Plasticity index (%)	Description
1	Less than 1	Non-plastic
2	1–7	Slightly plastic
3	7–17	Moderately plastic
4	17–35	Highly plastic
5	Over 35	Extremely plastic

which a soil can be rolled, without breaking, into a thread 3 mm in diameter, any further rolling causing it to crumble. Unfortunately, the inadequacy of control involved in the test means that the results obtained are not consistent for any particular clay. Turning to the liquid limit, this is defined as the minimum moisture content at which a soil will flow under its own weight. Clays may be classified according to their liquid limit as shown in Table 10.4.

The numerical difference between the liquid and plastic limits is referred to as the plasticity index (PI).

Table 10.6 Consistency of cohesive soils.

Description	Consistency index	Approximate undrained shear strength (kPa)	Field identification
Hard		Over 300	Indented with difficulty by thumbnail, brittle
Very stiff	Above 1	150–300	Readily indented by thumbnail, still very tough
Stiff	0.75–1	75–150	Readily indented by thumb, but penetrated only with difficulty. Cannot be moulded in the fingers
Firm	0.5–0.75	40–75	Can be penetrated to several centimetres by thumb with moderate effort, and moulded in the fingers by strong pressure
Soft	Less than 0.5	20–40	Easily penetrated to several centimetres by thumb, easily moulded
Very soft		Less than 20	Easily penetrated to several centimetres by fist, exudes between fingers when squeezed in fist

Table 10.7 Symbols used in the Casagrande Unified Soil Classification System soil classification.

Main soil type		Prefix
Coarse grained soils	Gravel	G
	Sand	S
Fine grained soils	Silt	M
	Clay	C
	Organic silts and clays	O
Fibrous soils	Peat	Pt

Subdivisions		Suffix
For coarse grained soils	Well graded, with little or no fines	W
	Well graded with suitable clay binder	C
	Uniformly graded with little or no fines	U
	Poorly graded with little or no fines	P
	Poorly graded with appreciable fines or well graded with excess fines	F
For fine grained soils	Low compressibility (plasticity)	L
	Medium compressibility (plasticity)	I
	High compressibility (plasticity)	H

This indicates the range of moisture content over which the material exists in a plastic condition. The plasticity index has been divided into five classes which are shown in Table 10.5. The liquidity index of a soil is defined as its moisture content in excess of the plastic limit, expressed as a percentage of the plasticity index. It describes the moisture content of a soil with respect to its index limits, and indicates in which part of its plastic range a soil lies, that is, its nearness to the liquid limit. The consistency index is the ratio of the difference between the liquid limit and natural moisture content to the plasticity index. It can be used to classify the different types of consistency of cohesive soils, as shown in Table 10.6.

Skempton (1953) defined the activity of a clay as

Table 10.8(a) Unified Soil Classification. Coarse grained soils. More than half of the material is larger than No. 200 sieve size.*

Field identification procedures (excluding particles larger than 76 mm and basing fractions on estimated weights)			Group symbols†	Typical names
Gravels. More than half of coarse fraction is larger than No. 7 sieve size‡	Clean gravels (little or no fines)	Wide range in grain size and substantial amounts of all intermediate particle sizes	GW	Well graded gravels, gravel–sand mixtures, little or no fines
		Predominantly one size or a range of sizes with some intermediate sizes missing	GP	Poorly graded gravels, gravel–sand mixtures, little or no fines
	Gravels with fines (appreciable amount of fines)	Non-plastic fines (for identification procedures, see ML in Table 10.8b)	GM	Silty gravels, poorly graded gravel–sand–silt mixtures
		Plastic fines (for identification procedures, see CL in Table 10.8b)	GC	Clayey gravels, poorly graded gravel–sand–clay mixtures
Sands. More than half of coarse fraction is smaller than No. 7 sieve size‡	Clean sands (little or no fines)	Wide range in grain sizes and substantial amounts of all intermediate particle sizes	SW	Well graded sands, gravelly sands, little or no fines
		Predominantly one size or a range of sizes with some intermediate sizes missing	SP	Poorly graded sands, gravelly sands, little or no fines
	Sands with fines (appreciable amount of fines)	Non-plastic fines (for identification procedures, see ML in Table 10.8b)	SM	Silty sands, poorly graded sand–silt mixtures
		Plastic fines (for identification procedures, see CL in Table 10.8b)	SC	Clayey sands, poorly graded sand–clay mixtures

* All sieve sizes in this table are US standard. The No. 200 sieve size is about the smallest particle visible to the naked eye.
† Boundary classifications. Soils possessing characteristics of two groups are designated by combinations of group symbols. For example, GW–GC, well-graded gravel–sand mixture with clay binder.
‡ For visual classification, the 6.3 mm size may be used as equivalent to the No. 7 sieve size.
Field identification procedures for fine grained soils or fractions. These procedures are to be performed on the minus No. 40 sieve size particles, approximately 0.4 mm. For field classification purposes, screening is not intended, simply remove by hand the coarse particles that interfere with the tests.
Dilatancy (reacting to shaking). After removing particles larger than No. 40 sieve size, prepare a pat of moist soil with a volume of about 1 cm³. Add enough water, if necessary, to make the soil soft but not sticky. Place the pat in the open palm of one hand and shake horizontally, striking vigorously against the other hand several times. A positive reaction consists of the appearance of water on the surface of the pat, which changes to a livery consistency and becomes glossy. When the sample is squeezed between the fingers, the water and gloss disappear from the surface, the pat stiffens and, finally, cracks and crumbles. The rapidity of appearance of water during shaking and of its disappearance during squeezing assist in identifying the character of the fines in a soil. Very fine clean sands give the quickest and most distinct reaction, whereas a plastic clay has no reaction. Inorganic silts, such as a typical rock flour, show a moderately quick reaction.
Dry strength (crushing characteristics). After removing particles larger than No. 40 sieve size, mould a pat of soil to the consistency of putty, adding water if necessary. Allow the pat to dry completely by oven, sun or air drying, and then test its

Information required for describing soils		Laboratory classification criteria	
Give typical name; indicate approximate percentages of sand and gravel; maximum size; angularity, surface condition and hardness of the coarse grains; local or geological name and other pertinent descriptive information; and symbols in parentheses	Use grain size curve in identifying the fractions as given under field identification	Determine percentages of gravel and sand from grain size curve. Depending on fines (fraction smaller than No. 200 sieve size), coarse-grained soils are classified as follows: less than 5%: GW, GP, SW, SP; more than 12%: GM, GC, SM, SC; 5% to 12%: borderline cases require use of dual symbols	$C_u = \dfrac{D_{60}}{D_{10}}$ Greater than 4
			$C_e = \dfrac{(D_{30})^2}{D_{10} \times D_{60}}$ Between 1 and 3
			Not meeting all gradation requirements for GW
For undisturbed soils, add information on stratification, degree of compactness, cementation, moisture conditions and drainage characteristics			Atterberg limits below A-line, or *PI* less than 4 / Atterberg limits above A-line with *PI* greater than 7
			Above A-line with *PI* between 4 and 7 are *borderline* cases requiring use of dual symbols
Example: Silty sand, gravelly; about 20% hard, angular gravel particles; 12.5 mm maximum size; rounded and subangular sand grains coarse to fine, about 15% non-plastic fines with low dry strength; well compacted and moist in place; alluvial sand; (SM)			$C_u = \dfrac{D_{60}}{D_{10}}$ Greater than 6
			$C_e = \dfrac{(D_{30})^2}{D_{10} \times D_{60}}$ Between 1 and 3
			Not meeting all gradation requirements for SW
			Atterberg limits below A-line with *PI* less than 5 / Atterberg limits above A-line with *PI* greater than 7
			Above A-line with *PI* between 4 and 7 are *borderline* cases requiring use of dual symbols

strength by breaking and crumbling between the fingers. This strength is a measure of the character and quantity of the colloidal fraction contained in the soil. The dry strength increases with increasing plasticity. High dry strength is characteristic for clays of the CH group. A typical inorganic silt possesses only very slight dry strength. Silty fine sands and silts have about the same slight dry strength, but can be distinguished by the feel when powdering the dried specimen. Fine sand feels gritty, whereas a typical silt has the smooth feel of flour.

Toughness (consistency near plastic limit). After removing particles larger than the No. 40 sieve size, a specimen of soil, about 1 cm³ in size, is moulded to the consistency of putty. If too dry, water must be added and, if sticky, the specimen should be spread out in a thin layer and allowed to lose some moisture by evaporation. Then the specimen is rolled out by hand on a smooth surface or between the palms into a thread about 3 mm in diameter. The thread is then folded and re-rolled repeatedly. During this manipulation, the moisture content gradually is reduced, and the specimen stiffens, finally it loses its plasticity and crumbles when the plastic limit is reached. After the thread crumbles, the pieces should be lumped together and a slight kneading action continued until the lump crumbles. The tougher the thread near the plastic limit and the stiffer the lump when it finally crumbles, the more potent is the colloidal clay fraction in the soil. Weakness of the thread at the plastic limit and a quick loss of coherence of the lump below the plastic limit indicate either inorganic clay of low plasticity, or materials such as kaolin-type clays and organic clays which occur below the A-line. Highly organic clays have a very weak and spongy feel at the plastic limit.

Table 10.8(b) Unified Soil Classification. Fine-grained soils. More than half of the material is smaller than No. 200 sieve size.*

Identification procedures on fraction smaller than No. 40 sieve size				Group symbols†	Typical names
	Dry strength (crushing characteristics)	Dilatancy (reaction to shaking)	Toughness (consistency near plastic limit)		
Silts and clays, liquid limit less than 50	None to slight	Quick to slow	None	ML	Inorganic silts and very fine sands, rock flour, silty or clayey fine sands with slight plasticity
	Medium to high	None to very slow	Medium	CL	Inorganic clays of low to medium plasticity, gravelly clays, sandy clays, silty clays, lean clays
	Slight to medium	Slow	Slight	OL	Organic silts and organic silt–clays of low plasticity
Silts and clays, liquid limit greater than 50	Slight to medium	Slow to none	Slight to medium	MH	Inorganic silts, micaceous or diatomaceous fine sandy or silty soils, clastic silts
	High to very high	None	High	CH	Inorganic clays of high plasticity, fat clays
	Medium to high	None to very slow	Slight to medium	OH	Organic clays of medium to high plasticity
Highly organic soils	Readily identified by colour, odour, spongy feel and, frequently, by fibrous texture			Pt	Peat and other highly organic soils

See footnotes to Table 10.8(a).

$$\text{Activity} = \frac{\text{Plasticity index}}{\% \text{ by mass finer than } 0.002 \, \text{mm}} \quad (10.3)$$

He suggested three classes of activity, namely, active, normal and inactive, which he further subdivided into five groups as follows:

1 inactive with activity less than 0.5;
2 inactive with activity range 0.5–0.75;
3 normal with activity range 0.75–1.25;
4 active with activity range 1.25–2; and
5 active with activity greater than 2.

It appears that there is only a general correlation between the clay mineral content of a deposit and its activity, that is, kaolinitic and illitic clays usually are inactive, whereas montmorillonitic clays range from inactive to active. Usually, active clays have a relatively high water holding capacity and a high cation exchange capacity. They are also highly thixotropic, have a low permeability and a low resistance to shear.

10.6 Soil classification

Any system of soil classification involves the grouping of the different soil types into categories possessing similar properties and, in so doing, provides a systematic method of soil description. Such a classification should provide a means by which soils can be identified quickly. Although soils include materials of various origins, for purposes of engineer-

Information required for describing soils	Laboratory classification criteria

Give typical name; indicate degree and character of plasticity; amount and maximum size of coarse grains; colour in wet condition; odour if any; local or geological name and other pertinent descriptive information; and symbol in parentheses

For undisturbed soils, add information on structure, stratification, consistency in undisturbed and remoulded states, moisture and drainage conditions

Example:
Clayey silt; brown; slightly plastic; small percentage of fine sand; numerous vertical root holes; firm and dry in place; loess; (ML)

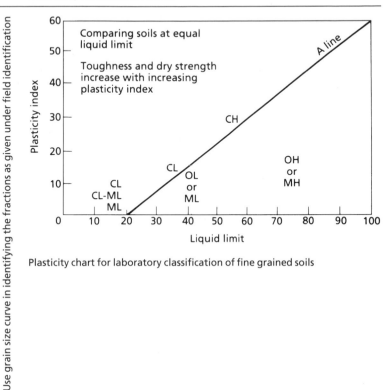

Plasticity chart for laboratory classification of fine grained soils

ing classification, it is sufficient to consider their simple index properties, which can be assessed easily, such as their particle size distribution or consistency limits.

Casagrande (1948) advanced one of the first comprehensive engineering classifications of soil. In the Casagrande system, the coarse grained soils are distinguished from the fine on the basis of particle size. Gravels and sands are the two principal types of coarse grained soil, and in this classification, both are subdivided into five subgroups on the basis of grading (Table 10.7). Well graded soils are those in which the particle size distribution extends over a wide range without excess or deficiency in any particular sizes, whereas in uniformly graded soils, the distribution

extends over a very limited range of particle size. In poorly graded soils, the distribution contains an excess of some particle sizes and a deficiency of others.

Each of the main soil types and subgroups is given a letter, a pair of letters combining to give the group symbol, the former being the prefix and the latter the suffix. A plasticity chart (Table 10.8) is also used when classifying fine grained soils. On this chart, the plasticity index is plotted against the liquid limit. The A-line is taken as the boundary between organic and inorganic soils, the latter lying above the line. Subsequently, the Unified Soil Classification (Table 10.8) was developed from the Casagrande system.

Table 10.9 Mixed soil types.

Term	Composition of the coarse fraction
Slightly sandy GRAVEL	Up to 5% sand
Sandy GRAVEL	5–20% sand
Very sandy GRAVEL	Over 20% sand
GRAVEL/SAND	About equal proportions of gravel and sand
Very gravelly SAND	Over 20% gravel
Gravelly SAND	5–20% gravel
Slightly gravelly SAND	Up to 5% gravel

The British Soil Classification for engineering purposes (Anon., 1981a) also uses particle size as a fundamental parameter. Boulders, cobbles, gravels, sands, silts and clays are distinguished as individual groups, each group being given the following symbol and size range:

1 boulders (B), over 200 mm;
2 cobbles (Cb), 60–200 mm;
3 gravel (G), 2–60 mm;
4 sand (S), 0.06–2 mm;
5 silt (M), 0.002–0.06 mm; and
6 clay (C), less than 0.002 mm.

The gravel, sand and silt ranges may be further divided into coarse, medium and fine categories. Mixed gravel–sand soils can be described as shown in Table 10.9.

These major soil groups are further divided into subgroups on the basis of grading in the case of cohesionless soils and on the basis of plasticity in the case of fine material. Granular soils are described as well graded (W) or poorly graded (P). Two further types of poorly graded granular soils are recognized, namely uniformly graded (Pu) and gap graded (Pg). Silts and clays are subdivided according to their liquid limits (*LL*) into low (under 35%), intermediate (35–50%), high (50–70%), very high (70–90%) and extremely high (over 90%) subgroups. Each subgroup is given a combined symbol, in which the letter describing the predominant size fraction is written first (e.g. GW represents well graded gravels and CH represents clay with a high liquid limit). Any group may be referred to as organic if it contains a significant proportion of organic matter, in which case the letter O is suffixed to the group symbol (e.g. CVSO represents organic clay of very high liquid limit with sand). The symbol Pt is given to peat. The British Soil

Classification can be made either by rapid assessment in the field or by full laboratory procedure (Tables 10.10 and 10.11, respectively).

The proportions of boulders and cobbles are recorded separately in the British Soil Classification. Their presence should be recorded in the soil description, a plus sign being used in symbols for soil mixtures, for example, G + Cb for gravel with cobbles. Very coarse deposits should be classified as follows:

1 *Boulders.* Over half of the very coarse material is of boulder size (over 200 mm). May be described as cobbly boulders if cobbles are an important second constituent in the very coarse fraction.

2 *Cobbles.* Over half of the very coarse material is of cobble size (200–60 mm). May be described as bouldery cobbles if boulders are an important second constituent in the very coarse fraction.

Mixtures of very coarse material and soil can be described by combining the terms for the very coarse constituent and the soil constituent as in Table 10.12.

10.7 Shear strength of soil

The shear strength of a soil is the maximum resistance which it can offer to shear stress. When this maximum has been reached, the soil is regarded as having failed, its strength having been fully mobilized. However, the shear strength value determined experimentally is not a unique constant characteristic of the material, but varies with the method of testing. Shear displacement also continues to take place after the shear strength has been exceeded. Shear displacements occur across a well defined single plane of rupture or across a shear zone.

The stress on any plane surface can be resolved into the normal stress σ_n, which acts perpendicular to the surface, and the shear stress τ, which acts along the surface, the magnitude of the resistance being given by Coulomb's equation

$$\tau = c + \sigma_n \tan\phi \qquad (10.4)$$

where ϕ is the angle of shear resistance and c is the cohesion.

The stress that controls the changes in the volume and strength of a soil is known as the effective stress (σ'). When a load is applied to a saturated soil, it is carried by either the pore water, which gives rise to an increase in the pore water pressure (u), or the soil skeleton, or both. The effect that a load has on a soil therefore is affected by the drainage conditions, however, it has been shown that, for most practical cases, the effective stress is equal to the intergranular stress, and can be determined from the equation

$$\sigma' = \sigma - u \qquad (10.5)$$

where σ is the total stress. Hence, the shear strength depends upon the effective stress and not the total stress. Accordingly, Coulomb's equation must be modified in terms of the effective stress, and becomes

$$\tau = c' + \sigma' \tan \phi' \qquad (10.6)$$

The internal frictional resistance of a soil, for example, as developed in a sand, is generated by friction, when the grains in the shear zone are caused to slide, roll and rotate against each other. Local crushing may occur at the points of contact which suffer the highest stress. The total resistance to rolling is the sum of the behaviour of all the particles, and is influenced by the confining stress, the coefficient of friction and the angles of contact between the minerals, as well as their surface roughness. However, the angle of internal friction does not depend solely on the internal friction between the grains, because a proportion of the shear stress on the plane of failure is utilized in overcoming interlocking, that is, it is also dependent upon the initial void ratio or density of a given soil. Furthermore, it is influenced by the size and shape of the grains. The larger the grains, the wider the zone which is affected. The more angular the grains, the greater the frictional resistance to their relative movement. Electrical forces of attraction and repulsion may also be involved in the shear resistance.

In clay soils, the cohesion which is developed by the molecular attractive forces between the minute soil particles mainly is responsible for the resistance offered to shearing. Because molecular attractive forces depend, to a large extent, on the mineralogical composition of the particles, and on the type and concentration of electrolytes present in the pore water, the magnitude of the true cohesion also depends on these factors.

The shear strength of an undisturbed clay generally found to be greater than that obtained when it is remoulded and tested under the same conditions and the same moisture content. The ratio of the undisturbed to the remoulded strength at the same moisture content is defined as the sensitivity of a clay.

Skempton and Northey (1952) proposed the following grades of sensitivity:
1 insensitive clays, under 1;
2 low-sensitive clays, 1–2;
3 medium-sensitive clays, 2–4;
4 sensitive clays, 4–8;
5 extra-sensitive clays, 8–16; and
6 quickclay, over 16.

Clays with high sensitivity values have little strength after being disturbed. Indeed, if they suffer slight disturbance, this may cause an initially fairly strong material to behave as a viscous fluid. The high sensitivity seems to result from the metastable arrangement of equidimensional particles. The strength of undisturbed clay chiefly is due to the strength of the framework developed by the particles and the bonds between their points of contact. If the framework is destroyed by remoulding, the clay loses most of its strength, and any subsequent regain in strength due to thixotropic hardening does not exceed a small fraction of its original value. Sensitive clays generally possess high moisture contents, frequently with liquidity indices well in excess of unity. A sharp increase in moisture content may cause a great increase in sensitivity, sometimes with disastrous results. The effect of remoulding on clays of various sensitivities is illustrated in Fig. 10.6.

Some clays with moderate to high sensitivity show a regain in strength when, after remoulding, they are allowed to rest under unaltered external conditions. Such soils are thixotropic. Thixotropy is the property of a material which allows it to undergo an isothermal gel-to-sol-to-gel transformation upon agitation and subsequent rest. This transformation can be repeated indefinitely without fatigue, and the gelation time under similar conditions remains the same. The softening and subsequent recovery of thixotropic soils appears to be due, first, to the destruction and, second, to the rehabilitation of the molecular structure of the adsorbed layers of the clay particles.

10.8 Consolidation

The theory of consolidation enables a determination to be made of the amount and rate of settlement which is likely to occur when structures are erected on cohesive soils. When a layer of soil is loaded, some of the pore water is expelled from its voids, moving slowly away from the region of high stress, as a result

Table 10.10 Field identification and description of soils. (After Anon., 1981a.)

	Basic soil type	Particle size (mm)	Visual identification	Particle nature and plasticity	Composite soil types (mixtures of basic soil types)
Very coarse soils	Boulders		Only seen complete in pits or exposures	Particle shape:	Scale of secondary constituents with coarse soils
		— 200			
	Cobbles		Often difficult to recover from boreholes	Angular	Term % of clay or silt
		— 60		Subangular	
Coarse soils (over 65% sand and gravel sizes)	Gravels	Coarse	Easily visible to naked eye; particle shape can be described; grading can be described	Subrounded Rounded Flat Elongate	Slightly clayey ⎫ GRAVEL* or Under 5 Slightly silty ⎭ SAND
		20	Well graded: wide range of grain sizes, well distributed. Poorly graded: not well graded (may be uniform: size of most particles lies between narrow limits; or gap graded: an intermediate size of particle is markedly underrepresented)		Clayey ⎫ GRAVEL or 5 to 15 Silty ⎭ SAND
		Medium			
		6			
		Fine		Texture:	Very clayey ⎫ GRAVEL or 15 to 35 Very silty ⎭ SAND
		— 2			
	Sands	Coarse	Visible to naked eye; very little or no cohesion when dry; grading can be described	Rough Smooth Polished	Sandy GRAVEL ⎫ Sand or gravel as important second Gravelly SAND ⎭ constituent of the coarse fraction
		0.6			
		Medium	Well graded: wide range of grain sizes, well distributed. Poorly graded: not well graded (may be uniform: size of most particles lies between narrow limits; or gap graded: an intermediate size of particle is markedly underrepresented)		For composite types described as: clayey: fines are plastic, cohesive silty: fines non-plastic or of low plasticity
		0.2			
		Fine			
		— 0.06			
Fine soils (over 35% silt and clay sizes)	Silts	Coarse	Only coarse silt barely visible to naked eye; exhibits little plasticity and marked dilatancy; slightly granular or silky to the touch. Disintegrates in water; lumps dry quickly; possesses cohesion, but can be powdered easily between fingers	Non-plastic or low plasticity	Scale of secondary constituents with fine soils Term % of sand or gravel Sandy ⎫ CLAY or 35 to 65 Gravelly ⎭ SILT
		0.02			
		Medium			
		0.006			
		Fine			
		0.002			
	Clays		Dry lumps can be broken, but not powdered, between the fingers; they also disintegrate under water but more slowly than silt; smooth to the touch; exhibits plasticity, but no dilatancy; sticks to the fingers and dries slowly; shrinks appreciably on drying usually showing cracks. Intermediate- and high-plasticity clays show these properties to a moderate and high degree, respectively	Intermediate plasticity (lean clay)	CLAY : SILT Under 35
					Examples of composite types
					(Indicating preferred order for description)
				High plasticity (fat clay)	Loose, brown, subangular, very sandy, fine to coarse gravel with small pockets of soft grey CLAY
					Medium dense, light brown, clayey, fine and medium SAND
Organic soils	Organic clay, silt or sand	Varies	Contains substantial amounts of organic vegetable matter		Stiff, orange brown, fissured sandy CLAY
					Firm, brown, thinly laminated silt and CLAY
	Peats	Varies	Predominantly plant remains, usually dark brown or black in colour, often with distinctive smell; low bulk density		Plastic, brown, amorphous PEAT

* Principal soil type given in capital letters.

Compactness/strength		Structure		Interval scales	Colour
Term	**Field test**	**Term**	**Field identification**		
Loose	By inspection of voids and particle packing	Homogeneous	Deposit consists essentially of one type	**Scale of bedding spacing**	Red
Dense		Interstratified	Alternating layers of varying types or with bands or lenses of other materials. Interval scale for bedding spacing may be used	Term — Mean spacing (mm)	Pink / Yellow / Brown / Olive
				Very thickly bedded — Over 2000	Green
				Thickly bedded — 2000–600	Blue
				Medium bedded — 600–200	White
Loose	Can be excavated with a spade; 50 mm wooden peg can be easily driven	Heterogeneous	A mixture of types	Thinly bedded — 200–60	Grey / Black etc.
				Very thinly bedded — 60–20	Supplemented as necessary with:
Dense	Requires pick for excavation; 50 mm wooden peg hard to drive	Weathered	Particles may be weakened and may show concentric layering	Thickly laminated — 20–6	Light / Dark / Mottled etc.
				Thinly laminated — Under 6	
Slightly cemented	Visual examination; pick removes soil in lumps which can be abraded				
Soft or loose	Easily moulded or crushed in the fingers				and
Firm or dense	Can be moulded or crushed by strong pressure in the fingers	Fissured	Break into polyhedral fragments along fissures. Interval scale for spacing of discontinuities may be used		Pinkish / Reddish / Yellowish / Brownish etc.
Very soft	Exudes between fingers when squeezed in hand	Intact	No fissures		
Soft	Moulded by light finger pressure	Homogeneous	Deposit consists essentially of one type	**Scale of spacing of other discontinuities**	
Firm	Can be moulded by strong finger pressure	Interstratified	Alternating layers of varying types. Interval scale for thickness of layers may be used	Term — Mean spacing (mm)	
				Very widely spaced — Over 2000	
Stiff	Cannot be moulded by fingers. Can be indented by thumb	Weathered	Usually has crumb or columnar structure		
Very stiff	Can be indented by thumbnail			Widely spaced — 2000–600	
				Medium spaced — 600–200	
Firm	Fibres already compressed together			Closely spaced — 200–60	
Spongy	Very compressible and open structure	Fibrous	Plant remains recognizable and retain some strength	Very closely spaced — 60–20	
Plastic	Can be moulded in hand, and smears fingers	Amorphous	Recognizable plant remains absent	Extremely closely spaced — Under 20	

Table 10.11 British Soil Classification System for engineering purposes. (After Anon., 1981a.)

Soil groups*		Subgroups and laboratory identification				
GRAVEL and SAND may be qualified sandy gravel and gravelly sand, etc., where appropriate		Group symbol†,‡	Subgroup symbol†	Fines (% less than 0.06 mm)	Liquid limit (%)	Name
Coarse soils, less than 35% of the material is finer than 0.06mm	GRAVELS More than 50% of coarse material is of gravel size (coarser than 2 mm) — Slightly silty or clayey GRAVEL	G — GW	GW	0 to 5		Well graded GRAVEL
	Slightly silty or clayey GRAVEL	GP	GPu GPg			Poorly graded/uniform/gap-graded GRAVEL
	Silty GRAVEL	G–F — GM	GWM GPM	5 to 15		Well graded/poorly graded silty GRAVEL
	Clayey GRAVEL	G–C	GWC GPC			Well graded/poorly graded clayey GRAVEL
	Very silty GRAVEL	GF — GM	GML, etc.	15 to 35		Very silty GRAVEL; subdivide as for GC
	Very clayey GRAVEL	GC	GCL GCI GCH GCV GCE			Very clayey GRAVEL (clay of low, intermediate, high, very high, extremely high plasticity)
	SANDS More than 50% of coarse material is of sand size (finer than 2 mm) — Slightly silty or clayey SAND	S — SW	SW	0 to 5		Well graded SAND
	Slightly silty or clayey SAND	SP	SPu SPg			Poorly graded/uniform/gap-graded SAND
	Silty SAND	S–F — S–M	SWM SPM	5 to 15		Well graded/poorly graded silty SAND
	Clayey SAND	S–C	SWC SPC			Well graded/poorly graded clayey SAND
	Very silty SAND	SF — SM	SML, etc.	15 to 35		Very silty SAND; subdivided as for SC
	Very clayey SAND	SC	SCL SCI SCH SCV SCE			Very clayey SAND (clay of low, intermediate, high, very high, extremely high plasticity)

Fine soils	Soil description	(F)	Group symbol	Subgroup symbol	% fines	
FINE SOILS, more than 35% of the material is finer than 0.06 mm	Gravelly or sandy SILTS and CLAYS 35% to 65% fines	FG	MG	MLG, etc.		Gravelly SILT; subdivide as for CG
	Gravelly SILT					
	Gravelly CLAYS		CG	CLG	<35	Gravelly CLAY of low plasticity
				CIG	35 to 50	of intermediate plasticity
				CHG	50 to 70	of high plasticity
				CVG	70 to 90	of very high plasticity
				CEG	>90	of extremely high plasticity
	SILTS and CLAYS 65% to 100% fines	FS	MS	MLS, etc.		Sandy SILT; subdivide as for CG
	Sandy SILTS		CS	CLS, etc.		Sandy CLAY; subdivide as for CG
	Sandy CLAY					
	SILT (M soil)	F	M	ML, etc.		SILT; subdivide as for C
	CLAY‖,¶		C	CL	<35	CLAY of low plasticity
				CI	35 to 50	of intermediate plasticity
				CH	50 to 70	of high plasticity
				CV	70 to 90	of very high plasticity
				CE	>90	of extremely high plasticity
Organic soils	Descriptive letter 'O' suffixed to any group or subgroup symbol					Organic matter suspected to be a significant constituent. Example MHO: organic SILT of high plasticity
Peat	Pt					Peat soils consist predominantly of plant remains which may be fibrous or amorphous

* The name of the soil group should always be given when describing soils, supplemented, if required, by the group symbol, although for some additional applications (e.g. longitudinal sections) it may be convenient to use the group symbol alone.

† The group symbol or subgroup symbol should be placed in parentheses if laboratory methods have not been used for identification, e.g. (GC).

‡ The designation FINE SOIL, or FINES, F, may be used in place of SILT, M, or CLAY, C, when it is not possible or not required to distinguish between them.

§ Gravelly if more than 50% of coarse material is of gravel size. Sandy if more than 50% of coarse material is of sand size.

‖ SILT (M soil), M, is material plotting below the A-line, and has a restricted plastic range in relation to its liquid limit, and relatively low cohesion. Fine soils of this type include clean silt-sized materials and rock flour, micaceous and diatomaceous soils, pumice and volcanic soils, and soils containing halloysite. The alternative term 'M soil' avoids confusion with materials of predominantly silt size, which form only a part of the group. Organic soils also usually plot below the A-line on the plasticity chart, when they are designated ORGANIC SILT, MO.

¶ CLAY, C, is material plotting above the A-line, and is fully plastic in relation to its liquid limit.

of the hydrostatic gradient created by the load. The void ratio accordingly decreases and settlement occurs. This is termed primary consolidation. Further settlement, usually of a minor degree, may occur due

Table 10.12 Mixtures of very coarse materials and soil.

Term	Composition
BOULDERS (or COBBLES) with a little finer material*	Up to 5% finer material
BOULDERS (or COBBLES) with some finer material*	5–20% finer material
BOULDERS (or COBBLES) with much finer material*	20–50% finer material
FINE MATERIAL* with many boulders (or cobbles)	50–20% boulders (or cobbles)
FINE MATERIAL* with some boulders (or cobbles)	20–5% boulders (or cobbles)
FINE MATERIAL* with occasional boulders (or cobbles)	Up to 5% boulders (or cobbles)

*The name of the finer material is given in parentheses when it is a minor constituent, for example, cobbly BOULDERS with some fine material (sand with some fines).

to the rearrangement of the soil particles under stress, this being referred to as secondary consolidation. However, in reality, primary and secondary consolidation are not distinguishable.

Terzaghi (1943) showed that the relationship between the unit load and the void ratio for a soil can be represented by plotting the void ratio e against the logarithm of the unit load p (Fig. 10.7). The shape of the e log$p$ curve is related to the stress history of a clay. In other words, the e log$p$ curve for a normally consolidated clay is linear, and is referred to as the virgin compression curve. On the other hand, if a clay is overconsolidated, the e log$p$ curve is not straight, and the preconsolidation pressure can be derived from the curve. The preconsolidation pressure refers to the maximum overburden pressure to which a deposit has been subjected. Overconsolidated clay is appreciably less compressible than normally consolidated clay.

The compressibility of a clay can be expressed in terms of the compression index (C_c) or the coefficient of volume compressibility (m_v). The compression index is the slope of the linear section of the e log$p$

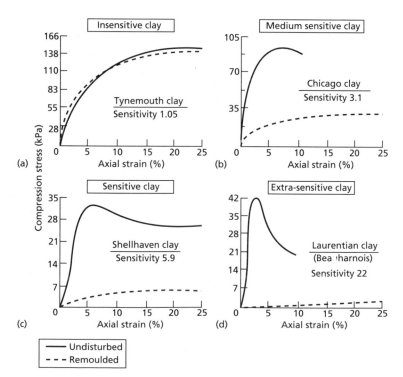

Fig. 10.6 Stress–strain curves of clay so with different sensitivities.

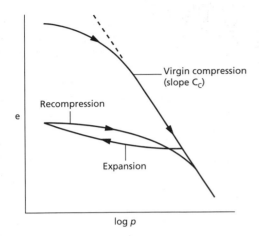

Fig. 10.7 Results of consolidation test: e log p curve.

curve and is dimensionless. The value of C_c for cohesive soils ranges between about 0.075 for silty clays to more than 1.0 for highly colloidal bentonic clays. Some examples of the degree of compressibility are as follows:

Range of C_c	Degree of compressibility	Soil type
Over 0.3	Very high	Soft clay
0.3–0.15	High	Clay
0.15–0.075	Medium	Silty clay
Less than 0.075	Low	Sandy clay

Hence the compressibility index increases with increasing clay content.

The coefficient of volume compressibility is defined as the volume change per unit volume per unit increase in load; its units are the inverse of pressure $m^2 MN^{-1}$, i.e. the strain per unit stress increase in one-dimensional compression). The value of m_v for a given soil depends upon the stress range over which it is determined. Anon. (1990) recommends that it should be calculated for a pressure increment of 100 kPa in excess of the effective overburden pressure on the soil at the depth in question.

The compressibility of a soil is dependent on the average rate of compression, and the soil structure has a substantial time-dependent resistance to compression. At the instant at which a load p on a layer of clay suddenly is increased by Δp, the thickness of the layer remains unchanged. Hence, the application of the load Δp produces an equal increase Δu in the hydrostatic pressure of the pore water. As time proceeds, the excess pore water pressure gradually is dissipated, and finally disappears, the grain-to-grain pressure simultaneously increases from an initial value p to $p + \Delta p$. The ratio between the decrease in the void ratio Δe at time t and the ultimate decrease Δe_1 represents the degree of consolidation U at time t:

$$U = 100 \frac{\Delta e}{\Delta e_1} \qquad (10.7)$$

With a given thickness H of a layer of clay, the degree of consolidation at time t depends exclusively on the coefficient of consolidation c_v, defined as

$$c_v = \frac{k}{\gamma_w m_v} \qquad (10.8)$$

where k is the coefficient of permeability and γ_w is the unit weight of water. The coefficient of consolidation, which determines the rate at which settlement takes place, is calculated for each load increment, and either a mean value or that value appropriate to the pressure range in question is used (c_v is measured in $m^2 y^{-1}$). With increasing values of p, both k and m_v decrease. The coefficient of consolidation decreases for normally consolidated clays from about $3.15 \, m^2 y^{-1}$ for very lean clays to about $0.03 \, m^2 y^{-1}$ for highly colloidal clays. At any value of c_v, the time t at which a given degree of consolidation U is reached increases in simple proportion to the square of the thickness H of the layer concerned.

10.9 Soil erosion

The loss of soil due to erosion by water or wind is a natural process. Soil erosion removes the topsoil, which contains a high proportion of the organic matter and the finer mineral fractions that provide nutrient supplies for plant growth. Unfortunately, soil erosion can be accelerated by the activity of man, witness the disasters in the Dust Bowl and Tennessee Valley of the USA prior to the Second World War. It is difficult, however, to separate natural from man-induced changes in erosion rates. Although accelerated, man-induced soil erosion does not figure as prominently amongst environmental issues now as it did then, despite the fact that soil erosion in some countries is as serious as it ever was, if not more so.

In the case of soil erosion by water, it is most active where a large part of the rainfall finds it difficult to infiltrate into the ground, so that most of it flows over the surface and, in so doing, removes soil. The intensity of rainfall is at least as important in terms of soil erosion as the total amount of rainfall and, as rainfall becomes more seasonal, the total amount of erosion tends to increase. This run-off can take the form of sheet flow or can be concentrated into rills and gullies. Such conditions are met most frequently in semi-arid and arid regions. Nevertheless, soil erosion occurs in many different climatic regions where vegetation has been removed, being at a maximum when intense rainfall and the vegetation cover are out of phase, that is, when the surface is bare.

Sheet flow may cover up to 50% of the surface of a slope during heavy rainfall, but erosion does not take place uniformly across a slope. The depth of sheet flow, up to 3 mm, and the velocity of flow are such that both laminar and turbulent flow take place. Erosion due to turbulent flow only occurs where flow is confined in linear concentrations within sheet flow. Hence, the flow elsewhere is laminar and non-erosive. The velocity of sheet flow ranges between 15 and 300 mm s^{-1}. Velocities of 160 mm s^{-1} are required to erode soil particles of 0.3 mm in diameter, but velocities as low as 20 mm s^{-1} will keep these particles in suspension.

Rills and gullies begin to form when the velocity of flow increases to rates in excess of 300 mm s^{-1}, and flow is turbulent. Whether they form depends on the soil factors, as well as the velocity and depth of water flow. Rills and gullies remove much larger volumes of soil per unit area than sheet flow. Severe soil erosion, associated with the formation of gullies (Fig. 10.8), can give rise to mass movements on the steepened slopes at the sides of these gullies.

Soil erosion by wind, like erosion by water, depends upon the force that the wind can exert upon soil particles. This is influenced by the roughness of the surface over which the wind blows. Where the surface is rough, for example, due to the presence of boulders, plants or other obstacles, this reduces the wind speed immediately above the ground. Obviously, particle size is important, as wind can only remove particles of a certain size, silt-sized particles being especially prone to wind erosion. Nonetheless,

Fig. 10.8 Gully erosion in northern Lesotho.

wind erosion is effective in lowering landscapes, for example, it is estimated that, during the period in which parts of Kansas formed the Dust Bowl, wind erosion accounted for soil losses of up to 10 mm per year at certain locations. By comparison, in some semi-arid regions, water may remove around 1 mm per year.

Chepil and Woodruff (1963) recognized four types of soil aggregation that produce textures which are more resistant to wind erosion, that is, primary aggregates, secondary aggregates or clods, fine material amongst clods and surface crusts. The primary aggregates are bound together by clay and colloids. Clods are cemented together in the dry state by fine particles, and their cohesion is also provided by silt and clay particles. Surface crusts are compacted by raindrop impact. If these aggregations are to be moved by the wind, they must be broken down by weathering, raindrop impact, flowing water or wind abrasion. The resistance of these aggregations to wind abrasion depends upon the soil type and its cohesion.

In most soils, according to Chepil and Woodruff (1963), the nature and stability of the textural aggregations govern the erodibility of the soil. Hence, in a well textured soil, the number of soil particles which are small enough to be moved may be very low, so that the abrasivity of the wind is also low. The stability of the textural aggregations depends upon the soil moisture content, organic content, cation exchange capacity, soil suction and cement bonds, on the one hand, and the processes responsible for breakdown on the other.

The smaller, more easily removed grains of soil and organic matter are readily subjected to wind erosion. This action leads to a reduction in the fertility of the soil, as well as lowering its water retention capacity. Erosion of soil by wind can give rise to dust storms and sand storms, which may choke small water courses, accumulate against buildings, bury small obstacles, block highways and even strip paint from cars. People and animals affected may suffer from respiratory problems.

Wind erosion is most effective in arid and semi-arid regions, where the ground surface is relatively dry and vegetation is absent or sparse. The problem is most acute in those regions where land-use practices are inappropriate and rainfall is unreliable. This means that the ground surface may be left exposed. Nonetheless, soil erosion by wind is not restricted to dry regions. It also occurs, although on a smaller scale, in humid areas.

Of paramount importance as far as soil erosion is concerned is vegetation cover. Generally, an increase in erosion occurs with increasing rainfall, and erosion decreases with increasing vegetation cover. However, the growth of natural vegetation depends on rainfall, so producing a rather complex variation of erosion with rainfall. Although the vegetation cover depends primarily upon rainfall, agriculture, especially where irrigation occurs, can mean that vegetation becomes more or less independent of rainfall. Hence, farming practices have had and do have an influence on soil erosion, because they alter the nature of the vegetation cover, the total rainfall and its intensity during times of low cover being most important. Conse-

Table 10.13 Beneficial and adverse effects of vegetation. (From Coppin & Richards, 1990.)

Hydrological effects		Mechanical effects	
Foliage intercepts rainfall causing:		Roots bind soil particles and permeate the soil, resulting in:	
1 Absorptive and evaporative losses, reducing rainfall available for infiltration	B	**1** Restraint of soil movement reducing erodibility	B
2 Reduction in kinetic energy of raindrops and thus erosivity	B	**2** Increase in shear strength through a matrix of tensile fibres	B
3 Increase in drop size through leaf drip, thus increasing localized rainfall intensity	A	**3** Network of surface fibres creates a tensile mat effect, restraining underlying strata	B
Stems and leaves interact with flow at the ground surface, resulting in:		Roots penetrate deep strata, giving:	
1 Higher depression storage and higher volume of water for infiltration	A/B	**1** Anchorage into firm strata, bonding soil mantle to stable subsoil or bedrock	B
2 Greater roughness on the flow of air and water, reducing its velocity, but	B	**2** Support to upslope soil mantle through buttressing and arching	B
3 Tussocky vegetation may give high localized drag, concentrating flow and increasing velocity	A	Tall growth of trees, so that:	
Roots permeate the soil, leading to:		**1** Weight may surcharge the slope, increasing normal and downslope force components	A/B
1 Opening up of the surface and increasing infiltration	A	**2** When exposed to wind, dynamic forces are transmitted into the ground	B
2 Extraction of moisture, which is lost to the atmosphere in transpiration, lowering pore water pressure and increasing soil suction, both increasing soil strength	B	Stems and leaves cover the ground surface, so that:	
		1 Impact of traffic is absorbed, protecting soil surface from damage	B
3 Accentuation of desiccation cracks, resulting in higher infiltration	A	**2** Foliage is flattened in high-velocity flows, covering the soil surface and providing protection against erosive flows	B

A, adverse effect; B, beneficial effect.

quently, land is most vulnerable to erosion during periods when ploughing and harvesting take place. In addition, overgrazing reduces the amount of vegetation cover, and so can lead to an increase in the rate of soil erosion. Semi-arid regions appear to be the most sensitive, in terms of the rate of soil erosion, to changes in the amount of rainfall, and therefore vegetation cover.

The overall influence of vegetation as far as soil erosion is concerned depends upon the relationship between beneficial and adverse influences (Table 10.13). The properties of vegetation which influence its engineering function behaviour and value may vary with the stage of vegetation growth, and so may change with the season. Although vegetation can control the amount of erosion which may occur, conversely, erosion may produce an environment which cannot sustain vegetation. Most vegetation is self-generating, but interference by man and animals may adversely affect the natural cycles of plant growth.

The influence of vegetation on evapotranspiration can be expressed in terms of the E_t/E_o ratio, where E_t is the evapotranspiration rate for the vegetation cover and E_o is the evaporation rate from open water. Some typical values of the E_t/E_o ratio are provided in Table 10.14. Evapotranspiration is not limited by the supply of water, in that it takes place at the potential rate (E_{tp}). Nonetheless, where the rate of evapotranspiration is high, the top layers of the soil dry out rapidly, so that plants find it increasingly difficult to abstract water from the soil. Consequently, in order to avoid dehydrating, plants reduce their transpiration. At this point, actual evapotranspiration (E_{ta}) becomes less than potential evapotranspiration. The E_{ta}/E_{tp} ratio is governed by the soil moisture deficit, that is, the difference between the reduced level of soil moisture and the field capacity.

Rainfall which is interrupted by vegetation is referred to as being intercepted. Some intercepted rainfall remains on the plants and, subsequently, is evaporated; the rest (i.e. temporarily intercepted throughfall) reaches the ground either as leaf drainage or stem flow. Interception storage varies widely, but it may account for 15–25% of annual rainfall in temperate deciduous forests, and for 20–25% of annual rainfall in tropical rain forests. The interception storage therefore reduces the quantity of rainfall reaching the ground surface. Not only

Table 10.14 E_t/E_o ratios for selected plant covers. (From Morgan, 1995.)

Plant (crop) cover	E_t/E_o ratio
Wet (padi) rice	1.35
Wheat	0.59–0.61
Maize	0.67–0.70
Barley	0.56–0.60
Millet/sorghum	0.62
Potato	0.70–0.80
Beans	0.62–0.69
Groundnut	0.50–0.87
Cabbage/Brussels sprouts	0.45–0.70
Banana	0.70–0.77
Tea	0.85–1.00
Coffee	0.50–1.00
Cocoa	1.00
Sugar cane	0.68–0.80
Sugar beet	0.73–0.75
Rubber	0.90
Oil palm	1.20
Cotton	0.63–0.69
Cultivated grass	0.85–0.87
Prairie/savannah grass	0.80–0.95
Forest/woodland	0.90–1.00

does vegetation affect the volume of rain reaching the ground, it also affects its local intensity and drop size distribution. Hence, the energy of rainfall, which is available for splash erosion under a vegetation cover, is governed by the proportions of rain falling as direct throughfall and as leaf drainage, as well as the height of the canopy, which determines the energy of leaf drainage.

The presence of vegetation can increase the infiltration rate of the soil by maintaining a continuous pore system due to root growth, by the presence of organic matter and by enhancing biological activity. This also leads to an increase in the moisture storage capacity of the soil and a decrease in run-off. Vegetation also increases the time taken for run-off to occur, and so heavier rainfall is required to produce a critical amount of run-off.

As commented above, soil erosion is a natural process. Hence, ideally, as soil is being removed, it should be replaced by newly formed soil. If this does not occur, the soil cover gradually is removed. This is accompanied by an increase in run-off by overland flow which, in turn, reduces the amount of water which infiltrates into the ground. This means that chemical breakdown of rock material becomes less effective, so adversely affecting soil formation.

The exhaustion of organic matter and nutrients in the soil is closely allied to soil erosion. Organic matter in the soil fulfils a similar role to clays in holding water and both organic and inorganic nutrients. The organic matter in the soil is important in terms of the aggregation of soil particles, and a good vegetation cover tends to increase biological activity and the rate of aggregate formation. Furthermore, increasing aggregate stability leads to increasing permeability and infiltration, as well as a moister soil. High permeability and aggregate strength minimize the risk of overland flow. The loss of organic matter depends largely on the vegetation cover and its management. The partial removal of vegetation or wholesale clearance prevents the addition of plant debris as a source of new organic material for the soil. Over a period of years, this results in a loss of plant nutrients and, in a dry climate, there is a significant reduction in soil moisture. The process can turn a semi-arid area into a desert in less than a decade. This organic depletion leads to a lower infiltration capacity and increased overland flow, with consequent erosion on slopes of more than a few degrees.

The strength of the soil is increased by the presence of plant roots. Grasses and small shrubs can have a notable strengthening effect to depths of up to 1.5 m below the surface, whereas trees can enhance soil strength to depths of 3 m or more. Roots have their greatest effects in terms of soil reinforcement close to the soil surface, where the root density is highest. The reinforcing effect is limited with shallow rooting vegetation, where roots fail by pullout before their peak tensile strength is attained. In the case of trees, because their roots penetrate soil to greater depths, they may be ruptured when the forces placed upon them exceed their tensile strength.

10.10 Estimation of soil loss

Soil loss refers to the amount of soil moved from a particular area, and it is transported elsewhere in the form of sediment yield. Techniques used to predict the amount of soil loss have been developed over many years as the processes of soil erosion have become better understood. The most widely used method for predicting soil loss is the Universal Soil

Loss Equation (USLE). Several empirical methods are available for estimating the sediment yield. These relate sediment concentrations to flow stage, or sediment yield to watershed or hydrological parameters.

Soil erosion by water involves the separation of soil particles from the soil mass and their subsequent transportation. There are two agents responsible for this, namely raindrops and flowing water. Erosion by raindrops involves the detachment of particles from the soil by impact and their movement by splashing. Run-off removes soil by the action of turbulent water flowing either as sheets or in rills or gullies. The nature of soil erosion depends on the relationship between the erosivity of raindrops and run-off and the erodibility of the soil. Splash and sheet erosion combined are referred to as interrill erosion (Watson & Laflen, 1986; Liebenow *et al.*, 1990).

10.10.1 The Universal Soil Loss Equation

As noted above, the most frequently used method to determine the soil loss is the USLE, developed by the United States Soil Conservation Service:

$$A = RKLSCP \qquad (10.9)$$

where A is the average annual loss of soil, R is the rainfall erosivity, K is the soil erodibility factor, L is the slope length factor, S is the slope gradient factor, C is the cropping management factor and P is the erosion control practice factor. The method was developed as a means to predict the average annual soil loss from interrill and rill erosion. However, the equation is not universally applicable. Its primary purpose is to predict losses from arable or non-agricultural land by sheet and rill erosion, and to provide guides for the selection of adequate erosion control practices (Wischmeier & Smith, 1978; Mitchell & Bubenzer, 1980). As such, the USLE may be used to determine how conservation practices may be applied or altered to permit more intensive cultivation, to predict the change in soil loss that results from a change in cropping or conservation practices, and to provide soil loss estimates for the use of conservationists in determining conservation needs. Lastly, the USLE provides a means of estimating the yield of sediments from watersheds.

The rainfall erosivity factor in the USLE was developed by Wischmeier (1959). Wischmeier showed that storm losses from fallow plots were highly correlated with the product of the total kinetic energy and the maximum 30-min rainfall intensity. This is expressed in terms of the EI_{30} index, which shows how much soil is lost due to rainfall, where E is the kinetic energy and I_{30} is the greatest intensity of rainfall in a 30-min period (Smith & Wischmeier, 1962). The rainfall erosivity factor R is obtained by dividing EI_{30} by 173.6. The rainfall erosivity indices can be summed for any time period to provide the erosivity of rainfall for that period. Long-term records of rainfall provide average annual values of the erosivity index, and these can be used to produce maps of rainfall erosivity (Fig. 10.9). Unfortunately, this index does not always correlate well with soil erosion outside the USA.

The soil erodibility factor K in the USLE is a quantitative description of the inherent erodibility of a particular soil. It reflects the fact that different types of soil are eroded at different rates as a result of several factors. Many of these factors have been discussed above, such as the infiltration capacity, saturation, permeability, dispersion, splash and transport of soil particles. Other factors of importance include the texture or crumb structure of the soil, the stability of the mineral aggregates, the strength of the soil and its chemical and organic components.

The effects of slope length and gradient are represented separately in the USLE, but they frequently are evaluated as a single topographical factor LS. In this instance, the slope is defined as the distance from the point at which overland flow begins to the position at which there is a sufficient decrease in the gradient of the slope for deposition to occur or where run-off enters a defined channel. The gradient of the slope usually is expressed as a percentage, and is the field or segment slope. The LS factor is used to predict the ratio of soil loss per unit area of a given field slope to that from a standard plot length.

The cropping management factor C represents the ratio of soil loss from a specific cropping cover to the soil loss from a fallow condition for the same soil type, slope and rainfall. This factor takes into account the nature of the cover, crop sequence, productivity level, length of growing season, tillage practices, residue management and expected time distribution of erosion events. Hence, the evaluation of this factor frequently proves to be difficult.

The erosion control practice factor P is the ratio of soil loss using a specific practice to that which takes

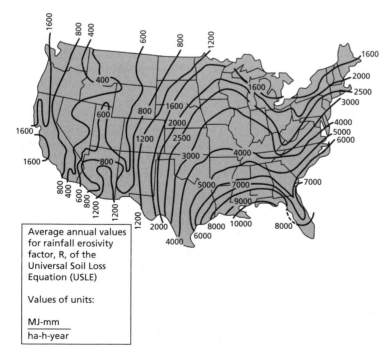

Average annual values for rainfall erosivity factor, R, of the Universal Soil Loss Equation (USLE)

Values of units:

$$\frac{MJ\text{-}mm}{ha\text{-}h\text{-}year}$$

Fig. 10.9 Rainfall and run-off erosivity index R by geographical location. (From Foster *et al.*, 1981.)

place when ploughing up and down a slope. This factor therefore takes into account erosion control practices, such as contouring, contour-strip cropping and terracing.

The aim of soil erosion control is to reduce the soil loss in order that soil productivity can be maintained economically. In this context, the soil loss tolerance is defined as the maximum rate of soil erosion which allows a high level of productivity to be sustained. Tolerances vary from place to place, because soil types and rates of erosion vary. In the USA, deep, medium-textured, moderately permeable soils have been assigned tolerance losses of around $1.1\,kg\,m^{-2}$ per year (Mitchell & Bubenzer, 1980). Soils with a shallow root zone or other detrimental characteristics were given lower tolerance values. Kirkby (1980) suggested that tolerances of $0.251–1.255\,kg\,m^{-2}$ per year generally may be suitable.

10.10.2 Soil erosion by wind

Surface winds which are capable of initiating soil particle movement are turbulent. As far as wind erosion is concerned, the most important characteristics of soil particles are their size and density. In other words, there is a maximum size for a wind of a given velocity. In terms of the wind itself, the force of the wind at the ground surface is the main factor affecting erosivity. Once particles have started to move, they exert a drag effect on the air flow, which then alters the velocity profile of the wind. The velocity of wind blowing over a surface increases with the height above the surface, the roughness of the surface retarding the speed of the wind immediately above it. As more material is picked up, the surface wind speed is reduced. Hence, the more erodible a soil, the greater the proportion of grains involved in saltation, and the greater the reduction in the wind speed (Chepil & Milne, 1941). Soil particles are also moved in suspension and by traction.

The Wind Erosion Equation is similar to the USLE, and takes into account the major factors involved in wind erosion (Woodruff & Siddaway, 1965). It is written as follows:

$$E = f(IKCLV) \tag{10.10}$$

where E is the potential erosion measured in tonnes per hectare per year, I is a soil erodibility index, K is a soil ridge roughness factor, C is a local wind erosion

climatic factor (i.e. wind speed and duration), L is the median unsheltered field length along the prevailing wind erosion direction and V is the equivalent vegetation cover. The wind erodibility I is a function of the soil aggregates with diameters exceeding $0.84\,mm$, and refers to the soil loss from a wide, unsheltered field which has a bare, smooth surface. The erodibility can be decreased by increasing the proportion of non-erodible clods (above $0.84\,mm$ in diameter) at the surface of the soil. This can be brought about in some soils by tillage. The roughness factor K is a measure of the effect of ridges made by tillage and planting machinery on the erosion rate. Ridges absorb and deflect wind energy, and trap moving particles of soil. On the other hand, too much roughness causes turbulence, which may accelerate the movement of particles. The climatic factor C is based on the principle of erosion varying directly as the cube of the wind speed and inversely as the square of the soil moisture (Chepil et al., 1962). The unsheltered distance is taken as the length from a sheltered edge of a field, parallel to the direction of the prevailing wind, to the end of the unsheltered field. The effect of vegetative cover in the Wind Erosion Equation is expressed in terms of the type, density and orientation of vegetative matter relative to its equivalent of small grain residue (i.e. flattened wheat straw defined as stalks, $254\,mm$ in height, lying flat on the soil in rows normal to the wind direction). Because the relationships between the variables in the equation are complex, wind erosion cannot be estimated just by multiplying the variables, and either nomograms and tables or computer programs are needed to obtain answers (Hagen, 1991). The Wind Erosion Equation tends to overestimate wind erosion and to give higher values of soil loss tolerance levels. Obviously, this must be taken into account if the equation is to be used for the planning of erosion control.

10.11 Erosion control and conservation practices

Satisfactory erosion control implies the achievement of the maximum sustained level of production from a given area of land and, at the same time, the maintenance of soil loss below a given threshold level. In other words, this implies the maintenance of equilibrium in the soil system, so that the rate of loss does not exceed the rate at which new soil is formed.

However, this state of equilibrium is difficult to predict and therefore to attain in practice. Accordingly, less rigorous targets generally are pursued, so that the rate of erosion is controlled to maintain soil fertility over a period of 20–25 years. Nonetheless, there are doubts regarding whether this target can maintain soil resources adequately.

As noted above, tolerable rates of soil loss obviously vary from region to region, depending upon the type of climate, vegetation cover, soil type and depth, slope and farming practices. Arnoldus (1977) proposed a range of values for tolerable mean annual soil loss from $0.2 \, \mathrm{kg \, m^{-2}}$ per year for soils in which the rooting depth is up to 250 mm to $1.1 \, \mathrm{kg \, m^{-2}}$ per year when the rooting depth exceeds 1.5 m. Nevertheless, where erosion rates are very high, higher threshold values may have to be adopted.

Conservation measures must protect the soil from raindrop impact, increase the infiltration capacity of the soil to reduce the volume of run-off, improve the aggregate stability of the soil to increase its resistance to erosion and, lastly, increase the roughness of the ground surface to reduce the run-off and wind velocity. In fact, soil conservation practices can be grouped into three categories, namely, erosion prevention practices which include agronomic measures, mechanical measures and soil management practices (Table 10.15). Sound soil management is vital to the success of any soil conservation scheme, because it helps to maintain the texture and fertility of the soil.

Land capability classification normally is used for the evaluation of land in soil conservation planning. The classification may be based upon that developed by the United States Soil Conservation Service (Klingebiel & Montgomery, 1966), or a modified version thereof, the rationale of the classification system being that the correct use of land represents the best method of controlling soil erosion. The system makes an assessment of the limitations of agricultural land use, placing special emphasis on erosion risk, soil depth and climate. The land is divided into capability classes according to the severity of these limiting factors (see Section 10.12).

10.11.1 Conservation measures for water erosion

As far as water erosion is concerned, conservation is directed mainly at the control of overland flow, because rill and gully erosion will be reduced effectively if overland flow is prevented. There is a critical

Table 10.15 Soil conservation practices. (After Morgan, 1980.)

| Practice | Control over | | | | | |
| | Rainsplash | | Run-off | | Wind | |
	D	T	D	T	D	T
Agronomic measures						
Covering soil surface	*	*	*	*	*	*
Increasing surface roughness	–	–	*	*	*	*
Increasing surface depression storage	+	+	*	*	–	–
Increasing infiltration	–	–	+	*	–	–
Mechanical measures						
Contour ridging	–	+	+	*	+	
Terraces	–	+	+	*	–	
Shelter belts	–	–	–	–	*	
Waterways	–	–	+	*	–	
Soil management						
Fertilizers, manures	+	+	+	*	+	
Subsoiling, drainage	–	–	+	*	–	

–, No control; +, moderate control; *, strong control; D, detachment; T, transport.

slope length at which erosion by overland flow is initiated. Hence, if the slope length is reduced by, for example, terracing, overland flow should be controlled. In other words, terraces break the original slope into shorter units, and have been used for agricultural purposes in certain parts of the world for centuries. Terraces may be regarded as small detention reservoirs, which contain water long enough to help it to infiltrate into the ground. The terraces follow the contours, and are raised on the outer side so as to form a channel. The terrace width is related primarily to the ground slope (Anon., 1989), but may also be influenced by climate (i.e. intensity and duration of rainfall) and soil type. Terraces help to conserve soil moisture, and should remove run-off in a controlled manner. Therefore, in order to remove excess water, drainage ditches must be incorporated into the terrace system. The build-up of excess pore water pressures in a slope is one of the principal reasons for slope failure. Consequently, it is most important that the terraced slopes are drained adequately.

Contour farming involves following the contours to plant rows of crops. As with terraces, this interrupts overland flow, thereby reducing its velocity and helping to conserve soil moisture. Contour farming probably is most successful in deep, well drained soils, where the slopes are not too steep. Alternating strips of crops and grass, planted around the contours of slopes, form the basis of strip cropping. Not only do strips reduce the velocity of overland flow, they protect the soil from raindrop impact and aid infiltration.

Strip cropping may be used on slopes which are too steep for terracing. According to Morgan (1995), the steepest slope on which strip cropping can be practised is around 8.5°. The width of the strip is reduced with increasing slope, for example, on a slope of 3°, the width could be 30m, but this is reduced to half that on a 10° slope. The widths also vary with the erodibility of the soil. The crop in each strip normally is rotated, so that fertility is preserved. However, grass strips may remain in place where the risk of erosion is high.

Multiple cropping refers to either the growth of two or more crops in sequence in the same year on the same plot of land or the growth of two or more crops simultaneously. In these ways, maximum canopy cover is available throughout the year. It is important that crops with different growth rates are involved, in order to take maximum advantage of vegetation cover over as long a period as possible.

Soil or crop management practices include crop rotation, where a different crop is grown on the same area of land in successive years in a 4- or 5-year cycle. This avoids exhausting the soil, and can improve the texture of the soil. The use of mulches, that is, covering the ground with plant residues, affords protection to the soil against raindrop impact, and reduces the effectiveness of overland flow. When mulches are ploughed into the soil, they increase the organic content and thereby improve the texture and fertility of the soil.

Reafforestation of slopes, where possible, also helps to control the flow of water and to reduce the impact of rainfall. The tree roots also help to improve the infiltration capacity of the soil.

Gullies can be dealt with in a number of ways. Most frequently, some attempt is made at revegetation by grassing or planting trees. Small gullies can be filled or partially filled, or traps or small dams can be constructed to collect and control the flow of sediment. A gully can be converted into an artificial channel with appropriate dimensions to convey water away.

Geotextiles can be used to simulate a vegetation cover so as to inhibit erosion processes, and thereby reduce soil loss (Rickson, 1995). They also can help to maintain soil moisture and promote the growth of seed. In fact, some geotextile mats contain seeds. Geotextiles used for erosion prevention can be made from either natural or synthetic materials to form an erosion blanket or mat. In the former case, the geotextile is biodegradable and so its role is temporary. Synthetic geotextiles remain in place to provide reinforcement for a soil. If the vegetation will control erosion when fully established, temporary geotextiles can be used. However, the establishment of vegetation may not coincide with the time at which the geotextile biodegrades. Temporary geotextiles are made from woven fibres of jute, coir, cotton or sisal. Others consist of loose mulch materials (e.g. paper strips, wood shavings or chips, straw, cotton waste or coconut fibres) held within a lightweight degradable polypropylene mesh. The synthetic geotextiles normally are made of polyethylene, and are either three-dimensional meshes or geowebs. Both types of geotextile are used for slope protection; they are rolled downslope and pegged into place. Seeding takes place prior to the positioning of temporary geo-textiles, whereas synthetic geotextiles tend to be placed first, then seeded and covered with topsoil.

10.11.2 Conservation measures for wind erosion

The problem of wind erosion can be mitigated by reducing the velocity of the wind and/or altering the ground conditions, so that particle movement becomes more difficult or particles are trapped. Windbreaks or shelter belts normally are placed at right angles to the prevailing wind. On the other hand, if winds blow from several directions, a her-ringbone or grid pattern of windbreaks may be more effective in reducing wind speed. Windbreaks inter-rupt the length over which the wind blows unimpeded, and thereby reduce the velocity of the wind at the ground surface. The spacing, height, width, length and shape of windbreaks influence their effec-tiveness. When trees or bushes act as windbreaks, as they are fairly permanent, they should be planned with care. For example, their spacing should be related to the amount of shelter which each wind-break offers. As wind velocities may increase at the ends of windbreaks, as a consequence of funnelling, they should be made longer than shorter. The recom-mended length of a windbreak is about 12 times its height for prevailing winds at right angles. However, in order to allow for deviations in the direction of the wind, this value may be doubled. Furthermore, wind-breaks should be permeable to avoid the creation of erosive vortices on their leeside. The shape of a veg-etative windbreak can be influenced by the careful selection of the bushes or trees involved. They prefer-ably should be triangular shaped in cross-section. A windbreak consisting of trees and shrubs may need to be up to 9m in width, comprising two rows of trees and three of bushes.

Geotextiles can interrupt the flow of wind over the ground surface, and so reduce its velocity, particularly if coarse fibred products are used. By reducing the initial erosion of particles, further erosion is reduced, because there are fewer particles in saltation, suspension and traction. In addition, because of the moisture retention characteristics, especially of natural geotextiles, the soil remains relatively moist, so that it is less susceptible to wind erosion.

Field cropping practices may be simpler and cheaper than windbreaks, and sometimes more effec-tive (Hudson, 1981). The effectiveness of the protec-tion against wind erosion offered by plant cover is affected by the degree of ground cover provided, and therefore by the height and density of the plant cover. It has been suggested that plants should cover over 70% of the ground surface to afford adequate protec-tion against erosion (Evans, 1980). Cover crops can be grown to protect the soil during the period when it usually would be exposed, or strip cropping may be practised. Vegetative cover helps to trap soil particles, and is particularly important in terms of reducing the amount of saltating particles. Strip cropping helps to inhibit soil avalanching. Although the choice of crop may be limited by the availability of water, it would appear that small grain crops, legumes and grasses are reasonably effective in reducing the effects of wind erosion. Mulches may be used to cover the soil, and should be applied at a minimum rate of $0.5\,kg\,m^{-2}$. As well as providing organic matter for the soil, mulches also trap soil particles and conserve soil moisture. Stubble may be left between strips of ploughed soil; the strip should be wide enough to prevent saltating particles from leaping across it. Because wind erosion is adversely affected by the surface roughness, the furrows produced by plough-ing, ideally should be orientated at right angles to the prevailing wind.

As far as arid and semi-arid regions are concerned, crop yields generally are low near desert margins, so that it is rarely worth spending large sums of money on conservation. Good management is necessary in these regions and, on low slopes, this involves con-trolling yields at low, sustainable levels, and varying production with wet and dry years. On steep slopes, or where there are migrating dunes, vegetation must be established and grazing discouraged. However, such policies are difficult to put into effect.

10.12 Assessment of soil erosion

Morgan (1995) referred to the assessment of soil erosion hazard as a special form of land resource evaluation, the purpose of which was to identify areas of land where the maximum sustained productivity from a particular use was threatened by the excessive loss of soil. Such an assessment subdivides a region into zones in which the type and degree of erosion hazard are similar. This can be represented in map form, and used as the basis for planning and conservation within a region.

Maps of erosivity using the rainfall erosion index R have been produced for the USA (Fig. 10.9), and Stocking and Elwell (1976) provided a map of Rhodesia (now Zimbabwe) showing the mean annual erosivity (Fig. 10.10). Stocking and Elwell (1973) also devised a rating system for the evaluation of erosion risk. Each of the parameters in the rating system, namely, the erosivity, erodibility, slope, ground cover and human occupation, has five grades. The last parameter includes the type and density of settlement. Grades range from 1 for a low risk of erosion to 5 for a high risk, and the ratings for each of the grades are summed to provide a group rating (Fig. 10.11). The group rating enables an area to be classified as low, average or high risk.

Land capability classification, as referred to above, was originally developed by the United States Soil Conservation Service for farm planning (Klingebiel & Montgomery, 1966). In other words, it provides a means of assessing the extent to which limitations, such as erosion risk, soil depth, wetness and climate,

hinder the agricultural use of land (Table 10.16). Three categories are recognized, namely, capability classes, subclasses and units. A capability class is a group of capability subclasses which has the same relative degree of limitation or hazard. Classes I to IV can be used for cultivation, whereas classes V to VIII cannot because of permanent limitations, such as wetness or steeply sloping land. The risk of erosion increases through the first four classes progressively, reducing the choice of crops which can be grown, and requiring more expensive conservation practices. A capability subclass is a group of capability units which possesses the same major conservation problems or limitations (e.g. erosion hazard, excess water, limited depth of soil or excessive stoniness). Lastly, a capability unit is a group of soil mapping units that has the same limitations and management responses. Soils within a capability unit can be used for the same crops and require similar conservation treatment.

Terrain evaluation is concerned with the analysis, classification and appraisal of a region (see Chapter 16). The initial interpretation of the form of the land can be made from large-scale maps and aerial

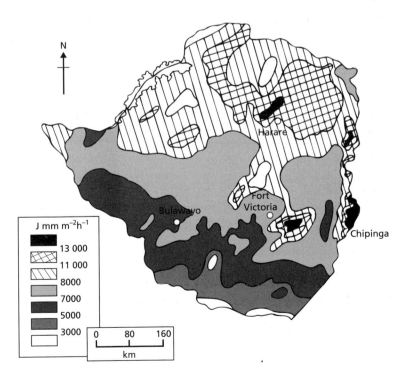

Fig. 10.10 Mean annual erosivity in Zimbabwe (formerly Rhodesia). (After Stocking & Elwell, 1976.)

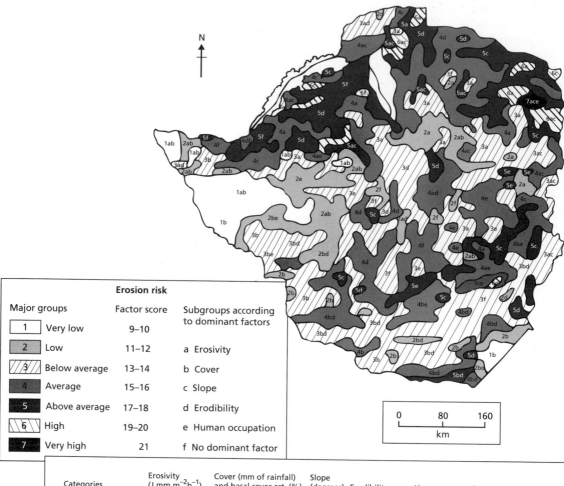

N

Erosion risk

Major groups		Factor score	Subgroups according to dominant factors
1	Very low	9–10	
2	Low	11–12	a Erosivity
3	Below average	13–14	b Cover
4	Average	15–16	c Slope
5	Above average	17–18	d Erodibility
6	High	19–20	e Human occupation
7	Very high	21	f No dominant factor

0	80	160

km

Categories		Erosivity (J mm m^{-2}h^{-1})	Cover (mm of rainfall) and basal cover est. (%)	Slope (degrees)	Erodibility	Human occupation*
Low	I	Below 5000	Above 1000 7–10	0–2	Orthoferralitic regosoils	Extensive large-scale commercial ranching, National Parks or Unreserved
Below average	II	5000–7000	800–1000 5–8	2–4	Paraferralitic	Large-scale commercial farms
Average	III	7000–9000	600–800 3–6	4–6	Fersiallitic	Low-density CLs (below 5 p.p.km^2) and SCCF
Above average	IV	9000–11 000	400–600 1–4	6–8	Siallitic vertisols lithosols	Moderately settled CLs (5–30 p.p.km^2)
High	V	Above 11 000	Below 400 0–2	Above 8	Non-calcic hydromorphic sodic	Densely settled CLs (above 30 p.p.km^2)

(Note: Cover, Erodibility and Human occupation are only tentative and cannot as yet be expressed on a firm quantitative basis)
*p.p.km^2, persons per square kilometre; CL, communal lands; SCCF, small-scale commercial farms.

Fig. 10.11 Erosion survey of Rhodesia (now Zimbabwe). (After Stocking & Elwell, 1973.)

Table 10.16 Land-use alternatives of capability classes. Based on the definitions in the original US version of the system (Kingebiel & Montgomery, 1966.)

Capability class	Management under cultivation			Capability						Characteristics and recommended land use
	Limitations	Choice of crops	Conservation practices	Cultivation	Pasture (improved)	Range (unimproved pasture)	Woodland	Wildlife food and cover	Recreation water supply aesthetic	
I	Few	Any	None	√	√	√	√	√	√	Deep, productive soils, easily worked, on nearly level land; not subject to overland flow; no or slight risk of damage when cultivated; use of fertilizers and lime, cover crops, crop rotations required to maintain soil fertility and soil structure
II	Some	Reduced or	Moderate	√	√	√	√	√	√	Productive soils on gentle slopes; moderate depth; subject to occasional overland flow; may require drainage; moderate risk of damage when cultivated; use of crop rotations, water control systems or special tillage practices to control erosion

Continued on p. 320

Table 10.16 *Continued.*

Capability class	Limitations	Management under cultivation		Capability						Characteristics and recommended land use
		Choice of crops	Conservation practices	Cultivation	Pasture (improved)	Range (unimproved pasture)	Woodland	Wildlife food and cover	Recreation water supply aesthetic	
III	Severe	Reduced	and/or special	√	√	√	√	√	√	Soils of moderate fertility on moderately steep slopes; subject to more severe erosion; subject to severe risk of damage, but can be used for crops provided that plant cover is maintained; hay or other sod crops should be grown instead of row crops
IV	Very severe	Restricted and/ or	Very careful management	√	√	√	√	√	√	Good soils on steep slopes; subject to severe erosion; very severe risk of damage, but may be cultivated if handled with great care; keep in hay or pasture, but a grain crop may be grown once in 5 or 6 years
V	Other than erosion				√	√	√	√	√	Land is too wet or stony for cultivation, but of nearly level slope; subject to only slight erosion if properly managed; should

Class	Limitations					Suitable uses
						be used for pasture or forestry, but grazing should be regulated to prevent plant cover from being destroyed
VI	Severe	√	√	√	√	Shallow soils on steep slopes; use for grazing and forestry; grazing should be regulated to preserve plant cover; if plant cover is destroyed, use should be restricted until cover is re-established
VII	Very severe	√	√	√	√	Steep, rough, eroded land with shallow soils; also includes droughty or swampy land; severe risk of damage even when used for pasture or forestry; strict grazing or forest management must be applied
VIII	Very severe	√				Very rough land; not suitable even for woodland or grazing; reserve for wildlife, recreation or watershed conservation

Classes I–IV denote soils suitable for cultivation.
Classes V–VIII denote soils unsuitable for cultivation.

photographs. The observation of relief should give particular attention to the direction and angle of the maximum gradient, the maximum relief amplitude and the proportion of the total area occupied by bare rock or slopes. An assessment of the risk of erosion (especially the location of slopes which appear to be potentially unstable) and the risk of excess sedimentation of water-borne or windblown material should be made. The land system, land facet and land element, in decreasing order of size, are the principal units used. A land systems map shows the subdivision of a region into areas with common physical attributes which differ from those of adjacent areas. Many of the factors considered in land systems analysis are of importance in relation to soil erosion, so that terrain evaluation is useful in the evaluation of the risk of soil erosion.

The first types of soil erosion surveys consisted of mapping sheet wash, rills and gullies within an area, frequently from aerial photographs. Indices, such as the gully density, were used to estimate the erosion hazard. Subsequently, sequential surveys of the same area were undertaken, in which mapping, again generally from aerial photographs, was performed at regular time intervals. Hence, changes in those factors responsible for erosion could be evaluated. They could also be evaluated in relation to other factors, such as changing agricultural practices or land use. Any data obtained from imagery should be checked and supplemented with extra data from the

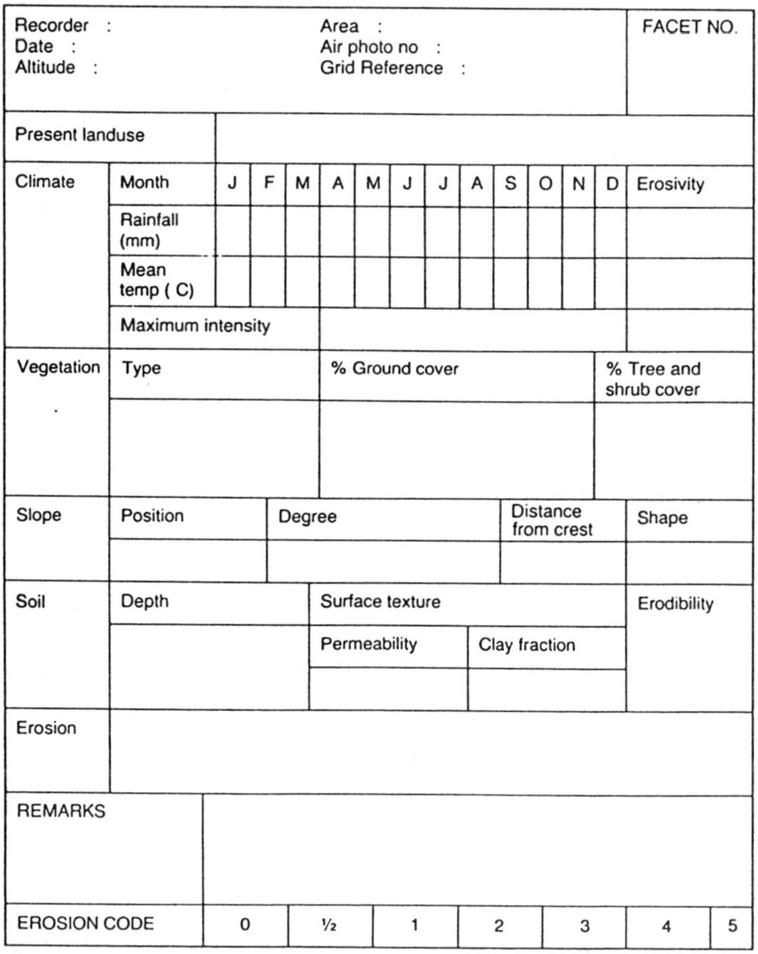
Recorder :
Date :
Altitude :
Area :
Air photo no :
Grid Reference :
FACET NO.
Present landuse
Climate
Month
J F M A M J J A S O N D
Erosivity
Rainfall (mm)
Mean temp (C)
Maximum intensity
Vegetation
Type
% Ground cover
% Tree and shrub cover
Slope
Position
Degree
Distance from crest
Shape
Soil
Depth
Surface texture
Erodibility
Permeability
Clay fraction
Erosion
REMARKS
EROSION CODE
0 ½ 1 2 3 4 5

Fig. 10.12 Pro forma for recording soil erosion in the field. (From Morgan, 1995.)

Table 10.17 Coding system for soil erosion appraisal in the field. (After Morgan, 1995.)

Code	Indicators
0	No exposure of tree roots; no surface crusting; no splash pedestals; over 70% plant cover (ground and canopy)
$^1/_2$	Slight exposure of tree roots; slight crusting of the surface; no splash pedestals; soil level slightly higher on upslope or windward sides of plants and boulders; 30–70% plant cover
1	Exposure of tree roots, formation of splash pedestals, soil mounds protected by vegetation, all to depths of 1–10 mm; slight surface crusting; 30–70% plant cover
2	Tree root exposure, splash pedestals and soil mounds to depths of 10–50 mm; crusting of the surface; 30–70% plant cover
3	Tree root exposure, splash pedestals and soil mounds to depths of 50–100 mm; 2–5 mm thickness of surface crust; grass muddied by wash and turned downslope; splays of coarse material due to wash and wind; less than 30% plant cover
4	Tree root exposure, splash pedestals and soil mounds to depths of 50–100 mm; splays of coarse material; rills up to 80 mm deep; bare soil
5	Gullies; rills over 80 mm deep; blow-outs and dunes; bare soil

field. It may be possible to rate the degree of severity of soil erosion as seen in the field. A pro forma for recording data is illustrated in Fig. 10.12, and a coding system suggested by Morgan (1995) is given in Table 10.17.

10.13 Soil surveys and mapping

Soil surveys are undertaken to assess the properties of soils, and are of value in relation to rural and urban planning, especially in terms of agriculture and forestry, and for feasibility and design studies in land development projects and engineering works (Davidson, 1980). One of the principal objectives of a soil survey is the production of a soil map (Fig. 10.13) and report, which involves the delineation of soil types in the field in the case of a general purpose soil survey. The mapping unit for soil mapping is based primarily on soil profiles, although features of landforms may be included in the descriptions of mapping units. The report not only describes the soils and their properties, but should provide background information on environmental factors, land potential, probable responses to various alternative forms of land management and, sometimes, land-use and management recommendations (Dent & Young, 1981). However, soil surveys can also be carried out for a special purpose, such as for the development of a scheme for irrigation, where the drainage properties

and degree of salinity may be of particular importance to demarcate areas prone to flooding or for land reclamation.

Soil surveys can prove to be difficult in that the boundaries between different soils usually are not readily defined, but grade into one another. The changes which do occur are not necessarily reflected in the surface expression. This means that soils must be examined from auger holes or small pits. In a general purpose survey, the soils generally are mapped according to the properties which are observed in the field, whereas in a special purpose survey, the relevant properties to be mapped are chosen before the survey commences. General purpose surveys involve the production of a pedological map which may be used in land evaluation. Such surveys are useful in underdeveloped regions, where little is known about the potential land use or the physical environment. Hence, basic information on soils must be obtained in such areas prior to decisions being made on land use. In developed countries, farmers tend only to be interested in soil surveys when significant changes in land use are contemplated. Nonetheless, soil surveys allow the results of experience of soil use to be communicated to those concerned with agriculture, in order to bring about the optimum cropping system and management of soils. Soil surveys may investigate soil erodibility, depth of soil, steepness of slope, shrink–swell poten-

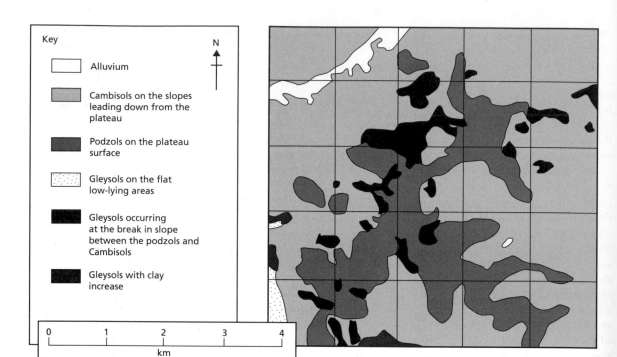

Fig. 10.13 Soil map.

tial, frequency of rock outcrops, possibility of salinization, fertilizer–plant response and crop variety trials, especially new crops. The most suitable soil conservation practices may be introduced with the aid of soil surveys. In cases where major changes in the type of land use are contemplated, an assessment must be made of the response of the soil to any proposed changes. Land suitability evaluation, giving the suitability of areas for each of several uses, may be undertaken in such cases. Accordingly, soil surveys often now form part of an environmental impact assessment.

Soil surveys may be used to assist in avoiding hazards, such as flooding, salinization due to irrigation or crop failure due to drought. The data from such surveys are used to adopt appropriate measures to counteract the hazard. Alternatively, they may form the basis of a decision not to use the land for the proposed purpose.

Data from surveys can be of value in the initial planning stages of construction projects, because they may indicate where certain problem soils occur, such as expansive clays, collapsible soils and peaty soils. They may also be of value in relation to the location of construction materials, such as gravels and sands, and suitable supplies of soils for embankments. Such data may be particularly useful in terms of trench construction, low-cost housing, light industrial units, road construction and recreational areas.

10.14 Soils as a construction material

10.14.1 Gravels and sands

Gravels and sands may occur as river deposits, river terrace deposits, beach deposits, raised beach deposits and fluvio-glacial deposits. Sand may also accumulate as windblown deposits. Gravels and sands represent sources of coarse and fine aggregate respectively.

A gravel deposit consists of a framework of pebbles between which are voids. The voids are rarely empty, being occupied by sand, silt or clay material. River and fluvio-glacial gravels notably are bimodal, the principal mode being in the gravel grade, the secondary in the sand grade. Marine gravels, however, are often unimodal, and tend to be more uniformly sorted than fluvial types of similar grade size.

The shape and surface texture of the pebbles in a gravel deposit are influenced by the agent responsible for their transportation and the length of time taken in transport, although the shape is also dependent on the initial shape of the fragment, which, in turn, is controlled by the fracture pattern within the parent rock. The shape of gravel particles usually is classified as rounded, irregular, angular, flaky and elongated (Anon., 1975b). Anon. (1975b) also defines a flakiness index, an elongation index and an angularity number. The flakiness index of an aggregate is the percentage of particles, by weight, whose least dimension (thickness) is less than 0.6 times their mean dimension. The elongation index of an aggregate is the percentage, by weight, of particles whose greatest dimension (length) is greater than 1.8 times their mean dimension. The angularity number is a measure of the relative angularity based on the percentage of voids in the aggregate. The least angular aggregates are found to have about 33% voids, and the angularity number is defined as the amount by which the percentage of voids exceeds 33%. The angularity number ranges from zero to about 12.

The composition of a gravel deposit reflects not only the type of rocks in the source area, but also the agent(s) responsible for its formation and the climatic regime in which it was or is being deposited. Furthermore, relief influences the character of a gravel deposit, for example, under low relief, gravel production is small, and the pebbles tend to be chemically inert residues, such as vein quartz, quartzite, chert and flint. By contrast, high relief and rapid erosion yield coarse, immature gravels. Nevertheless, a gravel achieves maturity much more rapidly than a sand under the same conditions. Gravels which consist of only one type of rock fragment are called oligomictic. Such deposits are usually thin and well sorted. Polymictic gravels usually consist of a varied assortment of rock fragments, and occur as thick, poorly sorted deposits.

Gravel particles generally possess surface coatings that may be the result of weathering or may represent mineral precipitates derived from circulating groundwater. The latter type of coating may be calcareous, ferruginous, siliceous or, occasionally, gypsiferous. Clay may also form a coating about pebbles. Surface coatings generally reduce the value of gravels for use as concrete aggregate; thick and/or soft and loosely adhering surface coatings are particularly suspect. Clay and gypsum coatings, however, can often be removed by screening and washing. Siliceous coatings tend to react with the alkalis in high-alkali cements, and therefore are detrimental to concrete.

The textural maturity of sand varies appreciably. A high degree of sorting, coupled with a high degree of rounding, characterizes a mature sand. The shape of sand grains, however, is not greatly influenced by the length of transport. Maturity is also reflected in the chemical or mineralogical composition, and it has been argued that the ultimate sand is a concentration of pure quartz. This is because the less stable minerals disappear due to mechanical or chemical breakdown during erosion and transportation, or even after the sand has been deposited.

Sands are used for building purposes to give bulk to concrete, mortars, plasters and renderings. For example, sand is used in concrete to lessen the void space created by the coarse aggregate. A sand consisting of a range of grade sizes gives a lower proportion of voids than one in which the grains are of uniform size. Indeed, grading probably is the most important property as far as the suitability of a sand for concrete is concerned (Anon., 1981b). In any concrete mix, consideration should be given to the total specific surface of the coarse and fine aggregates, because this represents the surface which has to be lubricated by the cement paste to produce a workable mix. Poorly graded sands can be improved by adding the missing grade sizes to them, so that a high-quality material can be produced with correct blending.

Generally, a sand with rounded particles produces a slightly more workable concrete than one consisting of irregularly shaped particles.

Sands used for building purposes usually are siliceous in composition and should be as free from impurities as possible. They should contain no significant quantity of silt or clay (less than 3% by weight), because these require a high water content to produce a workable concrete mix. This, in turn, leads to shrinkage and cracking on drying. Furthermore, clay and shaly material tend to retard setting and hardening, or they may spoil the appearance of the finished product. If sand particles are coated with clay, they form a poor bond with cement, and produce a weaker and less durable concrete. The presence of feldspars in sands used in concrete has sometimes given rise to hair cracking, and mica and particles of shale adversely affect the strength of concrete. Organic impurities may adversely affect the setting and hardening properties of cement by retarding hydration, and thereby reduce its strength and durability. Organic and coaly matter also cause popping, pitting and blowing. If iron pyrite occurs in sand, it gives rise to unsightly rust stains when used in concrete. The salt content of marine sands is unlikely to produce any serious adverse effects in good quality concrete, although it will probably give rise to efflorescence. Salt can be removed by washing sand.

High-grade quartz sands are also used for making silica bricks employed for refractory purposes. Glass sands must have a silica content of over 95% (over 96% for plate glass). The amount of iron oxides present in glass sands must be very low, in the case of clear glass under 0.05%. The uniformity of grain size is another important property, as this means that the individual grains melt in the furnace at approximately the same temperature.

10.14.2 Clay deposits, refractory materials and bricks

The principal clay minerals belong to the kandite, illite, smectite, vermiculite and palygorskite families. The kandites, of which kaolinite is the chief member, are the most abundant clay minerals. Deposits of kaolin or china clay are associated with granite masses which have undergone kaolinization. Kaolin is used for the manufacture of white earthenwares and stonewares, in white Portland cement and for special refractories.

Ball clays are composed almost entirely of kaolinite and, as between 70 and 90% of the individual particles are below 0.01 mm in size, these clays have a high plasticity. Their plasticity at times is enhanced by the presence of montmorillonite. They contain a low percentage of iron oxide and, consequently, when burnt give a light cream colour. They are used for the manufacture of sanitary ware and refractories.

If a clay or shale can be used to manufacture refractory bricks, it is termed a fireclay. Such material should not fuse below 1600 °C and should be capable of taking a glaze. Ball clays and china clays are in fact fireclays, fusing at 1650 °C and 1750 °C respectively,

however, they are too valuable, except for making special refractories. Most fireclays are highly plastic and contain kaolinite as their predominant material. Some of the best fireclays are found beneath coal seams. Indeed, in the UK, fireclays are almost entirely restricted to strata of Coal Measures age. The material in a bed of fireclay which lies immediately beneath a coal seam is often of better quality than that found at the base of the bed. Because fireclays represent fossil soils which have undergone severe leaching, they consist chiefly of silica and alumina, and contain only minor amounts of alkalis, lime and iron compounds. This accounts for their refractoriness (alkalis, lime, magnesia and iron oxides in a clay deposit tend to lower its temperature of fusion and act as fluxes). Very occasionally, a deposit contains an excess of alumina and, in such cases, it possesses a very high refractoriness. After the fireclay has been quarried or mined, it usually is left to weather for an appreciable period of time to allow it to break down before it is crushed. The crushed fireclay is mixed with water and moulded. Bricks, tiles and sanitary ware are made from fireclay.

Bentonite is formed by the alteration of volcanic ash, the principal clay mineral being either montmorillonite or beidellite. When water is added to bentonite, it swells to many times its original volume to produce a soft gel. Bentonite is markedly thixotropic, and this, together with its plastic properties, has given the clay a wide range of uses. For example, it is added to poorly plastic clays to make them more workable and to cement mortars for the same purpose. In the construction industry, it is used as a material for clay grouting, drilling mud, slurry trenches and diaphragm walls.

The suitability of a raw material for brick making is determined by its physical, chemical and mineralogical character, and the changes which occur when it is fired. The unfired properties, such as the plasticity, workability (i.e. the ability of the clay to be moulded into shape without fracturing, and to maintain its shape when the moulding action ceases), dry strength, dry shrinkage and vitrification range, are dependent upon the source material, but the fired properties, such as the colour, strength, total shrinkage on firing, porosity, water absorption, bulk density and tendency to bloat, are controlled by the nature of the firing process. The price that can be charged for a brick depends largely upon its attractiveness, that is its colour and surface appearance. The ideal raw material should possess moderate plasticity, good workability, high dry strength, total shrinkage on firing of less than 10% and a long vitrification range. However, the suitability of a clay for brick manufacture can be determined only by running it through a production line or by pilot plant firing tests.

The mineralogy of the raw material influences the behaviour of a clay during the brick making process, and hence the properties of the finished product (Bell, 1992). Mudrocks consist of clay minerals and non-clay minerals, mainly quartz. The clay mineralogy varies from one deposit to another. Although bricks can be made from most mudrocks, the varying proportions of the different clay minerals have a profound effect on the processing and on the character of the fired brick. Those clays which contain a single predominant clay mineral have a shorter temperature interval between the onset of vitrification and complete fusion than those consisting of a mixture of clay minerals. This is more true of montmorillonitic and illitic clays than those composed chiefly of kaolinite. In addition, those clays which consist of a mixture of clay minerals do not shrink as much when fired as those composed predominantly of one type of clay mineral. Clays containing significant amounts of disordered kaolinite tend to have moderate to high plasticity, and therefore are easily workable. They produce lean clays which undergo little shrinkage during brick manufacture. They also possess a long vitrification range and produce a fairly refractory product. However, clays containing appreciable quantities of well ordered kaolinite are poorly plastic and less workable. Illitic clays are more plastic and less refractory than those in which disordered kaolinite is dominant, and fire at somewhat lower temperatures. Smectites are the most plastic and least refractory of the clay minerals. They show high shrinkage on drying, because they require high proportions of added water to make them workable. As far as the unfired properties of the raw materials are concerned, the non-clay minerals present act mainly as a diluent, but may be of considerable importance in relation to the fired properties. The non-clay material may also enhance the working properties, for example, colloidal silica improves the workability by increasing the plasticity.

The presence of quartz, in significant amounts, gives strength and durability to a brick. This is because, during the vitrification period, quartz combines with the basic oxides of the fluxes released from the clay minerals on firing to form glass, which improves the strength. However, as the proportion increases, the plasticity of the raw material decreases.

The accessory minerals in clays play a significant role in brick making. The presence of carbonates particularly is important, and can influence the character of the bricks produced. When heated above 900°C, carbonates break down, yielding carbon dioxide, and leave behind reactive basic oxides, particularly those of calcium and magnesium. The escape of carbon dioxide can cause lime popping or bursting if large pieces of carbonate, for example, shell fragments, are present, thereby pitting the surface of a brick. To avoid lime popping, the material must be finely ground to pass a 20-mesh sieve. The residual lime and magnesia form fluxes which give rise to low-viscosity silicate melts. The reaction lowers the temperature of the brick and, hence unless additional heat is supplied, lowers the firing temperature and shortens the range over which vitrification occurs. The reduction in temperature can result in inadequately fired bricks. If excess oxides remain in the brick, they will hydrate on exposure to moisture, thereby destroying the brick. The expulsion of significant quantities of carbon dioxide can increase the porosity of bricks, reducing their strength. Engineering bricks must be made from a raw material which has a low carbonate content.

Sulphate minerals in clays are detrimental to brick making. For instance, calcium sulphate does not decompose within the range of the firing temperature of bricks. It is soluble and, if present in trace amounts in the fired brick, causes efflorescence when the brick is exposed to the atmosphere. Soluble sulphates dissolve in the water used to mix the clay. During drying and firing, they often form a white scum on the surface of a brick. Barium carbonate may be added to render such salts insoluble, and so prevent scumming.

Iron sulphides, such as pyrite and marcasite, frequently occur in clays. When heated in oxidizing conditions, the sulphides decompose to produce ferric oxide and sulphur dioxide. In the presence of organic matter, oxidation is incomplete, yielding ferrous compounds, which combine with silica and basic oxides, if present, to form black glassy spots. This may lead to a black vitreous core being present in some bricks, which can reduce the strength significantly. If the vitrified material forms an envelope around the ferrous compounds, and heating continues until this decomposes, the gases liberated cannot escape, causing bricks to bloat and distort. Under such circumstances, the rate of firing should be controlled in order to allow gases to be liberated prior to the onset of vitrification. Too high a percentage of pyrite or other iron-bearing minerals gives rise to rapid melting, which can lead to difficulties on firing.

The strength of a brick depends largely on the degree of vitrification. Theoretically, the strengths of bricks made from clays containing fine grained clay minerals, such as illite, should be higher than those of bricks made from clays containing the coarser grained kaolinite. Illitic clays, however, vitrify more easily, and there is a tendency to underfire, particularly if they contain fine grained calcite or dolomite. Kaolinitic clays are much more refractory and can stand harder firing, greater vitrification therefore is achieved.

The permeability also depends on the degree of vitrification. Clays containing a high proportion of clay minerals produce less permeable products than clays with a high proportion of quartz, but the former types of clays may give a high drying shrinkage and high moisture absorption.

The colour of the clay prior to burning gives no indication of the colour it will have after leaving the kiln. Indeed, a chemical analysis can only offer an approximate guide to the colour of the finished brick. The iron content of a clay, however, is important in this respect. For instance, as there is less scope for iron substitution in kaolinite than in illite, this often means that kaolinitic clays give a whitish or pale yellow colour on firing, whilst illitic clays generally produce red or brown bricks. More particularly, a clay containing about 1% of iron oxides, when burnt, tends to produce a cream or light yellow colour, one containing about 2–3% gives a buff colour and one containing about 4–5% produces a red colour. The presence of other constituents, notably calcium, magnesium or aluminium oxides, tends to reduce the colouring effect of iron oxide, whereas the presence of titanium oxide enhances it. High original carbonate content tends to produce yellow bricks.

Organic matter commonly occurs in clay. It may be concentrated in lenses or seams, or be finely disseminated throughout the clay. Incomplete oxidation of the carbon upon firing may result in black coring or bloating. Even minute amounts of carbonaceous material can give black coring in dense bricks if it is not burnt out. Black coring can be prevented by ensuring that all carbonaceous material is burnt out below the vitrification temperature. This means that, if a raw material contains much carbonaceous matter, it may be necessary to admit cool air into the firing chamber to prevent the temperature rising too quickly. On the other hand, the presence of oily material in a clay can be an advantage as it can reduce the fuel costs involved in brick making. For instance, the Lower Oxford Clay in parts of England contains a significant proportion of oil, so that, when it is heated above approximately 300°C, it becomes almost self-firing.

Sufficient quantities of suitable raw material must be available at a site before a brickfield can be developed. The volume of suitable clay must be determined, as well as the amount of waste, that is, the overburden and unsuitable material within the sequence that is to be extracted. The first stages of the investigation are topographical and geological surveys, followed by a drill-hole programme. This leads to a lithostratigraphical and structural evaluation of the site. It should also provide data on the position of the water table and the stability of the slopes which will be produced during excavation of the brick pit.

10.14.3 Fills and embankments

The engineering properties of soils used in fills and embankments, such as their shear strength, consolidation characteristics and permeability, are influenced by the amount of compaction they have undergone. Therefore, the desired amount of compaction is established in relation to the engineering properties required for the fill to perform its design function. A specification for compaction needs to indicate the type of compactor, mass, speed of travel and any other factors influencing the performance, such as the frequency of vibration, thickness of individual layers to be compacted and number of passes of the compactor (Table 10.18).

Fills and embankments are mechanically compacted by laying and rolling soil in thin layers. The soil particles are packed more closely due to a reduc

Table 10.18 Typical compaction characteristics for soils used in earthwork construction. (After Anon., 1981c.) The information in this table should be taken only as a general guide. When the material performance cannot be predicted, it may be established by earthwork trials. This table is applicable only to fill placed and compacted in layers. It is not applicable to deep compaction of materials *in situ*.

Soil	Major divisions	Subgroups	Suitable type of compaction plant	Minimum number of passes for satisfactory compaction	Maximum thickness of compacted layer (mm)	Remarks
Coarse soils	Gravel sand gravelly soils	Well graded gravel and gravel/sand mixtures; little or no fines Well graded gravel/sand mixtures with excellent clay binder Uniform gravel; little or no fines Poorly graded grave and gravel/sand mixtures; little or no fines Gravel with excess fines, silty gravel, clayey gravel, poorly graded gravel/sand/clay mixtures	Grid roller over 540 kg per 100 mm of roll Pneumatic-tyred roller over 2000 kg per wheel Vibratory plate compactor over 1100 kg/m^2 of baseplate Smooth-wheeled roller Vibratory roller Vibro-rammer Self-propelled tamping roller	3–12 depending on type of plant	75–275 depending on type of plant	
	Sands and Sandy soils	Well graded sands and gravelly sands; little no fines Well graded sands with excellent clay bincer				
	Uniform sands and gravels	Uniform gravels; little or no fines Uniform sands; little or no fines Poorly graded sands; little or no fines Sands with fines, silty sands, clayey sands poorly graded sand/clay mixtures	Smooth-wheeled roller below 500 kg per 100 mm of roll Grit roller below 540 kg per 100 mm Pneumatic-tyred roller below 1500 kg per wheel Vibratory roller Vibrating plate compactor Vibro-tamper	3–16 depending on type of plant	75–300 depending on type of plant	

Continued on p. 330

Table 10.18 *Continued.*

Soil	Major divisions	Subgroups	Suitable type of compaction plant	Minimum number of passes for satisfactory compaction	Maximum thickness of compacted layer (mm)	Remarks
Fine soils	Soils having low plasticity	Silts (inorganic) and very fine sands, silty or clayey fine sands with slight plasticity Clayey silts (inorganic) Organic silts of low plasticity	Sheepsfoot roller Smooth-wheeled roller Pneumatic-tyred roller Vibratory roller over 70 kg per 100 mm of roll Vibratory plate compactor over 1400 kg m^{-2} of baseplate Vibro-tamper Power rammer	4–8 depending on type of plant	100–450 depending on type of plant	If moisture content is low it may be preferable to use a vibratory roller Sheepsfoot rollers are best suited to soils at a moisture content below their plastic limit
	Soils having medium plasticity	Silty and sandy clays (inorganic) of medium plasticity Clays (inorganic) of medium plasticity Organic clays of medium plasticity				Generally unsuitable for earthworks
	Soils having high plasticity	Fine sandy and silty soils, plastic silts Clay (inorganic) of high plasticity, fat clays Organic clays of high plasticity				Should only be used when circumstances are favourable Should not be used for earthworks

Note. If earthwork trials are carried out, the number of field density tests on the compacted material should be related to the variability of the soils and the standard deviation of the results obtained. Compaction of mixed soils should be based on that subgroup requiring most compactive effort.

tion in the volume of the void space, resulting from the momentary application of loads, such as rolling, tamping or vibration. Compaction involves the expulsion of air from the voids without the moisture content being changed significantly. Hence, the degree of saturation is increased. However, all the air cannot be expelled from the soil by compaction, so that complete saturation is not achievable. Nevertheless, compaction does lead to a reduced tendency for changes in the moisture content of the soil to occur. The method of compaction used depends upon the soil type, including its grading and moisture content at the time of compaction, the total quantity of material, layer thickness and rate at which it is to be compacted and the geometry of the proposed earthworks.

The soil is stiff, and therefore more difficult to compact, when the moisture content is low. As the moisture content increases, it enhances the interparticle repulsive forces, thus separating the particles and causing the soil to soften and become more workable. This gives rise to higher dry densities and lower air contents. As saturation is approached, however, pore water pressure effects counteract the effectiveness of the compactive effort. Each soil therefore has an optimum moisture content at which the soil has a maximum dry density. This optimum moisture content can be determined by the standard compaction test (Anon., 1990). Unfortunately, however, this relationship between the maximum dry density and optimum moisture content varies with the compactive effort, and so test results can only indicate how easily a soil can be compacted. Field tests are needed to assess the actual density achieved by compaction on site.

Most specifications for the compaction of cohesive soils require that the dry density which is achieved represents a certain percentage of some laboratory standard. For example, the dry density required in the field can be specified in terms of the relative compaction or the final air void percentage achieved. The ratio between the maximum dry densities obtained *in situ* and derived from the standard compaction test is referred to as the relative compaction. If the dry density is given in terms of the final air void percentage, a value of 5–10% usually is stipulated, depending upon the maximum dry density determined from the standard compaction test. In most cases of compaction of cohesive soils, a 5% variation in the value

specified by either method is allowable, provided that the average value attained is equal to or greater than that specified.

Specifications for the control of compaction of cohesionless soils either require a stated relative density to be achieved or stipulate the type of equipment, thickness of layer and number of passes required. Field compaction trials should be carried out in order to ascertain which method of compaction should be used, including the type of equipment. Such tests should take into account the variability in grading of the material used and its moisture content, the thickness of the individual layers to be compacted and the number of passes per layer.

The compaction characteristics of clay largely are governed by its moisture content. For instance, a greater compactive effort is necessary as the moisture content is lowered. It may be necessary to use thinner layers and more passes by heavier compaction plant than required for granular materials. The properties of cohesive fills also depend to a much greater extent on the placement conditions than do those of a coarse grained fill.

The shear strength of a given compacted cohesive soil depends on the density and the moisture content at the time of shear. The pore water pressures developed while the soil is being subjected to shear are also of great importance in determining the strength of such soils. For example, if a cohesive soil is significantly drier than the optimum moisture content, it has a high strength, due to the high negative pore water pressures developed as a consequence of capillary action. The strength declines as the optimum moisture content is approached, and continues to decrease on the wet side of optimum. Increases in pore water pressures produced by volume changes coincident with shearing tend to lower the strength of a compacted cohesive soil.

Furthermore, there is a rapid increase in pore water pressure as the moisture content approaches the optimum. Nevertheless, the compaction of cohesive soil with a moisture content which is slightly less than optimum frequently gives rise to an increase in strength. This is because the increase in the value of friction more than compensates for the change in pore water pressure.

The compressibility of a compacted cohesive soil also depends on its density and moisture content at the time of loading. However, the placement moisture content tends to have a larger effect than the dry density on the compressibility. If a soil is compacted significantly dry of optimum, and then is saturated, extra settlement occurs on loading. This does not occur when soils are compacted at optimum moisture content or on the wet side of optimum.

A sample of cohesive soil compacted on the dry side of the optimum moisture content swells more, at the same confining pressure, when given access to moisture, than a sample compacted on the wet side of optimum. This is because the former type has a greater moisture deficiency, a lower degree of saturation and a more random arrangement of particles than the latter. Even at the same degree of saturation, a soil compacted on the dry side of optimum tends to swell more than one compacted on the wet side of optimum. Conversely, a sample compacted on the wet side of optimum shrinks more on drying than a soil, at the same density, which has been compacted on the dry side of optimum.

When expansive clays must be used as compaction materials, they should be compacted as wet as is practicable, consistent with compressibility requirements. Certainly, it would appear that compaction on the wet side of the optimum moisture content gives rise to lower amounts of swelling and swelling pressure. However, this produces fill with a lower strength and a higher compressibility. Therefore the choice of moisture content and dry density should not be based solely on lower swellability. Consideration should be given to the addition of various salts to reduce the swelling potential by increasing the ion concentration in the pore water.

A minimum permeability occurs in a compacted cohesive soil which is at or slightly above optimum moisture content, after which a slight increase in permeability takes place. The noteworthy reduction in permeability which occurs with an increase in moulding water on the dry side of optimum is brought about by an improvement in the orientation of the soil particles, which probably increases the tortuosity of flow, and by the decrease in the size of the largest flow channels. Conversely, on the wet side of the optimum moisture content, the permeability increases slightly, because the effects of decreasing dry density more than offset the effects of improved particle orientation. Increasing the compactive effect lowers the permeability because it increases both the

orientation of the particles and the compacted density.

The moisture content has a great influence on both the strength and compaction characteristics of silty soils. For example, an increase in moisture content of 1 or 2%, together with the disturbance due to spreading and compaction, can give rise to very considerable reductions in shear strength, making the material impossible to compact.

Because granular soils are relatively permeable, even when compacted, they are not affected significantly by their moisture content during the compaction process. In other words, for a given compactive effort, the dry density obtained is high when the soil is dry and high when the soil is saturated, with somewhat lower densities occurring when the soil contains intermediate amounts of water (Hilf, 1990). Moisture can be forced from the pores of granular soils by compaction equipment, and so a high standard of compaction can be obtained even if the material initially has a high moisture content. Normally, granular soils are easy to compact. However, if granular soils are uniformly graded, a high degree of compaction near the surface of the fill may prove difficult to obtain, particularly when vibratory rollers are used. This problem usually is resolved when the succeeding layer is compacted in that the loose surface of the lower layer is also compacted. According to Anon. (1981c), improved compaction of uniformly graded, granular material can be brought about by maintaining as high a moisture content as possible by intensive watering and by making the final passes at a higher speed using a non-vibratory smooth wheel roller or grid roller. When compacted, granular soils have a high load bearing capacity and are not compressible, and are not usually susceptible to frost action unless they contain a high proportion of fines. Unfortunately, however, if granular material contains a significant amount of fines, high pore water pressures can develop during compaction if the moisture content of the soil is high.

It is important to provide an adequate relative density in granular soil that may become saturated and subjected to static or, more particularly, dynamic shear stresses. For example, a quick condition might develop in granular soils with a relative density of less than 50% during ground accelerations of approximately 0.1g. On the other hand, if the relative density is greater than 75%, liquefaction is unlikely to occur for most earthquake loadings. Consequently, it has been suggested that, in order to reduce the risk of liquefaction, granular soils should be densified to a minimum relative density of 85% in the foundation area, and to at least 70% within the zone of influence of the foundation.

The most critical period during the construction of an embankment is just before it is brought to grade or shortly thereafter. At this time, pore water pressures, due to consolidation in the embankment and foundation, are at a maximum. The magnitude and distribution of pore water pressures developed during construction depend primarily on the construction water content, the properties of the soil, the height of the embankment and the rate at which dissipation by drainage can occur. Water contents above the optimum can cause high construction pore water pressures, which increase the danger of rotational slips in embankments. Referring to earth dams, Sherard *et al.* (1967) maintained that well graded clayey sands and sand–gravel–clay mixtures develop the highest construction pore water pressures, whereas uniform silts and fine silty sands are the least susceptible.

References

Anon. (1971) *Guide for Interpreting Engineering Uses of Soils.* United States Department of Agriculture, Washington, DC.

Anon. (1975a) *Soil Taxonomy. Agricultural Handbook No.* **436**. United States Department of Agriculture, Washington, DC.

Anon. (1975b) *Methods for Sampling and Testing Mineral Aggregates, Sands and Fillers* BS 812. British Standards Institution, London.

Anon. (1979) Classification of rocks and soils for engineering geological mapping. Part 1—Rock and soil materials. *Bulletin of the International Association of Engineering Geology* **19**, 364–371.

Anon. (1980) *Soil Map of the World*, Vol. 1, Legend. FAO–UNESCO, Paris.

Anon. (1981a) *Code of Practice on Site Investigation* BS 5930. British Standards Institution, London.

Anon. (1981b) *Speciation for Aggregates from Natural Sources for Concrete* BS 882. British Standards Institution, London.

Anon. (1981c) *Code of Practice on Earthworks* BS 6031. British Standards Institution, London.

Anon. (1989) *Design, Layout, Construction and Maintenance of Terrace Systems.* American Society of Agricultural Engineers, Standard S268.3. American Society of Agricultural Engineers, St Joseph, MI.

Anon. (1990) *Methods of Test for Soils for Civil Engineering Purposes* BS 1377. British Standards Institution, London.

Arnoldus, H.M.S. (1977) Predicting soil losses due to sheet and rill erosion. *Food and Agriculture Organization Conservation Guide.* Food and Agriculture Organization, New York, pp. 99–124.

Bell, F.G. (1992) An investigation of a site in Coal Measures for brickmaking materials: an illustration of procedures. *Engineering Geology* **32**, 39–52.

Casagrande, A. (1948) Classification and identification of soils. *Transactions of the American Society of Civil Engineers* **113**, 901–902.

Chepil, W.S. & Milne, R.A. (1941) Wind erosion of soil in relation to roughness of surface. *Soil Science* **52**, 417–431.

Chepil, W.S. & Woodruff, N.P. (1963) The physics of wind erosion and its control. *Advances in Agronomy* **15**, 211–302.

Chepil, W.S., Siddaway, F.H. & Armbrust, D.V. (1962) Climate factor for estimating wind erodibility of farm fields. *Journal of Soil and Water Conservation* **17**, 162–165.

Coppin, N.J. & Richards, I.G. (1990) *Use of Vegetation in Civil Engineering.* Construction Industry Research and Information Association (CIRIA), Butterworths, London.

Davidson, D.A. (1980) *Soils and Land Use Planning.* Longman, Harlow, Essex.

Dent, D. & Young, A. (1981) *Soil Survey and Land Evaluation.* Allen and Unwin, London.

Evans, R. (1980) Mechanics of water erosion and their spatial and temporal controls: an empirical viewpoint. In *Soil Erosion*, Kirby, M.J. & Morgan, R.P.C. (eds). Wiley, Chichester, pp. 109–128.

Foster, G.R., Lane, L.J., Nowlin, J.D., Lafler, J.M. & Young, R.A. (1981) Estimating erosion and sediment yield on field-sized areas. *Transactions of the American Society of Agricultural Engineers* **24**, 1253–1264.

Hagen, L.J. (1991) A wind erosion prediction system to meet user needs. *Journal of Soil and Water Conservation* **56**, 106–111.

Hilf, J.W. (1990) Compacted fill. In *Foundation Engineering Handbook*, Fang, H.S. (ed). Chapman and Hall, London.

Hudson, N.W. (1981) *Soil Conservation.* Batsford, London.

Kirkby, M.J. (1980) The problem. In *Soil Erosion*, Kirkby, M.J. & Morgan, R.P.C. (eds). Wiley, Chichester, pp. 1–16.

Klingebiel, A.A. & Montgomery, P.H. (1966) *Land Capability Classification. Soil Conservation Service Agricultural Handbook* **210**. United States Department of Agriculture, Washington, DC.

Liebenow, A.M., Elliot, W.J., Laflen, J.M. & Kohl, K.D.

(1990) Interrill erodibility: collection and analysis of data from cropland soils. *Transactions of the American Society of Agricultural Engineers* 33, 1882–1888.

Mitchell, J.K. & Bubenzer, G.D. (1980) Soil loss estimation. In *Soil Erosion*, Kirkby, M.J. & Morgan, R.C.P. (eds). Wiley, Chichester, pp. 17–62.

Morgan, R.P.C. (1980) Implications. In *Soil Erosion*, Kirkby, M.J. & Morgan, R.P.C. (eds). Wiley, Chichester, pp. 253–301.

Morgan, R.P.C. (1995) *Soil Erosion and Conservation*, 2nd edn. Longman, Harlow, Essex.

Rickson, R.J. (1995) Simulated vegetation and geotextiles. In *Slope Stabilization and Erosion Control: A Bioengineering Approach*, Morgan, R.P.C. & Rickson, R.J. (eds). Spon, London, pp. 95–131.

Sherard, J.L., Woodward, R.L., Gizienski, S.F. & Clevenger, W.A. (1967) *Earth and Earth Rock Dams*. Wiley, New York.

Skempton, A.W. (1953) The colloidal activity of clays. In *Proceedings of the Third International Conference on Soil Mechanics and Foundation Engineering, Zurich* 1, 57–60.

Skempton, A.W. & Northey, R.D. (1952) The sensitivity of clays. *Geotechnique* 2, 30–53.

Smith, D.D. & Wischmeier, W.H. (1962) Rainfall erosion. *Advances in Agronomy* 14, 109–148.

Stocking, M.A. & Elwell, H.A. (1973) Soil erosion hazard in Rhodesia. *Rhodesian Agricultural Journal* 70, 93–101.

Stocking, M.A. & Elwell, H.A. (1976) Rainfall erosivity over Rhodesia. *Transactions of the Institute of British Geographers, New Series* 1, 231–245.

Terzaghi, K. (1943) *Theoretical Soil Mechanics*. Wiley, New York.

Wagner, A.A. (1957) The use of the Unified Soil Classification System for the Bureau of Reclamation. In *Proceedings of the Fourth International Conference on Soil Mechanics and Foundation Engineering, London* 1, 125–134.

Watson, D.A. & Laflen, J.M. (1986) Soil strength, slope and rain intensity effects on interrill erosion. *Transactions of the American Society of Agricultural Engineers* 29, 98–102.

Wischmeier, W.H. (1959) A rainfall erosion index for a universal soil loss equation. *Proceedings of the Soil Science Society of America* 23, 246–249.

Wischmeier, W.H. & Smith, D.D. (1978) *Predicting Rainfall Erosion Losses. Agricultural Research Service Handbook* 537. United States Department of Agriculture, Washington, DC.

Woodruff, N.P. & Siddaway, F.H. (1965) A wind erosion equation. *Proceedings of the Soil Science Society of America* 29, 602–608.

11　Problem soils

Certain soils, because of their geotechnical properties, which are commonly related to their fabric and mineralogy, can be problematic or even regarded as hazards. These problems include, for example, the swelling and shrinkage of some clay soils, dispersivity in clay soils with particular chemical characteristics and the collapse potential of certain silty soils. Other soils, such as quicksands, quickclays and peat, lack strength; in particular, the first two, when disturbed, can give rise to major problems. The effects of these hazards, although unspectacular and rarely causing loss of life, can result in considerable financial loss. Swell–shrink in clay soils, for example, has caused losses of up to two billion pounds in recent years in the UK. The use of dispersive soils in southern Africa, Australia and the USA has led to the failure of dams and road embankments.

11.1 Quicksands

As water flows through silts and sands and loses head, its energy is transferred to the particles past which it is moving which, in turn, creates a drag effect on the particles. If the drag effect is in the same direction as the force of gravity, the effective pressure is increased and the soil is stable. Indeed, the soil tends to become more dense. Conversely, if water flows towards the surface, the drag effect is counter to gravity, thereby reducing the effective pressure between the particles. If the velocity of upward flow is sufficient, it can buoy up the particles, so that the effective pressure is reduced to zero. This represents a critical condition in which the weight of the submerged soils is balanced by the upward-acting seepage force. A critical condition sometimes occurs in silts and sands. If the upward velocity of flow increases beyond the critical hydraulic gradient, a quick condition develops.

Quicksands, if subjected to deformation or disturbance, can undergo a spontaneous loss of strength. This loss of strength causes them to flow like viscous liquids. Terzaghi (1925) explained the quicksand phenomenon in the following terms. First, the sand or silt concerned must be saturated and loosely packed. Second, on disturbance, the constituent grains become more closely packed, which leads to an increase in pore water pressure, reducing the forces acting between the grains. This brings about a reduction in strength. If the pore water can escape very rapidly, the loss in strength is momentary. Hence, the third condition requires that pore water cannot escape readily. This is fulfilled if the sand or silt has a low permeability and/or the seepage path is long. Casagrande (1936) demonstrated that a critical porosity exists above which a quick condition can be developed. He maintained that many coarse grained sands, even when loosely packed, have porosities approximately equal to the critical condition, whereas medium and fine grained sands, especially if uniformly graded, exist well above the critical porosity when loosely packed. Accordingly, fine sands tend to be potentially more unstable than coarse grained varieties. It must also be remembered that finer sands have lower permeabilities.

Quick conditions brought about by seepage forces frequently are encountered in excavations made in fine sands which are below the water table, for example, in coffer-dam work. As the velocity of the upward seepage force increases further from the critical gradient, the soil begins to boil more and more violently (Fig. 11.1). At such a point, structures fail by sinking into the quicksand. The liquefaction of potential quicksands may be brought about by sudden shocks caused by the action of heavy machinery (notably pile driving), blasting and earthquakes (see Chapter 3). Such shocks increase the stress carried by the water, the neutral stress, and give rise to

Fig. 11.1 Liquefaction of sand due to Tangshan earthquake (July 28, 1976) near Dougtian, Fengnan County, China.

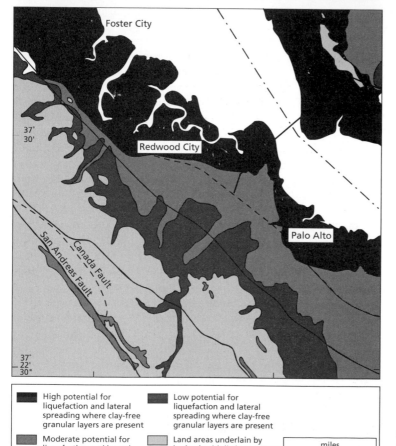

	High potential for liquefaction and lateral spreading where clay-free granular layers are present		Low potential for liquefaction and lateral spreading where clay-free granular layers are present
	Moderate potential for liquefaction and lateral spreading where clay-free granular layers are present		Land areas underlain by bedrock with little potential for liquefaction

miles
0   1   2   3

0 1 2 3 4
km

Fig. 11.2 Map of the relative potential for liquefaction in terms of the response of ground conditions to seismic loading in the area around Palo Alto, California. (From Blair & Spangle, 1979.)

a decrease in the effective stress and shear strength of the soil. There is also a possibility of a quick condition developing in a layered soil sequence, where the individual beds have different permeabilities. Hydraulic conditions particularly are unfavourable where water initially flows through a very permeable horizon with little loss of head, which means that flow takes place under a large hydraulic gradient. Maps showing the degree of liquefaction hazard have been produced, for example, for the San Francisco Bay area. The zones are based on the likely response of the surface materials to seismic loading (Fig. 11.2). Restriction on the location of certain types of building within particular zones can mean that severe damage due to liquefaction can be avoided.

There are several methods which may be employed during construction to avoid the development of quick conditions. One of the most effective techniques is to prolong the length of the seepage path, thereby increasing the frictional losses and so reducing the seepage force. This can be accompanied by placing a clay blanket at the base of an excavation where seepage lines converge. If sheet piling is used in excavation with critical soils, the depth to which it is sunk determines whether or not quick conditions will develop. Consequently, it should be sunk deep enough to avoid a potential critical condition occurring at the base level of the excavation. The hydrostatic head can also be reduced by means of relief wells, and seepage can be intercepted by a wellpoint system placed about the excavation. Furthermore, a quick condition may be prevented by increasing the downward-acting force. This may be brought about by laying a load on the surface of the soil where seepage

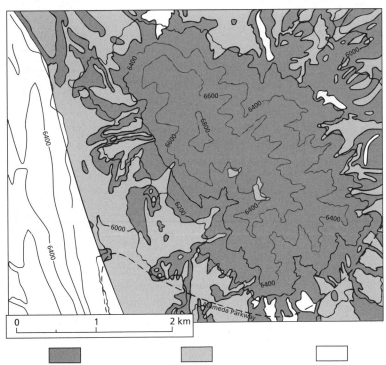

Fig. 11.3 Map showing areas of swelling clay in the Morrison Quadrangle, Colorado. (Courtesy of the United States Geological Survey, Washington, DC.)

Areas underlain by geological units that contain clays having swelling pressures higher than 2500 lb/ft² (120 kPa) PVC (potential volume change). Preconstruction investigations in these areas should include engineering laboratory swell–shrink tests

Area underlain by 5 or more feet of non-swelling surficial deposits which in turn are underlain by geological units which may have swelling pressures higher than 2500 lb/ft² (120 kPa) PVC. For any construction involving foundation excavation through the surficial deposits, preconstruction investigations should include engineering laboratory swell–shrink tests

Areas underlain by geological units having no or slight swelling pressures. Preconstruction swell–shrink tests unnecessary

is discharging. Gravel filter beds may be used for this purpose. Suspect soils can also be densified, treated with stabilizing grouts or frozen.

Subsurface structures should be designed to be stable with regard to the highest groundwater level that is likely to occur. Structures below groundwater level are acted upon by uplift pressures. If the structure is weak, this pressure can break it and, for example, cause a blow-out of a basement floor or the collapse of a basement wall. If the structure is strong, but light, it may be lifted, that is, subjected to heave. Uplift can be taken care of by adequate drainage or by resisting the upward seepage force. Continuous drainage blankets are effective, but should be designed with filters to function without clogging. The entire weight of structure can be mobilized to resist uplift if a raft foundation is used. Anchors, grouted into bedrock, can provide resistance to uplift.

11.2 Expansive clays

Some clay soils undergo slow volume changes, which occur independent of loading and are attributable to swelling or shrinkage (see also Vertisols, e.g. black cotton soils, Section 11.7). These volume changes can give rise to ground movements, which may result in damage to buildings. Low-rise buildings particularly are vulnerable to such ground movements, because they generally do not have sufficient weight or strength to resist. In addition, shrinkage settlement of embankments can lead to cracking and break-up of the roads they support. Maps delineating areas of expansive clay can be produced in order to facilitate land-use planning (Fig. 11.3).

Problems caused by expansive soils in the USA give rise to costs of over two billion dollars each year. This frequently is twice the cost of flood damage or damage caused by landslides, and more than 20 times the cost of earthquake damage. The principal cause of expansive clays is the presence of swelling clay minerals, such as montmorillonite. Construction damage especially is notable where expansive clay forms the surface cover in regions which experience alternating wet and dry seasons, leading to swelling and shrinkage of these soils (Fig. 11.4). Cyclic movements of the ground surface can be between 100 and 200 mm.

Differences in the period and amount of precipitation and evapotranspiration are the principal factors influencing the swell–shrink response of a clay soil

beneath a building. Poor surface drainage or leakage from underground pipes can also produce concentrations of moisture in clay. Trees with high water demand and uninsulated hot process foundations may dry out clay, causing shrinkage. Cold stores may also cause desiccation of clay soil. The depth of the active zone in expansive clays (i.e. the zone in which swelling and shrinkage occur in wet and dry seasons, respectively) varies. It may extend to over 6 m in depth in some semi-arid regions of South Africa, Australia and Israel. Many soils in temperate regions, such as the UK, especially in southeast England, possess the potential for significant volume changes due to changes in moisture content. However, owing to the damp climate in most years, volume changes are restricted to the upper 1.0–1.5 m in clay soils.

Fig. 11.4 Break-up of a road in Austin, Texas, due to expansive clay.

The potential for volume changes in clay soil is governed by the initial moisture content, initial density or void ratio, microstructure and vertical stress, as well as the type and amount of clay minerals present. Cemented and undisturbed expansive clay soils often have a high resistance to deformation, and may be able to absorb significant amounts of swelling pressure. Remoulded expansive clays therefore tend to swell more than their undisturbed counterparts. For example, Schmertmann (1969) maintained that some clays increase their swelling behaviour when they undergo repeated large shear strains due to mechanical remoulding. He introduced the term swell sensitivity for the ratio of the remoulded swelling index to the undisturbed swelling index.

Expansive clay minerals take water into their lattice structure. In less dense soils, they tend to expand initially into zones of looser soil before a volume increase occurs. However, in densely packed soil with a low void space, the soil mass has to swell more or less immediately to accommodate the volume change. Hence, clay soils with a flocculated fabric swell more than those which possess a preferred orientation. In the latter, the maximum swelling occurs normal to the direction of clay particle orientation.

Because expansive clays normally possess extremely low permeabilities, moisture movement is slow, and a significant period of time may be involved in the swelling–shrinking process. Accordingly, moderately expansive clays with a smaller potential to swell, but with higher permeabilities than clays having a greater swell potential, may swell more during a single wet season than more expansive clays.

Expansive clays often are heavily fissured due to seasonal changes in volume, which produce shrinkage cracks and shear surfaces. Consequently, near-vertical fissures frequently are found at shallow depth, with diagonal fissures at greater depth. Sometimes, the soil is so desiccated that the fissures are wide open and the soil is shattered or microshattered. The presence of desiccation cracks enhances evaporation from the soil.

The swell–shrink behaviour of a clay soil under a given state of applied stress in the ground is controlled by changes in soil suction. The relationship between soil suction and water content depends on the proportion and type of clay minerals present, their microstructural arrangement and the chemistry of the pore water. Changes in soil suction are brought about by moisture movement through the soil due to evaporation from its surface in dry weather, transpiration from plants or recharge consequent upon precipitation. The climate governs the amount of moisture available to counteract that which is removed by evapotranspiration (i.e. the soil moisture deficit). In semi-arid climates, there are long periods of high soil moisture deficit, alternating with short periods when precipitation balances or exceeds evapotranspiration.

The volume changes which occur due to evapotranspiration from clay soils can be conservatively predicted by assuming the lower limit of the soil moisture content to be the shrinkage limit. Desiccation beyond this value cannot bring about further volume change.

Transpiration from vegetative cover is a major cause of water loss from soils in semi-arid regions. Indeed, the distribution of soil suction in soil primarily is controlled by transpiration from vegetation, and represents one of the most significant changes made in loading (i.e. to the state of stress in a soil). The behaviour of root systems is exceedingly complex, and is a major factor in the intractability of swelling and shrinking problems. The spread of root systems depends on the type of vegetation, soil type and groundwater conditions. The suction induced by the withdrawal of water fluctuates with the seasons, reflecting the growth of vegetation, and probably varies between 100 and 1000 kPa (equivalent to pF = 3 and pF = 4, respectively). The complete depth of the active clay profile usually does not become fully saturated during the wet season in semi-arid regions. Nonetheless, changes in soil suction may occur over a depth of 2.0 m or so between the wet and dry seasons. The suction pressure associated with the onset of cracking is approximately pF = 4.6.

The extent to which the vegetation is able to increase the suction to the level associated with the shrinkage limit obviously is important. In fact, the moisture content at the wilting point exceeds that of the shrinkage limit in soils with a high content of clay, and is less in soils possessing low clay contents. This explains why settlement resulting from the desiccating effects of trees is more notable in low to moderately expansive soils than in expansive soils.

When vegetation is cleared from a site, its desiccating effect is also removed. Hence, the subsequent regain of moisture by clay soils leads to swelling. Swelling movements on expansive clays in South Africa, associated with the removal of vegetation and the subsequent erection of buildings, have amounted to about 150 mm in many areas, although movements over 350 mm have been recorded (Williams, 1980).

Methods of predicting volume changes in soils can be grouped into empirical methods, soil suction methods and oedometer methods. Empirical methods make use of the swelling potential as determined from the void ratio, natural moisture content, liquid and plastic limits, and activity. However, because the determination of plasticity is carried out on remoulded soil, it does not consider the influence of soil texture, moisture content, soil suction or pore water chemistry, which are important factors in relation to the volume change potential. Too much reliance on the results of such tests therefore must be avoided. Consequently, empirical methods should be regarded as simple swelling indicator methods and nothing more.

Soil suction methods use the change in suction from the initial to the final conditions to obtain the degree of volume change. Soil suction is the stress which, when removed, allows the soil to swell. In other words, the value of soil suction in a saturated, fully swollen soil is zero. O'Neill and Poormoayed (1980) quoted the United States Army Engineers Waterways Experimental Station (USAEWES) classification of swell potential (Table 11.1), which is based on the liquid limit, plasticity index and initial (*in situ*) suction. The latter is measured in the field by a psychrometer.

The oedometer methods of determining the potential expansiveness of clay soils represent more direct methods. As most expansive clays are fissured, this means that lateral and vertical strains develop locally within the ground. Even when the soil is intact, swelling or shrinkage is not truly one dimensional. The effect of imposing zero lateral strain in the oedometer therefore is likely to give rise to overpredictions of heave, and the greater the degree of fissuring, the greater the overprediction.

Effective and economic foundations for low-rise buildings on swelling and shrinking soils have proved to be difficult to achieve. This partly is because the cost margins on individual buildings are low. Obviously, detailed site investigation and soil testing are out of the question for individual dwellings. Similarly, many foundation solutions which are appropriate for major structures are too costly for small buildings. Nonetheless, the choice of foundation is influenced by the subsoil and site conditions, estimates of the amount of ground movement and the cost of alternative designs. In addition, different building materials have different tolerances to deflections. Hence, materials which are more flexible can be used to reduce potential damage due to differential movement of the structure.

Various methods can be adopted when choosing a design solution for building on expansive soils. The isolation of foundation and structure has been widely adopted for 'severe' and 'very severe' ground conditions. Straight-shafted bored piles can be used in conjunction with suspended floors for severe conditions. The piles are sleeved over the upper part. The use of stiffened rafts for low-rise buildings is fairly commonplace (Bell & Maud, 1995).

A simple method of reducing or eliminating ground movements due to expansive soil is to replace or partially replace it with non-expansive granular soils. A geomembrane should surround the granular material to prevent water entering the soil. If expansive soil is allowed to swell by wetting prior to construction, and if the soil moisture content is then maintained, the soil volume should remain relatively constant and no

Table 11.1 USAEWES classification of swell potential. (From O'Neill & Poormoayed, 1980.)

Liquid limit (%)	Plastic limit (%)	Initial (*in situ*) suction (kN m^{-2})	Potential swell (%)	Classification
Less than 50	Less than 25	Less than 145	Less than 0.5	Low
50–60	25–35	145–385	0.5–1.5	Marginal
Over 60	Over 35	Over 385	Over 1.5	High

Fig. 11.5 Lime slurry pressure injection. (Courtesy of Hayward Baker, Fort Worth, Texas.)

heave should take place. Williams (1980) described a case in which severe damage due to swelling was corrected by controlled wetting.

Many attempts have been made to reduce the expansiveness of clay soil by chemical stabilization. For example, lime stabilization of expansive soils, prior to construction, can minimize the amount of shrinkage and swelling that they undergo. In the case of light structures, lime stabilization may be applied immediately below strip footings. However, significant SO_4 content (i.e. in excess of $5000\,mg\,kg^{-1}$) in clay soils can mean that they react with CaO to form ettringite, with resultant expansion (Forster *et al.*, 1995). The treatment is better applied as a layer beneath a raft so as to overcome differential movement. The lime-stabilized layer is formed by mixing 4–6% lime with the soil. A compacted layer, $150\,mm$ in thickness, usually gives a satisfactory performance. Premix or mix-in-place methods can be used (Bell, 1993). Alternatively, lime treatment can be used to form a vertical cutoff wall at or near the footings in order to minimize the movement of moisture. The lime slurry pressure injection method has also been used to minimize differential movements beneath structures, although it is more expensive (Fig. 11.5). The method involves pumping hydrated lime slurry under pressure into the soil, the points of injection being spaced about $1.5\,m$ apart. The lime slurry forms a network of horizontal sheets, often interconnected by vertical veins. Cement stabilization has much the

same effect on expansive soils as lime treatment, although the dosage of cement needs to be greater for heavy expansive clays.

11.3 Dispersive soils

Dispersion occurs in soils found in semi-arid regions when the repulsive forces between clay particles exceed the attractive forces, thus bringing about deflocculation, so that, in the presence of relatively pure water, the particles repel each other. In non-dispersive soil, there is a definite threshold velocity below which flowing water causes no erosion. The individual particles cling to each other, and are only removed by water flowing with a certain erosive energy. By contrast, there is no threshold velocity for dispersive soil; the colloidal clay particles go into suspension even in quiet water, and therefore dispersive soils are highly susceptible to erosion and piping. For a given eroding fluid, the boundary between the flocculated and deflocculated states depends on the value of the sodium adsorption ratio (SAR) (see Chapter 7), the salt concentration, the pH value and the mineralogy.

Nonetheless, there are no significant differences in the clay fractions of dispersive and non-dispersive soils, except that soils with less than 10% clay particles may not have enough colloids to support dispersive piping. Dispersive soils contain a higher content of dissolved sodium (up to 12%) in their pore water than ordinary soils. The pH value of dispersive soils generally ranges between pH 6 and pH 8.

The *SAR* is used to quantify the role of sodium where free salts are present in the pore water. There is a relationship between the electrolyte concentration of the pore water and the exchangeable ions in the adsorbed layers of clay particles. This relationship is dependent upon the pH value, and may also be influenced by the type of clay minerals present. Hence, it is not necessarily constant. An *SAR* value above six suggests that the soil is sensitive to leaching. However, in Australia, Aitchison and Wood (1965) regarded soils in which the *SAR* value exceeded two to be dispersive. Subsequently, Bell and Maud (1994) suggested that the latter value seems to be more appropriate.

The presence of exchangeable sodium is the main chemical factor contributing towards dispersive behaviour in soils. This is expressed in terms of the exchangeable sodium percentage (*ESP*)

$$ESP = \frac{\text{Exchangeable sodium}}{\text{Cation exchange capacity}} \times 100 \qquad (11.1)$$

where the units are given in milli-equivalents per 100g of dry clay. A threshold value of *ESP* of 10%

has been recommended, above which soils that have their free salts leached by the seepage of relatively pure water are prone to dispersion. Soils with *ESP* values above 15% are highly dispersive (Bell & Maud, 1994). Those with low cation exchange values (15 meq per 100g of clay) have been found to be completely non-dispersive at *ESP* values of 6% or below. Similarly, soils with high cation exchange capacity (*CEC*) values and a plasticity index greater than 35% swell to such an extent that dispersion is not significant. High *ESP* values and piping potential generally exist in soils in which the clay fraction is composed largely of smectite and other 2:1 clays. Some illites are highly dispersive. On the other hand, high values of *ESP* and high dispersibility are rare in clays composed largely of kaolinite.

Another property which has been claimed to govern the susceptibility of clayey soils to dispersion is the total content of dissolved salts (*TDS*) in the pore water. In other words, the lower the content of dissolved salts in the pore water, the greater the susc-

Fig. 11.6 Potential dispersivity chart of Sherard *et al.* (1976), with examples of soil from Natal. (After Bell & Maud, 1994.)

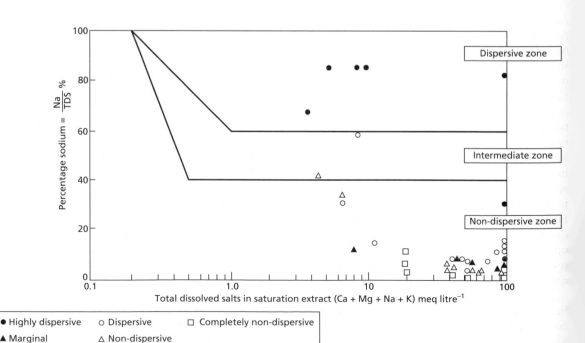

• Highly dispersive ○ Dispersive □ Completely non-dispersive
▲ Marginal △ Non-dispersive

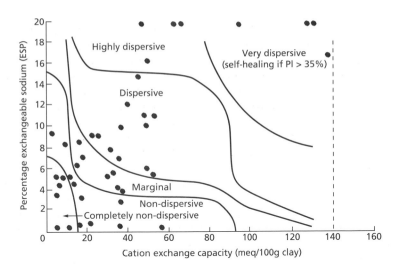

Fig. 11.7 Procedure for the classification of dispersive soils according to Gerber and von Harmse (1987), showing examples of soil from Natal. (After Bell & Maud, 1994.)

eptibility of sodium-saturated clays to dispersion. Sherard *et al.* (1976) regarded the total dissolved salts for this specific purpose as the total content of calcium, magnesium, sodium and potassium in milliequivalents per litre. They designed a chart in which the sodium content was expressed as a percentage of *TDS*, and was plotted against *TDS* to determine the dispersivity of soils (Fig. 11.6). However, Craft and Acciardi (1984) showed that this chart exhibited poor overall agreement with the results of physical tests. In South Africa, the determination of the dispersive potential frequently involves the use of a chart designed by Gerber and von Harmse (1987), which plots *ESP* against *CEC* (Fig. 11.7).

Unfortunately, dispersive soils cannot be differentiated from non-dispersive soils by routine soil mechanics testing, for example, they cannot be identified by their plasticity or activity. Although a number of tests have been used to recognize dispersive soils, no single test can be relied upon completely to identify them. For example, Gerber and von Harmse (1987) showed that the crumb test, the double hydrometer test and the pin-hole test were unable to identify dispersive soils when free salts were present in solution in the pore water, which frequently is the case with sodium-saturated soils.

Nonetheless, various tests show that the boundary between deflocculated and flocculated states varies considerably between different soils, so that the transition between dispersive and non-dispersive soils is wide. Because there is no clearly defined boundary between dispersive and non-dispersive soils, it is wise to perform several types of test on at least a proportion of all samples. Some properties of dispersive soils from Natal, South Africa, are given in Table 11.2.

Serious piping damage to embankments and failures of earth dams have occurred when dispersive soils have been used in their construction. Severe erosion damage can also form deep gullies on earth embankments after rainfall. Indications of piping take the form of small leakages of muddy-coloured water from an earth dam after initial filling of the reservoir. The pipes become enlarged rapidly, and this can lead to the failure of a dam (Fig. 11.8). Dispersive erosion may be caused by initial seepage through an earth dam in areas of higher soil permeability, especially areas in which compaction may not be so effective, such as around conduits, against concrete structures and at the foundation interface, and through desiccation cracks, cracks due to differential settlement or cracks due to hydraulic fracturing (Wilson & Melis, 1991).

In many areas in which dispersive soils are found, there is no economic alternative other than to use these soils for the construction of earth dams. Experience, however, indicates that, if an earth dam is built with careful construction control and incorporates filters, it should be safe enough even if it is constructed with dispersive clay. Alternatively, hydrated lime, pulverized fly ash, gypsum and aluminium sul-

Table 11.2 Some physical and chemical properties of dispersive soils from Natal. (From Bell & Maud, 1994.)

Location	Clay (%)	LL (%)	PI (%)	A	LS (%)	pH	K (meq l⁻¹)	Ca (meq l⁻¹)	Mg (meq l⁻¹)	Na (meq l⁻¹)	TDS (meq l⁻¹)	ESP (%)	SAR	CEC (meq per 100 g clay)	EC (mS m⁻¹)	Dispersivity (%)	Dispersivity potential
Makatini	30	28	14	0.47	6.0	8.6	2.6	9.0	8.0	97.6	117.2	39.4	33.4	97.9	900		HD
Makatini	29	43	24	0.83	8.0	8.6	2.6	181.9	78.9	115.9	379	30	10.1	132.86			HD
Paddock	40	31	12	0.30	5.3	8.6	1.6	95.6	147.2	23.5	268.2	89	2.1	66.3		44	HD
Paddock	44	33	14	0.32	6.0	8.6	0.02	0.3	0.4	4.4	5.1	16.6	7.5	60.2	52		HD
Winterton	31	27	13	0.42	4.0	8.6	0.03	0.8	0.4	7.7	8.9	24.4	9.9	50.9	95	71.9	HD
Winterton	44	35	10	0.23	5.3	8.95	0.04	0.8	0.5	8.3	9.6	16.1	10.2	36.6	110	53.2	HD
Winterton	37	28	13	0.35	5.3	7.5	0.04	1.9	1.5	4.8	8.2	6.3	3.7	53.5	81	35.4	D
Rietspruit	36	47	24	0.67	8.0	6.5	2.4	71.6	46.9	13.1	134	9.3	1.7	38.8			D
Rietspruit	60	66	34	0.57	10.0	7.55	1.8	160.4	139.1	17.1	318.4	5.3	1.4	53.8			D
Tala	22	45	6	0.27	0.7	5.05	5.3	33.7	4.8	4.8	48.6	8.8	1.1	25.9			D
Weenen	51	34	16	0.31	7.0	8.3	0.9	12.7	5.3	2.3	21.2	9.1	1.1	25.1			D
Ramsgate	18	29	8	0.44	5	5.2	0.9	22.0	46.7	5.4	75	7	0.9	38	77.5		D

LL, liquid limit; PI, plasticity index; A, activity; LS, linear shrinkage; K, potassium; Ca, calcium; Mg, magnesium; Na, sodium; TDS, total dissolved solids; ESP, exchangeable sodium percentage; SAR, sodium adsorption ratio; CEC, cation exchange capacity; EC, electrical conductivity; HD, highly dispersive; D, dispersive (according to the method of Gerber & von Harmse, 1987).

Fig. 11.8 Failure of small earth dam constructed of dispersive soil, which was subjected to piping, near Ramsgate, Natal, South Africa. Note the pipes on both the right and left flanks of the dam.

phate have been used to treat dispersive clays used in earth dams. The type of stabilization undertaken depends on the properties of the soil, especially the *ESP* and *SAR* values.

Dispersive soils can also present problems in earthworks, such as those required for roads on both the fill and cut slopes. In the case of embankment fills, dispersive soil can be used provided that it is covered by an adequate depth of better class material. Care must be exercised in the placement and compaction of the fill layers, so that no layer is left exposed during construction for such a period that it can shrink and crack, and thus weaken the fill. As in the case of the construction of earth dams, the dispersive soil material should be placed and compacted at 2% above its optimum moisture content to inhibit shrinkage and cracking. In some instances, to reduce erosion, it is prudent to stabilize the outer 0.3 m or so of dispersive soil material in an embankment with lime, the soil material and lime being mixed in bulk prior to placement, rather than being mixed *in situ*.

11.4 Collapsible soils

Soils such as loess (Fig. 11.9), brickearth and certain windblown silts may possess the potential to collapse. These soils generally consist of 50–90% silt particles, and sandy, silty and clayey types have been recognized by Clevenger (1958), with most falling into the silty category (Fig. 11.10). Collapsible soils possess porous textures with high void ratios and relatively low densities. They often have sufficient void space in their natural state to hold their liquid limit moisture at saturation. At their natural low moisture content, these soils possess high apparent strength, but they are susceptible to large reductions in void ratio upon wetting. In other words, the metastable texture collapses as the bonds between the grains break down when the soil is wetted. Hence, the collapse process represents a rearrangement of soil particles into a denser state of packing. Collapse on saturation normally only takes a short period of time, although the more clay such a soil contains, the longer the period tends to be.

The fabric of collapsible soils generally takes the form of a loose skeleton of grains (generally quartz) and microaggregates (assemblages of clay or clay and silty clay particles). These tend to be separate from each other, being connected by bonds and bridges, with uniformly distributed pores. The bridges are formed of clay-sized minerals, consisting of clay minerals, fine quartz, feldspar or calcite. Surface coatings of clay minerals may be present on coarser grains. Silica and iron oxide may be concentrated as cement at the grain contacts, and amorphous overgrowths of silica occur on the grains. As the grains are not in contact, the mechanical behaviour is governed by the structure and quality of the bonds and bridges.

Fig. 11.9 Loess escarpment of the Loess Plateau, Shaanxi Province, China.

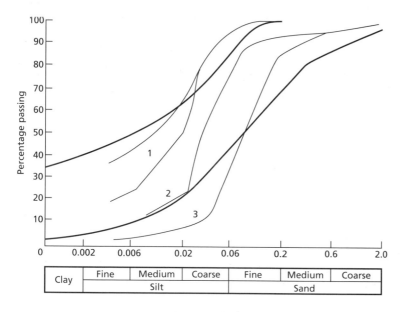

Fig. 11.10 Particle size distribution of brickearth from Essex compared with clayey (1), silty (2) and sandy (3) loess as recognized by Clevenger (1958). (After Northmore et al., 1996.)

The structural stability of collapsible soils is not only related to the origin of the material, its mode of transport and the depositional environment, but also to the amount of weathering undergone. For instance, Gao (1988) pointed out that the weakly weathered loess of the northwest of the Loess Plateau in China has a high potential for collapse, whereas the weathered loess of the southeast of the Loess Plateau is relatively stable. Moreover, in more finely textured loess deposits, the high capillary potential plus high perched groundwater conditions have caused loess to collapse naturally through time, thereby reducing its porosity. This reduction in porosity, combined with the high liquid limit, makes the possibility of collapse less likely. Gao concluded that, usually, highly collapsible loess occurs in regions near the source of the loess, where its thickness is at a maximum, and where the landscape and/or the climate are not conducive to the development of long-term saturated conditions within the soil. This view was supported by Grabowska-Olszewska (1988), who found that, in Poland, collapse is most frequent in the youngest loess, and is restricted almost exclusively to loess which contains slightly more than 10% of clay-sized particles. Such soils are characterized by a random texture and a carbonate content of less than 5%. They are more or less unweathered, and possess a pronounced pattern of vertical jointing. The size of the pores is all important, Grabowska-Olszewska maintaining that collapse occurs as a result of pore space reduction taking place in pores greater than 1 μm in size and, more especially, in those exceeding 10 μm in size. Phien-Wej *et al.* (1992) reported a pore size in loess in northeast Thailand frequently varying from 200 to 500 μm.

Popescu (1986) maintained that there is a limiting value of pressure, defined as the collapse pressure, beyond which the deformation of soil increases appreciably. The collapse pressure varies with the degree of saturation. He defined truly collapsible soils as those in which the collapse pressure is less than the overburden pressure. In other words, such soils collapse when saturated, because the soil fabric cannot support the weight of the overburden. When the saturation collapse pressure exceeds the overburden pressure, soils are capable of supporting a certain level of stress on saturation, and Popescu defined these soils as conditionally collapsible soils. The maximum load which such soils can support is the difference between the saturation collapse and overburden pressures.

Several collapse criteria have been proposed for predicting whether a soil is liable to collapse upon saturation and loading. For instance, Gibbs and Bara (1962) suggested the use of the dry unit weight and the liquid limit as criteria to distinguish between collapsible and non-collapsible soil types. Their method is based on the assumption that a soil which has enough void space to hold its liquid limit moisture content at saturation is susceptible to collapse on wetting. This criterion only applies if the soil is uncemented and the liquid limit is above 20%. When the liquidity index in such soils approaches or exceeds unity, collapse may be imminent. As the clay content of a collapsible soil increases, the saturation moisture content becomes less than the liquid limit, so that such deposits are relatively stable. However, North-

more *et al.* (1996) concluded that this method did not provide a satisfactory means of identifying the potential metastability of brickearth.

Collapse criteria have been proposed, which depend upon the void ratios at the liquid limit (e_l) and plastic limit (e_p) and the natural void ratio (e_o). For example, Fookes and Best (1969) proposed a collapse index (i_c), which involved these void ratios as follows:

$$i_c = \frac{e_o - e_p}{e_l - e_p} \tag{11.2}$$

Previously, Feda (1966) had proposed the following collapse index:

$$i_c = \frac{m/S_r - PL}{PI} \tag{11.3}$$

in which m is the natural moisture content, S_r is the degree of saturation, PL is the plastic limit and PI is the plasticity index. Feda also proposed that the soil must have a critical porosity of 40% or above, and an imposed load must be sufficiently high to cause structural collapse when the soil is wetted. He suggested that, if the collapse index was greater than 0.85, this was indicative of metastable soils. However, Northmore *et al.* (1996) suggested that a lower critical value of the collapse index, that is, 0.22, was more appropriate for the brickearths of south Essex. Derbyshire and Mellors (1988) also referred to a lower collapse index for the brickearths of Kent. This may be due to the greater degree of sorting of brickearths than loess.

The oedometer test can be used to assess the degree of collapsibility. The test involves the loading of an undisturbed specimen at natural moisture content in the oedometer up to a given load. At this point, the specimen is flooded, and the resulting collapse strain, if any, is recorded. The specimen is then subjected to further loading. The total consolidation upon flooding can be described in terms of the coefficient of collapsibility (C_{col}), given by Feda (1988) as

$$C_{col} = \Delta h / h \tag{11.4}$$
$$= \frac{\Delta e}{1 + e}$$

in which Δh is the change in height of the specimen after flooding, h is the height of the specimen before flooding, Δe is the change in void ratio of the specimen upon flooding and e is the void ratio of the speci-

Table 11.3 Collapse percentage as an indication of potential severity. (After Jennings & Knight, 1975.)

Collapse (%)	Severity of problem
0–1	No problem
1–5	Moderate trouble
5–10	Trouble
10–20	Severe trouble
Above 20	Very severe trouble

Table 11.4 Methods of treating collapsible foundations. (Based on Clemence & Finbarr, 1981.)

Depth of subsoil treatment (m)	Foundation treatment
	Current and past methods
0–1.4	Moistening and compaction (conventional extra heavy impact or vibratory rollers)
1.5–10	Overexcavation and recompaction (earth pads with or without stabilization by additives such as cement or lime). Vibroflotation (free-draining soils). Vibroreplacement (stone columns). Dynamic compaction. Compaction piles. Injection of lime. Lime piles and columns. Jet grouting. Ponding or flooding (if no impervious layer exists). Heat treatment to solidify the soils in place
Over 10	Any of the aforementioned or combinations of the aforementioned, where applicable. Ponding and infiltration wells, or ponding and infiltration wells with the use of explosive
	Possible future methods
	Ultrasonics to produce vibrations that will destroy the bonding mechanics of the soil. Electrochemical treatment. Grouting to fill pores

men prior to flooding. Table 11.3 provides an indication of the potential severity of collapse. This table indicates that those soils which undergo more than 1% collapse can be regarded as metastable. However, in China, a value of 1.5% is taken (Lin & Wang, 1988) and, in the USA, values exceeding 2% are regarded as indicative of soils susceptible to collapse (Lutenegger & Hallberg, 1988).

From the above, it may be concluded that significant settlements can take place beneath structures in collapsible soils after they have been wetted (in some cases of the order of metres; Feda *et al.*, 1993). These have led to foundation failures (Clevenger, 1958; Phien-Wej *et al.*, 1992). Clemence and Finbarr (1981) recorded a number of techniques which could be used to stabilize collapsible soils. These are summarized in Table 11.4. Evstatiev (1988) also provided a survey of methods which could be used to improve the behaviour of collapsible soils.

Another problem which may be associated with loess soils is the development of pipe systems. Extensive pipe systems, which have been referred to as loess karst, may run subparallel to a slope surface, and the pipes may have diameters up to 2.0 m. Pipes tend to develop by the action of weathering and widening along the joint systems in loess. The depths to which pipes develop may be inhibited by changes in permeability associated with the occurrence of palaeosols.

11.5 Quickclays

The material which makes up quickclays is predominantly smaller than 0.002 mm, but many deposits seem to be very poor in clay minerals, containing a high proportion of ground down, fine quartz. For instance, it has been shown that quickclay from St Jean Vienney, Canada, consists of very fine quartz and plagioclase. Indeed, the examination of quickclays with a scanning electron microscope has revealed that they do not possess clay-based structures, although such work has not lent unequivocal support to the view that non-clay particles govern the physical properties.

Cabrera and Smalley (1973) suggested that such deposits owe their distinctive properties to the predominance of short-range interparticle bonding forces, which they maintained were characteristic of deposits in which there was an abundance of glacially produced, fine, non-clay minerals. In other words, they contended that ice sheets had supplied abundant ground quartz in the form of rock flour for the formation of quickclays. Certainly, quickclays have a restricted geographical distribution, occurring in certain parts of the northern hemisphere, which were subjected to glaciation during Pleistocene times.

Gillott (1979) has shown that the fabric and mineralogical composition of sensitive soils from Canada, Alaska and Norway are qualitatively similar. He pointed out that they possess an open fabric, high moisture content and similar index properties. An examination of the fabric of these soils revealed the presence of aggregations. Granular particles, whether aggregations or primary minerals, are rarely in direct contact, being linked generally by bridges between the particles. Clay minerals usually are non-orientated, and clay coatings on primary minerals tend to be uncommon, as are cemented junctions. Networks of platelets occur in some soils. Primary minerals, particularly quartz and feldspar, form a higher than normal proportion of the clay-sized fraction, and illite and chlorite are the dominant phyllosilicate minerals. Gillott noted that the presence of swelling clay minerals varies from almost zero to significant amounts.

The open fabric which is characteristic of quickclays has been attributed to their initial deposition, when colloidal particles interacted to form loose aggregations by gelation and flocculation. Clay minerals exhibit strongly marked colloidal properties, and other inorganic materials, such as silica, behave as colloids when sufficiently fine grained. Gillott (1979) suggested that the open fabric may have been retained during very early consolidation, because it remained a near-equilibrium arrangement. Its subsequent retention to the present day may be due to the mutual interference between particles and the buttressing of junctions between granules by clay and other fine constituents, the precipitation of cement at particle contacts, low rates of loading and low load increment ratio.

Quickclays often exhibit little plasticity, their plasticity indices at times varying between 8 and 12%. Their liquidity index normally exceeds unity, and their liquid limit often is less than 40%. Quickclays usually are inactive, their activity frequently being less than 0.5. The most extraordinary property possessed by quickclays is their very high sensitivity (Fig. 11.11). In other words, a large proportion of their undisturbed strength permanently is lost following shear. The liquidity index can be used to show sensitivity increases in clay (Fig. 11.12). The small fraction of the original strength regained after remoulding quickclays may be attributable to the development of a different form of interparticle

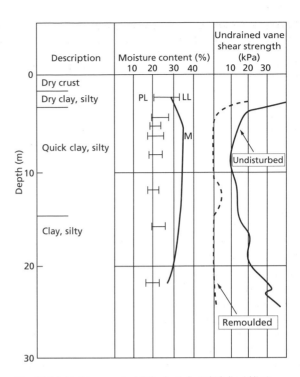

Fig. 11.11 Moisture content (M), plastic limit (PL), liquid limit (LL) and sensitivity of quickclay, near Trondheim, Norway.

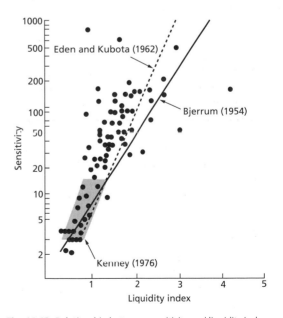

Fig. 11.12 Relationship between sensitivity and liquidity index of sensitive clays. (After Lutenegger & Hallberg, 1988.)

Table 11.5 Geotechnical properties of sensitive clays from eastern Canada. (After Locat *et al.*, 1984.)

Site	c_u (kPa)	c_{ur} (kPa)	S_t	S (g l⁻¹)	m (%)	LL (%)	PL (%)	PI (%)	LI	<2 μm (%)	SS (m² g⁻¹)	CEC (meq per 100 g)	P (%)	<2 μm (%)
Grande-Baleine	32	0.1	282	0.7	58	36.0	24.0	12.0	2.8	74	26	14.4	8.5	65.5
Grande-Baleine	24	0.2	120	—	46	35.8	23.0	12.8	1.8	66	27	—	15	51.0
Grande-Baleine	23	0.2	114	0.7	58	36.9	25.0	11.9	2.7	70	24	7.5	—	—
Olga	23	1.0	19	0.3	83	73.4	26.0	47.4	1.2	94	85	44.4	34.5	59.5
Olga	23	0.9	25	0.3	85	66.7	28.0	38.7	1.5	82	48	10.0	—	—
Olga	28	1.19	24	—	88	67.6	27.8	39.8	1.5	88	55	—	31.3	56.7
St Marcel	81	0.8	24	<2	81	64.0	28.0	36.0	1.5	85	67	21.9	26.7	58.3
St Marcel	—	1.21	—	<2	82	61.8	25.8	36.0	1.6	85	58	—	24.9	60.0
St Léon	80	3.4	24	12	58	61.0	23.5	37.5	0.9	74	56	14.1	32.2	41.8
St Alban	80	2.2	37	2.0	40	36.0	21.0	15.0	1.3	43	46	11.3	33.2	9.8
St Barnabé	157	3.1	50	4	48	43.0	20.0	23.0	1.2	50	40	8.5	19.4	30.6
Shawinigan	109	1.8	62	0.6	33	34.0	18.0	16.0	0.9	36	29	6.1	12.8	23.2
Chicoutimi	245	0.5	532	—	33	28.0	18.0	10.0	1.5	55	25	22.5	14.0	41.0
Outardes	109	0.6	181	—	64	34.0	21.5	12.5	3.4	50	30	7.3	—	—
Outardes	130	<0.07	>10³	—	35	25.0	17.0	8.0	2.3	47	23	—	—	—

c_u, undrained shear strength; c_{ur}, remoulded undrained shear strength; S_t, sensitivity; S, pore water salinity; m, natural moisture content; LL, liquid limit; PL, plastic limit; PI, plasticity index; LI, liquidity index; SS, specific surface area; CEC, cation exchange capacity; P, amount of phyllosilicates and amorphous minerals.

bonding; because the rate at which it develops is so slow, only a small fraction of the original strength can ever be recovered.

With regard to the physical properties of sensitive clays from eastern Canada, which were examined by Locat *et al.* (1984), with one exception, all samples fell above the A-line on the plasticity chart, indicating inorganic material. The natural moisture contents, plastic limits, liquid limits and plasticity indices of the soils are given in Table 11.5, from which it can be seen that the natural moisture contents always exceed the plastic limits and commonly exceed the liquid limits. In such cases, the liquidity indices are greater than unity. The strength decreases and the sensitivity increases dramatically as the liquidity index increases. This is illustrated by the fall cone strengths, the strength more or less disappearing on remoulding, giving sensitivity values varying from 10 to over 1000. Values of the specific surface area, cation exchange capacity and salinity of pore water are also provided in Table 11.5. The variation in the geotechnical properties of these soils primarily was attributed to their differences in mineralogy and texture.

Quickclays are associated with several serious engineering problems. Their bearing capacity is low, settlement is high and the prediction of the consolidation of quickclays by standard methods is unsatisfactory. Slides in quickclays sometimes have proved disastrous and, unfortunately, the results of slope stability analyses often are unreliable.

Quickclays can liquefy on sudden shock. This has been explained by the fact that, if quartz particles are small enough, having a very low settling velocity, and if the soil has a high water content, the solid–liquid transition can be achieved.

11.6 Soils of arid regions

Most arid deposits consist of the products of the physical weathering of bedrock formations. The process gives rise to a variety of rock and mineral fragments, which may then be transported and deposited under the influence of gravity, wind or water. Arid conditions may give rise to a variety of evaporite deposits and cemented layers due to the precipitation of soluble minerals.

A number of problems arise with Quaternary arid deposits. Some of these are common features of the materials themselves and others occur due to the climatic conditions. The shortage of surface water, together with the harsh conditions, inhibits biological and chemical activity, such that, in most arid regions, there is only a sparse growth, or perhaps absence, of vegetation. This makes the surface regolith of upland areas and slopes highly vulnerable to intense denudation and redistribution by gravitational or aeolian processes, or by the action of ephemeral water. Weathering activity tends to be dominated by physical breakdown, in places intense, of rock masses into poorly sorted assemblages of fragments, ranging in size down to silts. Many of the deposits within alluvial plains and covering hillsides are poorly consolidated. As such, they may undergo large settlements, especially if subjected to vibration due to earthquakes or cyclic loading. Much of the gravel-sized material is liable to consist of relatively weak, low-durability materials. Many arid areas are dominated by the presence of large masses of sand, which often are poorly graded. In the absence of downward leaching, surface deposits become contaminated with precipitated salts, particularly sulphates and chlorides. Alluvial plain deposits often contain gypsum particles and cement, and fragments of weak, weathered rock and clay.

Most beach materials in arid regions are carbonate sands with smooth, well-rounded, porous, chemically unstable particles. Contamination with evaporative salt can cause problems. Low-lying coastal zones and inland plains with shallow water tables are areas in which sabkha conditions commonly develop (see Chapter 7). Highly developed sabkhas tend to retain a greater proportion of soil moisture than moderately developed sabkhas. In addition, the higher the salinity of the groundwater, the greater the amount of water retained by the sabkha at a particular drying temperature (Sabtan *et al.*, 1995). The height to which water may rise from the water table is a function of the size and continuity of the pore spaces in the soil, and capillary rises of up to 3 m can occur in some fine grained deposits.

Within coastal sabkhas, the dominant minerals are calcite, dolomite and gypsum, with lesser amounts of anhydrite, magnesite, halite and carnallite, together with various other sulphates and chlorides. For example, James and Little (1994) described the fine grained soils of the sabkhas of Jubail on the Gulf Coast of Saudi Arabia as being partially cemented with sodium chloride and calcium sulphate (i.e. gypsum). In fact, they noted that little carbonate was present. James and Little also referred to the highly saline nature of the groundwater, which, at times, contained up to 23% NaCl, and occurred close to ground level. In such aggressively saline conditions, precautions have to be taken when using concrete. James and Little recommended the use of concrete with a high density and low water to cement ratio made from sulphate-resisting cement. The sodium chloride content of the water is also high enough to represent a corrosion hazard to steel reinforcement in concrete and to steel piles. Indeed, James and Little suggested that the thickness of steel may be reduced by one-half within 15 years. In such instances, reinforcement requires protective sheathing.

Minerals that are precipitated from groundwater in arid deposits also have high solution rates, so that flowing groundwater may lead quickly to the development of solution features. Problems, such as increased permeability, reduced density and settlement, are liable to be associated with engineering works or natural processes which result in a decrease in the salt concentration of groundwater. Hence, care needs to be exercised with irrigation, bridge, tunnel and harbour works that may lead to the removal of soluble minerals from the ground. Changes in the hydration state of minerals, such as swelling clays and calcium sulphate, also cause significant volume changes in soils. In the case of inland sabkhas, the minerals precipitated within the soil are much more variable than those of coastal sabkhas, because they depend on the composition of the local groundwater. The same applies to inland drainage basins and salt playas that are subjected to periodic desiccation.

Sabkha soils frequently are characterized by low strength (Table 11.6). Furthermore, some surface clays which normally are consolidated or lightly overconsolidated may be sensitive to highly sensitive (Hossain & Ali, 1988). The low strength is attributable to the concentrated salt solutions in sabkha brines, the severe climatic conditions under which sabkha deposits are formed (e.g. large variations in temperature and excessive wetting–drying cycles), which can give rise to instability in sabkha soils, and the ready solubility of some minerals which act as cements in these soils. As a consequence, the bearing

Property	Range of values for		
	Top crust	Main body of soft clay	Stiff brown clay
Natural water content (%)	15–40	40–78	25–31
Liquid limit (%)	20–50	36–60	50–56
Plastic limit (%)	16–40	18–30	25–33
Plasticity index (%)	5–20	20–32	24–30
Undrained shear strength by field vane shear test (kPa)	20–60	12–45	> 50
Sensitivity	3.5–22	1.3–11	1–2
M value from Mackintosh probing	15–140	3–40	50–300
Cone resistance from CPT (kg cm^{-2})	7–12	3–5	15–60
N value from SPT	2	1	9–24
Coefficient of volume compressibility (m^2 MN^{-1})	0.5–0.7	0.52–1.47	0.17–0.22
Compression index	0.37–0.42	0.40–0.88	0.17–0.31
Overconsolidation ratio	4–37	1.3–2.4	6–11

Table 11.6 Summary of geotechnical characteristics of clayey soil units in sabkhas. (After Hossein & Ali, 1988.)

capacity of sabkha soils and their compressibility frequently do not meet routine design requirements. Various ground improvement techniques therefore have been used in relation to large construction projects, such as vibroreplacement, dynamic compaction, compaction piles and underdrainage. Soil replacement and preloading have been used when highway embankments have been constructed. Al-Amoudi et al. (1995) suggested that sabkha soils can be improved by stabilization with cement and with lime. However, a high water to lime ratio is not satisfactory, and lime should not be used for the stabilization of sulphate-rich soils.

A number of silty deposits formed under arid conditions are liable to undergo considerable volume reduction or collapse when wetted. Such metastability arises due to the loss of strength of interparticle bonds, resulting from increases in water content. Thus, the infiltration of surface water, including that applied during the course of irrigation, by leakage from pipes and from a rise in the water table, may cause large settlements to occur. On the other hand, high rates of evaporation, leading to the precipitation of minerals in the capillary fringe, can give rise to ground heave.

A common feature of arid deposits is the cementa-tion of sediments by the precipitation of mineral matter from the groundwater. The species of salt held in solution, and those precipitated, depend on the source of the water, as well as on the prevailing temperature and humidity conditions. This process may lead to the development of various crusts or cretes, in which unconsolidated deposits or bedrock are cemented by gypsum, calcite, silica, iron oxides or other compounds. Thus cementation by gypsum would give rise to gypcrust or gypcrete, respectively. Cretes and crusts may form continuous sheets or isolated patch-like masses at the ground surface, where the groundwater table is at, or near, this level, or at some other position within the ground profile (Fig. 11.13).

Well cemented crusts and cretes may provide adequate bearing capacity for structures. Care must be taken that the underlying uncemented material is not overloaded. In addition, possible changes in the engineering behaviour of the material with any changes in the water conditions must be borne in mind.

11.7 Tropical soils

Engineers from countries with temperate climates often have experienced difficulties when dealing with

ig. 11.13 Calcrete in the north of
Iamib-Nankluft Park, Namibia.

ertain tropical soils, because it has been assumed
nat they will behave in a similar manner to soils in
he temperate zone. Consequently, methods of testing
nd engineering classification schemes have been used
hich were not designed to cope with the different
onditions found in soils formed in tropical environ-
nents (Gidigasu, 1988). Of course, it would be
rong to assume that all tropical soils are different
om those found in other climatic zones. Vargas
985) pointed out that many soils, such as alluvial
ays and sands or organic clays, behave in the same
anner and have similar geotechnical properties
gardless of the climatic conditions of the area of
position.

Drying initiates two important effects in tropical
sidual soils, namely, cementation by the sesqui-
ides and aggregate formation, on the one hand,
d loss of water from hydrated clay minerals on the
 her. In the case of halloysite, the latter causes an
eversible transformation to metahalloysite. Some
nsequences of these changes are illustrated in Table
.7(a). Drying can cause almost total aggregation of
 y-sized particles into silt- and sand-sized ranges,
 d a reduction or loss of plasticity (Table 11.7b).
e unit weight, shrinkage, compressibility and shear
ength can also be affected. Hence, classification
ts should be applied to the soil with as little drying
possible, at least until it can be established from
mparative tests that drying has no effect on the
ults.

Table 11.7 The effect of drying on index properties.

(a) Liquid limits (*LL*) and plastic limits (*PL*). (From Anon.,
1990.)

Soil type and location	Natural *LL*:*PL*	Air-dried *LL*:*PL*	Oven-dried *LL*:*PL*
Laterite (Costa Rica)	81:29		56:19
Laterite (Hawaii)	245:35	NP	
Red clay (Kenya)	101:70	77:61	65:47
Latosol (Dominica)	93:56	71:43	
Andosol (New Guinea)	145:75		NP
Andosol (Java)	184:146		80:74

NP, non-plastic.

(b) Effect of air drying on the particle size distribution of a
hydrated laterite clay from Hawaii. (From Gidigasu, 1988.)

Index properties	Wet (at natural moisture content)	Moist (partial air drying)	Dry (complete air drying)
Sand content (%)	30	42	86
Silt content (%) (0.05–0.005 mm)	34	17	11
Clay content (%) (<0.005 mm)	36	41	3

Conventional classifications of non-tropical soils are based principally on grading and plasticity. Once a soil has been classified, it may then be possible to infer some of the ways in which it will behave from an engineering point of view. This is not the case with many tropical soils. For some soils, the tests used are inadequate to determine the property being measured. For example, the liquid and plastic limit values obtained from 'standard' tests for some tropical red clay soils are dependent upon the precise test method employed and, in particular, on the amount of energy used in mixing the soil prior to carrying out the test. Consequently, two quite different values of the liquid limit can be obtained by two operators testing the same soil. Different results will also be obtained

depending upon whether the soil was predried prior to testing or kept close to its natural moisture content.

One of the most recent classifications of tropical residual soils has been proposed by Anon. (1990). The classification is pedologically based, and the main groupings are given in Table 11.8.

Duricrusts are not considered in detail here as they are discussed under soils of arid climates, although ferricretes (includes laterites) and alcretes (bauxites) are common in tropical regions. The term 'laterite' has been applied to soft, clay-rich horizons showing marked iron segregation or mottling, and also to gravelly materials comprising mainly iron oxide concretions or pisoliths. McFarlane (1976) considered that these non-indurated materials formed part of a sequence of lateritic weathering, ultimately resulting in the formation of an indurated surface or near surface sheet of duricrust.

Vertisols are characterized by the presence of clay minerals of the smectite group, which typically have high swell and shrink potential, and possess contraction cracks and slickensides. Anon. (1990) indicate that these soils are prone to erosion and dispersion, as well as to swell–shrink problems. Generally, the clay fraction in these soils exceeds 50%, the silty material

Table 11.8 Classification of tropical residual soils.

Mature soils	Duricrusts
Vertisols	Silcrete
Fersiallitic andosols	Calcrete
Fersiallitic (*sensu stricto*)	Gypcrete
Ferruginous (*sensu stricto*)	Ferricrete
Ferrallitic	Alcrete (Alicrete)

Table 11.9 Some enginnering properties of tropical residual soils. (From Bell, 1992.)

Soil	Location	Natural moisture content (%)	Plastic limit (%)	Liquid limit (%)	Clay content (%)	Activity	Void ratio	Unit weight (kN m⁻³)
1. Laterite	Hawaii		135	245	36			15.2–1
2. Laterite	Nigeria	18–27	13–20	41–51				
3. Laterite	Sri Lanka	16–49	28–31	33–90	15–45			
4. Laterite gravel	Cameroon							
5. Red clay	Brazil		26	42	35	0.46		
6. Red clay	Ghana		24	48	31	0.77		
7. Red clay	Brazil		22	28	27	0.3		
8. Red clay	Hong Kong							
9. Latosol	Brazil		30	47	35	0.51		
10. Latosol	Ghana		29	52	23	1.0		
11. Latosol	Java	36–38			80		1.7–1.8	
12. Black clay	Cameroon		21	75				
13. Black clay	Kenya		36	103				
14. Black clay	India		41	132	48			
15. Andosol	Kenya	62	73	107				11.5
16. Andosol	Java	58–63			29		2.1–2.2	

varies between 20 and 40% and the remainder is sand. Black cotton soils probably are the most common type, and are highly plastic, silty clays. The swell–shrink behaviour of these soils is a problem in many regions which experience alternating wet and dry seasons. However, the volume changes frequently are confined to an upper critical zone in the soil, often less than 1.5 m thick. Below this, the moisture content remains more or less constant.

The 'red' soils of tropical regions include the fersiallitic, ferruginous and ferrallitic soils. Each type relates to a broad set of climatic conditions. Fersiallitic soils are found in subtropical or Mediterranean climates. Smectite is the main clay mineral, but on older surfaces, well drained sites or silica-poor parent rocks, kaolinite may be present. Young volcanic ashes produce fersiallitic andosols characterized by allophane, which alters to imogolite or halloysite on weathering.

Ferruginous soils occur in either more humid (without dry season) or slightly hotter regions than Mediterranean, and are more strongly weathered than fersiallitic soils. Kaolinite is the dominant clay mineral and smectite is subordinate.

Ferrallitic soils develop in the hot, humid tropics. All the primary minerals, except quartz, are weath-

ered, with much silica and bases being removed in solution. Any residual feldspar is converted to kaolinite. Gibbsite may be present. Soils can be divided into ferrites and allites depending upon whether iron oxides or aluminium oxides dominate.

With this form of classification, the different groupings should be seen as part of a weathering continuum from fersiallitic soils through to ferrallitic soils. Nonetheless, this can lead to practical problems in that, for example, fersiallitic soils can be found in a climatic environment conducive to the formation of ferrallitic soils when the parent material has not been exposed to the climatic conditions for a long enough period of time. In the same way, different soils may occur within the soil profile.

Further practical problems may arise in the use of this classification, because of the difficulty in identifying the variation of climate, at a particular location, with time. In other words, the present environmental conditions are not necessarily a good indicator of the soil to be expected if the climate has changed since the formation of the soil.

Some values of the natural moisture content for lateritic soils are given in Table 11.9, values frequently falling within the range 6–22%. Usually, at or near the surface, the liquid limits of laterites do not exceed

Dry density (Mg m^{-3})	Specific gravity	Shear strength		Compressibility			Permeability (m s^{-1})
		ϕ'	c' (kPa)	C_c	m_v (m^2 MN^{-1})	c_v (m^2 y^{-1})	
		27–57	48–345				1.5×10^{-6}–5.4×10^{-9}
		28–35	466–782	0.0186	13	262	
	2.89–3.14	37	0–40				
	2.73						
	2.8						
	2.75				0.466–0.075	2.32–232	5×10^{-6}–2×10^{-6}
	2.86						
	2.83						
	2.74	21	26	0.03–0.06	0.995–0.01		4.7–5×10^{-8}
	2.64	16–20	47–58				
1.27–1.75							
	2.72						
		37	25				
		18	46	0.01–0.04	0.358–0.084		4×10^{-8}
							2.8×10^{-8}–5.6×10^{-9}

60%, and the plasticity indices are less than 30% (Table 11.9). Consequently, such laterites are of low to medium plasticity. The activity of laterites may vary between 0.5 and 1.75. Red clays and latosols have high plastic and liquid limits (Table 11.9).

Obviously, the cementation between particles of residual soils has a significant influence on their shear strength, as does the widely variable nature of the void ratio and the partial saturation which, as noted, can occur to appreciable depths. A bonded soil structure exhibits a peak shear strength, unrelated to density and dilation, which is destroyed by yield as large strains develop. According to Mitchell and Sitar (1982), the main sources of data on the strength characteristics of both lateritic soils and andosols (see below) are from tests performed on remoulded specimens. The reported range of the friction angle is from 10 to 41°, with the majority of the results falling in the range 28–38°. The values of cohesion range from zero to in excess of 48 kPa. Data on the undisturbed shear strength characteristics of residual soils are fairly limited. Andosols are characterized by fairly high friction angles, 27–57°, as opposed to 23–33° for lateritic soils. The cohesion for laterites can vary from zero to over 210 kPa, whilst that quoted for andosols ranges from 22 to 345 kPa. The moisture content affects the strength of andosols significantly, as the degree of saturation can have an appreciable effect on cementation.

Because of the presence of a hardened crust near the surface, the strength of laterite may decrease with increasing depth. For example, Nixon and Skipp (1957) quoted shear strength values of 90 and 25 kPa, derived from undrained triaxial tests, for samples of laterite taken from the surface crust and from a depth of 6 m, respectively. In addition, the variation of the shear strength with depth is influenced by the mode of formation, type of parent rock, depth of water table and its movement, degree of laterization and mineral content, and amount of cement precipitated.

Many tropical residual soils behave as if they are overconsolidated, in that they exhibit a yield stress at which there is a discontinuity in the stress–strain behaviour and a decrease in stiffness. This is the apparent preconsolidation pressure. The degree of overconsolidation depends on the amount of weathering. The cementation of soils formed in regions with a distinct dry season can be weakened by saturation, and this leads to a collapse of the soil structure.

Table 11.10 Engineering properties of a lateritic soil before and after leaching. (After Ola, 1978.)

Property	Before leaching	After leaching
Natural moisture content (%)	14	—
Liquid limit (%)	42	53
Plastic limit (%)	25	21
Relative density	2.7	2.5
Angle of shearing resistance ϕ' (degrees)	26.5	18.4
Cohesion c' (kPa)	24.1	45.5
Coefficient of compressibility* ($m^2 MN^{-1}$)	12	15
Coefficient of consolidation* ($m^2 y^{-1}$)	599	464

*For a pressure of 215 kPa.

As a consequence, the apparent preconsolidation pressure decreases and the compressibility of the soil increases.

The effects of leaching on lateritic soils have been investigated by Ola (1978). As mentioned, the cementing agents in lateritic soils help to bond the finer particles together to form larger aggregates. However, as a result of leaching, these aggregate break down, which is shown by the increase in the liquid limit after leaching. Moreover, the removal of cement by leaching gives rise to an increase in compressibility of more than 50%. Again, this mainly is due to the destruction of the aggregate structure. Conversely, there is a decrease in the coefficient of consolidation by some 20% after leaching (Table 11.10). The changes in the effective angle of the shearing resistance and the effective cohesion before and after leaching can be similarly explained. Prior to leaching, the larger aggregates in the soil cause it to behave as a coarse grained, weakly bonded particulate material. The strongly curvilinear form of the Mohr failure envelopes can also be explained as a result of the breakdown of the larger aggregates.

Some residual soils have high void ratios and large macropores, and therefore are associated with high permeability. For instance, the *in situ* permeabilities of clay-type lateritic soils may be as high as 10^- $10^{-4} m s^{-1}$. Both in the saprolite horizon and the underlying rock, the permeability is governed by discontinuities. Brand (1984) considered that the saprolitic soils in Hong Kong generally were relatively permeable, and that therefore drained conditions normally occurred. Water tables often are quite low in such soils. Blight (1990) noted that water tables often are deeper than 5–10 m. If evapotranspiration exceeds infiltration, deep desiccation of the soil profile is likely. As a result, residual soils may be cracked and fissured, which further increases the permeability.

Expandable clay minerals frequently are present in tropical residual clay soils of medium to high plasticity. When these clays occur above the water table, they can undergo a high degree of shrinkage on drying. By contrast, when wetted they swell, exerting high pressures. Indeed, highly expansive clay minerals may develop sufficient swelling pressure to break the bonded structure of the soil, and thereby give rise to large heave movements. Large heave movements can occur when desiccated black clay soils are wetted (see Section 11.2).

Allophane-rich soils or andosols are developed from basic volcanic ashes in high-temperature, high-rainfall regions. The soils have very high moisture contents (60–250%) and liquid limits (80–300%), with corresponding high plasticity indices (20–100%). The plasticity values vary considerably when the soils are tested with or without air drying prior to testing, values of the plasticity index and liquid limit being reduced by more than 50% in the air-dried state. The plasticity values obtained are also dependent upon the degree of working of the soil during pretest mixing, the plasticity decreasing with increased working. The soils are also characterized by very low dry densities and high void ratios (sometimes as high as six).

Soils containing halloysite, or its partially dehydrated form metahalloysite, have high moisture contents (30–65%) and liquid limits (40% to more than 100%). The plasticity indices can also be high (10–50%). Air drying of the soils causes a reduction in plasticity, approximately parallel to the A-line, but by a much smaller amount than for allophane-rich soils. The plasticity increases with mixing. Clay contents generally are high (50% to more than 80%), and predrying appears to have little effect on the particle size determination provided that complete dispersion of the particles is achieved. These soils are susceptible to collapse, that is, rapid consolidation on flooding with water at constant load. However, the phenomenon does not appear to occur universally, and the collapse susceptibility seems to be dependent on factors such as the sample depth and disturbance, void ratio, degree of saturation and test duration prior to flooding.

11.8 Peat

Peat represents an accumulation of partially decomposed and disintegrated plant remains, which have been preserved under conditions of incomplete aeration and high water content. It accumulates wherever the conditions are suitable, that is, in areas in which there is an excess of rainfall and the ground is poorly drained, irrespective of latitude or altitude. Nonetheless, peat deposits tend to be most common in those regions with a comparatively cold, wet climate. Physicochemical and biochemical processes cause this organic material to remain in a state of preservation over a long period of time. In other words, waterlogged, poorly drained conditions not only favour the growth of particular types of vegetation, but help to preserve the plant remains. Decay is most intense at the surface, where aerobic conditions exist; it occurs at a much slower rate throughout the mass of the peat.

Landva and Pheeney (1980) suggested that the proper identification of peat should include a description of the constituents, because the fibres of some plant remains are much stronger than others. The degree of humification and water content should also be taken into account. They proposed a modified form of the Von Post (1922) classification of peat (Table 11.11).

The void ratio of peat ranges between nine, for dense amorphous granular peat, up to 25, for fibrous types with high contents of sphagnum. It usually tends to decrease with depth within a peat deposit. Such high void ratios give rise to a phenomenally high water content. The water content of peats varies from a few hundred percentage dry weight (e.g. 500% in some amorphous granular peats) to over 3000% in some coarse fibrous varieties. Put another way, the water content may range from 75 to 98% by volume of peat. Moreover, changes in the amount of water content can occur over very small distances. The pH value of bog peat usually is less than pH 4.5 and may be less than pH 3.

Gas is formed in peat as plant material decays, and

Table 11.11 Classification of peat. (From Landva & Pheeney, 1980.)

| (1) Genera | | (2) Designation | H | (3) Degree of humification (H) | |
				Decomposition	Plant structure
Bryales (moss)	B	With few exceptions, peats	H_1	None	Easily identified
Carex (sedge)	C	consist of a mixture of two or	H_2	Insignificant	Easily identified
Equisetum (horsetails)	Eq	more genera. These are listed	H_3	Very slight	Still identifiable
Eriophorum (cotton grass)	Er	in decreased order of content,	H_4	Slight	Not easily identified
Hypnum (moss)	H	i.e. the principal component	H_5	Moderate	Recognizable, but vague
Lignidi (wood)	W	first, e.g. ErCS	H_6	Moderately strong	Indistinct
Nanolignidi (shrubs)	N		H_7	Strong	Faintly recognizable
Phragmites	Ph		H_8	Very strong	Very indistinct
Scheuchzeria (aquatic herbs)	Sch		H_9	Nearly complete	Almost unrecognizable
Sphagnum (moss)	S		H_{10}	Complete	Not discernible

(4) Water content (B)	(5) Fine fibres (F)	(6) Coarse fibres (R)	(7) Wood (W) and shrub remnant
Estimated from a scale of 1 (dry) to 5 (very high), and designated B_1, B_2, etc. Landva and Pheeney suggested the following ranges: B_2 less than 500% B_3 500–1000% B_4 1000–2000% B_5 over 2000%	These are fibres and stems less than 1 mm in diameter: F_0 = nil; F_1 = low content; F_2 = moderate content; F_3 = high content	These fibres have a diameter exceeding 1 mm: R_0 = nil; R_1 = low content; R_2 = moderate content; R_3 = high content	Wood and shrub are similarly graded: W_0 = nil; W_1 = low content; W_2 = moderate content; W_3 = high content

this tends to take place from the centre of stems, so that gas is held within stems. The volume of gas in peat varies; values of around 5–7.5% have been quoted (Hanrahan, 1954). At this degree of saturation, most of the gas is free, and so has a significant influence on the initial consolidation, rate of consolidation, pore pressure under load and permeability.

The bulk density of peat is both low and variable, being related to the organic content, mineral content, water content and degree of saturation. Peats frequently are not saturated, and may be buoyant under water due to the presence of gas. Except at low water contents (less than 500%), with high mineral contents, the average bulk density of peats is slightly lower than that of water. However, the dry density is a more important engineering property of peat, influencing its behaviour under load. Hanrahan (1954) recorded dry densities of drained peat within the range 65–120kgm⁻³. The dry density is influenced by the mineral content, and higher values than those quoted can be obtained when peats possess high mineral residues.

Peat undergoes significant shrinkage on drying. Nonetheless, the volumetric shrinkage of peat increases up to a maximum and then remains constant, the volume being reduced almost to the point of complete dehydration. The amount of shrinkage which can occur ranges between 10 and 75% of the original volume. The change in peat is permanent, in that it cannot recover all the water lost when wet conditions return. Hobbs (1986) noted that the more highly humified peats, even though they have lower water contents, tend to shrink more than less humified, fibrous peat.

Apart from its moisture content and dry density the shear strength of a peat deposit appears to be influenced by its degree of humification and its mineral content. As both of these factors increase, so does the shear strength. Conversely, the higher the moisture content of peat, the lower its shear strength. As the effective weight of 1 m³ of undrained peat is approximately 45 times that of 1 m³ of drained peat the reason for the negligible strength of the latter becomes apparent.

Consolidation of peat takes place when water is expelled from the pores, and the particles undergo some structural rearrangement. Initially, the two processes occur at the same time but, as the pore water pressure is reduced to a low value, the expulsion of water and structural rearrangement occur as a creep-like process (Berry & Poskitt, 1972). In other words, the initial stage of drainage can be regarded as primary consolidation, whereas the stage of continuing creep represents secondary compression. It must be remembered that primary and secondary consolidation are empirical divisions of a continuous compression process, both of which occur simultaneously during part of that process. In fact, an accurate prediction of the amount and rate of settlement of peat cannot be derived directly from laboratory tests. Hence, large-scale field trials seem to be essential for important projects.

Peat has a high coefficient of secondary compression, the latter being the dominant process in terms of the settlement of peat; the strain is virtually independent of the water content and degree of saturation. The short phase of primary consolidation is responsible for the slight distortion.

Differential and excessive settlement is the principal problem associated with peaty soil. When a load is applied to peat, settlement occurs because of the low lateral resistance offered by the adjacent unloaded peat. Serious shearing stresses are induced even by moderate loads. Worse still, should the loads exceed a given minimum, settlement may be accompanied by creep, lateral spread or, in extreme cases, rotational slip and upheaval of adjacent ground. At any given time, the total settlement in peat due to loading involves settlement with and without volume change. Settlement without volume change is the more serious, as it can give rise to the types of failure mentioned. Moreover, it does not enhance the strength of peat.

When peat is compressed, the free pore water is expelled under excess hydrostatic pressure. Because the peat initially is quite pervious, and the percentage of pore water is high, the magnitude of settlement is large, and this period of initial settlement is short (a matter of days in the field). The magnitude of initial settlement is directly related to the peat thickness and applied load. The original void ratio of a peat soil also influences the rate of initial settlement. The excess pore water pressure is almost entirely dissipated during this period. Settlement subsequently continues at a much slower rate, which is approximately linear with the logarithm of time. This is because the permeability of peat is reduced significantly due to the large decrease in volume. During this period, the effective consolidating pressure is transferred from the pore water to the solid peat fabric. The latter is compressible, and will only sustain a certain proportion of the total effective stress, depending on the thickness of the peat mass.

The use of precompression by surcharge loading in the construction of embankments across peatlands involves the removal of the surcharge after a certain period of time. This gives rise to some swelling in the compressed peat. Uplift or rebound can be quite significant, depending on the actual settlement and surcharge ratio (i.e. the mass of the surcharge in relation to the weight of the fill once the surcharge has been removed). The rebound is also influenced by the amount of secondary compression induced prior to unloading. The swelling index C_s is related to the compression index C_c, in that an average C_s value is around 10% of C_c, within a range of 5–20%. Rebound undergoes a marked increase when the surcharge ratio is greater than about three. Normally, rebound in the field is between 2 and 4% of the thickness of the compressed layer of peat before surcharge has been removed. Hence, the compressibility of preconsolidated peat is reduced greatly. With few exceptions, improved drainage has no beneficial effect on the rate of consolidation. This is because efficient drainage only accelerates the completion of primary consolidation which, anyway, is completed rapidly.

References

Aitchison, G.D. & Wood, C.C. (1965) Some interactions of compaction, permeability and post-construction deflocculation affecting the probability of piping failure in small earth dams. In *Proceedings of the Sixth International Conference in Soil Mechanics and Foundation Engineering, Montreal* 2, 442–446.

Al-Amoudi, O.S.B., Asi, I.M. & El-Naggar, Z.R. (1995) Stabilization of arid, saline sabkha using additives. *Quarterly Journal of Engineering Geology* 28, 369–379.

Anon. (1990) Tropical residual soils. Working Party Report. *Quarterly Journal of Engineering Geology* 23, 1–101.

Bell, F.G. (1992) *Engineering Properties of Soils and Rocks*, 3rd edn. Butterworth–Heinemann, Oxford.

Bell, F.G. (1993) *Engineering Treatment of Soils*. Spon, London.

Bell, F.G. & Maud, R.R. (1994) Dispersive soils: a review from a South African perspective. *Quarterly Journal of Engineering Geology* 27, 195–210.

Bell, F.G. & Maud, R.R. (1995) Expansive clays and construction, especially of low rise structures: a view point from Natal, South Africa. *Environmental and Engineering Geoscience* 1, 41–59.

Berry, P.L. & Poskitt, T.J. (1972) The consolidation of peat. *Geotechnique* 22, 27–52.

Blair, M. H. & Spangle, W. E. (1979) Seismic safety and land-use planning—selected examples from California. *United States Geological Survey Professional Paper* **G41-B**, 82 pp.

Blight, G.E. (1990) Construction in tropical soils. In *Proceedings of the Second International Conference on Geomechanics in Tropical Soils, Singapore* 2. Balkema, Rotterdam, pp. 449–468.

Brand, E.W. (1985) Geotechnical engineering in tropical residual soils. In *Proceedings of the First International Conference on Geomechanics in Tropical Lateritic and Saprolitic Soils, Brasilia, Brazilian Society for Soil Mechanics* 1, pp. 235–251.

Cabrera, J.G. & Smalley, I.J. (1973) Quick clays as products of glacial action: a new approach to their nature, geology, distribution and geotechnical properties. *Engineering Geology* 7, 115–133.

Casagrande, A. (1936) Characteristics of cohesionless soils affecting the stability of slopes and earth fills. *Journal of the Boston Society of Civil Engineers* 23, 3–32.

Clemence, S.P. & Finbarr, A.O. (1981) Design considerations for collapsible soils. *Proceedings of the American Society of Civil Engineers, Journal Geotechnical Engineering Division* 107, 305–317.

Clevenger, W.A. (1958) Experience with loess as foundation material. *Proceedings of the American Society of Civil Engineers, Journal Soil Mechanics Foundations Division* 85, 151–180.

Craft, D.C. & Acciardi, R.G. (1984) Failure of pore water analyses for dispersion. *Proceedings of the American Society of Civil Engineers, Journal Geotechnical Engineering Division* 110, 459–472.

Derbyshire, E. & Mellors, T.W. (1988) Geological and geotechnical characteristics of some loess and loessic soils from China and Britain: a comparison. *Engineering Geology* 25, 135–175.

Evstatiev, D. (1988) Loess improvement methods. *Engineering Geology* 25, 341–366.

Feda, J. (1966) Structural stability of subsidence loess from Praha-Dejvice. *Engineering Geology* 1, 201–219.

Feda, J. (1988) Collapse of loess on wetting. *Engineering Geology* 25, 263–269.

Feda, J., Bohac, J. & Herle, I. (1993) Compression of collapsed loess: studies on bonded and unbonded soils. *Engineering Geology* 34, 95–103.

Fookes, P.G. & Best, R. (1969) Consolidation characteristics of some late Pleistocene periglacial metastable soils of east Kent. *Quarterly Journal of Engineering Geology* 2, 103–128.

Forster, A., Culshaw, M.G. & Bell, F.G. (1995) The regional distribution of sulphate rocks and soils of Britain. In *Engineering Geology and Construction, Engineering Geology Special Publication No. 10*, Eddleston, M., Walthall, S., Cripps, J.C. & Culshaw, M.G. (eds). Geological Society, London, pp. 95–104.

Gao, G. (1988) Formation and development of the structure of collapsing loess in China. *Engineering Geology* 25, 235–245.

Gerber, A. & von Harmse, H.J.M. (1987) Proposed procedure for identification of dispersive soils by chemical testing. *The Civil Engineer in South Africa* 29, 397–399.

Gibbs, H.H. & Bara, J.P. (1962) Predicting surface subsidence from basic soil tests. *American Society for Testing and Materials (ASTM), Special Technical Publication* 322, 231–246.

Gidigasu, M.D. (1988) Potential application of engineering pedology in shallow foundation engineering on tropical residual soils. In *Proceedings of the Second International Conference on Geomechanics in Tropical Soils, Singapore*, 1. Balkema, Rotterdam, pp. 17–24.

Gillott, J.E. (1979) Fabric, composition and properties of sensitive soils from Canada, Alaska and Norway. *Engineering Geology* 14, 149–172.

Grabowska-Olszewsla, B. (1988) Engineering geological problems of loess in Poland. *Engineering Geology* 25, 177–199.

Hanrahan, E.T. (1954) An investigation of some physical properties of peat. *Geotechnique* 4, 108–123.

Hobbs, N.B. (1986) Mire morphology and the properties and behaviour of some British and foreign peats. *Quarterly Journal of Engineering Geology* 19, 7–80.

Hossain, D. & Ali, K.M. (1988) Shear strength and consolidation characteristics of Obhor sabkha, Saudi Arabia. *Quarterly Journal of Engineering Geology* 21, 347–359.

James, A.N. & Little, A.L. (1994) Geotechnical aspects of sabkha at Jubail, Saudi Arabia. *Quarterly Journal of Engineering Geology* 27, 83–121.

Jennings, J.E. & Knight, K. (1975) A guide to construction on or with materials exhibiting additional settlement due to collapse of grain structure. In *Proceedings of the Sixth African Conference on Soil Mechanics Foundation Engineering, Durban*, pp. 99–105.

Landva, A.O. & Pheeney, P.E. (1980) Peat fabric and structure. *Canadian Geotechnical Journal* 17, 416–435.

Lin, Z.G. & Wang, S.J. (1988) Collapsibility and deformation characteristics of deep-seated loess in China. *Engineering Geology* 25, 271–282.

Locat, J., Lefebvre, G. & Ballivy, G. (1984) Mineralogy,

chemistry and physical properties interrelationships of some sensitive clays from eastern Canada. *Canadian Geotechnical Journal* 21, 530–540.

Lutenegger, A.J. & Hallberg, G.R. (1988) Stability of loess. *Engineering Geology* 25, 247–261.

McFarlane, M.J. (1976) *Laterite and Landscape*. Academic Press, London.

Mitchell, J.K. & Sitar, N. (1982) Engineering properties of tropical residual soils. In *Proceedings of the Speciality Conference on Engineering and Construction in Tropical and Residual Soils, Honolulu*. American Soil Civil Engineers, Geotechnical Engineering Division, pp. 30–57.

Nixon, I.K. & Skipp, B.O. (1957) Airfield construction on overseas soils. Part 5—Laterite. *Proceedings of the Institution of Civil Engineers* 36, 253–275.

Northmore, K.J., Bell, F.G. & Culshaw, M.G. (1996) The engineering properties and behaviour of the brickearth of south Essex. *Quarterly Journal of Engineering Geology* 29, 147–161.

Ola, S.A. (1978) Geotechnical properties and behaviour of stabilized lateritic soils. *Quarterly Journal of Engineering Geology* 11, 145–160.

Ola, S.A. (1980) Mineralogical properties of some Nigerian residual soils in relation with building problems. *Engineering Geology* 15, 1–13.

O'Neill, M.W. & Poormoayed, A.M. (1980) Methodology for foundations on expansive clays. *Proceedings of the American Society of Civil Engineering, Journal Geotechnical Engineering Division* 106, 1345–1367.

Phien-Wej, N., Pientong, T. & Balasubramanian, A.S. (1992) Collapse and strength characteristics of loess in Thailand. *Engineering Geology* 32, 59–72.

Popescu, M.E. (1986) A comparison of the behaviour of swelling and collapsing soils. *Engineering Geology* 23, 145–163.

Sabtan, A., Al-Saify, M. & Kazi, A. (1995) Moisture retention characteristics of coastal sabkhas. *Quarterly Journal of Engineering Geology* 28, 37–46.

Schmertmann, J.H. (1969) Swell sensitivity. *Geotechnique* 19, 530–533.

Sherard, J.L., Dunnigan, L.P. & Decker, R.S. (1976) Identification and nature of dispersive soils. *Proceedings of the American Society of Civil Engineers, Journal Geotechnical Engineering Division* 102, 287–301.

Terzaghi, K. (1925) *Erdbaumechanik auf Boden Physikalischer Grundlage*. Deuticke, Vienna.

Vargas, M. (1985) The concept of tropical soils. In *Proceedings of the First International Conference on Geomechanics of Tropical Lateritic and Saprolite Soils, Brasilia, Brazilian Society for Soil Mechanics* 3, 101–134.

Von Post, L. (1922) Sveriges Geologiska Undersokings torvimventering och nogra av dess hittels vunna resultat (SGU peat inventory and some preliminary results). *Svenska Mosskulturfoereningens Tidskrift, Jonkoping, Sweden* 36, 1–37.

Williams, A.A.B. (1980) Severe heaving of a block of flats near Kimberley. In *Proceedings of the Seventh Regional Conference for Africa on Soil Mechanics and Foundation Engineering, Accra* 1, pp. 301–309.

Wilson, C. & Melis, L. (1991) Breaching of an earth dam in the Western Cape by piping. In *Geotechnics in the African Environment*, Blight, G.E., Fourie, A.B., Luker, I., Mouton, D.J. & Scheurenburg, R.J. (eds). Balkema, Rotterdam, pp. 301–312.

12 Rock masses, their character, problems and uses

12.1 Rock types

12.1.1 Igneous rocks

Rocks are aggregates usually of two or three major minerals. They are divided according to their origin into three groups, igneous, metamorphic and sedimentary rocks. Igneous rocks are formed when hot, molten, rock material, that is, magma, solidifies. Magmas are developed either within the Earth's crust or in the uppermost layers of the mantle. They comprise hot solutions of several liquid phases, the chief of which is a complex silicate phase. Hence, igneous rocks are composed primarily of silicate minerals. Moreover, of the silicate minerals, six families, namely the olivines, pyroxenes, amphiboles, micas, feldspars and silica minerals, are quantitatively by far the most important constituents. Figure 12.1 shows the approximate distribution of the minerals in the commonest igneous rocks.

Igneous rocks may be divided into intrusive or extrusive types, depending on their mode of occurrence. In the former type, the magma crystallizes within the Earth's crust, whereas in the latter, it solidifies at the surface, having been erupted as lavas or pyroclasts from a volcano (see Chapter 2). The intrusions may be further subdivided, on the basis of their size, into major and minor categories according to their mode of occurrence. The former are developed in a plutonic (deep-seated) environment, whereas the latter occur in a hypabyssal environment. Batholiths represent major intrusions which have been exposed at the surface by erosion (Fig. 12.2). Dykes and sills are the most frequently occurring types of minor intrusion (Fig. 12.3). Most of the plutonic intrusions have a granitic–granodioritic composition, whereas most of the extrusives are basaltic in composition.

An igneous rock may be composed of an aggregate of crystals, of natural glass or of crystals and glass in varying proportions. This depends on the rate of cooling, composition of the magma and the environment in which the rock was formed. If igneous rocks cool quickly, as in volcanic and hypabyssal environments, they may be glassy or fine grained, whereas slow cooling in a plutonic environment allows crystals to grow and develop a granular texture.

12.1.2 Metamorphic rocks

Metamorphic rocks are derived from pre-existing rock types, and have undergone mineralogical, textural and structural changes. These alterations have been brought about by changes taking place in the physical and chemical environments in which the rocks exist. The processes responsible for the changes give rise to progressive transformation in the solid state. The changing conditions of temperature and/or pressure are the primary agents causing metamorphic reactions in rocks. Individual minerals are stable over limited temperature–pressure conditions, which means that, when these limits are exceeded, mineralogical adjustment has to be made to establish equilibrium with the new environment.

Two major types of metamorphism may be distinguished on the basis of geological setting. One type is of local extent, whereas the other extends over a regional area. The first type includes thermal or contact metamorphism, and the latter refers to regional metamorphism. Thermal metamorphism occurs around igneous intrusions, so that the principal factor controlling the reactions is the temperature. The encircling zone of metamorphic rock around an igneous intrusion is referred to as the contact aureole (Fig. 12.4). Within a contact aureole there usually is a sequence of mineralogical changes which increase in their degree of alteration from the country rocks towards the intrusion. Regional meta-

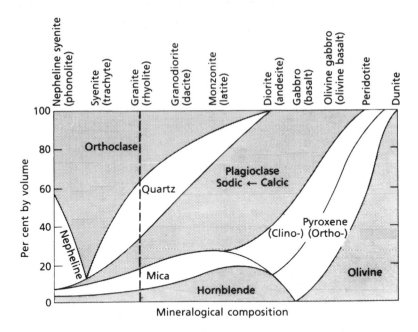

Fig. 12.1 Approximate mineralogical compositions of the more common types of igneous rock, e.g. granite is approximately 40% orthoclase, 33% quartz, 13% plagioclase, 9% mica and 5% hornblende. Plutonic types without parentheses; volcanic equivalents in parentheses.

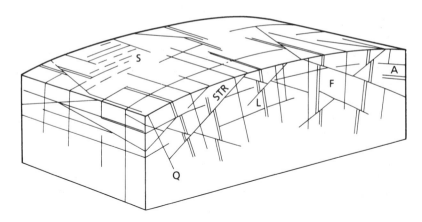

Fig. 12.2 Block diagram showing the types of structure in a batholith: Q, cross-joints; S, longitudinal joints; L, flat-lying joints; STR, planes of stretching; F, linear flow structure; A, aplite dykes.

morphic rocks occur in the Precambrian shields and the eroded roots of fold mountains. As such, they extend over hundreds or even thousands of square kilometres. Regional metamorphism involves both the processes of changing temperature and stress. In fact, regional metamorphism may involve temperatures up to a maximum of 800 °C, and confining pressures generally are in excess of 3 kbar. Moreover, temperatures and pressures conducive to regional metamorphism may have been maintained over millions of years. Regional metamorphism is a progressive process, that is, in any given terrain formed

initially of rocks of similar composition, zones of increasing grade may be defined by different mineral assemblages. The boundaries between the zones can be regarded as isograds, lines of equal metamorphic conditions.

Most deformed metamorphic rocks possess some kind of preferred orientation, commonly exhibited as planar structures, which allow the rock to split more easily in one direction than another. One of the most familiar examples of such a structure is cleavage, which is characteristic of slate (Fig. 12.5). When recrystallization occurs under conditions which

(a)

(b)

Fig. 12.3 (a) Basalt dyke, Isle of Skye, Scotland; (b) Great Whin Sill, Northumberland, England.

include shearing stress, a directional element is imparted to the newly formed rock. Minerals are arranged in parallel layers along the direction normal to the shearing stress. Slaty cleavage reflects a highly developed preferred orientation of mineral boundaries, particularly of those of plate-like habit, such as mica and chlorite. The most important minerals responsible for the development of schistosity are those which possess an acicular, plate-like or tabular habit, the micas again being the principal family involved (Fig. 12.6). Schists develop at higher temperature–stress conditions than slates. Foliation, which is characteristic of gneiss, consists of parallel bands formed of contrasting mineral assemblages, such as quartz–feldspar and biotite–hornblende (Fig. 12.7). Foliation appears to be related to the same stress system as that responsible for the development of schistosity in the area. At higher temperatures, the influence of stress becomes less important, and so schistosity tends to disappear in rocks of high-grade metamorphism, foliation becoming a more significant feature.

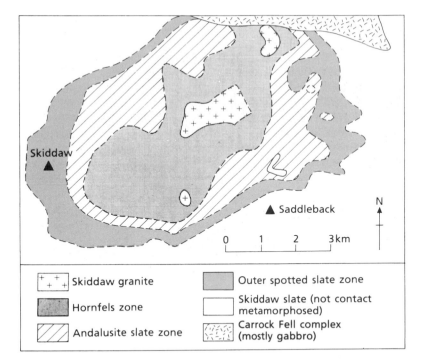

Fig. 12.4 Geological sketch map of the Skiddaw granite and its contact aureole.

Fig. 12.5 Cleavage in slate, near Llanberis, North Wales.

12.1.3 Sedimentary rocks

Most sedimentary rocks consist of detrital material derived from the breakdown of pre-existing rocks. Indeed, these clastic types may form between 80%

and 95% of all sedimentary rocks. The non-clastic sedimentary rocks are the products of chemical and biochemical precipitation, whilst others are of organic origin. One feature common to all sedimentary rocks is that they were deposited, and this gives

Fig. 12.6 Dalradian schist, folded with quartz rod, Isle of Arran, Scotland.

Fig. 12.7 Foliated gneiss, Valley of One Thousand Hills, Natal, South Africa.

rise to their most noteworthy characteristic, that is, their bedding or stratification (Fig. 12.8). Although sedimentary rocks may comprise only about 5% of the Earth's crust, they cover three-quarters of the continental areas and most of the ocean floor.

The breakdown processes involve weathering and erosion, and each cycle of erosion is accompanied by a cycle of sedimentation. The particles involved usually have undergone a varying amount of transportation, which has an effect on their size and

shape. The composition of clastic sedimentary rocks depends upon the composition of the parent material, the stability of its component minerals, the type of action to which it has been subjected and how long it has suffered such action. Quartz is one of the commonest minerals in clastic sedimentary rocks, because it is one of the most stable, many of the others breaking down to form clay minerals. Hence, sandstones and shales are common sedimentary rocks. In order to convert an unconsolidated sediment into a solid

Fig. 12.8 Bedding in sandstone, northwest of Nelson, South Island, New Zealand.

rock, it must undergo lithification. Lithification involves consolidation and cementation.

The clastic sedimentary rocks consist of conglomerates, breccias, sandstones, siltstones and mudrocks (i.e. shales and mudstones). Mudrocks are the commonest of these rock types. Limestones are polygenetic in origin, some being clastic, others chemical or biochemical precipitates and yet others organic, such as algal and coral limestones. Evaporitic deposits are formed by precipitation from saline waters. Organic plant residues accumulate as peaty deposits or as sapropelic deposits. In the latter case, organic material accumulates together with silt and mud in still bodies of water. Both can give rise to coal eventually.

12.1.4 Deformation and strength of rocks

The factors which influence the deformation characteristics and failure of a body of rock include the intrinsic properties of the rock. The most important of these, as far as the intact rock is concerned, are the mineralogical composition and texture. These two parameters are governed by the origin of the rock. Few rocks are composed of only one mineral species and, even when they are, the properties of that species vary slightly from mineral to mineral. Consequently, few rocks can be regarded as homogeneous, isotropic materials. As far as the texture is concerned, the degree of interlocking of the grains is important in terms of the physical behaviour. Fracture is more likely to take place along grain boundaries (intergranular fracture) than through grains (transgranular fracture), and therefore irregular boundaries make fracture more difficult. The bond between grains in many sedimentary rocks is provided by the cement and/or matrix, rather than by the interlocking of grains. In this case, the amount and, to a lesser extent, the type of cement/matrix are important, influencing not only the strength and elasticity, but also the density and porosity. Grain orientation in a particular direction facilitates failure in that direction, and this applies to all fissile rocks. The presence of moisture in rocks adversely affects their engineering behaviour, the moisture content increasing the strain velocity and lowering the fundamental strength.

The strength of most fresh, intact rocks normally is satisfactory as far as their engineering performance is concerned (Table 12.1). However, rock masses are intersected, to a greater or lesser extent, by discontinuities, which frequently represent planes of weakness. Furthermore, weathering leads to a decrease in

Table 12.1 Some examples of the physical properties of rocks.

	Specific gravity	Dry density (Mg m⁻³)	Porosity (%)	Unconfined compressive strength, dry (MPa)	Unconfined compressive strength, saturated (MPa)	Point load strength (MPa)	Young's modulus (E_{t50}, GPa)
Granite (Dalbeattie)*	2.67			147.8		10.3	41.1
Basalt (Derbyshire)	2.91			321.0		16.9	93.6
Slate† (North Wales)	2.67			96.4		7.9	31.2
Slate‡ (North Wales)				72.3		4.2	
Fell Sandstone (Rothbury)	2.69	2.25	9.8	74.1	52.8	4.4	32.7
Sherwood Sandstone (Edwinstowe)	2.68	1.87	25.7	11.6	4.8	0.7	6.4
Carboniferous Limestone (Buxton)	2.71	2.58	2.9	106.2	83.9	3.5	66.9
Bath Stone (Corsham)	2.7	2.3	15.3	15.6	9.3	0.9	16.1
Anhydrite (Sandwith)	2.93	2.82	2.9	97.5		3.7	63.9
Gypsum (Sherburne-in-Elmet)	2.36	2.19	4.6	27.5		2.1	24.8

*Scotland, all the others, except slate, are from England.
†Tested normal to cleavage.
‡Tested parallel to cleavage.

the bulk density and an accompanying increase in the porosity of rock masses. This, in turn, means that their strength is reduced. Ultimately, weathering can turn a competent rock mass into a soil. Accordingly, the mechanical behaviour of rock masses is very much influenced by the presence and character of discontinuities, on the one hand, and the degree of weathering on the other.

12.2 Description of rocks and rock masses

Description is the initial step in an assessment of rocks and rock masses. It should therefore be both uniform and consistent in order to gain acceptance. The data collected on rocks and rock masses should be recorded on data sheets for subsequent processing.

A data sheet for the description of rock masses is shown in Fig. 12.9.

The complete specification of a rock mass requires descriptive information on the nature and distribution in space of both the materials that constitute the mass (rock, soil, water and air-filled voids) and the discontinuities which divide it. The intact rock may be considered as a continuum or polycrystalline solid consisting of an aggregate of minerals or grains, whereas a rock mass may be looked upon as a discontinuum of rock material transected by discontinuities. The properties of the intact rock are governed by the physical properties of the materials of which it is composed, and the manner in which they are bonded to each other. The parameters which may be used in the description of intact rock therefore include the

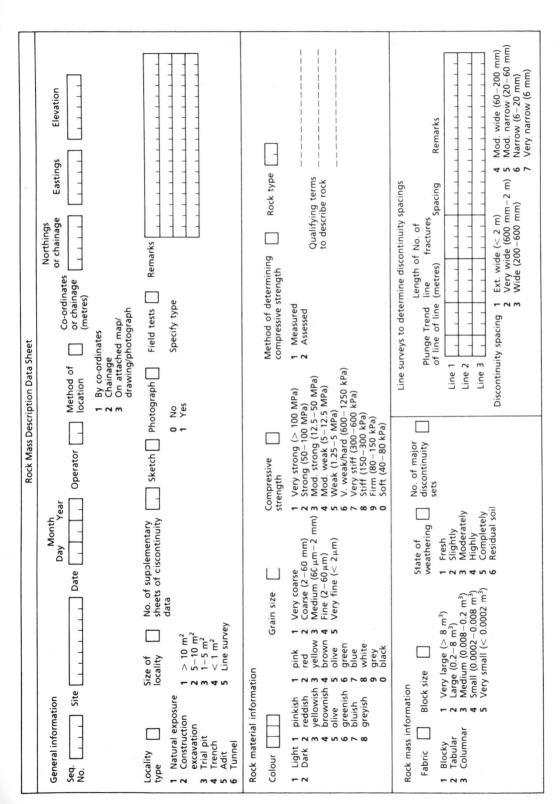

Fig. 12.9 Rock mass data description sheet. (After Anon., 1977.)

Table 12.2 Rock type classification. (After Anon., 1979.)

				Detrital sedimentary		Pyroclastic	Chemical/organic
Genetic/group				Detrital sedimentary		Pyroclastic	Chemical/organic
Usual structure				Bedded			
Composition				Grains of rock, quartz, feldspar and clay minerals	At least 50% of grains are of carbonate	At least 50% of grains are of fine grained igneous rock	
				Grains are of rock fragments			
Grain size (mm)	Group	Grade					
60	Rudaceous	Very coarse grained	Boulders / Cobbles	Rounded grains: conglomerate	Carbonate gravel — Calcirudite	Rounded grains: agglomerate. Angular grains: volcanic breccia, lapilli tuff	Saline rocks, Halite, Anhydrite
2	Rudaceous	Coarse grained	Gravel	Angular grains: breccia			Gypsum
	Arenaceous	Medium grained	Sand	Grains are mainly mineral fragments — Sandstone: grains are mainly mineral fragments. Quartz arenite: 95% quartz, voids empty or cemented. Arkose: 75% quartz, up to 25% feldspar: voids empty or cemented. Greywacke: 75% quartz, 15% fine detrital material: rock and feldspar fragments	Carbonate sand — Calcarenite	Tuff / Volcanic ash	Limestone / Dolomite
0.06	Argillaceous	Fine grained	Silt	Siltstone: 50% fine grained particles (Mudstone, shale: fissile mudstone; Mudstone; Marlstone)	Carbonate silt — Calcisiltite, chalk	Fine grained tuff	Chert
0.002	Argillaceous	Very fine grained	Clay	Claystone: 50% very fine grained particles	Carbonate mud — Calcilutite	Very fine grained tuff	Flint
		Glassy / Amorphous					Peat, Lignite, Coal

Limestone and dolomite (undifferentiated)

Genetic/group		Metamorphic		Igneous			
Usual structure		Foliated		Massive			
Composition		Quartz, feldspars, micas, acicular dark minerals		Light-coloured minerals are quartz, feldspar, mica		Dark and light minerals	Dark minerals
Grain size (mm)				Acid rocks	Intermediate	Basic rocks	Ultrabasic
60	Very coarse grained			Pegmatite			
2	Coarse grained	Gneiss (ortho-, para-). Alternate layers of granular and flaky minerals	Marble / Granulite	Granite	Diorite	Gabbro	Pyroxenite and peridotite
	Medium grained	Migmatite / Schist	Quartzite / Hornfels / Amphibolite	Microgranite	Microdiorite	Dolerite	Serpentinite
0.06	Fine grained	Phyllite		Rhyolite	Andesite	Basalt	
0.002	Very fine grained	Slate / Mylonite					
	Glassy			Obsidian and pitchstone		Tachylyte	
	Amorphous			Volcanic glasses			

petrological name, mineralogical composition, colour, texture, minor lithological characteristics, degree of weathering or alteration, density, porosity, strength, hardness, intrinsic or primary permeability, seismic velocity and modulus of elasticity. Swelling and slake durability can be taken into account where appropriate, such as in the case of argillaceous rocks.

The behaviour of a rock mass is, to a large extent, determined by the type, spacing, orientation and characteristics of the discontinuities present. As a consequence, the parameters which ought to be used in a description of a rock mass include the nature and geometry of the discontinuities, as well as its overall strength, deformation modulus, secondary permeability and seismic velocity. It is not necessary, however, to describe all the parameters for either a rock mass or intact rock.

Intact rock may be described from a geological or engineering point of view. In the first case, the origin and mineral content of a rock are of prime importance, as are the texture and any change which has occurred since its formation. In this respect, the name of a rock provides an indication of its origin, mineralogical composition and texture (Table 12.2). A useful system of petrographical description was provided by Dearman (1974) and was further developed by Anon. (1979). Only a basic petrographical description of the rock is required when describing a rock mass.

The micropetrographical description of rocks for engineering purposes includes the determination of all parameters which cannot be obtained from a macroscopic examination of a rock sample, such as the mineral content, grain size (Table 12.3) and texture, and which have a bearing on the mechanical behaviour of the rock or rock mass. In particular, a

microscopic examination should include a modal analysis, determination of microfractures and secondary alteration, determination of grain size and, where necessary, fabric analysis. The International Society of Rock Mechanics (ISRM) (Hallbauer *et al.*, 1978) recommends that the report of a petrographical examination should be confined to a short statement on the origin, classification and details relevant to the mechanical properties of the rock concerned. Wherever possible, this should be combined with a report on the mechanical parameters.

The colour of a rock has a composite character attributable to the different minerals from which it is formed, the size of these minerals and, in the case of sedimentary rocks, the type and amount of cement present. Hence, the overall colour should be assessed by reference to a colour system or chart, because it is difficult to make a quantitative assessment by the eye alone. For example, the Munsell colour system evaluates colour in terms of the hue, value and chroma. The hue refers to the basic colour or a mixture of basic colours. The chroma indicates the intensity, strength or degree of departure of a particular hue from a neutral grey of the same value. The value indicates the degree of lightness or darkness of a colour in relation to a neutral grey. A simple subjective scheme has been suggested by Anon. (1977), which involves the choice of colour from column 3 below, supplemented, if necessary, by a term from column 2 and/or column 1:

1	2	3
Light	Pinkish	Pink
Dark	Reddish	Red
	Yellowish	Yellow
	Brownish	Brown
	Olive	Olive
	Greenish	Green
	Bluish	Blue
		White
	Greyish	Grey
		Black

The texture of a rock refers to its component grains and their mutual arrangement or fabric. It is dependent upon the relative sizes and shapes of the grains, and their positions with respect to one another, and the groundmass or matrix, when present. The grain size, in particular, is one of the most important

Table 12.3 Description of grain size.

Term	Particle size	Equivalent soil grade
Very coarse grained	Over 60 mm	Boulders and cobbles
Coarse grained	2–60 mm	Gravel
Medium grained	0.06–2 mm	Sand
Fine grained	0.002–0.06 mm	Silt
Very fine grained	Less than 0.002 mm	Clay

Table 12.4 Description of dry density and porosity. (After Anon., 1979.)

Class	Dry density (Mg m⁻³)	Description	Porosity (%)	Description
1	Less than 1.8	Very low	Over 30	Very high
2	1.8–2.2	Low	30–15	High
3	2.2–2.55	Moderate	15–5	Medium
4	2.55–2.75	High	5–1	Low
5	Over 2.75	Very high	Less than 1	Very low

Table 12.5 Rock strength.

(a) Description of unconfined compressive strength.

Geological Society (Anon., 1970)		IAEG (Anon., 1979)		ISRM (Anon., 1981)	
Strength (MPa)	Description	Strength (MPa)	Description	Strength (MPa)	Description
Less than 1.25	Very weak	1.5–15	Weak	Under 6	Very low
1.25–5.00	Weak	15–50	Moderately strong	6–20	Low
5.00–12.50	Moderately weak	50–120	Strong	20–60	Moderate
12.50–50	Moderately strong	120–230	Very strong	60–200	High
50–100	Strong	Over 230	Extremely strong	Over 200	Very high
100–200	Very strong				
Over 200	Extremely strong				

IAEG, International Association of Engineering Geology.

(b) Estimation of the strength of intact rock. (After Anon., 1977.)

Description	Approximate unconfined compressive strength (MPa)	Field estimation
Very strong	Over 100	Very hard rock, more than one blow of geological hammer required to break specimen
Strong	50–100	Hard rock, hand-held specimen can be broken with a single blow of hammer
Moderately strong	12.5–50	Soft rock, 5 mm indentations with sharp end of pick
Moderately weak	5.0–12.5	Too hard to cut by hand
Weak	1.25–5.0	Very soft rock—material crumbles under firm blows with the sharp end of a geological hammer

aspects of texture, in that it exerts an influence on the physical properties of a rock. It now generally is accepted that the same descriptive terms for grain size ranges should apply to all rock types, and should be the same as those used to describe soils (Table 12.3). Other aspects of texture include the relative grain size and the grain shape. Anon (1979) suggested three types of relative grain size, namely, uniform, non-uniform and porphyritic. The grain shape was described in terms of the angularity (angular, subangular, subrounded and rounded), form (equidimensional flat, elongated, flat and elongated, and irregular) and surface texture (rough and smooth).

Anon. (1979) grouped the dry density and porosity of rocks into five classes. These are given in Table 12.4. A determination of the strength and deformability of intact rock is obtained with the aid of various laboratory tests. There are several scales of unconfined compressive strength. Three are given in Table 12.5(a). If the strength of a rock is not obtained by testing, it can be estimated as shown in Table 12.5(b), but this can only be very approximate.

Table 12.6 Description of the point load strength. (After Franklin & Broch, 1972.)

	Point load strength index (MPa)	Approximate uniaxial compressive strength (MPa)
Extremely high strength	Over 10	Over 160
Very high strength	3–10	50–160
High strength	1–3	15–50
Medium strength	0.3–1	5–15
Low strength	0.1–0.3	1.6–5
Very low strength	0.03–0.1	0.5–1.6
Extremely low strength	Less than 0.03	Less than 0.5

Table 12.7 Description of deformability. (After Anon., 1979.)

Class	Deformability (GPa)	Description
1	Less than 5	Very high
2	5–15	High
3	15–30	Moderate
4	30–60	Low
5	Over 60	Very low

Table 12.8 Estimation of secondary permeability from the discontinuity frequency. (After Anon., 1977.)

Rock mass description	Term	Permeability k (m s^{-1})
Very closely to extremely closely spaced discontinuities	Highly permeable	10^{-2}–1
Closely to moderately widely spaced discontinuities	Moderately permeable	10^{-5}–10^{-2}
Widely to very widely spaced discontinuities	Slightly permeable	10^{-9}–10^{-5}
No discontinuities	Effectively impermeable	Less than 10^{-9}

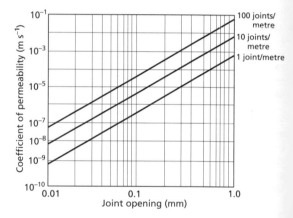

Fig. 12.10 Influence of joint opening and spacing on the coefficient of permeability in the direction of a set of smooth parallel joints in a rock mass. (After Hoek & Bray, 1981.)

Rocks have a much lower tensile than compressive strength. Unfortunately, however, the determination of the direct tensile strength has frequently proved to be difficult, because it is not easy to grip the specimen without introducing bending stresses. Hence, most values of the tensile strength quoted have been obtained by indirect methods of testing. One of the most popular of these methods is the point load test. Franklin and Broch (1972) suggested a scale for the point load strength, which is shown in Table 12.6. As far as the deformability is concerned, five classes have been proposed by Anon. (1979), and are shown in Table 12.7.

The permeability of intact rock (primary permeability) usually is several orders less than the *in situ* permeability (secondary permeability), as most water normally flows via discontinuities in rock masses. Although the secondary permeability is affected by the openness of discontinuities, on the one hand, and the amount of infilling on the other, a rough estimate of the permeability can be obtained from the frequency of discontinuities (Table 12.8; see also Fig.

12.10). Admittedly, such estimates must be treated with caution, and cannot be applied to rocks which are susceptible to solution.

Classifications of intact rock are based upon certain selected mechanical properties. The specific purpose for which a classification is developed obviously plays an important role in determining which mechanical properties of the intact rock are chosen. The object of the classification is to provide a reliable basis for assessing the rock quality. In fact, classification of intact rock for engineering purposes should be relatively simple, being based on significant mechanical properties, so that it has a wide appli-

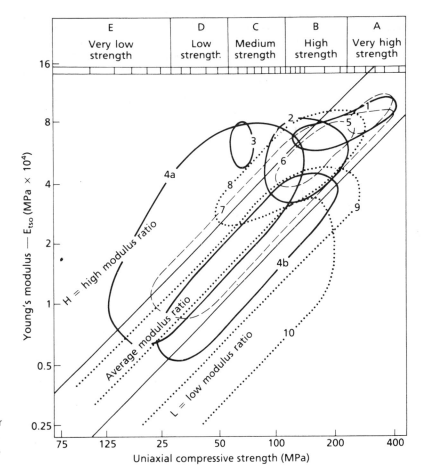

Fig. 12.11 Engineering classification of intact rock based on uniaxial compressive strength and modulus ratio. Fields are shown for igneous, sedimentary and metamorphic rocks. Metamorphic: 1, quartzite; 2, gneiss; 3, marble; 4a, schist, steep foliation; 4b, schist, flat foliation. Igneous: 5, diabase; 6, granite; 7, basalt and other flow rocks. Sedimentary: 8, limestone and dolomite; 9, sandstone; 10, shale. (After Deere & Miller, 1966.)

cation. For example, Deere and Miller (1966) based their engineering classification of intact rock on the unconfined compressive strength and Young's modulus (Fig. 12.11).

12.3 Discontinuities

A discontinuity represents a plane of weakness within a rock mass, across which the rock material is structurally discontinuous. Although discontinuities are not necessarily planes of separation, most in fact are, and they possess little or no tensile strength. Discontinuities vary in size from small fissures to huge faults. The most common discontinuities are joints and bedding planes (Fig. 12.12). Other important discontinuities are planes of cleavage and schistosity.

Joints are fractures along which little or no dis-

placement has occurred, and are present within all types of rock. At the ground surface, joints may open as a consequence of denudation, especially weathering, or the dissipation of residual stress.

Joints are formed through failure in tension, in shear or some combination of both. Rupture surfaces formed by extension tend to be clean and rough with little detritus. They tend to follow minor lithological variations. Simple surfaces of shearing generally are smooth and contain considerable detritus. They are unaffected by local lithological changes.

Joints are also formed in other ways. For example, joints develop within igneous rocks when they initially cool down, and in wet sediments when they dry out. The most familiar of these are the columnar joints in lava flows, sills and some dykes (Fig. 12.13). It has been noted that the frequency of sheet jointing is related to the depth of overburden; in other words,

Fig. 12.12 Two systems of diagonally opposed joints in argillaceous limestone of Carboniferous age exposed on the north coast of Northumberland, England.

Fig. 12.13 Columnar jointing in basalt, Giant's Causeway, Northern Ireland.

the thinner the rock cover, the more pronounced the sheeting. This suggests a connection between the removal of the overburden by denudation and the development of sheeting. Indeed, such joints often have developed suddenly during quarrying operations. It may well be that some granitic intrusions contain considerable residual strain energy, and that, with the gradual removal of load, the associated residual stresses are dissipated by the formation of sheet joints.

12.3.1 Description of discontinuities in rock masses

The shear strength of a rock mass and its deformability are very much influenced by the discontinuity pattern, its geometry and how well it is developed. The observation of discontinuity spacing, whether in a field exposure or in a core stick, aids in the appraisal of the rock mass structure. In sedimentary rocks, bedding planes usually are the dominant discontinuity, and the rock mass can be described as shown in

Table 12.9 Description of bedding plane and joint spacing.

Description of bedding plane spacing	Description of joint spacing	Limits of spacing
Very thickly bedded	Extremely wide	Over 2 m
Thickly bedded	Very wide	0.6–2 m
Medium bedded	Wide	0.2–0.6 m
Thinly bedded	Moderately wide	60 mm–0.2 m
Very thinly bedded	Moderately narrow	20–60 mm
Laminated	Narrow	6–20 mm
Thinly laminated	Very narrow	Under 6 mm

Table 12.9. The same boundaries can be used to describe the spacing of joints.

As joints generally represent surfaces of weakness, the larger and more closely spaced they are, the more influential they become in reducing the effective strength of a rock mass. The persistence of a joint plane refers to its continuity. This is one of the most difficult properties to quantify, because joints frequently continue beyond the rock exposure and, con-

sequently, in such instances, it is impossible to estimate their continuity.

The block size provides an indication of how a rock mass is likely to behave, because the block size and interblock shear strength determine the mechanical performance of a rock mass under given conditions of stress. The following descriptive terms have been recommended for the description of rock masses, in order to convey an impression of the shape and size of the blocks of rock material (Barton, 1978):

1 Massive—few joints or very wide spacing.
2 Blocky—approximately equidimensional.
3 Tabular—one dimension considerably shorter than the other two.
4 Columnar — one dimension considerably larger than the other two.
5 Irregular—wide variations of block size and shape.
6 Crushed—heavily jointed to 'sugar cube'.

The block size may be described using the terms given in Table 12.10.

Discontinuities, especially joints, may be open or closed. How open they are (Table 12.11) is of

Table 12.10 Block size and equivalent discontinuity spacing. (After Anon., 1977; Barton, 1978.)

Term	Block size	Equivalent discontinuity spacing in blocky rock	Volumetric joint count (J_v) (joints m^{-3})
Very large	Over 8 m^3	Extremely wide	Less than 1
Large	0.2–8 m^3	Very wide	1–3
Medium	0.008–0.2 m^3	Wide	3–10
Small	0.0002–0.008 m^3	Moderately wide	10–30
Very small	Less than 0.0002 m^3	Less than moderately wide	Over 30

Table 12.11 Description of the aperture of discontinuity surfaces

Anon. (1977)		Barton (1978)		
Description	Width of aperture	Description		Width of aperture
Tight	Zero	Closed	Very tight	Less than 0.1 mm
Extremely narrow	Less than 2 mm		Tight	0.1–0.25 mm
Very narrow	2–6 mm		Partly open	0.25–0.5 mm
Narrow	6–20 mm	Gapped	Open	0.5–2.5 mm
Moderately narrow	20–60 mm		Moderately wide	2.5–10 mm
Moderately wide	60–200 mm		Wide	Over 10 mm
Wide	Over 200 mm	Open	Very wide	10–100 mm
			Extremely wide	100–1000 mm
			Cavernous	Over 1 m

Table 12.12 Assessment of seepage from open and filled discontinuities. (After Barton, 1978.)

| Seepage rating | Open discontinuities | Filled discontinuities |
	Description	Description
1	The discontinuity is very tight and dry, water flow along it does not appear to be possible	The filling material is heavily consolidated and dry, significant flow appears to be unlikely due to very low permeability
2	The discontinuity is dry with no evidence of water flow	The filling materials are damp, but no free water is present
3	The discontinuity is dry, but shows evidence of water flow, i.e. rust staining, etc.	The filling materials are wet, occasional drops of water
4	The discontinuity is damp, but no free water is present	The filling materials show signs of outwash, continuous flow of water (estimate l min^{-1})
5	The discontinuity shows seepage, occasional drops of water, but no continuous flow	The filling materials are washed out locally, considerable water flow along outwash channels (estimate l min^{-1} and describe pressure, i.e. low, medium, high)
6	The discontinuity shows a continuous flow of water (estimate l min^{-1} and describe pressure, i.e. low, medium, high)	The filling materials are washed out completely, very high water pressures are experienced, especially on first exposure (estimate l min^{-1} and describe pressure)

importance in relation to the overall strength and permeability of a rock mass, and this often depends largely on the amount of weathering which the rocks have suffered. Some joints may be partially or completely filled. The type and amount of filling influence not only the effectiveness with which the opposing joint surfaces are bound together, thereby affecting the strength of the rock mass, but also the permeability. If the infilling is sufficiently thick, the walls of the joint will not be in contact, and hence the strength of the joint plane will be that of the infill material.

The nature of the opposing joint surfaces also influences the rock mass behaviour, the smoother they are, the more easily movement can take place along them. However, joint surfaces are usually rough and may be slickensided. Hence, the nature of a joint surface may be considered in relation to its waviness, roughness and the condition of the walls. The waviness and roughness differ in terms of scale and their effect on the shear strength of the joint. Waviness refers to first-order asperities, which appear as undulations of the joint surface, and are not likely to shear off during movement. Therefore the effects of waviness do not change with displacements along the joint surface. Waviness modifies the apparent angle of dip but not the frictional properties of the discontinuity. On the other hand, roughness refers to second-order asperities, which are sufficiently small to be sheared off during movement. An increased roughness of the discontinuity walls results in an increased effective friction angle along the joint surface. These effects diminish or disappear when infill is present.

The compressive strength of the rock comprising the walls of a discontinuity is a very important component of the shear strength and deformability, especially if the walls are in direct rock to rock contact. Weathering (and alteration) frequently is concentrated along the walls of discontinuities, thereby reducing their strength.

The seepage of water through rock masses usually takes place via the discontinuities, although, some sedimentary rocks, seepage through the pores may also play an important role. The prediction the groundwater level, probable seepage paths and approximate water pressures frequently provide an indication of stability or construction problems. Barton (1978) suggested that seepage from open filled discontinuities could be assessed according the descriptive scheme shown in Table 12.12.

12.3.2 Strength of discontinuous rock masses and its assessment

Discontinuities in a rock mass reduce its effective shear strength, at least in a direction parallel to discontinuities. Hence, the strength of discontinuous rock masses is highly anisotropic. Discontinuities offer no resistance to tension, whereas they offer high resistance to compression. Nevertheless, they may deform under compression if crushable asperities, compressible filling or apertures are present along the discontinuity, or if the wall rock is altered.

Barton (1976) proposed the following empirical expression to derive the shear strength τ along joint surfaces:

$$\tau = \sigma_n \tan\left[JRC \log_{10}\left(JCS/\sigma_n \right) + \phi_b \right] \quad (12.1)$$

where σ_n is the effective normal stress, JRC is the joint roughness coefficient, JCS is the joint wall compressive strength and ϕ_b is the basic friction angle. According to Barton, the values of the joint roughness coefficient range from zero to 20, from the smoothest to the roughest surface (Fig. 12.14). The joint wall

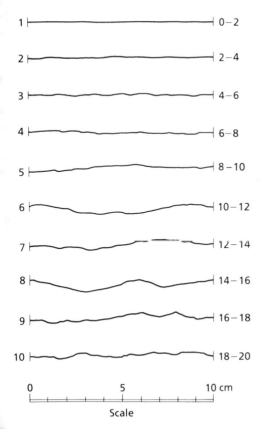

Fig. 12.14 Roughness profiles and corresponding range of JRC values. (After Barton, 1976.)

compressive strength is equal to the unconfined compressive strength of the rock if the joint is unweathered. This may be reduced by up to 75% when the walls of the joints are weathered. Both of these factors are related, as smooth-walled joints are less affected by the value of JCS, because the failure of asperities plays a less important role. The smoother the walls of the joints, the more significant is the part played by the mineralogy (ϕ_b). The experience gained from rock mechanics indicates that, under low effective normal stress levels, such as those that occur in engineering, the shear strength of joints can vary within relatively wide limits. The maximum effective normal stress acting across joints, considered to be critical for stability, lies, according to Barton, in the range 0.1–2.0 MPa.

Subsequently, Hoek and Brown (1980) proposed that the peak triaxial compressive strengths of a wide range of rock materials could be described by the expression

$$\sigma_1 = \sigma_3 \left(m\sigma_c\sigma_3 + s\sigma_c^2 \right)^{1/2} \quad (12.2)$$

where σ_1 is the major principal stress at failure, σ_3 is the minor principal stress (the confining pressure in the case of a triaxial test), σ_c is the uniaxial compressive strength of the intact rock and m and s are dimensionless constants, which are approximately analogous to the angle of friction and cohesion. The constant m varies with the rock type, ranging from about 0.001 for rock masses containing heavily weathered joint sets to about 25 for certain hard, intact rock (Table 12.13). For intact rock, s is unity. In 1983, Hoek suggested that the shear strength τ along a discontinuous surface could be obtained from

$$\tau = \left(\cot\phi_i - \cos\phi_i \right)\frac{m\sigma_c}{8} \quad (12.3)$$

where ϕ_i is the instantaneous angle of friction at given values of τ and σ_1.

12.3.3 Discontinuities and rock quality indices

Several attempts have been made to relate the numerical intensity of fractures to the quality of unweathered rock masses, and to quantify their effect on deformability. For example, the concept of rock quality designation (RQD) was introduced by Deere (1964). It is based on the percentage core recovery when drilling rock with NX (57.2 mm) or larger

Table 12.13 Approximate relationship between rock mass quality and material constants. (After Hoek, 1983.)

Empirical failure criterion
$\sigma_1 = \sigma_3 + (m\sigma_c\sigma_3 + s\sigma_c^2)^{1/2}$
σ_1 = major principal stress
σ_3 = minor principal stress
σ_c = uniaxial compressive strength of intact rock
m, s = empirical constants

	Carbonate rocks with well-developed crystal cleavage, e.g. dolostone, limestone and marble	Lithified argillaceous rocks, e.g. mudstone, siltstone, shale and slate (tested normal to cleavage)	Arenaceous rocks with strong crystals and poorly developed crystal cleavage, e.g. sandstone and quartzite	Fine grained polymineralic igneous crystalline rocks, e.g. andesite, dolerite, diabase and rhyolite	Coarse grained polymineralic igneous and metamorphic crystalline rocks, e.g. amphibolite, gabbro, gneiss, granite, norite and quartz diorite
Intact rock samples Laboratory-sized samples free from pre-existing fractures	$m = 7$ $s = 1$	$m = 10$ $s = 1$	$m = 15$ $s = 1$	$m = 17$ $s = 1$	$m = 25$ $s = 1$
Very good quality rock mass Tightly interlocking, undisturbed rock with rough, unweathered joints spaced at 1–3 m	$m = 3.5$ $s = 0.1$	$m = 5$ $s = 0.1$	$m = 7.5$ $s = 0.1$	$m = 8.5$ $s = 0.1$	$m = 12.5$ $s = 0.1$
Good quality rock mass Fresh to slightly weathered rock, slightly disturbed with joints spaced at 1–3 m	$m = 0.7$ $s = 0.004$	$m = 1$ $s = 0.004$	$m = 1.5$ $s = 0.004$	$m = 1.7$ $s = 0.004$	$m = 2.5$ $s = 0.004$
Fair quality rock mass Several sets of moderately weathered joints spaced at 0.3–1 m disturbed	$m = 0.14$ $s = 0.0001$	$m = 0.20$ $s = 0.0001$	$m = 0.30$ $s = 0.0001$	$m = 0.34$ $s = 0.0001$	$m = 0.50$ $s = 0.0001$
Poor quality rock mass Numerous weathered joints at 30–500 mm with some gouge. Clean, compacted rockfill	$m = 0.04$ $s = 0.00001$	$m = 0.05$ $s = 0.00001$	$m = 0.08$ $s = 0.00001$	$m = 0.09$ $s = 0.00001$	$m = 0.13$ $s = 0.00001$
Very poor quality rock mass Numerous heavily weathered joints spaced at 50 mm with gouge. Waste rock	$m = 0.007$ $s = 0$	$m = 0.010$ $s = 0$	$m = 0.015$ $s = 0$	$m = 0.017$ $s = 0$	$m = 0.025$ $s = 0$

Table 12.14 Classification of rock quality in relation to the incidence of discontinuities.

Quality classification	RQD (%)	Fracture frequency per metre	Velocity ratio (V_{cf}/V_{cl})	Mass factor (j)
Very poor	0–25	Over 15	0.0–0.2	—
Poor	25–50	15–8	0.2–0.4	Less than 0.2
Fair	50–75	8–5	0.4–0.6	0.2–0.5
Good	75–90	5–1	0.6–0.8	0.5–0.8
Excellent	90–100	Less than 1	0.8–0.10	0.8–0.10

diameter diamond core drills. Assuming that a consistent standard of drilling can be maintained, the percentage of solid core obtained depends on the strength and degree of discontinuities in the rock mass concerned. The *RQD* is the sum of the core sticks in excess of 100 mm expressed as a percentage of the total length of core drilled. However, the *RQD* does not take into consideration the joint opening and condition; a further disadvantage is that, with fracture spacings greater than 100 m, the quality is excellent irrespective of the actual spacing (Table 12.14). This particular difficulty can be overcome by using the fracture spacing index as suggested by Franklin *et al.* (1971). This simply refers to the frequency per metre with which fractures occur within a rock mass (Table 12.14).

The effect of discontinuities in a rock mass can be estimated by comparing the *in situ* compressional wave velocity with the laboratory sonic velocity of an intact core sample obtained from the rock mass. The difference in these two velocities is caused by the discontinuities which exist in the field. The velocity ratio, V_{cf}/V_{cl}, where V_{cf} and V_{cl} are the compressional wave velocities of the rock mass *in situ* and of the intact specimen respectively, was first proposed by Onodera (1963). For a high-quality massive rock with only a few tight joints, the velocity ratio approaches unity. As the degree of jointing and fracturing becomes more severe, the velocity ratio is reduced (Table 12.14). The sonic velocity is determined for the core sample in the laboratory under an axial stress equal to the computed overburden stress at the depth from which the rock material was taken, and at a moisture content equivalent to that assumed for the *in situ* rock. The field seismic velocity preferably is determined by up-hole or cross-hole seismic measurements in drill-holes (McCann, 1992) or test adits, because, by using these measure-ments, it is possible to explore individual homogeneous zones more precisely than by surface refraction surveys.

An estimate of the numerical value of the deformation modulus of a jointed rock mass can be obtained from various *in situ* tests. The values derived from such tests are always smaller than those determined in the laboratory from intact core specimens, and the more heavily the rock mass is jointed, the larger the discrepancy between the two values. Thus, if the ratio between these two values of the deformation modulus is obtained from a number of locations on a site, the rock mass quality can be evaluated. Accordingly, the concept of the rock mass factor (j) was introduced by Hobbs (1975), who defined it as the ratio of the deformability of a rock mass to the deformability of the intact rock (Table 12.14).

12.3.4 Discontinuity surveys

One of the most widely used methods of collecting discontinuity data is simply by direct measurement on the ground. A direct survey can be carried out subjectively, in that only those structures which appear to be important are measured and recorded. In a subjective survey, the effort can be concentrated on the apparently significant joint sets. Nevertheless, there is a risk of overlooking sets which might be important. Conversely, in an objective survey, all discontinuities intersecting a fixed line or area of the rock face are measured and recorded.

Several methods have been used to carry out direct discontinuity surveys. In the fracture set mapping technique, all discontinuities occurring in 6 m × 2 m zones, spaced at 30 m intervals along the face, are recorded. Alternatively, the use of a series of line scans is a satisfactory method of joint surveying. The technique involves extending a metric tape

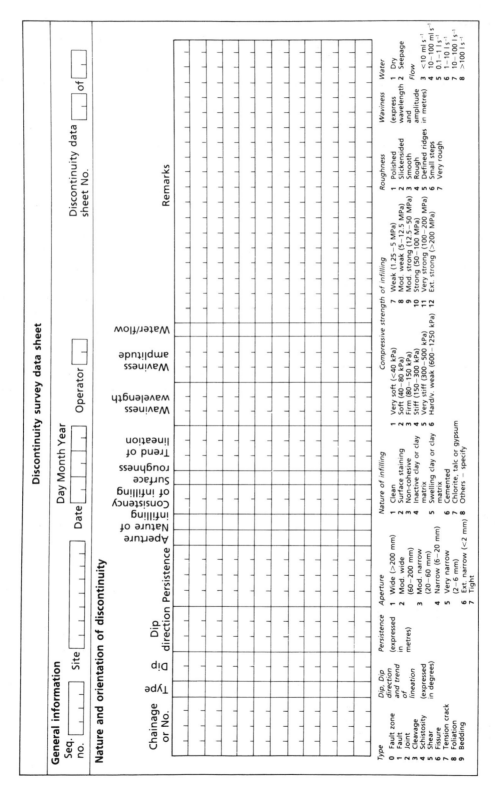

Fig. 12.15 Discontinuity survey data sheet. (After Anon., 1977.)

across an exposure, levelling the tape and then securing it to the face. Two other scan lines are set out as near as possible at right angles to the first, one more or less vertical and the other horizontal. The distance along a tape at which each discontinuity intersects is noted, as is the direction of the pole to each discontinuity (this provides an indication of the dip direction). The dip of the pole from the vertical is recorded, as this is equivalent to the dip of the plane from the horizontal. The strike and dip directions of discontinuities in the field can be measured with a compass, and the amount of dip with a clinometer. The measurement of the length of a discontinuity provides information on its continuity. Measurements should be taken over distances of about 30 m and, to ensure that the survey is representative, the measurements should be continuous over that distance. A minimum of at least 200 readings per locality is recommended to ensure statistical

reliability. A summary of the other details which should be recorded about discontinuities is given in Fig. 12.15.

The value of data on discontinuities gathered from orientated cores from drill-holes depends, in part, on the quality of the rock concerned, in that poor quality rock is likely to be lost during drilling. However, it is impossible to assess the persistence, degree of separation or nature of the joint surfaces. Moreover, infill material, especially if it is soft, is not recovered by the drilling operation.

The core orientation can be achieved by using a core orientator (Fig. 12.16a). In the Craelius core orientator, the teeth clamp the instrument in position inside the core barrel. The housing contains a soft aluminium ring, against which a ball bearing is indented by pressure from the cone when it reaches the bottom of the hole. The cone is released by pressure against the core stub and, when released, locks the probe in position, and releases the clamping teeth to allow the instrument to ride up inside the barrel ahead of the core.

In integral sampling, a drill-hole (diameter D) is drilled to a depth at which the integral sample is to be obtained; another hole (diameter D_1) is then drilled

Fig. 12.16 (a) Details and method of operation of the Craelius core orientator. The probes take up the profile of the core stub left by the previous drilling run, and are locked in position when the spring-loaded cone is released. (b) Stages of the integral sampling method.

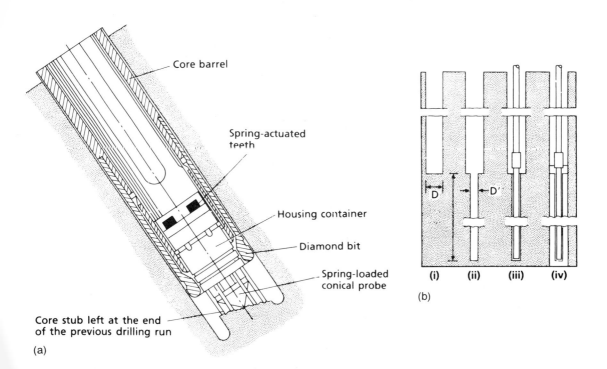

Core barrel

Spring-actuated teeth

Housing container

Diamond bit

Spring-loaded conical probe

Core stub left at the end of the previous drilling run

(a)

D D'

(i)   (ii)   (iii)   (iv)

(b)

coaxial with the former and with the same length as the required sample, and a reinforcing bar is placed in this hole (Fig. 12.16b). The bar is then grouted to the rock mass. The integral sample is obtained by over-drilling with diameter D. The method has been used with success in all types of rock mass, from massive to highly weathered varieties, and provides information on the spacing, orientation, opening and infilling of discontinuities.

Drill-hole inspection techniques include the use of drill-hole periscopes, drill-hole cameras or closed-circuit television. The drill-hole periscope affords direct inspection, and can be orientated from outside the hole. However, its effective use is limited to about 30 m. The drill-hole camera also can be orientated prior to photographing a section of the wall of a drill-hole. The television camera provides a direct view of the drill-hole, and a recording can be made on video-tape. These three systems are limited, in that they require relatively clear conditions, and so may be of little use below the water table, particularly if the water in the drill-hole is murky. The televiewer pro-duces an acoustic picture of the drill-hole wall. One of its advantages is that drill-holes need not be flushed prior to its use.

Many data relating to discontinuities can be obtained from photographs of exposures. Pho-tographs may be taken looking horizontally at the rock mass from the ground, or they may be taken from the air looking vertically or, occasionally, obliquely down at the outcrop. These photographs may or may not have survey control. Uncontrolled photographs are taken using hand-held cameras. Stereopairs are obtained by taking two photographs of the same face, from positions about 5% of the dis-tance of the face apart, along a line parallel to the face. Delineation of major discontinuity patterns and preliminary subdivision of the face into structural zones can be made from these photographs. Unfortu-nately, data cannot be transferred with accuracy from them onto maps and plans. Conversely, discontinuity data can be accurately located on maps and plans by using controlled photographs. Controlled pho-tographs are obtained by aerial photography with complementary ground control, or by ground-based phototheodolite surveys. Aerial photographs, with a suitable scale, have proved to be useful in the investi-gation of discontinuities. Photographs taken with a phototheodolite can also be used with a stereocom-parator which produces a stereoscopic model. Mea-surements of the locations or points in the model can be made with an accuracy of approximately 1 in 5000 of the mean object distance. As a consequence, a point on a face photographed from 50 m can be located to an accuracy of 10 mm. In this way, the fre-quency, orientation and continuity of discontinuities can be assessed. Such techniques prove to be particu-larly useful when faces which are inaccessible or unsafe have to be investigated.

Data from a discontinuity survey usually are plotted on a stereographic projection. The use of spherical projections, commonly the Schmidt or Wulf net, means that traces of the planes on the surface of the 'reference sphere' can be used to define the dips and dip directions of discontinuity planes. In other words, the inclination and orientation of a particular plane can be represented by a great circle or a pole, normal to the plane, which is traced on an overlay placed over the stereonet. The method whereby great circles or poles are plotted on a stereogram has been explained by Hoek and Bray (1981). When recording field observations of the direction and amount of dip of discontinuities, it is convenient to plot the poles rather than the great circles. The poles can then be contoured in order to provide an expression of the orientation concentration. This affords a qualitative appraisal of the influence of the discontinuities on the engineering behaviour of the rock mass concerned (Fig. 12.17).

12.4 Weathering

Weathering of rocks is brought about by physical dis-integration, chemical decomposition and biological activity. The agents of weathering, unlike those of erosion, do not themselves provide for the trans-portation of debris from a rock surface. Therefore, unless this rock waste is otherwise removed, it even-tually acts as a protective cover, preventing further weathering taking place. If weathering is to be con-tinuous, fresh rock exposures must be revealed con-stantly, which means that the weathered debris must be removed by the action of gravity, running water or wind.

The rate at which weathering takes place depends not only on the vigour of the weathering agent(s), but also on the durability of the rock itself. The latter is governed by the mineralogical composition, texture

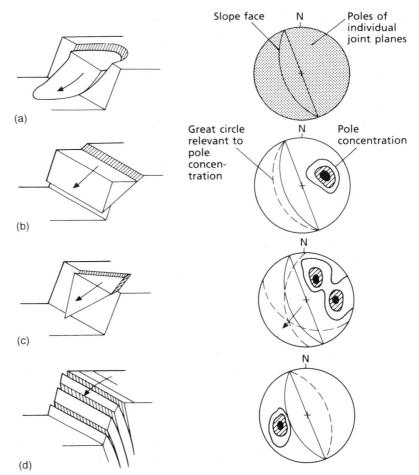

Fig. 12.17 Representation of structural data concerning four possible slope failure modes plotted on equal-area stereonets as poles, which are contoured to show the relative concentration, and great circles: (a) circular failure in heavily jointed rock with no identifiable structural pattern; (b) plane failure in highly ordered structure such as slate; (c) wedge failure on two intersecting sets of joints; (d) toppling failure caused by steeply dipping joints. (After Hoek & Bray, 1981.)

and porosity of the rock, and the incidence of discontinuities within the rock mass.

12.4.1 Mechanical weathering

Mechanical or physical weathering is effective particularly in climatic regions which experience significant diurnal changes in temperature. This does not necessarily imply a large range of temperature, as frost and thaw action can proceed where the range is limited.

As far as the frost susceptibility is concerned, the porosity, pore size and degree of saturation all play an important role. When water turns to ice, it increases in volume by up to 9%, thus giving rise to an increase in pressure within the pores. This action is enhanced further by the displacement of pore water away from the developing ice front. Once ice has formed, the ice pressures rapidly increase with decreasing temperature, so that, at approximately −22°C, ice can exert a pressure of up to 200 MPa. Usually, coarse grained rocks withstand freezing better than fine grained types. The critical pore size for freeze–thaw durability appears to be about 0.005 mm. In other words, rocks with larger mean pore diameters allow outward drainage and escape of fluid from the frontal advance of the ice line, and therefore are less frost susceptible. Alternate freeze–thaw action causes cracks, fissures, joints and some pore spaces to be widened. As the process advances, angular rock debris gradually is broken from the parent body.

The mechanical effects of weathering are well displayed in hot deserts, where wide diurnal ranges of

Fig. 12.18 Weathering in granite, near Grunau, Namibia.

temperature cause rocks to expand and contract. Because rocks are poor conductors of heat, these effects mainly are localized in their outer layers, where alternate expansion and contraction create stresses which eventually rupture the rock. In this way, flakes of rock break away from the parent material, the process being termed exfoliation. The effects of exfoliation are concentrated at the corners and edges of rocks, so that their outcrops gradually become rounded (Fig. 12.18).

12.4.2 Chemical and biological weathering

Chemical weathering leads to mineral alteration and the solution of rocks. Alteration is caused principally by oxidation, hydration, hydrolysis and carbonation, whereas solution is brought about by acidified or alkalinized waters. Chemical weathering also aids rock disintegration by weakening the rock fabric, and by emphasizing any structural weaknesses, however slight, that it may possess. When decomposition occurs within a rock, the altered material frequently occupies a greater volume than that from which it was derived, and, in the process, internal stresses are generated. If this swelling occurs in the outer layers of a rock, it causes them to peel off from the parent body.

In dry air, rocks decay very slowly. The presence of moisture hastens the rate tremendously, first, because water is itself an effective agent of weathering and,

second, because it holds in solution substances which react with the component minerals of the rock. The most important of these substances are free oxygen, carbon dioxide, organic acids and nitrogen acids.

Free oxygen is an important agent in the decay of all rocks containing oxidizable substances, iron and sulphur being particularly suspect. The rate of oxidation is quickened by the presence of water; indeed, it may enter into the reaction itself as, for example, in the formation of hydrates. However, its role is chiefly that of a catalyst. Carbonic acid is produced when carbon dioxide is dissolved in water, and it may possess a pH value of about pH 5.7. The principal source of carbon dioxide is not the atmosphere, but the air contained in the pore spaces in the soil, where its proportion may be a hundred or so times greater than it is in the atmosphere. An abnormal concentration of carbon dioxide is released when organic material decays. Furthermore, humic acids are formed by the decay of humus in soil waters; they ordinarily have pH values between pH 4.5 and pH 5.0, but occasionally they may be under pH 4.0.

The simplest reactions which take place on weathering are the solution of soluble minerals and the addition of water to substances to form hydrates. Solution commonly involves ionization, for example, this takes place when gypsum and carbonate rocks are weathered. Hydration and dehydration take place in some substances, a common example being gypsum and anhydrite. When anhydrite is hydrated to form gypsum, an increase in volume of approximately 6% occurs and, accordingly, causes the enclosing rocks to be wedged further apart.

Iron oxides and hydrates are conspicuous products of weathering. Usually, the oxides are a shade of red and the hydrates yellow to dark brown.

Sulphur compounds are oxidized readily by weathering. Because of the hydrolysis of the dissolved metal ion, solutions formed from the oxidation of sulphides are acidic. For instance, when pyrite initially is oxidized, ferrous sulphate and sulphuric acid are formed. Further oxidation leads to the formation of ferric sulphate. Very insoluble ferric oxide or hydrated oxide is formed if highly acidic conditions are produced.

Perhaps the most familiar example of a rock prone to chemical attack is limestone. Limestones are composed chiefly of calcium carbonate. They are subjected to acid attack because CO_3 readily combines with H to form the stable bicarbonate, HCO_3. In water with a temperature of 25 °C, the solubility of calcium carbonate ranges from 0.01 to $0.05\,gl^{-1}$, depending upon the degree of saturation with carbon dioxide. Dolostone is somewhat less soluble than limestone. When a limestone is subjected to dissolution, any insoluble material present in it remains behind.

Weathering of silicate minerals primarily is a process of hydrolysis. Much of the silica which is released by weathering forms silicic acid but, when it is liberated in large quantities, some of it may form colloidal or amorphous silica. Mafic silicates usually decay more rapidly than felsic silicates and, in the process, they release magnesium, iron and lesser amounts of calcium and alkalis. Olivine particularly is unstable, decomposing to form serpentine, which, on further weathering, forms talc and carbonates. Chlorite is the commonest alteration product of augite and of hornblende. When subjected to chemical weathering, feldspars decompose to form clay minerals; the latter are, consequently, the most abundant residual products. The process is effected by the hydrolysing action of weakly carbonated water, which leaches the bases out of the feldspars and produces clays in colloidal form. The alkalis are removed in solution as carbonates from orthoclase (K_2CO_3) and albite (Na_2CO_3), and as bicarbonate from anorthite ($Ca(HCO_3)_2$). Some silica is hydrolysed to form silicic acid. The colloidal clay eventually crystallizes as an aggregate of minute clay minerals.

Plants and animals play an important role in the breakdown and decay of rocks. Indeed, their part in soil formation is of major significance. Tree roots penetrate cracks in rocks, and gradually wedge the sides apart, whereas the adventitious root system of grasses breaks down small rock fragments to particles of soil size. Burrowing rodents also bring about mechanical disintegration of rocks. The action of bacteria and fungi largely is responsible for the decay of dead organic matter. Other bacteria are responsible, for example, for the reduction of iron or sulphur compounds.

12.4.3 Engineering classification of weathering

Several attempts have been made to devise an engineering classification of weathered rock. The problem can be tackled in two ways. One method is to attempt to assess the grade of weathering by reference to some simple index test. Such methods provide a quantitative, rather than qualitative, assessment. When coupled with a grading system, this means that the disadvantages inherent in these simple index tests largely are overcome.

For instance, Iliev (1967) developed a coefficient of weathering (K) for granitic rock, based upon the ultrasonic velocities of the rock material, according to the expression

$$K = (V_u - V_w)/V_u \qquad (12.4)$$

where V_u and V_w are the ultrasonic velocities of the fresh and weathered rock, respectively. A quantitative index indicating the grade of weathering, as determined from the ultrasonic velocity and the corresponding coefficient of weathering, is shown in Table 12.15.

Table 12.15 Ultrasonic velocity and grade of weathering.

Grade of weathering	Ultrasonic velocity ($m\,s^{-1}$)	Coefficient of weathering
Fresh	Over 5000	0
Slightly weathered	4000–5000	0–0.2
Moderately weathered	3000–4000	0.2–0.4
Strongly weathered	2000–3000	0.4–0.6
Very strongly weathered	Under 2000	0.6–1.0

Table 12.16 Weathering indices for granite. (After Irfan & Dearman, 1978a.)

Type of weathering	Quick absorption (%)	Bulk density (Mg m⁻³)	Point load strength (MPa)	Unconfined compressive strength (MPa)
Fresh	Less than 0.2	Over 2.61	Over 10	Over 250
Partially stained*	0.2–1.0	2.56–2.61	6–10	150–250
Completely stained*	1.0–2.0	2.51–2.56	4–6	100–150
Moderately weathered	2.0–10.0	2.05–2.51	0.1–4	2.5–100
Highly/completely weathered	Over 10	Less than 2.05	Less than 0.1	Less than 2.5

*Slightly weathered.

After an extensive testing programme, Irfan and Dearman (1978a) concluded that the quick absorption, Schmidt hammer and point load strength tests were reliable field tests for the determination of a quantitative weathering index for granite (Table 12.16). This index can be related to the various grades of weathering recognized by visual determination and given in Table 12.17.

As the mineral composition and texture influence the physical properties of a rock, petrographical techniques can be used to evaluate the successive stages in mineralogical and textural changes brought about by weathering. Accordingly, Irfan and Dearman (1978b) also developed a quantitative method of assessing the grade of weathering of granite in terms of its megascopic and microscopic petrography. The megascopic factors included an evaluation of the amount of discoloration, decomposition and disintegration shown by the rock. The microscopic analysis involved the assessment of the mineral composition and degree of alteration by modal analysis, and a microfracture analysis.

Another method of assessing the grade of weathering is based on a simple description of the geological character of the rock concerned, as seen in the field, the description incorporating different grades of weathering which are related to engineering performance (Table 12.17). This approach was first developed by Moye (1955), who proposed a grading system for the degree of weathering found in granite at the Snowy Mountains Scheme in Australia.

Similar classifications subsequently were advanced, and were based mainly on the degree of chemical decomposition exhibited by a rock mass. They were directed primarily towards weathering in granitic

rocks. Dearman (1974) suggested descriptions which could be used to establish the grade of mechanical weathering and of solution weathering of relatively pure carbonate rock (Table 12.17). Others, working on different rock types, have proposed modified classifications of weathering grade. For example, Lovegrove and Fookes (1972) made slight variations in their identification of grades of weathering of volcanic tuffs and associated sediments in Fiji. Classifications of weathered chalk and weathered marl (mudstone) have been developed by Ward et al. (1968) and Chandler (1969), respectively. The latest review of the description and classification of weathered rocks has been provided by Anon. (1995).

Usually, the grades will lie one above the other in a weathered profile developed from a single rock type, the highest grade being at the surface. However, this is not necessarily the case in complex geological conditions. Even so, the concept of the grade of weathering can still be applied. Such a classification can be used to produce maps or sections showing the distribution of the grade of weathering at a particular site (Fig. 12.19). The dramatic effect of weathering upon the strength of rock is illustrated, according to the grade of weathering, in Fig. 12.20.

12.5 Igneous and metamorphic rocks

The plutonic igneous rocks are characterized by a granular texture, massive structure and relatively homogeneous composition. In their unaltered state, they essentially are sound and durable, with adequate strength for any engineering requirement (Table 12.1). In some instances, however, intrusives may be highly altered by weathering or hydrothermal attack

Table 12.17 Engineering grade classifications of weathered rock and their relation to engineering behaviour.

Grade	Degree of decomposition	Field recognition (After Little, 1969, Fookes et al., 1972; Dearman, 1974)			Engineering behaviour		
		Rocks (mainly chemical decomposition)	Rocks (physical disintegration)*	Carbonate rocks (solution)	After Little (1969)	After Hobbs (1975)	After Martin and Hencher (1986)*
VI	Residual soil	The rock is discoloured and is completely changed to a soil, in which the original fabric of the rock is completely destroyed. There is a large volume change	The rock is changed to a soil by granular disintegration and/or grain fracture. The structure of the rock is destroyed and the soil is a residuum of minerals unaltered from the original rock	Grades V and VI cannot occur. These grades can be applied to interbedded, soluble and insoluble rocks. Void size should be recorded	Unsuitable for important foundations. Unstable on slopes when vegetation cover is destroyed, and may erode easily unless hard cap is present. Requires selection before use as fill	In completely weathered rock and residual soil, it may be possible to obtain fair quality samples, depending upon the parent rock type and the consistency of the product.	A soil mixture with the original texture of the rock completely destroyed
V	Extremely weathered	The rock is discoloured and is wholly decomposed and friable, but the original fabric is mainly preserved. The properties of the rock mass depend in part on the nature of the parent rock. In granite rocks, feldspars are completely kaolinized	The rock is changed to a soil by granular disintegration and/or fracture. The structure of the rock is preserved		Cannot be recovered as cores by ordinary rotary drilling methods. Can be excavated by hand or ripping without the use of explosives. Unsuitable for foundations of concrete dams or large structures. May be suitable for foundations of earth dams and for fill. Unstable in high cuttings at steep angles. New joint patterns may have formed. Requires erosion protection	Generally, the samples will tend to be less disturbed than when taken in the highly weathered state. The bearing capacity and settlement characteristics of rock in these extreme states can be assessed using the usual methods for testing soils	No rebound from Schmidt hammer; slakes readily in water; geological pick easily indents when pushed into surface; rock is wholly decomposed, but rock texture preserved

Continued on p. 390

Table 12.17 Continued.

Grade	Degree of decomposition	Field recognition (After Little, 1969; Fookes et al., 1972; Dearman, 1974)			Engineering behaviour		
		Rocks (mainly chemical decomposition)	Rocks (physical disintegration)*	Carbonate rocks (solution)	After Little (1969)	After Hobbs (1975)	After Martin and Hencher (1986)*
IV	Highly weathered†	The rock is discoloured; discontinuities may be open and have discoloured surfaces (e.g. stained by limonite), and the original fabric of the rock near the discontinuities is altered; alteration penetrates deeply inwards, but corestones are still present. The rock mass is partially friable. Less than 50% rock	More than 50% and less than 100% of the rock is disintegrated by open discontinuities or spheroidal scaling spaced at 60 mm or less and/or by granular disintegration. The structure of the rock is preserved	More than 50% of the rock has been removed by solution. A small residuum may be present in the voids	Similar to Grade V. Sometimes recovered as core by careful rotary drilling. Unlikely to be suitable for foundations of concrete dams. Erratic presence of boulders makes it an unreliable foundation for large structures	In highly weathered rock, difficulties generally will be encountered in obtaining undisturbed samples for testing. If samples are obtained, the strength and modulus generally will be underestimated, frequently by large margins, even with apparently undisturbed samples. In such rocks, in situ tests with either the Menard pressure meter or the plate should be carried out to determine the bearing capacity and settlement characteristics. The greatest difficulties in assessing the bearing capacity and	Schmidt hammer rebound value 0–25; does not slake readily in water; geological pick cannot be pushed into surface; hand penetrometer strength index >250 kPa; rock weakened so that large pieces broken by hand; individual grains plucked from surface

| III | Moderately weathered† | The rock is discoloured: discontinuities may be open and have greater discoloration with the alteration penetrating inwards; the intact rock is noticeably weaker, as determined in the field, than the fresh rock. The rock mass is not friable. 50–90% rock | Up to 50% of the rock is disintegrated by open discontinuities or by spheroidal scaling spaced at 60 mm or less and/or by granular disintegration. The structure of the rock is preserved | Up to 50% of the rock has been removed by solution. A small residuum may be present in the voids. The structure of the rock is preserved. | Possessing some strength; large pieces (e.g. NX drill core) cannot be broken by hand. Excavated with difficulty without the use of explosives. Mostly crushes under bulldozer tracks. Suitable foundations of small concrete structures and rockfill dams. May be suitable for semi-pervious fill. Stability in cuttings depends on structural features, especially joint attitudes | In moderately weathered rock, the intact modulus and strength can be very much lower than in the fresh rock, and thus the j value will be higher than in the fresh state, unless the joints and fractures have been opened by erosion or softened by the accumulation of weathering products. The intact modulus and strength can be measured in the laboratory and the bearing capacity assessed in the same way as for fresh rock. Triaxial tests may be more appropriate than uniaxial tests, and it would be advisable to adopt conservative values for the factor of safety | settlement are likely to be encountered in highly weathered rocks, in which the rock fabric becomes increasingly disintegrated or increasingly more plastic

Schmidt hammer rebound value 25–45; considerably weathered, but possessing strength such that pieces of 55 mm in diameter cannot be broken by hand; rock material not friable |

Continued on p. 392

Table 12.17 *Continued.*

Grade	Degree of decomposition	Field recognition (After Little, 1969; Fookes et al., 1972; Dearman, 1974)			Engineering behaviour		
		Rocks (mainly chemical decomposition)	Rocks (physical disintegration)*	Carbonate rocks (solution)	After Little (1969)	After Hobbs (1975)	After Martin and Hencher (1986)*
II	Slightly weathered	The rock may be slightly discoloured, particularly adjacent to discontinuities, which may be open and have slightly discoloured surfaces; the intact rock is not noticeably weaker than the fresh rock. Some decomposed feldspar in granites. Over 90% rock	100% rock; discontinuities open and spaced at more than 60 mm	100% rock; discontinuity surfaces open. Very slight solution etching of discontinuity surfaces may be present	Requires explosives for excavation. Suitable for concrete dam foundations. Highly permeable through open joints. Often more permeable than the zones above or below. Questionable as concrete aggregate	In faintly and slightly weathered rock, it is possible that the j value, owing to the reduction in stiffness of the joints as a result of penetrative weathering alone, will show a fairly sharp decrease compared with that of the same rock in the fresh state.	Schmidt rebound value >45; more than one blow of geological hammer to break specimen; strength approaches that of fresh rock
I	Fresh rock	The parent rock shows no discoloration, loss of strength or other effects due to weathering	100% rock; discontinuities closed	100% rock; discontinuities closed	Staining indicates water percolation along joints; individual pieces may be loosened by blasting or stress relief, and support may be required in tunnels and shafts	The intact modulus, by definition, is unaffected by penetrative weathering. The safe bearing capacity therefore is not affected by faint weathering, and may be only slightly affected by slight weathering	No visible signs of weathering

* Discontinuity spacing should be recorded.
† The ratio of the original rock to the altered material should be estimated where possible.

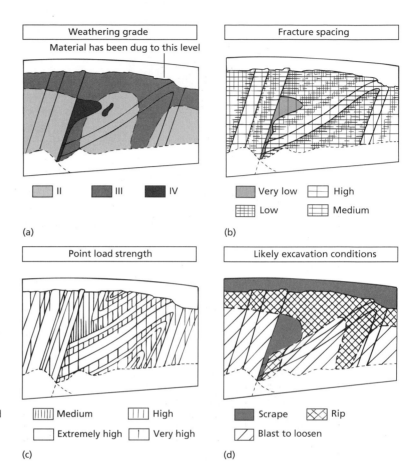

Fig. 12.19 Geotechnical assessment of the quarry face exposed at Knowle Quarry, Okehampton, UK: (a) weathering zones defined in terms of grade; (b) discontinuity spacing variation–fracture spacing classes; (c) strength zones determined by point load strength tests; (d) engineering appraisal in terms of likely excavation conditions. (From Dearman, 1991.)

(Fig. 12.21). Furthermore, fissure zones are by no means uncommon in granites. The rock mass may be very fragmented along such zones; indeed, it may be reduced to sand-sized material (Terzaghi, 1946), and may have undergone varying degrees of kaolinization.

In humid regions, valleys carved in granite may be covered with residual soils which extend to depths often in excess of 30 m. Fresh rock may only be exposed in valley bottoms which have actively degrading streams. At such sites, it is necessary to determine the extent of weathering and the properties of the weathered products. Generally, the weathered product of plutonic rocks has a large clay content, although that of granitic rocks sometimes is porous, with a permeability comparable with that of medium grained sand.

Joints in plutonic rocks often are quite regular, steeply dipping structures in two or more intersecting sets. Sheet joints tend to be approximately parallel to the topographical surface. Consequently, they may introduce a dangerous element of weakness into valley slopes.

Generally speaking, the older volcanic deposits do not prove to be problematical, ancient lavas having strengths frequently in excess of 200 MPa (Table 12.1). However, volcanic deposits of geologically recent age at times can prove to be treacherous. This is because they often represent markedly anisotropic sequences, in which lavas, pyroclasts and mudflows are interbedded. Hence, foundation problems in volcanic sequences arise because weak beds of ash, tuff and mudstone occur within lava piles, giving rise to problems of differential settlement and sliding. In

Weathering description

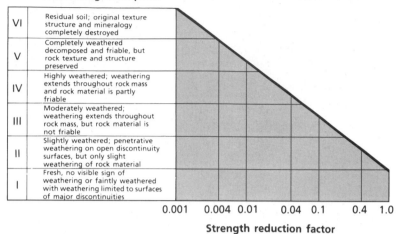

VI	Residual soil; original texture structure and mineralogy completely destroyed	
V	Completely weathered decomposed and friable, but rock texture and structure preserved	
IV	Highly weathered; weathering extends throughout rock mass and rock material is partly friable	
III	Moderately weathered; weathering extends throughout rock mass, but rock material is not friable	
II	Slightly weathered; penetrative weathering on open discontinuity surfaces, but only slight weathering of rock material	
I	Fresh, no visible sign of weathering or faintly weathered with weathering limited to surfaces of major discontinuities	

0.001 0.004 0.01 0.04 0.1 0.4 1.0

Strength reduction factor

Fig. 12.20 Strength reduction as a function of weathering. (After Stacey & Page, 1986.)

Fig. 12.21 Highly weathered granite, decomposed to clayey soil near the surface, near Hillcrest, Natal, South Africa.

addition, weathering during periods of volcanic inactivity may have produced fossil soils, these being of much lower strength.

Clay minerals are formed within basaltic rocks when they are altered or weathered. Their presence can mean that basalt breaks down rapidly once exposed (Fig. 12.22). The disintegration is a result of the swelling of expansive clay minerals when they absorb water. These expansive clay minerals form when basic volcanic glass, or olivine, pyroxene or plagioclase, is subjected to deuteric alteration. This breakdown process has been referred to as slaking (Higgs, 1976). Disintegration can also be brought about in basalts by the absorption of water by zeolites, specifically leonhardite, resulting in expansion. Zeolites commonly are present in amygdales. However, the breakdown of basalts is also dependent on their texture, in that water must have access to those minerals which swell when water is absorbed. In other words, breakdown is governed by the textural accessibility to swelling clay of a basalt.

Fig. 12.22 Highly weathered basalt, northern Lesotho. Note 'onion skin' weathering.

The disintegration of some basalts from Lesotho, according to Haskins and Bell (1995), may take the form of crazing. Crazing is extensive microfracturing that develops in some basalts on exposure to the atmosphere or moisture. These microfractures have a random orientation, tend to be interconnected and vary from hundreds of nanometres up to 1 mm in width. They expand with time, causing the basalt to disintegrate into gravel-sized fragments. The microfractures may form a radial pattern around some amygdales, suggesting that the amygdales play an important role in the development of this disintegration process. For example, crazing fractures may extend from areas containing zeolites, either in vesicles or veins. The microfractures also exploit mineralogical and structural weaknesses in the rock, such as grain boundaries and altered minerals.

Individual lava flows may be thin and transected by a polygonal pattern of cooling joints. They may also be vesicular or contain pipes, cavities or even tunnels (Fig. 2.14). Pyroclasts usually give rise to extremely variable ground conditions due to the wide variations in strength, durability and permeability. Their behaviour very much depends upon their degree of induration. For instance, ashes invariably are weak and often highly permeable. One particular hazard concerns ashes, not previously wetted, which are metastable, and exhibit a significant decrease in their void ratio on saturation. Tuffs and ashes frequently are prone to sliding. Montmorillonite is not an uncommon constituent in the weathered products of basic ashes.

Slates, phyllites and schists are characterized by textures which have a marked preferred orientation. Not only do cleavage and schistosity adversely affect the strength of such metamorphic rocks, they also make them more susceptible to decay. Generally speaking, however, slates, phyllites and schists weather slowly, but the areas of regional metamorphism in which they occur have suffered extensive folding, so that, in places, rocks may be fractured and highly deformed. Nonetheless, the quality of some schists, slates and phyllites is suspect. For instance, talc, chlorite and sericite schists are weak rocks containing planes of schistosity only 1 mm or so apart. Some schists become slippery upon weathering, and therefore fail under a moderately light load.

12.6 Mudrocks

Shale is the commonest sedimentary rock and is characterized by its lamination. Sedimentary rock of similar size range and composition, but which is not laminated, usually is referred to as mudstone. In fact, there is no sharp distinction between shale and mudstone, one grading into the other.

Shale frequently is regarded as an undesirable material to work in. Certainly, there have been many failures of structures founded on slopes in shales (Bell & Maud, 1996). Nonetheless, shales do vary in engineering behaviour, this being largely dependent upon their degree of compaction and cementation. Cemented shales invariably are stronger and more durable. The degree of packing and, hence, the porosity, void ratio and density of shale depend on its mineral composition, grain size distribution, mode of sedimentation, subsequent depth of burial, tectonic history and the effects of diagenesis. When the natural moisture content of shales exceeds 20%, they frequently are suspect, as they tend to develop potentially high pore water pressures. Generally, shales with a cohesion of less than 20 MPa and an apparent angle of friction of less than 20° are likely to present problems.

The higher the degree of fissility possessed by a shale, the greater the anisotropy with regard to strength, deformation and permeability. For instance, the influence of lamination on the behaviour of clay shale has been discussed by Wichter (1979). He noted that, in triaxial testing, the compressive strength parallel to the laminations was some 1.5–2 times less than that obtained at right angles to it, for confining pressures up to 1 MPa. The influence of the fissility on the Young's modulus can be illustrated by two values quoted by Chappel (1974), 6000 and 7250 MPa for cemented shale tested parallel and normal to the lamination respectively. Previously, Zaruba and Bukovansky (1965) found that the values of the Young's modulus were up to five times greater when shale was tested normal rather than parallel to the direction of lamination.

Argillaceous materials are capable of undergoing appreciable suction before pore water is removed, drainage commencing when the necessary air entry suction is achieved (about $pF = 2$). Under increasing suction pressure, the incoming air drives out the water from a shale, and some shrinkage takes place in the fabric before air can offer support. Generally, as the natural moisture content and liquid limit increase, so the effectiveness of soil suction declines.

The greatest variation found in the engineering properties of mudstones can be attributed to the effects of weathering. Weathering reduces the amount of induration or removes it completely, leading to an increase in moisture content and a decrease in density. The plasticity of mudrocks increases with increasing degree of weathering (Bell *et al.*, 1997). Weathering

ultimately returns mudstone to a normally consolidated, remoulded condition by destroying the bonds between particles. For instance, Spears and Taylor (1972) referred to a difference of 37% in the value of the effective angle of friction and of 93% in the effective cohesion between fresh and weathered samples of mudstone tested in triaxial conditions.

The breakdown of mudstones starts with exposure, which leads to the opening and development of fissures as the residual stress is dissipated, and to an increase in moisture content and softening. Taylor and Spears (1970) maintained that mudstones degrade rapidly to a dominantly gravel-sized aggregate, this being facilitated by the presence of polygonal fracture patterns, joints and bedding. They considered that physical disintegration was a much more important breakdown process than chemical weathering. This subsequently was reiterated by Wetzel and Einsele (1991). The lithological factors which govern the durability of mudstones include the degree of induration, degree of fracturing, grain size distribution and mineralogical composition, especially the nature of the clay mineral fraction. As discontinuities open on exposure and weathering, water finds access more easy, thereby facilitating chemical degradation. The anisotropic behaviour of mudrocks primarily is due to the presence of laminations. It was maintained by Dick and Shakoor (1992) that, in those mudstones which contain less than 50% clay-sized particles, the influence of clay minerals on slaking diminishes and microfractures become the dominant characteristic controlling the durability. In such cases, slaking is initiated along microfractures.

Depending upon the relative humidity, many shales slake almost immediately when exposed to air (Kennard *et al.*, 1967). The desiccation of shale, following exposure, leads to the creation of negative pore water pressures and, consequently, tensile failure of the weak, intercrystalline bonds. This leads to the breakdown of shale to particles of coarse sand or fine gravel size. Alternate wetting and drying causes a rapid breakdown of compaction shales. Low-grade compaction shales, in particular, undergo complete disintegration after several cycles of drying and wetting. On the other hand, well cemented shales are fairly resistant to slaking. The slaking behaviour of mudstones often is dominated by a tendency to breakage along irregular fracture patterns, which, when well developed, can mean that these rocks disintegrate within one or two cycles of wetting and drying. If mudstones undergo desiccation, air is drawn into the outer pores and capillaries as high suction pressures develop. On saturation, entrapped air is pressurized as water is drawn into the rock by capillarity. Slaking therefore causes the fabric of the rock to be stressed. According to Taylor (1988), disintegration takes place as a consequence of air breakage after a sufficient number of cycles of wetting and drying. The size of the pores is far more important than the volume of the pores, as far as the development of the capillary pressure is concerned. The capillary pressure is proportional to the surface tension.

The slake durability test estimates the resistance to wetting and drying of a rock sample, particularly mudrocks. The sample, which consists of 10 pieces of rock, each weighing about 40g, is placed in a test drum, oven dried and weighed. After this, the drum, with sample, is half immersed in a tank of water and attached to a rotor arm, which rotates the drum for a period of 10min at 20revmin⁻¹ (Fig. 12.23). The cylindrical periphery of the drum is formed of 2mm sieve mesh, so that broken down material can be lost whilst the test is in progress. After slaking, the drum and the material retained are dried and weighed. The slake durability index is then obtained by dividing the weight of the sample retained by its original weight, and expressing the answer as a percentage. The following scale of slake durability is used:

Very low	Under 25%
Low	25–50%
Medium	50–75%
High	75–90%
Very high	90–95%
Extremely high	Over 95%

Taylor (1988) suggested that durable mudrocks were more easily distinguished from non-durable types on the basis of the unconfined compressive strength and three-cycle slake durability index (i.e. those mudrocks with a compressive strength exceeding 3.6MPa and a three-cycle slake durability index above 60% were regarded as durable). Bell *et al.* (1997) showed that the results of one-cycle slake durability testing of mudstones did not compare favourably with those obtained by cyclic wetting and drying. They therefore tended to agree with Taylor.

The swelling properties of certain shales have proved to be extremely detrimental to the integrity of many civil engineering structures. Swelling, especially in clay shales, is attributable to the absorption of free water by particular clay minerals, notably montmorillonite, in the clay fraction of a shale. Highly fissured, overconsolidated shales have greater swelling tendencies than poorly fissured, clayey shales, the fissures providing access for water.

ig. 12.23 Slake durability apparatus.

The failure of poorly cemented mudrocks occurs during saturation, when the swelling pressure or internal saturation swelling stress σ_s, developed by capillary suction pressures, exceeds the tensile strength. An estimate of σ_s can be obtained from the modulus of deformation (E)

$$E = \sigma_s / \xi_D \qquad (12.5)$$

where ξ_D is the free swelling coefficient. The latter is determined by a sensitive dial gauge, which records the amount of swelling of an oven-dried core specimen along the vertical axis during saturation in water for 12 h; ξ_D is obtained as follows:

$$\xi_D = \frac{\text{Change in length after swelling}}{\text{Initial length}} \qquad (12.6)$$

Olivier (1979) proposed the geodurability classification for mudrocks, which is based on the free swelling coefficient and uniaxial compressive strength (Fig. 12.24).

Uplift frequently occurs in excavations in shales, and is attributable to swelling and heave. Rebound on unloading of shales during excavation is attributed to heave due to the release of stored strain energy. Shale relaxes towards a newly excavated face, and sometimes this occurs as offsets at weaker seams in the shale. The greatest amount of rebound occurs in heavily overconsolidated compaction shales, for example, at Garrison Dam, North Dakota, just over 0.9 m of rebound was measured in the deepest excavation in the Fort Union Clay Shales (Smith & Redlinger, 1953).

The problem of settlement in shales generally can be resolved by reducing the unit bearing load by widening the base of structures or using spread footings. In some cases, appreciable differential settlements are provided for by designing articulated structures capable of accommodating differential

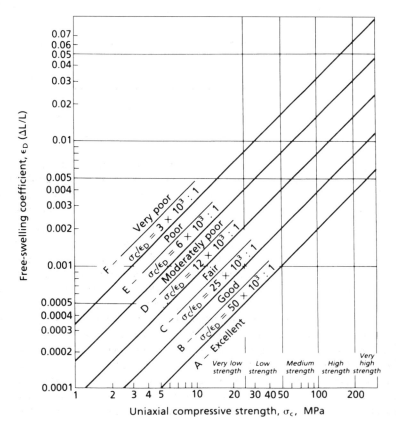

Fig. 12.24 Geodurability classification of intact rock material. Note: (i) ξ_D determined from oven-dried (105 °C) to 24-h saturation condition; (ii) ξ_D plotted as the range and mean of the test results; (iii) strength ratings according to Deere and Miller (1966) as modified by Bieniawski (1974). (After Olivier, 1979.)

movements of individual sections without damage to the structure. Severe settlements may take place in low-grade compaction shales. However, compaction shales contain fewer open joints or fissures which can be compressed beneath heavy structures than do cemented shales.

When a load is applied to an essentially saturated shale foundation, the void ratio in the shale decreases, and the pore water attempts to migrate to regions of lesser load. Because of the shale's relative impermeability, water becomes trapped in the voids and can only migrate slowly. As the load is increased, there comes a point at which it partly is transferred to the pore water, resulting in a build-up of pore water pressure. Depending on the permeability of the shale and the rate of loading, the pore water pressure can more or less increase in value, so that it equals the pressure imposed by the load. This greatly reduces the shear strength of the shale, and failure of structures can occur. Pore water pressure problems are not so important in cemented shales.

The stability of slopes in excavations can be a major problem in shale, both during and after construction. This problem becomes particularly acute in dipping formations and in formations containing expansive clay minerals.

Sulphur compounds frequently are present in argillaceous rocks. An expansion in volume large enough to cause structural damage can occur when sulphide minerals, such as pyrite and marcasite, suffer oxidation, and give rise to anhydrous and hydrous sulphates. Penner *et al.* (1973) quoted a case of heave in a black shale of Ordovician age in Ottawa, which caused displacement of the basement floor of a three-storey building. The maximum movement totalled some 107mm, the heave rate being almost 2mm per month. The heave was attributable to the breakdown of pyrite to produce sulphur compounds, which combined with calcium to form gypsum and jarosite. The latter minerals formed in the fissures and between the laminae of the shales in the altered zone.

The decomposition of sulphur compounds also gives rise to aqueous solutions of sulphate and sulphuric acid, which react with tricalcium aluminate in Portland cement to form calcium sulpho-aluminate or ettringite. This reaction is accompanied by expansion. The rate of attack is very much influenced by the permeability of the concrete or mortar and the position of the water table. For example, sulphates can only continue to reach cement by movement of their solutions in water. Thus, if a structure is permanently above the water table, it is unlikely to be attacked. By contrast, below the water table, movement of water may replenish the sulphates removed by reaction with cement, thereby continuing the reaction. Concrete with a low permeability is essential to resist sulphate attack, hence, it should be fully compacted. Sulphate-resistant cements, that is, those in which the tricalcium aluminate concentration is low, can also be used for this purpose (Anon., 1975a). Foundations can be protected by impermeable membranes or bituminous coatings.

12.7 Carbonate rocks

Carbonate rocks contain more than 50% carbonate minerals. The term limestone is applied to those rocks in which the carbonate fraction exceeds 50%, over half of which is calcite or aragonite. If the carbonate material is made up chiefly of dolomite, the rock is named dolostone.

Thick-bedded, horizontally lying limestones, relatively free from solution cavities, afford excellent foundations. On the other hand, thin-bedded, highly folded or cavernous limestones are likely to present serious foundation problems. A possibility of sliding may exist in highly bedded, folded sequences. Similarly, if beds are separated by layers of clay or shale, especially when inclined, these may serve as sliding planes and result in failure.

Limestones commonly are transected by joints. These generally have been subjected to various degrees of dissolution, so that some may gape (Fig. 12.25). Rain water usually is weakly acidic, and further acids may be taken into solution from carbon dioxide or organic or mineral matter in the soil. The degree of aggressiveness of water to limestone can be assessed on the basis of the relationship between the dissolved carbonate content, the pH value and the temperature of the water. At any given pH value, the cooler the water, the more aggressive it is. If solution continues, its rate slackens, and it eventually ceases when saturation is reached. Hence, solution is greatest when the bicarbonate saturation is low. This occurs when water is circulating, so that fresh supplies with low lime saturation continually are made

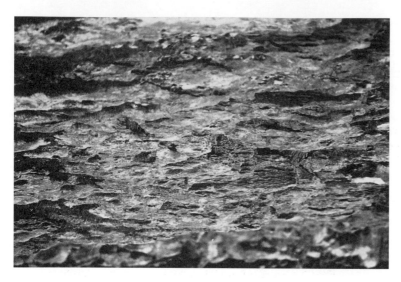

Fig. 12.25 Limestone pavement, the Burren, Ireland.

available. Fresh water can dissolve up to 400 mg l⁻¹ of calcium carbonate.

Sinkholes may develop where joints intersect, and these may lead to an integrated system of subterranean galleries and caverns (Fig. 12.26). The latter are characteristic of thick, massive limestones. The progressive opening of discontinuities by dissolution leads to an increase in mass permeability. Sometimes dissolution produces a highly irregular pinnacled surface or limestone pavement. The size, form, abundance and downward extent of the aforementioned features depend upon the geological structure and the presence of interbedded impervious layers. Solution cavities present numerous problems in the construction of large foundations.

An important effect of solution in limestone is the enlargement of the pores, which enhances the water circulation, thereby encouraging further solution (Sowers, 1975). This brings about an increase in stress within the remaining rock framework, which reduces the strength of the rock mass, and leads to increasing stress corrosion. On loading, the volume of the voids is reduced by fracture of the weakened cement between the particles, and by the re-orientation of the intact aggregations of rock that become separated by the loss of bonding. Most of the resultant settlement takes place rapidly within a few days of the application of load.

Rapid subsidence can take place due to the collapse of holes and cavities within limestone which has been subjected to prolonged solution, this occurring when the roof rocks are no longer thick enough to support themselves. It must be emphasized, however, that the solution of limestone is a very slow process; contemporary solution therefore is very rarely the cause of collapse. For instance, Kennard and Knill (1968) quoted mean rates of surface lowering of limestone areas in the UK which ranged from 0.041 to 0.099 mm annually.

Nevertheless, solution may be accelerated by man-made changes in the groundwater conditions, or by a

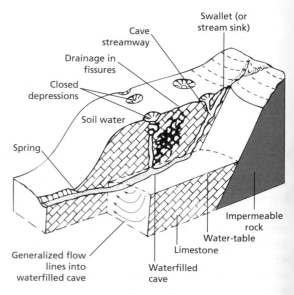

Fig. 12.26 Subsurface drainage through limestone terrain.

Table 12.18 Solution of soluble rocks. (After James & Kirkpatrick, 1980.)

Rock	(a) Solubility (c_s) in pure water ($kg\,m^{-3}$ at $10\,°C$)*	(b) Solution rate constant (K) at $10\,°C$ (flow velocity, $0.05\,m\,s^{-1}$) ($m\,s^{-1} \times 10^{-5}$)†	m^4 ($kg\,s^{-1} \times 10^{-6}$)
		$\theta = 1$	$\theta = 2$
Halite	360.0	0.3	
Gypsum	2.6	0.2	
Anhydrite	2.0		0.8
Limestone	0.015	0.4	

* c_s is dependent upon the temperature and the presence of other dissolved salts.
† K is dependent on the temperature, flow velocity and other dissolved salts.
θ, order of dissolution reaction.

(c) Limiting widths of fissures in massive rock*

Rock	(i) Upper limit for stable inlet face retreat	(ii) For a rate of retreat of 0.1 m per year (e.g. cavern formation)	(iii) Maximum safe lugeon value†
Halite	0.05	0.05	
Gypsum	0.2	0.3	50
Anhydrite	0.1	0.2	7
Limestone	0.5	1.5	700

* These values are for pure water at a fissure spacing in massive rock of one per metre and a hydraulic gradient of 0.2. For water containing $300\,mg\,l^{-1}$ of CO_2, the stable limit width of fissure becomes 0.4 mm in limestone.
† One lugeon unit is equal to a flow of $1\,l\,min^{-1}$ at a pressure of 1 MPa (it is approximately equal to a coefficient of permeability of $10^{-7}\,m\,s^{-1}$).

(d) Solution of particulate deposits*

Rock	(i) Limiting seepage velocity ($m\,s^{-1}$)	(ii) Width of solution zone (m)
Halite	6.0×10^{-9}	0.0002
Gypsum	1.4×10^{-6}	0.04
Anhydrite	1.6×10^{-6}	0.09
Limestone	3.0×10^{-4}	2.8

* Rate of movement of solution zone, 0.1 m per year. Mineral particles of 50 mm in diameter. Pure water.

...nge in the character of the surface water that ...ins into limestone. For instance, James and Kirk...rick (1980), in a consideration of the location of ...draulic structures on soluble rocks, stated that, if ...h dry, discontinuous rocks are subjected to sub...ntial hydraulic gradients, they will undergo disso...on along these discontinuities, hence leading to ...idly accelerating seepage rates. From experimen...work, carried out on the Portland Limestone, they ...nd that the values of the solution rate constant (K) ...eased appreciably at flow velocities which corre-

sponded to a transitional flow regime. They showed that such a flow regime occurred in joints about 2.5 mm in width which experienced a hydraulic gradient of 0.2 (Fig. 12.27a). According to these two authors, solution takes place along a small joint by retreat of the inlet face due to the removal of soluble material (Table 12.18; Fig. 12.27b). Dissolution of larger joints gives rise to long tapered enlargements, which enable seepage rates to increase rapidly and runaway situations to develop.

According to Sowers (1975), ravelling failures are

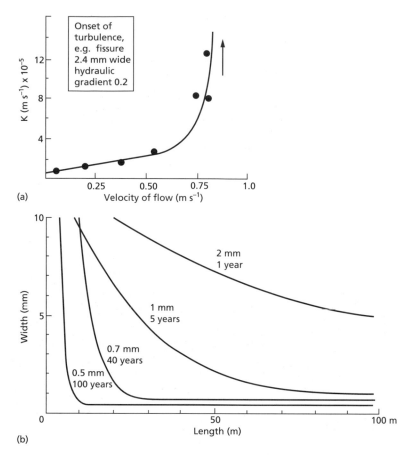

Fig. 12.27 (a) Rate of solution (K) of Portland Limestone and its dependence on the flow velocity. (b) Enlargement of fissures in calcium carbonate rock by pure flowing water. (After James & Kirkpatrick, 1980.)

the most widespread and probably the most dangerous of all the subsidence phenomena associated with limestones. Ravelling occurs when solution-enlarged openings extend upwards to a rock surface overlain by soil (Fig. 12.28). The openings should be interconnected, and lead into channels through which the soil can be eroded by groundwater flow. Initially, the soil arches over the openings but, as they are enlarged, a stage is reached at which the soil above the roof can no longer support itself, and so it collapses. A number of conditions accelerate the development of cavities in the soil and initiate collapse. Rapid changes in moisture content lead to aggravated slabbing or roofing in clays and flow in cohesionless sands. A lowering of the water table increases the downward seepage gradient and accelerates downward erosion. It also reduces capillary attraction and increases the instability of flow through narrow openings, and

gives rise to shrinkage cracks in highly plastic clay which weaken the mass in dry weather, and produce concentrated seepage during rains. Increased infiltration often initiates failure, particularly when it follows a period when the water table has been lowered.

Areas underlain by highly cavernous limestone possess most dolines, hence, the doline density has been proven to be a useful indicator of potential subsidence, as has the sinkhole density. As there is preferential development of solution voids along zones of high secondary permeability, because these concentrate groundwater flow, data on fracture orientation and density, fracture intersection density and the total length of fractures have been used to model the presence of solution cavities in limestone. Therefore the location of areas of high risk of cavity collapse has been estimated by using the intersection of lineaments

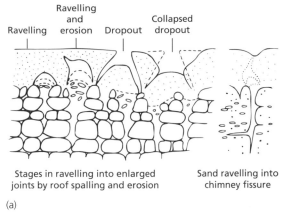

Stages in ravelling into enlarged
joints by roof spalling and erosion

Sand ravelling into
chimney fissure

(a)

Fig. 12.28 (a) Mechanisms of ravelling;
(b) appearance of a sinkhole at the
surface, Pretoria, South Africa.

(b)

formed by fracture traces and lineated depressions (dolines). Aerial photographs have proved to be particularly useful in this context.

Brook and Alison (1986) described the production of subsidence susceptibility maps of a covered karst terrain in Dougherty County, southwest Georgia. These have been developed using a geographical information system which incorporates many of the data referred to in the previous paragraph. The county was partitioned into 885 cells, each of 1.16 km² in area. Five cell variables were used in modelling, namely, the sinkhole density, sinkhole area, fracture density, fracture length and fracture intersection density. Broadly similar subsidence susceptibility models were developed from cell data by intersection and separately by linear combination. In the intersection technique, cells having specified values for all variables were located and mapped. In the linear combination technique, a map value, $MV = W_1 r_1 + \ldots W_n r_n$, where W is an assigned variable weight and r is an assigned weight value, was calculated for each cell.

Throughout the southeast USA, there are thousands of sinkholes of many different sizes and shapes. They may range from 1 or 2 m in diameter to more than 3 km. Depths of a few metres are common, although the largest sinkholes exceed 40 m in depth. Whereas it takes thousands of years to create natural sinkholes, those created by humans largely have occurred since the early 1900s. For instance, more than 4000 sinkholes have been catalogued in Alabama as being caused by human activities, with the great majority of these developing since 1950. Indeed, in Shelby County, Alabama, more than 1000 sinkholes developed between 1958 and 1973 in a 26 km^2 area. The largest was called the 'December Giant', because it suddenly developed in December 1972. It measures 102 m in diameter and 26 m in depth. Sinkholes are particularly dangerous because they form instantaneously by collapse, and often occur in significant number within a short time span. Sinkholes have resulted in costly damages to a variety of structures, and serve as a major local source of groundwater pollution. They have been largely caused by continuous dewatering projects in carbonate rocks for wells, quarry and mining operations, and drainage changes.

Chalk, being a relatively pure form of limestone, is also subjected to dissolution along discontinuities. However, subterranean solution features tend not to develop in chalk, because it is usually softer than limestone, and so collapses as solution occurs. Nevertheless, solution pipes and swallow holes are present in chalk, being commonly found near the contact of chalk with the overlying Tertiary and drift deposits. High concentrations of water, such as run-off from roads, can lead to the reactivation of swallow holes and the formation of small pipes within a few years. Moreover, voids gradually can migrate upwards through chalk due to material collapsing. Lowering of the chalk surface beneath overlying deposits due to solution can occur, disturbing the latter deposits and lowering their degree of packing. Hence, the chalk surface may be extremely irregular in places.

Chalk, especially when weathered, may suffer frost heave during cold weather, ice lenses up to 25 mm thick being developed along bedding planes. Higginbottom (1965) suggested that a probable volume increase of some 20–30% of the original thickness of the ground ultimately may result.

12.8 Evaporitic rocks

Evaporitic deposits quantitatively are unimportant as sediments. They are formed by precipitation from the saline waters of inland seas or lakes in arid areas. Evaporitic rocks exhibit varying degrees of plastic deformation before failing. For example, in rock salt, the yield strength may be as little as one-tenth of the ultimate compressive strength, whereas anhydrite undergoes comparatively little plastic deformation prior to rupture. Creep may account for anything between 20 and 60% of the strain at failure when these evaporitic rocks are subjected to incremental creep tests (Bell, 1981, 1994). Rock salt is most prone to creep. Langer (1982) proposed a rheological model for the mechanical behaviour of rock salt with time. Initially, primary creep occurs as rock salt is subjected to loading, the creep rate decreasing with time. During the stage of secondary creep, the creep rate is related to the amount of stress and the temperature conditions. Under a constant load, failure occurs when the stress, creep rate and temperature are combined in a given manner. The resistance to failure increases with increasing confining conditions.

Gypsum is more readily soluble than limestone (2100 mg l^{-1} vs. 400 mg l^{-1} can be dissolved in non-saline waters). Sinkholes and caverns therefore can develop in thick beds of gypsum more rapidly than they can in limestone (Eck & Redfield, 1965). Indeed in the USA, such features have been known to form within a few years, where beds of gypsum are located beneath dams. Extensive surface cracking and subsidence have occurred in certain areas of Oklahoma and New Mexico due to the collapse of cavernous gypsum. The problem is accentuated by the fact that gypsum is weaker than limestone, and therefore collapses more readily. Yuzer (1982) described karstic features found in the Sivas area of Turkey, where some 180 m of gypsum beds occur in Oligocene Miocene deposits.

Cooper (1988) has described subsidences which have occurred in the Ripon district, Yorkshire, UK due to the solution of gypsum. Wherever beds of gypsum approach the surface, their presence can be traced by broad, funnel-shaped craters formed by the collapse of overlying marl (mudstone) into areas from which gypsum has been removed by solution. These craters can take a matter of minutes to appear at the surface. Foundation problems due to the dissolution of gypsum in Rapid City, South Dakota, have been reported by Rahn and Davis (1996). However, when gypsum effectively is sealed from the ingress of water by overlying impermeable strata, such as mudstone, dissolution does not occur.

The solution rate of gypsum or anhydrite princi

pally is controlled by the area of the surface in contact with water and the flow velocity of water associated with a unit area of the material. Hence, the amount of fissuring in a rock mass and whether it is enclosed by permeable or impermeable beds are most important. The solution rate also depends on the subsaturation concentration of calcium sulphate in solution. According to James and Lupton (1978), the concentration dependence for gypsum is linear, whereas that for anhydrite is a square law. The salinity of the water is also influential. For example, the rates of solution of gypsum and anhydrite are increased by the presence of sodium chloride, carbonate and carbon dioxide in solution. It therefore is important to know the chemical composition of the groundwater.

Massive deposits of gypsum usually are less dangerous than those of anhydrite, because gypsum tends to dissolve in a steady manner, forming caverns or causing progressive settlements. For instance, if small fissures occur at less than 1m intervals, solution usually takes place by the removal of gypsum as a front moving 'downstream' at less than 0.01m per year. However, James and Lupton (1978) showed that, if the rock temperature was 10°C, the water involved contained no dissolved salts and a hydraulic gradient of 0.2 was imposed, a fissure, 0.2mm in width and 100m in length, in massive gypsum, would, in 100 years, have widened by solution so that a block 1m^3 in size could be accommodated in the entrance to the fissure. In other words, a cavern would be formed. If the initial width of the fissure exceeds 0.6mm, large caverns will form, and a runaway situation will develop in a very short time. In long fissures, the hydraulic gradient is low, and the rate of flow is reduced, so that solutions become saturated and little or no material is removed. Indeed, James and Lupton implied that a flow rate of 10^{-3} ms^{-1} was rather critical, in that if it was exceeded, extensive solution of gypsum could take place. The solution of massive gypsum is not likely to give rise to an accelerating deterioration in a foundation if precautions, such as grouting, are taken to keep seepage velocities low.

Massive anhydrite can be dissolved to produce uncontrollable runaway situations in which seepage flow rates increase in a rapidly accelerating manner. Even small fissures in massive anhydrite can prove to be dangerous. If anhydrite is substituted for gypsum, under the same conditions as mentioned in the previous paragraph, not only is a cavern formed, but the fissure is enlarged as a long, tapering section. Within about 13 years, the flow rate increases to a runaway situation. However, if the fissure is 0.1mm in width, the solution becomes supersaturated with calcium sulphate, and gypsum is precipitated. This seals the outlet from the fissure and, from that moment, any anhydrite in contact with the water is hydrated to form gypsum. Accordingly, a width of 0.1mm seems to be a critical fissure size in anhydrite.

If soluble minerals occur in particulate form in the ground, their removal by solution can give rise to significant settlements. In such situations, the width of the solution zone and its rate of progress obviously are important as far as the location of hydraulic structures are concerned (James & Kirkpatrick, 1980; Table 12.18). Anhydrite is less likely than gypsum to undergo catastrophic solution in a fragmented or particulate form. Again, the solubility of the mineral and its solution rate constant are the most important parameters. Of lesser consequence is information relating to the volumetric proportion of soluble minerals in the ground and details of their size, distribution and shape. This does not mean to say that such data should not be obtained, for they obviously aid in the determination of the amount of settlement which might take place.

Another point which should be borne in mind, and this particularly applies to conglomerates cemented with soluble material, is that, when this material is removed by solution, the rock is reduced greatly in strength. A classic example of this is associated with the failure of the St Francis Dam in California in 1928. One of its abutments was founded in conglomerate cemented with gypsum, which was gradually dissolved; the rock lost strength and, ultimately, the abutment failed.

On the other hand, anhydrite, on contact with water, may become hydrated to form gypsum. In so doing, there is a volume increase of between 30 and 58%, which exerts pressures that have been variously estimated between 2 and 70MPa. However, Yuzer (1982) claimed that the theoretical maximum swelling pressure is not a realistic value, and maintained that such swelling pressures commonly are between 1 and 8MPa, and rarely exceed 12MPa. It is believed that no great length of time is required to bring about such hydration.

Salt is even more soluble than gypsum, and the evidence of slumping, brecciation and collapse structures in rocks which overlie saliferous strata bear witness to the fact that salt has gone into solution in past geological times. It generally is believed, however, that in areas underlain by saliferous beds, measurable surface subsidence is unlikely to occur, except where salt is being extracted (Bell, 1992a). Perhaps this is because equilibrium has been attained between the supply of unsaturated groundwater and the salt available for solution. Exceptional cases have been recorded of rapid subsidence, such as the 'Meade salt sink' in Kansas. Johnson (1901) explained its formation as being due to the solution of beds of salt located at depth. This area of water, about 60 m in diameter, formed as a result of rapid subsidence in March 1879.

12.9 Building, roofing and facing stones

Stone has been used as a construction material for thousands of years. One of the reasons for this is that it generally is readily available locally. Furthermore, stone requires little energy for extraction and processing. Indeed, stone is used more or less as it is found, except for the seasoning, shaping and dressing that is necessary before it is employed for building purposes. Yet other factors, as many ancient buildings testify, include its attractiveness, durability and permanence.

A number of factors determine whether a rock will be worked as a building stone. These include the volume of material that can be quarried, the amount of overburden requiring removal, the ease with which the rock can be quarried, the wastage consequent upon quarrying, the cost of transportation, and the appearance and physical properties of the rock.

As far as the volume is concerned, the life of a quarry should be at least 20 years. The amount of overburden that has to be removed affects the economics of quarrying, and there comes a point when its removal makes operations uneconomic. Weathered rock represents waste; therefore, the ratio of fresh to weathered rock is another factor of economic importance. The ease with which a rock can be quarried depends mainly upon the geological structure, notably the geometry of the joints and bedding planes, where present. Ideally, rock for building stone should be massive and free from closely spaced joints

or other discontinuities, as these control the block size. The stone should be free of fractures and other flaws. In the case of sedimentary rocks, where beds dip steeply, quarrying has to take place along the strike. Steeply dipping rocks can also give rise to problems of slope stability when excavated. However, if beds of rock dip gently, it is advantageous to develop the quarry floor along the bedding planes. The massive nature of igneous rocks, such as granite, means that a quarry can be developed in any direction, within the constraints of planning permission.

A uniform appearance generally is desirable in building stone. Its appearance largely depends upon its colour which, in turn, is determined by its mineral composition. Texture also affects the appearance of a stone, as does the way in which it weathers, for example, the weathering of certain minerals, such as pyrite, may produce stains. Generally speaking, rocks of light colour are used as building stone.

For usual building purposes, a compressive strength of 35 MPa is satisfactory, and the strength of most rocks used for building stone is well in excess of this value. In certain instances, the tensile strength is important, for example, tensile stresses may be generated in a stone subjected to ground movements. However, the tensile strength of a rock or, more particularly, its resistance to bending is a fraction of its compressive strength. Hardness is a factor of small consequence, except where a stone is subjected to continual wear, such as in steps or pavings.

The texture and porosity of a rock affect its ease of dressing, and the amount of expansion, freezing and dissolution it may undergo. Fine grained rocks are more easily dressed than coarse varieties. Water retention in a rock with small pores is greater than in a rock with large pores, and so the former is more prone to frost attack.

The durability of a stone is a measure of its ability to resist weathering, and so to retain its original size, shape, strength and appearance over an extensive period of time (Sims, 1991; Bell, 1992b, 1993); it is one of the most important factors which determines whether or not a rock will be worked for building stone. The amount of weathering undergone by a rock in field exposures or quarries affords some indication of its qualities of resistance. However, there is no guarantee that the durability is the same throughout a rock mass and, if it changes, it is far mor difficult to detect than, for example, a change in colour.

The stresses imposed upon masonry by expansion and contraction brought about by changes in temperature and moisture content can result in masonry between abutments spalling at the joints. Blocks may even be shattered and fall out of place.

Frost damage is one of the major factors causing deterioration in a building stone. Sometimes small fragments are separated by frost action from the surface of a stone, but the major effect is gross fracture. Frost damage is most likely to occur on steps, copings, sills and cornices, where rain and snow can collect. Damage to susceptible stone may be reduced if it is placed in a sheltered location. Most igneous rocks, and the better quality sandstones and limestones, are immune. As far as frost susceptibility is concerned, the porosity, tortuosity, pore size and degree of saturation all play an important role. As water turns to ice, it increases in volume, resulting in pressure increase within the pores. Usually, coarse grained rocks withstand freezing better than fine grained ones. As noted in Section 12.4.1, the critical pore size for freeze–thaw durability appears to be about 0.005 mm, that is, rocks with larger mean pore diameters allow the outward drainage and escape of fluid from the frontal advance of the ice line, and are therefore less frost susceptible.

Deleterious salts, when present in a building stone, generally are derived from the ground or the atmosphere, although soluble salts may occur in the pores of the parent rock. Their presence in a stone gives rise to different effects. They may cause efflorescence by crystallizing on the surface of a stone. In subfluorescence, crystallization takes place just below the surface, and may be responsible for surface scabbing. The pressures produced by the crystallization of salts in small pores can be appreciable (see Section 12.4). Crystallization caused by freely soluble salts can lead to the crumbling or powdering of the surface of a stone. Deep cavities may be formed in magnesian limestone when it is attacked by magnesium sulphate. Salt action can give rise to honeycomb weathering in some sandstones and porous limestones. Conversely, surface induration of a stone by the precipitation of salts may give rise to a protective hard crust, that is, case hardening.

Building stones derived from sedimentary rocks may undergo a varying amount of decay in urban atmospheres, where weathering is accelerated due to the presence of aggressive impurities, such as SO_2, SO_3, NO_3, Cl_2 and CO_2, in the air, which produce corrosive acids. Limestones are most suspect (Fig. 12.29). For instance, weak sulphuric acid reacts with the calcium carbonate of limestones to produce calcium sulphate. The latter often forms just below the surface of a stone, and the expansion which takes place upon crystallization causes slight disruption. If this reaction continues, the outer surface of the limestone begins to flake off. Sandstones with calcareous cements may react with weak acids in urban atmospheres, leading to the irregular flaking off of the surface of a stone or, in extreme cases, to crumbling. Building stones derived from granitic igneous rocks usually suffer negligible decay.

Rocks used for roofing purposes must possess a sufficient degree of fissility to allow them to split into

Fig. 12.29 Weathering of limestone, Seville Cathedral, Spain.

thin slabs, as well as being durable and impermeable. Consequently, slate is one of the best roofing materials available, and has been used extensively. Today, however, more and more tiles are being used for roofing, these being cheaper than stone.

The specific gravity of a slate is about 2.7–2.9, with an approximate density of 2.59 Mg m^{-3}. The maximum permissible water absorption of a slate is 0.37%. Calcium carbonate may be present in some slates of inferior quality, which may result in flaking and, eventually, crumbling upon weathering. Accordingly, a sulphuric acid test is used to test their quality. Top quality slates, which can be used under moderate to severe atmospheric pollution conditions, reveal no signs of flaking, lamination or swelling after the test. Slates may be blue, grey, purple or red in colour.

An increasingly frequent method of using stone at the present day is as relatively thin slabs, applied as a facing to a building to enhance its appearance. Facing stone also provides a protective covering. Various thicknesses are used, from 40 mm in the case of granite, marble and slate, in certain positions at ground floor level, to 20 mm at first-floor level or above. If granite or syenite is used as a facing stone, it should not be overdried, but should retain some quarry sap, otherwise it becomes too tough and hard to fabricate. Limestone and sandstone slabs are somewhat thicker, varying between 50 and 100 mm. Because of their comparative thinness, facing stones should not be too rigidly fixed, otherwise differential expansion, due to changing temperatures, can produce cracking.

When fissile stones are used as facing stone and are given a riven or honed finish, they are extremely attractive. Facing stones, however, usually have a polished finish; they are even more attractive. Such stones are almost self-cleansing.

Rocks used for facing stones have a high tensile strength in order to resist cracking. The high tensile strength also means that thermal expansion is not a great problem when slabs are spread over large faces.

12.10 Crushed rock

12.10.1 Concrete aggregate

Crushed rock is produced for a number of purposes, the chief of which are for concrete and road aggregate. Approximately 75% of the volume of concrete

consists of aggregate; therefore its properties have a significant influence on the engineering behaviour of concrete. Aggregate is divided into coarse and fine varieties. The former usually consists of rock material that passes through a 40 mm sieve and is retained on a 5 mm sieve. The latter passes through this sieve and is caught on a 100 mesh sieve. Fines passing through a 200 mesh sieve should not exceed 10% by weight of the aggregate.

The crushing strength of rock used for aggregate generally ranges between 70 and 300 MPa. Aggregates that physically are unsound lead to the deterioration of concrete, inducing cracking, popping or spalling. On drying, cement shrinks. If the aggregate is strong, the amount of shrinkage is minimized and the cement–aggregate bond is good.

The shape of aggregate particles is an important property, and mainly is governed by the fracture pattern within a rock mass. Rocks such as basalts, dolerites, andesites, granites, quartzites and limestones tend to produce angular fragments when crushed. However, argillaceous limestones, when crushed, produce an excessive amount of fines. The crushing characteristics of a sandstone depend upon the closeness of its texture and the amount and type of cement. Angular fragments may produce a mix which is difficult to work, that is, it can be placed less easily and offers less resistance to segregation. Nevertheless, angular particles are said to produce a denser concrete. Rounded, smooth fragments produce workable mixes. The less workable the mix, the more sand, water and cement that must be added to produce a satisfactory concrete. Fissile rocks, such as those which are strongly cleaved, schistose, foliated or laminated, have a tendency to split and, unless crushed to a fine size, give rise to tabular- or planar-shaped particles. Planar and tabular fragments not only make concrete more difficult to work, but they also pack poorly, and so reduce its compressive strength and bulk weight. Furthermore, they tend to lie horizontally in the cement, so allowing water to collect beneath them, which inhibits the development of a strong bond on their undersurfaces.

The surface texture of aggregate particles largely determines the strength of the bond between the cement and themselves (French, 1991). A rough surface creates a good bond, whereas a smooth surface does not.

As concrete sets, hydration takes place, and alkalis

(Na$_2$O and K$_2$O) are released. These react with siliceous material. Table 12.19 lists some of the reactive rock types. If any of these types of rock are used as aggregate in concrete made with high-alkali cement, the concrete is liable to expand and crack, thereby losing strength. Expansion due to alkali–aggregate reaction has also occurred when greywacke has been used as aggregate. When concrete is wet, the alkalis that are released are dissolved by the water content and, as the water is used up during hydration, so the alkalis are concentrated in the remaining liquid. This caustic solution attacks reactive aggregates to produce alkali–silica gels. The osmotic pressure developed by these gels as they absorb more water eventually may rupture the cement around reacting aggregate particles. The gels gradually occupy the cracks produced, and these eventually extend to the surface of the concrete. If alkali reaction is severe, a polygonal pattern of cracking develops on the surface. These problems can be avoided if a preliminary petrological examination is made of the aggregate, that is, material that contains over 0.25% opal, over 5% chalcedony or over 3% glass or cryptocrystalline acidic to intermediate volcanic rock, by weight, will be sufficient to produce an alkali reaction in concrete unless low-alkali cement is used. This contains less than 0.6% of Na$_2$O and K$_2$O. The deleterious effect of the alkali–aggregate reaction can be avoided if pozzolana is added to the mix to react with the alkalis.

The reactivity may also be related to the percentage of strained quartz that a rock contains. For instance, Gogte (1973) maintained that rock aggregates containing 40% or more of strongly undulatory or highly granulated quartz were highly reactive, whereas those with between 30 and 35% were moderately reactive. He also showed that basaltic rocks with 5% or more of secondary chalcedony or opal, or about 15% palagonite, showed deleterious reactions with high-alkali cements. Sandstones and quartzites containing 5% or more of chert behaved in a similar manner.

Certain argillaceous dolostones have been found to expand when used as aggregates in high-alkali cement, thereby causing failure of concrete. This phenomenon has been referred to as the alkali–carbonate rock reaction, and an explanation has been attempted by Gillott and Swenson (1969). They proposed that

Table 12.19 Varieties of potentially reactive silica and their geological occurrence. (After Smith & Collis, 1993.)

Vartiety of silica	Common geological occurrence
Opal	Vein material and vug filling in a variety of rock types, a constituent of some types of chert, a replacement for siliceous fossil material and a cementing material in some sedimentary rocks
Volcanic glass	A constituent of some igneous volcanic rocks ranging from acidic to basic composition. Volcanic glass devitrifies over geological time and devitrified glass may also be potentially reactive
Tridymite and cristobalite	High-temperature, metastable polymorphs of silica found as a minor constituent in some acidic and intermediate volcanic rocks
Microcrystalline and cryptocrystalline quartz	The principal constituent of most cherts and flint. Vein material and vug fillings in a variety of rock types, groundmass mineral in some igneous and metamorphic rocks, cementing material in some sedimentary rocks
Chalcedony	A fibrous variety of microcrystalline quartz found as a constituent of some cherts and flint. Vein material and vug fillings in a variety of rock types, cementing material in some sedimentary rocks
strained quartz	Found especially in metamorphic rocks, but also in some igneous rocks, subjected to high stresses. Also occurs as a detrital mineral in clastic sediments. Current opinion is that strained quartz itself is probably not reactive, and reactivity may be associated with poorly ordered silica at the highly sutured grain boundaries commonly associated with strained quartz

Note: opal is the most reactive variety of silica. The relative reactivity of the other varieties shown in this Table will be controlled by many petrographic factors, and the order of the minerals is not intended to rank them in terms of reactivity.

the expansion of such argillaceous dolostones in high-alkali cements was due to the uptake of moisture by the clay minerals. This was made possible by dedolomitization, which provided access for moisture. Moreover, they noted that expansion only occurred when the dolomite crystals were less than 75 μm.

It usually is assumed that shrinkage in concrete will not exceed 0.045%, this usually taking place in the cement. However, basalt, gabbro, dolerite, mudstone and greywacke have been shown to be shrinkable, that is, they have large wetting and drying movements of their own, so much so that they affect the total shrinkage of concrete. Shrinking of coarse aggregate in concrete may lead to crazing, thereby reducing its durability. In some cases, it is possible to derive a predictive relationship between the water absorption of the aggregate and shrinkage of the aggregate on drying. Granite, limestone, quartzite and felsite remain unaffected.

Certain aggregates may be microporous and, as such, may be frost susceptible, especially when they are close to the surface of concrete and therefore exposed to the weather. In such instances, this may lead to 'popouts' by causing failure in the surrounding cement.

12.10.2 Road aggregate

Aggregate constitutes the basic material for road construction, and forms the greater part of a road surface. Thus, it has to bear the main stresses imposed by traffic and must resist wear. The rock material

used should therefore be fresh and have a high strength (Fookes, 1991).

Aggregate used as road metal must, as well as having a high strength, have a high resistance to impact and abrasion, polishing and skidding, and frost action. It must also be impermeable, chemically inert and possess a low coefficient of expansion. The principal tests used to assess the value of a roadstone are the aggregate crushing test, the aggregate impact test, the aggregate abrasion test and the polished stone value test. Other tests are those for water absorption, specific gravity, density and aggregate shape tests (Anon., 1975b). Some typical values are given in Table 12.20.

The properties of an aggregate obviously are related to the texture and mineralogical composition of the rock from which it was derived. Most igneous and contact metamorphic rocks meet the requirements demanded of good roadstone, but many regional metamorphic rocks are either cleaved or schistose, and are therefore unsuitable for roadstone, because they tend to produce flaky particles when crushed. Such particles do not achieve a good interlock, and therefore impair the development of dense mixtures for surface dressing. The amount and type of cement and/or matrix material which binds the grains together in a sedimentary rock influence the roadstone performance.

The way in which alteration develops can strongly influence the roadstone durability. Weathering may reduce the bonding strength between grains to such an extent that they are easily plucked out. Chemical alteration is not always detrimental to the mechanical

Table 12.20 Some representative values of roadstone properties of common aggregates.

Rock type	Water absorption (%)	Specific gravity	Aggregate crushing value (%)	Aggregate impact value (%)	Aggregate abrasion value (%)	Polished stone value
Basalt	0.9	2.91	14	13	14	58
Dolerite	0.4	2.95	10	9	6	55
Granite	0.8	2.64	17	20	15	56
Microgranite	0.5	2.65	12	14	13	57
Hornfels	0.5	2.81	13	11	4	59
Quartzite	1.8	2.63	20	18	15	63
Limestone	0.5	2.69	14	20	16	54
Greywacke	0.5	2.72	10	12	7	62

properties. Indeed, a small amount of alteration may improve the resistance of a rock to polishing (see below). On the other hand, resistance to abrasion decreases progressively with increasing content of altered minerals, as does the crushing strength. The combined hardness of the minerals in a rock, together with the degree to which they are cleaved, and the texture of the rock influence its rate of abrasion. The crushing strength is related to the porosity and grain size—the higher the porosity and the larger the grain size, the lower the crushing strength.

One of the most important parameters of road aggregate is the polished stone value, which influences the skid resistance. A skid-resistant surface is one which is able to retain a high degree of roughness whilst in service. The rate of polish initially is proportional to the volume of traffic, and straight stretches of road are less subject to polishing than bends. The latter may polish up to seven times more rapidly. Stones are polished when fine detrital powder is introduced between the tyre and the surface. Investigations have shown that detrital powder on a road surface tends to be coarser during wet periods than during dry ones. This suggests that polishing is more significant when the road surface is dry rather than wet, the coarser detritus more readily roughening the surface of stone chippings. An improvement in skid resistance can be brought about by blending aggregates.

Rocks within the same major petrological group may differ appreciably in their polished stone characteristics. The best resistance to polish occurs in rocks containing a small proportion of softer alteration materials. A coarser grain size and the presence of cracks in individual grains also tend to improve the resistance to polish. In the case of sedimentary rocks, the presence of hard grains set in a softer matrix produces a good resistance to polish. Sandstones, greywackes and gritty limestones offer a good resistance to polish but, unfortunately, not all of them possess sufficient resistance to crushing and abrasion to render them useful in the wearing course of a road. Purer limestones show a significant tendency to polish. In igneous and contact metamorphic rocks, the good resistance to polish is due to the variation in hardness between the minerals present.

The petrology of an aggregate determines the nature of the surfaces to be coated, the adhesion attainable, depending on the affinity between the individual minerals and the binder, as well as the surface texture of the individual aggregate particles. If the adhesion between the aggregate and binder is less than the cohesion of the binder, stripping may occur. Insufficient drying and the non-removal of dust before coating are, however, the principal causes of stripping. Acidic igneous rocks generally do not mix well with bitumen, as they have a poor ability to absorb it. Basic igneous rocks, such as basalts, possess a high affinity for bitumen, as does limestone.

12.11 Lime, cement and plaster

Lime is made by heating limestone, including chalk, to a temperature of between 1100 and 1200°C in a current of air, at which point CO_2 is driven off to produce quicklime (CaO). Approximately 56 kg of lime can be obtained from 100 kg of pure limestone. Slaking and hydration of quicklime take place when water is added, giving calcium hydroxide. Carbonate rocks vary from place to place, both in chemical composition and physical properties, so that the lime produced in different districts varies somewhat in its behaviour. Dolostones also produce lime, however, the resultant product slakes more slowly than does that derived from limestones.

Portland cement is manufactured by burning pure limestone or chalk with suitable argillaceous material (clay, mud or shale) in the proportions 3:1. The raw materials first are crushed and ground to a powder, and then blended. They are then fed into a rotary kiln and heated to a temperature of over 1800°C. Carbon dioxide and water vapour are driven off, and the lime fuses with the aluminium silicate in the clay to form a clinker. This is ground to a fine powder, and less than 3% gypsum is added to retard setting. Lime is the principal constituent of Portland cement, but too much lime produces a weak cement. Silica constitutes approximately 20% and alumina 5%. Both are responsible for the strength of the cement, but a high silica content produces a slow-setting cement, whereas a high alumina content gives a quick-setting cement. The percentage of iron oxides is low, and in white Portland cement it is kept to a minimum. The proportion of magnesium oxide (MgO) should not exceed 4%, otherwise the cement will be unsound. Similarly, the sulphate content must not exceed 2.75%. Sulphate-resisting cement is made by the addition of a very small quantity of tricalcium aluminate to normal Portland cement.

When gypsum ($CaSO_4.nH_2O$) is heated to a temperature of $170\,°C$, it loses three-quarters of its water of crystallization, becoming calcium sulphate hemihydrate or plaster of Paris. Anhydrous calcium sulphate forms at higher temperatures. These two substances are the chief materials used in plasters. Gypsum plasters have now more or less replaced lime plasters.

References

Anon. (1970) The logging of cores for engineering purposes. *Quarterly Journal of Engineering Geology* 3, 1–24.

Anon. (1975a) *Concrete in Sulphate Bearing Soils and Groundwaters. Digest* 90. Building Research Establishment, Her Majesty's Stationery Office, Watford.

Anon. (1975b) *Methods of Sampling and Testing Mineral Aggregates, Sands and Fillers,* BS 812. British Standards Institution, London.

Anon. (1977) The description of rock masses for engineering purposes. Working Party Report. *Quarterly Journal of Engineering Geology* 10, 355–388.

Anon. (1979) Classification of rocks and soils for engineering geological mapping. *Bulletin of the International Association of Engineering Geology* 19, 364–371.

Anon. (1981) Basic geotechnical description of rock masses. International Society for Rock Mechanics, Commission on the Classification of Rocks and Rock Masses. *International Journal of Rock Mechanics and Mining Science and Geomechanical Abstracts* 18, 85–110.

Anon. (1995) The description and classification of weathered rocks for engineering purposes. Working Party Report. *Quarterly Journal of Engineering Geology* 28, 207–242.

Barton, N. (1976) The shear strength of rock and rock joints. *International Journal of Rock Mechanics and Mining Science and Geomechanical Abstracts* 13, 255–279.

Barton, N. (1978) Suggested methods for the quantitative description of discontinuities in rock masses. *International Journal of Rock Mechanics and Mining Science and Geomechanical Abstracts* 15, 319–368.

Bell, F.G. (1981) Geotechnical properties of evaporites. *Bulletin of the International Association of Engineering Geology* 24, 137–144.

Bell, F.G. (1992a) Salt mining and associated subsidence in mid-Cheshire, England, and its influence on planning. *Bulletin of the Association of Engineering Geologists* 29, 371–386.

Bell, F.G. (1992b) The durability of sandstone as building stone, especially in urban environments. *Bulletin of the Association of Engineering Geologists* 24, 49–60.

Bell, F.G. (1993) Durability of carbonate rock as building stone with comments on its preservation. *Environmental Geology* 21, 187–200.

Bell, F.G. (1994) A survey of the engineering properties of some anhydrite and gypsum from the north and midlands of England. *Engineering Geology* 38, 1–23.

Bell, F.G. & Maud, R.R. (1996) Landslides associated with the Pietermaritzburg Formation in the greater Durban area, South Africa. *Environmental and Engineering Geoscience* 2, 557–573.

Bell, F.G., Entwistle, D.C. & Culshaw, M.G. (1997) A geotechnical survey of some British Coal Measures mudstones, with particular emphasis on durability. *Engineering Geology* 46, 115–129.

Bieniawski, I.T. (1974) Geomechanics classification of rock masses and its application to tunnelling. *Proceedings of the Third Congress of the International Society of Rock Mechanics,* Denver, 1, pp. 27–32.

Brook, C.A. & Alison, T.L. (1986) Fracture mapping and ground subsidence susceptibility modelling in covered karst terrain: an example of Dougherty County, Georgia. In *Proceedings of the Third International Symposium on Land Subsidence, Venice, International Association of Hydrological Sciences, Publication No.* 151, pp. 595–606.

Chandler, R.J. (1969) The effect of weathering on the shear strength properties of Keuper Marl. *Geotechnique* 19, 321–334.

Chappel, B.A. (1974) Deformational response of differently shaped and sized test pieces of shale rock. *International Journal of Rock Mechanics and Mining Science* 11, 21–28.

Cooper, A.H. (1988) Subsidence resulting from the dissolution of Permian gypsum in the Ripon area; its relevance to mining and water abstraction. In *Engineering Geology of Underground Movements, Engineering Geology Special Publication No.* 5, Bell, F.G., Culshaw, M.G., Cripps, J.C. & Lovell, M.A. (eds). Geological Society, London, pp. 387–390.

Dearman, W.R. (1974) Weathering classification in the characterization of rock for engineering purposes in British practice. *Bulletin of the International Association of Engineering Geology* 9, 33–42.

Dearman, W.R. (1991) *Engineering Geological Mapping.* Butterworth–Heinemann, Oxford.

Deere, D.U. (1964) Technical description of cores for engineering purposes. *Rock Mechanics and Engineering Geology* 1, 18–22.

Deere, D.U. & Miller, R.P. (1966) *Engineering Classification and Index Properties for Intact Rock. Technical Report No.* AFWL-TR-65-115. Air Force Weapons Laboratory, Kirtland Air Base, New Mexico.

Dick, J.C. & Shakoor, A. (1992) Lithological controls of mudrock durability. *Quarterly Journal of Engineering Geology* 25, 31–46.

Eck, W. & Redfield, R.C. (1965) Engineering geology problems at Sanford Dam, Texas. *Bulletin of the Association of Engineering Geologists* 3, 15–25.

Fookes, P.G. (1991) Geomaterials. *Quarterly Journal of Engineering Geology* 24, 3–16.

Fookes, P.G., Dearman, W.R. & Franklin, J.L. (1972) Some engineering aspects of weathering with field examples from Dartmoor and elsewhere. *Quarterly Journal of Engineering Geology* 3, 1–24.

Franklin, J.L. & Broch, E. (1972) The point load strength test. *International Journal of Rock Mechanics, Mineral Science* 9, 669–697.

Franklin, J.L., Broch, E. & Walton, G. (1971) Logging the mechanical character of rock. *Transactions of the Institution of Mining and Metallurgy* 81, *Mining Section*, A1–A9.

French, W.J. (1991) Concrete petrography: a review. *Quarterly Journal of Engineering Geology* 24, 17–48.

Gillott, J.E. & Swenson, E.G. (1969) Mechanism of alkali–carbonate reaction. *Quarterly Journal of Engineering Geology* 2, 7–24.

Gogte, B.S. (1973) An evaluation of some common Indian rocks with special reference to alkali aggregate reactions. *Engineering Geology* 7, 135–154.

Hallbauer, D.K., Nieble, C., Berard, J., *et al.* (1978) Suggested methods for petrographic description. International Society Rock Mechanics, Commission on Standardization of Laboratory and Field Tests. *International Journal of Rock Mechanics and Mining Science and Geomechanical Abstracts* 15, 14–41.

Haskins, D.R. & Bell, F.G. (1995) Drakensberg basalts: their alteration, breakdown and durability. *Quarterly Journal of Engineering Geology* 28, 287–302.

Higginbottom, I.E. (1965) The engineering geology of the Chalk. In *Proceedings of the Symposium on Chalk in Earthworks*. Institution of Civil Engineers, London, pp. 1–14.

Higgs, N.B. (1976) Slaking basalts. *Bulletin of the Association of Engineering Geologists* 13, 151–162.

Hobbs, N.B. (1975) Factors affecting the prediction of settlement of structures on rocks with particular reference to the Chalk and Trias. In *Settlement of Structures*, British Geotechnical Society. Pentech Press, London, pp. 579–610.

Hoek, E. (1983) Strength of rock masses. *Geotechnique* 33, 187–223.

Hoek, E. & Bray, J.W. (1981) *Rock Slope Engineering*, 3rd edn. Institution of Mining and Metallurgy, London.

Hoek, E. & Brown, E.T. (1980) *Underground Excavations in Rock*. Institution of Mining and Metallurgy, London.

Iliev, I.G. (1967) An attempt to estimate the degree of weathering of intrusive rocks from the physico-mechanical properties. In *Proceedings of the First Congress of the International Society for Rock Mechanics, Lisbon* 1, pp. 109–114. Balkema, Rotterdam.

Irfan, T.Y. & Dearman, W.R. (1978a) Engineering classification and index properties of weathered granite.

Bulletin of the International Association of Engineering Geology 17, 79–90.

Irfan, T.Y. & Dearman, W.R. (1978b) The engineering petrography of a weathered granite in Cornwall, England. *Quarterly Journal of Engineering Geology* 11, 233–244.

James, A.N. & Kirkpatrick, I.M. (1980) Design of foundations of dams containing soluble rocks and soils. *Quarterly Journal of Engineering Geology* 13, 189–198.

James, A.N. & Lupton, A.R.R. (1978) Gypsum and anhydrite in foundations of hydraulic structures. *Geotechnique* 28, 249–272.

Johnson, W.D. (1901) The high plains and their utilization. *United States Geological Survey, 21st Annual Report, Part* 4. United States Geological Survey, Washington, DC, pp. 601–741.

Kennard, M.F. & Knill, J.L. (1968) Reservoirs on limestone, with particular reference to the Cow Green Scheme. *Journal of the Institution of Water Engineers* 23, 87–113.

Kennard, M.F., Knill, J.L. & Vaughan, P.R. (1967) The geotechnical properties and behaviour of Carboniferous shale at Balderhead Dam. *Quarterly Journal of Engineering Geology* 1, 3–24.

Langer, M. (1982) Geotechnical investigation methods for rock salt. *Bulletin of the International Association of Engineering Geology* 25, 155–164.

Little, A.L. (1969) The engineering classification of residual tropical soils. In *Proceedings of the Seventh International Conference on Soil Mechanics and Foundation Engineering, Mexico City* 1, pp. 1–10.

Lovegrove, G.W. & Fookes, P.G. (1972) The planning and implementation of a site investigation for a highway in tropical conditions in Fiji. *Quarterly Journal of Engineering Geology* 5, 43–68.

Martin, R.P. & Hencher, S.R. (1986) Principles for description and classification of weathered rock for engineering purposes. In *Site Investigation Practice: Assessing BS 5930. Engineering Geology Special Publication No. 2*, Hawkins A.B. (ed). Geological Society, London, pp. 299–308.

McCann, D.M. (1992) Rock mass assessment using geophysical methods. In *Engineering in Rock Masses*, Bell, F.G. (ed). Butterworth–Heinemann, Oxford, pp. 170–189.

Moye, D.G. (1955) Engineering geology for the Snowy Mountains Scheme. *Journal of the Institution of Engineers of Australia* 27, 287–298.

Olivier, H.G. (1979) A new engineering–geological rock durability classification. *Engineering Geology* 14, 255–279.

Onodera, T.F. (1963) Dynamic investigation of foundation rocks. In *Proceedings of the Fifth Symposium on Rock Mechanics, Minnesota*. Pergamon Press, New York, pp. 517–533.

Penner, E., Eden, W.J. & Gillott, J.E. (1973) Floor heave due to biochemical weathering of shale. In *Proceedings of the Eighth International Conference on Soil Mechanics and Foundation Engineering, Moscow* **2**, pp. 151–158. Balkema, Rotterdam.

Rahn, P.H. & Davis, A.D. (1996) Gypsum foundation problems in the Black Hills area, South Dakota. *Environmental and Engineering Geoscience* **2**, 213–223.

Sims, I. (1991) Quality and durability of stone for construction. *Quarterly Journal of Engineering Geology* **24**, 67–74.

Smith, C.K. & Redlinger, J.F. (1953) Soil properties of the Fort Union clay shale. In *Proceedings of the Third International Conference on Soil Mechanics, Foundation Engineering, Zurich* **1**, pp. 56–61.

Smith, M.R. & Collis, L. (eds) (1993) *Aggregates: Sand, Gravel and Crushed Rock Aggregates for Construction Purposes*, 2nd edn. *Engineering Geology Special Publication No. 9*. Geological Society, London.

Sowers, G.F. (1975) Failures in limestones in the humid subtropics. *Proceedings of the American Society of Civil Engineers, Journal Geotechnical Engineers Division* **10**, GT8, 771–787.

Spears, D.A. & Taylor, R.K. (1972) The influence of weathering on the composition and engineering properties of *in situ* Coal Measures rocks. *International Journal of Rock Mechanics and Mining Science* **9**, 729–756.

Stacey, T.R. & Page, C.H. (1986) *Practical Handbook for Underground Rock Mechanics*. Trans. Tech. Publications, Clausthal-Zelterfeld.

Taylor, R.K. (1988) Coal Measures mudrocks: composition, classification and weathering. *Quarterly Journal of Engineering Geology* **21**, 85–99.

Taylor, R.K. & Spears, D.A. (1970) The breakdown of British Coal Measures rocks. *International Journal of Rock Mechanics and Mining Science* **14**, 291–309.

Terzaghi, K. (1946) Introduction to tunnel geology. In *Rock Tunnelling with Steel Supports*, Procter, R. & White, T. (eds). Commercial Shearing and Stamping Co., Youngstown, OH, pp. 17–99.

Ward, W.H., Burland, J.B. & Gallois, R.W. (1968) Geotechnical assessment of a site at Mundford, Norfolk, for a large proton accelerator. *Geotechnique* **18**, 399–431.

Wetzel, A. & Einsele, G. (1991) On the physical weathering of various mudrocks. *Bulletin of the International Association of Engineering Geology* **44**, 89–100.

Wichter, L. (1979) On the geotechnical properties of a Jurassic clay shale. In *Proceedings of the Fourth International Congress on Rock Mechanics (ISRM), Montreux* **1**. Balkema, Rotterdam, pp. 319–328.

Yuzer, E. (1982) Engineering properties and evaporitic formations of Turkey. *Bulletin of the International Association of Engineering Geology* **25**, 107–110.

Zaruba, Q. & Bukovansky, M. (1965) Mechanical properties of Ordovician shales of Central Bohemia. In *Proceedings of the Sixth International Conference on Soil Mechanics and Foundation Engineering, Montreal* **3**, pp. 421–424.

13   The impact of mining on the environment

13.1 Introduction

Mining, alongside agriculture, represents one of the earliest activities of man, the two being fundamental to the development and continuation of civilization. Indeed, the origins of mining date back to the Stone Age, when our ancestors mined flints, and the dependence of man on minerals has increased as society has evolved. Today, civilization could not exist without the products of mining, and the demand for these products continues to grow.

However, unlike agriculture, where there is some choice in where and what to grow, mining can only take place where minerals are present and economically viable. Another important aspect of mining is that it is a robber economy, in that a mineral deposit is a finite resource, and so mining comes to an end when the deposit is exhausted. Deposits which are abandoned when they become uneconomic to work may be reworked at some time in the future, when mining technology or demand makes their exploitation once again worthwhile.

Mining and the associated mineral processing and beneficiation have an impact on the environment. Unfortunately, this frequently has led to serious consequences. The impact of mining depends on many factors, in particular the type of mining and the size of the operation. It can mean that land is disturbed, that the topography is changed and that the hydrogeological conditions are affected adversely. Mining also has a social impact on the environment. Communities grow up around mines. When the mines close, the communities can suffer and die.

In the past, the mining industry frequently showed a lack of concern for the environment. In particular, the disposal of waste has led to unsightly spoil heaps being left to disfigure the landscape, and to the pollution of surface streams and groundwater. Urban areas have suffered serious subsidence damage by under-

mining. Today, however, the greater awareness of the importance of the environment has led to tighter regulations being imposed by many countries to lessen the impact of mining. The concept of the rehabilitation of a site after mining operations have ceased has become entrenched in law. An environmental impact assessment is necessary prior to the development of any new mine, and an environmental management programme must be produced to show how the mine will operate.

The adverse impacts are part of the price which has to be paid for the benefits of mineral consumption. Although the adverse impacts on the environment must be minimized, some environmental degradation due to mining is inescapable. Some consolation perhaps may be obtained from the fact that mines are local phenomena, although they may have an impact beyond mine boundaries. Mines also account for only a small part of the land area of a country (e.g. the mining industry accounts for less than 1% of the total area of South Africa or the USA). Be that as it may, land which has become derelict by past mining activity can be restored, at a cost. Rehabilitated spoil heaps frequently become centres of social amenity, such as parklands, golf courses and even artificial ski slopes. Open pits, when they fill with water, can be used as marinas, for fishing or as wildlife reserves. Even some underground mines can be used, such as those in limestone at Kansas City, Missouri, which are employed as warehouses, cold storage facilities and offices. Mining therefore can be regarded as one of the stages in the sequential use of land.

13.2 Metalliferous mining and subsidence

13.2.1 Mining methods

There are several underground mining methods used for the extraction of metalliferous deposits. In partial extraction methods, solid pillars are left unmined to provide support to the underground workings. For example, sill pillars and shaft pillars are designed to protect important underground areas. Pillars may be variable in shape and size and, in many cases, because pillars often represent reserves of ore, they are extracted or reduced in size towards the end of the life of the mine. In modern practice, it is possible to design against the occurrence of surface subsidence. However, this procedure was not practised in earlier times. In such situations, subsidence of the surface is a possibility. The collapse of the roof or hanging wall of the workings between pillars can result in localized surface subsidence. The subsidence profile may be very severe, with large differential subsidence, tilts and horizontal strains. Pillar collapse can give rise to localized, or substantial, areas of subsidence. If variable-sized pillars are present, or if larger barrier pillars are provided at regular spacings, the area of subsidence is likely to be restricted. The resulting subsidence profile is likely to be irregular, with large differential settlements, tilts and horizontal strains at the perimeter of the subsidence area. The presence of faults can give rise to fault steps. Failure of the roof and/or floor leads to a similar type of subsidence behaviour. When mining inclined ore bodies, the steeper the dip, the more localized the subsidence is likely to be.

Mining of tabular stopes with substantial spans is practised in the gold and platinum mines of southern Africa. Extraction of one, two or more reefs has taken place, each typically with stoping widths of 1–2 m. When this type of mining occurs at shallow depth, detrimental subsidence frequently has taken place. The surface profile commonly is very irregular, and tension cracks often have been observed. The effects are dependent on the dips of the reefs, which vary from as little as 10° to vertical. The closure of stopes is more likely for flatter dips, and hence surface movements are more likely to occur under such conditions. Steeply dipping stopes tend to remain open and present a longer term hazard. Adverse subsidence usually is not observed when the mining depth exceeds 250 m. Most of the detrimental subsidence that has occurred has been associated with geological features, such as dyke contacts and faults (Stacey & Rauch, 1981).

Many metalliferous deposits occur in large disseminated ore bodies, and often the only way in which such ore bodies can be mined economically is by means of very high-volume production. When the rock mass is sufficiently competent, large, open stopes can be excavated. The sizes of open stopes are the subject of design, and some form of pillar usually separates adjacent open stopes. The method therefore is a partial extraction approach, but open stopes may have spans of between 50 and 100 m. In caving, material from above is allowed to cave into the opening created by the mined extraction. Because caving occurs above the ore body, it can progress through to

Fig. 13.1 Cratering due to block caving at a copper mine in central Chile. (Courtesy of Dr Dick Stacey.)

the surface to form a subsidence crater (Fig. 13.1). This crater usually has several scarps around its perimeter, and contains material which has subsided *en bloc*, overlying caved and rotated material and loosely consolidated ravelled material. Owing to the large volumes that are extracted during mining, the extent and depth of the subsidence crater can be very large.

13.2.2 Stabilization of old workings

Taking Johannesburg as an example, gold deposits were extensively worked within the central area, where three reefs were mined. Narrow, tabular, open stopes, supported by occasional pillars, were left behind, the first level occurring at around 30–40 m beneath the surface. Subsequent levels occur at depth intervals of approximately 30 m. Much of the Johannesburg area, where the outcrops occur, is zoned for commercial or light industrial development, so that low-rise, relatively low-cost structures can be constructed. The aim of any shallow ground stabilization method required is to develop a competent surface to promote arching across stopes and prevent collapses. The construction of flexible structures can then accommodate any subsequent minor surface deformation which occurs. Dynamic compaction has been used as a method of shallow stabilization, where ground consisting of very loose sandy gravel and fill occurs above old mine workings. The dynamic compaction presumably improves the characteristics of the soil down to bedrock, and closes any cavities due to shallow mine workings. In this way, a ground arch is formed spanning across the reef outcrops. Another, more common, form of shallow stabilization is to plug the stopes, normally with concrete, to provide the necessary local support, and then to place backfill above the plug, the former being compacted to form a stabilizing ground arch.

Deep stabilization of abandoned reef workings has to provide rigid support between footwall and hanging wall, so that, if collapse or instability occurs at greater depth, the near-surface rigid zone forms an arch. In stope mining, support can comprise a combination of dip and strike pillars. Dip pillars tend to have greater ability than strike pillars to withstand shear deformation in a vertical plane normal to the reefs. Strike pillars, especially those at the bases of dip pillars, retain fill in the stopes, so preventing slabbing from hanging wall and footwall. The pillars are constructed by sinking winzes in the stopes and backfilling them with mass concrete. The treatment is completed by constructing a concrete crown pillar across the stope.

13.2.3 Mining and sinkhole development

Dewatering associated with mining in the gold-bearing reefs of the Far West Rand, South Africa, which underlie dolostone and unconsolidated deposits, has led to the formation of sinkholes and produced differential subsidence over large areas. Hence, certain areas became unsafe for occupation and were evacuated. Subsidence initially was noticed in 1959, and the seriousness of the situation was highlighted in December 1962 when a sinkhole appeared at the West Driefontein Mine and engulfed a three-storey crusher plant (Fig. 13.2) with the loss of 29 lives. Then, in August 1964, two houses and parts of two others disappeared into a sinkhole in Blyvooruitzicht Township with the loss of five lives. Consequently, it became a matter of urgency that the areas which were subject to subsidence or to the occurrence of sinkholes be delineated.

Sinkholes are formed concurrently with the lowering of the water table in areas which formerly, in general, were free from sinkholes. Normally, the cover of unconsolidated material is less than 15 m, otherwise it chokes the cavity on collapse. Bezuidenhout and Enslin (1970) divided these areas into three groups. First, there are those areas in which the original water table is less than 15 m below and, less frequently, within 30 m of the surface. Second, sinkholes form in scarp zones which border deep, buried valleys. Third, sinkholes occur in narrow, buried valleys where the limited width means that initially unconsolidated material bridges sinks in the dolostone below. Bezuidenhout and Enslin found that sinkholes of larger dimensions than would normally be expected to occur develop in such valleys. The largest one they recorded had a diameter of 80 m.

Slow subsidence occurs as a consequence of consolidation taking place in unconsolidated material as the water table is lowered. The degree of subsidence which takes place reflects the thickness and proportion of unconsolidated deposits which have consolidated. The thickness of these deposits varies laterally, thereby giving rise to differential subsidence which,

Fig. 13.2 Collapse of three-storey crusher plant into a sinkhole at West Driefontein Mine. (Courtesy of Gold Fields of South Africa Ltd.)

Fig. 13.3 Pillared workings in the Bethaney Falls Limestone, Kansas City, Missouri.

in turn, causes large fissures to occur at the surface. In fact, the most prominent fissures frequently demarcate the areas of subsidence. The total subsidence varies from several centimetres to over 9 m. The time lag between the lowering of the water table and surface subsidence, where observed, is fairly short.

13.3 Pillar workings

Stratified deposits, such as coal, limestone, gypsum, salt, sedimentary iron ore, etc., have been and are worked by partial extraction methods, whereby pillars of the mineral deposit are left in place to support the roof of the workings (Fig. 13.3). In pillared workings, the pillars sustain the redistributed weight of the overburden, which means that they and the rocks immediately above and below are subjected to added compression. Stress concentrations tend to be located at the edges of pillars, and the intervening roof rocks tend to sag. The effects at ground level normally are insignificant. Although the intrinsic strength of a stratified deposit varies, the important

Fig. 13.4 Subsidence due to pillar collapse, Witbank coalfield, South Africa.

factor in the case of pillars is that their ultimate behaviour and failure are a function of the bed thickness to pillar width, the depth below ground and the size of the extraction area. The mode of failure also involves the character of the roof and floor rocks. Pillars in the centre of the mined out area are subjected to greater stress than those at the periphery. Individual pillars in dipping seams tend to be less stable than those in horizontal seams, because the overburden produces a shear force on the pillar.

Collapse in one pillar can bring about collapse in others in a sort of chain reaction, because increasing loads are placed on those remaining (Bryan *et al.*, 1964). Slow deterioration and failure of pillars may take place after mining operations have long since ceased (Fig. 13.4). Old pillars at shallow depth have occasionally failed near faults, and they may fail if they are subjected to the effects of subsequent mining. The yielding of a large number of pillars can bring about a shallow, broad subsidence trough over a large surface area. The ground surface in such a trough displaces radially inwards towards the area of maximum subsidence. This inward radial movement generates tangential compressive strain, and circumferential tension fractures frequently are developed. Ground movements depend on the mine layout, in particular the extraction ratio and geology, as well as the topographical conditions at the surface. They tend to develop rather suddenly, the major initial movements lasting, in some instances, for about a week, with sub-

sequent displacements occurring over varying periods of time. The initial movements can produce a steep-sided, bowl-shaped area. Nonetheless, the shape of the profile can vary appreciably and, because it varies with the mine layout and geological conditions, it can be difficult to predict accurately. Normally, the greater the maximum subsidence, the greater the likelihood of variation in the profile. Maximum profile slopes and curvatures frequently increase with increasing subsidence. The magnitude of surface tensile and compressive strains can range from slight to severe.

In the past, especially in coal mining, pillars frequently were robbed on retreat. The extraction of pillars during the retreat phase can simulate longwall conditions (see below), although it can never be assumed that all pillars have been removed. At moderate depths, pillars, especially pillar remnants, are probably crushed and the goaf (i.e. the worked out area) is compacted but, at shallow depths, lower crushing pressures may mean that the closure is variable. This causes foundation problems when large or sensitive structures are erected above.

The prediction of subsidence as a result of pillar failure requires accurate data regarding the layout of the mine. Such information frequently does not exist in the case of old, abandoned mines. On the other hand, when accurate mine plans are available, or in the case of a working mine, the method outlined by

Fig. 13.5 Void migration in shaly roof rocks above the Five Quarter seam at Pethburn opencast site, County Durham, England.

Goodman *et al.* (1980) may be used to evaluate the collapse potential.

Even if pillars are relatively stable, the surface may be affected by void migration (Fig. 13.5). This can take place within a few months of, or a very long period of years after, mining. Void migration develops when roof rock falls into the worked out areas. When this occurs, the material involved in the fall bulks, which means that migration eventually is arrested, although the bulked material never completely fills the voids. Nevertheless, the process can, at shallow depth, continue upwards to the ground surface, leading to the sudden appearance of a crown-hole. The factors which influence whether or not void migration will take place include the width of the unsupported span, the height of the workings, the nature of the cover rocks, particularly their shear strength and the incidence and geometry of the discontinuities, the thickness and dip of the

seam, the depth of overburden, and the groundwater regime.

It frequently is maintained that the maximum height of void migration is directly proportional to the height of the workings, and inversely proportional to the change in volume of the collapsed material. Various methods have been proposed to predict the collapse of roof strata above rooms or stalls (Piggott & Eynon, 1978; Garrard & Taylor, 1988). In fact, the maximum height of migration in exceptional cases might extend to 10 times the height of the original room, however, it generally is 3–5 times the room height. In a sequence of differing rock types, if a competent rock beam is to span an opening, its thickness should be equal to twice the span width, in order to allow for arching to develop. A bed of sandstone will usually arrest a void, especially if it is located some distance above the immediate roof of the working. Sandstones apart, however, most voids are bridged when the span decreases through corbelling to an acceptable width, rather than when a more competent bed is encountered. Chimney-type collapses can occur to abnormally high levels of migration in massive strata in which the joints diverge downwards.

Exceptionally, void migrations in excess of 20 times the worked height of a seam have been recorded. The self-choking process may not be fulfilled in dipping seams, especially if they are affected by copious quantities of water which can redistribute the fallen material. The redistribution of collapsed material can lead to the formation of supervoids, and their migration to rockhead then produces large-scale subsidence at ground level. Weak, superficial deposits may flow into voids which have reached rockhead, thereby forming features which may vary from a gentle dishing of the surface to inverted cone-like depressions of large diameter.

A site investigation for an important structure(s) requires the exploration and sampling of all strata likely to be significantly affected by the structural loading. The location of subsurface voids due to mineral extraction is of prime importance in this context. In other words, an attempt should be made to determine the number and depth of mined horizons, the extraction ratio, the pattern of the layout and the condition of the old room and pillar workings (Bell, 1986; Cripps *et al.*, 1988). The sequence and type of roof rocks may provide some clue as to whether void migration has taken place and, if so, its possible extent. Of particular importance is the state of the old workings; careful note should be taken of whether they are open, partially collapsed or collapsed, and the degree of fracturing and bed separation in the roof rocks should be recorded, if possible. This helps to provide an assessment of past and future collapse, which is obviously very important.

13.4 Old mine workings and hazard zoning

Assessments of mining hazards usually have been made on a site basis, regional assessments being much less common. Nonetheless, regional assessments can offer planners an overview of the problems involved, and can help them to avoid imposing unnecessarily rigorous conditions on developers in areas where they really are not warranted.

Hazard maps of areas in which old mine workings are present ideally should represent a source of clear and useful information for planners. In this respect, they should realistically present the degree of risk (i.e. the probability of the occurrence of a hazard event). Descriptive terms such as high, medium and low risk must be defined, and there are social and economic dangers in overstating the degree of risk. Ideally, numerical values should be assigned to the degree of risk. However, this is no easy matter, as numerical values can only be derived from a comprehensive record of events which, unfortunately, in the case of old, abandoned mine workings, is available only infrequently. Furthermore, many events in the past may not have been recorded, which throws into question the reliability of any statistical analysis of the data. Indeed, subsidence is affected by so many mining and geological variables that many regard it as, to all intents and purposes, a random process. The matter of risk assessment is further complicated in terms of its tolerance, for example, people are less tolerant in relation to loss of life than to loss of property. Hence, the likelihood of an event causing loss of life would be assigned a higher risk value than a loss of property (e.g. the probability of a 1 in 100 loss of life presumably would be regarded as very risky, whereas very risky in terms of a loss of property may be accepted as 1 in 10). Nonetheless, if planners can be provided with some form of numerical assessment of the degree of risk, they will be better able to make sensible financial decisions and provide more effective solutions to problems which arise.

An attempt at hazard zoning in an area of old mine workings was made by Price (1971). After a detailed investigation at a site in Airdrie, Scotland, underlain by shallow, abandoned mine workings, he was able to propose safe and unsafe zones (Fig. 13.6). In the safe zones, the cover rock was regarded as being thick enough to preclude subsidence hazards (about 10 m of rock or 15 m of till was regarded as being sufficient to ensure that crown-holes did not appear at the surface), and normal foundations could be used for the two-storey dwellings which were to be erected. On the boundaries between the safe and unsafe zones, the dwellings were constructed with reinforced foundations, or rafts, as an added precaution against unforeseen problems. Development was prohibited in zones designated as unsafe. In effect, Price produced a thematic mining information plan of the site to facilitate its development.

A zoning system based upon the depth of cover above old mine workings in Johannesburg, South Africa, has been referred to by Stacey and Bakker (1992). They described the building restrictions which are imposed on surface development related to these zones. Basically, these prohibit development where old mine workings occur at a depth of less than 90 m. They then limit the permissible heights of proposed buildings in accordance with the depth of mining below the site. These restrictions are relaxed progressively to a depth of 240 m, after which no restrictions apply. The restrictions are flexible, in that they also take into account the number and separation of the reefs worked, the extent of mining, the stoping widths, the type and amount of any artificial support and the dip of the reefs.

Early thematic maps, produced by the British Geological Survey, depicted areas of undermining assumed to be within 30 m of the surface, on the one hand, and at depths exceeding 30 m below the surface on the other (McMillan & Browne, 1987). This 30 m depth is based on limited information, and therefore is subject to interpretation. It assumes that bulking factors of 10–20% will affect the strata involved in void migration. Initially, known and suspected mining areas were not differentiated on these maps. However, only areas of mining shown on mine plans were represented on the map of the Glasgow district,

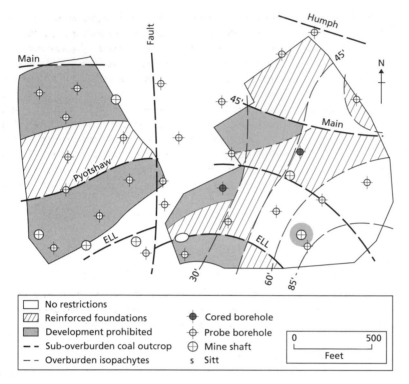

Fig. 13.6 Permissible foundation zones at a site in Airdrie. Normal foundations are permitted in unshaded areas (i.e. safe zones). (After Price, 1971.)

Legend:
- ☐ No restrictions
- ▨ Reinforced foundations
- ▦ Development prohibited
- – – Sub-overburden coal outcrop
- – – Overburden isopachytes
- ● Cored borehole
- ⊕ Probe borehole
- ⊕ Mine shaft
- s Sitt

0 — 500 Feet

and no areas of suspected mining were given. No attempt was made to infer the extent of working beyond the limits defined by mine plans, other than to plot relevant drill-hole data. The recognition of single and multiple seam working led to the requirement that areas of shallow working (i.e. less than 30 m below rockhead) should be identified in terms of the seams worked. Separate maps were prepared, illustrating areas of total known mining, current mining, known mining within 30 m of rockhead, together with the locations of shafts and drill-holes encountering old shallow workings, and mining for minerals other than coal and ironstone. An indication of the area in which old mine workings might be expected can be obtained by plotting all drill-holes which encounter spoil outside areas of workings known from abandonment plans. In regions in which mineral outcrops are reasonably well known, areas of suspected workings can be mapped as a separate category, although it is then necessary to assume that all workable beds have been exploited, at least in the near-surface area.

In an assessment of the degree of risk due to subsidence incidents associated with abandoned mine workings in South Wales, carried out by Statham

et al. (1987), it was found that, of the 388 events, 64% occurred in open land and so posed no threat to persons or property; 21% occurred when people were nearby or threatened property; the remainder caused damage to highways, buildings or other property, and only one event resulted in minor injuries. In the context of the coalfield of South Wales, this represents a low level of hazard. Assuming that a typical incident affects an area of 5 m², the probability of collapse occurring on any 25 m² plot is of the order of 10^{-7} per year. Even if the number of subsidence incidents which have remained undiscovered increased the above total by a factor of three, the overall risk would still be low. Statham *et al.* found that over 90% of the incidents occurred within 100 m of the outcrop of the seam concerned. They produced a development advice map for South Wales, which showed two zones inside the outcrops of worked seams, corresponding to migration ratios (thickness of rock cover divided by the extracted thickness) of 6 and 10, which were expected to contain 90% and 100%, respectively of the relevant subsidence incidents. This map makes a contribution towards regional planning by taking into account a possible development constraint.

However, it must be borne in mind that thematic maps which attempt to portray the degree of risk of a hazard event represent generalized interpretations of the data available at the time of compilation. Therefore, they cannot be interpreted too literally, and areas outlined as 'undermined' should not automatically be subject to planning blight. Obviously, there is a tendency to assume that the limits of old mine workings represented on a map indicate the full extent of the workings, but the interpretation of their location is based on scanty information and includes assumptions, some of which may be unfounded. It should be recognized that problems in areas of past mining only occur if buildings are not properly planned, designed and constructed with reference to the state of undermining. In addition, zoning based entirely upon the depth of cover above workings cannot be relied upon completely, because occasionally subsidences have occurred in zones designated as 'safe'.

13.5 Measures to reduce or avoid subsidence effects due to old mine workings

Where a site, which is proposed for development, is underlain by shallow, old mine workings, there are a number of ways in which the problem can be dealt with (Healy & Head, 1984). However, any decision regarding the most appropriate method should be based upon an investigation, which has evaluated the site conditions carefully (Bell, 1975) and examined the economic implications of the alternatives. One of the most difficult assessments to make is related to the possible effects of progressive deterioration of the workings and the associated potential subsidence risk. In addition, the placement of a new structure might alter a quasi-stable situation.

The first and most obvious method is to locate the proposed structure on sound ground away from old workings, or over workings proved to be stable. It generally is not sufficient to locate immediately outside the area undermined, as the area of influence should be considered. In such cases, the angle of influence or draw usually is taken as 25°; in other words, the area of influence is defined by projecting an angle of 25° to the vertical from the periphery and depth of the workings to the ground surface. Such a location, of course, is not always possible.

If old mine workings are at a very shallow depth, it

might be feasible, by means of bulk excavation, to found on the strata beneath. This is an economic solution at depths of up to 7 m, or on sloping sites.

Where the allowable bearing capacity of the foundation materials has been reduced by mining, it may be possible to use a raft. Rafts can consist of massive concrete slabs or stiff slab and beam cellular rafts. A raft can span weaker and more deformable zones in the foundation, thus spreading the weight of the structure well outside the limits of the building. However, rafts are expensive.

Reinforced bored pile foundations have also been resorted to in areas of abandoned mine workings. In such instances, the piles bear on a competent stratum beneath the workings. They should also be sleeved so that concrete is not lost into voids, and to avoid the development of negative skin friction if overlying strata collapse. Some authorities, however, have suggested that piling through old mine workings seems to be inadvisable for the following reasons: first, pile emplacement may precipitate collapse and, second, subsequent collapse at seam level possibly could lead to the buckling or shearing of piles (Price et al., 1969). There may also be a problem with the lateral stability of piles passing through collapsed zones above mine workings or through large remnant voids.

Where old mine workings are believed to pose an unacceptable hazard to development, and it is impracticable to use adequate measures in design or found below their level, the ground itself can be treated. Such treatment involves filling the voids in order to prevent void migration and pillar collapse. In exceptional cases in which, for example, the mine workings are readily accessible, barriers can be constructed underground, and the workings filled hydraulically with sand or pneumatically with some suitable material. Hydraulic stowing may also take place from the surface via drill-holes of sufficient diameter. Pneumatic or gravity stowing often is considered where large subsurface voids have to be filled.

Alternatively, grouting can be undertaken by drilling holes from the surface into the mine workings, on a systematic basis, and filling the remnant voids with an appropriate grout mix (Lloyd et al., 1995). If it has been impossible to obtain accurate details of the layout and extent of the workings, the zone beneath the intended structure can be subjected to consolidation grouting. The grouts used in these operations commonly consist of cement, fly ash and

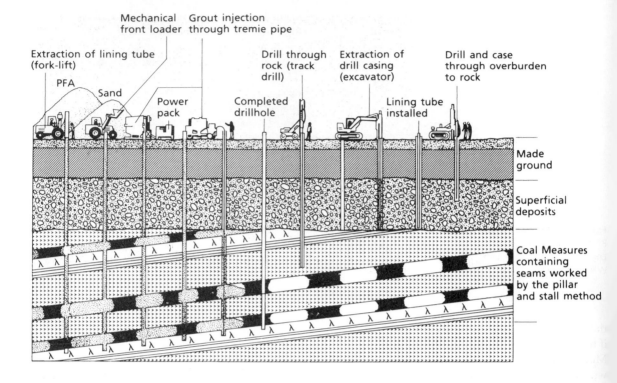

Extraction of lining tube
(fork-lift)

PFA

Sand

Mechanical   Grout injection
front loader   through tremie pipe

Power
pack

Completed
drillhole

Drill through
rock (track
drill)

Extraction of
drill casing
(excavator)

Lining tube
installed

Drill and case
through overburden
to rock

Made
ground

Superficial
deposits

Coal Measures
containing
seams worked
by the pillar
and stall method

Fig. 13.7 Sequence of grouting operations to fill large voids in coal seams.

sand mixes, economy and bulk being their important features. Gravel may be used as a bulk filler where a large amount of grout is required for treatment. Alternatively, foam grouts can be used. If the workings are still more or less continuous, there is a risk that grout will penetrate the bounds of the zone requiring treatment. In such instances, dams can be built by placing gravel down large-diameter drillholes around the periphery of the site. When the gravel mound has been formed, it is grouted. The area within this barrier is then grouted (Fig. 13.7).

Steel mesh reinforcement or geonets have been used in roads constructed over areas of potential mining subsidence. If a void should develop under the road, the reinforcing layer is meant to prevent the road from collapsing into the void. Any surface depression which occurs in the road suggests the presence of a void. It can be investigated and, if required, remedial measures can be taken.

13.6 Old mine shafts

Centuries of mining in many countries have left behind a legacy of old shafts. Unfortunately many, if not most, are unrecorded or are recorded inaccurately. In addition, there can be no guarantee of the effectiveness of their treatment, unless it has been carried out in recent years. The location of a shaft is of great importance as far as the safety of a potential structure is concerned because, although shaft collapse fortunately is an infrequent event, its occurrence can prove to be disastrous (Bell, 1988a). Moreover, from the economic point of view, the sterilization of land due to the suspected presence of a mine shaft is unrealistic.

The ground about a shaft may subside or, worse still, collapse suddenly (Fig. 13.8). The collapse of filled shafts can be brought about by a deterioration of the fill, which usually is due to adverse changes in groundwater conditions, or to the surrounding ground being subjected to vibration or overloading. Collapse at an unfilled shaft normally is due to the deterioration and failure of the shaft lining. Shaft collapse may manifest itself as a hole roughly equal to

Fig. 13.8 Shaft collapse along the Birmingham–Wolverhampton road, England. (Courtesy of *Express and Star*, Wolverhampton.)

the diameter of the shaft if the lining remains intact or if the ground around the shaft consists of solid rock. More frequently, however, shaft linings deteriorate with age to a point at which they are no longer capable of retaining the surrounding material. If superficial deposits surround a shaft which is open at the top, the deposits eventually collapse into the shaft to form a crown-hole at the surface. The thicker these superficial deposits are, the greater the dimensions of the crown-hole. Such collapses may affect adjacent shafts if they are interconnected.

For obvious reasons, the search for old shafts on land which is about to be developed must extend outside the site in question for a sufficient distance to find any shafts which could affect the site itself if a collapse occurred. An investigation should include a survey of plans of abandoned mines, geological records of shaft sinking, geological maps, all available editions of topographical maps, aerial photographs, and archival and other official records (Anon., 1976). The success of geophysical methods in locating old shafts depends on the existence of a sufficient contrast between the physical properties of the shaft and its surroundings to produce an anomaly. A shaft frequently will remain undetected when the top of the shaft is covered by more than 3 m of fill. The size, especially the diameter, of a shaft influences whether or not it is likely to be detected. The most effective geochemical method used so far in the loca-

tion of abandoned coal mine shafts is methane detection.

The confirmation of the existence of a shaft is accomplished by excavation, for example, with a mechanical boom-type digger which is anchored outside the search area. Alternatively, a rig, placed on a platform, can be used to drill exploratory holes.

If the depth is not excessive and the shaft is open, it can be filled with suitable granular material, the top of the fill being compacted. If, as is more usual, the shaft is filled with debris in which there are voids, these should be filled with gravel and grouted. Dean (1967) suggested that, if the exact positions of the mouthings in an abandoned open shaft are known, these areas should be filled with gravel, the rest of the shaft being filled with mine waste. Anon. (1982) supplied details concerning the concrete cappings needed to seal mine shafts.

Obviously, the easiest way to deal with old mine shafts is to avoid locating structures in their immediate vicinity. Because of the possible danger of collapse and cratering, it generally is recommended that the minimum distance for siting buildings from open or poorly filled shafts is twice the thickness of the superficial deposits up to a depth of 15 m, unless they are exceptionally weak. A consideration of the long-term stability of shaft linings is required, as well as the effects of increased lateral pressure due to the erection of structures nearby. Protective sheet piles or concrete walls can be constructed around shafts to counteract cratering, but such operations should be carried out with due care.

13.7 Longwall mining and subsidence

Longwall mining is used to work coal, and exposes a face of up to 220m in width between two parallel roadways (Whittacker & Reddish, 1989). The roof is supported only in and near the roadways and at the working face. After the coal has been won and loaded, the face supports are advanced, leaving the strata, in the areas in which coal has been removed, to collapse. Subsidence at the surface more or less follows the advance of the working face and may be regarded as immediate (Brauner, 1973a). Trough-shaped subsidence profiles associated with longwall mining develop tilt between adjacent points which

have subsided by different amounts, and curvature results from adjacent sections which are tilted by different amounts. Maximum ground tilts are developed above the limits of the area of extraction, and may be cumulative if more than one seam is worked up to a common boundary. Where movements occur, points at the surface subside downwards, and are displaced horizontally inwards towards the axis of the excavation (Fig. 13.9). Differential horizontal displacements result in a zone of apparent extension on the convex part of the subsidence profile (over the edges of the excavation), whereas a zone of compression develops on the concave section over the excavation itself. Differential subsidence can cause substantial damage, the tensile strains generated usually being the most effective in this respect.

The surface area affected by ground movement is greater than the area worked in the seam (Fig. 13.9).

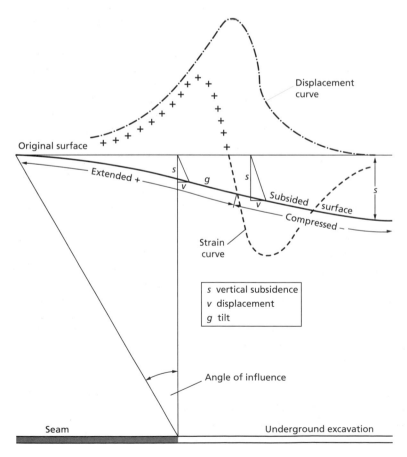

Fig. 13.9 Curve of subsidence showing tensile and compressive strains, vertical subsidence and tilt, together with the angle of influence or draw (not to scale)

The boundary of the surface area affected is defined by the limit angle of draw or angle of influence, which varies from 8 to 45° depending on the coalfield. It would seem that the angle of draw may be influenced by the depth, seam thickness and local geology, especially the location of the self-supporting strata above the coal seam.

Usually, there is an appreciable difference between the volume of mineral extracted and the amount of subsidence at the surface. For example, Orchard and Allen (1970) showed that, where the maximum subsidence of 90% of the seam thickness occurred, the volume of the subsided ground was only 70% of the coal extracted. This mainly is attributable to bulking.

One of the most important factors influencing the amount of subsidence is the width–depth relationship of the panel removed. In fact, in the UK, it usually has been assumed that the maximum subsidence generally begins at a width to depth ratio of 1.4:1 (this assumes an angle of draw of 35°; Anon., 1975). This is the critical condition above and below which maximum subsidence is and is not achieved respectively.

However, the width to depth ratio necessary to cause 90% subsidence usually can be achieved only in shallow workings, because with deeper workings, the critical area of extraction is made up of a number of panels, often with narrow pillars of coal left *in situ* to protect one or other of the roadways. These pillars reduce the subsidence and some coalfields, as a result, have no experience of subsidence in excess of 75–80% of maximum.

According to Brauner (1973b), in addition to the rate of advance and size of the critical area, the duration of surface subsidence depends on the geological conditions, depth of extraction, kinds of packings and previous extraction. For example, it lasts longer for a thickly bedded or stronger overburden and for complete caving. The depth, in particular, influences the rate of subsidence, first, because the diameter of the area of influence, and therefore the time taken for working with a given rate of advance to transverse it, increases with depth and, second, because, at greater depths, several workings may be necessary before the area of influence completely is worked out. Consequently, the time which elapses before subsidence is complete varies according to the circumstances.

Residual subsidence takes place at the same time as instantaneous subsidence, and may continue after the latter event for periods of up to 2 years. The magnitude of residual subsidence is proportional to the rate of subsidence of the surface, and is related to the mechanical properties of the rocks above the seam. For instance, strong rocks produce more residual subsidence than weaker ones. Residual subsidence rarely exceeds 10% of the total subsidence if the face is stopped within the critical width, and falls to 2–3% if the face has passed the critical width. Very occasionally values greater than 10% have been recorded.

Ground movements induced at the surface by mining activities are influenced by variations in the ground conditions, especially by the near-surface rocks and superficial deposits. However, the reactions of surface deposits to ground movements usually are difficult to predict reliably. Indeed, it has been suggested that 25% of all cases of mining subsidence undergo some measure of abnormal ground movement which, at least in part, is attributable to the near-surface strata.

In concealed coalfields, the strata overlying the Coal Measures often influence the basic movements developed by subsidence. In fact, abnormal subsidence behaviour and inconsistent movements are much more common in concealed than exposed coalfields. The occurrence of abnormally thick beds of sandstone can modify stratal movement due to mining. Such beds may resist deflection, in which case stratal separation occurs, and the effective movements at the surface are appreciably less than would otherwise be expected. The differences in behaviour disappear when the extraction becomes wide enough for the sandstone to collapse, when the subsidence behaviour reverts to normal.

The necessary readjustment in weak strata to subsidence usually can be accommodated by small movements along joints. However, as the strength of the surface rock and the joint spacing increase, so the movements tend to become concentrated at fewer points so that in massive limestones and sandstones, movements may be restricted to master joints. Well-developed joints or fissures in such rocks concentrate differential displacement. Tensile and compressive strains many times the basic values have been observed at such discontinuities (Bell & Fox, 1991). For example, joints may gape by anything up to 1 m

Fig. 13.10 A fault step is responsible for the dip in the road. The houses alongside were badly affected: windows dropped out, doors would not open and stonework was cracked. Elsecarr, South Yorkshire, England.

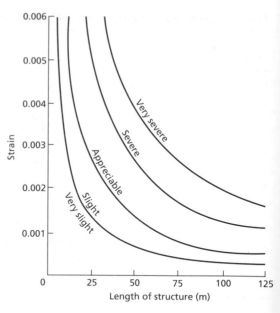

Fig. 13.11 Relationship of damage to length of structure and horizontal strain. (After Anon., 1975.)

in width at the surface. It is quite common for the total lateral movement caused by a given working to concentrate in such a manner. In such instances, no strain is measurable on either side of the discontinuity concerned.

Drift deposits often are sufficiently flexible to obscure the effects of movements at rockhead. In particular, thick deposits of till tend to obscure tensile effects. On the other hand, superficial deposits may allow movements to affect larger areas than otherwise.

Faults also tend to be locations at which subsidence movement is concentrated, thereby causing abnormal deformation of the surface. Whilst subsidence damage to structures located close to or on the surface outcrop of a fault can be very severe, in any particular instance, the areal extent of such damage is limited, often being confined to within a few metres

of the outcrop. Indeed, a subsidence step may occur at the outcrop of the fault (Fig. 13.10). The presence of a fault does not increase the amount of subsidence on the side of the fault nearer the workings compared with normal conditions, but subsidence on the opposite side is reduced considerably (by upwards of 50%). On the other hand, many faults have not reacted adversely when subjected to subsidence (Hellewell, 1988). The extent to which faults influence and modify subsidence movements cannot be quantified accurately. Dykes can have similar effects on subsidence.

Investigations carried out by the former National Coal Board (Anon., 1975) have revealed that typical mining damage starts to appear in conventional structures when they are subjected to effective strain of 0.5–1.0 mm m^{-1}, and such damage can be classified as negligible, slight, appreciable, severe and very severe (Fig. 13.11; Table 13.1). However, this relationship between damage and the change in length of a structure is only valid when the average ground strain produced by mining subsidence is equalled by the average strain in the structure. In fact, this commonly is not the case, the strain in the structure being less than it is in the ground. Hence, Fig. 13.11 only

Table 13.1 National Coal Board classification of subsidence damage. (After Anon., 1975.)

Change in length of structure (mm)	Class of damage	Description of typical damage
Up to 30	**1** Very slight or negligible	Hair cracks in plaster. Perhaps isolated slight fracture in the building, not visible on outside
30–60	**2** Slight	Several slight fractures showing inside the building. Doors and windows may stick slightly. Repairs to decoration probably necessary
60–120	**3** Appreciable	Slight fracture showing on outside of building (or main fracture). Doors and windows sticking, service pipes may fracture
120–180	**4** Severe	Service pipes disrupted. Open fractures requiring rebonding and allowing weather into the structure. Window and door frames distorted, floor sloping noticeable. Some loss of bearing in beams. If compressive damage, overlapping of roof joints and lifting of brickwork with open horizontal fractures
>180	**5** Very severe	As above, but worse, and requiring partial or complete rebuilding. Roof and floor beams lose bearing and need shoring up. Windows broken with distortion. Severe slopes on floors. If compressive damage, severe buckling and bulging of roof and walls

Table 13.2 Recommended damage criteria for buildings. (After Bhattacharya & Singh, 1985.)

Category	Damage level	Angular distortion (mm m^{-1})	Horizontal strain (mm m^{-1})	Radius of curvature (km)
1	Architectural	1.0	0.5	—
	Functional	2.5–3.0	1.5–2.0	20
	Structural	7.0	3.0	—
2	Architectural	1.3	—	—
	Functional	3.3	—	—
	Structural	—	—	—
3	Architectural	1.5	1.0	—
	Functional	3.3–5.0	—	—
	Structural	—	—	—

serves as a crude guide to the likely damaging effects of ground strains induced by mining subsidence. In addition, it does not take into account the design of the structure or construction materials. Nevertheless, it indicates that the larger the building, the more susceptible it is to differential vertical and horizontal ground movement.

More recent criteria for subsidence damage to buildings have been proposed by Bhattacharya and Singh (1985), who recognized three classes of damage, namely, architectural which is characterized by small-scale cracking of plaster and sticking of doors and windows; functional which is characterized by the instability of some structural elements, jammed doors and windows, broken window panes and restricted building services; and structural in which primary structural members are impaired, there is a possibility of collapse of members and complete or large-scale rebuilding is necessary. Their conclusions are summarized in Table 13.2. They observed that basements were the most sensitive parts of houses with regard to subsidence damage, and therefore usually suffered more damage than the rest of a building.

An important feature of subsidence due to longwall mining is its high degree of predictability. Empirical methods of subsidence prediction, such as those developed by the former National Coal Board, have been obtained by continuous study and analysis of survey data from British coalfields (Anon., 1975). Because they do not take into account the topography, the nature of the strata and the geological structure involved, or how the rock masses are likely to deform, such empirical relationships can be applied only under conditions similar to those in which the original observations were made. Theoretical methods of subsidence assessment assume that stratal displacement behaves according to the constitutive equations of continuum mechanics over most of its range. Numerical models permit a quantitative analysis of subsidence problems to be made, and are not subject to the same restrictive assumptions as theoretical methods. Most viable methods of subsidence prediction fall into the category of semi-empirical methods, because fitting field data to theory often results in good correlations between the predicted and actual subsidence. There are two principal methods of semi-empirical prediction, namely, profile function and influence function methods (Brauner, 1973b; Hood *et al.*, 1983). A review of the methods of subsidence prediction has been provided by Bell (1988b).

13.8 Measures to mitigate the effects of subsidence due to longwall mining

The contemporaneous nature of subsidence associated with longwall mining sometimes affords the opportunity to planners to phase long-term surface development in relation to the cessation of subsidence (Bell, 1987). However, the relationship between the future programming of surface development and that of subsurface working may be difficult to coordinate, because of the differences which may arise between the programmed intention and the performance achieved.

Damage attributable to subsidence due to longwall mining can be reduced or controlled by precautionary measures incorporated into new structures in mining areas, and preventive works applied to existing structures (Anon., 1977). Several factors must be considered when designing buildings for areas of active mining. First, where high ground strains are anticipated, the cost of providing effective rigid foundations may be prohibitive. Second, experience suggests that buildings with deep foundations, on which thrust can be exerted, suffer more damage than those in which the foundations are more or less isolated from the ground. Third, because of the relationship between the ground strain and the size of the structure, very long buildings should be avoided unless their long axes can be orientated normal to the direction of principal ground strain. Lastly, although tall buildings may be more susceptible to tilt rather than to the effects of horizontal ground strain, tilt can be corrected by using jacking devices.

The most common method of mitigating subsidence damage is by the introduction of flexibility into a structure. In flexible design, structural elements deflect according to the subsidence profile. The foundation therefore remains in contact with the ground as subsidence proceeds. Flexibility can be achieved by using specially designed rafts. Raft foundations should be as shallow as possible, preferably above ground, so that compressive strains can take place beneath them instead of transmitting direct compressive forces to their edges. They should be constructed on a membrane, so that they will slide as ground movements occur beneath them. For instance, reinforced concrete rafts, laid on granular material, reduce friction between the ground and the structure. Cellular rafts have been used for multi-storey buildings.

The use of piled foundations in areas of mining subsidence presents its own problems. The lateral and vertical components of ground movement, which occur as mining progresses, mean that the pile caps tend to move in a spiral fashion, and that each cap moves at a different rate and in a different direction according to its position relative to the mining subsidence. Such differential movements and rotations normally would be transmitted to the structure with a corresponding readjustment of the loadings on the pile cap. In order to minimize the disturbing influence of these rotational and differential movements, it is often necessary to allow the structure to move independently of the piles by the provision of a pin joint or roller bearing at the top of each pile cap. It may be necessary to include some provision for jacking the superstructure where severe dislevelment is likely to occur.

Preventive techniques frequently can be used to

reduce the effects of movements on existing structures. The type of technique depends upon the type of structure and its ability to withstand ground movements, as well as the amount of movement likely to occur. Again the principal objective is to introduce greater flexibility or to reduce the amount of ground movement which is transmitted to the structure. In the case of buildings longer than 18 m, damage can be reduced by cutting them into smaller, structurally independent units. The space produced should be large enough to accommodate deflection. In particular, items such as chimneys, lift shafts, machine beds, etc. can be planned independently of the other, generally lighter, parts of a building, and separated from the main structure by joints through foundations, walls, roof and floor, which allow freedom of movement. Extensions and outbuildings should be separated from the main structure by such joints.

Flexible joints can be inserted into pipelines to combat the effects of subsidence. Thick-walled plastics, which are more able to withstand ground movements, have been used for service pipes. Where pipes are already laid, they can be exposed, and flexible joints inserted, or pipes can be freed from contact with the surrounding ground. The latter reduces the ground strains transmitted to the pipeline. The zone around the pipe can be backfilled with gravel, so that the friction between it and the surrounding soil is lowered.

Damage to surface structures can also be reduced by adopting specially planned layouts of underground workings, which take into account the fact that surface damage to structures primarily is caused by ground strains. Thus, to minimize the risk of damage, underground extraction must be planned, so that the surface strain is reduced or eliminated. Unfortunately, it is seldom possible to arrange the direction and location of underground workings to suit the need to minimize surface movement in the neighbourhood of a particular surface structure.

Pillars of coal can be left in place to protect the surface structures above them. In panel and pillar mining, pillars are left between relatively long, but narrow panels. Adjacent panels are designed with a pillar of sufficient width in between so that the interaction of the ground movements results in flat subsidence profiles at the surface with low ground strains.

The resultant surface subsidence ranges from 3 to 20% of the thickness of the seam worked.

Maximum subsidence can also be reduced by packing the goaf. This subsidence factor, because it varies with the method, packing materials and depth, needs to be determined from actual observations of individual cases in each mining district. Pneumatic stowing can reduce subsidence by up to 50% (Anon., 1975).

13.9 Surface mining

Most surface mining methods are of large scale, involving the removal of massive volumes of broken material, either as ore or as overburden, to gain access to a mineral deposit. Huge amounts of waste can be produced during the process. For example, in gold mining, the ore only contains a few grams per tonne; hence, once the gold has been extracted, the volume of waste, due to processing, actually is larger than that of the ore mined.

Obviously, one of the factors which must be taken into account in the development of a surface mine is the stability of its slopes. This is influenced by the nature of the rock mass(es), the geological structure and the hydrogeological conditions. The latter determine whether dewatering is necessary. The slope stability will affect the limits of the excavation and the bench dimensions where benches are employed. Another important factor is the stripping ratio, that is, the ratio of overburden to ore. This can determine the limit of working in terms of the economic operation of a surface mine. A mineral occurring at a depth beyond the maximum stripping ratio will either have to remain unworked or be mined by underground methods.

13.9.1 Strip mining and opencasting

Stratified mineral deposits which occur at or near the surface, such as coal or sedimentary iron ore, in relatively flat terrain generally are mined by strip or opencast mining. The deposit is either horizontal or gently dipping, and normally is within 60 m of the surface. All the strata overlying the mineral deposit are removed, frequently by dragline, and placed in stockpiles. Bucket wheel excavators sometimes are used for stripping. When necessary, blasting is used to break rock above the mineral deposit. The mineral

(a)

(b)

Fig. 13.12 (a) Mining Athabasca Tar Sands, Fort McMurdo, Alberta. (b) Bison enjoying lunch on restored workings in the Athabasca Tar Sands, Alberta.

deposit itself may have to be drilled and blasted, once exposed, prior to its removal by conventional loading and haulage equipment. The deposition of spoil in mined out areas means that very little advanced stripping is necessary, and that mining activity can be located in a relatively small area. Because the face is exposed for a relatively short time, this allows it to have a steeper slope than otherwise. When the overburden is removed from the working face in long, parallel strips, and placed in long spoil heaps in the worked out area, this is referred to as strip mining

(Fig. 13.12a). Parallel rows of spoil are produced as the face moves on. Soil is removed before the overburden, and is placed in separate piles, so that it can be used in the rehabilitation process. The rehabilitation of the spoil heaps can proceed before the mining operation ceases, the spoil heaps being regraded to fit into the surrounding landscape (Fig. 13.12b). Hence, the short lifespan of the spoil heaps means that they can be maintained at their natural angle of repose. Once the regrading of spoil has been completed, it is covered with soil, fertilized if necessary and seeded.

In the UK, opencast coal mining involves the exploitation of shallow seams at relatively small sites (normally from 10 to 800 ha). This usually involves creating a box-cut to reveal the coal. Once the coal has been extracted, the box-cut is filled with overburden from the next cut. The maximum depths of working are about 100 m, with stripping ratios of up to 25:1. The progressive restoration of sites is undertaken, so that the handling of waste is limited. Draglines may be used for overburden removal, together with face shovels and dump trucks, and scrapers. The vast majority of opencast sites are restored to agriculture. Nevertheless, due to increasing pressures on land use and the proximity between some sites and urban areas, backfilled opencast sites have been used for industrial and housing developments. At those sites in which, at the planning stage, it has been suggested that buildings subsequently should be erected, the full extent of the backfill has been compacted. Nonetheless, varying amounts of settlement have been noted at restored opencast sites. Differential settlement can occur across a site and, especially, at the fill–solid interface. Such settlement can induce cracking in buildings, and can affect roads and services. Furthermore, Charles *et al.* (1984) reported the effects of a rising water table on opencast backfill. This can lead to a second phase of surface settlement after the initial phase, which occurs under the self-weight of the backfill. The latter generally is largely completed within 2 or 3 years.

13.9.2 Open pit mining

Open pit mining is used to work ore bodies which occur at or near the surface, where the ore body is steeply dipping or occurs in the form of a pipe. Initially, the overburden is stripped to expose the ore. Mining and stripping then continues in a carefully phased manner. The ore body is blasted in a series of benches, the walls of which are steeply dipping (Fig. 13.13). Rock surrounding the ore body may have to be removed in order to maintain stable slopes in the pit. This represents waste and, together with the waste produced by processing the ore, is disposed of in spoil heaps or tailings lagoons.

Because of the size of many open pits, for example, the Bingham Canyon copper mine in Utah is some 800 m in depth and covers nearly 8 km², it being one of the largest man-made excavations in the world, there normally is not enough waste material available to backfill them. Accordingly, the principal objectives after the cessation of mining operations generally are to ensure that the walls of the pit are stable, and that the waste dumps and tailings dams are rehabilitated. It is possible that an open pit could be used for the disposal of other waste if the hydrogeological conditions permit. Perhaps, open pits could be left to fill with water, and then could act as a source of supply and/or be used for recreational purposes. Bingham Canyon copper mine, although still a working mine, is also a tourist attraction.

Fig. 13.13 Bingham Canyon copper mine, Utah.

13.9.3 Blasting and vibrations

The three most commonly derived quantities relating to vibration are the amplitude, particle velocity and acceleration. Of the three, the particle velocity appears to be most closely related to damage in the frequency range of typical blasting vibrations. Over 35 years ago, Edwards and Northfield (1960) defined three categories of damage attributable to vibrations. Threshold damage refers to the widening of old cracks and the formation of new ones in plaster, and the dislodgement of loose objects. Minor damage is that which does not affect the strength of the structure; it includes broken windows, loosened or fallen plaster and hairline cracks in masonry. Major damage seriously weakens the structure; it includes large cracks, shifting of foundations and bearing walls and distortion of the superstructure caused by settlement and walls out of plumb. Edwards and Northfield proposed that a vibration level with a peak particle velocity of $50\,mm\,s^{-1}$ could be regarded as safe as far as structural damage was concerned, $50–100\,mm\,s^{-1}$ would require caution, and above $100\,mm\,s^{-1}$ would present a high probability of damage occurrence (Fig. 13.14a). The United States Bureau of Mines subsequently lent support to the idea that $50\,mm\,s^{-1}$ provides a reasonable safety margin from the possibility of damage (Nicholls *et al.*, 1971). However, Oriard (1972) maintained that no single value of the velocity, amplitude or acceleration could be used indiscriminately as a criterion for limiting blasting vibrations. In particular, low vibration levels may disturb sensitive machinery and, in this case, it is impossible to specify a limit of ground velocity; each instance should be separately assessed.

Moreover, blasting vibrations at $50\,mm\,s^{-1}$ peak particle velocity, in terms of human response, would

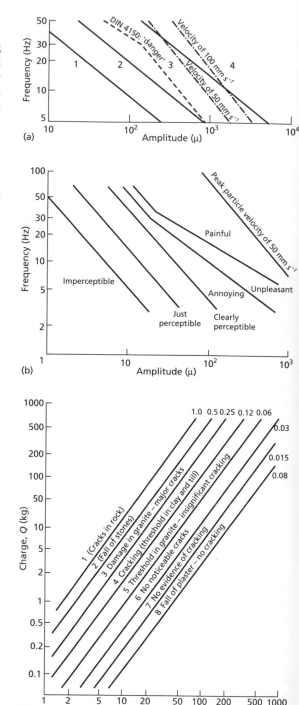

Fig. 13.14 (a) Possible damage to buildings for frequencies between 5 and 50 Hz, the range most frequently encountered in buildings: 1, no damage; 2, possibility of cracks; 3, possible damage to load-bearing structural units; 4, damage to load-bearing units. (b) Human sensitivity to vibration. If the peak particle velocity of $50\,mm\,s^{-1}$ is taken as the threshold of damage, the figure indicates that humans react strongly long before there is any reason to anticipate damage to property. (c) Damage as a function of distance for various charge levels. Numbers 3–8 inclusive describe damage expected in normal houses. (After Langefors & Kihlstrom, 1962.)

be regarded as highly unpleasant or intolerable (Nicholls *et al.*, 1971). Indeed, Crandell (1949) previously had concluded that the average person would consider a vibration to be 'severe' at about one-fifth of the level which might damage structures, and the threshold of subjective perception has been variously placed from as low as $0.5\,mm\,s^{-1}$ to $10\,mm\,s^{-1}$. Experience has shown that the great majority of complaints and even law suits against blasting operations are due to irritation, and that subjective response to vibrations normally leads a person to react strongly long before there is any likelihood of damage occurring to property (Fig. 13.14b). People will react more unfavourably to large-amplitude vibrations of the same intensity. It is probable that this sensitivity is increased in the low-frequency range of 3–10 Hz. The duration of the operation and the frequency of occurrence of the blasts are almost as important as the level of the physical effects.

When dealing with high levels of shock and vibration, the time history of the motion and the characteristic response of the structure concerned to the type of motion imposed become increasingly important. For instance, a structure with a slow response, such as a tall chimney, when subjected to vibrations of large amplitude, low frequency and long duration, would come closer to the resonant response of the structure. This would be more dangerous than vibrations of small amplitude, high frequency and short duration, even though both may have the same acceleration or velocity. Because of the dependence of the response on the frequency, conservative limits should be accepted when applying single values of velocity or acceleration as criteria for different types of structure subjected to different kinds of motion.

Vibrations associated with blasting generally fall within the frequency range of 5–60 Hz. The types of vibration depend on the size of the explosive charge, the volume of the ground set into vibration, the attenuation characteristics of the ground and the distance from the blast. A small explosive charge generates a low vibration with relatively high frequency and relatively low amplitude. By contrast, a large explosive charge produces a vibration with relatively low frequency and relatively high amplitude. The shock waves are attenuated with distance from the blast, the higher frequencies being maintained more readily in dense rock masses. In other words, these pulses rapidly are attenuated in unconsolidated

deposits which are characterized by lower frequencies.

Vibrographs can be placed in locations considered to be susceptible to blast damage in order to monitor the ground velocity. A record of the blasting effects, compared with the size of the charge and the distance from the point of detonation, normally is sufficient to reduce the possibility of damage to a minimum (Fig. 13.14c). A preblast survey informs the owners of buildings as to the condition of their buildings before the commencement of blasting, and offers some protection to the contractor against unwarranted claims for damage.

The effects of vibration due to blasting operations can be reduced by the following factors:
1 Time dispersion, which involves the use of delay intervals in ignition. Delay intervals mean that shock waves generated by individual blasts are mutually interfering, thereby reducing vibration.
2 Spatial dispersion, which concerns the pattern and orientation of the blast-holes.
3 The way in which the charge is distributed in the blast-hole, the diameter of the hole and the depth of lift.
4 The confinement of the charge, which involves the type of decking and the powder factor, as well as the width of burden and the spacing of the blast-holes.

13.9.4 Dredge mining

Dredge mining is used to recover minerals from alluvial, marine or aeolian (dune) deposits (Fig. 13.15a). It involves the underwater excavation of a deposit, usually carried out from a floating vessel, which may incorporate processing and waste disposal facilities. The body of water may be natural or man-made.

The deposits which contain the minerals must be diggable and able to hold water. For example, a pit usually is excavated in the deposit and filled with water, so forming a pond. The dredge excavates the deposit and pumps the material to the separation plant, which may be separate from the dredge. Heavy minerals (e.g. ilmenite, rutile, cassiterite, chromite or scheelite) are concentrated and pumped ashore for further processing. Others, such as gold or diamonds, are collected. The tailings from the separation plant are placed in the mined out area of the pond for subsequent rehabilitation. Hence, the dredge, separation

(a)

(b)

Fig. 13.15 (a) Dredge mining of heavy minerals, Richards Bay, Natal, South Africa. (Courtesy of Richards Bay Minerals Ltd.) (b) Some 20 years after the restoration of a dredge mined area at Richards Bay, Natal, South Africa. (Courtesy of Richards Bay Minerals Ltd.)

plant (if separate) and pond move on, excavating at one end of the pond and depositing waste at the other.

The flexibility of waste disposal allows different types of topography to be created after mining has passed on, so that restoration can be very successful (Fig. 13.15b). The topsoil is stripped prior to mining and placed initially in stockpiles for use during restoration. However, once mining is in progress, soil can be transported directly from the area ahead of the dredging operation to the waste which is being landscaped. The scope for establishing vegetation is governed by the moisture-holding capacity of the waste, which is usually sand or sand–gravel mixtures.

13.10 Subsidence associated with fluid abstraction

Subsidence of the ground surface occurs in areas where there is intensive abstraction of fluids, notably groundwater, oil or brine. In the case of groundwater, subsidence occurs when abstraction exceeds

Fig. 13.16 Land subsidence, compaction, water fluctuations and changes in effective stress near Pixley, California. (After Lofgren, 1979.)

the natural recharge, and the water table is lowered, the subsidence being attributed to the consolidation of the sedimentary deposits as a result of increasing effective stress (Bell, 1988c). The total overburden pressure in saturated deposits is borne by their granular structure and the pore water. When groundwater abstraction leads to a reduction in pore water pressure by draining water from the pores, this means that there is a gradual transfer of stress from the pore water to the granular structure. For instance, if the groundwater level is lowered by 1 m, this gives rise to a corresponding increase in the average effective overburden pressure of 10 kPa. As a result of having to carry this increased load, the fabric of the deposits affected may deform in order to adjust to the new stress conditions. In particular, the void ratio of the deposits concerned undergoes a reduction in volume, the surface manifestation of which is subsidence. Surface subsidence does not occur simultaneously with the abstraction of fluid from an underground reservoir, rather it occurs over a larger period of time than that taken for abstraction.

The amount of subsidence which occurs is governed by the increase in the effective pressure (i.e. the magnitude of the decline in the water table), the thickness and compressibility of the deposits

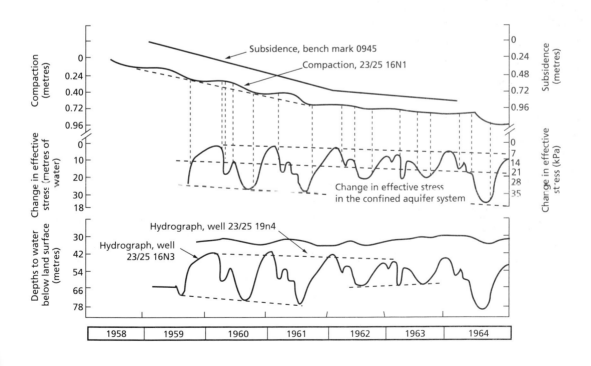

involved, the depth at which they occur, the length of time over which the increased loading is applied and, possibly, the rate and type of stress applied. For instance, Lofgren (1979) showed that 20 years of precise field measurements in the San Joaquin and Santa Clara Valleys of California indicated a close correlation between the hydraulic stresses induced by groundwater pumping and the consolidation of the water-bearing deposits (Fig. 13.16). The rate at which consolidation occurs depends on the thickness of the beds concerned, as well as the rate at which pore water can drain from the system which, in turn, is governed by its permeability. Thick, slow-draining, fine grained beds may take years or decades to adjust to an increase in applied stress, whereas coarse grained deposits adjust rapidly. However, the rate of consolidation of slow-draining aquitards reduces with time, and usually is small after a few years of loading.

Methods which can be used to arrest or control subsidence caused by groundwater abstraction include the reduction of the pumping draught, the artificial recharge of aquifers from the ground surface and the repressurization of the aquifer(s) involved via wells, or any combination thereof. The aim is to manage the rate and quantity of water withdrawal, so that its level in wells is either stabilized or raised somewhat, and thus the effective stress is not further increased.

Subsidence due to the abstraction of oil occurs for the same reason as does subsidence associated with the abstraction of groundwater. In other words, pore pressures are lowered in the oil-producing zones due to the removal of not only oil, but also gas and water, and the increased effective load causes consolidation in compressible beds. Such an explanation probably first was advanced to account for the spectacular and costly subsidence at Wilmington oilfield, California. This area is underlain by approximately 1800 m of Miocene to Holocene sediments, consisting largely of sands, silts and shales. Subsidence was noticed in 1940, after oil production had been underway for 3 years. Gilluly and Grant (1949) demonstrated that the area of maximum subsidence showed a remarkable coincidence with the productive area of the oilfield. They also indicated that there was a very close agreement between the relative subsidence of the various parts of the oilfield and the pressure decline, thickness of oil sand affected and mechanical properties of the oil sand (Fig. 13.17). By 1966, an elliptical area of over 75 km² had subsided by more than 8.8 m. By 1947, subsidence was occurring at a rate of 0.3 m per year, and had reached 0.7 m annually by 1951, when the maximum rate of withdrawal was attained. Because of the seriousness of the subsidence (and the realization that it was due to declines in fluid pressures), remedial action was taken in 1957 by injecting water into and thereby repressurizing the abstraction zones. By 1962, this had brought subsidence to a halt in most of the field.

Those sedimentary deposits which readily go into solution, notably salt, can be extracted by solution mining. For example, salt has been obtained by brine pumping in a number of areas in the UK, Cheshire being by far the most important. Consequently, subsidence due to salt extraction was an inhibiting factor

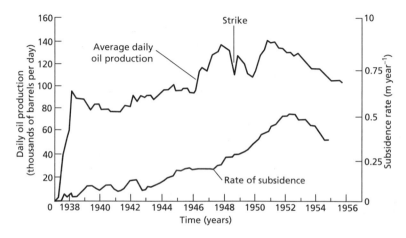

Fig. 13.17 Oil production and rate of subsidence at BM 700, Wilmington oilfield, California. (After Steinbrugge & Bush, 1958. Subsidence in Long Beach–Terminal Island–Wilmington, California. Pacific Fire Rating Bureau, San Francisco.)

Fig. 13.18 Severe cracking in the wall of a farm building due to subsidence caused by wild brine pumping, Cheshire, England.

in parts of Cheshire as far as major developments were concerned (Bell, 1992).

In the past, brine production primarily was by wild brine pumping, that is, pumping from the major natural brine runs. Such pumping accelerated the formation of solution channels. Active subsidence normally was concentrated at the head and sides of a brine run, where the fresh water first entered the system. Hence, serious subsidence could occur at considerable distances, anything up to 8 km, from the pumping centres.

Because the exact area from which salt was extracted was not known, the magnitude of subsidence developed could not be related to the volume of salt worked. Consequently, there was no accurate means of predicting the amount of ground movement or strain. However, the rate at which subsidence occurred appears to have been related to the rate at which brine was extracted. Although it is not easy to

tell, it seems that subsidence did not necessarily occur immediately after the removal of salt. For example, surface subsidence in some areas was found to continue for 6 months after the particular well thought to be responsible ceased to operate. Indeed, residual subsidence often continued for 1 or 2 years. The maximum strains developed by wild brine pumping could ruin buildings (Fig. 13.18). More serious, as far as structural damage was concerned, was the formation of tension scars (small faults) on the convex flanks of subsidence hollows. The vertical displacement along a scar usually was less than 1 m (Fig. 13.19). Subsidence due to wild brine pumping also gave rise to the formation of flashes, that is, water-filled linear hollows. They formed as a result of collapse above brine runs.

Today, wild brine pumping accounts for only 1% of the salt produced by solution mining in Cheshire. In other words, controlled solution mining has replaced wild brine pumping. Subsidence is not associated with this form of abstraction.

Deere (1961) described differential subsidence (resulting from the development of a concavo-convex subsidence profile), simultaneous horizontal displacements and surface faulting associated with the abstraction of sulphur from numerous wells in the Gulf region of Texas. Solution mining of sulphur and potash in Louisiana has given rise to vertical subsidence of up to 9 m and 6 m respectively.

Ground failure, that is, fissuring or faulting, often is associated with subsidence due to the abstraction of fluids. Fissures may appear suddenly at the surface, and the appearance of some may be preceded by the occurrence of minor depressions at the surface. Within a matter of a year or so, most fissures become inactive. In a few instances, new fissures have formed in close proximity to older fissures.

The longest fissure zone in the USA is 3.5 km in length, and lengths of hundreds of metres are typical (Holzer, 1986). Individual fissures commonly are not continuous, but consist of a series of segments with the same trend. Occasionally, a zone defined by several closely spaced parallel fissures occurs, the width of which may be of the order of 30 m. Fissure separations tend not to exceed a few centimetres, the maximum reported being 6.4 cm. In Arizona, they frequently have been enlarged by erosion to form gullies 1 or 2 m in width and 2 or 3 m in depth (Fig. 13.20).

Fig. 13.19 Tension scar on the flank of a subsidence trough caused by wild brine pumping. Note the flash in the top right-hand corner due to flooding in the trough. The road on which the car stands had to be made up periodically.

Fig. 13.20 An earth fissure in south-central Arizona, which was caused by subsidence due to groundwater abstraction. It has been enlarged by erosion. (Courtesy of Dr Thomas L. Holzer, United States Geological Survey.)

Those faults which are suspected of being related to groundwater withdrawal are much less common than fissures. They frequently have scarps more than 1 km in length and more than 0.2 m in height.

Obviously, it is necessary to determine the amount of subsidence which is likely to occur as a result of the withdrawal of fluids from the ground, as well as to estimate the rate at which it may take place. Unfortunately, a large number of prediction methods have been developed, some of which are relatively simple, whilst others are complex. One of the probable reasons for this is that stratal sequences are different in different areas where subsidence has occurred, and consequently different models have been devised.

Nonetheless, a number of steps can be taken in order to evaluate the subsidence likely to occur due to the abstraction of fluids from the ground. These include defining the *in situ* hydraulic conditions, computing the reduction in pore pressure due to the removal of a given quantity of fluid, converting the reduction in pore pressure to an equivalent increase in effective stress and estimating the amount of consolidation likely to take place in the formation affected from consolidation data and the increased effective load. In addition to the depth of burial, the ratio between the maximum subsidence and reservoir consolidation should take into account the lateral extent of the reservoir, in that small reservoirs which are deeply buried do not give rise to noticeable subsidence, even if undergoing considerable consolidation, whereas extremely large reservoirs may develop significant subsidence. The problem therefore is three dimensional, rather than one of simple vertical consolidation.

13.11 Waste materials from mining

Mine wastes result from the extraction of metals and non-metals. In the case of metalliferous mining, high volumes of waste are produced, because of the low or very low concentrations of metal in the ore. In fact, mine wastes represent the highest proportion of waste produced by industrial activity, billions of tonnes being produced annually. Wastes can be inert or contain hazardous constituents, but generally are of low toxicity. The chemical characteristics of mine waste and waters arising therefrom depend upon the type of mineral being mined, as well as the chemicals which are used in the extraction or beneficiation processes. Because of its high volume, mine waste historically has been disposed of at the lowest cost, often without regard to safety and often with considerable environmental impacts. Catastrophic failures of spoil heaps and tailings dams, although uncommon, have led to the loss of lives.

The character of waste rock from metalliferous mines reflects that of the rock hosting the metal, as well as the rock surrounding the ore body. The type of waste rock disposal facility depends on the topography and drainage of the site, and the volume of waste. Van Zyl (1993) referred to the disposal of coarse mine waste in valley fills, side-hill dumps and open piles. Valley fills normally commence at the upstream end

of a valley and progress downstream, increasing in thickness. Side-hill dumps are constructed by the placement of waste along hillsides or valley slopes, avoiding natural drainage courses. Open piles tend to be constructed in relatively flat-lying areas. Obviously, an important factor in the construction of a spoil heap is its stability, which includes its long-term stability. Acid mine drainage from spoil heaps is another environmental concern (see Section 13.12).

13.11.1 Spoil heaps and their restoration

Old spoil heaps represent one of the most notable forms of dereliction associated with subsurface mining, especially coal mining (Bell, 1996). They are particularly conspicuous, and can be difficult to rehabilitate into the landscape. For instance, the uppermost slopes of a spoil heap frequently are devoid of near-surface moisture, and so are often barren of vegetation. In fact, in order to support vegetation, a spoil heap should have a stable surface in which roots can become established, must be non-toxic and must contain an adequate and available supply of nutrients. Such derelict areas need to be restored to sufficiently high standards to create an acceptable environment. The restoration of such derelict land requires a preliminary reconnaissance of the site, followed by a site investigation. The investigation provides essential input for the design of remedial measures.

In the case of waste from coal mines, there are two types of discard, namely, coarse and fine. Coarse discard consists of run-of-mine material, and reflects the various rock types which are extracted during mining operations. It contains varying amounts of coal which have not been separated by the preparation process. Fine discard consists of tailings slurry from the washery, which is disposed of in lagoons (see below).

The configuration of a spoil heap depends upon the type of equipment used in its construction and the sequence of tipping the waste. The shape, aspect and height of a spoil heap affect the intensity of exposure, the amount of surface erosion which occurs, the moisture content in the surface layers and its stability.

The argillaceous content of colliery spoil influences regrading, although most colliery spoil heap material essentially is granular in the mechanical sense. Generally, as far as the particle size distribution of coarse discard is concerned, there is a wide variation. Often most material may fall within the sand range, but significant proportions of gravel and cobble range may also be present.

The shear strength of coarse discard within a spoil heap, and therefore its stability, is influenced by the pore water pressures developed within it. Excess pore water pressures in spoil heaps may be developed as a result of the increasing weight of material added during construction, or by seepage though the spoil of natural drainage. High pore water pressures usually are associated with fine grained materials, which have a low permeability and high moisture content. Thus the relationship between the permeability and the build-up of pore water pressures is crucial. It has been found that, in soils with a coefficient of permeability of less than $5 \times 10^{-9} \, m \, s^{-1}$, there is no dissipation of pore water pressures, whereas above $5 \times 10^{-7} \, m \, s^{-1}$ they are completely dissipated. The permeability of colliery discard depends primarily upon its initial grading, its degree of compaction and the amount of degradation undergone. Generally, coarse discard has a range of permeability varying from 1×10^{-4} to $5 \times 10^{-8} \, m \, s^{-1}$.

Pyrite frequently occurs in the shales and coaly material present in spoil heaps. When pyrite weathers, it gives rise to the formation of sulphuric acid, together with ferrous and ferric sulphates and ferric hydroxide, which lead to acidic conditions in the weathered material. The oxidation of pyrite within spoil heap waste is governed by the access of air which, in turn, depends upon the particle size distribution, amount of water saturation and degree of compaction. If highly acidic oxidation products develop, they may be neutralized by alkaline materials in the waste.

The leaching of the oxidation products of pyrite occasionally is associated with spoil heaps. Surface water run-off can leach out soluble salts, especially chlorides. This may result in the loss of up to 1 tonne of chloride per hectare of exposed spoil heap per annum under average (British) rainfall conditions. Significant concentrations of sulphates sometimes occur in low-volume seepages from older, more permeable, spoil heaps. Acidic drainage from such sources occasionally contains low concentrations (a few milligrams per litre) of copper, nickel and zinc, but other heavy metals rarely are found in concentrations greater than $0.1 \, mg \, l^{-1}$.

Spontaneous combustion of carbonaceous material, frequently aggravated by the oxidation of pyrite, is the most common cause of burning spoil. It can be regarded as an atmospheric oxidation (exothermic) process in which self-heating occurs. Coal and carbonaceous materials may be oxidized in the presence of air at ordinary temperatures, below their ignition point.

The oxidation of pyrite at ambient temperature in moist air is also an exothermic reaction. When present in sufficient amounts, and especially when in finely divided form, pyrite associated with coaly material increases the likelihood of spontaneous combustion. When heated, the oxidation of pyrite and organic sulphur in coal gives rise to the generation of sulphur dioxide. If there is insufficient air for complete oxidation, hydrogen sulphide is formed.

The moisture content and grading of colliery spoil are also important factors in spontaneous combustion (Cook, 1990). At relatively low temperatures, an increase in free moisture increases the rate of spontaneous heating. Oxidation generally takes place very slowly at ambient temperatures but, as the temperature rises, so oxidation increases rapidly. In material of large size, the movement of air can cause heat to be dissipated, whereas in fine material the air remains stagnant, however, this means that burning ceases when the supply of oxygen is consumed. Accordingly, ideal conditions for spontaneous combustion exist when the grading is intermediate between these two extremes, and hot spots may develop under such conditions, where temperatures up to 900 °C may be recorded (Fig. 13.21).

Spontaneous combustion may give rise to subsurface cavities in spoil heaps, the roofs of which may be incapable of supporting a person. Burnt ashes may also cover zones which are red-hot to appreciable depths. When steam comes into contact with red-hot carbonaceous material, watergas is formed and, when the latter is mixed with air, over a wide range of concentrations, it becomes potentially explosive. If a cloud of coal dust is formed near burning spoil when reworking a heap, this can also ignite and explode. Damping with a water spray may prove to be useful in the latter case.

Noxious gases are emitted from burning colliery spoil. These include carbon monoxide, carbon

Fig. 13.21 Earth moving over a 'hot spot' in colliery spoil, near Barnsley, South Yorkshire, England.

dioxide, sulphur dioxide and, less frequently, hydrogen sulphide. Each may be dangerous if breathed at certain concentrations, which may be present at fires on spoil heaps. The rate of evolution of these gases may be accelerated by disturbing a burning heap by excavating into or reshaping it. Carbon monoxide is the most dangerous, because it cannot be detected by taste, smell or irritation, and may be present in potentially lethal concentrations. By contrast, sulphur gases are readily detectable in the aforementioned ways, and usually are not present in high concentrations. Even so, when diluted, they still may cause distress to persons with respiratory ailments.

The problem of combustion in spoil material sometimes must be faced when restoring old spoil heaps. In such cases, Anon. (1973) recommended compaction, digging out, trenching, injection with noncombustible material and water, blanketing and water spraying as methods by which spontaneous combustion in spoil material may be controlled.

One of the most important factors influencing the cost of restoration of a spoil heap frequently is the volume of waste which may have to be removed and the haulage distances involved. Hence, in such instances, the cheapest schemes usually are those in which the quantities of 'cut and fill' are approximately equal, so that the movement of materials is restricted to the site and thereby kept to a minimum. Restoration is not only an exercise in large-scale earthmoving. Because it invariably involves spread-

ing the waste over a larger area, so that gradients on the existing site can be reduced by the transfer of spoil to adjacent land, this means that additional land beyond the site boundaries has to be purchased. Where a spoil heap is very close to the disused colliery, spoil may be spread over the latter area. This involves the burial or removal of derelict colliery buildings, and possibly the treatment of old mine shafts or shallow old workings (Johnson & James, 1990). Water courses may have to be diverted, as may services, notably roads.

After an old spoil heap has been regraded, the actual surface still needs to be restored. This is not so important if the area is to be built over (e.g. if it is to be used for an industrial estate), as it is if it is to be used as an amenity or for recreational purposes. When buildings are to be erected, however, the ground must be adequately compacted so that the bearing capacities are satisfactory and buildings are not subjected to adverse settlement. On the other hand, where the land is to be used as an amenity or for recreational purposes, the soil fertility must be restored, so that the land can be grassed and trees planted. This involves laying topsoil (where applicable) or substitute materials, the application of fertilizers and seeding (frequently by hydraulic methods). If the spoil is acidic, it can be neutralized by liming. Adequate subsoil drainage needs to be installed and should take into account erosion control during landscaping. A spoil heap which is burning, or which is so acidic that polluted waters are being discharged, requires special treatment (Cook, 1990). Sealing layers have been used to control spontaneous combustion and polluted drainage, the sealing layers being well compacted.

13.11.2 Waste disposal in tailings dams

Tailings are fine grained slurries which result from crushing rock containing ore or which are produced by the washeries at collieries. The water in tailings may contain certain chemicals associated with the metal recovery process, such as cyanide in tailings from gold mines and heavy metals in tailings from copper–lead–zinc mines. Tailings may also contain sulphide minerals, such as pyrite, which can give rise to acid mine drainage. The particle size distribution, permeability and resistance to weathering of tailings affect the process of acid generation. Acid drainage may also contain elevated levels of dissolved heavy metals. Accordingly, contaminants carried in the tailings represent a source of pollution for both groundwater and surface water, as well as soil.

Tailings are deposited as slurry generally in specially constructed tailings dams (Vick, 1983). Embankment dams are constructed to their full height before tailings are discharged (Fig. 13.22). They are constructed in a similar manner to earthfill dams used to impound reservoirs. They may be zoned, with a clay core, and may have filter drains.

Such dams are best suited to tailings impoundments with high water storage requirements. However, tailings dams usually consist of raised embankments, that is, the construction of the dam is staged over the life of the impoundment. Raised embankments consist initially of a starter dyke, which normally is constructed of earthfill from a borrow pit. This dyke may be large enough to accommodate the first 2 or 3 years of tailings production. A variety of materials can be used subsequently to complete such embankments, including earthfill, mine waste or tailings themselves. Tailings dams consisting of tailings can be constructed using the upstream, centreline or downstream method of construction (Fig. 13.23). In the upstream method of construction, tailings are discharged from spigots or small pipes to form the impoundment. This allows the separation of particles according to size, with the coarsest particles accumulating in the centre of the embankment beneath the spigots and the finer particles being transported down the beach. Alternatively, cycloning may be undertaken to remove coarser particles from tailings, so that they can be used in embankment construction. Centreline and downstream construction of embankments uses coarse particles separated by cycloning for the dam. The design of tailings dams must pay due attention to their stability, both in terms of static and dynamic loading. Failure of a dam can lead to catastrophic consequences. For example, failure of the tailings dam at Buffalo Creek in West Virginia after

Fig. 13.22 An embankment around tailings from china clay workings, Cornwall, England.

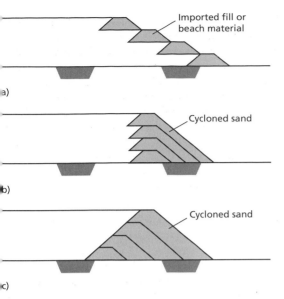

ig. 13.23 (a) Upstream, (b) centreline and (c) downstream
method of embankment construction for tailings dams.

eavy rain in February 1972 destroyed over 1500
ouses and cost 118 lives.

The deposition of tailings in a dam may lead to the
ormation of a beach or mudflat above the water
evel. When discharged, the coarser particles in tail-
ngs settle closer to the discharge point(s), with the
iner particles being deposited further away. The
mount of sorting which takes place is influenced by
he way in which the tailings are discharged. For
xample, a high-volume discharge from one point
roduces little sorting of tailings. On the other hand,
lischarge from multiple points at moderate rates
;ives rise to good sorting. After being deposited, the
;eotechnical properties of tailings, such as the mois-
ure content, density, strength and permeability, are
;overned by the amount and rate of consolidation.
The moisture content of deposited tailings can vary
rom 20% to over 60%, and the dry density from 1
o 1.3 Mg m⁻³. At the end of deposition, the density of
he tailings generally increases with depth due to
he increasing self-weight on the lower material. The
:oarser particles, which settle out first, also drain
nore quickly than the finer material, which accumu-
ates further down the beach, and so develop shear
trength more quickly than the latter. These varia-
ions in sorting also affect the permeability of the
material deposited.

The quantity of tailings which can be stored in a
dam of a given volume is dependent upon the density
which can be achieved. The latter is influenced by the
type of tailings, the method by which they are
deposited, whether they are deposited in water or
subaerially, the drainage conditions within the dam
and whether or not they are subjected to desiccation.
For example, Blight and Steffen (1979) referred to the
semi-dry or subaerial method of tailings deposition,
whereby, in semi-arid or arid climates, a layer of tailings is deposited in a dam and allowed to dry before
the next layer is placed. Because this action reduces
the volume of the tailings, it allows more storage to
take place within the dam. However, drying out can
give rise to the formation of desiccation cracks in the
fine discard which, in turn, can represent locations
where piping can be initiated. In fact, Blight (1988)
pointed out that tailings dams have failed as a result
of desiccation cracks and horizontal layering of fine
grained particles leading to piping failure. The most
dangerous situation occurs when the ponded water
on the discard increases in size, and thereby erodes
the cracks to form pipes, which may emerge on the
outer slopes of the dam.

The rate at which seepage occurs from a tailings
dam is governed by the permeability of the tailings
and the ground beneath the impoundment. The
climate and the way in which the tailings dam is
managed also have some influence on seepage losses.
In many instances, because of the relatively low permeability of the tailings compared with the ground
beneath, a partially saturated flow condition occurs
in the foundation. Nonetheless, the permeability of
tailings can vary significantly within a dam, depending on the nearness to discharge points, the degree
of sorting, the amount of consolidation which
has occurred, the development of stratification into
coarser and finer layers and the amount of desiccation
of the discard.

Seepage losses from tailings dams which contain
toxic materials can have an adverse effect on the environment. According to Fell et al. (1993), one of the
most cost-effective methods of controlling seepage
loss from a tailings dam is to cover the whole floor of
the impoundment with tailings from the start of the
operation (Fig. 13.24). This cover of tailings, provided it is of low permeability, will form a liner. Tailings normally have permeabilities between 10⁻⁷ and
10⁻⁹ m s⁻¹ or less. Nevertheless, Fell et al. mentioned

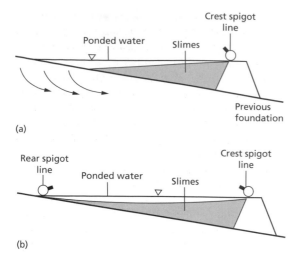

Fig. 13.24 Control of seepage by tailings spigotting procedures: (a) major seepage at water–foundation contact; (b) foundation sealing by near spigotting of tailings. (After Vick, 1983.)

that a problem could arise when using tailings to line an impoundment, that is, if a sandy zone develops near the point(s) of discharge, localized higher seepage rates will occur if water covers this zone. This can be avoided by moving the points of discharge or by placing fines. Alternatively, a seepage collection system can be placed beneath the sandy zone prior to its development. Clay liners also represent an effective method of reducing seepage from a tailings dam. The permeability of properly compacted clay soils usually varies between 10^{-8} and 10^{-9} m s^{-1}. However, clay liners may have to be protected from drying out, with the attendant development of cracks, by placement of a sand layer on top. On slopes, the sand may need to be kept in place by using geotextiles. Geomembranes have tended not to be used for lining tailings dams, primarily because of their cost. Filter drains may be placed at the base of tailings, with or without a clay liner. They convey water to collection dams. A toe drain may be incorporated into the embankment. This will intercept the seepage which emerges at that location. Where a tailings dam has to be constructed on sand or sand and gravel, slurry trenches may be used to intercept seepage water, but they are expensive.

The objectives of the rehabilitation of tailings impoundments include their long-term mass stability, long-term stability against erosion, prevention of environmental contamination and a return of the area to productive use. Normally, when the discharge of tailings comes to an end, the level of the phreatic surface in the embankment falls as water replenishment ceases. This results in an enhancement of the stability of the embankment slopes. However, where tailings impoundments are located on slopes, excess run-off into the impoundment may reduce the embankment stability, or overtopping may lead to failure by erosion of the downstream slope. The minimization of inflow due to run-off by judicious siting is called for when locating an impoundment which may be so affected. Diversion ditches can cater for some run-off, but have to be maintained, as do abandonment spillways and culverts.

The accumulation of water may be prevented by capping the impoundment, the capping sloping towards the boundaries. Erosion by water or wind can be impeded by placing rip-rap on slopes, and by the establishment of vegetation on the waste. The latter will also help to return the impoundment to some form of productive use. Where long-term potential for environmental contamination exists, particular precautions need to be taken. For example, as the water level in the impoundment declines, the rate of oxidation of any pyrite present in the tailings increases, reducing the pH and increasing the potential for heavy metal contamination. In the case of tailings from uranium mining, radioactive decay of radium gives rise to radon gas. Diffusion of radon gas does not occur in saturated tailings but, after abandonment, radon reduction measures may be necessary. In both of these cases, a clay cover can be placed over the tailings impoundment to prevent leaching of contaminants or to reduce the emission of radon gas.

Once the discharge of tailings ceases, the surface of the impoundment is allowed to dry. Drying of the decant pond may take place by evaporation and/or by drainage to an effluent plant. Desiccation and consolidation of the slimes may take a considerable time. Stabilization can begin once the surface is firm enough to support equipment. As mentioned above, this normally will involve the establishment of a vegetative cover.

13.12 The problem of acid mine drainage

The term acid mine drainage is used to describe the

drainage resulting from the natural oxidation of sulphide minerals which occur in mine rock or waste exposed to air and water. This is a consequence of the oxidation of sulphur in the mineral to a higher oxidation state and, if aqueous iron is present and unstable, the precipitation of ferric iron with hydroxide occurs. It can be associated with underground workings, open pit workings, spoil heaps, tailings ponds or mineral stockpiles (Brodie *et al.*, 1989).

Acid mine drainage is responsible for problems of water pollution in major coal and metal mining areas around the world. However, it will not occur if the sulphide minerals are non-reactive or if the rock contains sufficient alkaline material to neutralize the acidity. In the latter instance, the pH value of the water may be near neutral, but it may carry elevated salt loads, especially of calcium sulphate. The character and rate of release of acid mine drainage are influenced by various chemical and biological reactions at the source of acid generation. If acid mine drainage is not controlled, it can pose a serious threat to the environment, because acid generation can lead to elevated levels of heavy metals and sulphate in the water, which obviously have a detrimental effect on its quality (Table 13.3). This can have a notable impact on the aquatic environment, as well as the vegetation (Fig. 13.25). The development of acid mine drainage is time dependent and, at some mines, may evolve over a period of years.

Generally, acid mine drainage from underground mines occurs as point discharges. A major source of acid mine drainage may result from the closure of a mine. When a mine is abandoned and dewatering by pumping ceases, the water level rebounds. However, the workings often act as drainage systems, so that the water does not rise to its former level. Consequently, a residual dewatered zone remains that is subject to continuing oxidation. Groundwater may drain to the surface from old drainage adits, river bank mine mouths, faults, springs and shafts which intercept rock in which water is under artesian pressure. Nonetheless, it may take a number of years before this happens. Old adits often are unmapped and unknown, and even currently discharging ones are not always immediately evident. An examination of the catchment data is often the only way in which such discharges come to light. Discharges from old adits and mine mouths usually are gravity flows. Some of these are close to surface waters, and so acidified water can drain directly into them.

Table 13.3 Composition of acid mine water from a South African coalfield. (After Bullock & Bell, 1997.)

Determined parameter (mg l^{-1})	Sample 1	Sample 2	Sample 3	Sample 4	Sample 5	Sample 6
TDS	4844	2968	3202	2490	3364	3604
Suspended solids	33	10.4	12	10.0	7.6	
EC (mS m^{-1})	471	430	443	377	404	340
pH value	1.9	2.4	2.95	2.9	2.3	2.8
Turbidity as NTU	5.5	0.6	2.0	0.9	1.7	
Nitrate NO$_3$ as N	0.1	0.1	0.1	0.1	0.1	0.1
Chlorides as Cl	310	431	406	324	353	611
Fluoride as F	0.6	0.5	0.33	0.6	0.6	0.84
Sulphate as SO$_4$	3250	1610	1730	1256	2124	1440
Total hardness		484	411	576	585	377
Calcium hardness as CaCO$_3$		205	310	327	282	
Magnesium hardness as MgCO$_3$		199	101	249	305	
Calcium as Ca	173.8	114.0	124	131	113	84
Magnesium as Mg	89.4	48.4	49.5	60.5	49.3	31
Sodium as Na	247.0	326.0	311	278	267	399
Potassium as K	7.3	9.4	8.9	6.4	3.8	
Iron as Fe	248.3	128	140	87	89.9	193
Manganese as Mn	17.9	15	9.9	13.4	13.9	9.3
Aluminium as Al		124		112	204	84

Samples 1–5 from surface water courses; sample 6 from coal seam level.
NTU, nephelometric turbidity unit.

Fig. 13.25 Trees killed by acid mine drainage from pyrite discard at a tin mine in the Transvaal, South Africa.

The mine water quality is determined by the hydrogeological system and the geochemistry of the upper mine levels. Hence, in terms of a working mine, it is important that groundwater levels are monitored to estimate the rebound potential. In addition, records should be kept of the hydrochemistry of the water throughout the workings, drives and adits, so that the potential for acid generation on closure can be assessed.

The large areas of fractured rock exposed in opencast mines or open pits can give rise to large volumes of acid mine drainage. Even when abandoned, slope deterioration and failure can lead to fresh rock being exposed, allowing the process of acid generation to continue. Where the workings extend beneath the surrounding topography, the pit drainage system leads to the water table being lowered. Increased oxidation can occur in the dewatered zone.

Spoil heaps represent waste generated by the mining operation. As such they vary from waste produced by subsurface mining to waste produced by any associated smelting or beneficiation. Consequently, the sulphide content of the waste can vary significantly. Acid generation tends to occur in the surface layers of spoil heaps, where air and water have access to sulphide minerals (Bullock & Bell, 1995).

Tailings deposits which have a high content of sulphide represent another potential source of acid generation. However, the low permeability of many tailings deposits, together with the fact that they commonly are flooded, means that the rate of acid generation and release is limited. Consequently, the generation of acid mine drainage can continue to take place long after a tailings deposit has been abandoned.

Mineral stockpiles may represent a concentrated source of acid mine drainage. Major acid flushes commonly occur during periods of heavy rainfall after long periods of dry weather. Heap-leach operations at metalliferous mines include, for example, cyanide leach for gold recovery and acid leach for base metal recovery. Spent leach heaps can represent sources of acid mine drainage, especially those associated with low-pH leachates.

Certain conditions, including the correct combination of mineralogy, water and oxygen, are necessary for the development of acid mine drainage. Such conditions do not always exist. Consequently, acid mine drainage is not found at all mines with sulphide-bearing minerals. The ability of a particular mine rock or waste to generate net acidity depends on the relative contents of acid-generating minerals and acid-consuming or acid-neutralizing minerals. Acid waters produced by sulphide oxidation of mine rock or waste may be neutralized by contact with acid-consuming minerals. As a result, the water draining from the parent material may have a neutral pH value and negligible acidity despite continuing sulphide oxidation. If the acid-consuming minerals are dissolved, washed out or surrounded by other minerals, acid generation continues. Where neutralizing carbonate minerals are present, metal hydroxide sludges, such as iron hydroxides and oxyhydroxides, are formed. Sulphate concentrations generally are not affected by neutralization unless mineral saturation with respect to gypsum is attained. Hence, sulphate sometimes may be used as an overall indicator of the extent of acid generation after neutralization by acid-consuming minerals.

The oxidation of sulphide minerals may give rise to the formation of secondary minerals after a certain amount of pH neutralization, if the pH is maintained near neutral during oxidation. The secondary minerals which are developed may surround the sulphide minerals and, in this way, reduce the reaction rate.

The primary chemical factors which determine the rate of acid generation include the pH value, temperature, oxygen content of the gas phase if saturation is less than 100%, concentration of oxygen in the water phase, degree of saturation with water, chemical activity of Fe^{3+}, surface area of exposed metal sulphide and chemical activation energy required to initiate acid generation. In addition, the biotic microorganism *Thiobacillus ferrooxidans* may accelerate the reaction by its enhancement of the rate of ferrous iron oxidation. It may also accelerate the reaction through its enhancement of the rate of reduced sulphur oxidation. *Thiobacillus ferrooxidans* is most active in waters with a pH value around pH 3.2. If conditions are not favourable, the bacterial influence on acid generation will be minimal.

The presence of *Thiobacillus ferrooxidans* may also accelerate the oxidation of sulphides of antimony, arsenic, cadmium, cobalt, copper, gallium, lead, molybdenum, nickel and zinc.

Hence, the development of acid mine drainage is a complex combination of inorganic and, sometimes, organic processes and reactions. In order to generate severe acid mine drainage (pH < 3), sulphide minerals must create an optimum microenvironment for rapid oxidation and must continue to oxidize long enough to exhaust the neutralization potential of the rock.

An accurate prediction of acid mine drainage is required in order to determine how to bring it under control. The objective of acid mine drainage control is to satisfy environmental requirements using the most cost-effective techniques. The options available for the control of polluted drainage are greater at proposed rather than at existing operations, as control measures at working mines are limited by site-specific and waste disposal conditions. For instance, the control of acid mine drainage that develops as a consequence of mine dewatering, is helped by the approach taken towards the site water balance. In other words, the water resource management strategy developed during mine planning will enable mine water discharge to be controlled and treated, prior to release, or to be reused. The length of time over which the control measures are required to be effective is a factor which needs to be determined prior to the design of a system to control acid mine drainage.

The prediction of the potential for acid generation involves the collection of available data and the performance of static and kinetic tests. A static test determines the balance between potentially acid-generating and acid-neutralizing minerals in representative samples. One of the frequently used static tests is acid–base accounting. Acid–base accounting allows a determination to be made of the proportions of acid-generating and acid-neutralizing minerals present. However, static tests cannot be used to predict the quality of drainage waters or when acid generation will occur. If potential problems are indicated, the more complex kinetic tests should be used to obtain a better insight into the rate of acid generation. Kinetic tests involve the weathering of samples under laboratory or on-site conditions, in order to confirm the potential to generate net acidity, to determine the rates of acid formation, sulphide oxidation, neutralization and metal dissolution, and to test the control and treatment techniques. The static and kinetic tests provide data which may be used in various models to predict the effect of acid generation and control processes beyond the time frame of kinetic tests.

There are three key strategies in acid mine drainage management, namely, the control of the acid generation process, the control of acid migration, and the collection and treatment of acid mine drainage (Connelly *et al.*, 1995). The control of acid mine drainage may require different approaches, depending on the severity of potential acid generation, the longevity of the source of exposure and the sensitivity of the receiving waters. Mine water treatment systems installed during operation may be adequate to cope with both operational and long-term post-closure treatment with little maintenance. On the other hand, in many mineral operations, especially those associated with abandoned workings, the long-term method of treatment may be different from that used when the mine was operational. Hence, there may have to be two stages involved with the design of a system for the treatment of acid mine drainage, one during mine operation and another after closure.

Obviously, the best solution is to control acid generation, if possible. Source control of acid mine drainage involves measures to prevent or inhibit oxidation, acid generation or contaminant leaching. If acid generation is prevented, there is no risk of the contaminants entering the environment. Such control methods involve the removal or isolation of sulphide material, or the exclusion of water or air. The latter is much more practical and can be achieved by the placement of a cover over acid-generating material such as waste or air sealing of adits in mines.

Migration control is considered when acid generation is occurring and cannot be inhibited. Because water is the transporting medium, control relies on the prevention of water entry to the source of acid mine drainage. Water entry may be controlled by the diversion of surface water flowing towards the source by drainage ditches, the prevention of groundwater flow into the source by interception and isolation (this is very difficult to maintain over the long term) and the prevention of infiltration of precipitation into the source by the placement of cover materials (again the long-term integrity is difficult to ensure).

Release control is based on measures to collect and treat acid mine drainage. In some cases, especially at working mines, this is the only practical option available. Collection requires the collection of both groundwater and surface water polluted by acid mine drainage, and involves the installation of drainage ditches, collection trenches and wells. Treatment processes have concentrated on neutralization to raise the pH and precipitate metals. Lime or limestone commonly is used, although offering only a partial solution to the problem. More sophisticated processes (active treatment methods) involve osmosis (waste removal through membranes), electrodialysis (selective ion removal through membranes), ion exchange (ion removal using resins), electrolysis (metal recovery with electrodes) and solvent extraction (removal of specific ions with solvents).

Cambridge (1995) pointed out that conventional active treatment of mine water requires the installation of a treatment plant, continuous operation and maintenance. Hence, the capital and operational costs of active treatment are high. Alternatively, passive systems try to minimize the input of energy, materials and manpower, and so reduce operational costs. Acid mine water treated with active systems tends to produce a solid residue, which has to be disposed of in tailings lagoons. This sludge contains metal hydroxides. However, according to Cambridge, the long-term disposal of sludge in tailings lagoons is not appropriate. Alternatively, sludges could be placed in hazardous waste landfills, but such sites are limited.

Passive treatment involves engineering a combination of low-maintenance biochemical systems (e.g. anoxic limestone drains, aerobic and anaerobic wetlands and rock filters). Such treatment does not produce large volumes of sludge, the metals being precipitated as oxides or sulphides in the substrate materials.

Due to the impact on the environment of acid mine drainage, regular monitoring is required. The major objectives of a monitoring programme developed for acid mine drainage are, first, to detect the onset of acid generation, before acid mine drainage develops to the stage at which environmental impact occurs. If required, control measures should be put in place as quickly as possible. Second, it is necessary to monitor the effectiveness of the prevention–control–treatment techniques, and to detect whether the techniques are unsuccessful at the earliest possible time.

13.13 Waste waters and effluents from coal mining

Different types of waste waters and process effluent are produced as a result of coal mining. These may arise due to the extraction process, by the subsequent preparation of coal, from the disposal of colliery spoil or from coal stockpiles. The strata from which the groundwater involved is derived, the mineralogical character of the coal and the colliery spoil, and the washing processes employed all affect the type of effluent produced.

Generally, the major pollutants associated with coal mining are suspended solids, dissolved salt, acidity and iron compounds (Bell & Kerr, 1993). Elevated levels of suspended matter are associated with most coal mining effluents, with occasionally high values being recorded. Hence, the most common po lutant of a receiving stream associated with coal mining is the increased concentration of suspended solids. In particular, the blanketing effect of coal slurry particles on the bed of a river is unacceptable, both in terms of appearance and its influence on the flora and fauna in the stream. The turbidity of some streams into which colliery waste is discharged may restrict the penetration of light into the water, and so have adverse effects on the plant life therein. Moreover, reductions in both the abundance and diversity of invertebrates occur in reaches of streams where fine particles are deposited. However, where suspended solid concentrations are low or siltation sporadic, the community structure may remain unaffected. Where suspended solids concentrations averaged 110 mg l⁻¹ and occasionally exceeded 2000 mg l⁻¹ in a river receiving mine water, Edwards (1981) reported no change in the composition of the fauna, although a few species were virtually eliminated, and there was a 90% reduction in the overall abundance. In general, good fisheries are unlikely to be found where average concentrations of suspended solids exceed about 80 mg l⁻¹.

With respect to the disposal of spoil, solids are lost mainly from areas of current tipping, particularly during periods of heavy rainfall. Losses of solids from stable spoil heaps with a vegetation cover are not severe. However, in the nineteenth and early twentieth centuries, as there was no major demand for fine coal, much was tipped. Spoil disturbance, coupled with the use of inefficient washing procedures in the recovery of this fine coal, commonly is the cause of the high suspended solids concentrated in nearby rivers.

The character of drainage from coal mines varies from area to area and from coal seam to coal seam. Hence, mine drainage waters are liable to vary in both quality and quantity, sometimes unpredictably, as the mine workings develop. Although not all mine waters are ferruginous and, in fact, some are of the highest quality and can be used for potable supply, they commonly are high in iron and sulphates, low in pH and high in acidity. Ferruginous discharges can give rise to disastrous conditions in receiving streams, and can affect many kilometres of otherwise good quality water.

Although not all mine waters are highly mineralized, a high level of mineralization is typical of many coal mining discharges, and is reflected in the high values of the electrical conductivity. The range of dissolved salts encountered in mine water is variable, with electrical conductivity values up to 335 000 μS cm⁻¹ and chloride levels of 60 000 mg l⁻¹ being recorded (Woodward & Selby, 1981). Some average values for various coal mining effluents associated with the Nottinghamshire coalfield, England, are given in Table 13.4. The principal groups of salts in mine discharge waters are chlorides and sulphates

Table 13.4 Average quality characteristics of coal mining effluents in the Nottinghamshire coalfield. (From Bell & Kerr, 1993.)

Type of effluent	BOD (mg l⁻¹)	Suspended solids (mg l⁻¹)	Chloride (mg l⁻¹)	Electrical conductivity (μS cm⁻¹)	Minimum pH	Other potential contaminants
Mine waters	2.1	57	4900	14 000	3.5	Iron, barium, nickel, aluminium, sodium, sulphate
Drainage from coal stocking sites	2.6	128	600	2200	2.2	Iron, zinc
Spoil tip drainage	3.1	317	1600	4100	2.7	Iron, zinc
Coal preparation plant discharges	2.1	39	1500	4200	3.2	Oil froth flotation chemicals
Slurry lagoon discharges	2.4	493	2000	6100	3.8	

BOD, biological oxygen demand.

and, to a lesser extent, sodium and potassium salts. The chlorides occur in the waters lying in the confined aquifers between coal seams. Dissolved sulphates occur only in trace concentrations in waters of confined aquifers. Sulphates are either present in the waters lying in the more shallow, unconfined aquifers, or are generated in the workings by the action of atmospheric oxygen on pyrite. The more saline waters contain significant concentrations of barium, strontium, ammonium and manganese ions. All of these salts are released into the workings by mining operations. In general, the salinity increases with depth below the surface, and with distance from the outcrop or incrop.

The physical changes, such as delamination, bedding plane separation and fissuring of rock masses, caused by mining permit air to penetrate a much larger surface area than the immediate boundaries of the working faces and associated roadways. These changes also alter the hydrogeological conditions within a coalfield, and allow the wider movement of groundwater through the rock masses than existed prior to mining. These two factors mean that groundwater comes into contact with a large surface area of rock, which frequently contains pyrite and which is exposed to atmospheric oxidation. The sulphates and sulphuric acid which result react with clay and carbonate minerals to form secondary products, including manganese and aluminium sulphates. Further reactions with these minerals and incoming waters give rise to tertiary products, such as calcium and magnesium sulphates. Generally, the stratal waters are sufficiently alkaline to ensure that only the tertiary products appear in the discharge at the surface. Exceptionally, both the primary and secondary products may appear in waters from intermediate depths. The primary oxidation products tend to predominate in very shallow workings, liable to leaching by meteoric water.

The acidity of mine water discharges may affect the aquatic life in a receiving stream, but the precipitation of metal hydroxides as the acidity is neutralized by the alkalinity of the stream is more serious. For example, ferrous sulphate is oxidized in streams and precipitated as hydrated ferric oxide. When the pH exceeds pH 5, and the soluble metals are not toxic, the impact on the benthic fauna is similar to that of coal siltation, with a decline in the diversity of faunal species. Acidic ferruginous mine water may also

contain high concentrations of aluminium, which precipitates as hydroxide as the pH value rises on entering a receiving stream, giving a milky appearance to the water. Concentrations of heavy metals may be high in some acid waters and may have a toxic effect, however, their ecological significance is masked by the effects of acidity.

The high level of dissolved salts often present in mine waters represents the most intractable water pollution problem connected with coal mining. This is because dissolved salts are not readily susceptible to treatment or removal. The only treatment normally possible is dilution and dispersion in a receiving water course, which may involve transporting the discharge over considerable distances. In some situations, this option is not available.

The prevention of the pollution of groundwater by coal mining effluent is of particular importance. Movement of pollutants through strata often is very slow and difficult to detect. Hence, effective remedial action is either impractical or prohibitively expensive. Because of this, there are few successful recoveries of polluted aquifers. The pollution of an aquifer can be alleviated by lining the beds of influent streams which flow across the aquifer or by providing pipelines to convey mine discharge to less sensitive water courses which do not flow across the aquifer concerned.

13.14 Heap leaching

Heap leaching involves low-grade ores, such as finely disseminated gold deposits, from which the metal cannot be extracted by conventional methods. The deposits are placed on bases which have low permeability and are then sprayed with a solvent to extract the metal. For example, cyanide commonly is used to dissolve gold. As the complex formed between gold and cyanide is very strong, only a relatively weak solution is required. The solvent should percolate uniformly through the heap, so that it comes into contact with all the sources of metal in the ore. The solution and dissolved metal are collected in a plastic-lined pond, and are then treated to recover the metal.

The ores used in heap leaching can simply be run-of-mine material, the blasted ore being placed on the base for leaching without any prior preparation. Alternatively, the ore may be crushed. If there are excessive fines in the crushed ore, these can be bound together with Portland cement to form coarser particles.

Heap-leach projects can be developed on permanent bases, that is, the spent ore is left on the base after leaching, the latter acting as a liner. Conversely, if the leached ore is disposed of after the metal is extracted, the base can be used again. A single composite liner can provide environmental protection for permanent or reusable bases where the hydraulic head is low. A double composite liner may be necessary in the case of valley leach facilities, because of the higher hydraulic head which is maintained in the heap.

In order to reclaim the heaps after leaching, they are rinsed. Generally, a heap is regarded as having been successfully rinsed when, for instance, in the case of gold extraction, monitoring shows that the weak acid dissociable cyanide content is $0.5\,mg\,l^{-1}$ or less (Van Zyl, 1993).

The chemistry of cyanide is complex and, consequently, numerous forms of cyanide are present in the leached heaps and associated tailings dams at gold mines. The toxicity primarily arises from free cyanide and generally metal cyanide compounds are less toxic. The toxicity is also dependent upon the degree to which these compounds dissociate to release free cyanide. The different forms of cyanide vary widely in their rates of decay and potential toxicity. For example, free cyanide exists in solution as hydrocyanic acid and the cyanide anion. As the former has a low boiling point and high vapour pressure, it is reasonably volatile at atmospheric temperature–pressure conditions. Cyanide leaching of gold from ore usually is undertaken at pH 10.3 or above, so that most of the free cyanide in solution exists in the stable anion form; in this way, the loss of cyanide by volatilization is reduced to a minimum. When the waste is disposed of in a tailings dam, the pH value of the decant decreases to around pH 7, at which point most of the free cyanide occurs as hydrocyanic acid and is volatile. The greater the depth of water in the impoundment, the slower the rate of loss of free cyanide by volatilization.

Cyanide forms compounds with many metals, including cadmium, cobalt, copper, iron, mercury, nickel, silver and zinc. These cyanide compounds occur in the waste and effluents at gold mines which use the heap-leach process. Generally, the toxicity of these metal cyanide compounds is related to their sta-

bility, in that the more stable the compound, the less toxic it is, especially to aquatic life. For instance, zinc cyanide is weakly stable, copper cyanide and nickel cyanide are moderately stable and iron cyanide is very stable, the latter being more or less non-toxic. On the other hand, the weak and moderately stable metal cyanide compounds can break down to form highly toxic free cyanide forms.

In addition to copper cyanide and zinc cyanide, the principal weak acid dissociable metal cyanide species present in cyanide leach material are nickel, cadmium and silver cyanide. The breakdown of these metal cyanide species in an impoundment initially involves the dissociation of the cyanide ion and free metal ion. The hydrolysis of the cyanide ions then leads to the formation of hydrocyanic acid, which subsequently is lost by volatilization from the water in the impoundment. The general rate of decay of cyanide in the water in an impoundment therefore depends upon the rate at which the metal cyanide species dissociates and the rate of volatilization of free cyanide (Simovic et al., 1984).

Residual cyanide in the spent ore represents a potential source of cyanide release to the environment. However, the extent to which this represents a threat to the environment is debatable. For instance, Smith et al. (1984) found that cyanide concentrations at various depths within an impoundment in South Africa were very low. Subsequently, Miller et al. (1991) showed that natural decay and transformation processes within a tailings impoundment removed soluble cyanide from pore water at depth and that, as a result, the seepage which took place from the impoundment contained very low cyanide levels.

13.15 Other impacts

All coals spontaneously will ignite if the right conditions exist, although some coals ignite more easily than others. In other words, the oxidation of coal exposed to air can lead to spontaneous combustion. The factors which aid the spontaneous combustion of coal include the surface area of coal exposed, the moisture content of the coal, the rank of the coal and the temperature in the workings. The larger the surface area exposed, the greater the opportunity for oxidation of the coal. Michalski et al. (1990) indicated that, if a coal seam has a lower than normal moisture content and then the moisture content rises, this leads to the liberation of heat. High-rank coals are less prone to spontaneous combustion than those of low rank. The likelihood of spontaneous combustion increases as the temperature increases, so that, with increasing depth, and consequently increasing geothermal gradient, coal has a greater tendency to ignite. Michalski *et al.* also mentioned that, if the pyrite content in coal exceeds 2%, this helps spontaneous combustion, because the oxidation of pyrite is also an exothermic reaction.

When air gains access to shallow coal workings via surface cracks or due to partial collapse of the workings, conditions conducive to self-ignition may exist. Where the workings are old and abandoned, the sides of the pillars may be highly fractured or weakened due to weathering, and fine coal may be strewn in the roadways. Hence, a large surface area of coal is available for oxidation and, with the existence of a limited amount of air, the exothermic reaction produces a rise in temperature which eventually becomes self-generating. The flow of air through partially collapsed workings is unlikely to have a sufficiently high velocity to convey away the heat generated. If such occurrences are not detected early and controlled adequately, large areas of coal can be destroyed by self-combustion and surface areas seriously affected. Partially burnt pillars can collapse, leading to subsidence and ground fissuring, which can further accentuate the problem by allowing greater access of air to the workings. Air can also gain access to workings via the development of crown-holes at the surface or via poorly sealed shafts. Gases, such as steam, carbon dioxide, carbon monoxide and sulphur dioxide, may escape from the fissures or crown-holes (Fig. 13.26).

If the coal seam affected by spontaneous combustion is at a shallow depth, the spread of a fire may be limited by excavating a trench into the coal seam and then backfilling. Obviously, the depth imposes a limitation on trenching. Old workings sometimes have been flooded to extinguish fires. This means that the water pumped into the area of the mine concerned must be impounded by dams, or pumped into the mine more quickly than it can be discharged. The water must remain in place for a sufficient length of time to cool the coal and surrounding strata, otherwise re-ignition will occur when the water level is lowered. However, neither of these techniques is always successful. Furthermore, the pillar stability may be affected in workings which have been flooded when the workings drain, as the pore water pressure within fractured pillars may cause sidewall scaling, which can result in a reduction of the pillar width.

When grouting is used to extinguish a burning coal seam, drill-holes initially are sunk to intercept the workings at the lowest area in which burning is occurring. Holes are drilled at closely spaced intervals, and systematic filling takes place, moving

Fig. 13.26 Steam emanating from a crown-hole above workings in the No. 2 seam, Witbank coalfield, South Africa, due to seam burning.

towards the higher levels of the mine. Inert gases such as nitrogen, carbon dioxide, steam and combustion by-products, have also been used in attempts to extinguish mine fires, as have foams. As in the case of grouts, they are introduced into the mine via drill-holes. Any fissures or crown-holes at the surface, which allow entry of air to a mine, should be sealed prior to treatment. Thermocouples are used to determine the extent of the fire, its subsequent movement and the success of the treatment.

Methane and carbon dioxide are generated by the breakdown of organic matter and are associated with the Coal Measures. In other words, methane is a by-product formed during coalification, that is, the process which changes peat into coal. Biodegradation takes place in the early stages of accumulation of plant detritus, with the evolution of some methane and carbon dioxide. The major phase of methane production, however, probably takes place at a later stage in the coalification process after the deposits have been buried. During this process, approximately $140\,m^3$ of methane is produced per tonne of coal. As a consequence, the quantity of methane produced during coalification exceeds the holding capacity of the coal, resulting in the migration of excess methane into reservoir rocks that surround or overlie the coal deposit. In addition, coal has a very large internal surface area ($93\,000\,000\,m^2$ per tonne of coal) and, as methane can exist as a tightly packed monomolecular layer adsorbed on the internal surfaces of the coal matrix, it is able to hold two to three times more gas than conventional reservoirs. Gas contents of $14–17\,m^3$ per tonne have been measured in the higher rank bituminous coals in the USA. Hence, methane in the Coal Measures is adsorbed on coal, may be trapped in gas pockets or dissolved in groundwater.

Gases may be dissolved in groundwater depending on the pressure, temperature and concentration of other gases or minerals in the water. Dissolved gases may undergo advection by groundwater, and only when the pressure is reduced and the solubility limit of the gas in water is exceeded do they come out of solution and form a separate gaseous phase. Such a pressure release occurs when coal is removed during mining, tunnelling or shaft sinking operations in strata so affected. It is essential that such degassing is not allowed to occur in confined spaces, where an explosive mixture could develop.

Methane can be oxidized during migration to form carbon dioxide. However, carbon dioxide can be generated both microbially and inorganically in a number of ways which do not involve methane.

When methane is detected in the ground, it is necessary to determine its source, so that the most appropriate remedial action can be taken to minimize any possible danger. Methane can have more than one source, it need not originate in Coal Measures strata, but could arise from a landfill. In major construction operations and in the case of domestic dwellings, the sources of methane need to be considered and measures taken to minimize the risk posed. Cases are on record of explosions having occurred in buildings due to the presence of methane (5–15% methane mixed with air is explosive).

Generally, the connection between a source of methane and the location in which it is detected can be verified by detecting a component of the gas which is specific to the source, or by establishing the existence of a migration pathway from the source to the location in which the gas is detected. An analysis of the gas obviously helps to identify the source. For example, methane from landfill gas contains a larger proportion of carbon dioxide (16–57%) than does coal gas (up to 6%). Analysis may also involve trace components or isotopic characterization using stable isotope ratios ($^{13}C/^{12}C$; $^{2}H/^{1}H$) or the radio-isotope ^{14}C. This again allows a distinction to be made between coal gas and landfill gas as, in the former, all the radiocarbon has decayed.

Gas problems are not present in every coal mine. Nonetheless, gas may accumulate in abandoned mines, and methane, in particular, because it is lighter than air, may escape from old workings via shafts, and via crown-holes where the workings are at shallow depth.

Methane can move through coal by diffusion, which is relatively slow, however, as coal seams are likely to be fractured, the diffusion rate is increased. Gas also migrates through rocks via intergranular permeability or, more particularly, along discontinuities. Where strata have been disturbed by mining subsidence, the gas permeability is enhanced, as is that of the groundwater holding gas.

The determination of the migration pathway of gas involves both geological and hydrogeological investigation. The groundwater flow needs to be determined if gases are dissolved in groundwater, particular note being taken of areas of discharge. Hydrogeological assessments require an accurate measurement of the piezometric pressure, and sampling and chemical analysis of water. If methane is dissolved in groundwater, samples should be obtained at *in situ* pressure, and must not be allowed to degas as they equilibrate to atmospheric pressure.

References

Anon. (1973) *Spoil Heaps and Lagoons. Technical Handbook*. National Coal Board, London.

Anon. (1975) *Subsidence Engineer's Handbook*. National Coal Board, London.

Anon. (1976) *Reclamation of Derelict Land: Procedure for Locating Abandoned Mine Shafts*. Department of the Environment, London.

Anon. (1977) *Ground Subsidence*. Institution of Civil Engineers, London.

Anon. (1982) *Treatment of Disused Mine Shafts and Adits*. National Coal Board, London.

Bell, F.G. (1975) *Site Investigation in Areas of Mining Subsidence*. Newnes–Butterworths, London.

Bell, F.G. (1986) Location of abandoned workings in coal seams. *Bulletin of the International Association of Engineering Geology* 30, 123–132.

Bell, F.G. (1987) The influence of subsidence due to present day coal mining on surface development. In *Planning and Engineering Geology, Engineering Geology Special Publication No. 4*, Culshaw, M.G., Bell, F.G., Cripps, J.C. & O'Hara, M. (eds). Geological Society, London, pp. 359–368.

Bell, F.G. (1988a) Land development: state-of-the-art in the search for old mine shafts. *Bulletin of the International Association of Engineering Geology* 37, 91–98.

Bell, F.G. (1988b) The history and techniques of coal mining and the associated effects and influence on construction. *Bulletin of the Association of Engineering Geologists* 24, 471–504.

Bell, F.G. (1988c) Subsidence associated with the abstraction of fluids. In *Engineering Geology of Underground Movements, Engineering Geology Special Publication No. 5*, Bell, F.G., Culshaw, M.G., Cripps, J.C. & Lovell, M.A. (eds). Geological Society, London, pp. 363–376.

Bell, F.G. (1992) Salt mining and associated subsidence in mid-Cheshire, England, and its influence on planning. *Bulletin of the Association of Engineering Geologists* 22, 371–386.

Bell, F.G. (1996) Dereliction: colliery spoil heaps and their restoration. *Environmental and Engineering Geoscience* 2, 85–96.

Bell, F.G. & Fox, R.M. (1991) The effects of mining subsidence on discontinuous rock masses and the influence on foundations: the British experience. *The Civil Engineer in South Africa* 33, 201–210.

Bell, F.G. & Kerr, A. (1993) Coal mining and water quality with illustrations from Britain. In *Environmental Management, Geowater and Engineering Aspects*, Chowdhury, R.N. & Sivakumar, S. (eds). Balkema, Rotterdam, pp. 607–614.

Bezuidenhout, C.A. & Enslin, J.F. (1970) Surface subsidence and sinkholes in dolomitic areas of the Far West Rand, Transvaal, Republic of South Africa. In *Proceedings of the First International Symposium on Land Subsidence, Tokyo, International Association of Hydrological Sciences, UNESCO Publication No. 88 (2)*. UNESCO, Paris, pp. 482–495.

Bhattacharya, S. & Singh, M.M. (1985) *Development of Subsidence Damage Criteria. Office of Surface Mining, Department of Interior, Contract No. J5120129*. Engineering International Inc., Washington, DC, 226 pp.

Blight, G.E. (1988) Some less familiar aspects of hydraulic fill structures. In *Hydraulic Fill Structures. American Society of Civil Engineers, Geotechnical Special Publication No. 21*, Van Zyl, D.J.A. & Vick, S.G. (eds). American Society of Civil Engineers, New York, pp. 1000–1064.

Blight, G.E. & Steffen, O.K.H. (1979) Geotechnics of gold mining waste disposal. In *Current Geotechnical Practice in Mine Waste Disposal*. American Society of Civil Engineers, New York, pp. 1–52.

Brauner, G. (1973a) *Subsidence due to Underground Mining: Part II, Ground Movements and Mining Damage*. Bureau of Mines, Department of the Interior, US Government Printing Office, Washington, DC, 53 pp.

Brauner, G. (1973b) *Subsidence due to Underground Mining: Part I, Theory and Practice in Predicting Surface Deformation*. Bureau of Mines, Department of the Interior, US Government Printing Office, Washington, DC, 56 pp.

Brodie, M.J., Broughton, L.M. & Robertson, A. (1989) A conceptional rock classification system for waste management and a laboratory method for ARD prediction from rock piles. *British Columbia Acid Mine Drainage Task Force, Draft Technical Guide* 1, 130–135.

Bryan, A., Bryan, J.G. & Fouche, J. (1964) Some problems of strata control and support in pillar workings. *Mining Engineer* 123, 238–266.

Bullock, S.E.T. & Bell, F.G. (1995) An investigation of surface and ground water quality at a tin mine in the north west Transvaal, South Africa. *Transactions of the Institution of Mining and Metallurgy* 104, Section A, Mining Industry, A125–A133.

Bullock, S.E.T. & Bell, F.G. (1997) Some problems associated with past mining at a mine in the Witbank coalfield, South Africa. *Environmental Geology* 23, 61–71.

Cambridge, M. (1995) Use of passive systems for treatment of mine outflows and seepages. *Minerals Industry International, Bulletin of the Institution of Mining and Metallurgy* 1024, 35–42.

Charles, J.A., Hughes, D.B. & Burford, D. (1984) The effect of a rise of water table on the settlement of backfill at

Horsley restored opencast coal mining site, 1973–1983. In *Proceedings of the Third International Conference on Ground Movements and Structures, Cardiff*, Geddes, J.D. (ed). Pentech Press, London, pp. 423–442.

Connelly, R.J., Harcourt, K.J., Chapman, J. & Williams, D. (1995) Approach to remediation of ferruginous discharge in the South Wales coalfield and its application to closure planning. *Minerals Industry International, Bulletin of the Institution of Mining and Metallurgy* **1024**, 43–48.

Cook, B.J. (1990) Coal discard—rehabilitation of a burning heap. In *Reclamation, Treatment and Utilization of Coal Mining Wastes*, Rainbow, A.K.W. (ed). Balkema, Rotterdam, pp. 223–230.

Crandell, F.J. (1949) Ground vibrations due to blasting and its effect on stress meters. *Journal of the Boston Society of Civil Engineers* **36**, 222–225.

Cripps, J.C., McCann, D.M., Culshaw, M.G. & Bell, F.G. (1988) The use of geophysical methods as an aid to the detection of abandoned shallow mine workings. In *Minescape '88, Proceedings of the Symposium on Mineral Extraction, Utilization and Surface Environment, Harrogate*. Institution of Mining Engineers, Doncaster, pp. 281–289.

Dean, J.W. (1967) Old mine shafts and their hazard. *Mining Engineer* **127**, 368–377.

Deere, D.U. (1961) Subsidence due to mining—a case history from the Gulf region of Texas. In *Proceedings of the Fourth Symposium on Rock Mechanics. Bulletin Mining Industries Experimental Station, Engineering Series*, Hartman, H.L. (ed). Pennsylvania State University, Pennsylvania, pp. 59–64.

Edwards, A.T. & Northfield, R.D. (1960) Experimental studies of the effects of blasting on structures. *The Engineer* **210**, 539–546.

Edwards, R.W. (1981) The impact of coal mining on river ecology. In *Proceedings of the Symposium on Mining and Water Pollution, Nottingham*. Institution of Water Engineers and Scientists, London, pp. 31–38.

Fell, R., Miller, S. & de Ambrosio, L. (1993) Seepage and contamination from mine waste. In *Geotechnical Management of Waste and Contamination*, Fell, R., Phillips, A. & Gerrard, C (eds). Balkema, Rotterdam, pp. 253–311.

Garrard, G.E.C. & Taylor, R.K. (1988) Collapse mechanisms of shallow coal mine workings from field measurements. In *Engineering Geology of Underground Movements, Engineering Geology Special Publication No. 5*, Bell, F.G., Culshaw, M.G., Cripps, J.C. & Lovell, M.A. (eds). Geological Society, London, pp. 181–192.

Gilluly, J. & Grant, U.S. (1949) Subsidence in the Long Beach area, California. *Bulletin of the Geological Society of America* **60**, 461–560.

Goodman, R.E., Korbay, S. & Buchignani, A. (1980) Evaluation of collapse potential over abandoned room and pillar mines. *Bulletin of the Association of Engineering Geologists* **17**, 27–37.

Healy, P.R. & Head, J.M. (1984) *Construction over Abandoned Mine Workings, Construction Industry Research and Information Association, Special Publication* **32**. Construction Industry Research and Information Association, London.

Hellewell, F.G. (1988) The influence of faulting on ground movement due to coal mining. The UK and European experience. *Mining Engineer* **147**, 334–337.

Holzer, T.L. (1986) Ground failure caused by groundwater withdrawal from unconsolidated sediments. In *Proceedings of the Third Symposium on Land Subsidence, Venice, International Association of Hydrological Sciences, Publication No.* **151**, pp. 747–756.

Hood, M., Ewy, R.T. & Riddle, R.L. (1983) Empirical methods of subsidence prediction—a case study from Illinois. *International Journal of Rock Mechanics and Mining Science and Geomechanical Abstracts* **20**, 153–170.

Johnson, A.C. & James, E.J. (1990) Granville Colliery land reclamation/coal recovery scheme. In *Reclamation, Treatment and Utilization of Coal Mining Wastes*, Rainbow, A.K.M. (ed). Balkema, Rotterdam, pp. 193–202.

Langefors, U. & Kihlstrom, B. (1962) *The Modern Technique of Rock Blasting*. Wiley, New York.

Lloyd, P., Cripps, J.C. & Bell, F.G. (1995) The estimation of grout take for small scale developments in areas of shallow abandoned coal mining: some examples from the East Pennine coalfield. In *Engineering Geology and Construction, Engineering Geology Special Publication No. 10*, Culshaw, M.G., Cripps, J.C. & Walthall, S. (eds). Geological Society, London, pp. 135–141.

Lofgren, B.N. (1979) Changes in aquifer system properties with groundwater depletion. In *Evaluation and Prediction of Subsidence, Proceedings of the Speciality Conference of the American Society of Civil Engineers, Gainsville*, Saxena, S.K. (ed), New York, pp. 26–46.

McMillan, A.A. & Browne, M.A.E. (1987) The use and abuse of thematic information maps. In *Planning and Engineering Geology, Engineering Geology Special Publication No. 4*, Culshaw, M.G., Bell, F.G., Cripps, J.C. & O'Hara, M. (eds). Geological Society, London, pp. 237–246.

Michalski, S.R., Winschel, L.J. & Gray, R.E. (1990) Fires in abandoned coal mines. *Bulletin of the Association of Engineering Geologists* **27**, 479–495.

Miller, S.D., Jeffery, J.J. & Wong, J.W.C. (1991) In pit identification and management of acid forming rock waste at Golden Cross Gold Mine, New Zealand. In *Proceedings of the Second International Conference on the Abatement of Acid Drainage, Montreal*, pp. 125–132.

Nicholls, H.R., Johnson, C.F. & Duvall, W.I. (1971) *Blasting Vibrations and their Effects on Structures. United States Bureau of Mines, Bulletin* **656**. United States Bureau of Mines, Washington, DC.

Orchard, R.J. & Allen, W.S. (1970) Longwall partial extraction systems. *Mining Engineer* **129**, 523–535.

Oriard, L.L. (1972) Blasting operations in the urban environment. *Bulletin of the Association of Engineering Geologists* 9, 27–46.

Piggott, R.J. & Eynon, P. (1978) Ground movements arising from the presence of shallow abandoned mine workings. In *Proceedings of the First International Conference on Large Ground Movements and Structures, Cardiff*, Geddes, J.D. (ed). Pentech Press, London, pp. 749–780.

Price, D.G. (1971) Engineering geology in the urban environment. *Quarterly Journal of Engineering Geology* 4, 191–208.

Price, D.G., Malkin, A.B. & Knill, J.L. (1969) Foundations of multi-storey blocks on Coal Measures with special reference to old mine workings. *Quarterly Journal of Engineering Geology* 1, 271–322.

Simovic, L., Snodgrass, W.J., Murphy, K.L. & Schmidt, J.W. (1984) Development of a model to describe the natural degradation of cyanide in gold with effluents. In *Proceedings of the Conference on Cyanide and the Environment, Tucson*, pp. 413–430.

Smith, A., Dehomann, A. & Pullen, R. (1984) The effects of cyanide bearing gold tailings disposal on water qualities in the Witwatersrand, South Africa. In *Proceedings of the Conference on Cyanide in the Environment, Tucson*, pp. 221–229.

Stacey, T.R. & Bakker, D. (1992) The erection or construction of buildings and other structures on undermined ground. In *Proceedings of the Symposium on Construction over Mined Areas, Pretoria*. South African Institution of Civil Engineers, Yeoville, pp. 282–288.

Stacey, T.R. & Rauch, H.P. (1981) A case history of subsidence resulting from mining at considerable depth. *Transactions of the South African Institution of Civil Engineers* 23, 55–58.

Statham, I., Golightly, C. & Treharne, G. (1987) The thematic mapping of the abandoned mining hazard—a pilot study for the South Wales Coalfield. In *Planning and Engineering Geology, Engineering Geology Special Publication No. 4*, Culshaw, M.G., Bell, F.G., Cripps, J.C. & O'Hara, M. (eds). Geological Society, London, pp. 255–268.

Van Zyl, D.J.A. (1993) Mine waste disposal. In *Geotechnical Practice for Waste Disposal*, Daniel, D.E. (ed). Chapman and Hall, London, pp. 269–287.

Vick, S.G. (1983) *Planning, Design and Analysis of Tailings Dams*. Wiley, New York.

Whittacker, B.N. & Reddish, D.J. (1989) *Subsidence: Occurrence, Prediction and Control*. Elsevier, Amsterdam.

Woodward, G.M. & Selby, K. (1981) The effect of coal mining on water quality. In *Proceedings of the Symposium on Mining and Water Pollution, Nottingham*. Institution of Water Engineers and Scientists, London, pp. 11–19.

14 Waste and its disposal

14.1 Introduction

With increasing industrialization, technical development and economic growth, the quantity of waste has increased immensely. In addition, in developed countries, the nature and composition of waste have changed over the decades, reflecting industrial and domestic practices. Many types of waste material are produced by society, of which domestic, commercial, industrial, mining and radioactive wastes probably are the most notable. Over and above this, waste can be regarded as non-hazardous and hazardous. Waste may take the form of solids, sludges, liquids and gases, or any combination thereof. Depending on the source of generation, some of these wastes may degrade into harmless products, whereas others may be non-degradable and/or hazardous, thus posing health risks and environmental problems if not managed properly. A further problem is the fact that deposited waste can undergo changes through chemical reaction, resulting in the production of dangerous substances. Solving the waste problem is one of the fundamental tasks of environmental protection.

As waste products differ considerably from one another, the storage facilities they require also differ. Despite increased efforts at recycling wastes and avoiding their production, many different kinds of special waste are produced, which must be disposed of in special ways. Wastes that do not decompose within a reasonable time, mainly organic and hazardous wastes, and liquids that cannot be otherwise disposed of ideally should be burnt. All the organic materials are removed during burning, leading to a less hazardous form and leaving an inorganic residue. Solid, unreactive, immobile inorganic wastes can be disposed of at above-ground disposal sites. It sometimes is necessary to treat these wastes prior to disposal. To provide long-term isolation from the environment, high-toxicity, non-degradable wastes should be disposed of underground if they cannot be burnt.

The best method of disposal is determined on the basis of the type and amount of waste, on the one hand, and the geological conditions of the waste disposal site on the other. In terms of the location of a site, initially a desk study is undertaken (Sara, 1994). The primary task of the site exploration which follows is to determine the geological and hydrogeological conditions. Their evaluation provides the basis of the models used to test the reaction of the system to engineering activities. The chemical analysis of groundwater, together with the mineralogical analysis of rocks, may help to yield information about its origin, and hence about the future development of the site. At the same time, the leaching capacity of the water is determined, which allows a prediction to be made of the reactions between wastes and soil or rock. If groundwater must be protected, or highly mobile toxic or very slowly degradable substances are present in wastes, impermeable liners may be used to inhibit the infiltration of leachate into the surrounding ground.

In terms of waste disposal by landfill, a landfill is environmentally acceptable if it is correctly engineered. Unfortunately, if it is not constructed to sufficiently high standards, a landfill may have an adverse impact on the environment. Surface water or groundwater pollution may result. Consequently, a physical separation between waste, on the one hand, and groundwater and surface water on the other, as well as an effective surface water diversion drainage system, are fundamental to design.

14.2 Domestic refuse and sanitary landfills

Domestic refuse is a heterogeneous collection of almost anything (i.e. waste food, garden rubbish, paper, plastic, glass, rubber, cloth, ashes, building

Fig. 14.1 Disposing of domestic waste at a landfill in Kansas City, Missouri.

waste, metals, etc.; Fang, 1995), much of which is capable of reacting with water to give a liquid rich in organic matter, mineral salts and bacteria, namely, leachate. Leachate is formed when rainfall infiltrates a landfill and dissolves the soluble fraction of the waste, including the soluble products obtained as a result of the chemical and biochemical processes occurring within decaying wastes. The organic carbon content of waste especially is important, because this influences the growth potential of pathogenic organisms.

Matter exists in the gaseous, liquid and solid states in landfills, and all landfills comprise a delicate and shifting balance between the three states. Any assessment of the state of a landfill and its environment must take into consideration the substances present in a landfill, their mobility now and in the future, the potential pathways along which pollutants can travel and the targets potentially at risk from the substances involved.

Although domestic waste is disposed of in a number of ways, quantitatively the most important method is emplacement in a landfill (Fig. 14.1). As far as the location of a landfill is concerned, a decision must be made on the site selection, project extent, finance, construction materials and site rehabilitation. The major requirement in planning a landfill site is to establish exactly, by survey and analysis, the types, nature and quantities of waste involved. A waste survey is undertaken, after which future trends

are forecast. These forecasts form the basis of the decision to use a potential site for a landfill. Initially, the potential life of a site is estimated, and its distance from the proposed waste catchment area is assessed. The design of a landfill is influenced by the character of the material which it has to accommodate.

The selection of a landfill site for a particular waste or a mixture of wastes involves a consideration of the economic and social factors, as well as the geological and hydrogeological conditions. The ideal landfill site should be hydrogeologically acceptable, posing no potential threat to water quality when used for waste disposal, be free from running or static water and have a sufficient store of material suitable for covering each individual layer of waste. It should also be situated at least 200 m away from any residential development. As far as the hydrogeological conditions are concerned, most argillaceous sedimentary, massive igneous and metamorphic rock formations have a low intrinsic permeability, and therefore are likely to afford the most protection to the water supply. By contrast, the least protection is provided by rocks intersected by open discontinuities or in which solution features are developed. Granular materials may act as filters, leading to dilution and decontamination. Hence, sites for the disposal of domestic refuse can be chosen where decontamination has the maximum chance of reaching completion and where groundwater sources are located far enough away to enable dilution to be effective. The

Table 14.1 Classification of landfill sites based upon their hydrogeology. (After Barber, 1982.)

Designation	Description	Hydrogeology
Fissured site or site with rapid subsurface liquid flow	Material with well-developed secondary permeability features	Rapid movement of leachate via fissures, joints or through coarse sediments. Possibility of little dispersion in the groundwater or attenuation of pollutants
Natural dilution, dispersion and attenuation of leachate	Permeable materials with little or no significant secondary permeability	Slow movement of leachate into the ground through an unsaturated zone. Dispersion of leachate in the groundwater, attenuation of pollutants (sorption, biodegradation, etc.) probable
Containment of leachate	Impermeable deposits, such as clays or shales, or sites lined with impermeable materials or membranes	Little vertical movement of leachate. Saturated conditions exist within the base of the landfill

position of the water table is important, as it determines whether wet or dry tipping is involved. Generally, unless waste is inert, wet tipping should be avoided. The position of the water table also determines the location at which flow is discharged. The hydraulic gradient determines the direction and velocity of the flow of leachates when they reach the water table, and also influences the amount of dilution that leachates undergo. Aquifers which contain potable supplies of water must be protected. If they are close to a proposed landfill, a thorough hydrogeological investigation is necessary to ensure that site operations will not pollute them. If pollution is a possibility, the site must be designed to provide some form of artificial protection, otherwise the proposal should be abandoned.

Barber (1982) identified three classes of landfill site based upon hydrogeological criteria (Table 14.1). When assessing the suitability of a site, two of the principal considerations are the ease with which the pollutant can be transmitted through the substrata and the distance it is likely to spread from the site. Consequently, the primary and secondary permeabilities of the formations underlying a potential landfill area are of major importance. It is unlikely that the first type of site mentioned in Table 14.1 would be considered to be suitable. There would also be grounds for an objection to a landfill site falling within the second category of Table 14.1, if the site was located within the area of diversion to a water supply well. Generally, the third category, in which

the leachate is contained within the landfill area, is preferred. Because all natural materials possess some degree of permeability, total containment only can be achieved if an artificial, impermeable lining is provided over the bottom of the site. However, there is no guarantee that clay, soil–cement, asphalt or plastic linings will remain impermeable permanently. Thus the migration of materials from a landfill site into the substrata will occur eventually, although the length of time before this happens may be subject to uncertainty. In some instances, the delay will be sufficiently long for the pollution potential of the leachate to be greatly diminished. One of the methods of tackling the problem of pollution associated with landfills is by the dilution and dispersal of the leachate. Otherwise, leachate can be collected by internal drains within the landfill and conveyed away for treatment.

One of the difficulties in predicting the effect of leachate on groundwater is the continual change in the characteristics of the leachate as the landfill ages. Leachate may be diluted where it gains access to run-off or groundwater, but this depends on the quantity and chemical characteristics of the leachate, as well as the quantity and quality of the receiving water.

14.2.1 Design considerations for a landfill

The design of a landfill site is influenced by the physical and biochemical properties of the wastes. The need for the control of leachate production at a par-

Fig. 14.2 Cellular construction of a landfill in Kansas City, Missouri.

ticular landfill is dependent on the extent of the possible pollution problems at that site. Site selection therefore has an important influence on the need for leachate control. Nonetheless, the use of leachate control and/or treatment methods may permit unsuitable sites to be employed for the disposal of solid wastes. Leachate control should be planned during landfill development, especially if control techniques are to be installed beneath the waste.

The quantity of leachate produced is influenced by the amount of groundwater in a landfill (leachate percolation = net percolation – water absorbed by waste + liquid disposal into landfill). The quantity of water absorbed by waste depends on the age of emplacement of the waste. Initially, the water-absorbing capacity of the waste exceeds the net percolation, and the leachate flow is zero. At the other extreme, if the waste is totally saturated, the leachate flow equals the net percolation plus the groundwater or subsurface water flow into the landfill plus the liquid disposal into the landfill.

The most common means of controlling the leachate is to minimize the amount of water infiltrating the site by encapsulating the waste in impermeable material. Hence, well designed landfills usually possess a cellular structure, as well as a lining and a cover, that is, the waste is contained within a series of cells formed of clay (Fig. 14.2). The cells are covered at the end of each working day with a layer of soil and compacted.

At the present time, there is no standard rule as to how waste should be dumped or compacted. Nonetheless, compaction is important, because it reduces the settlement and hydraulic conductivity, whilst increasing the shear strength and bearing capacity. Furthermore, the smaller the quantity of air trapped within landfill waste, the lower the potential for spontaneous combustion. Wherever possible, waste should be uniformly distributed in thin layers prior to compaction. Locally available soil can be mixed with the waste. As far as the landfill stability is concerned, the potential for slope failure in a landfill is related to compaction control during disposal, and the heavier the roller used for compaction the better. Even so, conventional compaction techniques do not always achieve effective results, especially with highly non-uniform waste. In such instances, dynamic compaction has been used with good results.

Mitchell (1986) maintained that a properly designed and constructed liner and cover offer long-term protection for groundwater and surface water. Landfill liners are constructed from a wide variety of materials (Fig. 14.3). Adequate site preparation is necessary if a lining system is to perform satisfactorily. Nonetheless, no liner system, even if perfectly designed and constructed, will prevent all seepage losses. For instance, no liner, no matter how rigid or highly reinforced, can withstand large differential settlement without eventually leaking or possibly failing completely. If extreme concern is warranted, an

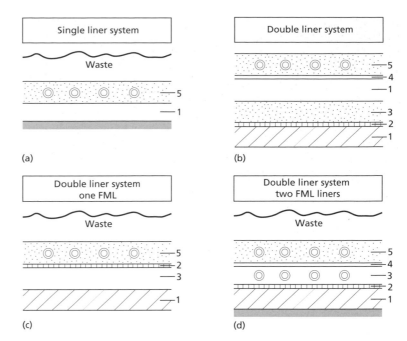

Fig. 14.3 Liner systems: 1, compacted, low-permeability clay; 2, flexible membrane liner (FML); 3, leachate collection/detection system; 4, FML; 5, primary leachate collection system; ◎, collection pipes.

underdrainage system can be placed beneath the primary liner to collect any leakage passing through, which can either be treated or recirculated back to the containment area. A secondary liner can be placed beneath the underdrainage system.

Clay, bentonite, geomembranes, soil–cement or bitumen–cement can be placed beneath a landfill to inhibit the movement of leachate into the soil. Clay liners are suitable for the containment of many wastes, because of the low hydraulic conductivity of clay and its ability to adsorb some wastes (Daniel, 1993). Clay liners compacted on the wet side of optimum moisture content are less permeable than those compacted on the dry side of optimum. Those leachates consisting of organic solvents, and those containing high levels of dissolved salts, acids or alkalis, can give rise to cracking and the development of pipes in clay. Clays, however, can be treated with polymers, which reduce their sensitivity to potential contaminants.

When geomembranes are used as liners, the chemical compatibility between the waste material and the geomembrane should be assessed. The geomembrane must possess sufficient thickness and strength to avoid failure due to physical stresses (Koerner, 1993). The foundation beneath the liner must be able to

support the liner and to resist pressure gradients. If the support system settles, compresses or lifts, the liner may rupture. Underliners and covers may be used to protect the geomembranes from puncture, tearing, abrasion and ultraviolet rays. For example, a bedding course of sand may be placed below the liner, and another layer of sand may be placed on top. Composite liners incorporate both clay blankets and geomembranes.

Soil–cement liners make use of local soils. The soil ideally should contain less than 20% silt/clay fraction and, when mixed with 3–12% cement by weight, should perform adequately as a liner. Lining slopes with soil–cement, however, can present difficulties. In addition, soil–cement can be degraded in acidic conditions, which tend to attack the cement. Furthermore, the likelihood of shrinking and cracking of soil–cement leads to an increase in its permeability.

Bitumen–cement and bituminous concrete have been used to line landfills. However, a major consideration is the chemical compatibility between the contents of the landfill and the bituminous material.

The double liner system is intended to prevent leakage. The primary liner of such a system is the upper geomembrane. Leachate should be properly collected above the primary liner at the bottom of the landfill, and disposed of in accordance with accepted environmental principles. A perforated pipe collector system is located below the primary liner, and is bedded within a crushed stone or sand drainage blanket. A secondary geomembrane liner occurs beneath the drainage blanket, which only need function if leakage from the primary liner is found in the underdrainage system. If leakage from the primary liner does occur, a downstream well monitoring system must be deployed to check for possible leakage from the secondary liner.

Leachate drainage systems should be designed to collect the anticipated volume of leachate likely to be produced by the landfill (Jessberger *et al.*, 1995). This will vary during the life of the site. One of their functions is to prevent the level of leachate rising to such an extent that it can drain from the landfill and pollute nearby water courses. Ideally, a leachate drainage collection system should occur over the whole base of a landfill and, if below ground level, should extend up the sides. The drainage blanket, consisting of granular material, should be at least 300 mm thick, with a minimum hydraulic conductivity of $1-10 \, \text{m s}^{-1}$. The perforated collection pipes within the drainage blanket convey leachate to collection points or wells, from which it can be removed by pumping.

If lining is considered to be too expensive, drainage is a relatively inexpensive alternative. For instance, a drainage ditch can be combined with a layer of free-draining granular material overlying a low-permeability base. The granular material is graded down towards the perimeter of the landfill, so that leachate can flow into the drainage ditches. The leachate can be pumped or can flow away from the ditches. Synthetic drainage layers may be used to achieve the drainage capacity. A filter medium may be used between the waste and the drainage layer to allow leachate to pass through, but to prevent the migration of fines.

The principal function of a cover is to minimize the infiltration of precipitation into a landfill, and hence minimize the formation of leachate. The nature of the soil used to cover the waste materials is very important. At many sites, however, the quality of the borrow material is less than the ideal, and some blending with imported soil may be required.

According to Daniel (1995), most cover systems in the USA consist of a number of components. However, not all the components shown in Fig. 14.4 need be present in a cover, and some layers may be combined.

Geosynthetic clay liners (GCLs) are thin layers of dry bentonite (approximately 5 mm thick) attached to one or more geosynthetic materials. In other words, the bentonite is sandwiched between an upper and lower sheet of geotextile, or the bentonite is mixed with adhesive and glued to a geomembrane. The primary purpose of the geosynthetic component(s) is to hold the bentonite together in a uniform layer. When the bentonite is wetted, it swells and provides a cover with a low permeability.

The key geotechnical factors in cover design are its stability and resistance to cracking. Cracking of a clay cover may be brought about by desiccation or by the build-up of gas pressure beneath, if a venting system is not functional or is not installed. It may also be difficult to maintain the integrity of the cover if large differential settlements occur in the landfill. A hydraulic conductivity of $10^{-9} \, \text{m s}^{-1}$ usually is specified for clay covers, but perhaps never attained because of cracking. Reinforcement with a high-strength geotextile on the top and bottom of a clay liner may help to reduce the likelihood of cracking. Differential settlement can be reduced by the dynamic compaction of refuse. Controlled percolation may not be critical if the treatment of leachate is regarded as more economically feasible than an expensive cover.

The slope of the surface of the landfill influences infiltration. Water tends to collect on a flat surface and subsequently infiltrates, whereas water tends to run off steeper slopes. However, when the surface slope exceeds about 8%, the possibility of surface run-off eroding the cover exists. Surface water should be collected in ditches and routed from the site.

The major function of all cutoff systems is to provide the isolation of wastes from the surrounding environment, so offering protection to soil and groundwater from contamination. Hence, the hydraulic and gas conductivities of cutoff systems are of paramount importance. Vertical cutoff walls may surround a site, but up-gradient and down-gradient walls may also be used. Where an impermeable horizon exists beneath the site, the cutoff wall should be keyed into it. A very deep cutoff is required where

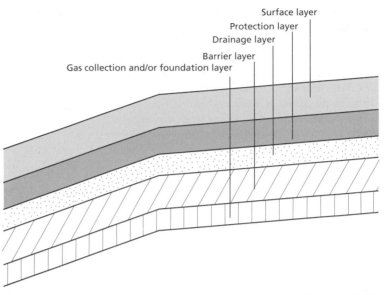

Surface layer
Protection layer
Drainage layer
Barrier layer
Gas collection and/or foundation layer

Layer	Description of layer	Typical materials	Typical thickness (m)
1	Surface layer	Topsoil; geosynthetic erosion control layer; cobbles; paving material	0.15
2	Protection layer	Soil; recycled or reused waste material; cobbles	0.3–1
3	Drainage layer	Sand or gravel; geonet or geocomposite	0.3
4	Barrier layer	Compacted clay; geomembrane; geosynthetic clay liner; waste material; asphalt	0.3–1
5	Gas collection layer and/or foundation layer	Sand or gravel; soil; geonet or geotextile; recycled or reused waste material	0.3

Fig. 14.4 Basic components of a final cap.

no impermeable stratum exists. If leachate flows from an existing landfill, the construction of a seepage cutoff system provides a solution.

Steel sheet piles, geosynthetic sheet piles and secant piles are used as cutoff walls, as are grout curtains. However, cutoff walls are more likely to be constructed by jet grouting than by injection. Barriers can also be constructed as trenches excavated under bentonite slurry and filled with soil–bentonite or cement–bentonite. Plastic concrete cutoff walls are similar to cement–bentonite cutoffs, except that they contain aggregate. Mitchell (1994) pointed out that the greater strength and stiffness of plastic concrete cutoff walls mean that they are better suited to situations where ground stability and movement control are important. Concrete diaphragm cutoff walls can also be used where high structural strength is required (Bell & Mitchell, 1986). A geomembrane sheeting-enclosed cutoff wall consists of a geomembrane that is fabricated to form a U-shaped envelope which fits the dimensions of a trench. Ballast is placed within the geomembrane in order to sink it into the slurry trench. After initial submergence into the trench, the envelope is filled with wet sand.

Leakage from a landfill in which the water table is high and there is no impermeable layer beneath can be treated by jet grouting. The grout pipe is inserted successively into a series of holes at predetermined centres at the base of the landfill, and rotated through 360°. This allows the formation of interlocking discs of grouted soil. Alternatively, inclined drilling can be used for the injection of grout or jet grouting, provided that the overall width of the site is not too great. Inclined drilling from opposite sides of a landfill forms an interlocking V-shaped barrier, which can be used together with vertical cutoff walls.

14.2.2 Degradation of waste in landfills

Leachate is formed by the action of liquids, primarily water, within a landfill. The generation of leachate occurs once the absorbent characteristics of the refuse are exceeded. The waste in a landfill site generally has a variety of origins. Many of the organic components are biodegradable. Initially, the decomposition of waste is aerobic. Bacteria flourish in moist conditions, and waste contains varying amounts of liquid, which may be increased by the infiltration of precipitation. Once decomposition starts, the oxygen in the waste rapidly becomes exhausted, so that the waste becomes anaerobic.

There are basically two processes by which the anaerobic decomposition of organic waste takes place. Initially, complex organic materials are broken down into simpler organic substances, typified by various acids and alcohols. The nitrogen present in the original organic material tends to be converted into ammonium ions, which are soluble readily and may give rise to significant quantities of ammonia in the leachate. The reducing environment converts oxidized ions to their reduced form, for example, ferric salts are converted to the ferrous state. Ferrous salts are more soluble, and therefore iron is leached from the landfill. The sulphate in the landfill may be reduced biochemically to sulphides. Although this may lead to the production of small quantities of hydrogen sulphide, the sulphide tends to remain in the landfill as highly insoluble metal sulphides. In a young landfill, the dissolved salt content may exceed $10\,000\,\mathrm{mg\,l^{-1}}$, with relatively high concentrations of sodium, calcium, chloride, sulphate or iron, whereas as a landfill ages, the concentration of inorganic materials usually decreases (Table 14.2). Suspended particles may be present in the leachate due to the washout of fine material from the landfill.

The second stage of anaerobic decomposition involves the formation of methane. In other words, methanogenic bacteria use the end products from the first stage of anaerobic decomposition to produce methane and carbon dioxide. Methanogenic bacteria prefer neutral conditions, so that if the acid formation in the first stage is excessive, their activity can be inhibited.

14.2.3 Attenuation of leachate

Physical and chemical processes are involved in the attenuation of leachate in soils. These include precipitation, ion exchange, adsorption and filtration. In the reducing zone, where organic pollution is greatest, insoluble heavy metal sulphides and soluble iron sulphide are formed. In the area between the reducing and oxidizing zones, ferric and inorganic hydroxides are precipitated. Other compounds may be precipitated, especially with ferric hydroxide. The reduction of nitrate may yield nitrite, nitrogen gas or, possibly, ammonia, although ammonia usually is present due to the biodegradation of nitrogen-bearing organic material. The ion exchange and adsorption properties of a soil or rock primarily are attributable to the presence of clay minerals. Consequently, a soil with a high clay content has a high ion exchange capacity. The humic material in soil also has a high ion exchange capacity. Adsorption is susceptible to changes in pH value, as the pH affects the surface charge on a colloid particle or molecule. Hence, at low pH values, the removal rates due to adsorption can be reduced significantly. However, the removal of pollutants by ion exchange and adsorption can be reversible. For example, after high-level pollution from a landfill subsides, and more dilute leachate is produced, the soil or rock may release the pollutants back into the leachate. Nevertheless, some ions will be irreversibly adsorbed or precipitated.

The degree of filtration brought about by a soil or rock mass depends on the size of the pores. In many porous soils or rocks, the filtration of suspended matter occurs within short distances. However, a rock mass containing open discontinuities may transmit leachate for several kilometres (Hagerty & Pavori, 1973). Pathogenic microorganisms, which may be found in a landfill, usually do not travel far in

Table 14.2 Leachate composition.

(a) Typical composition of leachates from recent and old domestic wastes at various stages of decomposition. (After Anon., 1986.)

Parameter	Leachate from recent wastes	Leachate from old wastes
pH	6.2	7.5
Chemical oxygen (mg l^{-1})	23 800	1160
Biochemical oxygen (mg l^{-1})	11 900	260
Total organic carbon (mg l^{-1})	8 000	465
Fatty acids (mg l^{-1})	5 688	5
Ammoniacal-N (mg l^{-1})	790	370
Oxidized (mg l^{-1})	3	1
Orthophosphate (mg l^{-1})	0.73	1.4
Chloride (mg l^{-1})	1315	2080
Sodium (Na) (mg l^{-1})	960	1300
Magnesium (Mg) (mg l^{-1})	252	185
Potassium (K) (mg l^{-1})	780	590
Calcium (Ca) (mg l^{-1})	1820	250
Manganese (Mn) (mg l^{-1})	27	2.1
Iron (Fe) (mg l^{-1})	540	23
Nickel (Ni) (mg l^{-1})	0.6	0.1
Copper (Cu) (mg l^{-1})	0.12	0.3
Zinc (Zn) (mg l^{-1})	21.5	0.4
Lead (Pb) (mg l^{-1})	8.4	0.14

(b) Chemical analyses of leachate from a sump at a landfill in Durban, South Africa, taken over a 5-year period from 1986 to 1991. (After Bell *et al.*, 1996b.)

	pH	Suspended solids (mg l^{-1})	Total dissolved solids (mg l^{-1})	Conductivity (mS m^{-1})	Oxygen absorption (mg l^{-1})	Chemical oxygen demand (mg l^{-1})
Max	8.9	6044	45 041	8080	6200	70 900
Min	6.6	200	900	450	145	11.2
Mean	7.7	965	17 029	1174	649	13 546
MPL	5.5–9.5		500	300	10	75

	Sodium (mg l^{-1})	Potassium (mg l^{-1})	Calcium (mg l^{-1})	Magnesium (mg l^{-1})	Sulphate (mg l^{-1})	Ammonium (mg l^{-1})
Max	8249	2646	1236	759	2237	3530
Min	80	355	60	70	6.5	92
Mean	1393	613	146	109	837	1093
MPL	400	400	200	100	600	2

MPL, South African maximum permissible limit for domestic water insignificant risk (Anon., 1993).

the soil, because of the changed environmental conditions. In fact, pathogenic bacteria normally are not present within a few tens of metres of a landfill.

14.2.4 Surface and groundwater pollution

Leachates contain many contaminants that may have a deleterious effect on surface water. If, for example, leachate enters a river, oxygen is removed from the river by bacteria as they break down the organic compounds in the leachate. In cases of severe organic pollution, the river may be completely depleted of oxygen, with disastrous effects on aquatic life. The net effect of oxygen depletion in a river is that its ecology changes and, at dissolved oxygen levels below $2 \, mg \, l^{-1}$, most fish cannot survive. The oxygen balance is affected by several factors. For instance, the rates of chemical and biochemical reactions increase with temperature, whereas the maximum dissolved oxygen concentration decreases as the temperature rises. At high river flows, bottom mud may be resuspended and exert an extra oxygen demand, but the extra turbulence also increases aeration. At low flows, organic solids may settle out and reduce the oxygen demand of the river.

The principal inorganic pollutants which may cause problems with leachate are ammonia, iron, heavy metals and, to a lesser extent, chloride, sulphate, phosphate and calcium. Ammonia can be present in landfill leachates at up to several hundred milligrams per litre, whereas unpolluted rivers have a very low content of ammonia. The discharge of leachate high in ammonia into a river exerts an oxygen demand on the receiving water. In addition, ammonia is toxic to fish (lethal concentration is in the range $2.5–25 \, mg \, l^{-1}$) and, as ammonia is a fertilizer, it may alter the ecology of the river. Leachate containing ferrous iron is particularly objectionable in a river, because ochreous deposits are formed by chemical or biochemical oxidation of the ferrous compounds to ferric compounds. The turbidity caused by the oxidation of ferrous iron can reduce the amount of light, and so decrease the number of flora and fauna. Heavy metals can be toxic to fish at relatively low concentrations. They can also affect the organisms on which fish feed.

Physically, leachate affects the river quality in terms of the suspended solids, colour, turbidity and temperature. The suspended solids, colour and turbidity reduce the light intensity in a river. This can affect the food chain, and the lack of photosynthetic activity by plants reduces the oxygen replacement in a river. The suspended solids may settle on the bed of a river in significant quantities. This can destroy plant and animal life.

If a leachate enters the phreatic zone, it mixes and moves with the groundwater (Fig. 14.5; Bell *et al.*, 1996b). The organic carbon content in the leachate leads to an increase in the biochemical oxygen demand (*BOD*) in the groundwater, which may increase the potential for the reproduction of pathogenic organisms. Organic matter is stabilized slowly, because the oxygen demand may deoxygenate the water rapidly, and usually no replacement oxygen is available. If anaerobic conditions develop, metals such as iron and manganese may dissolve in the water causing further problems. If the groundwater has a high buffering capacity, the effects of mixing of acidic or alkaline leachate with the groundwater are reduced. The worst situation occurs in discontinuous rock masses, where groundwater movement is dominated by fissure flow, or where groundwater movement is slow with a shallow water table so that little dilution occurs. The velocity of groundwater flow is important, because a high velocity gives rise to more dispersion in the direction of flow and relatively less laterally, so that the leachate plume forms a narrow cone in the direction of groundwater flow. Low groundwater velocity leads to a wider plume.

Ideally, a groundwater monitoring programme should be established prior to tipping, and should continue for anything up to 20 years after completion of the site. The number, location and depth of monitoring wells depend on the particular site (see Chapter 9). Monitoring should include sampling from the wells and analysis for pH, chlorides, dissolved solids, organic carbon and the concentration of any particular hazardous waste that has been tipped in the landfill.

The most serious effect of a leachate on groundwater is mineralization. Much of the organic carbon and organic nitrogen is biodegraded as it moves through the soil. Heavy metals may be attenuated due to ion exchange onto clay minerals. Mineralization is brought about by inorganic ions, such as chloride. The concentrations may only be reduced by dilution. Continuing input of inorganic ions into potable

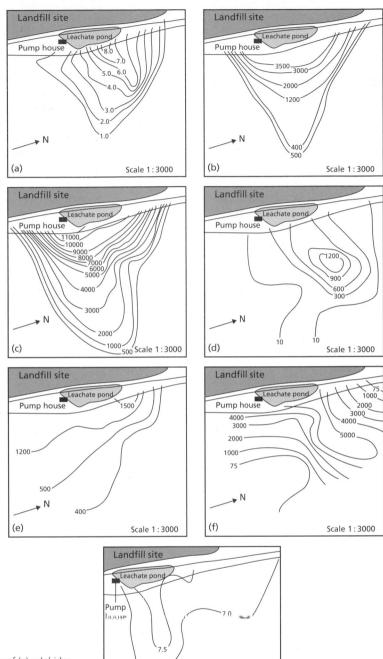

Fig. 14.5 Isopleths showing the distribution of (a) sulphides, (b) chlorides, (c) total dissolved solids, (d) oxygen absorption, (e) conductivity, (f) chemical oxygen demand and (g) pH value of the groundwater. All in milligrams per litre, except for the conductivity (mS m⁻¹) and pH value. (After Bell *et al.*, 1996b.)

groundwater eventually means that groundwater becomes undrinkable.

14.2.5 Landfill and gas formation

The biochemical decomposition of domestic and other putrescible refuse in a landfill produces gas consisting primarily of methane, with smaller amounts of carbon dioxide and volatile organic acids. In the initial weeks or months after emplacement, the landfill is aerobic, and gas production involves mainly CO_2, but also contains O_2 and N_2. As the landfill becomes anaerobic, the evolution of O_2 declines to almost zero and N_2 to less than 1%. The principal gases produced during the anaerobic stage are CO_2 and CH_4, with CH_4 production increasing slowly as methanogenic bacteria establish themselves.

The factors which influence the rate at which gas is produced include the character of the waste and the moisture content, temperature and pH of the landfill. The concentrations of salts, such as sulphate and nitrate, may also be important.

If the refuse is pulverized, the microbial activity in the landfill is higher which, in turn, may give rise to a higher production rate of gases. However, the period of time over which gas is produced may be reduced. The compaction or baling of refuse may decrease the rate of water infiltration into the landfill, retarding the bacterial degradation of the waste, with gas being produced at a lower rate over a longer period. If toxic chemicals are present in a landfill, the bacterial activity, in particular methanogenesis, may be inhibited. A moisture content of 40% or higher is desirable for optimized gas production. Generally, the rate of gas production increases with the temperature. The pH value of a landfill should be around pH 7.0 for optimum production of gas, methanogenesis tending to cease below pH 6.2. The amount of gas produced by domestic refuse varies appreciably, and a site investigation is required to determine the amount if such information is needed. Nonetheless, it has been suggested that between 2.2 and 250 litres per kilogram dry weight may be produced (Oweis & Khera, 1990).

Methane production can constitute a dangerous hazard because it is combustible and, in certain concentrations explosive (5–15% by volume in air). It is also asphyxiating. Appropriate safety precautions must be taken during site operation. In many instances, landfill gas is able to disperse safely to the atmosphere from the surface of a landfill. However, when a landfill is completely covered with a soil capping of low permeability in order to minimize leachate generation, the potential for gas to migrate along unknown pathways increases, and there have been cases of hazard arising from methane migration. Furthermore, there are unfortunately cases on record of explosions occurring in buildings due to the ignition of accumulated methane derived from nearby landfills or landfills on which the buildings were erected (Williams & Aitkenhead, 1991). The source of the gas should be identified so that remedial action can be taken. The identification of the source can involve a drilling programme and an analysis of the gas recovered. For instance, in a case referred to by Raybould and Anderson (1987), distinction had to be made between household gas and gas from sewers, old coal mines and a landfill before remedial action could be taken to eliminate the gas hazard affecting several houses. Accordingly, planners of residential developments should avoid landfill sites. Proper closure of a landfill site can require gas management to control methane gas by passive venting, power operated venting or the use of an impermeable barrier. The identification of possible migration pathways in the assessment of a landfill site is highly important, especially where old mine workings may be present, or where there is residential property nearby. Raybould and Anderson described the grouting of old mine workings which had acted as a conduit for the migration of methane from a landfill to residential properties. Although methane is not toxic to plant life, the generation of significant quantities can displace oxygen from the root zone and so suffocate plant roots. Large concentrations of carbon dioxide or hydrogen sulphide can produce the same result.

The monitoring of landfill gas is an important aspect of safety. Instruments usually monitor methane, as this is the most important component of landfill gas. Gas may be sampled from monitoring wells in the fill or from areas into which the gas has migrated. Leachate monitoring wells can act as gas collectors. Due caution must be taken when sampling.

Measures to prevent the migration of the gas include impermeable barriers (clay, bentonite,

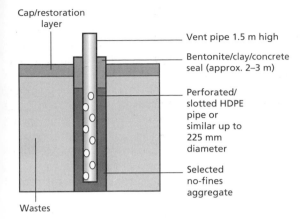

Cap/restoration layer

Vent pipe 1.5 m high

Bentonite/clay/concrete seal (approx. 2–3 m)

Perforated/ slotted HDPE pipe or similar up to 225 mm diameter

Selected no-fines aggregate

Wastes

Fig. 14.6 Diagrammatic sketch of a passive gas well. (From Attewell, 1993.)

geomembranes or cement) and gas venting. An impermeable barrier should extend to the base of the fill or to the water table, whichever is the highest. A problem, however, is to ensure that the integrity of the barrier is maintained during its installation and subsequent operation.

Venting either wastes the gas into the atmosphere or facilitates its collection for utilization. Passive venting involves venting gases to locations where they can be released to the atmosphere or burnt. The vents are placed at locations at which the gas concentrations and/or pressures are high (Fig. 14.6). This usually occurs towards the lower sections of a landfill. Perimeter wells are used where there are no geological constraints preventing the lateral migration of gas. Alternatively, the gas may be intercepted by a trench filled with coarse aggregate. Where atmospheric venting is insufficient to control the discharge of gas, an active or forced venting system is used. Active ventilation involves the connection of a vacuum pump to the discharge end of the vent.

A sand–gravel drainage layer located above the liner of a landfill can be used to collect carbon dioxide. Alternatively, a geocomposite of adequate transmissivity, together with a perforated pipe collection system, can be used to collect the gas.

14.3 Hazardous wastes

A hazardous waste can be regarded as any waste or combination of wastes of inorganic or organic origin which, because of its quantity, concentration, physical, chemical, toxicological or persistence properties, may give rise to acute or chronic impacts on human health and/or the environment when improperly treated, stored, transported or disposed. Such waste can be generated from a wide range of commercial, industrial, agricultural and domestic activities, and can take the form of liquid, sludge or solid. The characteristics of the waste not only influence its degree of hazard, but are also of great importance in the choice of a safe and environmentally acceptable method of disposal. Hazardous wastes may involve one or more risks, such as explosion, fire, infection, chemical instability or corrosion, or acute toxicity (in particular is the waste carcinogenic, mutagenic or teratogenic?). An assessment of the risk posed to health and/or the environment by hazardous waste must take into consideration its biodegradability, persistence, bioaccumulation, concentration, volume production, dispersion and potential for leakage into the environment. Hazardous wastes therefore require special treatment, and cannot be released into the environment, added to sewage or be stored in a situation which is either open to the air or from which aqueous leachate could emanate.

Some of the primary criteria for the identification of hazardous wastes include the type of hazard involved (flammability, corrosivity, toxicity and reactivity), the origin of the products, including industrial origins (e.g. medicines, pesticides, solvents, electroplating, oil refining), and the presence of specific substances or groups of substances (e.g. dioxin, lead compounds, polychlorinated biphenyls (PCBs)). These criteria and others are used alone or in combination, but in different ways in different countries. The compositional characteristics of the waste may or may not be quantified and, where levels of substance concentration are set, they again vary from country to country.

The assessment of the suitability of a site for hazardous waste disposal is a complex matter, which involves the use of models to predict the chemical behaviour of the waste in the ground, and the potential for mobilization and migration in groundwater. The form and rate of release of the waste to the environment, together with the time and place at which release occurs, can be produced with reasonable degrees of confidence. To translate these results into risk assessment requires a parallel prediction of the consequences of a release of waste. The risk associated with the disposal of wastes involves an estimation of the probability of an event or process, which leads to a release occurring within a given time period, multiplied by the consequences of that release (or, alternatively, the probability that an individual or group will be exposed to a pollutant multiplied by the probability that this exposure will give rise to a serious health effect). Given adequate epidemiological data on the health effects of toxic substances, the risk can be calculated.

The quantity of waste involved, the manner and conditions of use and the susceptibility of humans or other living things to a certain waste can be used to determine its degree of hazard. Hazard ratings can be categorized as extreme, high, moderate or low. Waste which contains significant concentrations of extremely hazardous material, including certain carcinogens, teratogens and infectious substances, is of primary concern. The low category of hazardous waste contains potentially hazardous constituents, but in concentrations which represent only a limited threat to health or the environment. If the hazard rating is less than the low category, the waste can be regarded as non-hazardous, and can be disposed of as general waste.

The minimum requirements for the treatment and disposal of hazardous waste involve ensuring that certain classes of waste are not disposed of without pretreatment. The objective of treating a waste is to reduce or destroy the toxicity of the harmful components, in order to minimize the impact on the environment. In addition, waste treatment can be used to recover materials during waste minimization programmes. The method of treatment chosen is influenced by the physical and chemical characteristics of the waste, that is, is it gaseous, liquid, in solution, sludge or solid, is it inorganic or organic and what are the concentrations of the hazardous and non-hazardous components? Physical treatment methods are used to remove, separate and concentrate hazardous and toxic materials. Chemical treatment is used in the application of physical treatment methods, and to lower the toxicity of a hazardous waste by changing its chemical nature. This may produce essentially non-hazardous materials. In biological treatments, microbial activity is used to reduce or destroy the toxicity of a waste. The principal objective of processes such as immobilization,

solidification or encapsulation is to convert hazardous waste into an inert mass with very low leachability. Macroencapsulation involves the containment of waste in drums or other approved containers within a reinforced concrete cell that is stored within a landfill. Incineration can be regarded as both a means of treatment and disposal.

The safe disposal of hazardous waste is the ultimate objective of waste management, disposal being in a landfill, by burial, by incineration or marine disposal. When landfill is chosen as the disposal option, the capacity of the site to accept certain substances without exceeding a specified level of risk must be considered. The capacity of a site to accept waste is influenced by the geological and hydrogeological conditions, the degree of hazard presented by the waste, the leachability of the waste and the design of the landfill (Fig. 14.7). Certain hazardous wastes may be prohibited from disposal in a landfill, such as explosive wastes or flammable gases. Obviously, medium- to high-level radioactive waste cannot be disposed of in a landfill.

The protection of groundwater from the disposal of toxic waste in landfills can be brought about by containment. A number of containment systems have been developed which isolate wastes, and include compacted clay barriers, slurry trench cutoff walls, geomembrane walls, sheet piling, grout curtains and hydraulic barriers (Mitchell, 1986). A compacted clay barrier consists of a trench which has been backfilled with clay compacted to give a low hydraulic conductivity. Slurry trench cutoff walls are narrow trenches filled with soil–bentonite mixtures, which extend downwards into an impermeable layer. The use of diaphragm walls is restricted to situations in which high structural stability is required. Grout curtains may be used in certain situations. Extraction wells can be used to form hydraulic barriers, and are located so that the contaminant plume flows towards them.

The monitoring of hazardous waste repositories forms an inherent part of the safety requirements governing their operational and postoperational periods. Hence, the repository operators are required to conduct monitoring programmes to detect any failure in the waste containment systems so that remedial action can be taken. The nature and duration of a monitoring programme needed to ensure the continuing safe isolation of waste depend upon a number of

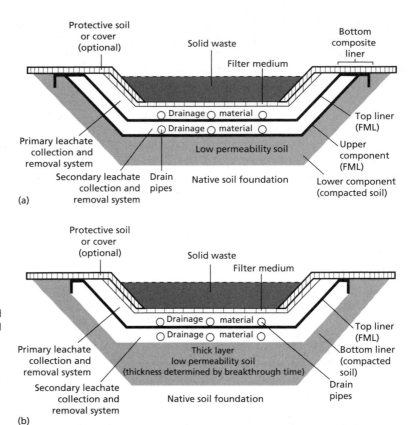

Fig. 14.7 Double liner systems proposed by the 1985 United States Environmental Protection Agency (USEPA) guidelines. The leachate collection layer is also considered to function as the geomembrane protection layer. (a) FML/composite double liner; (b) FML/compacted soil double liner. (After Mitchell, 1986.)

parameters, including the physical condition, composition and nature of the host formation. In addition, the regulatory authorities impose conditions and limitations on the disposal of hazardous wastes, in terms of the type and quantity of waste, and the level of activity for each particular repository.

Much lower level hazardous waste can be codisposed, that is, mixed, with very much larger quantities of 'non-hazardous' domestic waste, and buried in disused quarries, clay pits and other convenient holes in the ground (Chapman & Williams, 1987). The objectives of codisposal are to absorb, dilute and neutralize liquids, and to provide a source of biodegradable materials to encourage microbial activity. In other words, codisposal makes use of the attenuation processes in a landfill to minimize the impact of hazardous waste on the environment.

The disposal of liquid hazardous waste has also been undertaken by injection into deep wells located in rock below fresh water aquifers, thereby ensuring

that contamination or pollution of underground water supplies does not occur. In such instances, the waste generally is injected into a permeable bed of rock several hundreds or even thousands of metres below the surface, which is confined by relatively impervious formations. However, even where geological conditions are favourable for deep well disposal, the space for waste disposal frequently is restricted, and the potential injection zones usually are occupied by connate water. Accordingly, any potential formation into which waste can be injected must possess sufficient porosity, permeability, volume and confinement to guarantee safe injection. The piezometric pressure in the injection zone influences the rate at which the reservoir can accept liquid waste. A further point to consider is the fact that induced seismic activity has been associated with the disposal of fluids in deep wells. Two important geological factors relating to the cost of construction of a well are its depth and the ease with which it can be drilled.

Monitoring especially is important in deep well disposal involving toxic or hazardous materials. A system of observation wells sunk into the subsurface reservoir concerned, in the vicinity of the disposal well, allows the movement of liquid waste to be monitored. In addition, shallow wells sunk into fresh water aquifers permit the water quality to be monitored, so that any upward migration of the waste can be noted readily. Effective monitoring requires that the geological and hydrogeological conditions are accurately evaluated and mapped before the disposal programme is started.

14.4 Radioactive waste

Radioactive waste may be of low, intermediate or high level. Low-level waste contains small amounts of radioactivity, and so does not present a significant environmental hazard if properly dealt with. Intermediate-level waste comes from nuclear plant operations, and consists of items such as filters used to purify reactor water, discarded tools and replaced parts. This waste has to be stored for approximately 100 years. When reactors are closed down, decommissioning involves the safe disposal of high-level waste.

Although many would not agree, Chapman and Williams (1987) maintained that low-level radioactive waste can be disposed of safely by burying in carefully controlled and monitored sites where the hydrogeological and geological conditions severely limit the migration of radioactive material (Fig. 14.8). According to Rogers (1994), the most recent trend in low-level radioactive waste disposal in the USA is to provide engineering (concrete) barriers in the disposal facility, which prevent the short-term release of contamination. Such disposal facilities include below-ground vaults, above-ground vaults, modular concrete canisters placed in trenches, earth-mounded concrete bunkers, augered holes and mined cavities.

High-level radioactive waste unfortunately cannot be made non-radioactive, and so disposal must take into account the continuing emission of radiation. Furthermore, as radioactive decay occurs, the resulting daughter product is chemically different from the parent product. The daughter product may also be radioactive, but its decay mechanism may differ from that of the parent. This is of particular importance in the storage and disposal of radioactive material (Krauskopf, 1988). The half-life of a radioactive substance determines the time for which a hazardous waste must be stored for its activity to be reduced by half. However, if a radioactive material decays to form another unstable isotope, and if the half-life of the daughter product is longer than that of the parent, although the activity of the parent will decline with time, that of the daughter will increase because it decays more slowly. Consequently, the storage time of radioactive wastes must consider the half-lives of the products which result from decay.

Fig. 14.8 Waste disposal facility for low-level radioactive waste in southern Spain

In general, two types of high-level radioactive waste are being produced, namely, spent fuel rods from nuclear reactors and reprocessed waste. At present, both kinds of waste are isolated adequately from the environment in container systems. However, because much radioactive waste remains hazardous for hundreds or thousands of years, it should be disposed of far from the surface environment where it will require no monitoring. Baillieul (1987) reviewed various suggestions which have been advanced to meet this end, such as disposal in ice sheets, disposal in the ocean depths, disposal on remote islands and even disposal in space.

Ice sheets have been suggested as a repository for isolating high-level radioactive waste. The presumed advantages are disposal in a cold, remote area in a material which would entomb the wastes for many thousands of years. The high cost, adverse climate and uncertainties of ice dynamics are factors which do not favour such a means of disposal.

Disposal on the deep sea bed involves emplacement in sedimentary deposits at the bottom of the sea (i.e. thousands of metres beneath the surface). Such deposits have a sorptive capacity for many radionuclides that might leach from breached waste packages. In addition, if any radionuclides escaped, they would be diluted by dispersal. Currently, however, disposal of radioactive waste beneath the sea floor is prohibited by international convention.

The disposal of radioactive waste on a remote island involves the emplacement of waste within deep, stable, geological formations. It also relies upon the unique hydrological system associated with island geology. The remoteness, of course, is an advantage in terms of isolation.

The rock melt concept involves the direct emplacement of liquids or slurries of high-level wastes or dissolved spent fuel in underground cavities. After evaporation of the slurry water, the heat from radioactive decay would melt the surrounding rock. In about 1000 years, the waste–rock mixture would resolidify, trapping the radioactive material in a relatively insoluble matrix deep underground. The rock melt concept, however, is suitable only for certain types of waste. Moreover, because solidification takes about 1000 years, the waste is most mobile during the period of greatest fission product hazard.

The most favoured method, because it probably will give rise to the least problems subsequently, is disposal in chambers excavated deep within the Earth's crust in geologically acceptable conditions (Morfeldt, 1989). Deep disposal of high-level radioactive waste involves the multiple barrier concept, which is based upon the principle that uncertainties in performance can be minimized by conservation in design (Horseman & Volckaert, 1996). In other words, a number of barriers, both natural and man-made, exist between the waste and the surface environment. These can include encapsulation of the waste, waste containers, engineered barriers such as backfills and the geological host rocks which are of low permeability.

A deep disposal repository will consist of a large, underground system, located at least 200 m, and preferably 300 m or more, beneath the ground surface, in which there is a complex of horizontally connected tunnels for transportation, ventilation and the emplacement of high-level radioactive waste (Eriksson, 1989). It will also require a series of inclined tunnels and vertical shafts to connect the repository with the surface. Ideally, the waste should be so well entombed that none will reappear at the surface or, if it does so, the amounts will be minute enough to be acceptable.

The necessary safety of a permanent repository for radioactive wastes must be demonstrated by a site analysis, which takes into account the site geology and the type of waste, and their interrelationship (Langer, 1989). The site analysis must assess the thermomechanical load capacity of the host rock, so that disposal strategies can be determined. It must determine the safe dimensions of an underground chamber, and evaluate the barrier systems to be used. According to the multiple barrier concept, the geological setting for a waste repository must be able to make an appreciable contribution to the isolation of the waste over a long period of time. Hence, the geological and tectonic stability (e.g. mass movement or earthquakes), the load-bearing capacity (e.g. settlement or cavern stability) and the geochemical and hydrogeological development (e.g. groundwater movement and potential for dissolution of rock) are important aspects of safety assessment. Disposal would be best performed in a geological environment with little or no groundwater circulation, as groundwater is the most probable means of moving waste from the repository to the biosphere. The geological system should be complemented by multiple engineered barriers, such as the waste container, buffer materials and backfill.

If not already in solid form (e.g. spent fuel rods), waste should be treated to convert it into solid, ideally non-leachable, material. A variety of different solidification process materials have been proposed, including cement, concrete, plaster, glass and polymers. Currently, borosilicate glass is the most popular agent, as it can incorporate wastes of varying composition and has a low solubility in water. The glass would be placed within a metal canister and surrounded by cement or clay. The purpose of the container system is to provide a shield against radiation, and so it must be corrosion resistant.

The metal canisters would be stored in deep, underground caverns excavated in relatively impermeable rock types in geologically stable areas, that is, areas which do not experience volcanic activity, in which there is a minimum risk of seismic disturbance and which are not likely to undergo significant erosion (Fig. 14.9). Although earthquakes may represent a potential risk factor for rock chambers and tunnels, experience in mines in earthquake-prone regions has shown that vibrations in rock decrease with depth. Deep, structural basins are considered as possible locations.

Stress redistribution due to subsurface excavation and possible thermally induced stresses should not

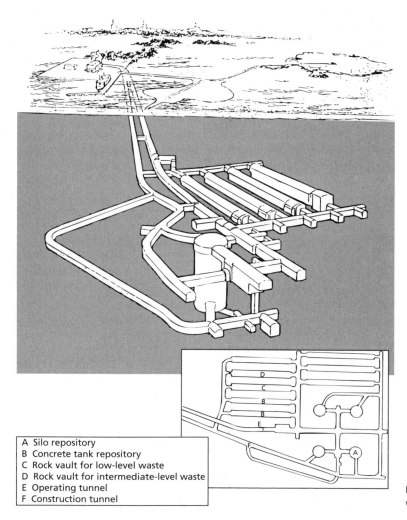

A Silo repository
B Concrete tank repository
C Rock vault for low-level waste
D Rock vault for intermediate-level waste
E Operating tunnel
F Construction tunnel

Fig. 14.9 Final repository for reactor waste. (After Morfeldt, 1989.)

endanger the state of equilibrium in the rock mass, and should not give rise to any inadmissible convergence or support damage during the operative period. The long-term integrity of the rock formations must be assured. Therefore, it is necessary to determine the distribution of stress and deformation in the host rock of the repository. This may involve a consideration of the temperature-dependent rheological properties of the rock mass, in order to compare them with its load-bearing capacity. Obviously, substantial strength is necessary for the engineering design of subsurface repository facilities, especially in maintaining the integrity of underground openings.

Completely impermeable rock masses are unlikely to exist, although many rock types may be regarded as practically impermeable, such as large igneous rock massifs, thick sedimentary sequences, metamorphic rocks and rock salt. The permeability of a rock mass depends mainly on the discontinuities present, their surfaces, width and amount of infill and their intersections. The repository needs to be watertight to prevent the transport of radionuclides by groundwater to the surface. Control and test pumping of the groundwater system and sealing by injection techniques may be necessary.

Rock types, such as thick deposits of salt or shale or granites or basalts at depths of 300–500 m, are regarded as the most feasible for the excavation of caverns for the disposal of high-level radioactive waste. Once a repository is fully loaded, it can be backfilled, with the shafts being sealed to prevent the intrusion of water. Once sealed, the system can be regarded as being isolated from the human environment.

As far as the disposal of high-level radioactive waste in caverns is concerned, thick deposits of salt have certain advantages. Salt has a high thermal conductivity, and so will rapidly dissipate any heat associated with high-level nuclear waste, it is 'plastic' at proposed repository depths, so that any fractures which may develop due to construction operations will 'self-heal', it possesses γ-ray protection similar to concrete, it undergoes only minor changes when subjected to radioactivity and it tends not to provide pathways of escape for fluids or gases (Langer & Wallner, 1988). Other attractive features of a deep salt deposit include the lack of water and the inability of water from an external source to move through it.

These advantages may be compromised if the salt contains numerous interbedded clay or mudstone horizons, open cavities containing brine or faults cutting the salt beds, so providing conduits for external water. The solubility of salt requires that unsaturated waters are totally isolated from underground openings in beds of salt by watertight linings, isolation seals or cutoffs and/or by collection systems. If suitable precautions are not taken in more soluble horizons in salt, any dissolution which occurs can lead to the irregular development of a cavern being excavated. Any water which does accumulate in salt will be a concentrated brine which, no doubt, will be corrosive to metal canisters. The potential for heavy groundwater inflows during shaft sinking requires the use of grouting or ground freezing. Rock salt is a visco-plastic material which exhibits short- and long-term creep. Hence, caverns in rock salt are subject to convergence as a result of plastic deformation of the salt. The rate of convergence increases with increasing temperature and stress in the surrounding rock mass. Because the temperature and stress increase with depth, convergence also increases with depth. If creep deformation is not restrained or not compensated for by other means, excessive rock pressures can develop on a lining system that may approach full overburden pressure.

Not all shales are suitable for the excavation of underground caverns, soft compacted shales present difficulties in terms of wall and roof stability. Caverns may also be subject to floor heave. Caverns could be excavated in competent cemented shales. Not only do these possess low permeability, they may also adsorb ions which move through them. A possible disadvantage is that, if temperatures in the cavern exceeded 100 °C, clay minerals could lose water and therefore shrink. This could lead to the development of fractures. In addition, the adsorption capacity of the shale would be reduced.

Granite is less easy to excavate than rock salt or shale, but is less likely to offer problems of cavern support. It provides a more than adequate shield against radiation, and will disperse any heat produced by radioactive waste. The quantity of groundwater in granite masses is small, and its composition generally is non-corrosive. However, fissure and shear zones occur within granites, along which copious quantities of groundwater can flow. Discontinuities tend to close with depth and faults may be sealed. For the location and design of a repository for spent nuclear fuel, the objective is to find 'solid blocks' which are large enough to host the tunnels and caverns of the repository. The Precambrian shields represent stable granite–gneiss regions.

The large thicknesses of lava flows in basalt plateaux mean that such successions could also be considered for disposal sites. Like granite, basalt can also act as a shield against radiation and can disperse heat. Frequently, the contact between flows is tight, and little pyroclastic material is present. Joints may not be well developed at depth and, strengthwise, basalt should support a cavern. However, the durability of basalts on exposure may be suspect, which could give rise to spalling from the perimeter of a cavern (Haskins & Bell, 1995). Furthermore, such basalt formations can be interrupted by feeder dykes and sills, which may be associated with groundwater. Groundwater generally is mildly alkaline with a low redox potential.

The emplacement of encapsulated nuclear waste in drill-holes as deep as 1000 m in stable rock formations cannot dispose of high volumes of waste. Similarly, the injection of liquid waste into porous or fractured strata, at depths from 1000 to 5000 m, can only accommodate limited quantities of waste. Such waste must be suitably isolated by relatively impermeable overlying strata, and disposal relies on the dispersal and diffusion of the liquid waste through the host rock. The limits of diffusion need to be well defined. Alternatively, thick beds of shale, at depths between 300 and 500 m, can be fractured by high-pressure injection, and waste, mixed with cement or clay grout, can be injected into the fractured shale and allowed to solidify in place. The fractures need to be produced parallel to the bedding planes. This requirement limits the depth of injection. The concept is applicable only to reprocessed wastes, or to spent fuel that has been processed in liquid or slurry form. This type of disposal, like that of deep well injection, can only dispose of limited amounts of waste.

14.5 Contaminated land

In many of the industrialized countries of the world, one of the legacies of the past two centuries is that land has been contaminated. The reason for this is that industry and society have tended to dispose of their waste with little regard for future consequences.

Hence, when such sites are cleared for the redevelopment of urban areas, they can pose problems. Contamination can take many forms and can be variable in nature across a site.

Contaminated land, because of its nature or former uses, is land which contains substances that, when present in sufficient quantities or concentrations, are likely to cause harm, directly or indirectly, to humans or the environment, or to give rise to hazards likely to affect a proposed form of development. The degree of hazard depends upon the mobility of the contaminant(s) within the ground, and different types of soil have different degrees of reactivity to compounds which are introduced. The risk is influenced by the future use of a site. However, at present, there is no consensus as to what is an acceptable risk.

The UK Government maintains a commitment to the 'suitable for use' approach to the control and treatment of existing contaminated land (Bell *et al.*, 1996a). This supports sustainable development by reducing damage from past activities and by permitting contaminated land to be kept in, or returned to, beneficial use wherever possible. Such an approach only requires remedial action when the contamination poses unacceptable actual or potential risks to health or the environment, and where there are appropriate and cost-effective means available to do so. Guideline values for concentrations of contaminants in soil may be used to indicate that a possibility of significant harm exists. For example, the Interdepartmental Committee on the Redevelopment of Contaminated Land (ICRCL, 1987) has proposed certain threshold limits (Table 14.3). However, there are no reference limits for certain organic contaminants in the ICRCL (1987) guidelines, and they contain no standards on groundwater quality. In the latter case, the practitioner has to fall back on drinking water abstraction standards. The ICRCL (1987) guidelines do attempt to relate hazards to land uses, thereby differing from assessment systems based only on concentration limits. In other words, they recognize lower thresholds for certain uses, such as residential developments with gardens, than for hard cover areas. Nonetheless, when using guideline values, it must be demonstrated that the values or assumptions underlying them are appropriate for use, and are characteristic of the land and the ecosystems or property in question (Anon., 1995). Even so, in any given circumstances, the hazards which arise from contaminated land will be peculiar to the site and will differ in significance. This is the principal reason why generally applicable criteria for the assessment of a contaminated site have not yet been developed.

In Europe, by comparison, the Dutch attitude towards contaminated land is more stringent. The Dutch recognize the possibility of a regular change in land use, and have insisted that, when contaminated land is redeveloped, the clean-up involved has to return the land to a standard which would allow any future use of the site in question. However, the Dutch have found that it has been impossible to organize, fund and execute all the clean-ups which have been deemed necessary (Cairney, 1993). Indeed, the concept of total clean-up is a standard of excellence which, in practice, usually is cost prohibitive. Hence, standards of relevance become a necessary prerequisite in order to avoid negative land values, which would mean that remedial action would not take place.

The presence of potentially harmful substances at a site may not necessarily require remedial action, if it can be demonstrated that they are inaccessible to living things or materials which may be detrimentally affected. However, consideration must always be given to the migration of soluble substances. The migration of soil-borne contaminants primarily is associated with groundwater movement, and the effectiveness of groundwater to transport contaminants is largely dependent upon their solubility. The quality of water can provide an indication of the mobility of contamination and the rate of dispersal. In an alkaline environment, the solubility of heavy metals becomes mainly neutral due to the formation of insoluble hydroxides. Provided that groundwater conditions remain substantially unchanged during the development of a site, the principal agent likely to bring about migration will be percolating surface water. On many sites, the risk of migration off site is of a very low order, because the compounds

Table 14.3 Guidance on the assessment and redevelopment of contaminated land. (After ICRCL, 1987.)

Contaminants	Use code	Reference value trigger concentrations (mg kg^{-1} air-dried soil)	
		Threshold	Action
Group A: selected inorganic contaminants that may pose hazards to health			
Arsenic	1	10	NS
	2	40	NS
Cadmium	1	3	NS
	2	15	NS
Chromium total	1	600	NS
	2	1000	NS
Chromium (hexavalent)*	1, 2	25	NS
Lead	1	500	NS
	2	2000	NS
Mercury	1	1	NS
	2	20	NS
Selenium	1	3	NS
	2	6	NS
Group B: contaminants that are phytotoxic, but not normally hazards to health			
Boron (water soluble)†	4	3	NS
Copper‡,‖	4	130	NS
Nickel‡,‖	4	70	NS
Zinc‡,‖	4	300	NS

Continued on p. 480

Table 14.3 *Continued.*

Contaminants	Use code	Reference value trigger concentrations (mg kg⁻¹ air-dried soil)	
		Threshold	Action
Contaminants associated with former coal carbonization sites			
Polyaromatic hydrocarbons¶, **,††	1	50	500
	3, 5, 6	1000	10 000
Phenols¶	1	5	200
	3, 5, 6	5	1000
Free cyanide¶	1, 3	25	500
	5, 6	25	500
Complex cyanides¶	1	250	1000
	3	250	5000
	5, 6	250	NL
Thiocyanate¶,††	All	50	NL
Sulphate¶	1, 3	2000	10 000
	5	2000‡‡	50 000‡‡
Sulphide	All	250	1000
Sulphur	All	500	20 000
Acidity (pH less than)	1, 3	pH 5	pH 3
	5, 6	NL	NL

Use codes: 1, domestic gardens and allotments; 2, parks, playing fields, open space; 3, landscaped areas; 4, any use where plants are grown (applies to contaminants that are phytotoxic, but not normally hazards to health); 5, buildings; 6, hard cover.

Conditions. (1) Tables are invalid if reproduced without the conditions and footnotes. (2) All values are for concentrations determined on 'spot' samples based on adequate site investigation carried out prior to development. They do not apply to the analysis of averaged, bulked or composited samples, nor to sites that have already been developed. (3) Many of these values are preliminary and will require regular updating. For contaminants associated with former coal carbonization sites, the values should not be applied without reference to the current edition of the report given by Anon. (1987). (4) If all sample values are below the threshold concentrations, the site may be regarded as uncontaminated as far as the hazards from these contaminants are concerned, and development may proceed. Above these concentrations, remedial action may be needed, especially if the contamination is still continuing. Above the action concentration, remedial action will be required or the form of development will need to be changed.

NS, not specified; NL, no limit set as the contaminant does not pose a particular hazard for this use.

* Soluble hexavalent chromium extracted by 0.1 M HCl at 37°C; solution adjusted to pH 1.0 if alkaline substances present.

† Determined by standard ADAS method (soluble in hot water).

‡ Total concentration (extraction by HNO₃/HClO₄).

‖ Total phytotoxic effects of copper, nickel and zinc may be additive. The trigger values given here are those applicable to the worst case phytotoxic effects that may occur at these concentrations in acid, sandy soils. In neutral or alkaline soils, phytotoxic effects are unlikely at these concentrations. The soil pH value is assessed to be about pH 6.5 and should be maintained at this value. If the pH falls, the toxic effects and uptake of these elements will be increased. Grass is more resistant to phytotoxic effects than most other plants, and its growth may be adversely affected at these concentrations.

¶ Many of these values are preliminary and will require regular updating. They should not be applied without reference to the current edition of the report given by Anon. (1987).

** Used here as a marker for coal tar, for analytical reasons (see Anon., 1987; Annex Al).

†† See Anon. (1987) for details of analytical methods.

‡‡ See also BRE Digest 250: *Concrete in Sulphate-Bearing Soils and Groundwater*. Building Research Establishment, Watford.

have low solubility and frequently most of their potential for leaching has been exhausted. Liquid and gas contaminants, of course, may be mobile. Obviously, care must be taken on site during working operations to avoid the release of contained contaminants (e.g. liquors in buried tanks) into the soil. Where methane has been produced in significant quantities, it can be oxidized by bacteria as it migrates through the ground with the production of carbon dioxide. However, this process is not necessarily continuous.

If any investigation of a site which is suspected of being contaminated is going to achieve its purpose, it must define its objectives and determine the level of data required, the investigation being designed to meet the specific needs of the project concerned. This is of greater importance when related to potentially contaminated land than to an ordinary site. According to Johnson (1994), all investigations of potentially contaminated sites should be approached in a staged manner. This allows for communication between interested parties, and helps to minimize costs and delays by facilitating the planning and progress of the investigation. After the completion of each stage, an assessment should be made of the degree of uncertainty and of acceptable risk in relation to the proposed new development. Such an assessment should be used to determine the necessity for, and type of, further investigation.

The first stage in any investigation of a site suspected of being contaminated is a desk study, which provides data for the design of the subsequent ground investigation. The desk study should be supplemented with a site inspection. This often is referred to as a land quality appraisal. The desk study should identify the past and present uses of the site and the surrounding area, and the potential for and likely forms of contamination. The objectives of the desk study are to identify any hazards and the primary targets likely to be at risk, to provide data for health and safety precautions for the site investigation and to identify any other factors which may act as constraints on development. Hence, the desk study should attempt to provide information on the layout of the site, including structures below ground, its physical features, the geology and hydrogeology of the site, the previous history of the site, the nature and quantities of the materials handled and the processing involved, health and safety records, and

the methods of waste disposal. It should allow a preliminary risk assessment to be made and the need for further investigation to be established.

The preliminary investigation should formulate objectives so that the work is cost effective. Such an investigation provides the data for the planning of the main field investigation, including the personnel and equipment needs, the sampling and analytical requirements, and the health and safety requirements. In other words, it is a fact finding stage, which should confirm the chief hazard and identify any additional ones, so that the main investigation can be carried out effectively. The preliminary investigation should also refer to any short-term or emergency measures that are required on the site before the commencement of full-scale operations.

As in a normal site investigation, the exploration of a site for contamination needs to determine the nature of the ground. In addition, it needs to assess the ability of the ground to transmit any contaminants either laterally or upward by capillary action. Permeability testing therefore is required. The investigation of contaminated sites frequently requires the use of a team of specialists and, without expert interpretation, many of the benefits of site investigation may be lost. Most sites require a careful interpretation of the ground investigation data. Sampling procedures are of particular importance, and the value of the data obtained therefrom is related to the representativeness of the samples. Some materials can change as a result of being disturbed when they are obtained or during handling. Hence, the sampling procedure should take into account the areas of the site which require sampling, the pattern, depth, types and numbers of samples to be collected, their handling, transport and storage, and the sample preparation and analytical methods. An assessment should be made as to whether protective clothing should be worn by site operatives.

The exploratory methods used in the main site investigation could include manual excavation, trenching and the use of trial pits, light cable percussion boring, power auger drilling, rotary drilling, and water and gas surveys. The excavation of pits and trenches is perhaps the most widely used technique for the investigation of contaminated land (Bell *et al.*, 1996a). Visually different materials should be placed in different stockpiles as a trench or pit is dug to facilitate sampling. The investigation must establish the location of perched water tables and aquifers, and any linkages between them, as well as the chemistry of the water on site.

Guidelines on how to conduct a sampling and testing programme are provided by the United States Environmental Protection Agency (USEPA, 1988) and Anon. (1988). A sampling plan should specify the objectives of the investigation, history of the site, analyses of any existing data, types of samples to be used, sample locations and frequency, analytical procedures and operational plan. Historical data should be used to ensure that any potential 'hot spots' are sampled satisfactorily. A sampling grid should be used in such areas (Anon., 1988; Sara, 1994).

Volatile contaminants or gas-producing material can be determined by sampling the soil atmosphere by using a hollow gas probe, inserted to the required depth. The probe is connected to a small vacuum pump and a flow of soil gas is induced. A sample is recovered by using a syringe. Care must be taken to determine the presence of gas at different horizons by the use of sealed response zones. Analysis usually is conducted on site by portable gas chromatography or photo-ionization. Standpipes can be used to monitor gases during the exploration work.

One of the factors which should be avoided is cross-contamination. This is the transfer of materials by the exploratory technique from one depth into a sample taken at a different depth. Consequently, cleaning requirements should be considered, and ideally a high specification of cleaning operation should be carried out on equipment between both sampling and borehole locations.

To ensure that the site investigation is conducted in the manner intended, and the correct data are recorded, the work should be carefully specified in advance. As the investigation proceeds, it may become apparent that the distribution of material about the site is not as predicted by the desk study or preliminary investigation. Hence, the site investigation strategy will need to be adjusted. The data obtained during the investigation must be accurately recorded in a manner whereby they can be understood subsequently. The testing programme should identify the types, distribution and concentration (severity) of contaminants, any significant variations or local anomalies. Comparisons with surrounding uncontaminated areas can be made.

Once completed, the site characterization process, when considered in conjunction with the development proposals, will enable the constraints on development to be identified. These constraints, however, cannot be based solely on the data obtained from the site investigation, but must take into account financial and legal considerations. If the hazard potential and associated risk are regarded as being too high, the development proposals will need to be reviewed. When the physical constraints and hazards have been assessed, a remediation programme can be designed, which allows the site to be economically and safely developed. It is at this stage when clean-up standards are specified in conjunction with the assessment of the contaminative regime on the surrounding area.

14.6 Remediation of contaminated land

Remedial planning and implementation are complex processes, which not only involve geotechnical methodology, but may be influenced by statutory and regulatory compliance. Indeed, it is becoming increasingly common for regulatory standards and guidance to provide very prescriptive procedural and technical requirements (Attewell, 1993; Holm, 1993).

A large number of technologies are available for the remediation of contaminated sites, and the applicability of a particular method will depend partly upon the site conditions, the type and extent of contamination and the extent of the remediation required. When the acceptance criteria are demanding and the degree of contamination is complex, it is important that the feasibility of the remediation technology is tested in order to ensure that the design objectives can be satisfied (Swane *et al.*, 1993). This can necessitate a thorough laboratory testing programme, together with field trials. In some cases, the remediation operation requires the employment of more than one method.

It obviously is important that the remedial works do not give rise to unacceptable levels of pollution, either on site or in the immediate surroundings. Hence, the design of the remedial works should include measures to control pollution during the operation. The effectiveness of the pollution control measures needs to be monitored throughout the remediation programme.

In order to verify that the remediation operation has complied with the clean-up acceptance criteria for the site, a further sampling and testing programme is required (USEPA, 1989). Swane *et al.* (1993) recommended that it is advisable to check that the clean-up standards are being attained as the site is being rehabilitated. In this way, any parts of the site which fail to meet the criteria can be dealt with there and then, so improving the construction schedule.

14.6.1 Soil remediation

Most of the remediation technologies differ in their applicability to treat particular contaminants in the soil. Landfill disposal and containment are the exceptions in that they are capable of dealing with most soil contamination problems. However, the removal of contaminated soil from a site for disposal in a special landfill facility transfers the problem from one location to another. Hence, the treatment of sites in place, where feasible, is the better course of action. Containment is used to isolate contaminated sites from the environment by the installation of a barrier system, such as a cutoff wall. In addition, containment may include a cover placed over the contaminated zone(s) to reduce the infiltration of surface water and to act as a separation layer between land users and the contaminated ground.

Soil washing involves the use of particle size fractionation, aqueous-based systems employing some type of mechanical and/or chemical process or countercurrent decantation, with solvents for organic contaminants and acids/bases or chelating agents for inorganic contaminants, to remove contaminants from excavated soils (Trost, 1993). Steam injection and stripping can be used to treat soils in the vadose zone contaminated with volatile compounds. One of the disadvantages is that some steam turns to water on cooling, which means that some contaminated water remains in the soil.

Solvents can be used to remove contaminants from the soil. For example, soil flushing makes use of water, water–surfactant mixtures, acids, bases, chelating agents, oxidizing agents and reducing agents to extract semi-volatile organic compounds, heavy metals and cyanide salts from the vadose zone of the soil. The technique is used in soils which are sufficiently permeable (not less than $10^{-7}\,\mathrm{m\,s^{-1}}$) to allow the solvent to permeate; the more homogeneous the soil, the better. The solvents are injected into the soil and the contaminated extractant is removed by pumping.

Vacuum extraction involves the removal of contaminants by the use of vacuum extraction wells. It can be applied to volatile organic compounds residing in the unsaturated soil or to volatile light non-aqueous phase liquids (LNAPLs) resting on the water table.

Fixation or solidification processes reduce the availability of contaminants to the environment by chemical reactions with additives (fixation) or by changing the phase (e.g. liquid to solid) of the contaminant by mixing with another medium. Such processes usually are applied to concentrated sources of hazardous wastes. Various cementing materials, such as Portland cement and quicklime, can be used to immobilize heavy metals.

Some contaminants can be removed from the soil by heating. For instance, soil can be heated to between 400 and 600 °C to drive off or decompose organic contaminants, such as hydrocarbons, solvents, volatile organic compounds and pesticides. Mobile units can be used on site, the soil being removed, treated and then returned as backfill.

Incineration, whereby wastes are heated to between 1500 and 2000 °C, is used for dealing with hazardous wastes containing halogenated organic compounds, such as PCBs and pesticides, which are difficult to remove by other techniques. Incineration involves the removal of the soil, which usually is then crushed and screened to provide fine material for firing. The ash which remains may require additional treatment, because heavy metal contamination may not have been removed by incineration. It is then disposed of in a landfill.

In situ vitrification transforms contaminated soil into a glassy mass. It involves the insertion of electrodes around the contaminated area and the application of a sufficient electric current to melt the soil (the required temperatures can vary from 1600 to 2000 °C). The volatile contaminants are either driven off or destroyed, and the non-volatile contaminants are encapsulated in the glassy mass when it solidifies.

Bioremediation involves the use of microbial organisms to bring about the degradation or transformation of contaminants so that they become harmless (Loehr, 1993). The microorganisms involved in the process either occur naturally or are introduced artificially. In the case of the former, the microbial action is stimulated by optimizing the conditions necessary for growth (Singleton & Burke, 1994). The principal use of bioremediation is in the degradation and destruction of organic contaminants, although it has also been used to convert some heavy metal compounds into less toxic states. Bioremediation can be carried out on ground *in situ*, or ground can be removed for treatment.

14.6.2 Groundwater remediation

Contaminated groundwater can either be treated *in situ* or can be abstracted and treated. The solubility in water and the volatility of the contaminants influence the selection of the remedial technique used. Some organic liquids are only slightly soluble in water and are immiscible. These are known as non-aqueous phase liquids (NAPLs). When dense they are referred to as DNAPLs or 'sinkers' and when light as LNAPLs or 'floaters'. Examples of the former include many chlorinated hydrocarbons, such as trichloroethylene and trichloroethane, whereas examples of the latter include petrol, diesel oil and paraffin. The permeability of the ground influences the rate at which contaminated groundwater moves, and therefore the ease and rate at which it can be extracted.

The pump and treat method is the most widely used means of the remediation of contaminated groundwater (Haley *et al.*, 1991). The latter is abstracted from the aquifer concerned by wells, trenches or pits and treated at the surface. It is then injected back into the aquifer. The pump and treat method proves to be most successful when the contaminants are highly soluble and are not readily adsorbed by clay minerals in the ground. In particular, LNAPLs can be separated from the groundwater either by using a skimming pump in a well or at the surface using oil–water separators. It usually is not possible to remove all light oil in this way, so that other techniques may be required to treat the residual hydrocarbons. Oily substances and synthetic organic compounds normally are much more difficult to remove from an aquifer. In fact, the successful removal of DNAPLs is impossible at present. They can be dealt with by containment. The methods of treatment used to remove contaminants which are dissolved in water, once it has been abstracted,

include standard water treatment techniques, air stripping of volatiles, carbon adsorption, microfiltration and bioremediation (Swane *et al.*, 1993).

Active containment refers to the isolation or hydrodynamic control of contaminated groundwater. The process makes use of pumping and recharge systems to develop 'zones of stagnation', or to alter the flow pattern of the groundwater. Cutoff walls are used in passive containment to isolate the contaminated groundwater.

Air sparging is a type of *in situ* air stripping, in which air is forced under pressure through an aquifer in order to remove volatile organic contaminants. It also enhances the desorption and bioremediation of contaminants in saturated soils. The air is removed from the ground by soil venting systems. The injection points, especially where contamination occurs at shallow depth, are located beneath the area affected (Waters, 1994). The air which is vented may have to be collected for further treatment as it may be hazardous.

In situ bioremediation makes use of microbial activity to degrade organic contaminants in the groundwater so that they become non-toxic. Oxygen and nutrients are introduced into an aquifer to stimulate activity in aerobic bioremediation, whereas methane and nutrients may be introduced in anaerobic bioremediation.

References

Anon. (1986) Please leave space for missing reference.

Anon. (1987) *Problems Arising from the Redevelopment of Gas Works and Similar Sites*, 2nd edn. Department of the Environment, London.

Anon. (1988) *Draft for Development, DD175: 1988, Code of Practice for the Identification of Potentially Contaminated Land and its Investigation*. British Standards Institution, London.

Anon. (1993) *South African Water Quality Guidelines*, Volume 1. Domestic use. Department of Water Affairs and Forestry, Pretoria.

Anon. (1995) *A Guide to Risk Assessment and Risk Management for Environmental Protection*. Department of the Environment, Her Majesty's Stationery Office, London.

Attewell, P.B. (1993) *Ground Pollution; Environmental Geology, Engineering and Law*. Spon, London.

Baillieul, T.A. (1987) Disposal of high level nuclear waste in America. *Bulletin of the Association of Engineering Geologists* **24**, 207–216.

Barber, C. (1982) *Domestic Waste and Leachate. Notes on Water Research No.* **31**. Water Research Centre, Medmenham.

Bell, F.G. & Mitchell, J.K. (1986) Control of groundwater by exclusion. In *Groundwater in Engineering Geology, Engineering Geology Special Publication No.* **3**, Cripps, J.C., Bell, F.G. & Culshaw, M.G. (eds). Geological Society, London, pp. 429–443.

Bell, F.G., Bell, A.W., Duane, M.J. & Hytiris, N. (1996a) Contaminated land: the British position and some case histories. *Environmental and Engineering Geoscience* **2**, 355–368.

Bell, F.G., Sillito, A.J. & Jermy, C.A. (1996b) Landfills and associated leachate in the greater Durban area: two case histories. In *Engineering Geology of Waste Disposal, Engineering Geology Special Publication No.* **11**, Bentley, S.F. (ed). Geological Society, London, pp. 15–35.

Cairney, T. (1993) International responses. In *Contaminated Land, Problems and Solutions*, Cairney, T. (ed). Blackie, Glasgow, pp. 1–6.

Chapman, N.A. & Williams, G.M. (1987) Hazardous and radioactive waste management: a case of dual standards? In *Planning and Engineering Geology, Engineering Geology Special Publication No.* **4**, Culshaw, M.G., Bell, F.G., Cripps, J.C. & O'Hara, M. (eds). Geological Society, London, pp. 489–493.

Daniel, D.E. (1993) Clay cores. In *Geotechnical Practice for Waste Disposal*, Daniel, D.E. (ed). Chapman and Hall, London, pp. 137–162.

Daniel, D.E. (1995) Pollution prevention in landfills using engineered final covers. In *Proceedings of the Symposium on Waste Disposal by Landfill, Green '93—Geotechnics Related to the Environment, Bolton*, Sarsby, R.W. (ed). Balkema, Rotterdam, pp. 73–92.

Eriksson, L.G. (1989) Underground disposal of high level nuclear waste in the United States of America. *Bulletin of the International Association of Engineering Geology* **39**, 35–62.

Fang, H.Y. (1995) Engineering behaviour of urban refuse, compaction control and slope stability analysis of landfill. In *Proceedings of the Symposium on Waste Disposal by Landfill, Green '93—Geotechnics Related to the Environment, Bolton*, Sarsby, R.W. (ed). Balkema, Rotterdam, pp. 47–62.

Hagerty, D.J. & Pavori, J.L. (1973) Geologic aspects of landfill refuse disposal. *Engineering Geology* **7**, 219–230.

Haley, J.L., Hanson, B., Enfield, C. & Glass, J. (1991) Evaluating the effectiveness of groundwater extraction systems. *Ground Water Monitoring Review* **12**, 119–124.

Haskins, D.R. & Bell, F.G. (1995) Drakensberg basalts: their alteration, breakdown and durability. *Quarterly Journal of Engineering Geology* **28**, 287–302.

Holm, L.A. (1993) Strategies for remediation. In *Geotechnical Practice for Waste Disposal*, Daniel, D.E. (ed). Chapman and Hall, London, pp. 289–310.

Horseman, S.T. & Volckaert, G. (1996) Disposal of radioactive wastes in argillaceous formations. In *Engineering Geology of Waste Disposal, Engineering Geology Special Publication No.* **11**, Bentley, S.P. (ed). Geological Society, London, pp. 179–191.

ICRCL (1987) *Guidelines on the Assessment and Redevelopment of Contaminated Land: Guidance Note 59/83. Interdepartmental Committee on the Redevelopment of Contaminated Land*, 2nd edn. Department of the Environment, Her Majesty's Stationery Office, London.

Jessberger, H.L., Manassero, M., Sayez, B. & Street, A. (1995) Engineering waste disposal (geotechnics of landfill design and remedial works). In *Proceedings of the Symposium on Waste Disposal by Landfill, Green '93—Geotechnics Related to the Environment, Bolton*, Sarsby, R.W. (ed). Balkema, Rotterdam, pp. 21–33.

Johnson, A.C. (1994) Site investigation for development on contaminated sites—how, why and when? In *Proceedings of the Third International Conference on Re-use of Contaminated Land and Landfills, London*, Forde, M.C. (ed). Engineering Technics Press, Edinburgh, pp. 3–7.

Koerner, R.M. (1993) Geomembrane liners. In *Geotechnical Practice for Waste Disposal*, Daniel, D.E. (ed). Chapman and Hall, London, pp. 164–186.

Krauskopf, K.B. (1988) *Radioactive Waste Disposal and Geology*. Chapman and Hall, London.

Langer, M. (1989) Waste disposal in the Federal Republic of Germany: concepts, criteria, scientific investigations. *Bulletin of the International Association of Engineering Geology* **39**, 53–58.

Langer, M. & Wallner, M. (1988) Solution-mined salt caverns for the disposal of hazardous chemical wastes. *Bulletin of the International Association of Engineering Geology* **37**, 61–70.

Loehr, R.C. (1993) Bioremediation of soils. In *Geotechnical Practice for Waste Disposal*, Daniel, D.E. (ed). Chapman and Hall, London, pp. 520–550.

Mitchell, J.K. (1986) Hazardous waste containment. In *Groundwater in Engineering Geology, Engineering Geology Special Publication No.* **3**, Cripps, J.C., Bell, F.G. & Culshaw, M.G. (eds). Geological Society, London, pp. 145–157.

Mitchell, J.K. (1994) Physical barriers for waste containment. In *Proceedings of the First International Congress on Environmental Geotechnics, Edmonton*, Carrier, W.D. (ed). BiTech Publishers Ltd, Richmond, BC, pp. 951–962.

Morfeldt, C.O. (1989) Different subsurface facilities for the geological disposal of radioactive waste (storage cycle) in Sweden. *Bulletin of the International Association of Engineering Geology* **39**, 25–34.

Oweis, I.S. & Khera, R.P. (1990) *Geotechnology of Waste Management*. Butterworths, London.

Raybould, J.G. & Anderson, J.G. (1987) Migration of landfill gas and its control by grouting—a case history. *Quarterly Journal of Engineering Geology* **20**, 78–83.

Rogers, V. (1994) Present trends in nuclear waste disposal.

In *Proceedings of the First International Congress on Environmental Geotechnics, Edmonton*, Carrier, W.D. (ed). BiTech Publishers Ltd, Richmond, BC, pp. 837–845.

Sara, M.N. (1994) *Standard Handbook for Solid and Hazardous Waste Facilities Assessment*. Lewis Publishers, Boca Raton, Florida.

Singleton, M. & Burke, G.K. (1994) Treatment of contaminated soil through multiple bioremediation technologies and geotechnical engineering. In *Proceedings of the Third International Conference on Re-use of Contaminated Land and Landfills, London*, Forde, M.C. (ed). Engineering Technics Press, Edinburgh, pp. 97–107.

Swane, I.C., Dunbavan, M. & Riddell, P. (1993) Remediation of contaminated sites in Australia. In *Proceedings of the Conference on Geotechnical Management of Waste and Contamination, Sydney*, Fell, R., Phillips, A. & Gerrard, C. (eds). Balkema, Rotterdam.

Trost, P.B. (1993) Soil washing. In *Geotechnical Practice for Waste Disposal*, Daniel, D.E. (ed). Chapman and Hall, London, pp. 585–603.

USEPA (1988) *Guidance for Conducting Remedial Investigations and Feasibility Studies under CERCLA*. Office of Emergency and Remedial Response, US Government Printing Office, Washington, DC.

USEPA (1989) *Methods for Evaluating the Attainment of Clean-up Standards*, Vol. 1: *Soils and Solid Media*. EPA/540/2–90/011. Office of Policy, Planning and Evaluation, US Government Printing Office, Washington, DC.

Waters, J. (1994) *In situ* remediation using air sparging and soil venting. In *Proceedings of the Third International Conference on Re-use of Contaminated Land and Landfills, London*, Forde, M.C. (ed). Engineering Technics Press, Edinburgh, pp. 109–112.

Williams, G.M. & Aitkenhead, N. (1991) Lessons from Loscoe: the uncontrolled migration of landfill gas. *Quarterly Journal of Engineering Geology* 24, 191–208.

15 Environmental geology and health

15.1 Introduction

The relationships between the environment, more specifically between the geochemistry of the environment, and the health of plants, animals and humans have been appreciated for a considerable period of time, and there is a growing awareness of the significance of such relationships. For instance, epidemiological studies have suggested that Balkan nephropathy, which is characterized by the progressive destruction of the kidneys, is not an infectious disease nor the result of a genetic fault. Cadmium, Pb, Ni and U have all, at one time or another, been considered as probable aetiological agents (Hopps, 1971). Excesses or deficiencies of certain elements within rocks, soils or water may be responsible for certain diseases in plants, animals and humans. Again, as an example, amyotrophic lateral sclerosis, according to Hopps (1971), is a progressive neurological disorder, which appears to be a disease that may reflect a deficiency of Mg or Mn. The decreased intake of Mg or Mn leads to a decreased ability to store and use thiamin (vitamin B_1). Hopps also noted that there is some evidence suggesting that Parkinson's disease may be causally related to Mn, but in this case, to an excess of Mn. Indeed, serious deficiencies or excesses of certain elements can result in crop failure or in the death of grazing animals. However, the relationships between the geochemical environment and human health are particularly complex.

Although geochemical anomalies which affect health occur naturally, humans themselves can adversely affect the environment by the disposal of waste, the indiscriminate use of fertilizers, pesticides and herbicides, chemical spillages, etc. The problems caused by humans tend to prevail in developed countries, whereas developing societies, which live more directly from the land, tend to suffer from different health problems. Hence, although developed countries may have virtually eliminated certain diseases, their peoples are more likely to suffer from diseases associated with the pollution of air, water or soil. Indeed, concern in industrial regions is focused on anthropogenic accumulations of potentially harmful elements, such as As, Cd, Hg and Pb, and on organic compounds, such as dichloro-diphenyl-trichloroethane (DDT), polychlorinated biphenyls (PCBs) and dioxins. Some of these chemicals can be classified as carcinogens, neurotoxins or irritants; others may cause reproductive failure or birth defects.

Nonetheless, the health problems associated with the geological environment are more acutely felt in the developing countries due to the added burden of poverty and malnutrition. Most of the people in these countries depend on local sources of food and water, so that any local geochemical anomaly can have a notable effect on the populace. Some examples of geochemical anomalies in soils and water, which can have a significant influence on the trace element uptake of developing and rural populations, are provided in Table 15.1. By contrast, the inhabitants of developed nations can include within their diets foods which come from different parts of the world.

Plants and animals, including humans, essentially are made up of 11 elements, namely, H, O, C, N, Ca, Mg, K, Na, P, S and Cl. In addition to these elements, certain others are required in trace amounts. Two main groups of trace elements are of particular importance as far as health is concerned. According to Mills (1996), those which are essential to animals are Fe, Mn, Ni, Co, Cr, Cu, Zn, V, Mo, Sn, Se, I and F. By contrast, potentially harmful elements, which have adverse physiological significance at relatively low levels, include As, Cd, Pb, Hg and some of the daughter products of U (Table 15.2). Aluminium can also have adverse physiological effects in trace amounts in plants and animals, especially in fish. In fact, all trace elements are toxic, or even lethal, if

Table 15.1 Typical geochemical and soil features associated with inorganic element anomalies causing nutritional diseases in humans and domesticated livestock. (After Mills, 1996.)

Syndrome	Environmental anomaly	Species affected*
Deficiencies		
Low cobalt	Soils intrinsically low in Co, e.g. extensively leached, acid arenaceous soils, or with Co immobilized with Fe/Mn hydroxide complexes	R
Low phosphorus	High Fe/Al parent materials with low pH and highly organic soils	R
Low selenium	Soils intrinsically low in Se, e.g. leached arenaceous soils, particularly when low in organic and argillaceous fractions. Fixation of Se in soils high in Fe	M, F
Low zinc	Calcareous parent materials and derived soils, especially when adventitious soil present in diets high in cereals and legumes. Arid arenaceous soils	M, F
Toxicities		
High arsenic	Waters from some hydrothermal sources or soils derived from detritus of mineral ore (especially Au) workings. Well waters or irrigation waters from sandstones high in arsenopyrite	
High fluoride	Waters from some aquifers, especially from rhyolite-rich rocks, black shales or coals; soils from F-containing residues of mineral or industrial deposits. Aggravated by high evaporative losses	M, F
High molybdenum	Mo from molybdeniferous shales or local mineralization, especially if drainage poor and soil pH > 6.5 (a significant cause of secondary Cu deficiency)	R
High selenium	Bioaccumulation of Se in organic-rich soil horizons; accumulation by high evaporative losses of high-pH groundwaters	M, F

* M, man; F, farm livestock, general; R, ruminant livestock, specific.

Table 15.2(a) Trace elements and humans.* (From Bowen, 1966.)

Element	Symbol	Intake level below which deficiencies occur	Normal intake	Intake level above which toxicity occurs	Lethal intake
Arsenic	As		0.1–0.3	5–50	100–300
Boron	B		10–30	4000	
Cadmium	Cd		0.5	3	
Chlorine	Cl	70	2400–4000		
Chromium	Cr		0.05	200	3000
Cobalt	Co		0.0002	500	
Copper	Cu		2–5	250–500	
Fluorine	F		0.5	20	2000
Iodine	I	0.015	0.2	1000	
Iron	Fe		12–15		
Lead	Pb		0.3–0.4		10000
Manganese	Mn		3–9		
Mercury	Hg		0.005–0.02		150–300
Molybdenum	Mo		0.5		
Selenium	Se	0.015	0.03–0.075	3.0	
Silver	Ag		0.06–0.08	60	1300
Sodium	Na	45	1600–2700		

All quantities in milligrams per day.
* Mean body weight, 70 kg; weight of dry diet, 750 g per day.

Table 15.2(b) Functions of the essential trace elements. (After Mertz, 1987.)

Element	Function	Deficiency signs		Occurrence of imbalances in humans
		Animals	Humans	
Fluorine	Structure of teeth, possibly of bones; possibly growth effect	Caries; possibly growth depression	Increased incidence of caries; possibly risk factor for osteoporosis	Deficiency and excess known
Silicon	Calcification; possibly function in connective tissue	Growth depression; bone deformities	Not known	Not known
Vanadium	Not known	Growth depression, change of lipid metabolism, impairment of reproduction	Not known	Not known
Chromium	Potentiation of insulin	Relative insulin resistance	Relative insulin resistance, impaired glucose tolerance, elevated serum lipids	Deficiency known in malnutrition, ageing, total parenteral alimentation
Manganese	Mucopolysaccharide metabolism, superoxide dismutase	Growth depression, bone deformities, β-cell degeneration	Not known	Deficiency not known; toxicity by inhalation
Iron	Oxygen, electron transport	Anaemia, growth retardation	Anaemia	Deficiencies widespread; excesses dangerous in haemochromatosis; acute poisoning
Cobalt	As part of vitamin B_{12}	Anaemia; growth retardation in ruminant species	Only as vitamin B_{12} deficiency	Inability to absorb vitamin B_{12}; low B_{12} intake from vegetarian diets
Nickel	Interaction with iron absorption	Growth depression, anaemia, ultrastructural changes in liver; impaired reproduction	Not known	Not known
Copper	Oxidative enzymes; interaction with iron; cross-linking of elastin	Anaemia, rupture of large vessels, disturbances of ossification	Anaemia, changes of ossification; possibly elevated serum cholesterol	Deficiencies in malnutrition, total parenteral alimentation
Zinc	Numerous enzymes involved in energy metabolism and in transcription and translation	Failure to eat, severe growth depression, skin lesions, sexual immaturity	Growth depression, sexual immaturity, skin lesions, depression of immunocompetence, change of taste acuity	Deficiencies in Iran, Egypt, in total parenteral nutrition, genetic diseases, traumatic stress
Arsenic	Not known	Impairment of growth, reproduction; sudden heart death in third generation lactating goats	Not known	Not known

Continued on p. 490

Table 15.2(b) *Continued.*

| Element | Function | Deficiency signs | | Occurrence of imbalances in humans |
		Animals	Humans	
Selenium	Glutathione peroxidase; interaction with heavy metals	Different, depending on species: muscle degeneration (ruminants), pancreas atrophy (chicken)	Endemic cardiomyopathy (Keshan disease) conditioned by selenium deficiency	Deficiency and excess in areas of China: one case resulting from total parenteral alimentation
Molybdenum	Xanthine, aldehyde, sulphide oxidases	Difficult to produce; growth depression	Not known	Excessive exposure in parts of Soviet Union associated with gout-like syndrome
Iodine	Constituent of thyroid hormones	Goitre, depression of thyroid function	Goitre, depression of thyroid function, cretinism	Deficiencies widespread; excessive intakes may lead to thyrotoxicosis

ingested or inhaled at sufficiently high levels for sufficiently long periods of time. A substance is toxic if it inhibits the growth or metabolism of an organism when present above a certain concentration. Cumulative toxins are substances which are retained by an organism, and so are particularly dangerous to deal with, Cd being an example. Synergistic interactions of toxic substances produce an effect which is much more notable than the effects of the individual substances. Selenium, F and Mo represent examples of elements which have relatively low concentration ranges between essential and toxic levels (i.e. around a few micrograms per gram). Yet other elements, such as B, are known to be essential to higher plants, but so far have not been shown to be necessary for animals (Thornton & Plant, 1980).

15.2 The occurrence of elements

The chemical form or speciation of elements affects their distribution, mobility and toxicity (Fig. 15.1). Some of the most important controls, especially on trace element speciation and mobility, according to Plant *et al.* (1996), include the pH, Eh, temperature, surface properties of solids, abundance and speciation of ligands, major cations and anions, presence or absence of dissolved and/or particulate organic matter, and biological activity. Of these, two of the most important factors which directly control the solubility and mobility are the pH and Eh (Fig. 15.2).

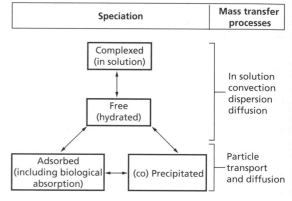

Fig. 15.1 Relationships between the speciation of chemical elements and mass transfer processes. (From Plant *et al.*, 1996.)

The solution chemistry of an element is affected profoundly by changes in the oxidation state, whereas dissolution reactions, including hydrolysis, inorganic complexation, complexation with smaller organic anions, such as oxalate, and sorption/desorption, are all pH controlled. Under high-pH conditions, anions and oxyanions (e.g. Te, Se, Mo, U, As, P and B) are more mobile, and cations (such as Cu, Pb, Hg and Cd) are less mobile. The converse usually is the case under low-pH conditions. Where organic matter is present, however, stable organometallic complexes are formed, increasing the trace element mobility.

Relative mobilities	Environmental conditions			
	Oxidizing	Acid	Neutral–alkaline	Reducing
Very high	I	I	I Mo *U* Se	I
High	Mo *U* Se F *Ra* Zn	Mo *U* Se F *Ra* Zn Cu Co Ni *Hg*	F *Ra*	F *Ra*
Medium	Cu Co Ni *Hg* *As Cd*	*As Cd*	*As Cd*	
Low	*Pb Be Bi Sb Tl*	*Pb Be Bi Sb Tl* Fe Mn	*Pb Be Bi Sb Tl* Fe Mn	Fe Mn
Very low to immobile	Fe Mn Al Cr	Al Cr	Al Cr Zn Cu Co Ni *Hg*	Al Cr Mo *U* Se Zn Cu Co Ni *Hg* *As Cd* *Pb Be Bi Sb Tl*

Fig. 15.2 General relationships between Eh, pH and the mobility of some essential and potentially toxic elements. Essential elements are shown in normal type and potentially hazardous elements in italic type. (From Plant *et al.*, 1996.)

The trace element mobility is governed by the physicochemical conditions and the interaction of other chemical constituents, but varies with each individual element (Edmunds & Smedley, 1996). For example, Al, Be, Cd, Fe, Mn and Pb are mobilized preferentially under acidic conditions. Fluoride is most mobile in alkaline conditions, given low dissolved Ca concentrations, and I may be a largely conservative element, depending mainly on the I concentrations derived from local sources. The complete reduction of oxygen in an aquifer is accompanied by a sharp decrease in the redox potential. Oxidizing conditions aid the mobility of some species, for example, NO_3, Se and U, whereas others such as Fe, Mn and As, possess increased mobility under reducing conditions.

A redox boundary is present at shallow depths, particularly in soils or aquifers with a high organic content or a high proportion of sulphide minerals, which act as the main substrate (electron donor) for the reduction of oxygen. On the other hand, oxidizing conditions may persist for thousands of years in sediments which are deficient in organic matter.

Acidic water does not pose a health risk, however, as many minerals are more soluble in acidic water, toxic trace metals (e.g. Al, Be, Cd and Pb) may be present at higher concentrations. In addition, most acidic waters are base-poor, soft waters, and their Ca and Mg deficiencies may also be implicated in various forms of heart disease (see Section 15.7).

All known natural elements are present in rocks, but the chemical compounds or minerals in which they occur differ between the major rock groups (Table 15.3). A variation of the elements obviously occurs from one rock type to another. For example, in Britain, Thornton and Plant (1980) indicated that areas underlain by acid igneous rocks or sandstones generally contain lower levels of essential trace elements, especially Fe, Mn, Ni, Co, V, Zn, Co and Cr, than areas underlain by basic and ultrabasic igneous rocks or shales. However, at times, these may contain sufficiently large concentrations of a potentially toxic element to be hazardous. As an illustration, Thornton and Plant mentioned levels of Ni and Cr in poorly drained soils derived from ultrabasic rocks in northeast Scotland, which give rise to toxicity in cereal crops, and black shales in Derbyshire, England, which may contain sufficiently high concentrations of Mo to cause disease in livestock. Increased levels of trace elements are associated with areas of metalliferous mineralization.

Most of the minerals which occur in igneous rocks, when subjected to chemical weathering, are altered considerably. Some elements may be lost during these changes, whereas others may be gained. However, the loss of specific elements in weathering and soil formation generally is incomplete, but mass movements,

Table 15.3 Average compositions of earth materials. (From Leckie & Parks, 1978.)

	Igneous rocks				Sedimentary rocks			Soils	Hydrosphere			Atmosphere
	Lithosphere	Acid	Basic	Ultrabasic	Shales	Sandstones	Limestones		Precipitation	Rivers	Oceans	
Major elements	(wt.%)								Major elements (p.p.m.)			Wt.%
Aluminium	8.2	7.7	8.8	0.45	8.0	3.2	0.9	4.5			0.01	Nil
Calcium	4.1	1.6	6.7	0.7	2.5	2.2	27.2	0.88		15	400	Nil
Iron	5.6	2.7	8.6	9.8	4.7	1.9	0.8	5.6		0.67	0.01	Nil
Magnesium	2.3	0.16	4.5	25.9	1.34	0.8	4.5	0.47	0.1	4.1	0.135	Nil
Potassium	2.1	3.3	0.83	0.03	2.3	1.3	0.2	1.2	0.05	2.3	380	Nil
Silicon	28.2	32.3	24.0	19.0	23.8	35.9	0.003				3	Nil
Sodium	2.4	2.8	1.9	0.57	0.66	0.4	0.04	0.4	0.5	6.3	10500	Nil
Titanium	0.57	0.23	0.9	0.03	0.45	0.2	0.04	0.25				Nil
Hydrogen	0.14	Nil	Nil	Nil	3.4	Nil	Nil				0.001	Variable
Oxygen	46.4	48.7	43.5	42.5	52.8	52.7	54.9					20.1 (O_2)
Selected minor elements												
	Parts per million (p.p.m.) except where indicated as parts per billion (p.p.b.)											Wt.%
Arsenic	1.8	1.5	2	0.5	6.6	1	0.9			0.013	0.003	Nil
Boron	10	15	5	1	100	90	16	26		1 to >10 p.p.b.	0.0006 p.p.b.	Nil
Cadmium	0.2	0.2	0.2	0.05	0.3	0.02	0.05	0.05–0.5		0.1 p.p.b.	0.11 p.p.b.	Nil
Carbon	200	300	100	100	1000	14000	114000		28	11	5.5	0.3 (CO_2)
Chromium	100	4	200	2000	100	120	7.1	37		0.1 to 10 p.p.b.	0.05 p.p.b.	Nil
Copper	55	10	100	20	57	15	4	18		10 p.p.b.	0.003 p.p.b.	Nil
Fluorine	625	850	400	100	500	220	112			<1	1.3	Nil
Lead	125	20	5	0.1	20	14	16	16		1 to 10 p.p.b.	0.03 p.p.b.	Nil
Manganese	950	400	1500	1500	850	392	842	340			0.002	Nil
Mercury	0.08	0.08	0.08	0.01	0.4	0.06	0.05	0.08		0.09 p.p.b.	0.03 p.p.b.	Nil
Molybdenum	1.5	2	1	0.2	2	0.5	0.8	2		<10	0.01	Nil
Nitrogen	20	20	20	6	60				0.3	0.23	0.5	78.1 (N_2)
Selenium	0.05	0.05	0.05	0.05	0.6	0.5	0.3				0.4 p.p.b.	Nil
Sulphur	260	270	250	100	220	945	4550		1	3.7	885	Nil
Uranium	2.7	4.8	0.6	0.003	3.2	1	2.2				0.003	Nil
Zinc	70	40	100	30	80	16	16	44		10	0.01	Nil

erosion and groundwater flow may spread traces of elements over areas many times larger than the zones of concentration in the parent rock(s).

The total trace element content of soils reflects, to varying degrees, that of the parent material. This relationship, however, can be modified by soil-forming processes, such as gleying, leaching, podzolization and the accumulation of organic matter. These processes, together with the soil pH, Eh and drainage, influence the forms and mobility of trace elements in soils and their availability to plants.

The climate has changed frequently in the geological past and, in some regions, has changed profoundly. Consequently, some regions have undergone a succession of different climates with different weathering and dispersion processes. At lower latitudes, especially in continental areas of low relief, the regolith may represent an expression of the cumulative effects of weathering throughout millions of years as, for example, in many parts of Africa. Some soils have been deeply leached, and generally tropical and subtropical soils, because they have been subjected to a number of weathering cycles, tend to be poorer, thinner and more fragile than those found in temperate regions. Nonetheless, weathering profiles may extend to depths in excess of 150 m in humid, tropical zones, so that the fresh supply of elements from rocks to the biosphere is limited. Laterites, including bauxites and ferricretes, and Fe and Al duricrusts have a homogeneous mineralogy and chemical composition, which usually bear little relationship to the underlying bedrock. The intense oxidation which is characteristic of such conditions frequently extends to the weathering front. This, in turn, means that organic matter is lacking, as are some of the major elements, such as N, P and K. Moreover, the formation of clay minerals, such as kaolinite, increases the Al to Si ratio. Such soils often are deficient in soluble ions, such as Na^+, Ca^{2+}, Cl^-, PO_4^{3-} and NO_3^-, but contain high levels of resistant oxides of Fe, Al, Ti and Mn. The soils typically are ferrallitic, with calcretes or silcretes occurring locally in areas of inland drainage affected by high evaporation rates and fluctuating water tables. Hardening of Al and Fe oxides to form cuirasses (hard pans), which reinforce the surface soil, is common in tropical regimes. Essential nutrients are leached further when natural vegetation is removed in these regions, and the topsoil easily is eroded. The soils in most desert regimes consist of detrital quartz, accessory minerals and clay minerals, with minimal organic matter. Surface waters are highly oxidizing, with variable but commonly high pH. Such waters may be saline, such that the chloride content is more important than that of bicarbonate.

Chemical elements tend to undergo more marked separation in tropical than in temperate regions, although this is governed by their chemical mobility (Plant *et al.*, 1996). Accordingly, areas in which a potential for deficiency or toxicity conditions exists generally are more common in the former rather than the latter regions. Tropical soils, in which deep leaching prevails, only retain the most inert elements, such as Fe, Mn, Zr, Hf and the rare earths, either as primary or secondary oxides. Conversely, the most mobile elements, namely, the alkali and alkaline earth elements, halogens and elements mobile in high-pH conditions, including anions and oxyanions such as those of B, Se, Mo, V and U, tend to accumulate in arid environments, in systems of inland drainage and near the base of weathered profiles. The variable oxidation states of the transition elements Co, Cr, Cu and Zn favour their removal from solution by ion exchange, precipitation and surface sorption. The lack of organic matter attributable to intense oxidation can mean that the total concentration of such elements in the regolith may be high, but they may be attached to Fe or Mn oxides. Consequently, their availability to plants may be exceptionally low which, in turn, can mean a potential deficiency as far as animals are concerned. This especially is the case in regions in which the surface has undergone deep oxidation.

Geochemical interaction between the lithosphere and hydrosphere, on the one hand, and the biosphere on the other depends partly on sorption processes and partly on chemical speciation (Thornton, 1983). Some elements, such as Al, Ti and Cr, are relatively poorly assimilated by plants, whereas others, for example, Cd, Se, Mo and Co, can readily cross from soil to plant and so enter the food chain. The sorption of elements by clay minerals and organic material in soils is the dominant fixing mechanism, with the soil pH controlling the sorption processes and metal solubility/bioavailability. Many metals become more soluble at low pH, induced by natural organic acids and root exudates.

Edmunds and Smedley (1996) noted that nearly all natural waters contain traces of most elements, but

TRACE ELEMENTS IN GROUNDWATER AND THEIR SIGNIFICANCE IN TERMS OF HEALTH AND ENVIRONMENTAL PROTECTION

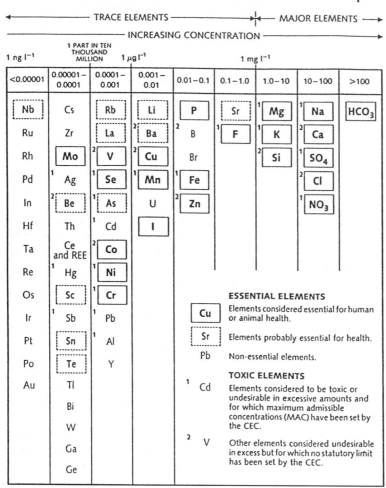

CONCENTRATIONS IN DILUTE, OXYGENATED GROUNDWATER AT pH 7

Fig. 15.3 Major and trace elements in groundwater and their significance in terms of health. Concentrations shown are those typical of dilute oxygenated groundwater at pH 7. Elements outlined are those considered to be essential (or probably essential) for health. Those elements which have guidelines or statutory limits set by the Commission of the European Community (CEC) or the World Health Organization (WHO) are indicated. (After Edmunds & Smedley, 1996.)

usually at extremely low or unquantifiable concentrations. Be that as it may, concentrations of naturally occurring mobile elements obviously can exceed the recommended maximum limits for potable waters, and/or may exceed the limits of general acceptability for domestic use. The typical concentrations of the elements in natural waters at pH 7 are given in Fig. 15.3. Those elements which, at present, are regarded as being essential for human health and metabolism are also indicated in Fig. 15.3, together with the elements which are considered to produce harmful effects if present above certain limits in drinking water (Commission of the European Community (CEC), 1980; World Health Organization (WHO), 1993). Nine major species (HCO_3, Na, Ca, SO_4, Cl, NO_3, Mg, K and Si) invariably account for more than 99% of the solute content of natural waters. However, the concentrations of minor and trace elements, below 1% of the total, can vary appreciably from those indicated, depending on the geochemical conditions. In particular, a reduction in the pH value of 1 may be responsible for an increase of more than

Table 15.4 Chemicals of health significance in drinking water. (From Edmunds & Smedley, 1996.)

	WHO (1993) guideline maxima (mg l⁻¹)	CEC (1980) guideline values (mg l⁻¹)	CEC (1980) maximum admissible concentrations (mg l⁻¹)
Antimony (Sb)	0.005 (P)		0.010
Arsenic (As)	0.01 (P)		0.050
Barium (Ba)	0.7	0.1	
Beryllium (Be)	NAD		
Boron (B)	0.3	1.0	
Cadmium (Cd)	0.003		0.005
Chromium (Cr)	0.05 (P)		0.050
Copper (Cu)	2 (P)	0.1	
Fluorine (F)	1.5		1.5*
Lead (Pb)	0.01		0.05
Manganese (Mn)	0.5 (P)	0.02	0.05
Mercury (Hg) total	0.001		0.001
Molybdenum (Mo)	0.07		
Nickel (Ni)	0.02		0.05
Nitrate (NO₃)	50	25	50
Selenium (Se)	0.01		0.01
Uranium (U)	NAD		

Data from WHO (1993), with values from CEC (1980) where appropriate. P, provisional value; NAD, no adequate data to permit recommendation of a health-based guideline value.
* Climatic conditions, volume of water consumed and intake from other sources should be considered when setting national standards.

one order of magnitude in the concentration of certain metals. A change from oxidizing to reducing conditions may have a similar effect on elements such as Fe. The currently agreed limits for inorganic constituents of importance in relation to health are summarized in Table 15.4. However, there are no agreed limits for some elements, such as Be and U, in spite of their effects on health.

The chemistry of surface waters and shallow groundwaters is influenced by the local geology. Reactions between rain water and bedrock, over a matter of days or months as infiltration and percolation occur, are responsible for the mineral content of groundwater. The extent to which reaction with the host rock proceeds is governed by the residence time of the water, which, in turn, may be influenced by the type of flow movement, that is, intergranular or fissure flow, and the mineralogy of the aquifer. The concentration of CO_2 in the soil influences the degree of reaction of carbonate or silicate minerals in a rock mass.

By contrast, deeper groundwaters can undergo notable changes in mineral composition with increas-

ing residence time. Accordingly, Edmunds *et al.* (1987) maintained that it therefore is necessary to consider the likely changes which take place between areas with different geology, together with the sequential changes that occur within an aquifer. They provided an example of health effects related to differences in water chemistry along flow gradients in aquifers by quoting the work of Lamont (1959). Lamont showed that changes in dental health could be related to groundwater drawn from the Lincolnshire Limestone, England, a confined aquifer. Once identified as a fluoride-related problem, the blending of high- and low-F waters was used to eliminate fluorosis and to provide an optimum concentration of F in the water supply.

The soil–plant relationship is affected by the plant species, stage of growth, season and the application of fertilizer and/or lime. Plants have a wide tolerance of many trace elements, and show marked species variability in the extent to which these elements are accumulated. The stage of maturity also affects the uptake of trace elements. For instance, the trace element compositions of cruciferous and leguminous crops generally fluctuate widely with different trace element contents of soils. On the other hand, the contents of trace elements in cereals, although frequently lower, are not affected so readily. However, the Se and, to a lesser extent, Zn contents of cereal crops are influenced significantly by the concentrations of these elements in the soil (Levander, 1986). In particular, maize concentrates Se to a greater extent than rice. In fact, in some regions of China, where maize residues have been used as fertilizer for crops grown on soils with a high content of Se, this has led to Se intoxication in animals and humans.

Investigations have shown that changes in soil and crop management can modify the uptake of trace elements by food crops, and so change trace element disease patterns. Aggett and Mills (1996) indicated that Cu can vary fourfold, Zn sevenfold, B 21-fold and Mo up to 46-fold depending on the soil and crop management techniques. In addition, excess soil moisture, arising from flooding or irrigation, in certain soil types, can enhance the uptake of Mo and Se. Mills (1995) pointed out that a high pH in soil can restrict the supply of Co, Fe and Zn, whilst aiding the accumulation of F, Mo and Se in plants.

The relationship between trace elements in plants and the amounts absorbed and utilized by animals is affected by the proportion of grass in the diet, the digestibility of the diet and the availability of ingested trace elements. Grazing animals may also ingest up to 10% of their dry matter intake as soil. In addition, the age and gender of animals exert a significant control on the uptake of trace elements.

The assimilation of elements in higher animals and humans occurs by the ingestion of nutrients and contaminants, by absorption through the skin and by respiration. Speciation in both the natural environment and in the gastrointestinal tract exerts a major influence on the uptake and assimilation of trace nutrients and potentially harmful elements.

Speciation studies, when used in relation to land degraded by mining, industrial activity or urbanization, can prove to be extremely useful in remediation schemes, and can lead to improved management practices. A knowledge of those factors which control speciation in different environments can aid the prediction of the potential for the absorption of potentially harmful elements and can act as a guide for the supplementation of trace elements for crops and animals.

15.3 Geochemical surveys and maps

Under favourable circumstances, geochemical mapping is capable of indicating changes in the chemical composition of bedrock, soils and water. It can also provide a cost-effective means of indirectly investigating the chemical composition of crops.

Ideally, comparable multielement data for rocks, soils and water should be determined during a geochemical survey. This is especially important, because new data on speciation, combined with computer modelling and geographical information systems (GISs), make possible the preparation of geochemical maps focused on environmental problems. The methods and requirements for the preparation of high-resolution geochemical baseline data have been reported by Darnley et al. (1995).

Geochemical reconnaissance by stream sediment sampling provides a ready means of mapping the distribution of elements on a regional scale. For example, Appleton (1992) demonstrated that, when levels in the fine fraction of stream sediments are considered on a regional basis, they can provide a good indication of likely soil concentrations.

Geochemical maps, together with geochemical data, can provide a source of ancillary information in epidemiological surveys, especially in rural areas and in the less well developed parts of the world (Webb, 1971). Such maps, according to Plant et al. (1996), have been used to identify areas of trace element deficiency or toxicity, thereby enabling agricultural, veterinary or medical investigations to be targeted more successfully (Fig. 15.4). Geochemical maps also provide basic data for investigations in areas contaminated by past and present mining and smelting operations, and from industrial and urban sources. Plant et al. also maintained that geochemical surveys of soil, stream sediments, and surface and groundwater are of considerable value in investigations which attempt to relate diet and health. The method is based on the assumption that stream sediment approximates to a composite sample of the products of weathering and erosion of the rocks and soils upstream of the sample. By analysing the trace element content of sediment samples, collected at densities ranging from 0.50 to 250 km^2, it often is possible to detect anomalous patterns in the distribution of a wide range of elements, and so to detect the incidence of certain trace elements which are respon-

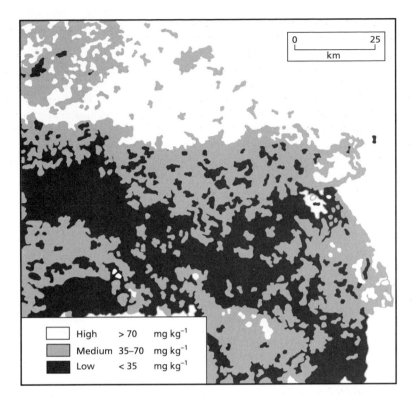

	High	> 70	mg kg^{-1}
	Medium	35–70	mg kg^{-1}
	Low	< 35	mg kg^{-1}

Fig. 15.4 Geochemical map of Zn in stream sediments in northeast Zimbabwe. (After Fordyce *et al.*, 1996.)

...ible for disorders in crops and livestock. Webb 1971) referred to the use of this technique to produce geochemical maps to show the relationship between soil geochemistry, Se toxicity and Mo-induced Cu deficiency in cattle in the southern Pennines of Derbyshire and north Staffordshire, England (Fig. 15.5).

The total concentration of elements indicated on geochemical maps may be of value in studies of the relationships between trace element levels and disease Appleton & Ridgway, 1993). However, the total amount of each element is less important than the amount which is bioavailable. In addition, there are difficulties in relating the biological and geochemical activities of trace elements to the total concentrations s represented on geochemical maps. For example, factors which affect the solubility and availability of trace elements include the mineral phases in which the elements are bound and the pH and Eh of surface and groundwaters. Moreover, the tempera-ure and total dissolved concentration and activity of

other dissolved species also affect the trace element mobility and sorption onto hydrous Fe and Mn oxides, and clay minerals limit the solubility of many trace elements. Accordingly, the concentration of ele-ments in surface waters only partly reflects the total concentration, which may be enhanced or diminished by mineral reactivity. In this way, increased quantities of elements may be released to surface and ground-waters in areas of metalliferous mineralization, because of the higher solubility of many ore minerals relative to silicate minerals. As an illustration, Al occurs in both bioavailable and inert forms. Hence, its potential toxicity is governed by its speciation. Aluminium can occur in a large number of dissolved species, particularly at either low or high pH, or in colloids with organic carbon or silica, which limit its toxicity. The toxicities of As and Sb also depend on their chemical form, they are most toxic in the M^{3+} gaseous state, their toxicity decreasing in the sequence $M^{3+} > M^{5+} >$ methyl As/Sb (Chen *et al.*, 1994).

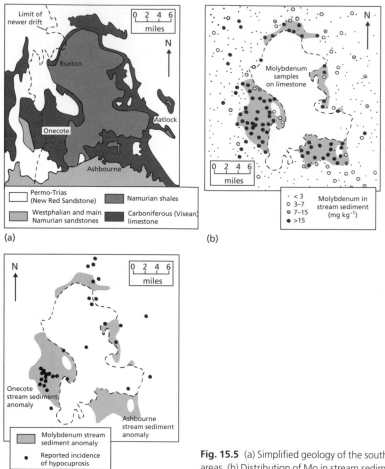

Fig. 15.5 (a) Simplified geology of the southern Pennines, England, and adjoining areas. (b) Distribution of Mo in stream sediment. (c) Incidence of hypocuprosis in cattle. (From Webb, 1971.)

15.4 Some trace elements and health

15.4.1 Arsenic

Arsenic is the main constituent of more than 200 mineral species, of which 60% are arsenates, 20% are sulphides and sulphosalts and the remaining 20% include arsenides, arsenites, oxides, silicates and elemental arsenic (Thornton, 1996). It is associated with many types of mineral deposit, especially those containing sulphide minerals. The common arsenic minerals are arsenopyrite, orpiment, realgar and enargite. Arsenic usually is present in chalcopyrite, iron pyrite and galena, and more rarely in sphalerite.

The average concentration of As in igneous and sedimentary rocks is approximately 2 mg kg^{-1}, and common concentrations in most rocks range from 0.5 to 2.5 mg kg^{-1}, although higher concentrations are found in finer grained mudrocks and phosphorites. Arsenic is concentrated in some reducing marine sediments, which may contain as much as 3000 mg kg^{-1}. Coal may contain up to 2000 mg kg^{-1} of As. The As contents of metamorphic rocks reflect those of the igneous and sedimentary rocks from which they were formed.

The weathering of rocks may mobilize As as salts of arsenous acid and arsenic acid. Under oxidizing conditions, arsenates are stable species. Under reducing conditions, arsenites are the predominant As compounds. Inorganic As compounds can be converted to methylated As species by microorganisms, plants, animals and humans. The oxidative methylation reactions act on trivalent As compounds, and produce monomethylarsonic acid, dimethylarsinic acid and trimethylarsine oxide. Under reducing conditions, these As compounds can be reduced to volatile and easily oxidized methylarsines (Thornton, 1996).

The average concentration of As in soil is around $56\,mg\,kg^{-1}$, and so generally is higher than that found in rocks. The lowest concentrations of As are found in sandy soils and those derived from granites, with higher levels in alluvial soils and those rich in organic matter. Arsenates of Fe and Al are the dominant phases in acid soils, and are less soluble than calcium arsenate, which dominates in many calcareous soils. The presence of clay minerals, Al, Fe and Mn oxides and organic matter, which can influence the sorption, solubility and rate of oxidation of As species, presumably complicates the position. The solubility of As is also controlled by adsorption reactions and biological activity. Soils close to or derived from sulphide ore deposits may contain up to $8000\,mg\,kg^{-1}$ of As.

Arsenic species in aqueous systems mainly involve arsenite and arsenate oxyanions, and are highly soluble over a wide range of pH and Eh conditions. However, under reducing conditions in the presence of sulphide, the mobility of As is reduced due to the precipitation as orpiment, realgar or arsenopyrite (although, at low pH, the aqueous species $HAsS_2$ may be present). The biomethylation of As may also occur, giving rise to monomethylarsonic acid (MMAA) and dimethylarsinic acid (DMAA). These usually are rare in natural waters compared with the organic forms, but may be present in relatively high concentrations in organic-rich waters.

Arsenic is strongly sorbed onto, or coprecipitated with, ferric hydroxide, the arsenate forms being more strongly sorbed than the arsenite forms. This produces potentially much higher concentrations of dissolved As under reducing conditions, not only because of the lower sorption affinity, but also because ferric hydroxide is more soluble at low Eh. However, As may also be present in oxidizing waters, particularly in groundwater environments where the oxidation of sulphide minerals occurs (Smedley et al., 1996). Furthermore, As readily is sorbed onto Al hydroxide, except at low pH, where Al is stable in dissolved ionic form.

The As concentration in unpolluted fresh waters typically ranges from 1 to $10\,\mu g\,l^{-1}$, rising to $100-5000\,\mu g\,l^{-1}$ in areas of sulphide mineralization and mining. WHO (1993) recently reduced its recommended limit for As in drinking water from 50 to 10 $\mu g\,l^{-1}$ in response to evidence from toxicological studies. Occurrences of high As concentrations in drinking water are relatively rare, most cases being associated with sources of natural sulphide minerals, such as pyrite and arsenopyrite; hence, high As levels may be produced by the disposal of associated mine wastes.

Arsenic contamination of the environment has arisen as a result of mining and smelting activities in several countries. However, a recent survey of the effects of As mining wastes and soils in southwest England by Mitchell and Barr (1995) suggested that the additional number of deaths arising from widespread contamination was small. On the other hand, As toxicity in cattle was positively identified in the form of dysentery and respiratory distress. In fact, it was suggested that the uptake of As by affected cattle may be as high as 50 mg per day due to grazing on contaminated pastures, with 60–70% of the As being present in accidentally ingested soil. Drainage from disused metalliferous mines and coal mines frequently is a source of both soluble and particulate As in surface river systems. The roasting of sulphide ores containing As and the burning of As-rich coal release As trioxide, which may react in air with basic oxides, such as alkaline earth oxides, to form arsenates. These inorganic As compounds can then be deposited onto soils, and may be leached into surface and groundwaters.

Arsenic is toxic, and as little as 0.1 g of arsenic trioxide may be lethal to humans (Jarup, 1992). In addition, at high doses, As is a human carcinogen. Skin cancer and a number of internal cancers, such as cancers of the bladder, liver, lung and kidneys, can result (North, 1992; Morton & Dunette, 1994). Hyperpigmentation, depigmentation, keratosis and peripheral vascular disorders are the most commonly reported symptoms of chronic As exposure. Neurological effects include tingling, numbness and peripheral neuropathy. Inhalation of As may give rise to respiratory cancer. As an illustration, Tseng (1977) showed that a relationship existed between high concentrations of As in drinking water and skin cancer, keratosis and Blackfoot disease (a type of gangrene) in southwest Taiwan. Subsequent investigations in Taiwan by Chen *et al.* (1992) established relationships between high As exposure and the internal cancers mentioned previously.

The toxicity depends on the form of As ingested, notably the state of oxidation and whether in organic or inorganic form. Reduced forms of As are more toxic than oxidized forms, with the order of toxicity decreasing from arsine, organo-arsine compounds, arsenite and oxides, arsenate, arsonium to native arsenic (Welch *et al.*, 1988). Arsenic intake by humans probably is greater from food (e.g. seafood) than from drinking water, however, As present in fish occurs as organic forms of low toxicity. Drinking water therefore represents by far the greatest hazard, because the species present in groundwater predominantly are the more toxic inorganic forms (Smedley *et al.*, 1996). In specific situations, this exposure may be added to through the inhalation of atmospheric particulates derived from industrial emission or from suspended As-rich soils and waste materials.

15.4.2 Cadmium

Cadmium is an uncommon element and, except for certain organic-rich shale, its content in rocks and soils is $0.2\,mg\,kg^{-1}$ or less. There are no mineral deposits which are composed predominantly of Cd, other than a few minor deposits of cadmium sulphide and cadmium carbonate. Cadmium, however, does occur as a minor constituent in sulphide ores, such as those of Zn and Pb. Cadmium enters the hydrosphere by leaching from rocks and soils, and by the disposal of domestic and industrial waste waters. Nonetheless, the concentration of dissolved Cd in groundwater and surface water ranges from $1\,\mu g\,l^{-1}$ to more than $10\,\mu g\,l^{-1}$. However, the concentration in bottom sediments of some water bodies is $1000–10\,000$ times greater than in the water above.

Cadmium is more volatile than most heavy metals, with a boiling point of $790\,°C$. Consequently, appreciable amounts of Cd, primarily in gaseous form, are released to the atmosphere when Zn and Pb ores are processed. The gas is oxidized rapidly, and is then deposited as fine particles over the surrounding areas. This represents a major source of Cd to the environment. Other significant sources include Cd in phosphate fertilizers and in sewage sludge.

Human exposure to Cd generally is greatest from food intake and the inhalation of tobacco smoke. Drinking water contains low concentrations of Cd unless the sources of water are affected by volcanic exhalations, leachate from landfills or mine waters. Cadmium solubility is limited by $CdCO_3$ (Hem, 1985). Hence, it occurs at higher concentrations in low-pH conditions. Cadmium may also be sorbed onto organic substances, such as humic and fulvic acids, and therefore organic-rich waters may have higher Cd concentrations given a local Cd source. The WHO (1993) limit for Cd in drinking water is $5\,\mu g\,l^{-1}$.

The average diet contains about $0.5\,mg$ per day of Cd, of which about 5% is absorbed through the intestinal wall. A cigarette contains about $1\,\mu g$ of Cd, but Cd is absorbed much more readily in lung tissue than in the intestines. Once Cd is in the bloodstream, it is transported to the kidneys and liver, where about two-thirds of the total body Cd is retained. In particular, Cd accumulates in the kidneys during childhood and adolescence, where it is bound to a protein (metallothionein) of unknown function. Cadmium accumulates slowly over the course of a lifetime, and there is no known mechanism for ridding the body of it.

Cadmium is an acute toxin, producing symptoms such as giddiness, vomiting, respiratory difficulties, cramps and loss of consciousness at high doses. Chronic exposure to the metal can lead to kidney disorders, anaemia, emphysema, anosmia (loss of sense of smell), cardiovascular diseases, renal problems and hypertension (Robards & Worsfold, 1991). Furthermore, there is evidence that increased ingestion of Cd can promote Cu and Zn deficiency in humans, and it may also be a carcinogen (Tebbutt, 1983).

Schroeder (1965) suggested that Cd may be implicated in hypertension, because humans exhibiting arterial hypertension pass appreciably more Cd in their urine than do normal individuals. Subsequently, Schroeder *et al.* (1968) related hypertension in rats to renal Cd/Zn ratios, rather than to absolute values of Cd. Moreover, they reported a reversal of Cd-induced hypertension by feeding rats a zinc chelate, which presumably bound Cd and released Zn, hence depleting renal Cd and repleting renal Zn. Perry (1971) also referred to a relationship between Cd and hypertension.

Itai-itai disease appears to be a Cd-related disease, which is very painful and causes the wastage and embrittlement of bones. The disease occurred along the River Zintsu in Japan in 1945. Mining wastes due to the processing of Pb and Zn, and containing Cd, were disposed of into the river system. The contaminated water was used for agricultural and domestic purposes (Pettyjohn, 1972). Although samples of water generally contained less than $1\,mg\,l^{-1}$ of Cd and $50\,mg\,l^{-1}$ of Zn, these elements were selectively concentrated in sediment, and yet more highly in plants. For example, the Cd content in rice averaged $125\,mg\,kg^{-1}$ and, in the roots, it was $1250\,mg\,kg^{-1}$.

15.4.3 Mercury

Mercury can exist in the solid, liquid and vapour phases. Under normal conditions, it is a liquid which evaporates and adds Hg vapour to the air. The abundance of Hg in the Earth's crust is estimated to be around $60-80\,\mu g\,kg^{-1}$. Although there are more than 12 Hg-bearing minerals, only a few are abundant in nature, cinnabar being the most important. It generally is found in mineral veins, as impregnations or having replaced quartz in rocks near recent volcanic or hot spring areas. Igneous rocks form the basic sources of Hg, but generally contain less than $0.02\,mg\,kg^{-1}$, averaging $0.01\,mg\,kg^{-1}$. The Hg content of soils and sedimentary rocks also averages around $0.01\,mg\,kg^{-1}$, and that of soils varies within relatively narrow limits. However, certain organic-rich shales contain concentrations exceeding $1\,mg\,kg^{-1}$.

Surface waters tend to contain tolerably small concentrations of Hg. Industrial, agricultural, medical and scientific uses of Hg and its compounds introduce further Hg into surface waters. Whatever the source, however, the concentration of Hg compounds, dissolved or suspended, is reduced rapidly by sorption and by complexing reactions with clays, plankton, colloidal proteins, humic materials and other organic and inorganic colloids (Hem, 1970).

Because of the tendency of Hg to vaporize, it enters the atmosphere in both gaseous and particulate forms. The rate of vaporization of Hg and certain of its inorganic compounds decreases in the sequence $Hg > Hg_2Cl_2 > HgO$ (Jenne, 1970). Gaseous and particulate Hg commonly occur in the exhaust fumes given off by various industrial and smelting processes. Most of the Hg in the air returns to the Earth's surface in rainfall, the average annual contribution being estimated to be about $1.2\,g\,ha^{-1}$.

Mercury in living tissues is largely organic and primarily involves methylmercury (CH_3Hg), the latter being soluble in water. It tends to concentrate in living tissue and, at critical concentrations, can be extremely toxic.

The toxic effects of water-borne Hg on humans were emphasized in 1953 when over 100 people in the vicinity of Minamata, Kyushu, Japan, developed strange symptoms, including tottering, jerky gaits, loss of manual dexterity, impairment of speech, and commonly, deafness and blindness (Minamata disease). Eventually, 50 of those afflicted died. Extensive investigations revealed that the deaths were caused by the consumption of Hg-contaminated fish and shellfish from Minamata Bay. The bay had received large amounts of methylmercury compounds in the waste effluents from a plastics factory (Kurland et al., 1960). Methylmercury can penetrate and erode brain tissue, particularly the areas that control sight, hearing and equilibrium.

As a result of these findings, a tentative upper limit of $5\,\mu g\,l^{-1}$ of Hg in drinking water was proposed by the United States Public Health Service (Greeson, 1970). This subsequently has been reduced to $1\,\mu g\,l^{-1}$ by WHO (1993). In addition, the United States Food and Drug Administration declared that fish and other foods which contained $0.05\,mg\,kg^{-1}$ or more Hg were unsafe for human consumption (Greeson, 1970).

15.4.4 Lead

The average abundance of Pb in the Earth's crust is approximately $15\,mg\,kg^{-1}$. Although Pb is a major constituent of more than 200 minerals, most of these are very rare, and only three are commonly found in sufficient abundance to form mineable Pb deposits. These are galena, anglesite and cerussite, a Pb sulphide, sulphate and carbonate respectively. Galena is a common primary constituent of sulphide ore deposits, whereas anglesite and cerussite normally form by the oxidation of galena close to the surface. Lead is present in trace amounts in feldspars, potash feldspar containing the most Pb of any of the common silicate minerals. The Pb content of clay minerals is very variable, but frequently is around $10-20\,mg\,kg^{-1}$.

The Pb contents of igneous rocks range from $4\,mg\,kg^{-1}$ for basalts to some $30\,mg\,kg^{-1}$ for granitic rocks

(Lovering, 1976). Metamorphic and sedimentary rocks also have Pb contents falling within this range. Unconsolidated, terrestrial sediments have a mean Pb content close to the average crustal abundance, but deep marine muds are appreciably richer in Pb, containing an average of $60\,mg\,kg^{-1}$. Ores of Pb and Zn often are closely associated in deposits formed by the replacement of limestone or dolostone. In addition, Pb ore frequently is present together with ores of Cu, Zn, Ag, As and Sb in complex vein deposits that are related genetically to granitic intrusions, but it may occur in a variety of host rocks.

The Pb content of young residual soils is influenced markedly by the parent rock from which they were derived. Such a relationship is unlikely to exist in mature soils developed on deeply weathered parent material. Generally, Pb tends to be more mobile in acid soils than in alkaline soils, tending to be leached out of the former and to form residual concentrations in the latter. Soil may be contaminated by mining waste from old Pb workings. Davies (1972) quoted an example from the Tamar Valley in southwest England, where some garden soils contained as much as $522\,mg\,kg^{-1}$ of Pb. Radishes were used to measure the Pb uptake, and contained up to $74\,mg\,kg^{-1}$ in their dry ash. However, a number of factors decrease the solubility of Pb from mine waste (Davis *et al.*, 1994). These include the mineral composition, the degree of encapsulation in pyrite of silicate matrices, the nature of the alteration rinds and the particle size. These factors decrease the bioavailability of Pb in soils derived from, or contaminated by, mining wastes, especially when compared with smelter-derived or urban soils (Fig. 15.6).

The concentration of dissolved Pb in most natural water systems, which normally contain dissolved carbon dioxide and have pH near 7, is very low. According to Hem (1985), it commonly is less than $10\,\mu g\,l^{-1}$. The low solubility of Pb is due mainly to its ability to combine readily with the carbonates, sulphates and hydroxides normally present in such waters to form compounds of low solubility. In particular, Pb solubility is controlled principally by $PbCO_3$ and low alkalinity, low-pH waters holding higher concentrations of Pb. The WHO (1993) maximum recommended concentration for Pb in drinking water has been reduced recently from 50 to $10\,\mu g\,l^{-1}$ because of concerns about chronic toxicity, although this concentration rarely is exceeded in natural waters.

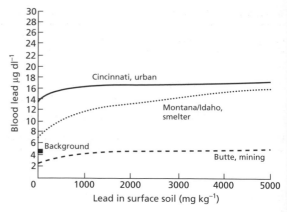

Fig. 15.6 Comparison of the blood Pb dose–response at mining, smelter and urban sites in the USA. (After Davis *et al.*, 1994.)

In the past, Pb has entered the atmosphere mainly in exhaust fumes from vehicles (the increasing use of Pb-free petrol is reducing this source), and in smoke from the large-scale industrial burning of coal. Hence, the average Pb content in the air of large urban areas may be around $2.5\,\mu g\,m^{-3}$ compared with $0.5\,\mu g\,m^{-3}$ in rural areas.

The amount of Pb introduced into the environment from natural sources is small compared with that made available by humans. Brinkman (1994) found levels of Pb in soil samples in Tampa, Florida, and Milwaukee, Wisconsin, which were considered to be hazardous, the Pb being derived from man-made sources. Lead also occurs in dusts in urban atmospheres. Although guidelines on acceptable levels of Pb in urban areas generally have not been established, Davies (1994) quoted the former Greater London Council, which adopted a guideline Pb concentration of $500\,\mu g\,g^{-1}$ as justifying further investigation and $5000\,\mu g\,g^{-1}$ as justifying control measures. A significantly lower level of control was associated with the Port Pirie lead abatement programme, South Australia (Table 15.5; Calder *et al.*, 1994). Moreover, Patterson (1965) estimated that the amount of Pb being absorbed by humans in certain urban areas is 30 times greater than that of the inferred natural rates. Cannon (1976) indicated that, although the quantity of Pb ingested by humans from food and water is much greater than that inhaled from urban air, inhaled Pb is much more readily absorbed.

Table 15.5 Port Pirie lead soil abatement protocols 1984–1994. (After Calder *et al.*, 1994.)

Category	Pb level (p.p.m.)	Action
A	<250	No action
B	250–1250	Home owners are advised to maintain a barrier between soil and children, and advice given to parents about behaviour
C	1250–5000	As for category B, plus assistance to cover contaminated soil with 50 mm of gravel or soil
D	>5000	Excavate contaminated soil and replace with clean fill

Indeed, some 20–40% of Pb absorbed probably is derived from the air, and the higher the atmospheric level, the higher the level of Pb in the blood.

Lead in animals, including humans, is concentrated largely in bone. Lead interferes with the normal maturation of erythroid elements in the bone marrow and inhibits haemoglobin synthesis in precursor cells. Studies conducted on animals have shown that Pb is an immunosuppressive agent at levels well below those causing overt toxicity. It also affects porphyrin metabolism and interferes with the activity of several enzymes (Cannon, 1976). Overt manifestations of acute Pb poisoning, plumbism, in children differ from those of adults, and their mortality rate is relatively high. Indeed, children less than 6 years of age are those at highest risk from exposure to environmental Pb (Lutz *et al.*, 1994). Symptoms include anaemia, gastrointestinal distress and encephalopathy. The latter may result in early death or permanent symptoms of brain damage, or recovery may take place. Adult or chronic Pb poisoning requires years to develop to a critical level, which is recognizable by symptoms such as headache, irritability, behavioural changes, constipation, impairment of mental ability, abdominal tenderness, colic, weight loss, fatigue and neuromuscular problems and muscle pains (Tebbutt, 1983). Renal lesions, which may be caused by Pb ingestion, can give rise to hypertension, and may predispose to gout as a result of defective urate secretion. Chronic exposures to Pb in water can result in increased miscarriages and stillbirths.

Yet another health problem which appears to be associated with Pb was mentioned by Anderson and Davies (1980). They reported the results of surveys of the prevalence of dental caries in children in the Tamar Valley, England, and in Ceredigion, Wales. Both are areas where past base metal mining has left a legacy of extensive heavy metal contamination of agricultural and garden soils. In both areas, a higher level of dental caries was associated with high levels of Pb in soils which were available to plants (Fig. 15.7).

15.4.5 Iodine

Iodine generally is not concentrated in primary minerals, having a relatively uniform content in all mineral groups. It is, however, relatively concentrated in the chlorine-bearing minerals eudialyte and sodalite, where it replaces the chloride ion. The low concentration of I in the primary minerals is reflected in the low and generally uniform levels in all igneous rock groups, the mean value being 0.24 mg kg⁻¹ (Fuge & Johnson, 1986). Iodine concentrations are low in metamorphic rocks, being similar to those found in igneous rocks. The distribution of I in sedimentary rocks is more variable, with higher concentrations in argillaceous than in arenaceous rocks. Some limestones possess higher concentrations of I. Iodine concentrations, however, only are increased significantly in organic-rich sediments which, at times, may have values exceeding 40 mg kg⁻¹. Marine sediments are richer in I than those of non-marine origin, with recent marine sediments being enriched particularly. Fuge and Johnson suggested mean values of 2.7, 2.3 and 0.8 mg kg⁻¹ for carbonates, shales and sandstones respectively, with recent sediments having a range of 5–200 mg kg⁻¹.

The I content of soils varies appreciably, from less than 0.1 to over 100 mg kg⁻¹. However, the concentration of I in soils generally is considerably higher than that in the parent rocks, the major supplier of I to the soil being the atmosphere, by way of precipitation. Therefore soils in coastal localities contain

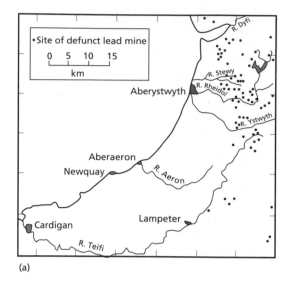

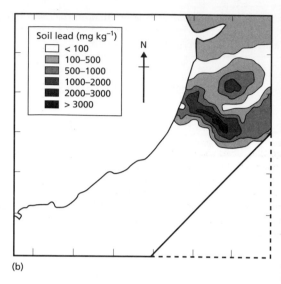

Fig. 15.7 Ceredigion district: (a) positions of old lead mines; (b) a computer-derived map of soil lead levels. (After Anderson & Davies, 1980.)

higher I concentrations than those of inland regions. Organic matter plays a major role in the retention of I in the soil. Iron oxide, Fe hydroxide and hydrated Al oxide are also important for the sorption of soil I. Furthermore, sorption of I by Al and Fe oxides is influenced markedly by the pH value, with greater sorption attributable to increasing acidic conditions and no sorption occurring under neutral conditions (Whitehead, 1979).

The oceans represent the most important source of I, the mean I content of sea water being estimated to be $58\,\mu g\,l^{-1}$. Iodine from the sea is transferred to the air above which, in turn, conveys it to the landmasses. In fact, appreciable quantities of I are released from the sea in volatile form, the volatilization to the atmosphere occurring as I gas, I_2 or methyl iodide, CH_3I.

Although the I content of surface waters is highly variable, it tends not to exceed $15\,\mu g\,l^{-1}$, with a mean of about $5\,\mu g\,l^{-1}$ (Fuge & Johnson, 1986). As expected, there is appreciable regional variation in relation to the proximity of the sea, with surface run-off being considerably enhanced in I in coastal regions. Iodine in rainfall over coastal areas generally

is higher than over inland areas (1.5–$2.5\,\mu g\,l^{-1}$ compared with $1\,\mu g\,l^{-1}$ or less respectively). Groundwater commonly is richer in I than surface water (Fuge, 1989). High natural baseline concentrations of I may occur in carbonate aquifers, probably being derived from the oxidation of organic matter. For example, this has been demonstrated by Edmunds and Smedley (1996) for the Chalk aquifer in the London Basin, where the median concentration of I is $32\,\mu g\,l^{-1}$ and the I to Cl ratio is 5.84×10^{-4}, which is about four times higher than for non-carbonate aquifers in England (Fig. 15.8).

According to Fuge (1996), I was the first element to be recognized as being essential to human health, with goitre, which occurs due to I deficiency, being the first endemic disease to be related to environmental geochemistry. The thyroid gland, at the base of the neck, requires I in order to function normally, I deficiency causing a tumorous condition involving the enlargement of the thyroid gland, that is, goitre. Unfortunately, despite this early recognition of the significance of I in relation to human health, the problems of goitre and other I deficiency disorders are still a cause for concern. For instance, Hetzel (1991) maintained that more than one billion people are potentially at risk from some form of iodine deficiency disease (IDD), of which some 200 million have goitre and up to 20 million have brain damage attributable to I deprivation during fetal development

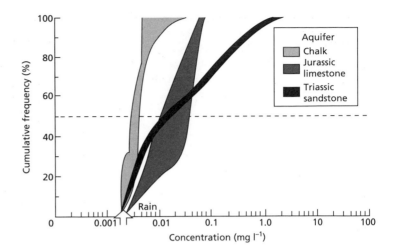

Fig. 15.8 Iodine concentrations in groundwater from UK aquifers (Chalk, Jurassic limestones, triassic sandstones). Note that only carbonate rocks contribute I from water–rock interaction, over and above rainfall inputs. (After Edmunds & Smedley, 1996.)

and infancy. Stewart and Pharoah (1996) remarked that endemic goitre, however, was only the tip of the clinical I deficiency iceberg. They went on to note that IDD includes stillbirths, abortions, congenital abnormalities, endemic cretinism (commonly characterized by mental deficiency, deaf-mutism, spastic diplegia and lesser degrees of neurological defect) and impaired mental function in children. Other more controversial effects of I deficiency are breast disorders, such as fibrocystic disease. Goitre itself rarely is a major health problem in a community. It is the fetal wastage and the impaired mental function associated with goitre that are so devastating to the individual. By far the most important of the disorders is endemic cretinism (Pharoah, 1985). There are two recognized types, the neurological and the hypothyroid. The former is due to I deficiency in early pregnancy, whereas I deficiency which occurs in late pregnancy or early babyhood is responsible for the latter type. Iodine is required by the thyroid gland to produce hormones. The thyroid hormones are associated with brain development. Impairment of the brain at this early stage means that the disorder is permanent.

Severe goitre and cretinism often are associated with highland regions, central continental regions and rain shadow areas. The geographical distribution of these diseases correlates well with the behaviour of I in the secondary environment, although there are anomalies within this general distribution (Fuge, 1996). Some of the anomalies have been related to goitrogenic substances in the diets of those affected.

In addition, it has been suggested that other elements may be involved in the aetiology of IDD. These include Ca, F, As, Zn, Mg, Mn, Co and Se (Thilly *et al.*, 1992). The possible link between Ca and goitre aetiology relates to the fact that areas underlain by limestone are prone to endemic goitre. For example, some areas of England which, in the past, were badly affected by endemic goitre included the Yorkshire Dales and the Peak District of Derbyshire. Both areas are underlain by limestone. However, the latter relationship is not straightforward. Fuge (1996), for instance, quoted several areas in the UK where goitre had occurred in the past, but where the soils were not deficient in I. He therefore proposed that the occurrence of IDD in such areas was due to the non-bioavailability rather than depletion. As many of these areas were underlain by limestone, such as the Peak District referred to above, he further suggested that rather than Ca being a goitrogen, its presence gives rise to soils of relatively high pH, which decrease the uptake of I by plants.

Some 20% of the daily requirement of I by humans probably is obtained from drinking water. The total I concentrations in drinking water range between 0.01 and $70\,\mu g\,l^{-1}$, depending on the location, topography and rainfall pattern. Concentrations much below this mean value frequently are associated with the occurrence of goitre. The problem today, however, tends to be restricted to rural areas of developing countries.

Nonetheless, goitre is still endemic, especially among the poor in certain areas of the USA. Remediation through dietary supplements generally is an effective treatment. For example, the use of iodized salt was found to reduce the incidence of goitre in Michigan.

15.4.6 Fluorine

Fluorine is a fairly common trace element in crustal rocks, the average content being 300 mg kg⁻¹. Fluorite is the most common F-bearing mineral, but F is also present in apatite and in trace quantities in amphibole, mica, sphene and pyroxene. Fluoride occurrence commonly is associated with volcanic activity (being especially high in volcanic gases), geothermal fluids and granitic rocks.

Most of the fluorine in soils is derived from the parent rock, but it can also be added by volcanic activity, which deposits F-rich volcanic ash on the land. For instance, the eruption of Hekla in Iceland on May 5, 1970 caused ash to fall over a large part of the island. The coarser ash that fell near Hekla contained 100 mg kg⁻¹ of F, whereas the finer particles of ash, which were deposited over areas 200 km or more to the north and northwest, contained up to 2000 mg kg⁻¹ of F. Some 100 000 sheep and many cattle were poisoned from eating grass on which the ash had settled. Industrial activity (the manufacture of Al and of bricks, and the smelting of F-bearing ores) and the application of phosphate fertilizers have also contributed locally to an increase in the concentration of F in soils.

The principal form of F occurring in water is free dissolved F, but at low pH, the species HF may be stabilized. This may be the dominant species at pH 3.5 and below (Hem, 1985). Fluorine readily forms complexes with Al, Be, Fe^{3+}, B and Si. Concentrations of F in water are limited by the solubility of fluorite, such that, in the presence of 10^3 M Ca, F is limited to 3.1 mg l⁻¹. Therefore, the absence of Ca in solution permits higher concentrations of F to be stable (Fig. 15.9). As a consequence, high F concentrations are likely to occur in groundwater in Ca-poor aquifers and in areas where F minerals (or F-substituted minerals, e.g. biotite) are common. The F concentration also increases in groundwater where cation exchange of Ca by Na takes place. Thermal, high-pH waters possess particularly high concentrations of F.

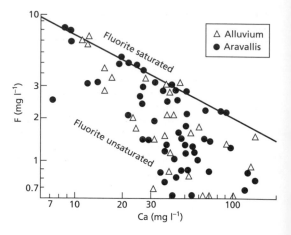

Fig. 15.9 Fluoride concentration in groundwaters from Rajasthan, India, showing solubility control by fluorite. (From Edmunds & Smedley, 1996.)

Where high-temperature processes are involved in manufacturing, such as in brick making, F is released as hydrogen fluoride or hydrofluoric acid. Hydrogen fluoride is extremely toxic. It can be taken out of the air, concentrated in plants and later ingested by animals. Concentrations as low as 1 p.p.b. are damaging to some plant species. A few parts per million in plants indicate that the local atmosphere is contaminated. Livestock ingesting high-F vegetation develop fluorosis. Fluorosis causes their teeth to mottle and wear more quickly than otherwise. Their bones become deformed and brittle, and break easily.

Fluorine is an important trace element, as calcium fluoride increases the crystallinity of hydroxyapatite in microcrystals in the enamel of teeth. In this way, it helps to prevent tooth decay by facilitating the growth of larger, more perfect crystals. Dental caries is the localized destruction of tooth enamel. The same process occurs in bones, where calcium fluoride assists in the development of a more perfect bone structure. Dental caries may result at low concentrations of total F, that is, less than about 0.5 mg l⁻¹, whereas at higher concentrations, chronic exposure can result in dental fluorosis (mottled enamel) or skeletal fluorosis. Concentrations above which these factors become problematic are around 2 mg l⁻¹ and 4 mg l⁻¹ respectively (Table 15.6). Dean (1945) originally recommended 1.0 mg l⁻¹ as the optimal level of fluoride in drinking water, based on the relation

Table 15.6 Impact of fluoride in drinking water on health. (From Dissanayake, 1991.)

Concentration of fluoride (mg l⁻¹)	Impact on health
Nil	Limited growth and fertility
0–0.5	Dental caries
0.5–1.5	Promotes dental health resulting in healthy teeth, prevents tooth decay
1.5–4.0	Dental fluorosis (mottling of teeth)
4.0–10.0	Dental fluorosis, skeletal fluorosis (pain in back and neck bones)
>10.0	Crippling fluorosis

between caries inhibition and the severity of dental fluorosis. Subsequently, Myers (1978) showed that, in temperate climates, fluoride levels of up to $1.5\,mg\,l^{-1}$ in drinking water produce only mild fluorosis of no public health significance. This therefore is the desirable upper limit recommended by WHO (WHO, 1993). However, recent work by Warnakulasuriya *et al.* (1992) has suggested that in hot dry climates, where water consumption is higher, dental fluorosis could occur in areas where the fluoride content in groundwater is less than $0.3\,mg\,l^{-1}$, as in parts of Sri Lanka. Their studies showed that, of those consuming drinking water with less than $1.0\,mg\,l^{-1}$ fluoride, 32% of the children had mild forms and 9% severe forms of dental fluorosis. They therefore recommended an upper limit of fluoride in drinking water of $0.8\,mg\,l^{-1}$ for those living in hot dry climates. Poor nutrition is also an important contributory factor. Fluoride levels in drinking water above the recommended limits can be reduced by exposing the water to solid alumina or bone ash. The effects of fluorosis are permanent and incurable. Moreover, high F concentrations in drinking water have been linked with cancer (Marshall, 1990).

As an illustration, Dissanayake (1991) found that high concentrations of F in groundwater in the Dry Zone of Sri Lanka, that is, up to $10\,mg\,l^{-1}$, were associated with dental and skeletal fluorosis. By contrast, in the Wet Zone, intensive rainfall and the long-term leaching of F from rocks presumably are responsible for the low concentrations of F in groundwater, and consequently the incidence of dental caries is high.

Aquifers in F-bearing rocks or the irrigation of F-rich soils can increase the intake of F through drink-

ing water or crops, and so lead to health problems attributable to the adverse effects of F intoxication on skeletal development. The increased consumption of water in arid environments may increase the risk of fluorosis if the F content of water is greater than $2\,mg\,l^{-1}$.

15.4.7 Selenium

Selenium occurs in nature in four oxidation states (2−, 0, 4+ and 6+). In its 2− state, Se occurs as H_2Se, a highly toxic and reactive gas, which readily oxidizes in the presence of oxygen. In elemental form (Se^0), Se is insoluble, and therefore non-toxic, because it is unavailable to plants and therefore to animals. Selenium occurs in the 4+ oxidation state as inorganic selenite (SeO_3^{2-}), which is also highly toxic. Under reducing and acidic conditions, however, selenite is reduced to elemental Se. Oxidizing and alkaline conditions favour the stability of the 6+ form, selenate (SeO_4^{2-}), which is highly soluble (Edmunds & Smedley, 1996). The mobility of selenium therefore is greater in oxidizing aquifers, although its dissolved concentration may be restricted due to its ease of sorption onto ferric hydroxide, which precipitates under such conditions.

One of the primary sources of Se is volcanic eruptions. For instance, Lakin (1973) maintained that the average concentration of Se in the soils of Hawaii, which are derived from volcanic material, is around 6–$15\,mg\,kg^{-1}$. This is markedly greater than the average content of Se in the Earth's crust, which is about $0.05\,mg\,kg^{-1}$. Selenium has a strong affinity for organic matter, and is incorporated readily into sulphide minerals. Consequently, it often is associated with sulphide-bearing hydrothermal veins, and is present in relatively high concentrations in uranium deposits. It may also form ferroselite ($FeSe_2$) if present in sufficiently high concentrations. Selenium also occurs in organic shales, coal and petroleum, its content tending to range between 0.5 and $1.0\,mg\,kg^{-1}$. In fact, the annual release of Se from the combustion of coal and oil in the USA is about 3.6 million kilograms. Fortunately, Se compounds formed during combustion are relatively insoluble, and so are more or less unavailable to plants.

Selenium in soils varies from about $0.1\,mg\,kg^{-1}$, in soils which are deficient, to as much as $1200\,mg\,kg^{-1}$ in organic-rich soils, such as the seleniferous soils, developed from black organic shales, which occur in certain areas of the Western Plains in the USA.

Selenium is an essential element in the diet of both animals and humans at concentrations between 0.04 and 0.1 mg per day. It is toxic above 4 mg per day. Selenium deficiency is more likely to occur in the diet of animals and humans than Se toxicity. Selenium deficiency may promote muscular degeneration, impeded growth, fertility disorders, anaemia and liver disease (Peereboom, 1985). Keshan disease, a chronic cardiomyopathy, is also thought to be related to Se deficiency, and is found in many parts of China. On the other hand, at high ingested concentrations of 10 mg per day or above, other problems, such as gastrointestinal ailments, skin discoloration and tooth decay, may occur. The WHO (1993) recommended limit for Se in drinking water is $10 \mu g l^{-1}$, however, concentrations in natural water rarely exceed $1 \mu g l^{-1}$. In concentrated amounts, Se is possibly one of the most toxic elements.

15.4.8 Zinc

The deficiency of Zn in soils is more common worldwide than that of any other trace element. Such deficiency results in a variety of diseases in crops. Pories *et al.* (1971) maintained that Zn deficiency in soils is attributable to three soil conditions, namely, the low content of total Zn, unavailability of bound Zn and poor soil management practices. Total Zn is low in highly leached soils, such as those found in many coastal areas. The element is made unavailable to plants by alkaline soils, by a high content of organic matter and by a high concentration of Mg or P, often found in clay soils. More recently, heavy fertilization of soils with phosphates and nitrates has contributed significantly to Zn deficiency in such soils.

Zinc is now recognized as an essential element for all animals. It is required only in minute concentrations, such as 20–100 p.p.m., yet even slight or moderate deficiencies can retard growth, lower feed efficiency and inhibit general well-being. Characteristic signs of Zn deficiency in animals include disorders of the bones, the joints, and the skin, delayed healing and loss of fertility. In severe form, Zn deficiency can lead to the death of animals.

Fordyce *et al.* (1996) found strong spatial and statistical correlations between sediment, soil and forage Zn in northeast Zimbabwe. These suggest that increased levels of Zn in soils give rise to increased uptake of Zn by plants. However, the ingestion of plants containing higher levels of Zn did not reveal an increase in the Zn levels found in cattle serum. Indeed, Zn in serum appeared to decrease slightly as Zn in forage increased. Fordyce *et al.* suggested that this could be due to the antagonistic relationships which exist between elements as they are absorbed during digestion by cattle. Furthermore, Lebdosoekojo *et al.* (1980) proposed that high levels of Fe and Mn may interfere with the metabolism of other trace elements in cattle. According to Mertz (1987), several studies in animals have identified mutually antagonistic relationships between Cu and Zn, and between Fe and Zn. This was borne out by Fordyce *et al.*, who found that, in northeast Zimbabwe, the area of high Zn in soils and pasture coincided with high Cu, Fe and Mn, and that therefore it was possible that the uptake of these elements was inhibiting the absorption of Zn in cattle. This may have wide-ranging implications in many tropical regions, where there is a preponderance of ferrallitic soils. Zinc concentrations in cattle serum in northeast Zimbabwe tended to decrease slightly as the Ca content of soils increased. Mertz showed that high Ca and P ingestion reduced Zn absorption in pigs, poultry and humans, but these relationships were less clear in cattle.

In humans, Zn deficiency retards growth and development, causes a loss of fertility and delays the healing of wounds and the repair of bones. Atherosclerosis is related to Zn deficiency, Zn being required for arterial repair. Mineral imbalances are additional factors in arterial disease. Deficiencies in Zn, Cu and V may interfere with arterial metabolism, whereas excesses of Ca, Cd and Co may be injurious to the arterial wall.

The addition of Zn supplements to soils deficient in Zn has resulted in spectacular gains in yields by plants. Similarly, the addition of supplemental Zn has produced substantial improvements in the health of livestock. Zinc sulphate therapy was referred to by Pories *et al.* (1971) as a means of accelerating the healing of wounds in humans.

15.4.9 Molybdenum

Plants do not appear to suffer from concentrations of

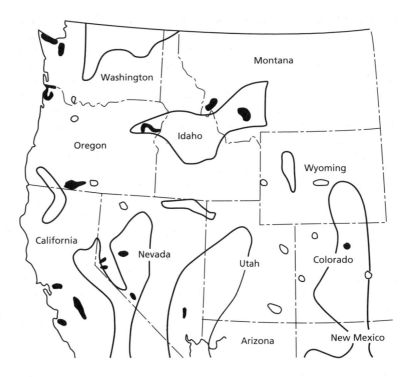

Fig. 15.10 Regions with rocks with a high content of Mo (shown outlined) and regions of occurrence of molybdenosis (shaded black) in western USA. (From Leckie & Parks, 1978.)

Mo up to 100 p.p.m., and some, notably legumes such as alfalfa and clover, accumulate high concentrations when grown in soils rich in Mo or in soils irrigated with water with a high content of Mo. Molybdenum may be added to fertilizers to correct for Mo deficiency in soil and water. Molybdenum deficiency detracts from the quality of pasture for livestock and reduces the quality of crops such as citrus fruit.

Molybdenum in trace amounts is essential in animal nutrition, but is poisonous in large doses. Livestock that consume pasture in which the Mo content exceeds $10–20\,mg\,kg^{-1}$ tend to develop molybdenosis, a disease characterized by a loss of appetite, diarrhoea, abnormalities in the joints and sometimes death. It can be corrected by supplemental copper. Figure 15.10 outlines the regions in the western USA where natural concentrations of Mo are above the average for the Earth's crust, together with areas where molybdenosis has occurred. Most reported cases of molybdenosis either have been within an area of natural high Mo, or downstream from such an area. Other occurrences were characterized by highly organic, wet soils which trapped metals

by binding them to organic soil particles. Such wet soils concentrate metals carried into the area by water, which itself may have only a slightly higher than normal content of Mo.

15.4.10 Some other trace elements and health

Aluminium is a major element in aluminosilicate minerals, and therefore is a common constituent of most rocks. The solubility of Al is strongly pH dependent, and significant concentrations only are found below pH 5.5, where the increasing concentrations are related to the solubility of microcrystalline gibbsite (Bache, 1996). The presence of inorganic ligands, notably F and SO_4, also increases the solubility of Al. At values in excess of pH 5, it is unlikely that labile, monomeric forms of Al are present in natural waters, although colloidal Al and other aluminosilicate colloids and particles may contribute to the total Al in waters (Edmunds & Smedley, 1996).

The occurrence of high Al in drinking water has been linked to the development of Alzheimer's disease (Martyn *et al.*, 1989). Aluminium is more likely to be toxic if present in the labile, monomeric form, in which it may make a disproportionate contribution to the amount absorbed from the gastrointestinal tract. The greater bioavailability of Al in drinking water therefore may render it a relatively more harmful source than food. The role of Al in Alzheimer's disease (let alone the role of Al in water) remains a controversial subject. The WHO (1993) maximum recommended concentration for Al in drinking water is $0.2\,mg\,l^{-1}$. However, this limit is a compromise between reducing aesthetic problems and retaining the efficacy of water treatment, and WHO concluded that at present it is not possible to derive a health-based guideline.

Manganese oxides are present in soils as nodules, coatings on peds and grains, and as disseminated black/grey pigment. They invariably are impure, being contaminated with clay minerals and organic matter, and often contain notable amounts of Fe. The total Mn content in soils often can be negatively correlated with the Cu and Co content. However, Lidiard et al. (1993) suggested that it may be the Al content of Mn oxides which is the controlling factor as far as their ability to concentrate Cu and Co is concerned. They found that the total Mn content appeared to be significant in the areas of Devon, England, where there were deficiencies in Cu and Co. Manganese therefore may act as a biochemical antagonist to Cu and Co.

Fordyce et al. (1996) studied the mineral status in cattle in northeast Zimbabwe. One of their findings was that there was a significant correlation between the Cu content of soil and forage. However, the Cu content of cattle serum showed no correlation with forage. They suggested that this lack of correlation may reflect antagonistic relationships between certain elements during uptake. The most significant negative correlations found by Fordyce et al. were those between serum Cu content, on the one hand, and Mn and Zn in stream sediment and forage, respectively, on the other. A similar correlation with Fe was slightly negative. Hence, these negative relationships with Mn and Zn, and possibly Fe, ingested in forage and soil, may mean that they are inhibiting the uptake of Cu by cattle. The inhibiting influence of Fe and Zn on Cu absorption by animals has been documented by Mertz (1987), and substantiates the findings of Lidiard et al. (1993) of Mn being associated with Cu deficiency.

Chromium deficiency in animals and humans is characterized by insulin resistance, manifested by the impairment of glucose tolerance in the presence of normal concentrations of insulin. Disorders of the central and peripheral nervous system may also be associated with Cr deficiency. Mertz (1987) suggested that Cr deficiency may be a risk factor in cardiovascular disease, and noted that the daily human requirement was around $50–200\,\mu g$. Both Cr and Cu deficiency are regarded as factors in cardiovascular disease in animals.

The daily turnover of Co as a constituent of vitamin B_{12} in an adult human varies between 100 and $150\,ng$. The average vitamin B_{12}-Co concentration in blood is approximately $5\,ng\,l^{-1}$. Deficiency of vitamin B_{12} causes pernicious anaemia.

Beryllium is substituted as a trace component in the silicate lattice of some rock-forming minerals. It is also present in the minerals beryl and bertrandite, and is concentrated in the residual deposits of silicic volcanic rocks. Beryllium is one of the most toxic metals, but only limited data are available on its occurrence and toxicity in the environment. Of particular concern is the possibility that Be might be mobilized, together with Al, under conditions of increasing acidity. In fact, Edmunds and Trafford (1993) maintained that Be is concentrated especially in acidic waters, being present as dissolved Be^{2+} at pH less than 5.5, but it may also be soluble as $Be(OH)$ complexes at higher pH. The solubility of Be may also be enhanced by the formation of F complexes. No guide level has been set by WHO (1993) for Be in drinking water, because of insufficient toxicological data. It is assumed that its concentration in drinking water is very low.

Barium occurs as a minor element in many rock types, being most abundant in acid igneous rocks. It is released readily by the interaction of water and rock, however, its solubility is controlled by the solubility of barite. Accordingly, concentrations of Ba in natural waters are inversely proportional to the sulphate concentration. High concentrations of Ba are present mainly where sulphate reduction has occurred and where sulphate concentrations are less than $10\,mg\,l^{-1}$. The concentration of Ba rarely exceeds $1\,mg\,l^{-1}$ in natural waters (Edmunds & Smedley, 1996). A maximum guideline value for drinking water of $0.7\,mg\,l^{-1}$ has been proposed by WHO

(1993). Brenniman *et al.* (1981) suggested that Ba has a possible association with cardiovascular disease.

15.5 Mineral dusts and health

As far as mineral dusts are concerned, they can take the form of fumes, particulates or fibres. The diseases which result from occupational exposure to dust are referred to as pneumoconioses.

Wagner (1980) described how the inhalation of mineral dusts can damage lung tissue, leading to illness and death. He showed that excessive exposure to mineral dusts, such as those produced in coal mining, slate quarrying and the processing of china clay, damage the lung, the damage consisting of little nodules of scar tissue. Once these nodules begin to coalesce, the disease becomes progressive, and is independent of further exposure. It is at this stage that breathlessness and ill health become apparent. This is associated with the disturbance of the exchange of gases, which ultimately leads to respiratory failure. In addition, the lungs become more sensitive to infections. These diseases are caused by the accumulation of isometric dust particles in the lung. Particles are retained in the lung if they are sufficiently small so that they do not fall onto the wall of the airway before they reach the alveolar (gas exchange) part of the lung. Consequently, isometric particles are retained if they are smaller than about 5 μm in diameter and larger than about 0.5 μm. Isometric particles of the size which are retained in the lung are removed by scavenger cells known as macrophages. Unless the particles damage the cells, as happens in the case of quartz dust, the dust is removed either to the lymphatic system or up the airways, leaving the lung undamaged.

Although exposure to quartz dust, that is, silicosis, produces severe scarring of the lungs, it rarely causes death by itself. The most common cause of death among individuals who contact silicosis is tuberculosis. An outline of the pathogenesis of silicosis has been provided by Wagner (1980).

Unlike quartz dust, coal dust has very little effect on the macrophages (macrophages engulf the dust in the air spaces of the lungs; quartz can kill these cells without any damage to itself). Hence, macrophages can ingest large quantities of coal dust with very little deleterious effect. It is considered that other materials in coal dust, such as mica, kaolinite and small amounts of quartz, may play a part in lungs which are so full of dust and engorged macrophages that the normal methods of clearance are overwhelmed completely.

A considerable quantity of dust is required in the lung to cause disease. For example, Elmes (1980) suggested that a coal miner needed to retain 100 g or more of dust, a slate quarryman around 10–15 g and a worker exposed to pure quartz dust around 5 g. The quantity of dust required depends partly on the amount of crystalline quartz it contains, the surface area of the particles and the presence of other compounds, such as Fe oxides, which may modify the effect of the quartz. However, dust-related diseases can be prevented by measures that suppress dust (wet processing), by adequate ventilation or by the use of masks which filter out dust.

Elongated particles and fibres have different aerodynamic properties from isometric particles, and tend to fall at a speed which is related to their cross-sectional diameter and is independent of their length. Length, however, becomes a limiting factor when the particles are longer than the diameter of the airways. Consequently, fibres are retained if they are less than 3 μm in diameter and less than 50–100 μm in length. The lower limit of retention may be around 0.1 μm in diameter and 5 μm in length. Unfortunately, elongated particles cannot be engulfed and removed by the macrophages. Fibres exceeding 10–12 μm in length remain in the alveolar part of the lung. Smaller fibres move into the lung tissue and, although some may enter the lymphatic system, others reach the surface of the lung and the pleural space. Indeed, the way in which the particle shape influences the clearance from the lung determines the relative particle potency with regard to the causation of the disease and the location at which it develops.

Wagner (1980) also described how fibrous minerals, such as asbestos, can cause extensive lung scarring (asbestosis), primary lung cancer and a cancer of the pleura or peritoneum called a mesothelioma. The difference between the effect of isometric and fibrous dust is qualitative, in that lung scarring is diffuse in the latter and nodular in the former. In addition, both types of cancer can occur when fibrous dusts are involved, whereas isometric dusts (unless radioactive or contaminated with chemical carcinogens) do not cause cancer. There is also a notable quantitative difference, in that severe disease can be associated with the retention of less than 1 g of asbestos, and mesothelioma with less than 1 mg.

The inhalation of asbestos dusts is associated with a variety of pathological changes. Some of these depend upon the type of asbestos to which the individual is exposed. Amphibole fibres are straight and stiff, and can split longitudinally. They can have an extremely fine diameter. Of the common commercial types of asbestos, crocidolite (which has a diameter of less than $0.1\,\mu m$) can penetrate the lung parenchyma through the conducting airways. Amphiboles do not undergo leaching or disintegrate in the tissue. The longer amphibole fibres become coated with an Fe-containing protein complex to form asbestos bodies which probably are inert. It is probably the smaller, uncoated fibres which cause the damage, and may continue to do so for a lifetime. By contrast, although chrysotile has a diameter less than that of crocidolite, because of its coiled configuration, it presents a far larger aerodynamic profile, and so most of these fibres tend to become caught and immobilized, and do not penetrate the smaller conducting airways. Consequently, only the shortest fibres can migrate through the lung and pleura. Once in the tissue, the Mg in the fibres is leached out, and the fibres eventually disintegrate.

In asbestosis, the scarring progresses until the alveoli along the respiratory bronchioles are replaced with a layer of scar tissue. The fibrous tissue then spreads further down the walls of the air sacs, so that more and more lung becomes involved, with scar tissue surrounding collapsed and useless air spaces. Breathlessness becomes evident once the scarring includes one-third of the lung parenchyma. This increases as the disease progresses, until the heart cannot cope, and goes into muscle fatigue and eventual failure. Excessive exposure to all types of asbestos fibre produces asbestosis.

The incidence of lung cancer is dose related and becomes frequent when exposure causes fibrosis. Elmes and Simpson (1977) showed that the risk of lung cancer developing in workers exposed to asbestos was enhanced significantly if they smoked. According to Elmes (1980), it would appear that mesothelioma is more likely to develop when a person is exposed to crocidolite. This is a rare cancer, and the latent period between first exposure and the development of cancer averages about 40 years.

15.6 Radon

Radon is a naturally occurring radioactive gas which is produced by the radioactive decay of U and Th. It is the only radioactive gas, and is colourless, odourless and tasteless. Uranium and Th are the parents of a series of radioactive daughter products, which ultimately decay to stable Pb isotopes. Three isotopes of radon are members of these series, namely, ^{219}Rn (actinon), ^{220}Rn (thoron) and ^{222}Rn (radon). The half-lives of the first two are only a matter of seconds, whereas that of the latter is 3.82 days. However, the immediate parent of ^{222}Rn is radium, ^{226}Ra. Its half-life is 1622 years. As far as release into the environment by weathering is concerned, Ra is less mobile than U. Radon decays to the solid daughter product ^{218}Po.

Although Rn frequently is present in notable amounts in areas where there is no U mineralization, it normally is associated with rocks with high concentrations of U (Ball et al., 1991). The range of concentration of ^{238}U in rocks and soil varies significantly. Rocks, such as sandstone, generally contain less than $1\,mg\,kg^{-1}$, whereas some carbonaceous shales, some rocks rich in phosphates and some granites may contain more than $3\,mg\,kg^{-1}$. For example, Ball and Miles (1993) noted that the granites in Cornwall and Devon, England, are characterized by moderate concentrations of U and are deeply weathered. Although the U in minerals in these granites may be relatively easily removed, Ra can remain. Radon therefore is easily emanated from the host rock, accounting for the high levels in soil gas. The granite areas are also characterized by high levels of U in stream sediments, groundwater and surface waters. Uranium can also be concentrated by weathering processes and resedimentation, and so can be redistributed in a form more amenable to Rn release (Bottrell, 1993). Such processes may give rise to local Rn anomalies in areas of otherwise low Rn exposure. Indeed, faults and shear zones often are enriched with U, and so are associated with elevated concentrations of Rn gas in the overlying soils. In addition, the presence of particularly high levels of Rn along a fault may be due to the increased rate of migration of Rn, because the permeability is higher along the fault, or to the presence of Ra- or Rn-bearing groundwater within the fault (Rn is the most soluble noble gas).

The amount of Rn which is emitted at the Earth's surface is related to the concentration of U in the rock and soil, as well as the efficiency of the transfer processes from the rock or soil to soil water and soil gas. In fact, most of the Rn produced in a mineral remains entrapped, but some may escape into voids in rock or soil. How much Rn is released depends upon the surface area, shape, degree of fracturing and other imperfections in the host mineral. The amount of Rn which escapes is greater in soils than in rocks. The movement of Rn in the pores depends upon the fluid flow through the rock and soil, because most Rn is transported by carrier gases or liquids, the movements of which are governed by the permeability of the rocks or soils. Jones (1995) found a significant association in Illinois between the chemical form of U occurring in soils and the basement Rn levels. Uranium atoms dispersed throughout the soil matrix were efficient emanators of Rn to the soil gas, whereas Rn which was trapped diffused very slowly from the several U-bearing minerals in the soils. Hence, it did not contribute significantly to basement Rn.

As Rn is moderately soluble in water, it can be transported over considerable distances, and therefore anomalous concentrations can occur far from the original sources of U or Th. Transport by fluids is especially rapid in limestones and along faults.

Radon represents a health hazard because it emits α particles. Outside the body, these do not present a problem, because their large size and relatively high charge mean that they cannot pass through the skin. However, when α particles are ingested or inhaled, they can damage tissue, because they are not penetrative, and therefore release energy over a relatively small volume of tissue. Indeed, the inhalation of Rn and its daughter products accounts for about one-half of the average annual exposure to ionizing radiation for the populations of the UK and USA. According to Bottrell (1993), exposure to Rn is mostly due to inhalation of indoor air (98%), but varies quite widely, with doses in much of Cornwall (from granites) being three times the UK average. Although the inhalation of Rn is the principal way in which α particles enter the human body, most is breathed out. The solid daughter products (e.g. ^{218}Po, which adheres to dust), however, are also α particle emitters, but are more dangerous, because they often

are retained in lung tissue where they increase the risk of lung cancer. The basis of Rn as an aetiological factor in lung cancer derives from the increased incidence of lung cancer in uranium miners. Jones (1995) found an association between dispersed U and the incidence of lung cancer in males and females in Illinois (Fig. 15.11). He further mentioned that concentration levels of basement Rn were more closely related to dispersed soil U than to total U, which is the sum of the dispersed and clustered forms associated with soil minerals bearing trace amounts of U. Some Rn may be dissolved in body fats, and its daughter products transferred to the bone marrow. The accumulated dose in older people can be high, and may give rise to leukaemia (Henshaw et al., 1990). Radon has also been linked with melanoma, cancer of the kidney and some childhood cancers. The United States Environmental Protection Agency (USEPA, 1986) has taken $4 pCil^{-1}$ as the limit beyond which Rn is considered to be a hazard. The average concentration of Rn outdoors is around $0.2 pCil^{-1}$ compared with approximately $1.0 pCil^{-1}$ indoors.

Radon and its daughter products accumulate in confined spaces, such as buildings. Soil gas is drawn into a building by the slight underpressure indoors, which is attributable to the rising of warmer air. A relatively small contribution is made by building materials. Radon can seep through concrete floors and foundations, drains, small cracks or joints in walls below ground level or cavities in walls. The emission of Rn from the ground can vary, for example, according to the barometric pressure and the moisture content of the soil. The accumulation of Rn is also affected by how well a building is ventilated. It can also enter a building via the water supply, particularly if a building is supplied by a private well. Public supplies of water, however, usually contain relatively little Rn, as the time over which the water is stored aids Rn release.

In some surveys of both homes and schools in the USA, some 8% registered Rn levels above $4 pCil^{-1}$ (Brenner, 1989; USEPA, 1989). For example, one school in Tennessee recorded a level of $136 pCil^{-1}$, and much higher levels have been found in some homes. Nero (1988) indicated that those who occupy homes for around 20 years where the average concentration of Rn is approximately $25 pCil^{-1}$ have a 2–3% chance of contracting lung cancer.

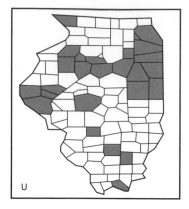

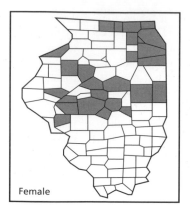

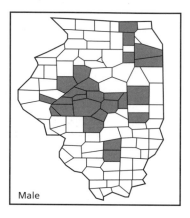

Fig. 15.11 Relationship between dispersed U and average annual incidence of lung cancer in females and males for counties in Illinois. (After Jones, 1995.)

In the UK, the National Radiological Protection Board (NRPB), in 1990, issued revised advice on Rn in homes, resulting in a new action level for Rn of 200 Bqm^{-3} (1 pCil^{-1} = 37 Bqm^{-3}). Parts of the country, in which 1% or more of the homes were above the action level, were regarded as 'affected areas'. The NRPB recommended that, within affected areas, the government should designate localities where precautions against Rn were required in new houses. The first affected area so defined was Cornwall and Devon. Ball and Miles (1993) produced maps showing the concentration of Rn in homes in this area (Fig. 15.12a,b). These maps bear a notable resemblance to the distribution of granite outcrops (Fig. 15.12c) and U mineralization (Fig. 15.12d).

The reduction of the potential hazard from Rn in homes involves the location and sealing of the points of entry of Rn, and improved ventilation by keeping more windows open or using extraction fans. Ventila-

tion systems can be built into a house during its construction, for example, a system can be installed beneath the house.

Ball *et al.* (1991) stated that Rn can be detected by zinc sulphide scintillation counting, liquid scintillation counting, α track registration, semiconductor detectors and absorbers (e.g. activated charcoal, silica gel or charged metal plates). When α particles interact with zinc sulphide, they give off pulses of light, which may be counted electronically. It is possible to calculate the activities of both radon and thoron, because of the different half-lives of radon, thoron and their immediate daughter products. The technique frequently is used for the measurement of Rn in soil gases. In liquid scintillation counting, an organic fluorescence agent is dissolved in an organic solvent, such as xylene or toluene. The solvent extracts Rn from gas or liquid phases or absorbers. The ionization of the solvent by α and β particles, and its subse-

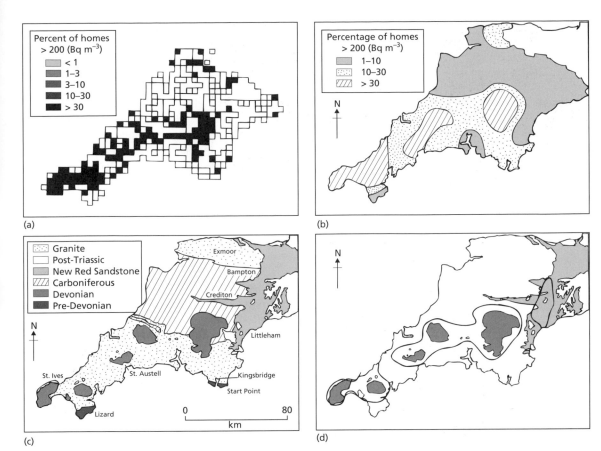

Fig. 15.12 (a) Summary of house data: percentage above action level in 5 km squares for Devon and Cornwall. (b) Estimated percentage of houses above the action level in Devon and Cornwall. (c) Simplified geological map of southwest England. (d) Simplified geological map of southwest England showing the granites (darker shading) and New Red Sandstone areas (lighter shading). Bold lines show the limit or uranium mineralization and stream sediment uranium anomalies. (After Ball & Miles, 1993.)

quent de-excitation, result in light emission, which is proportional to the Rn extracted. The α particles may be registered or detected by cellulose nitrate film, which is damaged by the particles, but is insensitive to light. The track density made by the α particles is proportional to the Rn concentration. Due to the lack of penetration of α particles, surface barrier semiconductor detectors are suitable for measuring Rn and its daughter products. Absorbers can concentrate Rn and/or its decay products. The build-up of the radioactivity of the daughter products may be determined, after a suitable time, by extraction of the Rn into a liquid scintillator.

Surveys of soil gas, based on geological information, can be used to estimate the Rn potential of an area. Such Rn surveys should be carried out in stable weather conditions. The resulting maps show the levels of Rn in the soil gas in relation to the rock type. Each rock formation should be tested several times, and some traverses should be repeated to test the variability. The best time to undertake a survey in temperate climatic regions is during the spring to autumn months, when the soils are less wet and thus are more permeable.

Radon in soil gas is measured by means of a hollow spike hammered into the ground and linked to a gas pump and detection unit. Detection of the Rn normally is by the zinc sulphide scintillation method. Account must be taken of the weather conditions and the permeability of the soil, as both can have a profound effect on Rn levels in soil gas. Alternative, passive methods employ detectors which are buried in the soil and recovered some time later, often up to a month. This procedure is used when long-term monitoring is required to overcome the problems of variation in the concentration of Rn due to changing weather conditions. Etched track methods are also used for long-term monitoring of soil Rn. Occasionally, semiconductor or absorber methods are employed.

15.7 Environmental geology and chronic disease

At present the western world, in particular, is concerned with cardiovascular disease, diseases of the central nervous system and cancers. When the incidences of these diseases are mapped within a region, variations are revealed, and the particular patterns may persist over many years. Environmental factors seem to be influencing the disease pattern, and it frequently has been suggested that geochemistry plays a part.

Many studies have indicated possible correlations between human health and geological setting. For instance, the chemistry of drinking water commonly has been cited as an important geological factor in cardiovascular disease. However, the situation is never straightforward, and other factors are involved which probably are more important, such as smoking, exercise and diet (Foster, 1992). It must also be admitted that geochemical associations may be coincidental or, at best, only partially correlated with disease incidence, and that other environmental or human constitutional factors can be important. In addition, there are considerable variations in dietary and culinary practices from one region to another, which can have a bearing on the validity of particular geochemical and disease relationships. Lastly, when vegetables are boiled, they absorb certain trace elements in water, whilst other elements are released to the water.

Nonetheless, the relationship between cardiovascular disease and water hardness has received particular attention. Masironi (1979), for example, indicated that it has long been suspected that there is an inverse relationship between water hardness (i.e. primarily dissolved Ca and Mg) and cardiovascular disease. Previously, Crawford *et al.* (1968), after

an examination of the constituents of hard water, reported that calcium carbonate appeared to be the principal component responsible for the inverse correlation of hard water with cardiovascular disease, and that there was an association between arteriosclerosis and soft water. A year earlier, Crawford and Crawford (1967) had carried out a study of individuals with coronary disease in Glasgow and London. Those in Glasgow had lower concentrations of Ca and Mg in their coronary arteries. Glasgow is located in a very soft water area as compared with London, which is in a very hard water area. Hence, the metals were accumulated in larger amounts in the coronary arteries in the hard water area, and this appeared to be beneficial in terms of heart disease.

Subsequently, Piispanea (1993) reported the results of an investigation undertaken in Finland, a country with one of the highest mortality rates of cardiovascular diseases. The incidence of cardiovascular disease differs markedly in the provinces of Vaasa and Northern Karelia. Piispanea showed that, in Vaasa, where the percentage of deaths attributable to cardiovascular disease was much lower, the hardness and Mg content (but not Ca content) of well water were significantly higher. However, Morton (1971) had suggested that rather than a simple correlation with water hardness, heart disease was related to specific ions and trace elements in water. Possible correlations seemed to exist with fluoride, sulphate, bicarbonate, carbonate, Ca and Mg, as well as pH. This was supported by Bain (1979), who carried out a survey in Ohio to determine whether any relationship existed between geological setting, water chemistry and disease. The concentrations of 11 ions, pH, total dissolved solids and hardness were compared with geological setting and with disease recorded in county death statistics for the years 1968–1971 inclusive. The survey found no correlation between death rates due to heart attack and hardness, Ca or Mg content and pH, as had been suggested by previous studies. The survey showed a slight positive correlation between death rates due to heart attack and sulphate concentration, and a slight negative correlation with bicarbonate content in water. Moreover, those counties in the southeast of Ohio, which were underlain by coal-bearing strata and where the water was sulphate rich, had high death rates due to heart attacks, whereas those counties which were covered with

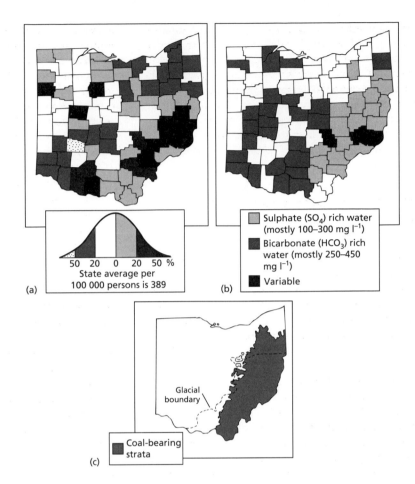

Fig. 15.13 Occurrence of (a) deaths by heart attack in individual counties, (b) water chemistry by counties and (c) simplified geology showing glacial drift covered areas and areas underlain by coal-bearing strata in Ohio. (After Bain, 1979.)

glacial deposits and had water which was bicarbonate rich recorded a low incidence of death by heart attack (Fig. 15.13).

More recently, the UK Committee on Medical Aspects of Food Policy (COMA, 1994) found a relationship between water hardness and cardiovascular disease mortality, but noted that the size of the effect was small and most clearly seen at levels of water hardness of less than $170\,mg\,l^{-1}$ (as $CaCO_3$). Several suggestions have been advanced to explain the connection with water. These include the potential for Ca and/or Mg to protect against some forms of cardiovascular disease; the fact that some trace elements, which are more prevalent in hard water, may be beneficial; and the greater solubility of many metals in soft water, which may promote cardiovascular disease.

Multiple sclerosis interferes with the autoimmune system of the body and may be viral in origin. It is possible that the virus may lie dormant until triggered by an environmental influence which causes a breakdown in the immunosuppressive system. Mims (1983), however, argued that there may be no specific virus, but that multiple sclerosis occurs when antibodies to other viruses enter the brain. It has been suggested that Pb plays a role in the aetiology of the disease. Campbell *et al.* (1950), for instance, investigated patients who were long-term residents of a village in Berkshire and another in Gloucestershire, England. In both cases, they found that garden soils were contaminated with Pb, and that the Pb content in the teeth of the patients was notably higher than that in the teeth of control groups. Nevertheless, it has not always been possible to demonstrate the link

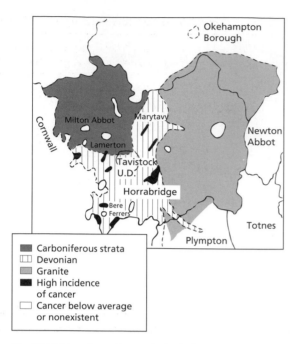

Fig. 15.14 Map of west Devon, England, showing generalized rock types and areas where the incidence of cancer is high or low. (From Allen-Price, 1960.)

with Pb. Ward *et al.* (1985), for instance, were unable to establish a relationship between multiple sclerosis sufferers and Pb, Zn and Mo.

Davies (1994) pointed out that little real progress has been made in the attempt to involve trace elements in the aetiology of cancer, and that some earlier promising associations have been found to be non-existent in later studies. Nonetheless, several reports have suggested that cancer may be related to environmental conditions. Furst (1971) considered that Ni, Cd and some Cr compounds were true metal carcinogens. He further mentioned that suggestive, but nonetheless inadequate, evidence existed for Co and Pb. Arsenic has been strongly indicated as a primary human carcinogen (North, 1992; Mitchell & Barr, 1995). Asbestos may prove to be a carrier for the carcinogenic metals, Ni and Cr. Selenium compounds are questionably carcinogenic.

The mechanisms of metal carcinogenesis are unknown, but Furst (1971) suggested that metals enter the cells by combining with a small protein. Once in the cell, the metal complex can dissociate, and the free ion can combine with S-containing units

in enzymes. This new metal derivative can inhibit the enzyme action completely or modify the kinetics of the enzyme. Under such conditions, cells may grow at different rates. The carcinogenic metals may also combine with nucleic acids, and alter the genetic information transferred to new cells. Metal toxicity can be a general manifestation of these phenomena or, in selected instances, the deranged metabolism may give rise to malignant new growth, that is, to cancer. The presence of an excess amount of a metal in an organ may result in the formation of a malignant tumour.

A number of studies were carried out in the UK in the 1960s concerning the relationship between cancer and the environment. Stocks and Davies (1960) carried out an analysis of soil elements in samples taken from 300 gardens in North Wales and Cheshire. They found that abnormal rates of mortality from stomach cancer could be correlated with high levels of Co, Zn and organic matter in the soil. These components were not related to the incidence of intestinal cancer, whereas it appeared that Cr was connected with the occurrence of both types of cancer. Vanadium and Fe showed inconclusive relationships, and Ni, Ti and Pb showed no connection. In the same year, Allen-Price (1960) published the results of a study undertaken in Devon. The study suggested that the incidence of stomach cancer possibly was associated with water supplies derived from highly mineralized Devonian strata (Fig. 15.14). Subsequently, Stocks and Davies (1964) reported that the average logarithm of the ratio of Zn to Cu in garden soils in the two areas studied was always higher where, after 10 years or more of residence, a person had just died of stomach cancer than it was in homes where an individual had died of a non-malignant cause. The effect was more pronounced and consistent for soils sampled from vegetable gardens, and was not found when the length of residence was less than 10 years.

The incidence of oesophageal cancer around the Caspian Sea in northern Iran has been reported by Kmet and Mahboubi (1972). The rates of oesophageal cancer in this region vary greatly, notable differences occurring within short distances. Kmet and Mahboubi maintained that there was a relationship between cancer, on the one hand, and climate, soils, vegetation and farming practices on the other, with the best correlation existing between rates of oeso-

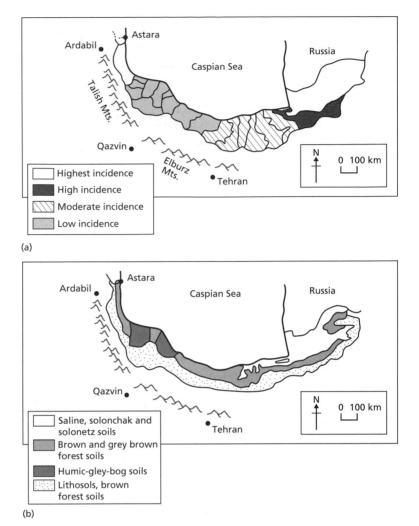

Fig. 15.15 (a) Incidence of oesophageal cancer in the Caspian littoral of Iran. (b) Types of soil in the Caspian littoral of Iran. (After Kmet & Mahboubi, 1972.)

phageal cancer and soil type (Fig. 15.15). The highest rates of the disease were associated with the saline soils of the eastern part of the region where the rainfall obviously is low. Rainfall increases towards the west, and so the leaching of salts in the soils becomes increasingly more significant. The lowest incidence of cancer occurs in the wetter western region.

References

Aggett, P.J. & Mills, C.F. (1996) Detection and anticipation of the risks of development of trace element-related disorders. In *Trace Elements in Human Nutrition and Health*. World Health Organization, Geneva, pp. 289–308.

Allen-Price, E.D. (1960) Uneven distribution of cancer in west Devon. *Lancet* **1**, 1235–1258.

Anderson, R.J. & Davies, B.E. (1980) Dental caries prevalence and trace elements in soil with special reference to lead. *Journal of the Geological Society* **137**, 547–558.

Appleton, J.D. (1992) *Review of the Use of Regional Geochemical Maps for Identifying Areas Where Mineral Deficiencies or Excesses May Affect Cattle Productivity in Tropical Countries. British Geological Survey, Technical Report* **WC/92/94**. British Geological Survey, Keyworth.

Appleton, J.D. & Ridgway, J. (1993) Regional geochemical mapping in developing countries and its application to environmental studies. *Applied Geochemistry, Supplement* **2** (8), 103–110.

Bache, B.W. (1996) Aluminium mobilization in soils and waters. *Journal of the Geological Society* **143**, 699–706.

Bain, R.J. (1979) Heart disease and geologic setting in Ohio. *Geology* 7, 7–10.

Ball, T.K. & Miles, J.C.H. (1993) Geological and geochemical factors affecting the radon concentration in homes in Cornwall and Devon, United Kingdom. *Environmental Geochemistry and Health* 15, 27–36.

Ball, T.K., Cameron, D.G., Colman, T.B. & Roberts, P.D. (1991) Behaviour of radon in the geological environment: a review. *Quarterly Journal of Engineering Geology* 24, 169–182.

Bottrell, S.H. (1993) Redistribution of uranium by physical processes during weathering and implications for radon production. *Environmental Geochemistry and Health* 15, 21–25.

Bowen, H.J.M. (1966) *Trace Elements in Biochemistry*. Academic Press, New York.

Brenner, D.J. (1989) *Radon, Risk and Remedy*. W.H. Freeman, New York.

Brenniman, G.R., Kojola, W.H., Levey, P.S., Carnow, B.W. & Nameteka, T. (1981) High barium levels in public drinking water and its association with elevated blood pressure. *Archives of Environmental Health* 36, 28–32.

Brinkman, R. (1994) Lead pollution in soils adjacent to homes in Tampa, Florida. *Environmental Geochemistry and Health* 16, 59–64.

Calder, I.C., Maynard, E.J. & Heyworth, J.S. (1994) Port Pirie Lead Abatement Program, 1992. *Environmental Geochemistry and Health* 16, 137–145.

Campbell, A.M.G., Herdann, G., Tatlow, W.F.T. & Whittle, E.G. (1950) Lead in relation to disseminated sclerosis. *Brain* 73, 52–71.

Cannon, H.L. (1976) Lead in the atmosphere, natural and artificially occurring lead, and the effects of lead on health. In *Lead in the Environment, United States Geological Survey, Professional Paper* 957, Lovering, T.G. (ed). United States Geological Survey, Washington, DC, pp. 75–80.

CEC (1980) *Directive Related to the Quality of Water for Human Consumption 80/778/EEC.* Commission of the European Community, Brussels.

Chen, C.J., Chen, C.W., Wu, M.M. & Kuo, T.L. (1992) Cancer potential in liver, lung, bladder and kidney due to ingested inorganic arsenic in drinking water. *British Journal of Cancer* 66, 888–892.

Chen, S.L., Dzeng, S.R., Yang, M.H., Chiu, K.H., Shieh, G.M. & Wai, C.M. (1994) Arsenic species in groundwater of the Blackfoot disease area, Taiwan. *Environmental Science and Technology* 28, 877–881.

COMA (1994) *Nutritional Aspects of Cardiovascular Disease No. 46.* Committee on Medical Aspects of Food Policy, Her Majesty's Stationery Office, London.

Crawford, M.D., Gardner, M.J. & Morris, J.N. (1968) Mortality and hardness of local water supplies. *Lancet* 1, 827–830.

Crawford, T. & Crawford, M.D. (1967) Prevalence and pathological changes of ischaemic heart disease in a hard water and in a soft water area. *Lancet* 1, 229–232.

Darnley, A.G., Bjorklund, A., Bolviken, B., Gustavsson, N. & Koval, P. (1995) *A Global Geochemical Database: Recommendations for International Geochemical Mapping. Final Report of ICGP Project 259.* UNESCO, Paris.

Davies, B.E. (1972) Occurrence and distribution of lead and other metals in two areas of unusual disease incidence in Britain. In *Proceedings of the International Symposium on Environmental Health Aspects of Lead, Amsterdam*, pp. 125–134.

Davies, B.E. (1994) Trace elements in the human environment: problems and risks. *Environmental Geochemistry and Health* 16, 97–106.

Davis, A., Ruby, M.V. & Bergstrom, P.D. (1994) Factors controlling lead bioavailability in the Butte mining district, Montana, USA. *Environmental Geochemistry and Health* 16, 147–157.

Dean, H.T. (1945) On the epidemiology of fluoride and dental caries. In *Fluoride in Dental Public Health*, Gies, W.J. (ed). Academy of Medicine, New York, pp. 19–30.

Dissanayake, C.B. (1991) The fluoride problem in the groundwater of Sri Lanka—environmental management and health. *International Journal of Environmental Studies* 38, 137–156.

Edmunds, W.M. & Smedley, P.L. (1996) Groundwater geochemistry and health. In *Environmental Geochemistry and Health with Special Reference to Developing Countries, Geological Society Special Publication No.* 113, Appleton, J.D., Fuge, R. & McCall, G.J.H. (eds). Geological Society, London, pp. 91–105.

Edmunds, W.M. & Trafford, J.M. (1993) Beryllium in river baseflow, shallow groundwaters and major aquifers in the U.K. *Applied Geochemistry, Supplementary Issue* 2 (8), 223–233.

Edmunds, W.M., Cook, J.M., Darling, W.G., Kinniburgh, D.G., Miles, D.L., Bath, A.H., Morgan-Jones, M. & Andrews, J.N. (1987) Baseline geochemical conditions in the Chalk aquifer, Berkshire, U.K.: a basis for groundwater quality management. *Applied Geochemistry* 2, 251–274.

Elmes, P.C. (1980) Fibrous minerals and health. *Journal of the Geological Society* 137, 525–535.

Elmes, P.C. & Simpson, M.J.C. (1977) Insulation workers in Belfast: a further study of mortality due to asbestos exposure (1940–1975). *British Journal of Industrial Medicine* 34, 174–180.

Fordyce, F.M., Masara, D. & Appleton, J.D. (1996) Stream sediment, soil and forage chemistry as indicators of cattle mineral status in north east Zimbabwe. In *Environmental Geochemistry and Health with Special Reference to Developing Countries, Geological Society Special Publication No.* 113, Appleton, J.P., Fuge, R. & McCall, G.J.H. (eds). Geological Society, London, pp. 23–37.

Foster, H.D. (1992) *Health, Disease and the Environment.* Belhaven, London.

Fuge, R. (1989) Iodine in waters: possible links with endemic goitres: a review. *Applied Geochemistry* **4**, 203–308.

Fuge, R. (1996) Geochemistry of iodine in relation to iodine deficiency diseases. In *Environmental Geochemistry and Health with Special Reference to Developing Countries, Geological Society Special Publication No. 113*, Appleton, J.D., Fuge, R. & McCall, G.J.H. (eds). Geological Society, London, pp. 201–211.

Fuge, R. & Johnson, C.C. (1986) The geochemistry of iodine: a review. *Environmental Geochemistry and Health* **8**, 34–54.

Furst, A. (1971) Trace elements related to cancer. In *Environmental Geochemistry in Disease and Health, Geological Society of America Memoir* **123**, Cannon, H.L. & Hopps, H.C. (eds). Geological Society of America, New York, pp. 109–130.

Greeson, P.E. (1970) Biological factors in the chemistry of mercury. In *Mercury in the Environment, United States Geological Survey, Professional Paper 713*. United States Geological Survey, Washington, DC, pp. 32–34.

Hem, J.D. (1970) Chemical behaviour of mercury in aqueous media. In *Mercury in the Environment, United States Geological Survey, Professional Paper 713*. United States Geological Survey, Washington DC, pp. 19–24.

Hem, J.D. (1985) *Study and Interpretation of the Chemical Characteristics of Natural Water. United States Geological Survey, Water Supply Paper* **2254**. United States Geological Survey, Washington, DC.

Henshaw, D.L., Eatough, J.P. & Richardson, R.B. (1990) Radon: a causative factor in the induction of myeloid leukaemia and other cancers in adults and children? *Lancet* **335**, 1008–1012.

Hetzel, B.S. (1991) The conquest of iodine deficiency: a special challenge to Australia from Asia. *Proceedings of the Nutrition Society of Australia* **16**, 68–78.

Hopps, H.C. (1971) Geographic pathology and the medical implications of environmental geochemistry. In *Environmental Geochemistry in Disease and Health, Geological Society of America Memoir* **123**, Cannon, H.L. & Hopps, H.C. (eds). Geological Society of America, New York, pp. 1–11.

Jarup, L. (1992) *Dose–Response Relations for Occupational Exposure to Arsenic and Cadmium.* National Institute for Occupational Health, Stockholm.

Jenne, E.A. (1970) Atmospheric and fluvial transport of mercury. In *Mercury in the Environment, United States Geological Survey, Professional Paper 713*. United States Geological Survey, Washington, DC, pp. 40–45.

Jones, R.L. (1995) Soil uranium, basement radon and lung cancer in Illinois, USA. *Environmental Geochemistry and Health* **17**, 21–24.

Kmet, J. & Mahboubi, F. (1972) Esophageal cancer in the Caspian littoral of Iran. *Science* **175**, 846–853.

Kurland, L.T., Faro, S.N. & Siedler, H.S. (1960) Minimata disease. *World Neurologist* **1**, 320–325.

Lakin, W.H. (1973) Selenium in our environment. In *Trace Elements in the Environment*, Kothny, E.L. (ed), *Advances in Chemistry Series* **123**. American Chemical Society, New York, pp. 96–111.

Lamont, P. (1959) A soft water zone in the Lincolnshire Limestone. *Journal of the British Water Association* **41**, 48–71.

Lebdosoekojo, S., Ammerman, C.B., Raun, N.S., Gomez, J. & Litell, R.C. (1980) Mineral nutrition of beef cattle grazing native pastures on the eastern plains of Colombia. *Journal of Animal Science* **51**, 1249–1260.

Leckie, J.O. & Parks, G.A. (1978) Geochemistry and environmental impact. In *Geology in Environmental Planning*, Howard, A.D. & Remson, I. (eds). McGraw–Hill, New York, pp. 276–288.

Levander, O.A. (1986) Selenium. In *Trace Elements in Human and Animal Nutrition*, 5th edn., Vol. 2, Mertz, W. (ed). Academic Press, Orlando, pp. 209–279.

Lidiard, H.M., Rae, J.E. & Parker, A. (1993) Identification of Mn-oxide minerals in some soils from Devon, U.K. and their varying capacity to absorb Co and Cu. *Environmetal Geochemistry and Health* **15**, 93–99.

Lovering, T.C. (1976) Summary. In *Lead in the Environment, United States Geological Survey Professional Paper* **957**, Lovering, T.G. (ed). United States Geological Survey, Washington, DC, pp. 1–4.

Lutz, P.M., Jayachandran, C., Gale, N.L., *et al.* (1994) Immunity in children with exposure to environmental lead: 1. Effects on cell numbers and cell-mediated immunity. *Environmental Geochemistry and Health* **16**, 167–177.

Marshall, E. (1990) The fluoride debate: one more time. *Science* **247**, 276–277.

Martyn, C.N., Barker, D.J.P., Osmond, C., Harris, E.C., Edwardson, J.A. & Lacey, R.F. (1989) Geophysical relations between Alzheimer's disease and aluminium in drinking water. *Lancet* **1**, 59–62.

Masironi, R. (1979) Geochemistry and cardiovascular diseases. *Philosophical Transactions of the Royal Society of London* **B288**, 193–203.

Mertz, W. (1987) *Trace Elements in Human and Animal Nutrition*. Academic Press, Orlando.

Mills, C.F. (1995) Trace element bioavailability and interactions. In *Trace Elements in Human Nutrition and Health*. World Health Organization, Geneva, pp. 22–46.

Mills, C.F. (1996) Geochemistry and trace element related diseases. In *Environmental Geochemistry and Health with Special Reference to Developing Countries, Geological Society Special Publication No. 113*, Appleton, J.P., Fuge, R. & McCall, G.J.H. (eds). Geological Society, London, pp. 1–5.

Mims, C. (1983) Multiple sclerosis—the case against viruses. *New Scientist* **98**, 938–940.

Mitchell, P. & Barr, D. (1995) The nature and significance of public exposure to arsenic: a review of its relevance to south west England. *Environmental Geochemistry and Health* **17**, 57–82.

Morton, W.E. (1971) Hypertension and drinking water constituents in Colorado. *American Journal of Public Health* **61**, 1371–1378.

Morton, W.E. & Dunette, D.A. (1994) Health effects of environmental arsenic. In *Arsenic in the Environment, Part II: Human Health and Ecosystem Effects*, Nriagu, J.O. (ed). Wiley, New York, pp. 159–170.

Myers, H.M. (1978) *Fluorides and Dental Fluorosis. Monograph on Oral Science*. Karger, Basle.

Nero, A.V. (1988) Controlling indoor air pollution. *Scientific American* **258**, No. 5, 42–48.

North, D.W. (1992) Risk assessment for ingested inorganic arsenic: a review and status report. *Environmental Geochemistry and Health* **14**, 59–62.

Patterson, C.C. (1965) Contaminated and natural lead environments of man. *Archives of Environmental Health* **11**, 344–360.

Peereboom, J.W.C. (1985) General aspects of trace elements and health. *Science of the Total Environment* **42**, 1–27.

Perry, H.M. (1971) Trace elements related to cardiovascular disease. In *Environmental Geochemistry in Disease and Health, Geological Society of America Memoir* **123**, Cannon, H.L. & Hopps, H.C. (eds). Geological Society of America, New York, pp. 179–195.

Pettyjohn, W.A. (1972) No thing is without poison. In *Man and His Physical Environment*, McKenzie, G.D. & Utgard, R.O. (eds). Burgess Publishing, Minneapolis, pp. 109–110.

Pharoah, P.O.D. (1985) The epidemiology of endemic cretinism. In *The Endocrine System and the Environment*, Follet, B.K., Ishil, S. & Chandola, A. (eds). Springer, Berlin, pp. 315–322.

Piispanen, R. (1993) Water hardness and cardiovascular mortality in Finland. *Environmental Geochemistry and Health* **15**, 201–208.

Plant, J.A., Baldock, J.W. & Smith, B. (1996) The role of geochemistry in environmental and epidemiological studies in developing countries: a review. In *Environmental Geochemistry and Health with Special Reference to Developing Countries, Geological Society Special Publication No.* **113**, Appleton, J.P., Fuge, R. & McCall, G.J.H. (eds). Geological Society, London, pp. 7–22.

Pories, W.W., Strain, W.H. & Rob, C.G. (1971) Zinc deficiency in delayed healing and chronic disease. In *Environmental Geochemistry in Disease and Health, Geological Society of America Memoir* **123**, Cannon, H.C. & Hopps, H.C. (eds). Geological Society of America, pp. 73–95.

Robards, K. & Worsfold, P. (1991) Cadmium toxicology and analysis: a review. *Analyst* **16**, 549–568.

Schroeder, H.A. (1965) Cadmium as a factor in hypertension. *Journal of Chronic Disease* **18**, 647–656.

Schroeder, H.A., Nason, A.P. & Mitchener, M. (1968) Action of a chelate of zinc on trace metals in hypertensive rats. *American Journal of Physiology* **214**, 796–805.

Smedley, P.L., Edmunds, W.M. & Pelig-Ba, K. (1996) Mobility of arsenic in groundwater in the Obuasi gold mining area of Ghana: some indications for human health. In *Environmental Geochemistry and Health with Special Reference to Developing Countries, Geological Society Special Publication No.* **113**, Appleton, J.P., Fuge, R. & McCall, G.J.H. (eds). Geological Society, London, pp. 223–230.

Stewart, A.G. & Pharoah, P.O.D. (1996) Clinical and epidemiological correlation of iodine deficiency disorders. In *Environmental Geochemistry and Health with Special Reference to Developing Countries, Geological Society Special Publication No.* **113**, Appleton, J.D., Fuge R., & McCall, G.J.H. (eds). Geological Society, London, pp. 223–230.

Stocks, P. & Davies, R.I. (1960) Epidemiological evidence from chemical and spectrographic analysis that soil is concerned in the causation of cancer. *British Journal of Cancer* **14**, 8–22.

Stocks, P. and Davies, R.I. (1964) Zinc and copper contents of soils associated with the incidence of cancer of the stomach and other organs. *British Journal of Cancer* **18**, 14–24.

Tebbutt, T.H.Y. (1983) *Relationship between Natural Water Quality and Health*. UNESCO, Paris.

Thilly, C.H., Vanderplas, J.B., Bebe, N., Ntambue, K. & Contempre, B. (1992) Iodine deficiency, other trace elements and goitrogenic factors in etiopathogeny of iodine deficiency disorders (IDD). *Biological Trace Elements Research* **32**, 229–243.

Thornton, I. (1983) *Applied Environmental Geochemistry*. Academic Press, London.

Thornton, I. (1996) Sources and pathways of arsenic in the geochemical environment: health implications. In *Environmental Geochemistry and Health with Special Reference to Developing Countries, Geological Society Special Publication No.* **113**, Appleton, J.P., Fuge, R. & McCall, G.J.H. (eds). Geological Society, London, pp 153–161.

Thornton, I. & Plant, J.A. (1980) Regional geochemical health and mapping in the United Kingdom. *Journal of the Geological Society* **137**, 575–586.

Tseng, W.P. (1977) Effects of dose response relationships on skin cancer and Blackfoot disease with arsenic. *Environmental Health Perspectives* **19**, 109–119.

USEPA (1986) *Citizen's Guide to Radon. EPA-86-004* United States Environmental Protection Agency, Washington, DC.

USEPA (1989) *Radon, Risk and Remedy. EPA-520/1-89 010*. United States Environmental Protection Agency, Washington, DC.

Wagner, J.C. (1980) The pneumoconioses due to mineral dusts. *Journal of the Geological Society* **137**, 537–545.

Ward, N.I., Bryce-Smith, D., Minski, M. & Matthews, W.B. (1985) Multiple sclerosis—a multi-element survey. *Biological Trace Elements Research* **7**, 153–159.

Warnakulasuriya, K.A.A.S., Balasuriya, S. & Perera, P.A.J. (1990) Prevalence of dental fluorosis in four selected schools from different areas of Sri Lanka. *Ceylon Medical Journal* **35**, 125–128.

Webb, J.S. (1971) Regional geochemical reconnaissance in medical geography. In *Environmental Geochemistry in Disease and Health, Geological Society of America Memoir* **123**, Cannon, H.C. & Hopps, H.C. (eds). Geological Society of America, New York, pp. 31–42.

Welch, A.H., Lico, M.S. & Hughes, J.L. (1988) Arsenic in groundwater of the western United States. *Ground Water* **26**, 333–347.

Whitehead, D.C. (1979) Iodine in the United Kingdom environment with particular reference to agriculture. *Journal of Applied Ecology* **16**, 269–279.

WHO (1993) *Guidelines for Drinking Water Quality.* World Health Organization, Geneva.

16 Land evaluation and site assessment

16.1 Introduction

Land evaluation and site assessment are undertaken to help to determine the most suitable use of land in terms of planning and development or for construction purposes. In the process, the impact on the environment of a particular project may have to be determined. This especially is the case as far as large projects are concerned. Obviously, there has to be a geological input into this process. This may be in terms of earth processes and geological hazards, mineral resources and the impact of mining, water supply and hydrogeological conditions, soil resources, ground conditions or the disposal of waste. The impact of the development of land is most acute in urban areas, where the human pressures on land are greatest. Urban development, together with the growth of industry and mining, in particular, has led to the spoilation and dereliction of land in the past and, no doubt, will in the future, but hopefully, to a much lesser extent. This is where planning has a vital role to play, and where the geologist must be involved.

Investigations in relation to land-use planning and development obviously can take place at various scales, from regional, to local, to site investigations. Regional investigations generally are undertaken on behalf of government authorities at, for example, state or county level, and may involve the location and use of mineral resources, the identification of hazards, the recognition of problems due to past types of land use, and land capability studies and zoning for future land use. In this context, it is necessary to recognize those geological factors which represent a resource or constitute a constraint. A constraint imposes a limitation on the use to which land can be put, so that a particular locality which is so affected is less suited to a specific activity than another. Local and site investigations tend to be undertaken for specific reasons, for example, the location of a suitable site for a landfill, the determination of a site for the construction of a reservoir and dam or the development of a brick pit and works. In such cases, investigations will be necessary to obtain the relevant information, including geological data, for the planning processes, which in many countries, will include a public inquiry. Regional investigations entail the production of engineering geomorphological and environmental geological maps, including hazard maps, with associated reports. Site investigations tend to involve the production of engineering geological (or geotechnical) maps and reports, although this again is not clearcut, e.g. road schemes frequently involve the production of engineering geomorphological maps as part of the site investigation.

Any investigation begins with the formulation of aims, what does the investigation want to achieve, and which type of information is of relevance to the particular project in question? Once the pertinent questions have been posed, the nature of the investigation can be defined, and the process of data collection can begin. The amount of detail required will depend largely on the purpose of the investigation. For instance, less detail is required for a feasibility study for a project than by engineers for the design and construction of that project. Various methodologies are employed in data collection. These may include the use of remote sensing imagery, aerial photography, existing literature and maps, fieldwork and mapping, subsurface exploration by boring and drilling, sample collection and geophysical surveying. In some instances, geochemical data, notably where water or ground is polluted or contaminated, may need to be gathered, or field testing or monitoring programmes may need to be carried out. Once the relevant data have been obtained, they must be interpreted and evaluated, and together with the conclusions, embodied in a report.

Aerial photographs and remote imagery have proved to be valuable aids in land and site evaluation, particularly in those underdeveloped regions of the world where good topographical maps do not exist. However, the amount of useful information obtainable from aerial photographs and imagery depends upon their characteristics, as well as the nature of the terrain they portray. Remote imagery and aerial photographs prove to be of most value during the planning and reconnaissance stages of a project. The information they provide can be transposed to a base map, and this is checked during fieldwork. This information not only allows the fieldwork programme to be planned much more readily, but should also help to shorten the period spent in the field.

16.2 Remote sensing

Remote sensing commonly represents one of the first stages of land assessment. It involves the identification and analysis of phenomena on the Earth's surface by using devices borne by aircraft or spacecraft. Most techniques used in remote sensing depend upon the recording of energy from part of the electromagnetic spectrum, ranging from γ-rays, through the visible spectrum, to radar. The scanning equipment used measures both emitted and reflected radiation, and the employment of suitable detectors and filters permits the measurement of certain spectral bands to be performed. Signals from several bands of the spectrum can be recorded simultaneously by multispectral scanners. The two principal systems of remote sensing are infrared line scanning (IRLS) and side-looking airborne radar (SLAR).

16.2.1 Infrared line scanning

Infrared line scanning is dependent upon the fact that all objects emit electromagnetic radiation, generated by the thermal activity of their component atoms. Emission is greatest in the infrared region of the electromagnetic spectrum for most materials at ambient temperature. The reflected infrared region ranges in wave length from 0.7 to 3.0 μm, and includes the photographic infrared band. This can be detected by certain infrared-sensitive films. The thermal infrared region ranges in wave length from 3.0 to 14.0 μm. The most effective wave band used for thermal IRLS for geological purposes is 8–14 μm.

Infrared line scanning involves the scanning of a succession of parallel lines across the track of an aircraft with a scanning spot. The spot travels forwards and backwards in such a manner that nothing is missed between consecutive passes. Because only an average radiation is recorded, the limits of resolution depend on the size of the spot. The diameter of the spot usually is around 2–3 mrad, which means that, if the aircraft is flying at a height of 1000 m, the spot measures 2–3 m across. The radiation is picked up by a detector, which converts it into electrical signals that, in turn, are transformed into visible light via a cathode ray tube, thereby enabling a record to be made on film or magnetic tape. The data can be processed in colour, as well as black and white, colours substituting for grey tones. Unfortunately, prints are distorted increasingly with increasing distance from the line of flight, which limits the total useful angle of scan to about 60° on either side. In order to reduce the distortion along the edges of the imagery, flight lines have a 50–60% overlap. According to Warwick et al. (1979), a temperature difference of 0.15 °C between objects of 500 mm in diameter can be detected by an aircraft at an altitude of 300 m. The spatial resolution is, however, much lower than that of aerial photographs, in which the resolution at this height would be 80 mm. At higher altitudes, the difference becomes more marked.

The use of IRLS depends on clear, calm weather. Moreover, some thought must be given to the fact that thermal emissions vary significantly throughout the day. The time of the flight therefore is important. From the geological point of view, predawn flying proves to be most suitable for thermal IRLS. This is because radiant temperatures are fairly constant, and reflected energy is not important, whereas during a sunny day, the radiant and reflected energy are roughly equal, so that the latter may obscure the former. In addition, because sun-facing slopes are warm and shaded slopes cool, rough topography tends to obliterate the geology in post-dawn imagery.

Although temperature differences of 0.1 °C can be recorded by IRLS, these do not represent differences in the absolute temperature of the ground, but in the emission of radiation. Careful calibration therefore is needed in order to obtain absolute values. The emitted radiation is determined by the temperature of the object and its emissivity, which can vary with the surface roughness, soil type, moisture content and vegetative cover.

A grey scale can be used to interpret the imagery, it being produced by computer methods from line scan data which have been digitized. This enables maps of isoradiation contours to be produced. Colour enhancement has also been used to produce isotherm contour maps, with colours depicting each contour interval. This method has been used in the preparation of maps of engineering soils.

The identification of grey tones is the most important aspect as far as the interpretation of thermal imagery is concerned, because these provide an indication of the radiant temperatures of a surface. Warm areas give rise to light and cool areas to dark tones. Relatively cold areas are depicted as purple and relatively hot areas as red on a colour print. Thermal inertia is important in this respect, because rocks with high thermal inertia, such as dolostone or quartzite, are relatively cool during the day and warm at night. Rocks and soils with low thermal inertia, for example, shale, gravel or sand, are warm during the day and cool at night. In other words, the variation in the temperature of materials with high thermal inertia during the daily cycle is much less than those with low thermal inertia. Because clay soils possess relatively high thermal inertia, they appear warm in predawn imagery, whereas sandy soils, because of their relatively low thermal inertia, appear cool. The moisture content of a soil influences the image produced, that is, soils which possess high moisture content may mask differences in soil types. Fault zones often are picked out because of their higher moisture content. Similarly, the presence of old landslides frequently can be discerned as their moisture content differs from that of their surroundings.

Texture can also aid interpretation. For instance, outcrops of rock may have a rough texture due to the presence of bedding or jointing, whereas soils usually give rise to a relatively smooth texture. However, where the soil cover is less than 0.5 m, the rock structure usually is observable on the imagery, because deeper, more moist soil occupying discontinuities gives a darker signature. Free-standing bodies of water usually are readily visible on thermal imagery, however, the high thermal inertia of highly saturated organic deposits may approach that of water masses. The two therefore may be difficult to distinguish at times.

16.2.2 Side-looking airborne radar

In SLAR, short pulses of energy, in a selected part of the radar wave band, are transmitted sideways to the ground from antennae on both sides of an aircraft (Sabins, 1996). The pulses of energy strike the ground along successive range lines, and are reflected back at time intervals related to the height of the aircraft above the ground. The reflected pulses are transformed into black and white photographs with the aid of a cathode ray tube. Returning pulses cannot be accepted from any point within 45° from the vertical, so that there is a blank space under the aircraft along its line of flight. In addition, the image becomes increasingly distorted towards the track of the aircraft. The belt covered by normal SLAR imagery varies from 2 to 50 km and, although the scanning is oblique, the system converts it to an image which is more or less planimetric.

There are some notable differences between SLAR images and aerial photographs. For instance, although variations in vegetation produce slightly different radar responses, an SLAR image depicts the ground more or less as it would appear on aerial photographs devoid of vegetation. Displacements of relief are to the side towards the imaging aircraft and not radial about the centre as in aerial photographs. Furthermore, radar shadows fall away from the flight line and are normal to it. The shadows on SLAR images form black areas which yield no information, whereas most areas of shadow on aerial photographs partially are illuminated by diffuse lighting. The subtle changes of tone and texture which occur on aerial photographs are not observable on SLAR images.

Because the wave lengths used in SLAR are not affected by cloud cover, imagery can be obtained at any time. This is particularly important in equatorial regions, which rarely are free of cloud. Consequently, this technique provides an ideal means of reconnaissance survey in such areas.

Typical scales for radar imagery, available commercially, are 1:100 000–1:250 000, with a resolution of between 10 and 30 m. Smaller objects than this can appear on the image if they are strong reflectors, and the original material can be enlarged. Mosaics are suitable for the identification of regional geological features and for the preliminary identification of terrain units. The lateral overlap of radar cover can give a stereoscopic image, which offers a more reliable assessment of the terrain. Furthermore, imagery recorded by radar systems can provide appreciable detail of landforms as they are revealed due to the low angle of incident illumination.

16.2.3 Satellite imagery

Small-scale space imagery provides a means of initial reconnaissance, which allows areas to be selected for further, more detailed, investigation, either by aerial and/or ground survey methods. Indeed, in many parts of the world, a Landsat image may provide the only form of base map available. The large areas of the ground surface covered by satellite images give a regional physiographical setting, and permit the distinction of various landforms to be made according to their characteristic photopatterns. Accordingly, such imagery can provide a geomorphological framework from which a study of the component landforms is possible. The character of the landforms may afford some indication of the type of material from which they are composed, and geomorphological data aid in the selection of favourable sites for field investigation on larger scale aerial surveys. Small-scale imagery may enable regional geological relationships and structures to be identified, which are not noticeable on larger scale imagery or mosaics.

The capacity to detect surface features and landforms from imagery obtained by multispectral scanners on satellites is facilitated by the recording of energy, reflected from the ground surface, within four specific wave length bands. These are visible green (0.5–0.6 μm), visible red (0.6–0.7 μm) and two invisible infrared bands (0.7–0.8 μm and 0.9–1.0 μm). The images are reproduced on photographic paper, and are available for the four spectral bands plus two false colour composites. The infrared band between 0.7 and 0.8 μm is probably the best for geological purposes. Because separate images within different wave lengths are recorded at the same time, the likelihood of recognizing different phenomena is enhanced significantly. Because the energy emitted and reflected from objects commonly varies according to the wave length, its characteristic spectral pattern or signature in an image is determined by the amount of energy transmitted to the sensor within the wave length range in which that sensor operates. As a consequence, a unique tonal signature frequently may be

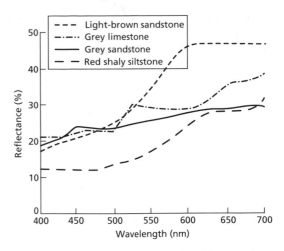

Fig. 16.1 Spectral reflectance curves for four different rock types. (From Beaumont, 1979.)

identified for a feature, if the energy which is being emitted and/or reflected from it is broken into specially selected wave length bands.

In Fig. 16.1, the reflectance curves for four different rock types illustrate the higher reflectance of brown sandstone at longer (orange–red) wave lengths and the lower reflectance of siltstone in the shorter (blue) wave lengths of the visible spectrum (Beaumont, 1979). This indicates that, if reflected energy from the shorter and longer ends of the visible spectrum are recorded separately, differentiation between rock types can be achieved. The ability to distinguish between different materials increases when imagery is recorded by different sensors outside the visible spectrum, the spectral characteristics then being influenced by the atomic composition and molecular structure of the materials concerned.

In addition to the standard photographs at a scale of 1:1 000 000, both transparencies (positive and negative) and enlargements, at scales of 1:250 000 and 1:500 000, are available, as are false colour composites. The latter often identify features not easily observable on black and white images.

Satellite images may be interpreted in a similar manner to aerial photographs, although the images do not come in stereopairs. Nevertheless, a pseudo-stereoscopic effect may be obtained by viewing two different spectral bands (band-lap stereo) of the same image, or by examining images of the same view taken at different times (time-lap stereo). There is also a certain amount of side-lap, which improves with latitude. This provides a true stereographic image across a restricted strip of a print, however, significant effects are only produced by large relief features. The interpretation of satellite data may also be accomplished by automated methods using digital data directly, or by using interactive computer facilities with visual display devices.

The value of space imagery is important where existing map coverage is inadequate. For example, it can be of use for the preparation of maps of terrain classification, for regional engineering soil maps, for maps used for route selection, for regional inventories of construction materials and for inventories of drainage networks and catchment areas (Beaumont, 1979). A major construction project is governed by the terrain, the optimum location requiring minimum disturbance of the environment. In order to assess the ground conditions, it is necessary to make a detailed study of all the photopattern elements that comprise the landforms on the satellite imagery. Important evidence relating to soil types or surface or subsurface conditions may be provided by erosion patterns, drainage characteristics or vegetative cover. Engineering soil maps frequently are prepared on a regional basis for both planning and location purposes, in order to minimize construction costs, the soils being delineated for the landforms within the regional physiographical setting.

16.3 Aerial photographs and photogeology

Aerial photographs generally are taken from an aeroplane which is flying at an altitude of between 800 and 9000 m, the height being governed by the amount of detail that is required. Photographs may be taken at different angles, ranging from vertical to low oblique (excluding horizon) to high oblique (including horizon). Vertical photographs, however, are the most relevant for photogeological purposes. Oblique photographs occasionally have been used for survey purposes, but because their scale of distortion from foreground to background is appreciable, they are not really suitable. Nevertheless, because they offer a graphic visual image of the ground, they constitute good illustrative material.

Normally, vertical aerial photographs have 60% overlap on consecutive prints on the same run, and adjacent runs have a 20% overlap or side-lap. As a result of tilt (the angular divergence of the aircraft from a horizontal flight path), no photograph is ever exactly vertical, but the deviation is almost invariably less than 1°. Scale distortion away from the centre of the photograph represents another source of error.

Not only does a study of aerial photographs allow the area concerned to be divided into topographical and geological units, but it also enables the geologist to plan fieldwork and to select locations for sampling. This should result in a shorter, more profitable, period in the field. When a detailed interpretation of aerial photographs is required, the photographs can be enlarged to approximately twice the scale of the final map to be produced (Rengers & Soeters, 1980).

The examination of consecutive pairs of aerial photographs with a stereoscope allows an observation to be made of a three-dimensional image of the ground surface. This is due to parallax differences brought about by photographing the same object from two different positions. The three-dimensional image means that heights can be determined and contours can be drawn, thereby producing a topographical map. However, the relief presented in this image is exaggerated, so that slopes appear steeper than they actually are. Nonetheless, this aids in the detection of minor changes in slope and elevation. Unfortunately, exaggeration proves to be a definite disadvantage in mountainous areas, as it becomes difficult to distinguish between steep and very steep slopes. A camera with a longer focal lens reduces the amount of exaggeration, and therefore may be preferable in such areas.

Aerial photographs may be combined in order to cover larger regions. The simplest type of combination is the uncontrolled print laydown, which consists of photographs, laid alongside each other, which have not been accurately fitted into a surveyed grid. Photomosaics represent a more elaborate type of print laydown, requiring more care in their production, and controlled photomosaics are based on a number of geodetically surveyed points. They can be regarded as having the same accuracy as topographical maps.

16.3.1 Types of aerial photograph

There are four main types of film used in normal aerial photography, namely, black and white, infrared monochrome, true colour and false colour. Black and white film is used for topographical survey work and for normal interpretation purposes. The other types of film are used for special purposes. For example, infrared monochrome film makes use of the fact that near-infrared radiation is strongly absorbed by water. Accordingly, it is of particular value when mapping shorelines, the depth of shallow underwater features and the presence of water on land, for example, in channels, at shallow depths underground or beneath vegetation. Furthermore, it is more able to penetrate haze than conventional photography. True colour photography displays variations of hue, value and chroma, rather than tone only, and generally offers much more refined imagery. As a consequence, colour photographs have an advantage over black and white ones as far as photogeological interpretation is con-cerned, in that there are more subtle changes in colour in the former than in grey tones in the latter; hence they record more geological information. However, colour photographs are more expensive, and it is difficult to reproduce slight variations in shade consistently in processing. Another disadvantage is the attenuation of colour in the atmosphere, with the blue end of the spectrum suffering a greater loss than the red end. Even so, at the altitudes at which photographs normally are taken, the colour differentiation is reduced significantly. Obviously, true colour is only of value if it is closely related to the geology of the area shown on the photograph. False colour is the term frequently used for infrared colour photography, because on reversed positive film, green, red and infrared light are recorded, respectively, as blue, green and red. False colour provides a

Table 16.1 Types of photogeological investigation.

Structural geology	Mapping and analysis of folding. Mapping of regional fault systems and recording of any evidence of recent fault movements. Determination of the number and geometry of joint systems
Rock types	Recognition of the main lithological types (crystalline and sedimentary rocks, unconsolidated deposits)
Soil surveys	Determination of the main soil type boundaries, relative permeabilities and cohesiveness; periglacial studies
Topography	Determination of relief and landforms. Assessment of stability of slopes; detection of old landslides
Stability	Slope instability (especially useful in detecting old failures which are difficult to appreciate on the ground) and rockfall areas, quickclays, loess, peat, mobile sand, soft ground, features associated with old mine workings
Drainage	Outlining of catchment areas, stream divides, surface run-off characteristics, areas of subsurface drainage, such as karstic areas, especially of cavernous limestone, as illustrated by surface solution features; areas liable to flooding. Tracing swampy ground, perennial or intermittent streams and dry valleys. Levées and meander migration. Flood control studies. Forecasting effect of proposed obstructions. Run-off characteristics. Shoals, shallow water, stream gradients and widths
Erosion	Areas of wind, sheet and gully erosion, excessive deforestation, stripping for opencast work, coastal erosion
Groundwater	Outcrops and structure of aquifers. Water-bearing sands and gravels. Seepages and springs; possible productive fracture zones. Sources of pollution. Possible recharge sites
Reservoirs and dam sites	Geology of reservoir site, including surface permeability classification. Likely seepage problems. Limit of flooding and rough relative values of land to be submerged. Bedrock gullies, faults and local fracture pattern. Abutment characteristics. Possible diversion routes. Ground needing clearing. Suitable areas for irrigation
Materials	Location of sand and gravel, clay, rip-rap, borrow and quarry sites with access routes
Routes	Avoidance of major obstacles and expensive land. Best graded alternatives and ground conditions. Sites for bridges. Pipe and power line reconnaissance. Best routes through urban areas
Old mine workings	Detection of shafts and shallow, abandoned workings, subsidence features

more sensitive means of identifying exposures of bare grey rocks than any other type of film. Lineaments, variations in water content in soils and rocks, and changes in vegetation, which may not readily be apparent on black and white photographs, often are depicted clearly by false colour. The choice of the type of photograph for a project is governed by the use that it will serve during the project. A summary of the types of geological information which can be obtained from aerial photographs is given in Table 16.1.

16.3.2 Photogeology

Allum (1966) pointed out that when stereopairs of aerial photographs are observed, the image perceived represents a combination of variations in both relief and tone. However, relief and tone on aerial photographs are not absolute quantities for particular rocks. For instance, relief represents the relative resistance of rocks to erosion, as well as the amount of erosion which has occurred. Tone is important, because small variations may be indicative of different types of rock. Unfortunately, tone is affected by light conditions, which vary with the weather, time of day, season and processing. Nevertheless, basic intrusions normally produce darker tones than acidic intrusions. Quartzite, quartz schist, limestone, chalk and sandstone tend to give light tones, whereas slates, micaceous schists, mudstones and shales give medium tones, and basalts, dolerites and amphibolites give dark tones.

The factors which affect the photographic appearance of a rock mass include the climate, vegetative cover, absolute rate of erosion, relative rate of erosion of a particular rock mass compared with that of the country rock, colour and reflectivity, composition, texture, structure, depth of weathering, physical characteristics, factors inherent in the type of photography and the conditions under which the photograph was obtained. Many of these factors are interrelated.

Regional geological structures frequently are easier to recognize on aerial photographs, which provide a broad synoptic view, than they are in the field.

Lineaments are alignments of features on an aerial photograph. The various types recognized include topographical, drainage, vegetative and colour alignments. Bedding is portrayed by lineaments, which usually are few in number and occur in parallel groups. If a certain bed is more resistant than those flanking it, it forms a clear topographic lineament. Even if bedding lineaments are interrupted by streams, they usually are persistent, and can be traced across the disruptive feature. Foliation may be indicated by lineaments. It often can be distinguished from bedding, because parallel lineaments, which represent foliation, tend to be both numerous and non-persistent.

Care must be exercised in the interpretation of the dip of strata from stereopairs of aerial photographs. For example, dips of 50 or 60° may appear to be almost vertical, and dips between 15 and 20° may seem to be more like 45° because of vertical exaggeration. However, with practice, dips can be estimated reliably in the ranges less than 10°, 10–25°, 25–45° and over 45°. Furthermore, the displacement of relief makes all vertical structures appear to dip towards the central or principal point of a photograph. Because relief displacement is much less in the central areas of photographs than at their edges, it obviously is wiser to use the central areas when estimating dips. It must also be borne in mind that the topographic slope need bear no relation to the dip of the strata composing the slope. However, scarp slopes do reflect the dip of rocks and, as dipping rocks cross interfluves and river valleys, they produce crescent- and V-shaped traces respectively. The pointed end of the V always indicates the direction of dip, and the sharper the angle of the V, the shallower the dip. If there are no dip slopes, it may be possible to estimate the dip from bedding traces. Vertical beds are independent of relief.

The axial trace of a fold can be plotted, and the direction and amount of plunge can be assessed when the direction and amount of dip of the strata concerned can be estimated from aerial photographs. Steeply plunging folds have well-rounded noses, and the bedding can be traced in a continuous curve. On the other hand, gently plunging folds occur as two bedding lineaments meeting at an acute angle (the nose) to form a single lineament. In addition, the presence of repeated folding may sometimes be recognized by plotting bedding plane traces on aerial photographs.

Straight lineaments which appear as slight negative

features on aerial photographs usually represent faults or master joints. In order to identify the presence of a fault, there should be some evidence of movement. Usually, this evidence consists of the termination or displacement of other structures. In areas of thick soil or vegetation cover, faults may be less obvious. Faults running parallel to the strike of strata may also be difficult to recognize. Joints, of course, show no evidence of displacement. Jointing patterns may assist the recognition of certain rock types, as for example, in limestone or granite terrains.

Dykes and veins also give rise to straight lineaments, which are, at times, indistinguishable from those produced by faults or joints. If, however, dykes or veins are wide enough, they may give a relief or tonal contrast with the country rock. They are then distinctive. Acid dykes and quartz veins often are responsible for light-coloured lineaments, and basic dykes for dark lineaments. Even so, because the relative tone depends very much on the nature of the country rock, positive identification cannot be made from aerial photographs alone.

If the area portrayed by the aerial photographs is subject to active erosion, it frequently is possible to differentiate between different rock masses, although it is not possible to identify the rock types. Normally, only general, rather than specific, rock types are recognizable from aerial photographs, for example, superficial deposits, sedimentary rocks, metamorphic rocks, intrusive rocks and extrusive rocks. Superficial deposits can be grouped into transported and residual categories. Transported superficial deposits can be recognized by their blanketing effect on the geology beneath, by their association with their mode of transport and with diagnostic landforms, such as meander belts, river terraces, drumlins, eskers, sand dunes, etc., and their relatively sharp boundaries. Residual deposits generally do not blanket the underlying geology completely and, in places, there are gradational boundaries with rock outcrops. Obviously, no mode of transport can be recognized. It usually is possible to distinguish between metamorphosed and unmetamorphosed sediments, as metamorphism tends to make rocks more similar as far as resistance to erosion is concerned. Metamorphism should also be suspected when rocks are tightly folded and associated with multiple intrusions. By contrast, rocks which are horizontally bedded or gently folded, and

are unaffected by igneous intrusions, are unlikely to be metamorphic. As noted above, acid igneous rocks give rise to light tones on aerial photographs and may display evidence of jointing. The recognition of volcanic cones indicates the presence of extrusive rocks.

16.4 Terrain evaluation

Terrain evaluation only is concerned with the uppermost part of the land surface of the Earth, that is, that which lies at a depth of less than 6 m, excluding permanent masses of water. Mitchell (1991) described terrain evaluation as involving the analysis (the simplification of the complex phenomena which make up the natural environment), classification (the organization of data in order to distinguish and characterize individual areas) and appraisal (the manipulation, interpretation and assessment of data for practical ends) of an area of the Earth's surface. There are two different approaches to terrain evaluation, namely, parametric evaluation and landscape classification. Parametric land evaluation refers to the classification of land on the basis of selected attribute values appropriate to the particular study, such as the class of slope or the extent of a certain kind of rock. The simplest form of parametric map is one which divides a single factor into classes. Landscape classification is based on the principal geomorphological features of the terrain.

In terrain evaluation, the initial interpretation of landscape can be made from large-scale maps and aerial photographs (Webster & Beckett, 1970). The observation of relief should give particular attention to the direction (aspect) and angle of maximum gradient, maximum relief amplitude and the proportion of the total area occupied by bare rock or slopes. In addition, an attempt should be made to interpret the basic geology and the evolution of the landscape. An assessment of the risk of erosion (especially the location of slopes which appear to be potentially unstable) and the risk of excess deposition of water-borne or windblown debris should also be made.

Terrain evaluation provides a method whereby the efficiency and accuracy of preliminary surveys can be improved. In other words, it allows a subsequent investigation to be directed towards the relevant problems. It also offers a rational means of correlating known and unknown areas, that is, of applying information and experience gained on one project to a subsequent project. This is based on the fact that landscape systems of terrain evaluation have indicated that landscapes in different parts of the world are sufficiently alike to make predictions from the known to the unknown.

The following units of classification of land have been recognized for purposes of terrain evaluation, in order of decreasing size, namely, land zone, land division, land province, land region, land system, land facet and land element (Brink *et al.*, 1966). The land system, land facet and land element are the principal units used in terrain evaluation (Lawrence, 1972, 1978; Anon., 1978).

A land systems map shows the subdivision of a region into areas with common physical attributes, which differ from those of adjacent areas. Land systems usually are recognized from aerial photographs, the boundaries between different land systems being drawn where there are distinctive differences between landform assemblages. For example, the character of land units can be largely determined from good stereopairs of photographs with an optimum scale of about 1:20000, depending on the complexity of the terrain. Fieldwork is necessary to confirm the landforms and to identify soils and bedrock.

In order to establish the pattern identified on the aerial photographs as a land system, it is necessary to define the geology and range of small topographic units referred to as land facets. A land system extends to the limits of a geological formation over which it is developed, or until the prevailing land-forming process gives way and another land system is developed. Land systems maps usually are prepared at scales of 1:500000 or 1:1000000. More detailed maps may be required in complex terrain. They provide background information which can be used in a preliminary assessment of the ground conditions in an area, and permit locations to be identified where detailed investigations may prove to be necessary.

A land system comprises a number of land facets. Each land facet possesses a simple form, generally being developed on a single rock type or superficial deposit. The soils, if not the same throughout the facet, at least vary in a consistent manner. An alluvial fan, a levée, a group of sand dunes or a cliff are examples of land facets. Indeed, geomorphology frequently provides the basis for the identification of land facets. Land facets occur in a given pattern

within a land system. They may be mapped from aerial photographs at scales between 1:10000 and 1:60000.

A land facet, in turn, may be composed of a small number of land elements, some of which deviate somewhat in a particular property, such as soils, from the general character. They represent the smallest unit of landscape that normally is significant. For example, a hillslope may consist of two land elements, an upper steep slope and a gentle lower slope. Other examples of land elements include small river terraces, gully slopes and small outcrops of rock.

Although nearly all terrain evaluation mapping is carried out at the land system level, the land region may be used in a large feasibility study for some projects. A land region consists of land systems which possess the same basic geological composition and have an overall similarity of landforms. Land regions usually are mapped at a scale between 1:1000000 and 1:5000000.

Most land system maps are accompanied by a report, which gives the basic information used to establish the classification of landforms within the area surveyed. The occurrence of land facets normally is shown on a block diagram (Fig. 16.2), cross-section or map; maps are more often used in areas where the relative relief is very small, such as alluvial plains. The descriptions of land facets include data on the slope and soil profile, with vegetation and water regime referred to where appropriate.

The scale of aerial photography used in terrain evaluation varies widely, but can be divided roughly into three ranges. Large-scale photographs have a scale greater than about 1:20000, and are used for the detailed interpretation of small features, such as beach ridges, river terraces, periglacial deposits, etc. Medium-scale photographs range from around 1:20000 to 1:50000. These most frequently are used in terrain study. Small-scale photographs range from approximately 1:50000 to 1:80000. They provide a method of cheap, rapid reconnaissance of low-value terrain, such as deserts.

16.5 Land capability studies

An important part of land-use planning involves the matching of an area of land with a land use which is most appropriate, taking into consideration the physical characteristics of the land, including any geologi-

Land region:	Lowland sediments
Climate (rainfall):	1750–2500 mm p.a.
Geology:	Shales and mudstones. A very small area of granite exists south-east of Masjid Tanah in Malacca (Alor Gajah Variant)
Landscape:	Low hills with uneven slopes and small gullies; frequent broad river valleys with well-developed terraces. All slopes are gentle, and the terraces are particularly advantageous for road location
Soils:	Clays and silty clays, often with laterite horizons (sometimes massive). BSCS: GPF (laterite gravels); I, E (clays and silty clays)
Vegetation:	Rubber plantations; padi in major valleys
Relief:	20–50 m
Altitude:	Few–90 m
Area:	847 km²

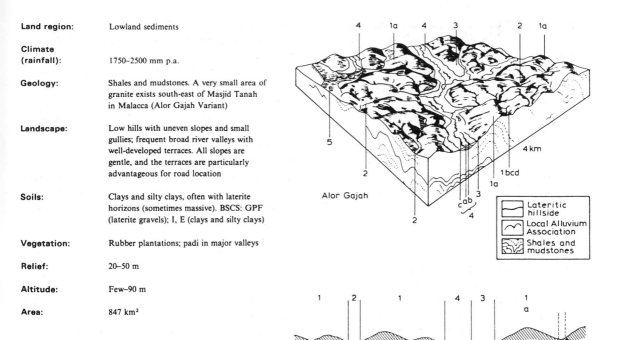

Fig. 16.2 Alor Gajah land system. (After Lawrence, 1978.)

cal constraints. Land capability analysis involves an evaluation of these physical characteristics in relation to different types of land use (Dent & Young, 1981). For example, areas threatened by landsliding can be allocated to recreation, forestry or grazing land. Hence, land capability analysis seeks to make geological data more understandable, and so more useful, to planners and politicians, and hopefully, makes planning more responsive to this information. This information can be used to relate geological factors to other environmental, social, political and economic factors, which can be expressed in monetary form, and so can contribute towards a land development decision. The geology of an area can either create opportunities for, or impose constraints on development. A land suitability study considers economic, social and political factors in addition to land capability factors.

Dearman (1991) provided examples of a number of land capability studies taken from the USA. He suggested that five steps usually are involved in a land capability study:

1 Identification of the types of land use for which the land capability is to be determined.
2 Determination of the natural factors which have a significant effect on the capability of the land to accommodate each use.
3 Development of a scale of values for rating each natural factor in relation to its effect on land capability.
4 Assignment of a weight to each natural factor indicating its importance relative to the other factors as a determinant of land capability.
5 Establishment of the land units, rating of each land unit for each factor, calculation of the weighted ratings for each factor and aggregation of the weighted ratings for each land unit.

One of the land capability studies, used by Dearman as an illustration, was that for Palo Alto, California. In this study, the area was divided into 330 cells, each cell being 20 acres (8.1 ha) in area. Twenty-three factors were chosen, 10 relating to geology and soils, four ecological, four visual and recreational, and five planning and market factors. Each cell was evaluated and rated for each factor, and the ratings were weighted, an illustra-

Table 16.2 Rating system for some land capability factors in the Palo Alto study, California, USA. (From Dearman, 1991.)

Geological and soil factors	Rating	Weight	Weighted rating
Average slope			
Over 50%	1 ⎫		10 ⎧
31–50%	2 ⎬	10	20
16–30%	4		40
0–15%	5 ⎭		50 ⎩
San Andreas fault line			
Within zone	1 ⎫	7	7 ⎧
Not within zone	5 ⎭		35 ⎩
Landslides			
Within slide area	1 ⎫	6	6 ⎧
Not within slide area	5 ⎭		30 ⎩
Natural slope stability			
Poor	1 ⎫		5 ⎧
Fair	3 ⎬	5	15
Good	5 ⎭		25 ⎩
Cut-slope stability			
Poor	1 ⎫		4 ⎧
Fair	3 ⎬	4	12
Good	5 ⎭		20 ⎩
Soil suitability as fill			
Poor	1 ⎫		2 ⎧
Fair	3 ⎬	2	6
Good	5 ⎭		10 ⎩
Soil erosion			
Severe	1 ⎫	6	6 ⎧
Moderate	5 ⎭		30 ⎩
Soil expansion			
High	1 ⎫		3 ⎧
Moderate	3 ⎬	3	9
Low	5 ⎭		15 ⎩

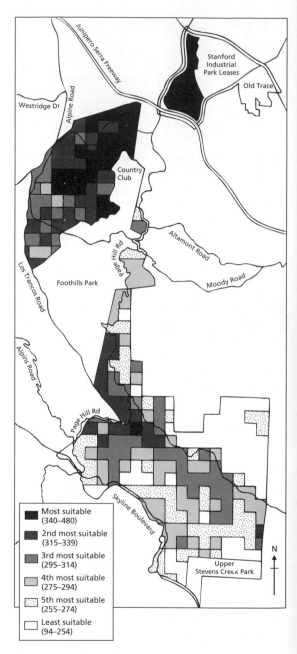

Fig. 16.3 Map of part of Palo Alto, California, showing land suitability for development ratings. (From Dearman, 1991.)

tion of the system being given in Table 16.2. The weighted ratings for all the factors were then summed for each cell to produce a final score. These final scores were divided into six groups of land suitability for development, ranging from least suitable to most suitable, and portrayed on a map (Fig. 16.3).

16.6 Site investigation

The general objective of a site investigation is to determine the suitability of a site for a proposed development. It primarily involves the gathering of data on ground conditions at and below the surface, and is a prerequisite for the successful and economic design of engineering structures and earthworks. A site investigation should attempt to foresee and provide against difficulties that may arise during construction operations due to ground and/or other local conditions, and should continue after construction begins. It is essential that the assessment of the ground conditions, which constitutes the basic design assumption, is checked as construction proceeds, and designs should be modified accordingly if conditions differ from those predicted. An investigation of a site for an important project requires both the exploration and sampling of strata likely to be affected. Data appertaining to the groundwater conditions, hazards, past land use, the extent of weathering and the discontinuity pattern in rock masses are also important. In certain areas, special problems need to be investigated, for example, potential subsidence in areas of shallow, abandoned mine workings.

16.6.1 Desk study and preliminary reconnaissance

Anon. (1981) recognized the desk study as the first stage in a site investigation. The objective of the desk study is to examine available records and data relevant to the area or site concerned to ascertain a general idea of the existing geological conditions prior to a field investigation. In addition, a desk study can be undertaken in order to determine the factors which affect a proposed development for feasibility assessment and project planning purposes. The terms of reference for a desk study, in both cases, need to be defined clearly in advance of the commencement of the study. The effort expended in a desk study depends upon the type of project, the geotechnical complexity of the area or site and the availability of relevant information. Over and above these factors is the budget allocated to the study, which affects the time which can be spent on it.

Therefore, a desk study for the planning stage of a project can encompass a range of appraisals, from the preliminary rapid response to the comprehensive. Nonetheless, there are a number of common factors throughout this spectrum which need to be taken into account. These are summarized in Table 16.3, from which it can be concluded that an appraisal report typically includes a factual and interpretative description of the surface and geological conditions, information on previous site usage, a preliminary assessment of the suitability of the site for the planned development, an identification of potential hazards, and provisional recommendations with regard to ground engineering aspects. However, a desk study should not be regarded as an alternative to a ground investigation for a construction project.

Detailed searches for information can be extremely time consuming, and may not be justified for small schemes at sites where the ground conditions are relatively simple or well known. In such cases, a study of the relevant topographical and geological maps and memoirs, and possibly aerial photographs, may suffice. On large projects, literature and map surveys may save time and thereby reduce the cost of the associated site exploration. The data obtained during such searches should aid in the planning of the subsequent site exploration, and should prevent the duplication of effort. A desk study can also reduce the risk of encountering unexpected ground conditions which could adversely affect the financial viability of a project. In some parts of the world, however, little or no literature or maps are available.

Topographical, geological and soil maps, together with remote sensing imagery or aerial photographs, can provide valuable information for use during the planning stage of a construction operation. Geological maps afford a generalized picture of the geology of an area. Generally, the stratum boundaries and positions of the structural features, especially faults, are interpolated. Consequently, their accuracy cannot always be trusted. Map memoirs may accompany maps, and these provide a detailed survey of the geology of the area concerned.

The preliminary reconnaissance involves a site walkover noting, where possible, the distribution of the soil and rock types present, the relief of the ground, the surface drainage and associated features, actual or likely landslip areas, ground cover and obstructions, and earlier uses of the site such as tipping or evidence of underground workings, etc. The inspection should not be restricted to the site, but should examine adjacent areas to see how they affect or will be affected by the development of the site in question.

Table 16.3 Summary of contents of engineering geological desk study appraisals. (After Herbert *et al.*, 1987.)

Item	Content and main points of relevance
Introduction	Statement of terms of reference and objectives, with indication of any limitations. Brief description of nature of project and specific ground-orientated proposals. Statement of sources of information on which appraisal is based
Ground conditions	Description of relevant factual information. Identification of any major features which might influence scheme layout, planning or feasibility
Site description and topography	Descriptions of existing surface conditions from study of topographical maps and actual photographs, and also from site walkover inspection (if possible)
Engineering history	Review of information on previous surface conditions and usage (if different from present) based on study of old maps, photographs, archival records and related to any present features observed during site walkover. Identification of features such as landfill zones, mine workings, pits and quarries, sources of contamination, old water courses, etc.
Engineering geology	Description of subsurface conditions, including any information on groundwater, from study of geological maps and memoirs, previous site investigation reports and any features or outcrops observed during site walkover. Identification of possible geological hazards, e.g. buried channels in alluvium, solution holes in chalk and limestone, swelling/shrinkable clays
Provisional assessment of site suitability	Summary of main engineering elements of proposed scheme, as understood. Comments on suitability of site for proposed development, based on existing knowledge
Provisional land classification	Where there is significant variation in ground conditions or assessed level of risk, subdivision of the site into zones of high and low risk, and any intermediate zones. Comparison of various risk zones with regard to the likely order of cost and scope of subsequent site investigation requirements, engineering implications, etc.
Provisional engineering comments	Statement of provisional engineering comments on aspects such as foundation conditions and which method(s) appears to be most appropriate for structural foundations and ground slabs, road pavement subgrade conditions, drainage, ease of excavation of soils and rocks, suitability of local borrow materials for use in construction, slope stability considerations, nature and extent of any remedial works, temporary problems during construction
Recommendations for further work	Proposals for phased ground investigation, with objectives, requirements and estimated budget costs

The importance of the preliminary investigation is that it should assess the suitability of the site for the proposed works, and if it is suitable, will form the basis upon which the site exploration is planned. The preliminary reconnaissance also allows a check to be made on any conclusions reached in the desk study.

16.6.2 Site exploration

The aim of a site exploration is to try to determine, and thereby understand, the nature of the ground conditions on and surrounding a site. The extent to which this is carried out depends upon the size and importance of the project. A report embodying

the findings of an investigation can be used for design purposes, and should contain geological plans of the site, with accompanying sections, thereby conveying a three-dimensional picture of the subsurface strata.

The scale of the mapping will depend on the particular requirement, the complexity of the geology and the staff and time available. For example, geological mapping frequently is required on a large and detailed scale for a large project.

Rock and soil types should be mapped according to their lithology and, if possible, presumed physical behaviour, that is, in terms of their engineering classification, rather than age. Geomorphological conditions, hydrogeological conditions, landslips, subsidences, borehole and field test information all can be recorded on maps. Particular attention should be given to the nature of the superficial deposits and, where present, made-over ground.

There are no given rules regarding the location of boreholes or drill-holes, or the depth to which they should be sunk. This depends upon two principal factors, the geological conditions and the type of project concerned. The information provided by the preliminary reconnaissance and any trial trenches should provide a basis for the initial planning and layout of the borehole programme. Holes should be located so as to detect the geological sequence and structure. Obviously, the more complex this is, the greater the number of holes needed. In some instances, it may be as well to start with a widely spaced network of holes. As information is obtained, further holes can be put down, if and where necessary.

Exploration should be carried out to a depth which includes all strata likely to be significantly affected by structural loading. Experience has shown that damaging settlement usually does not take place when the added stress in the soil due to the weight of a structure is less than 10% of the effective overburden stress. It would therefore seem logical to sink boreholes on compact sites to depths at which the additional stress does not exceed 10% of the stress due to the weight of the overlying strata. It must be borne in mind that, if a number of loaded areas are in close proximity, the effect of each is additive. Under certain special conditions, boreholes may have to be sunk more deeply, as, for example, when old voids due to mining operations are suspected, or when it is thought that there may be highly compressible layers, such as interbedded peats, at depth. If possible, boreholes should be taken through superficial deposits to rockhead. In such instances, adequate penetration of the rock should be specified to ensure that isolated boulders are not mistaken for the solid formation.

The results from a borehole or drill-hole should be documented on a log (Fig. 16.4). Apart from the basic information, such as the number, location, elevation, date, client, contractor and engineer responsible, the fundamental requirement of a borehole log is to show how the sequence of strata changes with depth. Individual soil or rock types are presented in symbolic form on a borehole log. The material recovered must

be adequately described, and in the case of rocks, frequently includes an assessment of the degree of weathering, fracture index and relative strength. The type of boring or drilling equipment should be recorded, the rate of progress made being a significant factor. The water level in the hole and any water loss, when it is used as a flush during rotary drilling, should be noted, as these reflect the mass permeability of the ground. If any *in situ* testing is performed during boring or drilling operations, the type(s) of test and the depth at which it/they were carried out must be recorded. The depths from which samples are taken must be noted. A detailed account of the logging of cores for engineering purposes has been given by Anon. (1970).

The direct observation of strata, discontinuities and cavities can be undertaken by cameras or closed-circuit television equipment, and drill-holes can be viewed either radially or axially. Remote focusing for all heads and rotation of the radial head through 360° are controlled from the surface. The television heads have their own light source. Colour changes in rocks can be detected as a result of the varying amount of light reflected from the drill-hole walls. Discontinuities appear as dark areas because of the non-reflection of light. However, if the drill-hole is deflected from the vertical, variations in the distribution of light may result in a certain lack of picture definition.

16.6.3 Subsurface exploration in soils

The simplest method whereby data relating to subsurface conditions in soils can be obtained is by hand augering. The two most frequently used types of auger are the posthole auger and the screw auger. These are used principally in cohesive soils.

Soil samples which are obtained by augering are badly disturbed and, invariably a certain amount of mixing of soil types occurs. Critical changes in the ground conditions therefore are unlikely to be located accurately. Even in very soft soils, it may be very difficult to penetrate more than 7 m with hand augers. The Mackintosh probe and Lang prospector are more specialized forms of hand tool.

Power augers are available as solid stem or hollow stem, both having an external continuous helical flight. The latter is used in those soils in which the borehole does not remain open. The hollow stem can

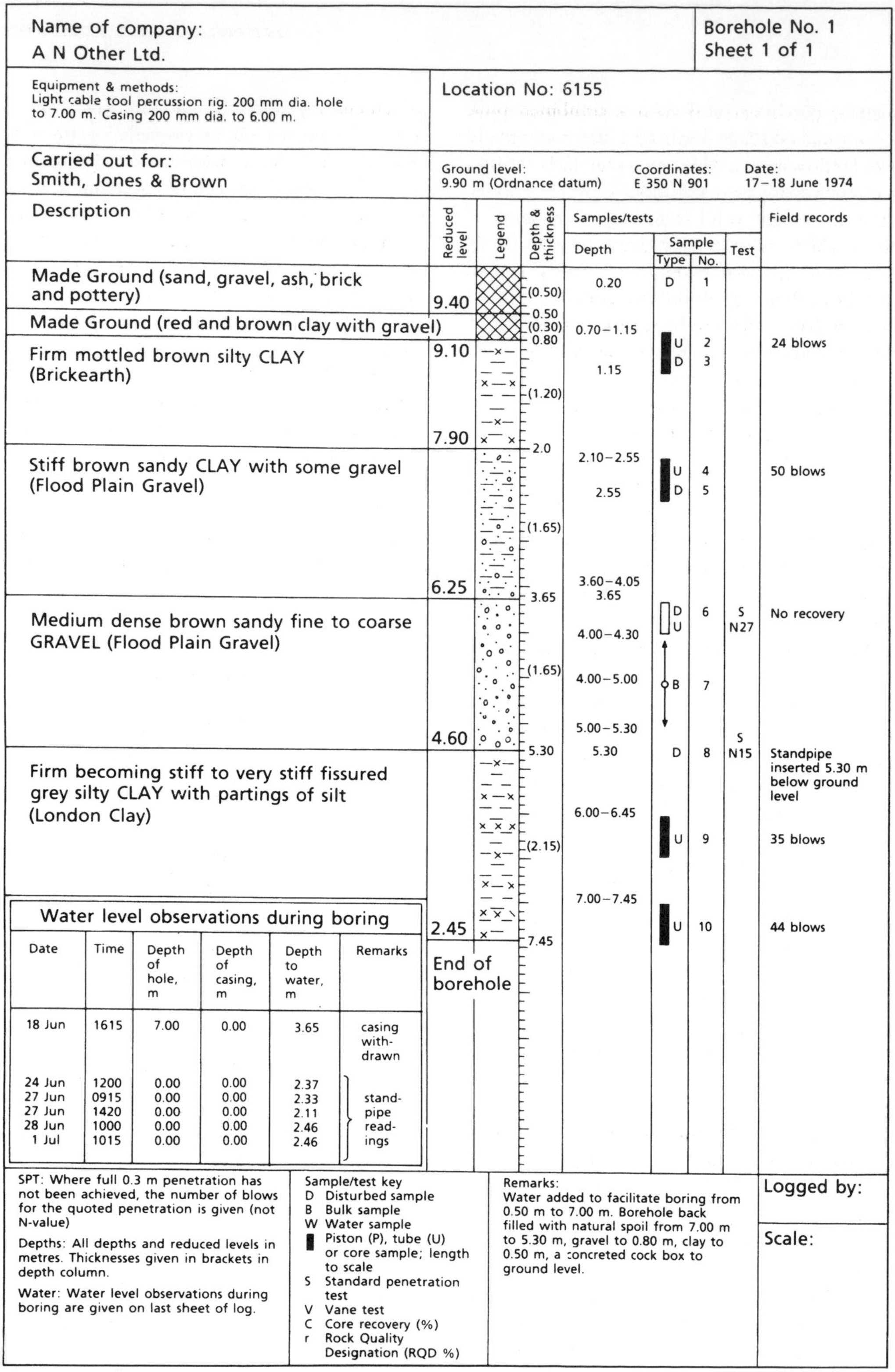

Name of company: A N Other Ltd.		Borehole No. 1 Sheet 1 of 1
Equipment & methods: Light cable tool percussion rig. 200 mm dia. hole to 7.00 m. Casing 200 mm dia. to 6.00 m.	Location No: 6155	
Carried out for: Smith, Jones & Brown	Ground level: 9.90 m (Ordnance datum) Coordinates: E 350 N 901 Date: 17–18 June 1974	

Description	Reduced level	Legend	Depth & thickness	Samples/tests Depth	Sample Type	No.	Test	Field records
Made Ground (sand, gravel, ash, brick and pottery)	9.40		(0.50) 0.50	0.20	D	1		
Made Ground (red and brown clay with gravel)			(0.30) 0.80	0.70–1.15				
Firm mottled brown silty CLAY (Brickearth)	9.10		(1.20)	1.15	U D	2 3		24 blows
	7.90		2.0					
Stiff brown sandy CLAY with some gravel (Flood Plain Gravel)			(1.65)	2.10–2.55 2.55	U D	4 5		50 blows
	6.25		3.65	3.60–4.05 3.65				
Medium dense brown sandy fine to coarse GRAVEL (Flood Plain Gravel)			(1.65)	4.00–4.30 4.00–5.00	D U B	6 7	S N27	No recovery
	4.60		5.30	5.00–5.30 5.30	D	8	S N15	Standpipe inserted 5.30 m below ground level
Firm becoming stiff to very stiff fissured grey silty CLAY with partings of silt (London Clay)			(2.15)	6.00–6.45	U	9		35 blows
	2.45		7.45	7.00–7.45	U	10		44 blows
			End of borehole					

Water level observations during boring					
Date	Time	Depth of hole, m	Depth of casing, m	Depth to water, m	Remarks
18 Jun	1615	7.00	0.00	3.65	casing with-drawn
24 Jun	1200	0.00	0.00	2.37	
27 Jun	0915	0.00	0.00	2.33	stand-pipe read-ings
27 Jun	1420	0.00	0.00	2.11	
28 Jun	1000	0.00	0.00	2.46	
1 Jul	1015	0.00	0.00	2.46	

SPT: Where full 0.3 m penetration has not been achieved, the number of blows for the quoted penetration is given (not N-value)

Depths: All depths and reduced levels in metres. Thicknesses given in brackets in depth column.

Water: Water level observations during boring are given on last sheet of log.

Sample/test key
D Disturbed sample
B Bulk sample
W Water sample
▮ Piston (P), tube (U) or core sample; length to scale
S Standard penetration test
V Vane test
C Core recovery (%)
r Rock Quality Designation (RQD %)

Remarks:
Water added to facilitate boring from 0.50 m to 7.00 m. Borehole back filled with natural spoil from 7.00 m to 5.30 m, gravel to 0.80 m, clay to 0.50 m, a concreted cock box to ground level.

Logged by:

Scale:

Fig. 16.4 Typical log of data from a light cable percussion borehole. (After Anon., 1981.)

be sealed at the lower end with a combined plug and cutting bit, which is removed when a sample is required. Hollow stem augers are useful for investigations in which the requirement is to locate bedrock beneath overburden. Solid stem augers are used in stiff clays which do not need casing. However, if an undisturbed sample is required, they must be removed. Disturbed samples taken from auger holes are often unreliable. In favourable ground conditions, such as firm and stiff homogeneous clays, auger rigs are capable of high output rates.

The development of large earth augers and patent piling systems has made it possible to sink boreholes of 1 m in diameter in soils more economically than previously. The ground conditions can be inspected directly from such holes. Depending on the ground conditions, the boreholes may be unlined, lined with steel mesh or cased with steel pipe. In the latter case, windows are provided at certain levels for inspection and sampling.

Trenches and pits allow the ground conditions in soils and highly weathered rocks to be examined directly, although they are limited as far as their depth is concerned. Trenches, to a depth of some 5 m, can provide a flexible, rapid and economic method of obtaining information (Hatheway & Leighton, 1979). The groundwater conditions and stability of the sides obviously influence whether or not they can be excavated, and safety must be observed at all times. This, at times, necessitates shoring the sides. Pits are expensive and should be considered only if the initial subsurface survey has revealed any areas of special difficulty. The soil conditions in pits and trenches can also be photographed. Undisturbed, as well as disturbed, samples can be collected. Such excavations are used to locate slip planes in landslides.

The light cable and tool boring rig is used for investigating soils (Fig. 16.5). The hole is sunk by repeatedly dropping one of the tools into the ground. A power winch is used to lift the tool, suspended on a cable wire, and by releasing the clutch of the winch, the tool drops and cuts into the soil. Once a hole is established, it is lined with casing, the drop tool operating within the casing. This type of rig usually is capable of penetrating about 60 m of soil. In doing so, the size of the casing in the lower end of the borehole is reduced. The basic tools are the shell and the clay-cutter, which are essentially open-ended steel tubes to

which cutting shoes are attached. The shell, which is used in granular soils, carries a flap valve at its lower end, which prevents the material from falling out on withdrawal from the borehole. The material is retained in the cutter by the adhesion of the clay.

For boring in stiff clays, the weight of the claycutter may be increased by adding a sinker bar. A little water often is added to assist boring progress in very stiff clays. This must be performed with caution, so as to avoid possible changes in the properties of the soil about to be sampled. In such clays, the borehole often can be advanced without lining, except for a short length at the top to keep the hole stable. If cobbles or small boulders are encountered in clays, particularly tills, these can be broken by using heavy chisels.

When boring in soft clays, although the hole may not collapse, it tends to squeeze inwards and prevents

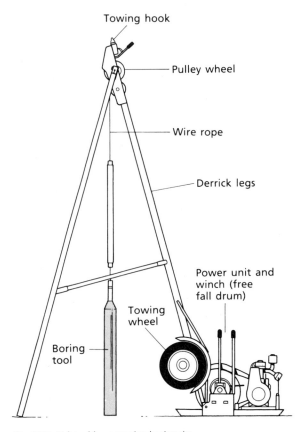

Fig. 16.5 Light cable percussion boring rig.

the cutter operating. The hole therefore must be lined. The casing is driven in and winched out, however, in difficult conditions, it may have to be jacked out. Casing tubes have internal diameters of 150, 200, 250 and 300 mm, the most commonly used sizes being 150 and 200 mm (the large sizes are used in coarse gravels).

Boreholes in sands or gravels almost invariably require lining. The casing should be advanced with the hole, or overshelling is likely to occur, that is, the sides collapse and prevent further progress. Because of the mode of operation of the shell, the borehole should be kept full of water, so that the shell may operate efficiently. Where cohesionless soils are water bearing, all that is necessary is for the water in the borehole to be kept topped up. If the flow of water occurs, it should be from the borehole to the surrounding soil.

Rotary attachments are available which can be used with light cable and tool rigs. However, they are much less powerful than normal rotary rigs, and tend to be used only for short runs, for example, to demonstrate rockhead at the base of a borehole.

In the wash boring method, the hole is advanced by a combination of chopping and jetting the soil or rock, the cuttings thereby produced being washed from the hole by the water used for jetting (Fig. 16.6). The method cannot be used for sampling, and therefore its primary purpose is to sink the hole between sampling positions. When a sample is required, the bit is replaced by a sampler. Some indication of the type of ground penetrated may be obtained from the cuttings carried to the surface by the wash water, from the rate of progress made by the bit or from the colour of the wash water.

Several types of chopping bit are used. Straight and chisel bits are used in sands, silts, clays and very soft rocks, whereas cross-bits are used in gravels and soft rocks. Bits are available with the jetting points either facing upwards or downwards. The former type are better at cleaning the base of the hole than the latter.

The wash boring method may be used in both cased and uncased holes. Casing obviously must be used in cohesionless soils to avoid the collapse of the sides of the hole. Although this method of boring commonly is used in the USA, it rarely has been employed elsewhere. This mainly is because wash boring does not lend itself to many of the ground conditions encountered, and also because it is difficult to identify strata with certainty.

Fig. 16.6 Wash boring rig: (a) driving the casing; (b) advancing the hole.

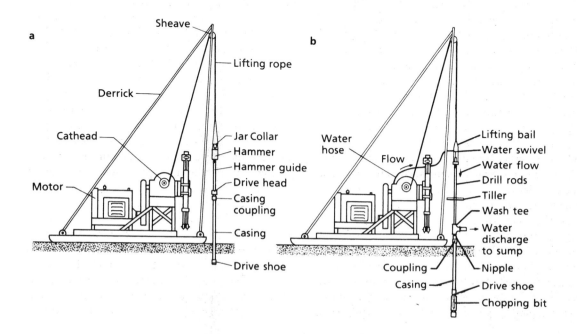

Cable and tool or churn drilling is a percussion drilling method which can be used in more or less all types of ground conditions, but the rate of progress tends to be slow. The rig can drill holes up to 0.6 m in diameter to depths of about 1000 m. The hole is advanced by raising and dropping heavy drilling tools which break the soil or rock. Different bits are used for drilling in different formations, and an individual bit can weigh anything up to 1500 kg, so that the total weight of the drill string may amount to several thousand kilograms. A slurry is formed from the broken material and the water in the hole. The amount of water introduced into the hole is kept to the minimum required to form the slurry. The slurry periodically is removed from the hole by means of bailers or sand pumps. In unconsolidated materials, casing should be kept near to the bottom of the hole in order to avoid caving.

Changes in the type of strata penetrated can again be inferred from the cuttings brought to the surface, the rate of drilling progress or the colour of the slurry. Sampling, however, has to be performed separately. Unfortunately, the ground to be sampled may be disturbed by the heavy blows of the drill tools.

16.6.4 Sampling in soils

As far as soils are concerned, samples may be divided into disturbed and undisturbed types. Disturbed samples can be obtained by hand, by auger or from the claycutter or shell of a boring rig. Samples of cohesive soil should be approximately 0.5 kg in weight, this providing a sufficient size for index testing. They are sealed in jars. A larger sample is necessary if the particle size distribution of granular material is required, and this may be retained in a tough plastic sack. Care must be exercised when obtaining such samples to avoid the loss of fines.

An undisturbed sample can be regarded as one which is removed from its natural condition without disturbing its structure, density, porosity, moisture content and stress condition. Although it must be admitted that no sample is ever totally undisturbed, every attempt must be made to preserve the original condition of such samples. Undisturbed samples may be obtained by hand from surface exposures, pits and trenches. Careful hand trimming is used to produce a regular block, normally a cube of about 250 mm. Block samples are waxed, together with reinforcing layers of thin cloth. Such samples are particularly useful when it is necessary to test specific horizons, such as shear zones.

The fundamental requirement of any undisturbed sampling tool is that, on being forced into the ground, it should cause as little displacement of the soil as possible. The amount of displacement is influenced by a number of factors, such as the cutting edge of the sampler. A thin cutting edge and sampling tube minimize displacement, but are easily damaged and cannot be used in gravelly and hard soils. The internal diameter of the cutting edge (D_i) should be slightly less than that of the sample tube, thus providing inside clearance, which reduces the drag effects due to friction. Similarly, the outside diameter of the cutting edge (D_o) should be from 1 to 3% larger than that of the sampler, again to allow for clearance. The relative displacement of a sampler can be expressed by the area ratio (A_r):

$$A_r = \frac{D_i^2 - D_o^2}{D_o^2} \times 100 \qquad (16.1)$$

This ratio should be kept as low as possible, for example, according to Hvorslev (1949), displacement is minimized by keeping the area ratio below 15%. It should not exceed 25%. Friction can also be reduced if the tube has a smooth inner wall. A coating of light oil may also prove to be useful in this respect.

The standard sampling tube for obtaining samples from cohesive soils is referred to as the U100 tube, having a diameter of 100 mm, a length of approximately 450 mm and walls 1.2 mm thick (Fig. 16.7). The cutting shoe should meet the requirements noted above. The upper end of the tube is fastened to a check valve, which allows air or water to escape during driving, and helps to hold the sample in place when it is being withdrawn. On withdrawal from the borehole, the sample is sealed in the tube with paraffin wax and the end caps are screwed on. In soft materials, two or three tubes may be screwed together to reduce disturbance in the sample.

A thin-walled piston sampler should be used to obtain clays with a shear strength of less than 50 kPa, because soft clays tend to expand into the sample tube. Expansion is reduced by a piston in the sampler, a thin-walled tube being jacked down over a stationary internal piston, which, when sampling is complete, is locked in place and the whole assembly is

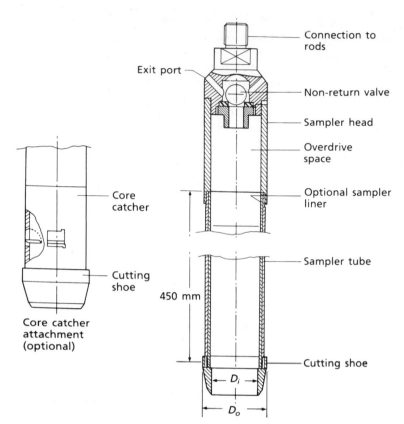

Exit port

Connection to rods

Non-return valve

Sampler head

Overdrive space

Core catcher

Optional sampler liner

Cutting shoe

450 mm

Sampler tube

Cutting shoe

Core catcher attachment (optional)

D_i

D_o

Fig. 16.7 The general purpose open-tube sampler (U100).

then pulled (Fig. 16.8). Piston samplers range in diameter from 54 to 250 mm. A vacuum tends to be created between the piston and the soil sample, and thereby helps to hold it in place.

Where continuous samples are required, particularly from rapidly varying or sensitive soils, a Delft sampler may be used (Fig. 16.9). This can obtain a continuous sample from ground level to depths of about 20 m. The core is retained in a self-vulcanizing sleeve as the sampler continuously is advanced into the soil.

The most difficult undisturbed sample to obtain is that from saturated sand, particularly when it is loosely packed. In such instances, the Bishop sand sampler, which makes use of compressed air and incorporates a thin-walled sampling tube, is employed (Fig. 16.10). The thin-walled sampling tube is housed in an outer tube. The inner tube is driven into the soil, and compressed air introduced into the outer tube expels the water. The sampling

tube is then retracted into the outer tube, the air pressure creating capillary zones which retain the soil.

16.6.5 Subsurface exploration in rocks

Rotary drilling rigs may be skid mounted, trailer mounted or, in the case of larger types, mounted on lorries (Fig. 16.11). They are used for drilling through rock, although they can, of course, penetrate soil.

Rotary percussion rigs are designed for rapid drilling in rock. The rock is subjected to rapid, high-speed impacts while the bit rotates, which brings about compression and shear in the rock. The technique is most effective in brittle materials, because it relies on the chipping of rock. The rate at which drilling proceeds depends upon the type of rock, particularly on its strength, hardness and fracture index, the type of drill and drill bit, the flushing medium and the pressures used, as well as the experience of the drilling crew. Drill flushings should be sampled at

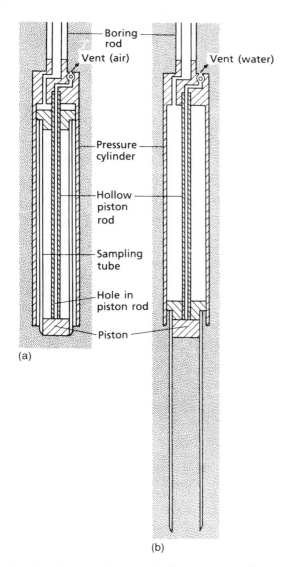

(a)

(b)

Fig. 16.8 Piston sampler of hydraulically operated type: (a) lowered to bottom of bore-hole, boring rod clamped in fixed position at ground surface; (b) sampling tube after being forced into soil by water supplied through boring rod.

regular intervals, when changes in the physical appearance of the flushings are observed and when significant changes in penetration rates occur. The interpretation of rotary percussion drill-holes should be related to a cored drill-hole nearby. This method of drilling sometimes is used as a means of advancing a hole at low cost and high speed between intervals where core drilling is required.

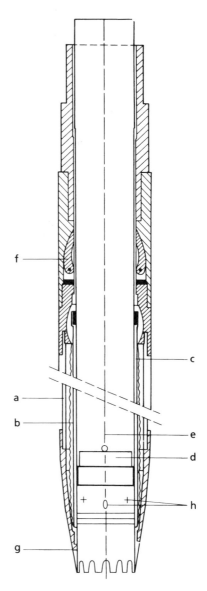

Fig. 16.9 Section through a 66 mm continuous sampling apparatus: (a) outer tube; (b) stocking tube over which precoated nylon stocking is slid; (c) plastic inner tube; (d) cap at top of sample; (e) steel wire to fixed point at ground surface (tension cable); (f) sample-retaining clamps; (g) cutting shoe; (h) holes for entry of lubricating fluid. (Courtesy of Delft Soil Mechanics Laboratory.)

For many exploration purposes a solid, and as near as possible continuous rock core is required for examination. The core is cut with a bit and housed in a core barrel (Fig. 16.12). The bit is set with dia-

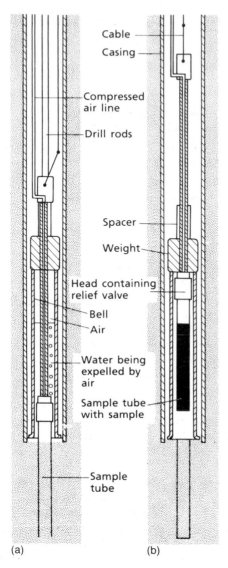

(a)   (b)

Fig. 16.10 Principle of Bishop sampler for sand below water table: (a) sampler is forced into sand by drill rods and water in bell is displaced by compressed air; (b) sampler is lifted by cable into air-filled bell.

Labels in figure: Cable, Casing, Compressed air line, Drill rods, Spacer, Weight, Head containing relief valve, Bell, Air, Water being expelled by air, Sample tube with sample, Sample tube

Fig. 16.11 Medium size, skid-mounted rotary drill. (Courtesy of Atlas Copco Ltd.)

monds or tungsten carbide inserts. The coarser surface set diamond and tungsten carbide tipped bits are used in softer formations. These bits generally are used with air rather than with water flush. Impregnated bits possess a matrix impregnated with diamond dust, and their grinding action is suitable for hard and broken formations. Tungsten bits are not suitable for drilling in very hard rocks. Core bits vary in size, and accordingly core sticks range between 17.5 and 165 mm in diameter—generally, the larger the bit, the better the core recovery.

A variety of core barrels are available for rock sampling. The simplest type of core barrel is the single tube but because it is suitable only for hard massive rocks, it rarely is used. In the single tube barrel, the barrel rotates the bit and the flush washes over the core. In double tube barrels, the flush passes between the inner and outer tubes. Double tubes may be of the rigid or swivel type. The disadvantage of the rigid barrel is that both the inner and outer tubes rotate together and in soft rock this can break the core as it enters the inner tube. It therefore is only suitable for hard rock formations.

In the double tube swivel-type core barrel, the outer tube remains stationary (Fig. 16.12). It is suitable for use in medium and hard rocks, and gives

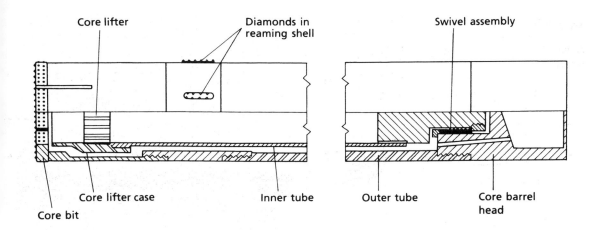

Fig. 16.12 Double tube swivel-type core barrel.

improved core recovery in soft friable rocks. The face-ejection barrel is a variety of the double tube swivel type, in which the flushing fluid does not affect the end of the core. This type of barrel is a minimum requirement for coring badly shattered, weathered and soft rock formations. Triple tube barrels have been developed for obtaining cores from very soft rocks, and from highly jointed and cleaved rock.

Both the bit and core barrel are attached by rods to the drill, by which they are rotated. The flush is pumped through the drill rods and discharged at the bit. The flushing agent serves to cool the bit and to remove the cuttings from the drill-hole.

Most rock cores should be removed by hydraulic extruders whilst the tube is held in a horizontal position. To reduce disturbance during extrusion, the inner tube of double core barrels can be lined with a plastic sleeve before drilling commences. On completion of the core run, the plastic sleeve containing the core is withdrawn from the barrel.

If casing is used for diamond drilling operations, it is drilled into the ground using a tungsten carbide or diamond tipped casing shoe with air, water or mud flush. The casing may be inserted down a hole, drilled to a larger diameter, to act as conductor casing when reducing and drilling ahead at a smaller diameter. Alternatively, it may be drilled or reamed with a larger diameter than the initial hole to allow continued drilling at the same diameter.

Many machines will core drill at any angle between vertical and horizontal. Unfortunately, inclined drill-holes tend to go off line, the problem being magnified in highly jointed formations. In deeper drilling, the sag of the rods causes the hole to deviate. Drill-hole deviation can be measured by an inclinometer.

The weakest strata generally are of the greatest interest, but these are the very materials which are most likely to be lost during drilling or to deteriorate after extraction. Consequently, Hawkins (1986) introduced the concept of the lithological quality designation (LQD), defining it as the percentage of solid core present greater than 100 mm in length within any lithological unit. The total core recovery and the maximum intact core length ($MICL$) should also be recorded. Shales and mudstones are particularly prone to deterioration once sampled, and some may disintegrate completely if allowed to dry. If samples are not properly preserved, they will dry out. Deterioration of suspect material may be reduced by wrapping the cores with aluminium foil or plastic sheeting. Core sticks may be photographed before they are removed from the site.

A simple, but nonetheless important, factor is labelling. This must record the site, the drill-hole number and the position in the drill-hole from which the material was obtained. The labels themselves must be durable and properly secured. When rock samples are stored in a core box, the depth of the top and bottom of the core contained and of the separate core runs should be noted both outside and inside the box. Zones of core loss should also be identified.

16.7 Geophysical exploration

A geophysical exploration can provide subsurface information over a large area at reasonable cost (Anon., 1988; McCann *et al.*, 1997). The information obtained may help to eliminate less favourable alternative sites, may aid in the location of testholes in critical areas and may prevent unnecessary repetitive drilling in fairly uniform ground. A geophysical survey may also help to detect variations in subsurface conditions between testholes. Testholes provide information about the strata in which they are sunk, but yield nothing about the ground in between. On the other hand, boreholes or drill-holes are required to aid in the interpretation and correlation of the geophysical measurements. Therefore an appropriate combination of direct and indirect methods often can yield a high standard of results.

Seismic and resistivity methods are more applicable to the determination of horizontal or near-horizontal changes or contacts, whereas magnetic and gravimetric methods generally are used to delineate lateral changes or vertical structures. Geophysical methods can also be used to provide an indirect assessment of the engineering properties of rock masses, particularly in terms of rock mass quality (McCann *et al.*, 1990), and can be employed in groundwater exploration.

In a geophysical survey, measurements of the variations in certain physical properties usually are taken in a traverse across the surface, although they may be made in order to log a hole. Anomalies in the physical properties measured generally reflect anomalies in the geological conditions. The ease of recognition and interpretation of these anomalies depends on the contrast in physical properties which, in turn, influences the choice of method employed.

The actual choice of method to be used for a particular survey may not be difficult to make. When dealing with layered rocks, provided that their geological structure is not too complex, seismic methods have a distinct advantage, because they give more detailed, precise and unambiguous information than any other method. Electrical methods may be preferred for small-scale work, where the geological structures are simple. On occasions, more than one method may be used to resolve the same problem.

16.7.1 Seismic methods

The sudden release of energy from the detonation of an explosive charge in the ground or the mechanical pounding of the surface generates shock waves, which radiate out in a hemispherical wave front from the point of release. The waves generated are compressional (P), dilational shear (S) and surface waves. The velocities of the shock waves generally increase with depth below the surface, because the elastic moduli increase with depth. Compressional waves travel faster, and are more easily generated and recorded than shear waves. They therefore are used almost exclusively in seismic exploration. The shock wave velocity depends on many variables, including the rock fabric, mineralogy and pore water. In general, velocities in crystalline rocks are high to very high (Table 16.4). Velocities in sedimentary rocks increase concomitantly with increasing consolidation and decreasing pore fluids, and with an increase in the degree of cementation and diagenesis. Unconsolidated sedimentary accumulations have maximum velocities varying as a function of the volume of the voids, either air filled or water filled, mineralogy and grain size.

When seismic waves pass from one layer to another, some energy is reflected back towards the surface, and the remainder is refracted. Thus, two methods of seismic surveying can be distinguished, namely, seismic reflection and seismic refraction. The measurement of the time taken from the generation of

Table 16.4 Velocities of compressional waves of some common rocks.

V_p (kms^{-1})		V_p (kms^{-1})	
Igneous rocks		*Sedimentary rocks*	
Basalt	5.2–6.4	Gypsum	2.0–3.5
Dolerite	5.8–6.6	Limestone	2.8–7.0
Gabbro	6.5–6.7	Sandstone	1.4–4.4
Granite	5.5–6.1	Shale	2.1–4.4
Metamorphic rocks		*Unconsolidated deposits*	
Gneiss	3.7–7.0	Alluvium	0.3–0.6
Marble	3.7–6.9	Sands and gravels	0.3–1.8
Quartzite	5.6–6.1	Clay (wet)	1.5–2.0
Schist	3.5–5.7	Clay (sandy)	2.0–2.4

the shock waves until they are recorded by detector arrays forms the basis of the two methods.

The seismic reflection method is the most extensively used of all geophysical techniques, its principal employment being in the oil industry. In this technique, the depth of investigation is large compared with the distance from the shot to detector array. This is to exclude refraction waves. Indeed, the method can record information from a large number of horizons down to depths of several thousands of metres.

In seismic refraction, one ray approaches the interface between two rock types at a critical angle, which means that, if the ray is passing from a low-velocity (V_0) to a high-velocity (V_1) layer, it will be refracted along the upper boundary of the latter layer. After refraction, the pulse travels along the interface with velocity V_1. The material at the boundary is subjected to oscillating stress from below. This generates new disturbances along the boundary, which travel upwards through the low-velocity rock and eventually reach the surface.

At short distances from the point at which the shock waves are generated, the geophones record direct waves, whilst at a critical distance, both the direct and refracted waves arrive at the same time. Beyond this, because the rays refracted along the high-velocity layer travel faster than those through the low-velocity layer above, they reach the geophones first. In refraction work, the object is to develop a time–distance graph, which involves plotting the arrival times against the geophone spacing (Fig. 16.13). Thus the distance between the geophones, together with the total length and arrangement of the array, must be carefully chosen to suit each particular problem.

The most common arrangement in refraction work is profile shooting. Here the shot points and geophones are laid out in long lines, with a row of geophones receiving refracted waves from the shots fired. The process is repeated at uniform intervals down the line. For many surveys, where it is required to determine the depth to bedrock, it may be sufficient to record from two shot point distances at each end of the receiving spread. By traversing in both directions, the angle of dip can be determined.

In the simple case of refraction by a single high-velocity layer at depth, the travel times for the seismic wave, which proceeds directly from the shot point to the detectors, and the travel times for the critical refracted wave to arrive at the geophones are plotted graphically against the geophone spacing (Fig. 16.13). The depth (Z) to the high-velocity layer can then be obtained from the graph using the expression

$$Z = \frac{X}{2}\frac{(V_1 - V_0)}{(V_1 + V_0)} \tag{16.2}$$

where V_0 is the speed in the low-velocity layer, V_1 is the speed in the high-velocity layer and X is the critical distance. The method also works for multilayered rock sequences, if each layer is sufficiently thick and transmits seismic waves at higher speeds than that above it. However, in the refraction method, a low-velocity layer underlying a high-velocity layer usually cannot be detected, because in such an inversion the pulse is refracted into the low-velocity layer. In addition, a layer of intermediate velocity between an underlying refractor and overlying layers can be masked as a first arrival on the travel–time curve. The latter is known as a blind zone. The position of faults can also be estimated from the time–distance graphs.

The porosity tends to lower the velocity of a shock wave through a material. In fact, the compressional wave velocity V_p is related to the porosity n of a normally consolidated sediment as follows:

$$\frac{1}{V_p} = \frac{n}{V_{pf}} + \frac{1-n}{V_{pl}} \tag{16.3}$$

where V_{pf} is the velocity in the pore fluid and V_{pl} is the compressional wave velocity for the intact material as determined in the laboratory. The compressional wave velocities may be increased appreciably by the presence of water. Because of the relationship between the seismic velocity and porosity, the seismic velocity broadly is related to the intergranular permeability of sandstone formations. However, in most sandstones, fissure flow makes a more important contribution to groundwater movement than intergranular or primary flow.

16.7.2 Resistivity methods

The resistivity of rocks and soils varies within a wide range. Because most of the principal rock-forming

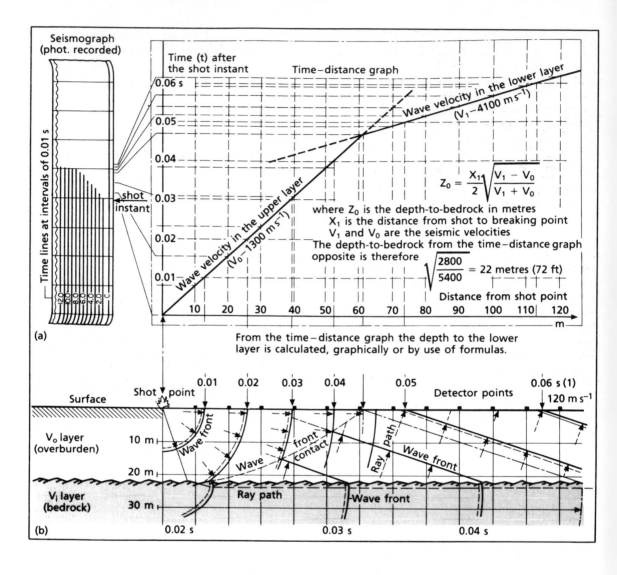

Fig. 16.13 Time–distance graphs for a theoretical single layer problem, with parallel interface. With non-parallel interfaces, both forward and reverse profiles must be surveyed.

minerals are practically insulators, the resistivity of rocks and soils is determined by the amount of conducting mineral constituents and the content of mineralized water in the pores. The latter condition is by far the dominant factor, and indeed most rocks and soils conduct an electric current only because they contain water. The widely differing resistivity values of the various types of impregnating water can cause

variations in the resistivity of rocks, ranging from a few tenths of an ohm metre to hundreds of ohm metres, as can be seen from Table 16.5.

In the resistivity method, an electric current is introduced into the ground by means of two current electrodes, and the potential difference between the two potential electrodes is measured. It is preferable to measure the potential drop or apparent resistance directly in ohms rather than to observe both current and voltage. The ohm value is converted to the apparent resistivity using a factor that depends on the particular electrode configuration in use (see below).

Table 16.5 Resistivity values of some types of natural water.

Type of water	Resistivity (Ωm)
Meteoric water, derived from precipitation	30–1000
Surface waters, in districts of igneous rocks	30–500
Surface waters, in districts of sedimentary rocks	10–100
Groundwater, in areas of igneous rocks	30–150
Groundwater, in areas of sedimentary rocks	Larger than 1
Sea water	About 0.2

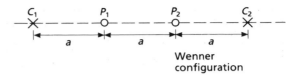

Wenner configuration

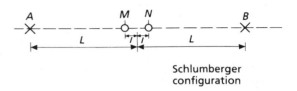

Schlumberger configuration

Fig. 16.14 Wenner and Schlumberger configurations.

Table 16.6 Resistivity values of some common rock types.

Rock type	Resistivity (Ωm)
Topsoil	5–50
Peat and clay	8–50
Clay, sand and gravel mixtures	40–250
Saturated sand and gravel	40–100
Moist to dry sand and gravel	100–3000
Mudstones, marls and shales	8–100
Sandstones and limestones	100–1000
Crystalline rocks	200–10000

The resistivity method is based on the fact that any subsurface variation in conductivity alters the pattern of current flow in the ground, and therefore changes the distribution of the electric potential at the surface. Because the electrical resistivities of superficial deposits and bedrock are different (Table 16.6), the resistivity method may be used in their detection, and to give their approximate thicknesses, relative positions and depths. The first step in any resistivity survey should be to conduct a resistivity depth sounding at the site of a borehole, in order to establish a correlation between the resistivity and lithological layers. If a correlation cannot be established, an alternative method is required.

The electrodes normally are arranged along a straight line, the potential electrodes being placed inside the current electrodes, and all four are symmetrically disposed with respect to the centre of the configuration. The configurations of the symmetrical type most frequently used are those introduced by

Wenner and by Schlumberger. Other configurations include the dipole–dipole and the pole–dipole arrays. In the Wenner configuration, the distances between all four electrodes are equal (Fig. 16.14). The spacings can be progressively increased, keeping the centre of the array fixed, or the whole array, with fixed spacings, can be shifted along a given line. In the Schlumberger arrangement, the potential electrodes maintain a constant separation about the centre of the station, whereas if changes with depth are being investigated, the current electrodes are moved outwards after each reading (Fig. 16.14). The expressions used to compute the apparent resistivity ρ_a for the Wenner and Schlumberger configurations are

$$\rho_a = 2\pi a R \qquad (16.4)$$

and

$$\rho_a = \frac{\pi\left(L^2 - l^2\right)}{2l} \times R \qquad (16.5)$$

respectively, where a, L and l are explained in Fig. 16.14 and R is the resistance reading.

The relation between the depth of penetration and the electrode spacing is given in Fig. 16.15, from which it can be seen that 50% of the total current passes above a depth equal to about one-half of the electrode separation and 70% flows within a depth equal to the electrode separation. An analysis of the variation in the value of the apparent resistivity with respect to the electrode separation enables inferences to be drawn about the subsurface formations.

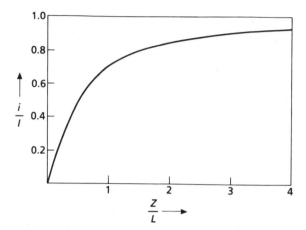

Fig. 16.15 Fraction of total current I which passes about a horizontal plane at depth Z as a function of the distance L between two current electrodes.

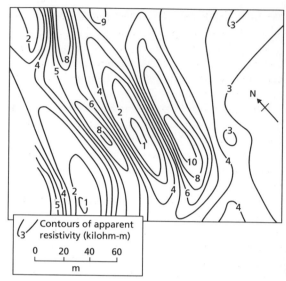

Fig. 16.16 Isoresistivity map.

Horizontal profiling is used to determine the variations in the apparent resistivity in a horizontal direction at a preselected depth. For this purpose, an electrode configuration, with fixed interelectrode distances, is moved along a straight traverse, resistivity determinations being made at stations at regular intervals. The length of the electrode configuration must be carefully chosen, because it is the dominating factor regarding the depth of penetration. The data of a constant separation survey, consisting of a series of traverses arranged in a grid pattern, may be used to construct a contour map of lines of equal resistivity (Fig. 16.16). These maps often are extremely useful in locating areas of anomalous resistivity, such as gravel pockets in clay soils, and the trend of buried channels. Even so, the interpretation of resistivity maps, as far as the delineation of lateral variations is concerned, is mainly qualitative.

Electrical sounding furnishes information concerning the vertical succession of different conducting zones and their individual thicknesses and resistivities (Fig. 16.17). For this reason, the method is particularly valuable for investigations on horizontally stratified ground. In electrical sounding, the midpoint of the electrode configuration is fixed at the observation station, and the length of the configuration is increased in a number of stages. As a result, the current penetrates deeper and deeper, the apparent resistivity being measured each time the

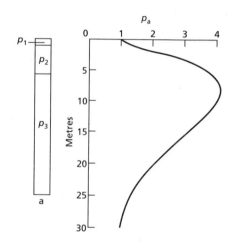

Fig. 16.17 Variation of the apparent resistivity with Wenner electrode separation for a three-layer earth.

current electrodes are moved outwards. The readings therefore become increasingly affected by the resistivity conditions at advancing depths. The Schlumberger configuration is preferable to the Wenner configuration for depth sounding. The data obtained usually are plotted as a graph of the apparent resistivity against the electrode separation, in the case of the Wenner array, or half the current electrode separation for the Schlumberger array. The electrode separation, at which inflection points occur in the graph, provides an idea of the depth of the interfaces. The apparent resistivities of the different parts of the curve provide some idea of the relative resistivities of the layers concerned.

If the ground approximates to an ideal condition, a quantitative solution, involving a curve-fitting exercise, should be possible. The technique requires a comparison of the observed curve with a series of master curves prepared for various theoretical models.

Generally, it is not possible to determine the depths to more than three or four layers. If a second layer is relatively thin, and its resistivity is much larger or smaller than that of the first layer, the interpretation of its lower contact will be inaccurate. For all depth determinations from resistivity soundings, it is assumed that there is no change in resistivity laterally. This is not the case in practice. Indeed, sometimes the lateral change is greater than that occurring with increasing depth, and therefore corrections must be applied for the lateral effects when depth determinations are made.

As the amount of water present is influenced by the porosity of a rock, the resistivity provides a measure of porosity. For example, in granular materials in which there are no clay minerals, the relationship between the resistivity ρ on the one hand and the density of the pore water ρ_w, the porosity n and the degree of saturation S_r on the other is as follows:

$$\rho = a\rho_w n^{-x} S_r^{-y} \qquad (16.6)$$

where a, x and y are variables (x ranges from 1.0 for sand to 2.5 for sandstone, and y is approximately 2.0 when the degree of saturation is greater than 30%). For those formations which occur below the water table and are therefore saturated, the above expression becomes

$$\rho = a\rho_w n^{-x} \qquad (16.7)$$

because $S_r = 1$ (i.e. 100%).

In a fully saturated sandstone, a fundamental empirical relationship exists between the electrical and hydrogeological properties, which involves the concept of the formation resistivity factor F_a, defined as

$$F_a = \frac{\rho_0}{\rho_w} \qquad (16.8)$$

where ρ_0 is the resistivity of the saturated sandstone and ρ_w is the resistivity of the saturating solution. In a clean sandstone, that is, one in which the electric current passes through the interstitial electrolyte during testing, with the rock mass acting as an insulator, the formation resistivity factor is closely related to the porosity. The formation resistivity factor is related to the true formation factor F by the expression

$$F = \frac{\rho_A F_a}{\rho_A - F_a \rho_w} \qquad (16.9)$$

where ρ_A is a measure of the effective resistivity of the rock matrix. In clean sandstone ρ_A is infinitely large, and consequently $F = F_a$. Generally, F is related to the porosity n by the equation

$$F = a/n^c \qquad (16.10)$$

where a and m are constants for a given formation.

The true formation factor in certain formations has also been shown to be broadly related to the intergranular permeability k_g by the expression

$$F = b/k_g^d \qquad (16.11)$$

where b and n are constants for a given formation. In sandstones in which intergranular flow is important, the above expressions can be used to estimate the hydraulic conductivity and, if the thickness of the aquifer is known, the transmissivity. The techniques are not useful in highly indurated sandstones, where the intergranular permeabilities are less than 1.0×10^{-7} m s^{-1}, and the flow is controlled by fissures. Similarly, these methods tend to be of little value in multilayered aquifers.

16.7.3 Electromagnetic methods

In most electromagnetic methods, electromagnetic energy is introduced into the ground by inductive coupling, and is produced by passing an alternating current through a coil. The receiver also detects the signal by induction, for example, the terrain conductivity meter measures the conductivity of the ground by such an inductive method. The conductivity meter is carried along traverse lines across a site and can provide a direct continuous readout. Hence, surveys can be carried out rapidly. Conductivity values are taken at positions set out on a grid pattern.

McCann *et al.* (1997) referred to the use of the very low-frequency (VLF) and electromagnetic pulse methods. One of the disadvantages of the former method is that wave penetration is limited. Hence, measurements of resistivity based on electromagnetic energy primarily are applied to soil cover and fill. The attenuation of electromagnetic energy is a problem in the electromagnetic pulse technique, but it has been used with success to investigate very dry sites and low-porosity rocks.

The ground probing radar method is based upon the transmission of pulsed electromagnetic waves in the frequency range 1–1000 MHz. In this method, the travel times of the waves reflected from subsurface interfaces are recorded as they arrive at the surface, and the depth (Z) to an interface is derived from

$$Z = vt/2 \qquad (16.12)$$

where v is the velocity of the radar pulse and t is its travel time. The conductivity of the ground imposes the greatest limitation on the use of radar probing, that is, the depth to which radar energy can penetrate depends upon the effective conductivity of the strata being probed. This, in turn, is governed chiefly by the water content and its salinity. The nature of the pore water exerts the most influence on the dielectric constant. Furthermore, the effective conductivity is also a function of the temperature and density, as well as the frequency of the electromagnetic waves being propagated. The least penetration occurs in saturated clayey materials, or where the moisture content is saline. For example, the attenuation of electromagnetic energy in wet clay and silt means that the depth of penetration frequently is less than 1 m. The technique appears to be reasonably successful in sandy soils and rocks in which the moisture content is non-saline. Rocks such as limestone and granite can be penetrated for distances of tens of metres and, in dry conditions, the penetration may reach 100 m. Dry rock salt is radar translucent, permitting penetration distances of hundreds of metres.

Grasmück and Green (1996) described a three-dimensional method of ground probing radar. According to them, this system is capable of producing vivid images of the subsurface up to depths of 50 m.

16.7.4 Magnetic and gravity methods

All rocks are magnetized to a lesser or greater extent by the Earth's magnetic field. Thus, in magnetic prospecting, accurate measurements are made of the anomalies produced in the local geomagnetic field by this magnetization. The intensity of magnetization, and hence the amount by which the Earth's magnetic field is changed locally depend on the magnetic susceptibility of the material concerned. In addition to the magnetism induced by the Earth's field, rocks possess a permanent magnetism that depends upon their history.

Rocks have different magnetic susceptibilities related to their mineral content. Some minerals, for example, quartz and calcite, are magnetized reversely to the field direction, and therefore have a negative susceptibility and are described as diamagnetic. Paramagnetic minerals, which are the majority, are magnetized along the direction of the magnetic field, so that their susceptibility is positive. The susceptibilities of ferromagnetic minerals, such as magnetite, ilmenite, pyrrhotite and haematite, are a very complicated function of the field intensity. However, because their susceptibilities are 10–10^5 times the order of susceptibility of the paramagnetic and diamagnetic minerals, the ferromagnetic minerals can be found by magnetic field measurements.

If the magnetic field ceases to act on a rock, the magnetization of paramagnetic and diamagnetic minerals disappears. However, in ferromagnetic minerals, the induced magnetization is diminished only to a certain value. This residuum is called the remanent magnetization, and is of great importance in rocks. All igneous rocks have a very high remanent magnetization, acquired as they cooled down in the Earth's magnetic field. In the geological past, during sedimentation in water, grains of magnetic materials were orientated by ancient geomagnetic fields, so that some sedimentary rocks show stable remanent magnetization.

The strength of the magnetic field is measured in nanoteslas (nT). The average strength of the Earth's magnetic field is about 50 000 nT, but the variations associated with magnetized rock formations are very much smaller than this.

Aeromagnetic surveying has almost completely supplanted ground surveys for regional reconnaissance purposes. An accurate identification of the plan position of the aircraft for the whole duration of the magnetometer record is essential. The object is to produce an aeromagnetic map, the base map with transcribed magnetic values being contoured at 5–10 nT intervals.

The aim of most ground surveys is to produce isomagnetic contour maps of anomalies to enable the form of the causative magnetized body to be estimated. Profiles are surveyed across the trend of linear anomalies, with stations, if necessary, at intervals of as little as 1 m. A base station is set up beyond the anomaly, where the geomagnetic field is uniform. The reading at the base station is taken as zero, and all subsequent readings are expressed as plus-or-minus differences. Corrections need to be made for the temperature of the instrument, as the magnets lose their effectiveness with increasing temperature. A planetary correction is also required, which eliminates the normal variation of the Earth's magnetic field with latitude. Large metallic objects, such as pylons, are a serious handicap to magnetic exploration, and must be kept at a sufficient distance, as it is difficult to correct for them.

A magnetometer may also be used for mapping geological structures, for example, in some thick sedimentary sequences it is sometimes possible to delineate the major structural features, because the succession includes magnetic horizons. These may be ferruginous sandstones or shales, tuffs or basic lava flows. In such circumstances, anticlines produce positive and synclines negative anomalies.

Faults and dykes are indicated on isomagnetic maps by linear belts of somewhat sharp gradient, or by sudden swings in the trend of the contours. However, in many areas the igneous and metamorphic basement rocks, which underlie the sedimentary sequence, are the predominant influence controlling the pattern of anomalies, because they usually are far more magnetic than the sediments above. Where the basement rocks are brought near the surface in structural highs, the magnetic anomalies are large and characterized by strong relief. Conversely, deep sedimentary basins usually produce contours with low values and gentle gradients on isomagnetic maps.

The Earth's gravitational field varies according to the density of the subsurface rocks, but at any particular locality, its magnitude is also influenced by the latitude, elevation, neighbouring topographical features and the tidal deformation of the Earth's crust. The effects of these factors must be eliminated in any gravity survey, where the object is to measure the variations in acceleration due to gravity precisely. This information can then be used to construct a contoured gravity map. In survey work, modern practice is to measure anomalies in gravity units (1 g.u. = 10^{-6} m s^{-2}). Formerly, the unit of measurement was the milligal (mGal), that is, 0.001 Gal, 980 Gal being the approximate acceleration at the Earth's crust due to gravity. Hence, 10 g.u. is equal to 1 mGal. Modern gravity meters used in exploration measure not the absolute value of the acceleration due to gravity, but the small differences in this value between one place and the next.

Gravity methods mainly are used in regional reconnaissance surveys to reveal anomalies which subsequently may be investigated by other methods. Because the gravitational effects of geological bodies are proportional to the contrast in density between them and their surroundings, gravity methods are particularly suitable for the location of structures in stratified formations. Gravity effects due to local structures in near-surface strata may be partly obscured or distorted by regional gravity effects caused by large-scale basement structures. However, regional deep-seated gravity effects can be removed or minimized in order to produce a residual gravity map showing the effects of shallow structures which may be of interest.

A gravity survey is conducted from a local base station, at which the value of the acceleration due to gravity is known with reference to a fundamental base, where the acceleration due to gravity has been accurately measured. The way in which a gravity survey is carried our largely depends on the objective in view. Large-scale surveys covering hundreds of square kilometres, carried out in order to reveal major geological structures, are performed by vehicle or helicopter, with a density of only a few stations per square kilometre. For more detailed work, such as the delineation of ore bodies or basic minor intrusions, or the location of faults, the spacing between stations may be as small as 20 m. Because gravity differences large enough to be of geological significance are produced by changes in elevation of several millimetres and of only 30 m in north–south distance, the location and elevation of stations must be established with very high precision.

16.7.5 Drill-hole logging techniques

Drill-hole logging techniques can be used to identify some of the physical properties of rocks. The electrical resistivity method makes use of various electrode configurations down the hole. As the instrument is raised from the bottom to the top of the hole, it provides a continuous record of the variations in resistivity of the wall rock. In the normal or standard resistivity configuration, there are two potential electrodes and one current electrode in the sonde. The depth of penetration of the electric current from the drill-hole is influenced by the electrode spacing. In a short normal resistivity survey, the spacing is about

400 mm, whereas in a long normal survey, the spacing generally is between 1.5 and 1.75 m. Unfortunately, in such a survey, because of the influence of thicker adjacent beds, thin resistive beds yield resistivity values which are much too low, whilst thin conductive beds produce values which are too high. The microlog technique may be used in such situations. In this technique, the electrodes are very closely spaced (25–50 mm), and are in contact with the wall of the drill-hole. This allows the detection of small lithological changes, so that much finer detail is obtained than with the normal electrical log (Fig. 16.18). A microlog is particularly useful in recording the position of permeable beds.

If, for some reason, the current tends to flow between the electrodes on the sonde instead of into the rocks, the laterolog or guard electrode is used.

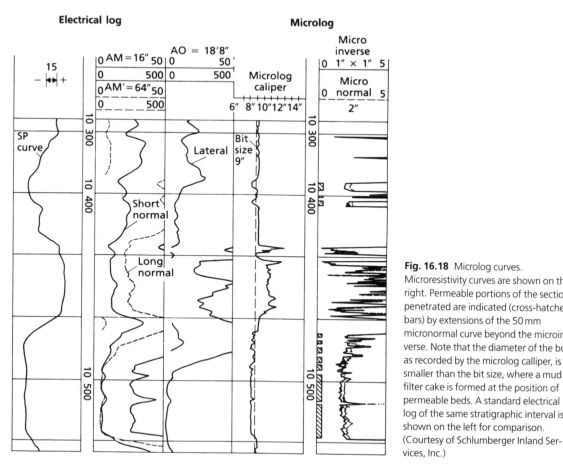

Fig. 16.18 Microlog curves. Microresistivity curves are shown on the right. Permeable portions of the section penetrated are indicated (cross-hatched bars) by extensions of the 50 mm micronormal curve beyond the microinverse. Note that the diameter of the bore, as recorded by the microlog calliper, is smaller than the bit size, where a mud filter cake is formed at the position of permeable beds. A standard electrical log of the same stratigraphic interval is shown on the left for comparison. (Courtesy of Schlumberger Inland Services, Inc.)

The laterolog 7 has seven electrodes in an array, and focuses the current into the strata of the drill-hole wall. The microlaterolog, a focused microdevice, is used in such a situation instead of the microlog.

A dipmeter is a three-, four- or six-arm side-wall microresistivity device which measures small changes in resistivity, thereby permitting the relative shifts of the characteristic patterns of variation on the traces to be used to determine the attitudes of discontinuities in contact with the drill-hole wall. In this way, a fracture log can be produced.

Induction logging may be used when an electrical log cannot be obtained. In this technique, the sonde sends electrical energy into the strata horizontally, and therefore only measures the resistivity immediately opposite the sonde, in contrast with normal electrical logging where the current flows between the electrodes. As a consequence, the resistivity is measured directly in an induction log, whereas in a normal electrical log, because the current flows across the stratal boundaries, it is measured indirectly from the electrical log curves. A gamma-ray log (see below) usually is run with an induction log in order to reveal the boundaries of stratal units.

A spontaneous potential (*SP*) log is obtained by lowering down a drill-hole a sonde which generates a small electrical voltage at the boundaries of permeable rock units and, especially, between such strata and less permeable beds. For example, permeable sandstones show large *SP* values, whereas shales typically are represented by low values. If sandstone and shale are interbedded, the *SP* curve has numerous troughs separated by sharp or rounded peaks, the widths of which vary in proportion to the thicknesses of the sandstones. *SP* logs frequently are recorded at the same time as resistivity logs. The interpretation of both sets of curves yields precise data on the depth, thickness and position in the sequence of the beds penetrated by the drill-hole. The curves also enable a semi-quantitative assessment to be made of the lithological and hydrogeological characteristics.

The sonic logging device consists of a transmitter–receiver system, the transmitter(s) and receiver(s) being located at given positions on the sonde. The transmitters emit short high-frequency pulses several times a second, and the differences in the travel times between receivers are recorded in order to obtain the velocities of the refracted waves. The velocities of the sonic waves propagated in sedimentary rocks are largely a function of the character of the matrix. Normally, beds with high porosities have low velocities, and dense rocks are typified by high velocities. Hence, the porosity of strata can be assessed. In the three-dimensional sonic log, one transmitter and one receiver are used at a time. This allows both compressional and shear waves to be recorded, from which, if density values are available, the dynamic elastic moduli of the beds concerned can be determined. As the velocity values vary independently of the resistivity or radioactivity, the sonic log permits a differentiation to be made between strata which may be less evident on the other types of log.

The televiewer emits pulses of ultrasonic energy from a piezoelectric transducer, which are then reflected by the fluid in the drill-hole wall rock to be picked up by the transducer, which also acts as a receiver. The transducer is orientated relative to the Earth's magnetic field by a down-the-hole magnetometer in the sonde, and is rotated within the drill-hole at $3\,\mathrm{revs^{-1}}$. In this way, data can be obtained relating to the orientation of discontinuities within the drill-hole wall (dip angles can be delineated down to 20° and dip directions down to values of 15°).

Radioactive logs include gamma-ray, natural gamma, gamma–gamma or formation density and neutron logs. They have the advantage of being obtainable through the casing in a drill-hole. On the other hand, the various electrical and sonic logs, with the exception of interborehole acoustic scanning, can only be used in uncased holes. The natural gamma log provides a record of the natural radioactivity or gamma radiation from elements such as potassium-40 and uranium and thorium isotopes in the rocks. This radioactivity varies widely among sedimentary rocks, being generally high for clays and shales and lower for sandstones and limestones. Evaporites give very low readings. The gamma–gamma log uses a source of gamma rays which are sent into the wall of the drill-hole. There they collide with electrons in the rocks and thereby lose energy. The returning gamma-ray intensity is recorded, a high value indicating a low electron density, and hence a low formation density. The neutron curve is a recording of the effects caused by bombardment of the strata with neutrons. As the neutrons are absorbed by atoms of hydrogen, which then emit γ rays, the log provides an indication of the quantity of hydrogen in the strata around the sonde. The amount of hydrogen is related to the water (or hydrocarbon) content, and therefore provides a means of estimating the porosity. Because carbon is a good moderator of neutrons, carbonaceous rocks are liable to yield spurious indications as far as porosity is concerned.

The calliper log measures the diameter of a drill-hole. Different sedimentary rocks show a greater or lesser ability to stand without collapsing from the walls of the drill-hole. For instance, limestones may present a relatively smooth face slightly larger than the drilling bit, whereas soft shale may cave to produce a much larger diameter. A calliper log is obtained together with other logs to help to interpret the characteristics of the rocks in the drill-hole.

16.7.6 Cross-hole seismic methods

The cross-hole seismic method is based on the transmission of seismic energy between drill-holes. Cross-hole seismic measurements are made between a seismic source in one drill-hole (i.e. a small explosive charge, an air-gun, a drill-hole hammer or an electric sparker) and a receiver at the same depth in an adjacent drill-hole. The receiver can either be a three-component geophone array clamped to the drill-hole wall or a hydrophone, as in interborehole acoustic scanning, where a hydrophone is used in a liquid-filled drill-hole to receive signals from an electric sparker in another drill-hole similarly filled with liquid. The choice of source and receiver is a function of the distance between the drill-holes, the required resolution and the properties of the rock mass. The best results are obtained with a high-frequency repetitive source (McCann *et al.*, 1986).

Generally, the source and receiver in the two drill-holes are moved up and down together. Drill-holes must be spaced closely enough to achieve the required resolution of detail and be within the range of the equipment. This is up to 400 m in some clays, 160 m in chalk and 80 m in sands and gravels. By contrast, because soft organic clay is highly attenuating, transmission is only possible over a few metres. These distances are for saturated material, and the effective transmission is reduced considerably in dry superficial layers.

Cross-hole seismic measurements provide a means by which the engineering properties of the rock mass between drill-holes can be assessed. For example, the dynamic elastic properties can be obtained from the values of the compressional and shear wave velocities, and the formation density. Other applications include the assessment of the continuity of lithological units between drill-holes, the identification of fault zones, the assessment of the degree of fracturing, and the detection of subsurface voids.

16.8 *In situ* testing

Under certain circumstances, especially when samples are difficult to obtain, data may be obtained directly by carrying out tests in the field. This particularly is the case with cohesionless soils and sensitive clays. In addition, the data obtained by direct testing in place are likely to give more reliable results than those obtained from laboratory tests.

There are two types of penetrometer test, that is, dynamic and static tests. Both methods measure the resistance to penetration of a conical point offered by the soil at any particular depth. Penetration of the cone creates a complex shear failure, and thus provides an indirect measure of the *in situ* shear strength of the soil.

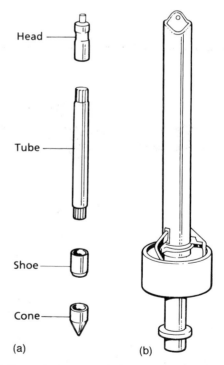

Head

Tube

Shoe

Cone

(a)   (b)

Fig. 16.19 Standard penetration test equipment: (a) split-spoon sampler; (b) trip hammer.

Table 16.7 Relative density and consistency of soil. (After Terzaghi & Peck, 1968; Sanglerat, 1972.)

(a) Relative density of sand and SPT values, and relationship to angle of internal friction

SPT (N)	Relative density (D_r)	Description of compactness	Angle of internal friction (ϕ)
4	0.2	Very loose	Under 30°
4–10	0.2–0.4	Loose	30°–35°
10–30	0.4–0.6	Medium dense	35°–40°
30–50	0.6–0.8	Dense	40°–45°
Over 50	0.8–1	Very dense	Over 45°

(b) N-values, consistency and unconfined compressive strengths of cohesive soils

N	Consistency	Unconfined compressive strength (kPa)
Under 2	Very soft	Under 20
2–4	Soft	20–40
5–8	Firm	40–75
9–15	Stiff	75–150
16–30	Very stiff	150–300
Over 30	Hard	Over 300

$$D_r = \frac{e_{max} - e}{e_{max} - e_{min}}; \; e \text{ is void ratio.}$$

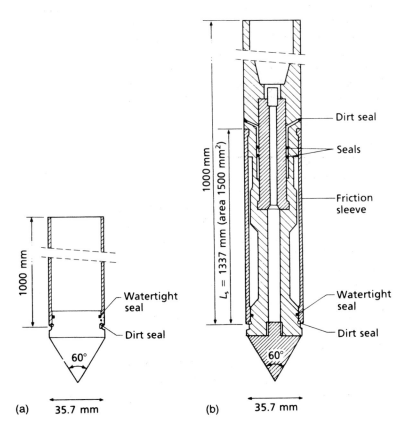

Fig. 16.20 An electric penetrometer tip: (a) without friction sleeve; (b) with friction sleeve.

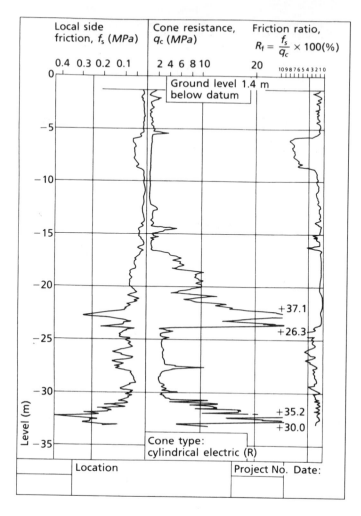

Fig. 16.21 Typical record of cone penetration test.

The most widely used dynamic method is the standard penetration test. This empirical test consists of driving a split-spoon sampler, with an outside diameter of 50 mm, into the soil at the base of a borehole. A trip hammer, weighing 63 kg, falls freely through a distance of 750 mm onto the drive head, which is fitted at the top of the rods (Fig. 16.19). Initially, the split-spoon is driven 150 mm into the soil at the bottom of the borehole. It is then driven a further 300 mm, and the number of blows required to drive this distance is recorded. The blow count is referred to as the N value, from which the relative density of the sand can be assessed (Table 16.7). Refusal is regarded as 100 blows. The results obtained from the standard penetration test provide an evaluation of the degree of compaction of cohesionless sands, and the N values may be related to the values of the angle of internal friction (ϕ) and the allowable bearing capacity. The standard penetration test can also be employed in clays (Table 16.7), weak rocks and in the weathered zones of harder rocks.

The most widely used static method employs the Dutch cone penetrometer (Fig. 16.20). It is particularly useful in soft clays and loose sands, where boring operations tend to disturb *in situ* values. In this technique, a tube and inner rod with a conical point at the base are hydraulically advanced into the ground, the reaction being obtained from pickets screwed into place. The cone has a cross-sectional area of 1000 mm² with an angle of 60°. At approximately every

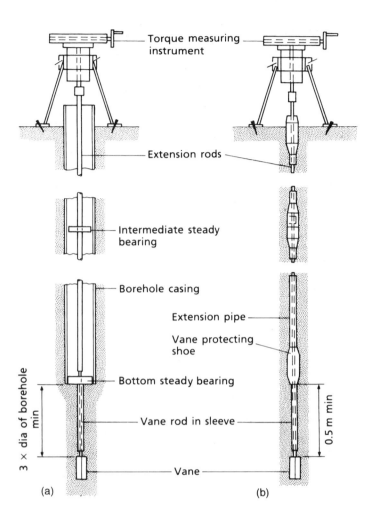

Fig. 16.22 Shear vane tests: (a) borehole vane test; (b) penetration vane test.

300 mm depth, the cone is advanced ahead of the tube by a distance of 50 mm, and the maximum resistance is noted. The tube is then advanced to join the cone after each measurement, and the process is repeated. The cone resistances are plotted against their corresponding depths so as to give a profile of the variation in consistency (Fig. 16.21). Recordings can also be taken continuously. One type of Dutch cone penetrometer has a sleeve behind the cone which can measure side friction. The ratio of sleeve resistance to that of cone resistance is higher in cohesive than in cohesionless soils, thus affording some estimate of the type of soil involved (Fig. 16.22).

In the piezocone, a cone penetrometer is combined with a piezometer, the latter being located between the cone and the friction sleeve. The pore water pressure is measured at the same time as the cone resistance and sleeve friction. Because of the limited thickness of the piezometer (the filter is around 5 mm), much thinner layers can be determined with greater accuracy than with a cone penetrometer. If the piezocone is kept at a given depth, so that the pore water pressure can dissipate with time, this allows an assessment to be made of the *in situ* permeability and consolidation characteristics of the soil.

Because soft clays may suffer disturbance when sampled, and therefore give unreliable results when tested for strength in the laboratory, a vane test often is used to measure the *in situ* undrained shear strength. Vane tests can be used in clays which have a consistency varying from very soft to firm. In its simplest form, the shear vane apparatus consists of four blades arranged in cruciform fashion and attached to the end of a rod (Fig. 16.22). To eliminate the effects of friction of the soil on the vane rods during the test, all rotating parts, other than the vane, are enclosed in guide tubes. The vane normally is housed in a protective shoe. The vane and rods are pushed into the soil from the surface or the base of a borehole to a point 0.5 m above the required depth. The vane is then pushed out of the protective shoe and advanced to the test position. It is then rotated at a rate of 6° to 12° per minute. The torque is applied to the vane rods by means of a torque measuring instrument, mounted at ground level and clamped to the borehole casing or rigidly fixed to the ground. The maximum torque required for rotation is recorded. When the vane is rotated, the soil fails along a cylindrical surface defined by the edges of the vane, as well as along the horizontal surfaces at the top and bottom of the blades. The shearing resistance is obtained from the following expression:

$$\tau = \frac{M}{\pi\left[\left(D^2 H/2\right) + \left(D^3/6\right)\right]} \qquad (16.13)$$

where τ is the shearing resistance, D and H are the diameter and height of the vane respectively, and M is the torque.

The plate load test provides valuable information from which the bearing capacity and settlement characteristics of a foundation can be assessed. It is carried out in a trial pit, usually at excavation base level. Plates vary in size from 0.15 to 0.61 m in diameter, the size of plate used being determined by the spacing of the discontinuities. The plate should be properly bedded, and the test carried out on undisturbed material, so that reliable results may be obtained. The load is applied by a jack, in increments, either of one-fifth of the proposed bearing pressure or in steps of 25–50 kPa (these steps are smaller in soft soils, i.e. where the settlement under the first increment of 25 kPa is greater than 0.002D, where D is the diameter of the plate). Successive increments should be made after settlement has ceased. The test generally is continued up to two or three times the proposed loading or, in clays until settlement equal to 10–20% of the plate dimension is reached or the rate of increase of settlement becomes excessive. Consequently, the ultimate bearing capacity at which settle-

ment continues without increasing the load rarely is reached.

Large plate bearing tests frequently are used to determine the value of Young's modulus of the foundation rock at large construction sites. Loading of the order of several meganewtons is required to obtain measurable deformation of representative areas. Tests usually are carried out in specially excavated galleries, in order to provide a sufficiently strong reaction point for the loading jacks to bear against.

The Menard pressuremeter (Fig. 16.23) is used to determine the *in situ* strength of the ground. It is particularly is useful in those soils from which undisturbed samples cannot be obtained readily. This pressuremeter consists essentially of a probe which is placed in a borehole at the appropriate depth and

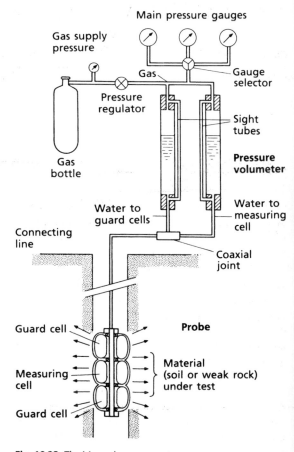

Fig. 16.23 The Menard pressuremeter.

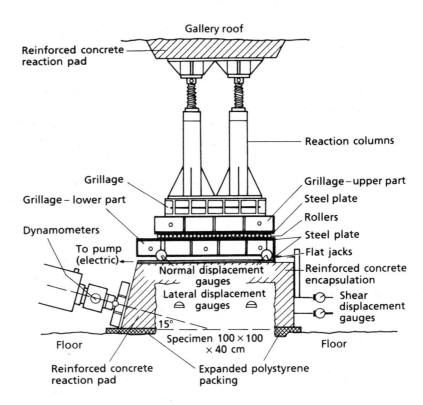

Fig. 16.24 The *in situ* shear test apparatus.

then expanded. Where possible the test is carried out in an unlined hole but, if necessary, a special slotted casing is used to provide support. The probe consists of a cylindrical metal body over which are fitted three cylinders. A rubber membrane covers the cylinders, and is clamped between them to give three independent cells. The cells are inflated with water, and a common gas pressure is applied by a volumeter located at the surface; thus a radial stress is applied to the soil. The deformations produced by the central cell are indicated on the volumeter. This test thus provides the ultimate bearing capacity of soils, as well as their deformation modulus. The test can be applied to any type of soil, and takes into account the influence of discontinuities. It can also be used in weathered zones of rock masses, and in weak rocks such as shales and marls. It provides an almost continuous method of *in situ* testing.

The major advantage of a self-boring pressuremeter is that a borehole is unnecessary. Hence, the interaction of the probe and the soil is improved. Self-boring is brought about either by jetting or by using a chopping tool.

A dilatometer is a type of pressuremeter, and is used in a drill-hole to obtain data relating to the deformability of a rock mass. These instruments range up to about 300 mm in diameter and over 1 m in length, and can exert pressures of up to 20 MPa on the drill-hole walls. Diametrical strains can be measured either directly along two perpendicular diameters, or by measuring the amount of liquid pumped into the instrument.

In situ shear tests usually are performed on blocks, 700×700 mm, cut in the rock. The block is sheared while a horizontal jack exerts a vertical load. It is advantageous to make the tests inside galleries, where reactions for the jacks are readily available (Fig. 16.24). The tests are performed at various normal loads, and give an estimate of the angle of shearing resistance and cohesion of the rock.

References

Allum, J.A.E. (1966) *Photogeology and Regional Mapping.* Pergamon Press, Oxford.

Anon. (1970) Logging of cores for engineering purposes.

Engineering Group Working Party Report. *Quarterly Journal of Engineering Geology* **5**, 1–24.

Anon. (1978) *Terrain Evaluation for Highway Engineering and Transport Planning. Transport Road Research Laboratory, Report* **SR448**. Department of the Environment, Crowthorne.

Anon. (1981) *Code of Practice on Site Investigations, BS 5930*. British Standards Institution, London.

Anon. (1988) Engineering geophysics. Engineering Group Working Party Report. *Quarterly Journal of Engineering Geology* **21**, 207–273.

Beaumont, T.E. (1979) Remote sensing for location and mapping of engineering construction materials. *Quarterly Journal of Engineering Geology* **12**, 147–158.

Brink, A.B.A., Mabbutt, J.A., Webster, R. & Beckett, P.H.T. (1966) *Report on the Working Group on Land Classification and Data Storage. Military Engineering Experimental Establishment, Report No. 940*. Military Engineering Experimental Establishment, Christchurch.

Dearman, W.R. (1991) *Engineering Geological Maps*. Butterworth–Heinemann, Oxford.

Dent, D. & Young, A. (1981) *Soil Survey and Land Evaluation*. Allen and Unwin, London.

Grasmück, M. & Green, A.G. (1996) 3-D georadar mapping: looking into the subsurface. *Environmental and Engineering Geosciences* **2**, 195–200.

Hatheway, A.W. & Leighton, F.B. (1979) Trenching as an exploration method. In *Geology in the Siting of Nuclear Power Plants, Reviews in Engineering Geology*, Vol. 4. Geological Society of America, Boulder, Colorado, pp. 169–195.

Hawkins, A.B. (1986) Rock descriptions. In *Site Investigation Practice: Assessing BS 5930, Engineering Geology Special Publication No. 2*, Hawkins, A.B. (ed). Geological Society, London, pp. 59–66.

Herbert, S.M., Roche, D.P. & Card, G.B. (1987) The value of engineering geological desk study appraisals in scheme planning. In *Planning in Engineering Geology, Engineering Geology Special Publication No. 4*, Culshaw, M.G., Bell, F.G., Cripps, J.C. & O'Hara, M. (eds). Geological Society, London, pp. 151–154.

Hvorslev, M.J. (1949) *Subsurface Exploration and Sampling for Civil Engineering Purposes, American Society of Civil Engineers Report*. American Society of Civil Engineers, Waterways Experimental Station, Vicksburg.

Lawrence, C.J. (1972) *Terrain Evaluation in West Malaysia, Part 7—Terrain Classification and Survey Methods. Transport Road Research Laboratory, Report* **506**. Department of the Environment, Crowthorne.

Lawrence, C.J. (1978) *Terrain Evaluation in West Malaysia, Part 2—Land Systems of South West Malaysia. Transport Road Research Laboratory, Report* **SR378**. Department of the Environment, Crowthorne.

McCann, D.M., Baria, R., Jackson, P.D. & Green, A.S.P. (1986) Application of crosshole seismic measurements to site investigation. *Geophysics* **51**, 914–925.

McCann, D.M., Culshaw, M.G. & Northmore, K. (1990) Rock mass assessments from seismic measurements. In *Field Testing in Engineering Geology, Engineering Geology Special Publication No. 6*, Bell, F.G., Culshaw, M.G., Cripps, J.C. & Coffey, J.R. (eds). Geological Society, London, pp. 257–266.

McCann, D.M., Culshaw, M.G. & Fenning, P.J. (1997) Setting the standard for geophysical surveys in site investigation. In *Modern Geophysics in Engineering Geology, Engineering Geology Special Publication No. 12*, McCann, D.M., Eddleston, M., Fenning, P.J. & Reeves, G.M. (eds). Geological Society, London, pp. 3–34.

Mitchell, C.W. (1991) *Terrain Evaluation*, 2nd edn. Longman, Harlow.

Rengers, N. & Soeters, R. (1980) Regional engineering geological mapping from aerial photographs. *Bulletin of the International Association of Engineering Geology* **21**, 103–111.

Sabins, F.F. (1996) *Remote Sensing—Principles and Interpretation*. Freeman, San Francisco.

Warwick, D., Hartopp, P.G. & Viljoen, R.P. (1979) Application of thermal infrared linescanning technique to engineering geological mapping in South Africa. *Quarterly Journal of Engineering Geology* **12**, 159–180.

Webster, R. & Beckett, P.H.T. (1970) Terrain classification and evaluation using air photography, *Photogrammetric Engineering* **26**, 51–75.

Index

Page numbers in *italic* type indicate figures; those in **bold** type indicate tables

ablation till   205–6, **207**, **215**
  large stones in   206
abrasion
  by stream load   115
  by wind   166
accelerographs   60, 73–4
accessory minerals   493
  role in brick making   327
accretion, lateral   126
acid mine drainage   441, 444, 446–50
  acid generation, determination of rate of   449
  control of   449
    migration control   450
    release control   450
    source control   450
  defined   446–7
  ferruginous, containing aluminium   452
  from colliery spoil heaps   442
  from mine workings   452
  heavy metals and sulphate a serious environmental threat   447, **447**
  key management strategies   449–50
  need for regular monitoring   450
  neutralized by contact with acid-consuming materials   448–9
  prediction of potential for   449
  treatment processes, active and passive   450
acid rain
  acidifying groundwater   278
  from volcanic gases   55
acid–base accounting   449
Admiralty charts, information on tidal currents   142–3
adsorption   466
aeolian deposits, due to desertification   188
aeolian processes   163
aeration, of soils   287

aerial photographs   17, 402–3, 525, 535
  aiding delineation of volcanic hazard zones   48
  black and white, for topographical surveys   529
  can provide base maps   19
  data for discontinuities   384
  false colour photography   529–30
  help in locating potential aquifers   255
  infrared monochrome film   529
  for landslide movement   101
  for mapping land facets   532
  for morphological mapping   13
  and photogeology   530–1
  in potential reservoir site investigations   235
  relief and tone   530
  for studying arid landscapes   176
  in terrain evaluation   532
  to monitor desertification   190
  true colour photography   529
  types of   528–9
  use of to geologists   528
aeromagnetic surveying   552–3
afforestation, to stabilize shallow slides   111
Africa
  development of basement aquifers for rural water supply   249
  leaching of soils   493
  see also South Africa; Zimbabwe
aftershocks, post-earthquake   59
Agadir, Morocco, buildings destroyed by earthquake   66
agglomerate   43
aggradation, upstream from reservoirs   235
aggregates
  for concrete   408–10
  microporous possibly frost susceptible   410
  reactive rock types   **403**, 409
  for road construction   410–11
    importance of petrology   411
  shape and surface texture of particles   408

used for base/sub-base courses, subject to salt weathering   185
agricultural land, purpose of Universal Soil Loss Equation   311–13
agriculture
  crop tolerance to salt concentrations   **193–4**
  elements necessary for healthy plant growth   288
  factors affecting plant growth   287–8
  nitrate pollution resulting from intensive cultivation   277
  practices affect soil erosion   309–10
  soil conservation measures   **314**, **315**
air sparging   484
airborne vertical laser sensing, use over forested landslides   101
Alaskan earthquakes
  1964, Anchorage   65, *65*, 69–70
  1969   69
alcretes (bauxite)   354, **354**, 493
alkali–aggregate reaction   409
alkali–carbonate reaction   409
alkali–silica gels   409
alluvial deposits, foundation problems   245
alluvial fans   **215**
  areas of high flood risk   171
  intermontane basins   170–1
alluvial stratigraphy, improving estimates of flood recurrence intervals   130
alluvial terraces   120, *120*
alluvium
  amplitude and acceleration of shock waves greater   63–4
  coarse grained, in ephemeral stream channels   174
  floodplain   126
α particle tracking   515
alteration
  affecting roadstone durability   410–11

through chemical weathering 386
aluminium 292
  in acid mine waters 452
  and health 509–10
  potential toxicity governed by speciation 497
aluminium hydroxide, arsenic sorbed onto 499
Alzheimer's disease, and aluminium in drinking water 509
ammonia 466
  toxicity of 468
amyotrophic lateral sclerosis, a deficiency disease 487
andisols 356, 357
andosols 354–5
  fersiallitic 354, 355
angle of influence
  longwall mining 427
  of undermining 423
anglesite 501
angularity number 325
anhydrite 180, 405
  hydration to form gypsum 405
  massive, dissolved to produce runaway situations 405
anhydrous calcium sulphate 411–12
animals
  absorption and utilization of trace elements 496
  zinc an essential element for 508
antimony (Sb) 497
The Apostles (stacks), Australia 148
apparent preconsolidation pressure 356
aquicludes 247
aquifer recharge 273
  see also groundwater recharge
aquifers 247, 473
  annual volume of recharge to 250
  basement 248–9
  coastal, problem of saline intrusion 275–7
  confined 282
    and unconfined 247–8, 247
  contamination of 262
  discharge by seepage and springs 251
  interaquifer flow, may spread pollution 273
  iodine concentration in, UK 504, 505
  potential, identification of 255
  protection against leachates 461
  storage coefficient/storativity defined 251–2
  suitability for artificial recharge 271
  tracers may go undetected 260
  transmissivity and safe yield 271

unconfined, storage coefficient 252
aquitards 247
aragonite 180
arêtes 203
argillaceous material
  causing valley bulges and cambering 83
  and suction pressure 396
arid regions
  accumulation of most mobile elements 493
  carbonate sand beach material 351
  cementation of sediments by precipitation 352
  soil erosion through wind action 163
  soils of 350–2
    problems with Quaternary deposits 351
arid and semi-arid regions
  deflation hollows and differential erosion 166
  desertification 188–90
  distribution of 163, 164
  irrigation 190–7
  necessity for good agricultural practices 316
  sabkha soil conditions 180–3
  salt weathering 183–8
  stream action in 169–74
  wind erosion in 309
arsenic 489, 490–500, 497
  biomethylation of 499
  in groundwater 264
  sorbed onto or co-precipitated with ferric hydroxide 499
  species in aqueous systems 499
  toxic and a carcinogen 499, 518
arsenic toxicity, in cattle 499
arsenites 498
arteriosclerosis, and soft water 416
artesian conditions 248, 248
artificial recharge 271–2, 281
  source of water for 271–2
  used dilute pollutants 279
asbestos dusts
  carcinogenic 511, 518
  inhalation causing many pathological changes 512
asbestos 511, 512
ash 243
  hazardous when not previously wetted 395
  lateral and vertical variation 43
  subject to hydrocompaction 43
ash falls, damage from 51
ash flow tuff 51
ash flows 39–40
ashfall, effects of 43
asphalt, cutback/asphalt emulsion, for stabilizing loose sand 179

Athabasca Tar Sands, strip mining 432
Atterberg limits see consistency limits
attrition 166
augers 537, 539
available water, defined 192–3
avalanche forecasting 220
avalanches 220, 221
  dry 220
  ice 220
  location prediction 220
  loose snow 220
  slab 220
  wet/slush 220
avalanching
  of sand, leads to increased surface abrasion 166
  of soil 316

backshore 139
backswamp deposits 126, 127
backwash 144
  tsunami waves, may destroy buildings 160
bacteria
  coliform 261, 265
  in landfills 466
  lifespan in groundwater 273
  methanogenic 466
  sulphate-reducing 263
bahadas 171, 172
Bahrain, northern, hazard intensity maps for aggressive ground conditions 186, 188, 188
ball clays 326
Balwin Hill Reservoir, Los Angeles, fault creep beneath 69
bank erosion 123, 174
  a self-sustaining process 122
bank scour and slumping 125
bank stability, reservoir 235
bankful discharge 120, 121
barium 264
  and health 510–11
barkhans 167, 168
  formation 168–9
  migration 169
  rate of movement 175
barrages, used to shorten coastlines 153
barrier systems, contaminated sites 483
basal till 207
basalt 362, 363
  could provide sites for deep disposal repositories 478
  disintegration of 393
    by microfracturing, Lesotho 394
  flood basalt 38

thick and massive, good dam sites 243
young may be permeable   243
basalt plateaux   38
northern Lesotho   *38*
base level, altered   118–20
basement aquifers   248–9
susceptible to surface pollution 248
basin method, artificial recharge   272
'basket of eggs' topogaphy *see* drumlin fields
batholiths   362, *363*
bauxite   354, **354**, 493
bay-bars   150, *150*
bay-head beaches   150–1
beach nourishment
maintains beaches and dunes 149, 153
material used   158
beach slope
influencing type of breaking wave 143–4
produced by interaction of swash and backwash   149
related to grain size   149
beach zones, within the nearshore current system   143–4
beaches   139, 145
bay-head   150–1
and longshore drift   147, 149–51
raised   220
sandy with high dunes   149, 153
shingle, storm waves constructive 149
bedding lineaments   530
bedding planes   377, **377**
affect marine erosion   146
bedforms, sand bed channels   124, *125*
bedload
contribution to total streamload 125
slow, intermittent movement 123–4
bedload transport, sand bed channels 124
beidellite   326
benches, acting as rock traps   106
benching   107
bentonite   326, 464
bentonite liners   463
berms (foreshore ridges)   149
beryllium, great toxicity of   510
bicarbonates   387, 400
binding agents   165
Bingham Canyon copper mine, Utah 433, *433*
bioavailability
of aluminium in drinking water 510
of lead   502

of trace elements   497
biochemical oxygen demand (BOD), leachate   274, 468
biochemical systems, for passive acid mine drainage treatment   450
biodegradation   455, 466
biological pretreatment, of hazardous waste   472
bioremediation
of contaminated soils   483–4
in situ, of groundwater   484
Bishop sand sampler   542, *544*
bits
chopping bits   540
diamond/tungsten carbide   543–4
bitumen, cutback/bitumen emulsion primers, perform better with aggressive ground conditions 186
bitumen–cement liners   463
black cotton soils   291–2, 354
blasting   431, 432
preblast survey for owners of buildings   435
and vibrations   434–5
blind zone   547
block and ash flows   51–2
block failure, active and passive   96
block lava   46
block size, described   377, **377**
blocks, accumulate as scree   88, 90
blockwork, igneous rocks   156
blow-holes   146, *147*
boiling mud pits   41, *42*
borehole extensometers, use of   98
boreholes *see* drill-holes
bornhardts *see* inselbergs
boron   264
borosilicate glass   476
boulder clay *see* till
boulders   300
braided rivers/streams   *117*, 118, 170
brain damage, and iodine deprivation 504–5
breaker zone   143, *145*
breakwaters   161
attached, cause accumulation of sand/shingle   154
detached offshore, effects of   154
rubble mound   154–5
breccia   347, **370**
brick making, raw materials for 326–8
brickearth   345, *346*
collapse index   347
bricks
colour of   328
strength depends on degree of vitrification   328
bridges, correct design of   134
brine pumping, Cheshire   438–9

effects of wild brine pumping   439
brines, groundwater
in formation of coastal sabkhas 180
in formation of continental sabkhas 182–3
British Geological Survey
early thematic maps   421–2
urban environmental survey, Wolverhampton area   30–1
British Soil Classification, for engineering purposes   300
field identification and description 302–3
full laboratory procedure   305–6
brittle fracture, in mines   80
brown earths   290
brush matting, temporary sand stabilization   178
building stone   406–8
colour and texture   406
decay in urban atmospheres   407
durability of   406–7
the ideal   406
presence of salts, effects of   407
buildings
accumulating radon   513
attacked by salt weathering   185
constructed to withstand tsunamis 161
damaged by avalanches   220
designing for areas of active mining 430
on expansive soils   340–1
on old spoil heaps   443
poorly constructed and unreinforced, damage from shaking   66, 67
possible damage by blasting vibration   434, *434*
relating dynamic characteristics to those of subsoil   64
susceptible, location away from fault traces   66
see also property; residential development
bulkheads, cantilevered or anchored 156–7
buried channels   220
burrows (crotovinas), in Chernozems 291

cable lashing and cable nets, to restrain loose rock blocks 106
cadmium   264, 500–1
an acute toxin   500
a more volatile heavy metal   500
in sewage sludge   277, 500
in sulphide ores   500
calcite   180

calcium
  possible link to goitre 505
  in sedimentary rocks 263
calcium bicarbonate, conversion to
    sodium bicarbonate 263
calcium carbonate 170, 173, 516
calcium deficiency 491
calcium fluoride, for tooth enamel and
    bone structure 506
calcium sulphate hemi-hydrate
    (plaster of Paris) 411–12
calcrete 170, 173, **354**, 493
calderas 37
  formation of 52
caliche deposits 170, 173
calliper logs 556
cambering 83
cambisols (altosols), brown earths
    290
canal banks, failure due to
    earthquakes 66
cancer
  and metal carcinogens 518–19
  oesophageal, Caspian Sea, linked to
    soil type 518–19, *519*
  stomach cancer linked to cobalt and
    zinc 518
  linked to zinc and copper, Devon
    518
canopies, rigid, protection from
    rockfalls 106–7, *107*
canyons 173, *173*
capability classes/subclasses 317
capability units 317
capillary fringe 183
capillary pressure 285
capillary rise 172, 173, 193, 285
  in hot arid climates 183, **184**,
    185
carbon dioxide 53, 443, 455, 470
  in soils 386–7
  soluble in water, leaching effect
    227
carbon monoxide, from burning
    colliery spoil 442, 443
carbonate rocks 400–4
carbonates 387
  and brick making 327
carbonic acid 262, 386
cardiovascular disease, and water
    hardness 516–17
Casagrande soil classification **295**,
    299
casing
  for diamond drilling 545
  of wells 265
cation exchange capacity, high 342
cation exchange reactions, importance
    increases with depth 261
caving, in ore body mining 416–17
cavity collapse risk estimation
    402–3

cement stabilization, of expansive soils
    341
cement–bentonite cutoffs 465
cementation
  by mineral precipitation, arid
    region soils 352
  in tropical soils 353
cemented layers 350
cereal crops
  toxicity caused by ultrabasic rocks
    491
  trace elements in 496
cerussite 501
chalk
  dissolution along discontinuities
    403–4
  problem of pipes and swallow holes
    227
channel avulsions 123
channel bars, transitory 126
channel form, changes during flood
    125
channel shifting, damage from 123
chemical oxygen demand (COD),
    leachates 274
chemical pretreatment, of hazardous
    waste 472
chemical stabilizers, for stabilizing
    loose sand 179
chernozems (mollisols), black earths
    290, 291
Chezy formula 121
China
  Khansu Province, landslide
    earthquake-induced 88
  Loess Plateau, potential for soil
    collapse variable 346
  sand liquefaction, Tangshan
    earthquake *336*
chloride(s) 261, 263
  in drinking water 265
  increased levels may indicate saline
    intrusion 276
  in mine water 452
chlorinated hydrocarbon compounds
    278
chlorite 364, 387
chopping bits 540
chromium 264, **489**, 518
chromium deficiency, and insulin
    resistance 510
chrysotile 512
chute channels 122–3
cinder 43
cinder cones 35, 36, *36*, 37
cinnabar 501
clapotis 140
clay barriers, compacted 472
clay cores, prevention of seepage
    155
clay deposits, refractory materials and
    bricks 326–8

clay illuviation 249
clay liners 463
  reduce seepage from tailings dams
    446
clay minerals 493
  formed in basaltic rocks 393, *395*
  increasing Al : Si ratio 493
  in soils 294
clay soils
  cohesion developed by molecular
    attractive forces 301
  dams usually earth dams 244
  deformation due to repeated
    loading 65
  potential for volume change 339
  problems if irrigation water is poor
    quality 195
  rotational slides 90
  soil failure 65
  stress–strain curves 301, *306*
  volume changes from
    evapotranspiration 339
clays
  activity of 195, 298
  for brickmaking 326–8
  classified according to liquid limit
    294, **294**
  compaction characteristics 331
  compressibility of 306–7
  frozen, and plastic failure 224
  laterite, Hawaii, effects of air drying
    353
  overconsolidated 306
    fissured, progressive slope failure
      94
    non-circular slips 91
    shear strength on a stable slope
      93
  silty *see* black cotton soils
  sodium-saturated 342–3
  soft, boring in 539–40
  stiff, boring in 539
  in tills 212
  undisturbed, shear strength of
    301
  unfissured, first time shear failure
    93–4
  in weathered plutonic rocks 393
  see also clay minerals; clay soils;
    expansive clays; varved clays
cleavage 364–5, *365*, 395
cliff units, in morphological mapping
    13, 14
climatic changes, noticeable in regolith
    composition 493
climatic factor, local wind erosion
    313
clinker 45, 46
coal dust 511
coal mining
  effluents from and waste waters
    450–2

UK, associated seismic events 81
waste from 441–4
see also longwall mining, and
    subsidence; opencast coal
    mining
coal slurry, blanketing effect of 451
coalfields
    concealed, influence of overlying
        strata on subsidence movement
        427–8
    effects of air penetration 452
coast 139
coastal areas, low-lying, liable to
    marine inundation 151
coastal current *145*
coastal defences, breaching and
    inundation 151
coastal embankments 153, 155–6,
    156, 161
coastal erosion 144–7
    rate dependent on nature of coast
        145, *146*
coastal flood forecasting and warning
    systems 152–3
coastal management
    socio-economic cost of 159
    for tackling coastal erosion
        158–9
coastlines
    drowned 220
    indented, wave refraction 141
    retreating 146–7
cobalt **489**
    and health 510
cobbles 300
coefficient of consolidation 307
    decrease in after leaching 356
coefficient of permeability 252, 258,
    307
coefficient of storage 258
coefficient of uniformity 293
coefficient of volume compressibility
    306, 307
coefficient of weathering, for granitic
    rock 387, **387**
cohesion 96
cold pyroclastic surges 52
coliform bacteria
    and acceptability of drinking water
        265
    in drinking water 261
collapse index 347
collapse pressure, collapsible soils
    347
collapsible soils 345–8, 357
    collapse criteria 347–8
    fabric of 345
    oedometer test assesses degree of
        collapsibility 347–8
    structural stability of 346
collector galleries 111
collector wells 249

colliery spoil heaps
    configuration of 441–2
    permeability of 442
    spontaneous combustion of
        442–3
    restoration 441–4
Colorado Plateau 173
Colorado River, Arizona 173
colour, of groundwater 261
columnar jointing 376, *376*
comminution till **207**
compaction
    and engineering properties of soils
        328
    of landfill, important 462
    mechanical 328, 331
    necessary for embankment soils
        246
composite volcanoes 35, 36
compressibility, degrees of 307
compression index 306–7
compressional wave velocities 546,
    546, 547
compressional waves *see* P waves
compressive strength
    building stone 406
    peak triaxial 380
concrete
    hydration and alkali release 409
    sand for 325
    shrinkage in 410
    sulphate-attack resistant 399
    for use in aggressively saline
        conditions 351
concrete aggregate 408–10
concrete dams
    arch 239, *240*
        importance of value of Young's
            modulus 242–3
        shales unsatisfactory for
            foundations 244
    buttress 239, *241*
    composite 239, *242*, 245
    gravity 239, *240*
    in postglacial rock-cut valleys
        245
    sliding at the base 239, *241*
    tensile stresses develop with
        deformation of foundations
        242
cone of depression 273–4, *274*
conglomerate 367, **370**
    cemented with soluble material,
        problems of 405
conjunctive use, of surface and
    groundwater 281–2
    advantages and disadvantages
        282
    design and operation of schemes
        281–2
connate water 246, 261, 275
conservation, involves reconciliation

    of differing views 5
consistency index **210**
    of a soil 295, **295**
consistency limits 294–8
    tills 211
consolidation, of soils 301, 306–7
    primary and secondary 306
    theory of 301, 306
constant head method of *in situ*
    permeability testing 257–8
constant-pumping-rate aquifer test
    258
construction
    in arid regions, avoidance of
        aggressive ground conditions
        185
    damage due to expansive clays
        338, *338*
    location of materials 324
    in permafrost regions 227–8
        active method 228
        prevention of heatflow to
            permafrost 227–8
    projects using satellite imagery
        528
construction operations, information,
    materials 535
construction/foundation stage plan
    29
contact aureole 362, *365*
containment, of contaminated sites
    483
containment systems, for hazardous
    waste 472
contaminants
    associated with coal carbonization
        sites **480**
    inorganic **479**
    liquid and gas, may be mobile
        481
    methods of removal from soil 483
    phytotoxic but not health hazards
        362, *365*
    soil-borne, migration of 479
    volatile, in soil atmosphere 482
contaminated land 478–82
    defined 478
    Dutch attitude 479
    geochemical maps aid
        investigations 496
    guidance on assessment and
        redevelopment of 478,
        479–80
    identification of development
        constraints 482
    remediation of 482–4, 496
    site investigation 481–2
    suitable for use approach 478
contamination
    of aquifers, reasons for 262
    bacteriological 278
    by arsenic 499

by tailings dams, prevention of
445–6
of drinking water   193
heavy metal   446
continental crust   34
continental shelf, modifies form of
tsunamis   160
contour farming/ploughing   133,
315
copper   264, **489**
and health   510
core barrels   543
double tube and single tube
544–5, *545*
face-ejection   545
triple tube   545
corrasion, lateral   115, *115*, 116
corries   201, *202*, 203
corrosion hazard, steel reinforcements
in saline sabkha groundwater
351
cost–benefit analysis   10
Costa Rica, cost of Limon-Telire
earthquake   6
Coulomb's equation   300, 301
counterfort drains   110
cover crops   316
covers
over contaminated zones   483
over landfill sites   464, *465*
geotechnical design factors   464
Craelius core orientator   383–4, *383*
crag and tail   200, *202*
crater lakes
drained to reduce hazard   56
hot lahars from   46–7
cratering   *416*, 417
creep   164
in ice   222
in rock salt   404
see also fault creep; hill creep; soil
creep; valley creep
cretinism, endemic, and iodine
deficiency   505
crevasse splays   126
crib systems, reasonably flexible
109–10
critical erosion velocity   123
critical slope angle, bedded and
jointed rock mass   94–5
critical tractive force   123
crocidolite   512
Cromer Till   **210**, 211
crop diseases, through zinc deficiency
508
crop management factor   312
crop rotation   195, 315
cross-contamination, avoidance of in
site investigations   482
cross-grading   114
cross-hole seismic methods, assesses
engineering properties of rock

mass between drill holes   556
crumb structure   291, 294, 312
crushed rock
concrete aggregate   408–10
road aggregate   410–11
crust/crete development   352
crustal deformation, regional   69
crustal strain   77
cryptodomes   38
cuestas   173
cuirasses *see* hard pans
currents, inshore and offshore
142–3
cuspate bars   150
cuspate forelands   150
Dungeness   *151*
cutoff drains, to reduce amount
surface run-off into landfill
sites   275
cutoff systems, for wastes   464–5
cutoff walls
concrete diaphragm   465, 472
geomembrane sheeting-enclosed
465
piles for   465
plastic concrete   465
protection against scouring
156–7
slurry trench   472
cutoffs, prevent water reaching zones
of potential instability   110
cutting shoes   541
cyanide   444
formation of compounds   453
used in heap leaching   452, 453

dam sites   239–43, **529**
and geology   243–6
in glaciated areas, difficult to
appraise   244–5
damage, from blasting   434
dams
height of   243, 405
horizontal pressure from reservoir
water   239
keyed into foundations   241
problems if underlain by gypsum
beds   404
see also concrete dams; earth dams;
tailings lagoons/dams
Darcy's law   252, 254
data sources
for geographical information
systems (GIS)   29
for landslide investigations   101
Davisian cycle of erosion   115, *116*
dead storage, reservoirs   235
débâcles, water release from moraine-
dammed lakes   221, **221**
debris avalanches, rapid progressive
failure   92

debris flows   170
cut V-shaped channels   92
debris slides   91
Deccan Plateau, flood basalt,
intermittent   38
decomposition
anaerobic, in landfills   466
chemical, of rocks   386–7, 388
of sulphur compounds   399
dedolomitization   410
deep disposal repositories, permanent
described   475
in impermeable rock in geologically
stable areas   476
lack of groundwater circulation
ideal   475
multibarrier concept   475
once loaded backfilling and sealing
brings isolation   477
permeability of host rock   477
safety shown by site analysis   475
stress/deformation in host rock
distribution determined
476–7
waste converted into solid, non-
leachable material   476
deep sea bed, for disposal of
radioactive waste   475
deflation   166
effects of   175
prevention of using gravel   178
deflocculation, of clays   341
deformability   374, **374**
deformation
elastic, in ice   222
of ice-contact deposits   215
in metamorphic rocks   363–4
plastic   245
before failure in evaporitic rocks
404
in ice   224
deformation till   **207**
Delft sampler   542, *543*
Delta Scheme, Netherlands   153
denitrification   272
dense non-aqueous phase liquids
(DNAPLs)   278, 484
dental caries, in children
associated with higher soil lead
levels   503, *504*
prevention using calcium fluoride
506
dentition, use of   107, *108*
deposition
coastal, extensive beaches   145
in lower reaches or arid region
streams   174
marine, straightening coastlines
151
in the plain tract   116
depth-to-water contours   256
depth-to-water table maps   256

desert margins, problems severe   163
desert pavements, creation of   165
desert varnish   170
desertification
  cost of reversal high   190
  distribution   189, *190*
  expansion of   189
  indicators of   188–9, **189**
  much attributable to unwise
    agricultural practices   189–90
deserts
  hot, mechanical weathering   386,
    386
  natural expansion and contraction
    189
desiccation
  of shale on exposure   396
  of soils   285–6
  of tropical soils   357
desiccation ground patterns, and
    capillary rise   173
desiccation structures, in playa
    sediments   172–3
'design flood', for structural design
    134
desk studies
  contaminated land site
    investigations   481
  engineering geological appraisal,
    contents   **536**
  groundwater resources assessment
    254–5
  in site investigation   535–6, **536**
  large projects   535
  for waste disposal sites   459
developed countries, disease
    associated with pollution of air,
    water and soil   487
developing countries
  development placing
    people/property at risk   6
  diseases associated with the
    geological environment   487
development, discouraged in
    hazardous areas   105
development advice map, South Wales
    422
diagenesis, in coastal sabkhas   180–1
dilatometer   561
dipmeters   555
disaster mitigation   7, **8**
disaster risk, reduction of   7
discontinuities   375–84, 460
  can increase flow to wells   270
  determine permeability   254
  flat-lying, in frost-shattered zones
    226
  movement along   80
  mudstones   396
  open or closed   377–8, **377**
  in rock masses   377–8
    and landsliding   94

and rock quality indices   380–1
  and rock slides   91
  and strength of rock masses
    378–80
  see also faults; fractures
discontinuity surveys   381–4
  data from photographs of
    exposures   384
data located on maps/plans using
    controlled photographs   384
  data plotted on stereographic
    projections   384, *385*
  direct   381
  objective   381, 383
  value of data from orientated cores
    383–4
discriminant analysis   103–4
diseases
  chronic, and environmental geology
    516–19
  crop diseases   508
  dust-related, prevention of   511
  and elemental excess/deficiency
    487
  related to high concentrations of
    arsenic   499–500
  see also Alzheimer's disease;
cardiovascular disease; iodine
deficiency diseases (IDDs); itai-itai
disease; Keshan disease; leukaemia;
lung cancer; Minimata disease
dispersion effects, increase duration of
    strong shaking   62
dispersive soils   341–5
  classification of   343, *343*
  problems in use of   335
  some physical and chemical
    properties of   **344**
dispersive–non-dispersive transition
    343
disseminated ore bodies, use of open
    stopes   416
dissolution, of limestone   399, *399*
doline density, and potential
    subsidence   402
dolines   402
dolomite   181, 400
  source of magnesium in
    groundwater   263
dolostone   387, 400, 411, 417
  alkali–carbonate reaction   409
domestic refuse
  leachate composition from   466,
    467
  and sanitary landfills   459–71
downcutting, by master streams   115
drainage   **529**
  adequate provision of in retaining
    walls   109
  beneficial to agriculture   288
  for improving slope stability
    110–11

for landslide control   106
  sub-surface, limestone   400, *400*
  subsurface   111
drainage basins
  characteristics changed by human
    activity   129
  size of and sediment yield to
    reservoirs   238
drainage blankets
  granular   464
  over soft alluvial clays   245
drainage density   128, 235
  inverse relationship with
    groundwater discharge   122
  relationship with mean annual
    flood discharge   122
drainage ditches, for removal of
    leachate   464
drainage diversion, by ice sheets   219
drainage galleries   111
drainage holes, subhorizontal   111
drainage patterns
  associated with high-magnitude
    flood peaks   126
  dendritic   114
  trellised   *115*
drainage systems
  development of   114–20
  for leachate drainage   464
dredge mining   435–7
  flexibility of waste disposal   437
drift deposits
  and mining subsidence movement
    428
  stratified   203, 213, 215–19
    beach deposits   216
    glacial lake deltas   216
    ice-contact deposits   213,
    215–16
    proglacial   213, 216–19
    terminal moraines in lakes   216
  unstratified see till
  varved sediments   217–19
drill-hole logging techniques   554–6
  calliper logs   556
  electrical resistivity method   554
  estimation of porosity   555–6
  induction logging   555
  laterolog (guard log), use of
    554–5
  microlaterolog   555
  microlog technique   554
  radioactive logs   555–6
  sonic logging devices   555
  spontaneous potential (SP) logs
    555
  televiewers   555
drill-holes
  cored   543–5
  data from orientated cores   383–4
  deviation   545
  direct observations through   537

inspection techniques 384
lining of, sands and gravels 540
location of 537
percussion drilling 539–43
for potential reservoir site
investigation 235
results documented on a log 537,
538
drinking water
acceptability of 265
arsenic limit 499
bacterial quality of 265
barium in 510
cadmium in 500
and cardiovascular disease 516
chemicals of health significance in
495, **495**
high aluminium and Alzheimer's
disease 509
iodine in 505
standards for 265, **266–8**
trace elements, permissible and
mandatory levels **269**
drinking water safety, standard tests
261
drumlin fields 201
drumlins 200–1, *202*
dry avalanches 220
dry densities
compacted soil 331
drained peat 358
and porosities of rocks 373, **373**
dry farming, on kastanozems 291
drying, in tropical residual soils 353,
353
Dubai, hazard maps for new airport
sites 177, *177*
dunes 149, 167
active, a threat 177
construction of 153
desert 166–9
development of 167–9
rate of movement 175
fixed 176
formed by onshore winds 149
hazard mapping of 177, *177*
migratory, arid regions 163
movement of, effects 175
removal/flattening not a complete
solution 177–8
stabilization by oiling 179, *179*
dunes and antidunes, sand bed
channels 124
duricrusts 170, 173
iron and aluminium 493
see also crust/crete development
dust
atmospheric, increase in affects
radiation balance 189
movement of 174–80
and sand, sources and stores 176,
176

Dust Bowl, USA 307, 308
dust storms 163, 165, 174–5, 309
dusts, isometric, not cancer-forming
511
Dutch cone penetrometer *557*,
558–9
dykes 553
affecting subsidence 428
and sills 362, *364*
and veins, as lineaments on aerial
photographs 531
dynamic compaction
in landfills 462
for shallow stabilization 417

earth dams 239, *241*
in areas of glacial deposition 245
careful construction control if
dispersive soils used 345–6
constructed on clay soils 244
construction materials from within
the dam basin 245–6
impervious membrane used as core
wall 246
earthfill dams, semi-plastic behaviour
239
earthflows 93
at spreading toes of rotational slides
93
earthquake damage, highly localized
77–8
earthquake design parameters,
selection of 74
earthquake foci 58
earthquake hazard
broad picture from seismic zoning
76
defined 74
recognition of areas exposed to
73
site specific level justifiable 75
earthquake hazard assessment,
importance of prediction of
seismic shaking 62
earthquake intensity scales 60
modified Mercalli **61**
earthquake maps 75
classification 76
earthquake precursors 70–1
earthquake prediction 70–2
earthquake regions, establishment of
risk to structures 60
earthquake risk
defined 74
shown by seismic microzoning
76
earthquake risk assessment 74–5
earthquake shock waves 58
speed and destructiveness of 58
see also L waves; P waves; S waves
earthquake swarms 59

prior to some volcanic eruptions
49
earthquakes 162
associated with edges of tectonic
plates 58, *59*
can trigger displacements on
subsidiary faults 62, 245
caused by faulting 58–9
causing landslides 87
classified 58
determination of probability of
occurrence 71–2
effects of 66–70
generation of other hazards 7
importance of duration of 62
intensity and magnitude of 60–3
largest reported 62
magnitudes of 60
maximum ground
accelerations/durations of
strong phase shaking 62,
62
triggered by induced change in
porewater pressures 78–81
see also tsunamis
ebb current 142
economic and social planning,
volcanic hazard maps for 56
Edale Shales, rotational slide at Mam
Tor *90*
eddies 124
Edinburgh, Castle Rock, crag and tail
202
effective stress 300, *301*
Eh 497
and trace element
solubility/mobility 490, *491*
elastic rebound theory of the cause of
earthquakes, use of 71
electrical conductivity mapping,
northern Bahrain 186, 188,
188
electrodes, resistivity surveying
549–51
electrodialysis 450
electrolysis 450
electromagnetic methods, geophysical
exploration 551–2
electromagnetic pulse technique 552
electromagnetic radiation 525
elements
excess or deficiency of, and disease
487
minor in earth materials **492**
occurrence of 490–6
solution chemistry of 490
Elk Valley, British Columbia, varved
clays, geotechnical properties
of 217–18, **218**
elongation index 325
embankment dams
to contain tailings slurry 444

zoned, safe at active fault sites
    245
embankments   328–33
    flood embankments   134
    most critical period   333
    pipe damage through use of
        dispersive soils   343, *345*
    raised, for tailings dams   444
    rotational slip in   333
    shrinkage settlement of   338
    use of dispersive soils as fills   345
    see also coastal embankments;
        dams
engineering behaviour
    of concrete, influenced by aggregate
        408–9
    of shale   396
engineering bricks   327
engineering classification
    mapped for rock and soil types
        356
    of weathering   387–8
engineering data, in seismic zoning
        maps   77
engineering geological desk appraisal
    536
engineering geological hazard
        mapping   11, *12*
engineering geological maps   23–9,
        524
    accompanied by cross-section,
        explanatory text and legend
        23
    analytical or comprehensive   23
    archive searching   23
    engineering information in tabular
        form   23, *25*, **26–7**
    information for planners and
        engineers   23
    special purpose or multipurpose
        23, *24*, *25*
    suite of, Leiden   *21*, 22
engineering geological plans   23, *28*
engineering geomorphological
        investigation, general
        procedure   **18**, 19
engineering geomorphological maps
        17–19, 524
    aims of survey for   18
    landform recognition by size and
        shape   17
    large scale maps/plans, local surveys
        19
    scale influenced by project
        requirements   19
engineering geotechnical maps *see*
        engineering geological maps
engineering performance, some rocks
        367–8, **368**
engineering problems
    with quickclays   350
    with sabkha soils   351–2

engineering properties
    mudstones   396–400
    rock masses, indirect assessment of
        546
engineering soil maps   528
englacial moraine   208
englacial till   **207**
environmental degradation,
        desertification   188–90
environmental evaluation   3
    aim of   3
    checklists   3, *102*
environmental geological maps
        19–23, 524
    comprehensive suites of in reports
        20, 22
    derivative maps   132–3
    limitations of   22–3
    for specialists or non-specialists
        19
environmental geology
    and chronic disease   516–19
    and health   487–519
environmental impact, of mining
        415–56
environmental impact assessment,
        includes soil surveys   324
environmental impact statements   3
environmental potential maps   20
environmental protection laws   2–3
epicentres   58, 75–6
    below or near large reservoirs   79
    indicating active faults   72
epidemiological surveys   496
epinephropathy   487
ergs   166
erodibility   174, 313
    field assessment of   174
erosion
    coastal   144–7
        dunes a natural defence against
            149
    down stream from new reservoirs
        235
    glacial   199–203
    interrill   311
    river   122–3
    vulnerability of arid and semi-arid
        regions   163
    see also glacial erosion; soil erosion;
        wind erosion
erosion control practice factor
        312–13
erosion risk, rating systems for   317,
        318
erosion risk assessment   322, 531
erosion velocity   123
erosivity maps   317, *317*
eruption plumes, measurement of
        movement   49
eruptions   34
    central, types of   38–40

determination of recurrence
        intervals   48
    evolution of   50
    fissure   34
    Hawaiian   38
    losses caused by can be minimized
        56
    Pelean   39–40
    Plinian   39
        effects of   52
    Strombolian   35, 38
    Vesuvian   39
    Vulcanian   35, 39
escarpments   173
Escherichia coli   261
eskers   215, *215*, *216*, 219
estuaries, protection for major ports
        against flooding   153
eudialyte   503
evacuation
    from tsunami areas   161
    from volcanic danger areas   50, 56
    permanent, areas of continuing
        slope failure   105
evaporation   233
    in desert areas   174
evaporative pumping   180
evaporite deposits   350, 367
evaporitic rocks   404–6
evapotranspiration   233
    actual   250
    exceeding precipitation   234, 291
    influence of vegetation on   310
    and irrigation   191, *191*
excess deposition/sedimentation risk
        322, 531
exchangeable sodium, and dispersive
        behaviour   342
exfoliation   386, *386*
Exmoor, devastating floods (1952)
        123
expansive clay minerals, swelling of
        393
expansive clays   337, 338–41
    compaction as wet as possible
        332
    design solutions for building on
        340–1
    fissures due to seasonal volume
        change   339
    low permeabilities give slow
        moisture movement   339
    Saudi Arabia, swelling potential
        183
    stabilization of prior to
        construction   341
    in tropical soils   357
    vegetation removal leads to swelling
        effect   340
expected intensity maps   76–7
explosion vents, produced by escaping
        gas   35, *36*

extraction barriers, abstracts
encroaching salt water   276
extraction wells   472

fabric
of collapsible soils   345
of quickclays   349
facing stone   408
factor of safety
selection of   97
sensitivity of   **94**, 95–6
failure
clays   93–4, 224
quickclays and varved clays   91
mudrocks   398
progressive, debris avalanches   92
ravelling failure, limestone   401
in rock masses   95
of soils   65, 93, 300
see also rock slopes; shear failure;
slope failure
falling head test   257
farming practices, poor   188,
189–90
fault creep   59–60
indicates fault activity, factor in
land-use planning   60
may weaken structures and cause
failure   69
fault planes, affect marine erosion
146
fault springs   249, *249*
fault steps, in mining subsidence   416
fault zones
displacements
causes of   59–60
migration associated with
earthquakes   72
and landsliding during dam
excavations   245
leakage along   237
picked out by infrared line scanning
(IRLS)   526
faults   255, 553
active
and dam building   245
geological indications of   73
little information on frequency of
movement   62
zones of geothermal energy
release   71
on aerial photographs   530–1
assessing relative activity of   72–3
due to fluid abstraction   439, 440
inactive   73
movement associated with
earthquakes   245
recent, investigation of   72
and subsidence   428
those to be regarded as potentially
active   73

see also tsunamis
feasibility studies   524, 532
feldspars
decomposition of   387
in sand, effect on concrete   326
fencing, for containment of small
rockfalls   106
ferralitic soils   354, 355, 493
ferrasols (oxisols)   *290, 292*
ferricrete   170, 173, 354, **354**, 493
ferroselite   507
ferruginous soils (*sensu stricto*)   354,
355
fersiallitic andosols   **354**, 355
fersiallitic soils (*sensu stricto*)   354,
355
fertilisers, nitrogenous   277
fetch, of wind   140
and storm surges   152
fibre particles, shape influences
clearance from lungs   511
fibrosis   512
field capacity   288
irrigation   191
field compaction trials   331
field cropping practices, protection
against wind erosion   316
field investigations   101
field mapping   19
in studying arid landscapes   176
field pumping tests   258–9
field reconnaissance   525
in potential reservoir site
investigation   235
field seismic velocity   381
fills
badly affected by earthquake
vibrations   66
and embankments   328–33
filter drains   246, 446
filtering action, soil and rock   260
filtration, and leachate attenuation
466, 468
Finland, incidence of cardiovascular
disease   516
fiords   203, *205*, 220
fireclays   326
Fish River, Namibia   169, 173
fissility   367
of roofing stone   408
of shale   396–7
fissure eruptions   34
fissure flow   99
fissure permeability, regolith   248–9
fissure volcanoes
monogenetic   35
pyroclastic   36
fissures
development in mudstone   396
due to differential subsidence   418
due to fluid abstraction   439
in expansive clays   339

fixation/solidification, of soil
contaminants   483
flakiness index   325
flame structures   205
flash floods   *128*, 129, 163
flashes, caused by wild brine pumping
439, *440*
flood barriers, movable   153
flood control   282
flood current   142
flood defences   134–5
temporary   134
flood diversion   136
flood embankments   134
flood forecasting
flood routeing   130, 136–7
importance of lag time   129
longer term, associated with snow
melt   129–30
problems related to run-off
predictions   130
flood frequency analysis   130
flood frequency curves   130, *130*
flood hazard   126
alluvial fans   171, *171*
of exceptional floods, reduced
135
factors to be considered   131
lessened by improving channel
hydraulic capacity   136
not reduced by financial assistance
134
rises with urbanization   129, 131
flood hazard maps   131, *132*
map units in central Arizona   **133**
flood hazard zones, and management
strategies   **133**
flood mitigation projects, assessment
of future effectiveness   134
flood potential, engineering
assessment of   131
flood prediction, for design purposes
129
flood routeing   130, 136–7
flood walls   134, 135–6
flood warning systems
success depends on hydrological
characteristics of the river
134
to alert communities   134
floodplains   115, *127*
apparent adjustment to flood
discharge   126
floodplain management plans
131
temporary sediment stores   126
zones, in land-use planning
131–3
floods/flooding   126–31
arid and semi-arid regions   169
discharge rate, used to predict
flooding magnitude   129

flood response, influenced by
    agricultural practices   133–4
flood-proofing measures   133
glacier   221, **221**
    importance of antecedent ground
        moisture quantity   131
    increases erosion in upper reaches
        123
    intensifying factors   126, 128
    lag time   129
    multiple event   129
    peaks vary in height   126
    reduction of peak discharge by
        temporary storage   137
    single event   129
    upstream from reservoirs   235
    a volcanic hazard   53
flow, in a porous medium   252, 254
flow duration curves   131, *131*
flow slides, due to ground liquefaction
    65
flow till   **207**
flow velocity   252
flowmeter logs   259
    identification of zones contributing
        to discharge   259
flows   91–3
    boundary with material beneath
        92
    dry (rock fragment)   92
    wet   92
fluid abstraction, subsidence
        associated with   437–41
    prediction of subsidence amount
        440–1
fluoride
    mobility of   491
    volcanic associations   506
fluorine   **489**, 506–7
    released in manufacturing processes
        506
    in water   506
fluorite   506
fluorosis
    dental   495, 506, 507
    skeletal   506
fluvial processes   120–6, 170
    river erosion   122–3
    river flow   120–2
    river transport   123–5
    sediment deposition   125–6
fluvio-glacial deposits, stratified drift
        213, 215–19
folds, axial traces of   530
foliation   364, 366, 530
footings
    placement below frost penetration
        depth   230
    spread   244
foreshocks, precede an earthquake
    59
foreshore   139

formal statements   3
formation resistivity factor   551
foundations   **348**
    on alluvial deposits, problems
        245
    for bulkheads   156–7
    dams
        concrete dams   242, 244
        heave under   242
        importance of geology   243
        rock surfaces roughened to
            prevent sliding   241
        stress acting upon   239, 242
    failure in collapsible soils   348
    in permafrost areas   228
    piled
        problems in mining subsidence
            areas   430
        reinforced and bored   423
    pressure from earthfill dams   239
    problems
        in alluvial deposits   245
        due to gypsum solution   404
        in volcanic sequences   243, 393
    on swelling and shrinking soils
        340
    variable conditions in pyroclastics
        243
    varied responses to earthquakes
        63–4
    see also footings; raft foundations
fracture mapping set technique   383
fracture spacing index   381
fracture/discontinuity systems   248
fracture(s)
    affected by frost shattering   225
    in sedimentary rocks   367
free waves   139
freeze–thaw action   385–6
    causing rockfall   90
    and rock slides   91
friction   95
frictional loss, dependent on channel
        roughness and shape   121
frost boil   222
frost churning   227
frost heave   222, 226, 228–30
    in chalk   404
    controlled by permeability below
        frozen zone   228
    estimation of   30
    factors necessary for occurrence
        228
    maximum rate of   229–30
    process of   230
frost sapping   227
frost shattering   225, 227, *227*
frost susceptibility   385–6
    of building stones   407
frost susceptibility system, US Army
        Corps of Engineers   228, *229*
Froude number   121, 124

frozen ground
    a flood intensifying factor   128
    impermeable, problems of thaw
        224–5
    settlement associated with thawing
        225
fumaroles   34
    associated with volcanic dormancy
        or senility   40
    composition of gases   41
    cool and hot   41
    gas emissions may change prior to
        eruptions   50
    new, when an eruption pending
        49
funnelling   142, 152
    of tsunamis   161
        causes great devastation   160

gabions   110
Gale Common, Yorkshire, varved
        clays   217
galena   501
gap grading, of tills   209, *209*
gases
    in abandoned mines, accumulation
        of   455
    dissolved
        affecting groundwater chemistry
            264
        in groundwater   455
    formed in peat   357–8
    from burning colliery spoil   442–3
    inert   455
        released before an earthquake
            71
    in landfills
        capped, dangers of migration
            from   470
        formation of   470–1
        from waste disposal   274
        rate of production   470
        venting of   470, 471
    migration pathways, determination
        of in mines   455–6
    see also radon; volcanic gases
geochemical anomalies
    affecting health   487
    affecting uptake of trace elements
        487, **488**
geochemical data   524
geochemical interaction, dependent on
        sorption processes and
        chemical speciation   493
geochemical maps   496–7, *497*
    showing total concentrations of
        elements   497
geochemical reconnaissance, by
        stream sediment sampling
        496
geochemical surveys   496–7

geodimeters and tiltmeters, measuring
    volcano uplift   49
Geographical Information Systems
    (GIS)   29–31
    advantages and disadvantages   29,
      30, **30**
    for evaluating landslide-controlling
      factors   101
    for vulnerability assessment   29,
      30
geohazards   6
    constraints imposed by   6
    control or prevention of   3–5
    data required to reduce losses from
      4
    losses attributable to, attract
      government compensation   8,
      134
    may impede economic development
      7
    one type may generate others   7
    and planning   6–8
    problem soils   335–59
    response to by risk evaluation   7
geological data
    collection methodologies   524
    for land-use planning investigations
      524
    provided for planners   2
geological maps/mapping   535
    for groundwater exploration   255
    for seismic investigations   72
    shortcomings of   19
geological plans, in site exploration
    report   536
geological risk assessment   4
geological risk reduction   4–5
geology
    and dam sites   243–6
    local, influence on surface waters
      and shallow groundwaters
      495
    the starting point for planning   2
    structural   **529**
geomembranes   463
    as part of double liner system
      463–4
geomorphological features, on
    morphological maps   13–14
geomorphological frameworks, from
    satellite imagery   527
geomorphological hazard maps, arid
    regions   177, *177, 178*
geomorphological maps   17
    site/situation of proposed bridge
      crossing, E Nepal   *16*
    slope angle classification for   **14**
geomorphology, the basis for land
    facets   532
geonets, for roads over potential
    mining subsidence   424
geophysical exploration   546–56

cross-hole seismic methods   556
drill-hole logging techniques
    554–6
electromagnetic methods   551–2
magnetic and gravity methods
    552–4
resistivity methods   547–51
seismic methods   546–8
geophysical observations, and
    eruption prediction   48–9
geoplan maps   22
geoscientific survey, multidisciplinary,
    Wolverhampton area   30–1
geosynthetic clay liners (GCLs)   464
geosynthetic sheet piles   465
geotechnical plans   29
geotextiles   446
    for drain linings   110
    mats containing seeds   315
    natural, used with seeding for sand
      stabilization   178
    synthetic   315
    temporary   315
    to inhibit soil erosion   315–16
    see also geomembranes; geonets
geysers   41, *42*, 250
Ghyben–Herzberg expression
    275–6, *276*
gibbsite   292, 355
GIS *see* Geographical Information
    Systems (GIS)
glacial deposits
    fluvio-glacial deposits   213,
      215–19
    tills and moraines   203, 205–9
glacial erosion   199–203
    variable rate   199–200
glacial hazards   220–3
glacier outbursts   221, **221**
glaciers
    acquisition, transportation and
      deposition of tills   *206*
    defined   199
    fluctuations and surges   221, **221**
    ice sheets/ice caps   199
    piedmont   199
    valley   199
glass sand   326
Glenrothes pilot study, suite of
    environmental geological maps
      in report   20, 22
gleysols (cryosols), tundra soils   289
glowing clouds   43
gneiss   243, 364, *366*, **371**
goitre, and iodine deficiency   504,
    505–6
grabens, at head of flow slides   65,
    65, 69–70
graded bedding, in volcanic ash   43
grading curves   293, *294*
grain size
    and occurrence of frost heave   228

    and rock texture   372–3, **372**
grain-to-grain pressure   307
grains, orientation and failure   367
Grand Canyon, Arizona   *173*
granite   363, *365*, **368**, **371**, 408,
    410
    assessment of weathering grade
      388
    coefficient of weathering for   387,
      387
    could provide areas for deep
      disposal repositories   477–8
    field tests for quantitative
      weathering index   388, **388**
    fissure zones in   388, 393
    a source of radon   512
    weathering in   *386*
granular materials, act as filters for
    landfills   460
grasslands   291
gravel   424
    hard to entrain   123
    oligomictic and polymictic   325
    particle shape classification   325
    river and fluvio-glacial   324–5
    surface coatings on particles   325
gravel and sand *see* sand and gravel
gravel sheets   174
gravitational field, of the Earth,
    variation in   553
gravity, generating movement on
    slopes   86
gravity maps, contoured   553
gravity meters   49
gravity surveys   553–4
Great Goose Neck, entrenched
    meander, Utah   *119*
Great Whin Sill   *364*
ground
    displaced during earthquakes
      68–9
    fissured or fractured by earthquakes
      69, *69*
    frost-shattered, stability problems if
      fissures silt/clay filled   225–6
    thermal radiation of   49
ground conditions, and seismicity
    63–6
ground conductivity profiling, to
    detect polluted water plumes
      281
ground contamination   5
ground failure *see* failure; faults;
    fissures
ground infiltration capacity   126,
    128
ground moraine   208
ground movement
    due to rock dilatancy before
      earthquakes   71
    Pacoima Dam, San Fernando
      earthquake (1971)   *75*

ground probing radar method 552
  limited by conductivity of strata
    probed 552
ground shaking
  importance of ground conditions
    66
  local, influenced by geological
    structure 63
  recorded by accelerographs 73–4,
    74, 75
  see also seismic shaking
groundwater 246–8, **529**
  aggressive, arid areas 184, 185,
    351
  aquifer, often pure 260
  in basement aquifers 248
  calcium in 263
  change in movement and chemistry
    before earthquakes 71
  chemical elements in 262–4, **262**
  classified according to total
    dissolved solids (TDS) 264
  and contaminant migration 479
  declining level indicating future
    problems 271
  deep, changed mineral composition
    with increasing residence time
    495
  direction of flow 255
  draining from abandoned mines
    447
  fluorine in 506, 506
  generally free from pathogenic
    bacteria and viruses 261
  and the hydrological cycle 233,
    234
  ionic content monitored in drill
    holes 281
  may become non-potable at depth
    261
  mineral content of 495
  movement of in coal workings
    452
  physical properties in relation to
    water supply 261
  rate of replenishment 246
  regime changed by construction
    operations 184
  sabkhas
    coastal, brine concentration
      180
    flowing, removal of soluble salts
      causing settlement 182, 184
    saline, precipitation of salts from
      183–4
  and slope stability 98–9
  subsidence when abstraction
    exceeds recharge 437–8, 437
  and surface water, conjunctive use
    281–2
  trace elements in and their health
    significance 494, 494

used to augment river flow 281
vulnerability evaluated using GIS
  30, 31
groundwater chemistry, affected by
  dissolved gases 264
groundwater conditions, on reservoir
  flanks 236, 236
groundwater contour maps 255,
  256
groundwater discharge 233, 250,
  255
  becomes base-flow of rivers 251
  maintaining water courses 250–1
groundwater exploration 254–7
  data required for investigations
    254
groundwater mining see overpumping
groundwater monitoring 98–9,
  279–81
  monitoring programme objectives
    279
  programmes established prior to
    tipping 468
groundwater mounds 255–6
groundwater pollution 5, 272–9
  biological pollution 272–3
  by coal mining effluent, prevention
    of 452
  by leachates 274–5, 468–70
  investigations 278
  through sinkholes 403
groundwater prediction maps 255
groundwater pressure 237
groundwater quality, various types of
  rock mass **262**
groundwater recharge 233, 236,
  250, 255, 270
groundwater regime, upset by
  reservoir formation 237–8
groundwater remediation 484
  active containment 484
  pump and treat method 484
  zones of stagnation 484
groundwater reservoirs, slow reaction
  to changed conditions 281
groundwater resources assessment
  254–5
  desk study 254–5
grout curtains 245, 465, 472
grouting
  foam grouts 424
  of old mine workings 423–4
  used to extinguish burning coal
    seams 454–5
groynes
  for arresting longshore drift
    157–8
  close spaced to control rip currents
    158
  length, determination of 158
  permeable 158
  spacing of 158

gullies 169, 170, 174, 311
  causing severe soil erosion 308
  fans at exits from 174
  formed from fissures 439, 440
  and soil conservation 315
'gulls' 83
gunite (shotcrete), used to seal rock
  face 108
gypcrete/gypcrust 170, 173, 352,
  354
gypsum 180
  dehydration in sabkha environment
    182
  massive deposits less dangerous
    than anhydrite 405
  and production plaster of Paris
    411–12
  readily soluble 405
    problem beneath dams 405

Hackensack Valley, New Jersey,
  overconsolidated varved clays
  217
halite 183, 263
  forming surface crust 180–1
halloysite/metahalloysite 353, 357
hand augering 537
hanging valleys 203
hard pans 290, 493
hard water 264–5
harmonic tremor, associated with rise
  of magma 49
Hawaii, selenium in soils 507
Hawaiian Islands 161
Hawaiian type eruptions 38, 39
Hayward fault, California, creep on
  69
hazard maps 10–12, 524
  aggressive salty ground conditions
    186–8
  areas with old mine workings 421
  disadvantages of 11
hazard risk mapping 10–12
  thematic maps 423
hazard risk reduction 104–5
hazard zone maps 10
  volcanic gases, Dieng Mountains,
    Java 53
hazard zoning, and old workings
  421–3
  Airdrie, Scotland, safe and unsafe
    zones 421, 422
hazardous waste 441, 471–4
  assessment of site suitability for
    471–2
  defined 471
  hazard ratings for 472
  identification of 471
  liquid, disposal of 473–4
    injection below freshwater
      aquifers 473

monitoring of 474
low-level, co-disposal 473
monitoring of repositories 472–3
pretreatment 472
safe disposal of 472
see also deep disposal repositories;
radioactive waste
hazards
avoidance of preferable 177
effects of 6
from contaminated land, site
peculiar 479
involve risk 6–7
natural and man-made 6
net impact of 7
rapid onset 7
unacceptable to development, old
mine workings 423
head see solifluction
health 495
effects related to water chemistry
differences along flow gradients
495
human, and iodine 504–5
impact of fluoride in drinking water
506, **507**
and mineral dusts 511–12
threatened by nitrate pollution
277
and trace elements 498–511
health hazard
from cemeteries/graveyards 278
radon and daughter products,
emitting α particles 513
Health and Safety Executive, UK,
definition of concept of
tolerability 9
heap leaching 452–3
development on permanent bases
453
sources of acid mine drainage 448
use of cyanide 452, 453
heart disease, possible relationship to
ions and trace elements in
water 516
heave
black shales in Ottawa 399
under dam foundations, must be
counteracted 242
heavy metal contamination 446
heavy metals 444, 479
toxicity of 468
heavy minerals, dredging for 435,
436
herringbone ditch drainage 110
Hessle Till 210, 211, 212
hill creep, accelerated by solifluction
227
histosols, marshy vegetation 289
Hjulstrom curves 123, 124
Holderness, Humberside tills
210–12

Basement Till 210
Hessle Till 210, 211, 212
Skipsea Till 210, 211
Withernsea Till 210, 211
hollow stem augers 537, 539
Hong Kong
shallow slides triggered by
proximate rainfall 88
terrain evaluation 101
hot avalanches 51–2
hot pyroclastic surges 52
hot springs 41, 249–50
temperature rise before imminent
eruption 49
used for domestic heating, Iceland
250
used for generation of electric
power 250
Huascaran, Mount, Peru, landslide
fatalities 69
humans
accelerating soil erosion 307
accumulation of cadmium in
500
affecting the environment 487
effects of zinc deficiency 508
intake of arsenic 500
humic acids 387
Hunstanton Till 210, 211
hurricanes, generating storm surges
152
Hurst Castle Spit, recurved laterals
150
hydrates 386, 387
hydration, and dehydration 387
hydraulic conductivity 252, 551
determination of in the field 260
variation in 258
hydraulic gradient 252
and leachate flow 461
hydrocompaction, on alluvial fans
171
hydrocyanic acid 453
hydrogen chloride, in volcanic gases
40
hydrogen sulphide 264, 443
in solfatara steam 40–1
hydrogeological acceptability, of
landfill sites 460–1, **461**
hydrogeological assessment, for gases
in groundwater 455–6
hydrological cycle 233–4, 234
hydrological records, for potential
reservoir site investigation
235
hydrolysis
by weakly carbonated water 387
of cyanide ions 453
of silicate minerals 387
hydroseeding 179
hydrosphere, average composition
492

hypertension
cadmium may be implicated 500
through renal lesions 503

ice
in mechanical weathering 385–6
pressure melting of 224
ice avalanches 220
ice caps 199
ice lenses 228
formation of in soils 222
ice mounds/hummocks 226
ice segregation, in soils 222
ice sheets 199
advance and retreat of 220
as repositories for high-level
radioactive waste 475
slow movement of 200
suppliers of ground quartz 348
ice wedges 222, 226
igneous intrusions 255
igneous rocks 362
acid, do not mix with bitumen
411
development of columnar jointing
376, 376
elemental composition **492**
elements altered by chemical
weathering 491, 493
high remanent magnetization 552
intrusive and extrusive 362
lead content of 501–2
and metamorphic rocks 388,
393–6
mineralogical composition 362,
363
quarrying in 406
ignimbrite 43–4, 51
associated with nuées ardentes 44
ignimbrite sheets, erupted from
fissures 36
illite 292, 326, 327
immunosupressive agents 503, 517
impervious linings, increase resistance
to seepage 237, 237
in situ testing 556–61
cone penetration tests 558–9, 558
in situ shear tests 561
penetrometer tests, dynamic and
static 556
plate load tests 560
shear vane tests 559–60, 559
standard penetration test 557,
558–9
incineration, of hazardous waste
472, 483
incised meanders 118–20
inclinometers, measurement of
subsurface horizontal
movement 98
induced seismicity 78–81, 473

induction logging   555
industrial processes, quality of water
    required variable   265
infiltration   233
    of artificial recharge, effects of
        272
    dependent on precipitation   246
    in desert regions   174
    ground infiltration capacity   126,
        128
    induced   273–4, *274*
    into fan deposits   170
    and landfill surface slope   464
    lessens with desertification   188
    rate increased by vegetation   310
infiltration capacity   246
    reduced   233
infiltration rate, effective and potential
    250
inflow tests   257
infrared line scanning (IRLS)   525–6
    imagery interpretation   526
    predawn flying most suitable   525
    texture and interpretation   526
infrastructure, routing may be unable
    to avoid active faults   69
inselbergs   172, *172*
insurance premiums, adjusted for
    flooding   134
Integrated Planning Decision Support
    System (IPDSS)   30
interblock shear strength   377
interception storage   310
interceptor drains   110
Interdepartmental Committee on the
    Redevelopment of
    Contaminated Land (ICRCL)
    478, **479–80**
interflow   233, 250
intergranular fracture   367
intergranular movements   92
intergranular permeability   551
intergranular stress   301
intermontane basins, arid/semi-arid
    regions, stream action
    169–73
internal friction, angle of   301
International Seismological Centre
    (Britain)   75
interparticle bonding, short-range
    348
interrill erosion   311
invertebrates, affected by deposition
    of fine particles   451
involutions   226
iodine   **490**, 503–6
iodine deficiency, and goitre   504,
    505–6
iodine deficiency diseases (IDDs)
    504–5
    geographical distributions   505
ion exchange   450

affects chemical nature of
    groundwater   264
and attenuation of leachate   466
and removal of sodium ions   263
ionization   386, 387
ions, inorganic, causing mineralization
    468, 470
IRLS *see* infrared line scanning (IRLS)
iron   292, **489**
    in groundwater   264
iron oxides   387
iron sulphides, black glassy spots/
    vitreous core in bricks   327–8
irrigation   190–7
    determination of economic
        feasibility of   191
    management of water usage   195
    scheduling according to water
        availability and crop need
        195, 197
irrigation systems, arid and semi-arid
    regions
    comparison of   **192**
    drip systems   191
    flooding   191
    need for proper planning/design
        and operation   190–1
    sprinkler systems   191
islands, remote, for disposal of
    radioactive waste   475
isomagnetic contour maps
    of anomalies   553
    faults and dykes on   553
isopach maps   255
isoradiation contours   526
isoresistivity lines   550, *550*
isoseismal lines, San Francisco
    Peninsula   *63*
isostatic rebound (recovery)   139,
    219–20
itai-itai disease   277
    a cadmium-related disease   500–1

jet grouting   245, 465, 466
Johannesburg, South Africa
    seismic tremors peaking around
        blasting time   80–1
    zoning system over old mine
        workings   421
joint inclination   95
joint surfaces, and rock mass
    behaviour   378
joint wall compressive strength   380
jointing patterns   531
joints   375–7
    accommodation of subsidence
        427–8
    affect marine erosion   146
    columnar jointing   376, *376*
    filled   378
    formation of   376

influence on coefficient of
    permeability   *374*
    in limestone   399, *399*, 400
    open and closed   377–8, **377**
    in plutonic rocks   393–4
    relict   96
    sandstone   244
    shear strength along   96
    and shear zones, unsound
        foundation areas   243–4
    sheet jointing   376–7
jökulhlaups   221, **221**
juvenile water   246

kame terraces   215, *215*
kames   215, *215*, 216
kaolinite   292, 326, 327, 355, 493
    disordered   327
kaolinization   326, 393
Kariba Reservoir, earthquakes and
    lake level   80, *80*
kastanozems (ustols)   291
Katse Reservoir, Lesotho, tremors
    associated with filling   80
Keshan disease   508
kettle holes   **215**
kinetic tests   449
knick points   118
Koyna Reservoir, Bombay, dam
    fissured by large earthquake
    79

L waves   58
lacustrine deposits
    interbedded with fluvio-glacial
        deposits   216
    proglacial lakes   219
lahar prediction   50
lahars   46–8
    cold   47
    from a danger zone   55
    generation by rainfall   46
    hot   47–8
        crater lake   46–7
    mitigation of   56
    movement in waves/pulses   48
    speed and distance of travel   52
lakes
    ice-dammed, drainage of   221
    proglacial   219
laminar flow   120, 171, 252, 308
    in lahars   46
    in lava streams   44
lamination, and clay behaviour
    396–7
land, grading according to potential
    uses and capabilities   2
land capability analyses   532–3
land capability classification   314,
    317, **319–21**

land capability studies 532–4
   Palo Alto 533–4, *534*
   steps in 533
land classifications 2
land degradation, through waste
   disposal 5
land elements 322, 532
land evaluation
   pedological maps for 323
   and site assessment 524–61
land facets 322, 532
land management, importance of
   slope steepness 14
land quality appraisal, contaminated
   sites 481
land reclamation 6
   see also remediation, of
   contaminated land
land regions *533*
   in large feasibility studies 532
land suitability, Palo Alto 534, *534*
land suitability evaluation 324
land suitability studies 533
land systems 322, 532
   Alor Gajah land system *533*
land systems maps 322, 532
land units 532
land use
   adjustment to microzoning
      recommendations 11
   most suitable, determination of
      524
land-use planning 1–2, 524
   and conservation 5
   establishment of acceptable criteria
      for location of activities 1
   maps for 19
   objective of 1
   for prevention and mitigation of
      geological hazard 4–5
   process of 2
   recognition of fault creep important
      60
   using floodplain zones 131–3
landfill
   assessment of state of 460
   cellular construction 462, *462*
   design considerations 461–6
   and gas formation 470–1
   and groundwater pollution
      274–5
   waste disposal by 459
landfill sites
   capability of to accept hazardous
      waste 472
   classification of 461, **461**
   distant from water supply sources
      275
   gas migration prevention 470–1
   landfill gas, monitoring of a safety
      aspect 470

planning requires a waste survey
   460
   selection of 460–1
   and leachate control 462
   venting, for methane control after
      closure 470
landscape classification 531
landslide hazard
   effective management 104
   investigation and mapping
      99–106
   recognition and reduction schemes
      104–5
   reduced by regulation 105–6
landslide hazard assessment 103,
   105
landslide hazard mapping 11, *12*
   scheme design 103
landslide hazard zoning maps 103
landslide investigations
   local and site specific 101
   use of checklists and inventory
      forms 101–3
landslide mapping, purpose of 103
landslide mitigation programmes
   100
landslide movement, derivation of
      magnitude of 101
landslide recurrence intervals 100–1
landslide risk maps 103
landslide susceptibility maps 103,
   104
   associate land characteristics with
      geomorphological units
      103–4
   multivariate 103, *105*
landslides
   active and inactive, indicative
      features **100**
   ancient, reactivated by human
      interference 170
   causes of 84–8
      triggered by earthquakes 69
      triggered by rainfall 86, 101
   classification of 88–93
      falls 88, 90, *90*
      flows 91–3
      slides 90–1
   coastal 87
      in unconsolidated material
      145, *146*
   complex slope movements 88
   correction and prevention 106
   costs direct and indirect 99
   deep-seated, safety improved by
      changed slope geometry 107
   defined 84
   due to sector collapse 52–3
   internal, caused by increased pore
      water pressure in slope material
      88

losses from increasing 99–100
   mapping of 101
   multiple movements 88
   not suitable areas for reservoirs
      245
   occurring after reservoir filling
      238
   in rock masses 94–7
   in soils 93–4
Lang prospector 537
lapilli 43
lateral blast, a volcanic hazard 51
lateral moraines 208, *208*
laterites/lateritic soils 354, *354*–5,
   355–6, 493
   effects of leaching on 356
lava
   aa 44–5, 45, 46, *46*
   basaltic 44
   blocky 44
   dacite and rhyolite 44
   pahoehoe 44, 45, *45*
   pillow 46, *47*
   viscosity affects type of eruption
      38–40
   viscous, blocking vents 37–8
lava flood eruptions 52
lava flows 44–6, 395
   ancient and recent 393
   damage to property 52
   recent, basaltic, not watertight
      237
   threat can be mitigated 56
lava streams 45–6
   laminar flow within 44
lava tunnels 45–6
lava volcanoes 37
   central monogenetic 37
   polygenetic 37
leachate drainage systems 464
leachate production, control of
   461–2
leachates
   affecting rivers 468
   amount influenced by groundwater
      462
   attenuation of in soils 466, 468
   control of 462
   in landfills
      formation of 466
      from domestic refuse 460
      from waste disposal in 274
   methods of tackling pollution by
      275
   surface and groundwater pollution
      by 468–70
leaching 290
   affecting coefficient of
      consolidation 356
   effects on lateritic soils 356
   removal of salts accumulated by

irrigation 195
of soils 493
lead 264, 501–3
acceptable levels in urban areas
502
effects of 503
entering the atmosphere 502
role in the development of multiple
sclerosis? 517–18
in water 502
lead poisoning (plumbism), in children
and adults 503
leukaemia, and radon 513
levées 127
along debris flow banks 170
described 134–5, 135
eliminate natural overflow basins
135
slope into flood basins 126
light cable and tool rigs 529–40,
539
rotary attachments 540
light non-aqueous phase liquids
(LNAPLs) 278
removal from soil 483
separation from groundwater
484
lime 411
lime popping 327
lime stabilization, of expansive soils
341
limestone 367, 399, 427
affected by salt weathering 184
dam sites vary in suitability 244
decays in urban atmospheres 408
ravelling failure in 401
subject to chemical attack 387
limestone pavement 399, 399, 400
lineaments
on aerial photographs 530–1
found by false colour 530
liners 445–6
for complete landfill containment
461
composite 463
double liner system 463–4, 473
for landfill 462–3, 463
permanent bases for heap leaching
453
see also geotextiles
liquefaction 91
with shock 332–3, 350
see also soil liquefaction
liquefaction hazard maps 336, 337
liquid limits 210, 211, 212, 294
laterite 355–6
Tees Laminated Clay 218
in tropical soils 354, 357
varved clays 217
liquid scintillation counting 514–15
liquidity index 210

quickclays 349, 349, 350
of a soil 295
literature surveys 19
lithification 367
lithological quality designation (LQD)
545
loads, effect on soil affected by
drainage conditions 300–1
local investigations 524
lodgement till 205, 205–6, 206,
207, 215
fissures in 206
Northumberland, geochemical
properties 211
loess 345, 346
collapse areas 346–7
loess karst 348
long-rain floods 129
longitudinal dunes 169
longshore currents 145
removal of landslide debris 145
longshore drift, stabilization of
157–9
longshore (littoral) drift 149
direction indicated by spit
orientation 149
longwall mining, and subsidence
426–30
area affected by ground movement
426–7, 426
influence of width–depth
relationship of extraction
427
measures to mitigate subsidence
effects 430–1
panel and pillar mining 431
preventative techniques 430–1
reduction of maximum
subsidence 431–3
predictability of subsidence 430
Love waves 58
Lower Oxford Clay, oily material,
almost self-firing 328
lung cancer 512
and radon 513
lung tissues, damaged by dust 511
luvisols (alfisols) 290–1

Macintosh probe 537
macroencapsulation, of hazardous
waste 472
magma 362
acidic 34
associated with explosive activity
34
basaltic 34, 35, 49
composition and viscosity of 34
eruption of 40
magma feeder channels, central vent
or fissure 34

magma formation
at constructive margins 34
at destructive margins 34
magnesite 181
magnesium 263
magnesium deficiency 491
magnesium oxide, in Portland cement
411
magnetic field
local, changed before an earthquake
71
strength of 552
magnetic and gravity methods,
geophysical exploration 546,
552–4
magnetic prospecting 552
magnetometers, for mapping
geological structures 553
manganese 489
in drinking water 265
and health 510
Manning formula 121
Manoun, Lake, carbon dioxide release
53
map memoirs 535
marble 243
marine arches 146, 148
marine inundation, and storm surges
151–3
mass movement 308
an accompaniment of earthquakes
69–70
varieties of 83
mass permeability 537
mass transport 139
mass wastage, solifluction 224–5
materials 529
susceptible to earthquake motions
87–8
materials (drift) maps 15
matrix system of analysis, acts as
super-check list 3
maximum intact core length (MICL)
545
Mead, Lake, Colorado, impounding
level and seismic activity 79
Meade salt sink, Kansas 406
mean recurrence intervals (MRIs),
used in volcanic hazard zoning
56
meanders 115, 116, 122
deformed and compressed
116–17
entrenched 118, 119
growth of in streams 122–3
incised 118–20
ingrown 118, 119
length and amplitude 117–18
medial moraines 208
megacities, future 1
meltwater, deposition of till 203

meltwater floods, seasonal   126
meltwater streams   213
    deposition of kame terraces   215,
        215
meltwater till   207
Menard pressuremeter, determines in
    situ strength of ground
    560–1
mercury   264
    enters atmosphere in gaseous and
        particulate forms   501
    and health   501
    in living tissue   501
mesas and buttes   166, 167, 173
metal carcinogenesis, mechanisms of
    unknown   518
metal hydroxide deposition, impact on
    benthic fauna   452
metal hydroxide sludges, formation of
    448
metal sulphides   466
metalliferous mineralizations,
        elements released to surface
        and groundwaters   497
metalliferous mining, and subsidence
    415–18
metamorphic aureole see contact
    aureole
metamorphic rocks   362–4
metamorphism
    and aerial photographs   531
    regional   362, 363
        rocks may not make good dam
            sites   243
    thermal/contact   11
methaemoglobinaemia   277
methane   466, 470
    a byproduct of coalification   455
    a dangerous hazard   470
    determination of source   455
    migration into reservoir rocks
        455
    moves through coal by diffusion
        455
    a problem in water   264
methanogenesis
    a dangerous hazard   470
    inhibited by toxic chemicals   470
methylation reaction, oxidative   499
methylmercury   501
Mexico city, earthquakes, affecting
    buildings   65, 67
mica   364
microfractures   395
microirrigation systems   191
microlaterolog   555
microlog techniques   554, 554
micropiracy see river capture
microzonation   11, 79
mine closure, leads to acid mine
    drainage   447
mine drainage waters, quality and

    quantity variable   451
mine shafts, old   424–6
    filling and capping   425
    locations unrecorded/inaccurate
        424
    protection against cratering
        425–6
    shaft linings   425
    sudden collapse   424–5, 425
mine wastes   441–6
    disposal of in tailings dams   444–6
    from metalliferous mines   441
    spoil heaps and their restoration
        441–4
mine water treatment systems
    449–50
mineral dusts, and health   511–12
mineral extraction
    causing water pollution and land
        degradation   5–6
    further uses of sites   5–6, 415,
        433, 443
mineralization, a serious effect of
    leachate on groundwater
    468, 470
minerals
    diamagnetic, paramagnetic,
        ferromagnetic   552
    exploitable resources   19
    fibrous, and health   511–12
    soluble
        in particulate form in the ground
            405
        solution of   387
    stockpiles a concentrated source of
        acid mine drainage   448
Minimata disease, toxic effects of
    mercury   501
minimum pool level   235
mining
    associated induced seismicity
        80–1
    partial extraction   418–21
    a robber economy   415
    and sinkhole development
        417–18
    stabilization of old workings   417
    waste materials from   441–6
    see also acid mine drainage; coal
mining; metalliferous mining; mine
shafts, old; old mine workings; surface
mining
Mississippi floodplain
    deposits of   127
    major embankments   136
mixed volcanoes see composite
    volcanoes
modulus of elasticity of rocks   242
Mohr failure envelope   356
moisture, in rocks   367
moisture content
    deposited tailings   445

natural
    tills   210, 211, 212
    varved clays, Ontario   217
    optimum   331, 332
moisture storage capacity   310
moisture zones
    dry lands   187
    and road construction   186, 186,
        187, 188
molecular attractive forces   301
molybdenosis   509
molybdenum   490
    excess causes disease in livestock
        491, 497, 498
    and health   508–9
    poisonous in large doses   509
molybdenum deficiency, and poor
    pasture   509
montmorillonite   291–2, 292, 326,
        327, 395
    absorption of free water   397
    enhances plasticity   326
    a swelling clay mineral   338
moraines   206, 208–9, 219
Mornos Reservoir, Greece, asphalt
    membrane   237
morphogenetic maps   14, 15
morphological maps   11–17
mountain catchments, effect on
    run-off   170
mudflows   46–8, 91, 92–3, 225
    coastal   145
mudrock   367, 395–400
    for brick making   327
    distinguishing durable from
        non-durable   397
    geodurability classification   397,
        398
    poorly cemented, failure of   398
mudstone, slaking behaviour of   396
mulches, use of   315, 316
multiple barrier concept, deep
    disposal, radioactive waste
    475
multiple cropping   315
multiple hazard–multiple purpose
    maps   11
multiple retrogressive slides   91
multiple sclerosis, lead may play a role
    517–18
multispectral scanners, satellite,
    detecting surface features and
    landforms   527

National Earthquake Information
    Service (USA)   75
National Radiological Protection
    Board (NRPB), advice on
    radon levels in homes   514,
    515
natural hazards   6

neap tides   142
Netherlands
    engineering geological maps, Leiden
        21, 22
    land reclamation   6
    protection from sandy beaches and
        high dunes   149
nickel   **489**, 518
Niigata, Japan, movements prior to
    earthquake   69, 71
Nile River, Egypt   169
nitrate pollution   277–9
    alleviation of   277
nitrate(s)   261
    dissolved, contaminate irrigation
        water   193
    may indicate a pollution source
        263–4
    reduction of   466
nitrification   272
nitrogen, converted to ammonia in
    leachate   466
nitrosamines   277
nivation hollows   201
non-aqueous phase liquids (NAPLs)
    278, 484
north Norfolk tills   209, *209*, **210**
    Contorted Drift   **210**
    Cromer Till   **210**, 211
    Hunstanton Till   **210**, 211
    Marly Drift   **210**
Northumberland lodgement till   *212*
    geochemical properties   **211**
    zones of weathering   212, *213*
notches, wave-cut   146
nuclear waste
    encapsulated, placed in deep drill
        holes   478
    reprocessed   475
nuées ardentes   39–40, 51–2, 55
Nyos, Lake, carbon dioxide release
    53

oceanic crust   34
oceans, important source of iodine
    504
offshore   139
offshore bars   151
oil, subsidence due to abstraction of
    438
old mine workings   **529**
    act as drainage systems   447
    and hazard zoning   421–3
    measures to reduce/avoid
        subsidence effects   423–4
    stabilization of   417
    state of, assessment of past and
        future collapse   421
    void filling to prevent migration
        and collapse   423
olivine, decomposition of   387

open pit mining   433
    waste disposal, spoil heaps or
        tailings lagoons   433
open pits, uses for   415, 433
opencast coal mining, UK   433
    restoration to agriculture   433
optimum moisture content   331, 332
orbital velocity, water particles   139
organic content, of soils   285
organic depletion, effects of   311
organic impurities, affecting cement
    326
organic matter   300
    decay products   386–7
    in gleysols (cryosols)   289
    importance of in soils   311
    incomplete oxidation and black
        coring in bricks   328
    lost by leaching   290
organic waste, anaerobic
    decomposition of   466
oscillatory waves   139
osmosis   450
outwash deposits   213
overbank deposits   *127*
overbank flow   126
overdraught   271
    reduced by artificial recharge   271
overflow channels   219
overgrazing   189–90, 310
overland flow   174, 310–11
    control of   314–16
overlays, use of with morphological
    maps   14
overprinting, use of   23
overpumping   275, 278
overstocking   188
oxbow lakes   116, *117*, 123
oxidation   386, 493
    of ablation till   206
    in coal workings   452
    in dewatered zones   447, 448
    of ferrous iron in rivers   468
    incomplete, of organic matter
        328
    of pyrite   387, 446, 454
    of sulphide minerals   447–50
oxidizing conditions   491
oxygen, free, and rock decay   386

P waves   58
    changes in record of   70–1
    and S waves, velocities decrease in
        ratio before major quakes   71
    used in seismic exploration   546
Pacific Ocean, majority of destructive
    tsunamis   161
Pacific Tsunami Warning System
    (PTWS)   161–2
packer tests   258
    double packer test   258

    evaluation of permeability from
        258
parasitic cones   35
Paricutin, Mexico   35, 36
Parkinson's disease   487
partial extraction methods   416,
    418–21
particle mobility/movement
    critical velocity for   163–4
    due to wind   313
    and wind erosion   164–5
particle size   287
    and frost heave   228, 229
particle size analysis   293, *294*
particle size distribution
    and maximum rate of frost heave
        229
    in soils   293–4, *293*, **293**
    and strength of frozen ground   223
    tills   211
particle velocity, and blasting
    vibration damage   434–5
peak triaxial compressive strength
    380
peat   300, 350, 357–9, 367
    classification of   **358**
    compression expels free pore water
        359
    consolidation of   359
    formation of   289
    high coefficient of secondary
        compression   359
    high water content   357
    permanent shrinkage on drying   358
peat soils   289
pebbles   325
pediment gaps   172
pediment passes   172
pediments   171–2, *172*
pediplains   172
pedological maps   323
peds   293–4
    laminar, lenticular, prismatic   294
penetrometers, for *in situ* testing
    556–9
perched water table   247
percolation   169, 233
    slow, delays nitrate concentration in
        groundwater   277
    of water through foundations of
        concrete dams   242
percussion drilling   539–43
periglacial environments, frozen
    ground phenomena   221–7
permafrost   199, *200*, 221–2, 289
    construction in permafrost regions
        227–8
permanent wilting point   192, 288
permeability   251
    of aquifers surrounding wells   270
    of bricks   328
    and compaction   332

and flow, assessment of in the field
257–60
governed by discontinuities,
tropical soils   356–7
of peat, reduced by compression
359
and pore water pressure in spoil
heaps   442
primary and secondary   252
relative values of   **253**
of a soil/rock type, defined   252
pesticides and herbicides, source of
groundwater pollution
277
petrographical descriptions, of rocks
372
pF value   191–2
pH value   342, 400, 497
bog peat   357
high in solonetz (naturargids) soils
291
for soils   288
and trace element solubility and
mobility   490, *491*
of vertisols   291–2
photogeology   530–1
types of investigation   **529**
photographs, uncontrolled, of
discontinuities   384
photography, in slope monitoring
98
photomosaics   528
phreatic water   246
phreatotypes   255
phyllite   **372**, 395
physical pretreatment, of hazardous
waste   472
piedmont glaciers   199
piezocones   559
piezometers, recording pore water
pressure   98
piezometric pressure   248
piles/piling   465
in permafrost regions   228
sleeving of   423
steel sheet and secant   465
pillar collapse   416, 419, *419*
pillar working   416, 417, 418–21
pillars
robbed as mining retreated   419
stability affected in flooded
workings   454
sustaining redistributed weight of
overburden   418–19
ultimate behaviour and failure
419
pillow lavas   46, *47*
pingos (hydrolaccoliths)   227
pipeline maintenance, affected by
dune movement   175
pipelines, flexible joints in   431
piping damage

caused by dispersive soils   343,
345
in collapsible soils   348
failure of tailings dams   445
piston samplers, thin-walled   541–2
pits, only used in areas of special
difficulty   539
plagioclase, source of sodium in
groundwater   263
plain tract   116
plane failure   96
planning   535
for avoidance of difficult ground
conditions   3
and geological hazards   6–8
maps for   19
sequential land planning concept
2
use of seismic zoning/microzoning
76–8
planning and development
ground characteristics for, Torbay
area   *20*
need for geological data   2
planning proposals, often
controversial   1
plant maturity, and trace element
uptake   495–6
plastic limit   210, 211, *212*, 294
Tees Laminated Clay   218, **218**
plasticity indices   210, 211, 294–5
and frost susceptibility   228
laterites   355–6
quickclays   349
tropical soils   357
plasticity, of soils   294
according to liquid limit   294
plate boundary earthquakes   71
plate load test   560
plate margins, constructive and
destructive   34
playa lakes   172, *172*
playas   172, *181*, 182–3
ground conditions   **182**
salt weathering attack on buildings
185
Pleistocene, glacial episodes   199
plunge line   144
plunge point   143–4
plunging breakers   143–4
removal of material from top of the
beach   144
plutonic rocks   362
for dam sites   243
weathering of   388, 393–4
pneumoconioses   511–12
podzols (spodosols), in coniferous
forest regions   289–90, *290*
point bar deposition   125
point bars   *127*
point discharge, acid mine drainage
447

point load strength   374, **374**
Poland, collapse of loess   347
polish, rate of   411
polished stone value, affects skid
resistance   411
pollutants
associated with coal mining
450–1
attenuation of   272
concentrated sources undesirable
273
inorganic, in rivers   468
potential, for groundwater   277–9
problem of slow travel in
underground strata   272
pollution
associated with landfill, tackling of
461
bacterial, maximum rate of travel
261
biological, in groundwater   272–3
of groundwater   272–9
nitrate   277–9
possible through induced
infiltration   274, *274*
susceptibility of shallow unconfined
water   260
of water supplies   271
'popouts'   410
population, increased in desert
margins   189
pore size distribution, and frost heave
230
pore space reduction
increases pore water pressure   87
and loess collapse   347
pore water, in stratified rocks of dam
foundations   241–2
pore water pressure   331, 399
development of in cohesive soils
93
and earthflows   93
embankments   333
excess   306
causing soil liquefaction   65
reduced by drainage   110
and slope failure   315
spoil heaps   442
in thawing of frozen ground
225
high, and failure in some clays   91
induced change in causes
earthquakes   78–81
influenced by precipitation   88
and landslides   87, 88
reduced by overabstraction of
groundwater   437
variations in affecting dams   241
porosity
of collapsible soils   345, 346
defined   251
of soils   292

Port Pirie, South Australia, lead abatement programme   502, **503**
Portland cement   399, 411
Portland Limestone, solution and flow velocity   401, *402*
posthole auger   537
potassium, in groundwater   263
potential dispersivity chart   *342*
power augers   537, 539
Precambrian shields   478
precipitation
    and attenuation of leachate   466
    see also rainfall
precompression, by surcharge loading, use in peat areas   359
preconsolidation pressure   306
pressuremeters, self-boring   561
pressure recorders   161
pressure transducers   98
process maps   14, *15*
profile shooting, seismic refraction work   547
property
    damage by lahars and lava flows   52
    hazards from pyroclastic falls   51
pumice   40
pumice cones   35
pumiceous pyroclastic flows   51
pumped storage reservoirs, for direct water supply   238–9
pumping, continuous   275
putty chalk   227
Puy de Sarcoui, Auvergne   37
pyramidal peaks   203, *203*
pyrite
    in colliery spoil heaps, weathering of   442
    oxidation of   387, 446, 454
pyroclastic fall deposits   51
pyroclastic flows   39–40, 51–2
    composition of   51
pyroclastic surges, volcanic hazard   52
pyroclastic volcanoes   35
pyroclastics   243
pyroclasts   41–4
    defined   41
    give very variable ground conditions   395
    size of variable   42–3

quarries, life of   406
quarrying process, ice erosion   199
quartz   366
    in brick making   327
    detrital   493
    fine, in quickclays   348
    in tills   205
    in varved clays   217

quartz dust   511
quartz sand   326
quick conditions, avoidance methods   337–8
quickclays   70, 348–50
    eastern Canada, geotechnical properties of   350, **350**
    high sensitivity of   349, *349*
quicklime, slaking and hydration of   411
quicksands   335–8

radioactive decay, daughter products and half-lives   474, 512
radioactive logs   555–6
radioactive tracers   260
radioactive waste   446, 474–8
    high-level
        deep disposal of   475–8
        disposal of   474–8
    low- and intermediate-level, disposal of   474
radon   446, 512–16
    amount released   513
    associated with rocks with high U concentrations   512
    and daughter products, emitting α particle   513
    detection of   514–15
    in homes   513, 514
    in soil gas, measurement of   515–16
    soluble in water   513
    transport of   513
radon hazard, potential, reduction of   514
raft foundations   244
    in mining areas   423, 430
railway tracks, fractured through fault creep   69
raindrop impact   174, 310, 311
rainfall
    in arid and semi-arid regions   169
    desert, erratic in time and space   173–4
    intensity, and run-off   174
    intercepted   310
    iodine in   504
    and soil erosion   308
    triggering landslides   86
rainfall erosivity factor   311–12
rainfall erosivity indices   312
rainfall erosivity maps   312, *312*
raised beaches   220
raisin cake structure   205
Rangely area, Colorado, seismic activity   78
ravelling   96–7, 402, *403*
ravelling failure, in limestone   401
Rayleigh waves   58
reafforestation   315

lowers intensity of flooding   133
recharge wells   272
reclaimed land, prone to inundation   151, *152*
reconnaissance surveys
    engineering geomorphological maps   17–18
    geochemical   496
    preliminary   535
    regional, use of magnetic survey methods   553
    see also field reconnaissance
recurrence interval, of floods   130
red clays   354–5, *356*
red soils, tropical   355
redevelopment, of older industrial and urban areas   3
redox boundary   491
reef workings, abandoned, deep stabilization of   417
refraction diagrams   141–2
refractory materials   326
regional investigations   524
regional physiographical settings, from satellite imagery   527
regolith
    of arid uplands, vulnerable to erosion   351
    thickness of   248
    throughflow channels in   248
rejuvenation, through land elevation   118–20
remanent magnetization   552
remediation, of contaminated land   482–4
    soil remediation   483–4
remote sensing   525–8
    infrared line scanning   525–6
    satellite imagery   527–8
    side-looking airborne radar (SLAR)   526–7
remote sensing imagery   525, 536
    for studying arid landscapes   176
    to monitor desertification   190
remote sensing surveys   76
reservoir basins, factors determining retention of water   235–6
reservoir design, capacity-yield relationship   235
reservoir leakage
    downstream from dams   236
    economics of   236
    resulting from cavernous limestone conditions   236–7
    through buried channels   237
reservoir sites   **529**
    potential, investigation of   235
reservoirs   234–9
    controlled and uncontrolled storage   137
    estimation of volume   235

flanks of
  conditions on   236, *236*
  slumping and sliding   238, *238*
for flood control   137
impermeability of floor important
    236
large
  attractive sites for   235
  earthquake epicentres below or
      near   79–80
  for irrigation   190
rapid fill   281
residential development, avoidance of
    landfill sites   470
residual factor, slip surfaces in clays
    94
resistivity
  decreasing before an earthquake   71
  a measure of porosity   551
  of saturated sandstone   551
    or limestone aquifer   280
resistivity depth soundings   549
resistivity methods, geophysical
    exploration   546, 547–51
  correction for lateral effects   551
  described   547–8
  determining boundaries of polluted
      groundwater plume   280
  good for investigations on
      horizontally stratified ground
      550–1, *550*
  laterolog (guard electrode), use of
      554–5
  microlog techniques   554, *554*
  penetration depth-electrode spacing
      relationship   549, *550*
  use of electrode configurations
      down hole   554
resistivity profiling   280, 550
retaining walls   109
  reinforced earth structures   109
retarding basins   137
return interval defined   270
Reuss Valley, Switzerland, cost of
    glacial flooding   220
revetments, to protect embankments
    against wave erosion   156
Reynolds number   252
  stream flow   120–1
Rheidol Valley, Wales, incised
    meanders   *119*
rias   220
ridge and swale topography   125–6
rills   114, 170, 174, 308, 311
rip currents   144, *145*
  control by groynes   158
rip-rap   135, 446
  for permeable sea walls   153
  as stone revetment   156, *156*
Ripon district, Yorkshire, subsidence
    due to gypsum solution   404
rising head test   257

risk   7
  acceptable level of   9
  actual and perceived   9
  from hazardous waste   471, 472
  potential, decreasing the level of   7
  and probability   9
risk analysis   9
risk assessment   8–10
  in areas of old mine workings   421
risk balancing   9–10
risk management
  problems inherent in   9–10
  requires a value system against
      which decisions are made   10
  in volcanic areas   56
risk probability, Bayesian approach to
    9
risk zoning maps   11
river capture   114, *115*
river channels
  factors causing adjustment   116
  unable to cope, arid/semi-arid
      regions   169
river erosion   122–3
river flow   120–2
  estimate of quantity   121
  prediction of   121–2
  recurrence interval, extreme events
      121–2
  reduction of   281
river terraces   118
river transport   123–5
  sediments not continually
      transported   125
rivers
  adjustment of regime to new
      reservoirs   235
  base-flow   233, 251
  competence and capacity of   123
  effects of inorganic pollutants
      468
  effects of oxygen depletion   468
  erosive power increased at flood
      times   129
  hazard zones in lower stage   131,
      132
  large, flowing through arid regions
      169
    typical landscapes   173
  longitudinal profile (long profile)
      116
  maturity   115
  organic pollution of   468
  as part of the hydrological cycle
      114
  regrading courses   118
  sediment discharge, defined   123
  sinuosity of   118
  sources of arsenic in   499
  see also streams
road aggregate   410–11
  desirable properties of   410, **410**

roads
  arid regions, behaviour of, influence
      of moisture zones   **186**
  avoidance of aggressive ground
      conditions   185–8
  and runways, construction of,
      permafrost regions   228
roches moutonnées   200, *201*, **215**
rock anchors, for major stabilization
    work   108
rock basins, glaciated valleys   203
rock bending, at faults, indicates
    increasing strain   72
rock blocks, rotational sliding after
    excavation   243
rock bolts, to enhance slope stability
    in rock masses   107–8, *108*
rock cores   545
  correct preservation and labelling of
      545
rock debris
  abrasion by   199
  variable assortment forming till
      205
rock fragment flows   92
rock mass factor concept   381
rock masses
  on aerial photographs   530
  bedded and jointed, critical slope
      angle   94–5
  behaviour determined by
      discontinuities   372
  discontinuous, strength of and its
      assessment   378–80
  estimation of deformation modulus
      of   381
  hard, liable to sudden and violent
      failure   95
  jointed, permeability of   254
  landsliding in   94–7
  rebound when load removed   242
  relationship between quality and
      material constants   **379**
  and rocks, description of   368–75
    data sheet for   *368*
    parameters used in description
        368, 372
  soft, failure by gradual sliding   95
  weathering   384–8
rock melt concept, for disposal of
    radioactive waste   475
rock pinnacles   166
rock quality designation (RQD)
    380–1
rock salt   404, 477
  dry, radar translucent   552
rock slides   91, 170
  small   96–7
rock slopes
  circular failure   96
  idealized failure mechanisms   96
  rotational failure   96

translational failure
  active and passive block failure
    96
  plane failure   96
  toppling failure   96
  wire meshing of   106
rock strength   373, **373**
rock traps   106
rock types   362–8, **529**
  classification   **370–1**, 372
  reflectance curves for   527, *527*
  strong influence on early stages of
    river development   114–15
rockbursts   80
rockfalls   88, 90, *90*, 170
  treatment of   106–7
rocks
  colour of   372
  common
    resistivity values of   549, **549**
    velocities of compressional waves
      546
  decay rate   386
  deformation and strength of
    367–8
  demagnetization of prior to
    eruptions   50
  description for engineering
    purposes   **370–1**, 372, **372**
  detectable pre-eruption volume
    increase   71
  disintegration
    aided by chemical weathering
      386
    through salt weathering
      184–5
  fractured and exposed, acid mine
    drainage from   448
  fresh metamorphosed, good dam
    sites   243
  intact
    engineering classification of
      375, *375*
    permeability of   374–5, *374*,
      374
  quarried, ease of dressing   406
  rapid intensity attenuation on   64
  and rock masses, description of
    368–75
    parameters for description of
      intact rock   368, 372
  role of plants and animals in
    breakdown of   387
  some physical properties of   **368**
  sub-surface exploration in   542–5
  texture of   372–3
  type and structure affect drainage
    systems   114
  weathering of   384–8
roof collapse, in mine workings   416
roofing stone   408
root zone water, estimation of   197

rotary percussion drilling rigs   542,
  544
roughness factor, and stream velocity
  121
run-off   *122*
  accelerated by loss of vegetative
    covering   87
  arid regions   170, 174
  basic components   233
  by overland flow   310–11
  causing soil erosion   308
  collection in drainage ditches at top
    of slopes   110
  diversion from landfill   275
  ephemeral   163
  from roads, a potential pollutant
    278
  intense   169
  moving sediments on slopes   174
  rapid from mountain catchments
    174
  and rehabilitation of tailings dams
    446
rupture, of faults   59
  causing ground displacement
    68–9
  total energy of   61

S waves   58, 71, 546
sabkhas
  coastal   180–2, 351
    Jubail, Saudi Arabia, fine grained
      soils   351
  dissolution of soluble salts with
    excessive moisture   182
    an engineering problem   351
  inland *see* playas; salinas
  salt weathering attack on buildings
    185
  soil conditions   180–3
  soils frequently low strength
    351–2, **352**
safe yield   270–1
  estimation of   271
  related to a return interval   270
  reservoirs   235
St Francis Dam, California, failure of
  405
St Helens, Mount
  a collapse caused by an earthquake
    52–3
  lateral blast damage   *51*
  mudflow   *47*
  nuée ardente   *40*
  side bulge before eruption   49
salinas   *172*, *173*, 182–3
  ground conditions   **182**
  salt weathering attack on buildings
    185
saline intrusion   275–7
  checked by maintaining fresh water

    gradient towards the sea   276
saline–fresh water interface   275
salinization
  avoidance of   195
  from inefficient irrigation   193
salt, solubility of   406
salt content, groundwater   260–1
salt crusts   *172*, *173*
salt crystallization, pressures
  produced by in small pores
  184
salt deposits, thick, suitable for deep
  disposal repositories   477
salt weathering   183–8
  damage to bituminous paved roads
    and built-over areas   185
saltation   124, 164, 168
salts
  accumulation by irrigation, removal
    by leaching   195
  dissolved, in mine water   451–2
  effect on plant growth   194
  soluble
    crystallization and hydration-
      dehydration thresholds
      184–5
    damage to roads and pavements
      185–6
sampling and investigation
  programme, contaminated land
  482
sampling tubes   541, *542*
San Francisco earthquake (1906),
  ground displacement   68
San Francisco fault, movement along
  68
San Francisco Peninsula
  geological map   **63**
  isoseismal lines   **63**
sand
  desert   166
    threshold velocity for   164
  and dust, aeolian
    potential sources of   177, *178*
    sources and stores   176, *176*
  fine, development of quick
    condition   335, 337
  frozen, initial deformation   224
  marine, effect of salt content   326
  mobile, methods of stabilization
    178–9
  more easily eroded   123
  movement of   174–80
    by saltation   165–6, 174
  saturated, undisturbed samples
    from   542, *544*
  textural maturity of   325
  used for building purposes   325–6
sand accumulations   166–7
  factors controlling form of   167
sand blast effects   166
sand boils   65

sand disturbance, at the plunge line
144
sand drifts 167
sand dunes *see* dunes
sand fences
  impound/divert moving sand
    179–80
  semi-permeable 180
sand and gravel
  coarse, good earth dam foundations
    245
  as construction materials 324–6
sand liquefaction, caused by shocks
    335, *335*, 337
sand seas 166
sand sheets 167
sand storms 163, 309
sand trapping 149
sands and clays 293
sandstone 366, 427
  as a foundation rock 244
  fully saturated, resistivity of 280,
    551
sapropelic deposits 367
satellite data, for monitoring volcanic
    activity 49
satellite imagery 527–8
  aiding delineation of volcanic
    hazard zones 48
  band-lap stereo and time-lap stereo
    527–8
  important where map coverage
    inadequate 528
  interpreted in similar manner to
    aerial photographs 527–8
saturation
  increased by compaction 331
  sandstone 280, 551
  of soil 292
  zone of 233, 237, 246
    water quality in 260
Scandinavia, isostatic recovery 220
scarp slopes 530
schist 364, *366*, **371**, 395
schistosity 364, *366*, 395
Schlumberger configuration,
    electrodes 549, *549*, *550*
scoria 43
scour-and-fill 123, 125
scours, erosion of 115
screw auger 537
sea level
  recent changes influenced present
    coastline 139
  rise in causes rivers to regrade
    courses 120
sea walls 153, 161
  designed to prevent wave erosion
    and overtopping 154
sector collapse, volcanoes 52
sediment load, increased during flood
    125

sediment yield 311
sedimentary rocks 365–7
  clastic, composition of 366–7
  deposition causes
    bedding/stratification 366,
      367
  distribution of iodine 503
  elemental composition **492**
  organic 365
  precipitated 365
sedimentation
  cycle of 366
  in reservoirs 238–9, **239**
sediments
  deposition of 125–6
  fine grained, soil moisture unfrozen
    at low temperatures 224
  marine
    reducing, arsenic in 498
    richer in iodine 503
  marine muds, rich in lead 502
  movement on slopes 174
  unconsolidated, coastal erosion
    145, *146*
seepage
  excessive, may damage dam
    foundations 242
  from discontinuities, assessment of
    378, **378**
  from reservoirs 235
    controlled by exclusion or
      drainage techniques 237
  from tailings dams 445–6
  of radon into buildings 513
seepage cutoff systems 465
seepage forces, and quick conditions
    335
seepage springs 249
seiches/seiching 70, 152
seif dunes 167, *168*
  development of 169
seismic hazard and risk 74–6
seismic intensity increments, for basic
    ground types 66
seismic investigation, methods of
    70–4
  accelerographs 73–4
  aseismic investigation 72–3
  assessment of movements along
    faults 72
  earthquake prediction 70–2
seismic methods, geophysical
    exploration 546–8
  cross-hole 556
  seismic reflection method 547
  seismic refraction method 547–8,
    548
seismic microzoning 76, 78
seismic microzoning maps
  Los Angeles County 79
  uses of 78
seismic moment 60–1

seismic risk, finding an acceptable
    level 74
seismic risk maps 77, *78*
seismic shaking, duration and
    intensity 62
seismic velocities, field seismic velocity
    381
seismic waves 58
  recording of 60
seismic zoning maps 76
  distinguishing zones of destruction
    77, *77*
  problem of parameter choice 77
seismic zoning and microzoning
    76–8
seismicity
  at reservoir sites 79–80
  and ground conditions 63–6
  important in tsunami hazard studies
    161
  induced 78–81
    by fluid disposal in deep wells
      473
seismo-tectonic maps 76
seismographs 161
seismological data, used to recognize
    active faults 73
seismometers 60
selenium **490**, 496
  from volcanic eruptions 507
  and health 507–8, 518
  in organic shale, coal and petroleum
    507
selenium deficiency 508
semi-arid regions
  encroachment of deserts 188
  hazardous though climatic
    variation 163
  moisture loss through transpiration
    339
  sensitive to soil erosion 310
semiconductor detectors and
    absorbers 515
sensitive clays, effect of disturbance
    301
sensitive soils 349
sequential land planning concept 2
settlement
  differential
    domestic refuse 462, 464
    and excessive, a problem with
      peat 359
    restored opencast sites 433
  in low-grade compaction shales
    398
  through solution of sabkha salts by
    groundwater 182, 184
sewage sludge, contaminants in 277,
    500
shale 366, 396
  characterized by laminations 396
  compaction shale 396

overconsolidated, rebound in 398
deep thick beds, reprocessed waste and cement injected   478
development of hydrostatic pressure on downstream abutment rocks   244
low grade compaction, settlement in   244
not all suitable for deep disposal repositories   477
organic-rich, containing mercury 501
overconsolidated, swelling tendencies   397
settlement problem   398
swelling properties detrimental to engineering structures   397
variable engineering behaviour 396
well cemented
few dam site problems   244
resistant to slaking   396
shallow ground stabilization, promotes arching across stopes 417
shear failure   245
resulting in slide movement   90
shear strength
along discontinuities   96
along joint surfaces   380
coarse material in spoil heaps   442
of a compacted cohesive soil   331
interblock   377
of peat   358
of soil   300–1
in situ indirect measurement of 556
tropical soils   356
undrained, of soft clays, use of shear vane tests   559–60
shear stress
dependent on effective stress   301
landslides in soils   93
on plain of failure   300, 301
shear tests, *in situ*   561, *561*
shear vane tests   559–60, *559*
sheet floods   169
sheet flow   170, 174, 308, 311
sheet joints/jointing   393–4
related to depth of overburden 376–7
sheet wash   171
shelter belts   316
shield volcanoes   35, 37
shingle crests, height of   144
shore   139
shorelines   139
erosion or accretion at change of direction   149
shrinkage, in soils subject to freezing 226

shrinkage limit   294
side-hill dumps, mine waste   441
side-looking airborne radar (SLAR) 526–7
differences between SLAR images and photographs   526
lateral overlap can give a stereoscopic image   426–7
not affected by cloud cover   526
silcrete   170, 173, *354*, 493
silica   387
colloidal   327
and viscosity of lava   44
silicate minerals, hydrolysis of   387
silicon   **489**
in groundwater   264
silicosis   511
sills *see* dykes, and sills
silts
and clays   300
lower transport velocity   123
frozen, moisture content   225
and sands, development of quick condition   335
silty soils
formed under arid conditions, collapsible   352
moisture content influences strength and compaction   332
single hazard–multiple purpose maps 11
sinkholes   400
density of and potential subsidence 402
in reef-mining areas   417, *418*
formed through lowering of the water table   417
southeast USA   403
sinter terraces   41
site assessment, and land evaluation 524–61
site exploration   536–7
mapping of rock and soil types 536
waste disposal sites   459
site investigations   524, 535–45
availability of data from   3
of contaminated sites   481–2
exploratory methods   481–2
sampling plan   482
use of specialists   481
desk studies and preliminary reconnaissance   535–6, **536**
exploration to include all strata affected by structural loading 537
mining areas, include location of subsurface voids   420–1
objectives   535
plans   29
sampling in soils   541–2

search for mine shafts on or near sites   425
by methane detection   425
site exploration   536–7
subsurface exploration
in rocks   542–5
in soils   537, 539–41
site restoration/rehabilitation   5
after cessation of mining   415
of dredge-mined area   436, *436*, 437
opencast coal sites   433
site selection
for landfill   274–5
for toxic waste disposal   275
skid resistance   411
Skipsea Till   **210**, 211
slab avalanches   220
slab slides   91
slake durability test   397
SLAR *see* side-looking airborne radar (SLAR)
slates   408
slaty cleavage   364
Sleepy Hollow oilfield, Nebraska, earth tremors   78–9
slide surfaces, quasi-planar   91
slides
rotational   90, 145, 244
safety improved by partial removal of material   107
translational, in stratified deposits 91
slippage, slow, differential *see* fault creep
slope, in morphological mapping *13*, 14
slope alteration, to reduce sand deposition   180
slope category maps   14
slope control and stabilization 106–11
alteration of slope geometry   107
drainage   110–11
restraining structures   108–10
rockfall treatment   106–7
slope reinforcement   107–8
slope extensometer   98
slope failure
delayed   88
potential for in landfills   462
in soils, caused by influence of water   93
slope length and gradient, Universal Soil Loss Equation   312
slope material
increased weight increases shearing stress   87
strength reduced by weathering 88
slope monitoring   97–9
automated processes   98

monitoring acoustic emissions (noise)   99
monitoring groundwater   98–9
monitoring movement   97–8
slope monitoring system, cost of   97
slope movement, demonstrated by distressed vegetation   87
slope stability   **529**
defined by a factor of safety   84
in excavations in shale   399
influenced by vegetation   86–7, **87**
investigation/assessment of   **99**, 100, **100**
a problem
in cuts during and after dam construction   244
with steeply dipping strata   406
use of checklists and inventory forms in stability investigations   101–3
slope stability calculations, sensitivity analysis for   *95*
slope stability classification, frequency and potential criteria   **99**
slope steepness, critical   14, **14**
slope units, in morphological maps   11
slopes
failure and potential for failure   100
oversteepened   227
protected with geotextiles   315–16
sediment movement on   174
stresses in   84
smectite   292, 326, 327, 355
snow-line   199
society, risk to of natural hazards   7
sodalite   503
sodium   263
dissolved, in dispersive soil pore water   341–2
sodium adsorption ratio (SAR)   195, 341, 342
sodium hazard, from irrigation   195, *196*
soft water   262
soil aggregations
importance of   311
in quickclays   349
resistant to wind erosion   308
and soil erodibility   308–9
stability of   314
soil classification   **295**, **296–9**, 298–300
according to erodibility   176
soil conservation measures   314, 324
for water erosion   314–16
for wind erosion   316
soil creep, causes   83
soil density   292–3
soil erodibility factor   312
soil erodibility index   313

soil erosion   7, 307–11
assessment of   316–23
by water   308, 311
by wind   308–9, 313
control of to maintain soil fertility   314
erosion control and conservation practices   313–16
field recording sheet   *322, 323*
reduced by good agricultural practices   133
soil erosion hazard assessment   316–17
soil erosion surveys   322–3
soil exhaustion   311
soil fertility   287–8
restoration of, old spoil heaps   443
soil flushing   483
soil formation, adversely affected by increased run-off   310–11
soil horizons   286–7
containing organic material   286
mineral horizons   286–7
soil liquefaction   64, 65, 69–70, 332–3
soil loss
estimation of   311–13
tolerable rates of   314
soil management   314, **314**
soil mapping units   317, 323
soil maps   323, *324*, 536
soil moisture
conserved by terraces   315
loss by evapotranspiration   287–8
and particle cohesion   165
retention of   287
and soil texture   **288**
soil moisture content   287–8, 292
assessment of   197
little changed by compaction   331
soil moisture deficits   234
soil nailing, with shotcrete-reinforced wire mesh   109
soil profiles   286, *286*
mature   287
soil remediation   483–4
soil ridge roughness factor   313
soil structure, and resistance to compression   307
soil suction   288, 340
controlling swell–shrink behaviour   339
and moisture content   191–2
sabkhas   183
soil surveys   **529**
and mapping   323–4
special purpose   323–4
soil washing   483
soil water tension   197
soil–cement liners   463
soil–plant relationship   495–6

soils
acting as filters   262
of arid regions   350–2
sabkha soils   351–2
assessment of relative densities   **557, 558**
basic properties of   292–300
changes in after accumulation   285–6
chemical changes in   286
coarse grained, arid regions, problem with   183
cohesionless
compacted by ground vibrations   64–5
specifications for control of compaction   331
cohesive
compacted, compressibility of   331–2
consistency of   294, **295**
development of pore water pressures   93
phase relations   292
slope failure   93
specifications for compaction of   331
compaction at optimum moisture content   332
compressibility of   307
concentration of arsenic in   499
as construction materials   324–33
in desert regions   493
fluorine in   506
frost action in   222
frost-susceptible, addition of chemicals to reduce heave   230
frozen
creep strength defined   224
description and classification of   222, **223**
granular, influence of relative density   223–4
mechanical properties of   222–3
strength of   223
granular   300
compaction of   332
conversion of water to ice   224, *225*
failure of   93
need for adequate relative density   332–3
relative density   292–3
highly saline or alkaline   291
internal frictional resistance of   301
investigation of, light cable and tool boring rig   539–40
landslides in   93–4
lead in   502

lost by desertification   188
morainic   209, *209*
origin of   285–6
pedological types   288–92
prediction of volume changes
   oedometer method   340
   soil suction methods   340
problem soils   335–59
radon concentrations in   512
retention of water in   246
and rock, act as purifying agents
   260
saline, classified   195, *197*
saline-sodic   195
salinization of   193, 195
salt-bearing, frequently hygroscopic
   183
sampling in   541–2
   disturbed and undisturbed
     samples   541
shear strength of   300–1
sodic   195
softened by compaction   331
strength increased by plant roots
   311
subsurface exploration in   537,
   539–41
thixotropic   301
trace element content reflects parent
   rock   493
tropical   352–7, 493
used in earthworks, compaction
   characteristics   328, **329–30**
variation in iodine content   503–4
zinc deficiency in   508
see also collapsible soils; dispersive
   soils; expansive clays;
   quicksands
solfataras   34
   associated with volcanic dormancy
     or senility   40
   super-heated steam from   40–1
solid stem augers   539
solidification process, radioactive
   waste   476
solifluction   83, 224–5
solifluction sheets   225
solonchaks (salorhids), potentially
   useful for agriculture   291
solonetz (naturargids) soils   291
solution
   of gypsum/anhydrite   405
   in limestone   401
solution cavities/features   460
   in limestone
     modelled   402–3
     problems in construction of large
       dams   244
solution channels, formed by wild
   brine pumping   439
solution mining   438–9
   controlled, of brine   439

solution pipes   403
solvent extraction   450
solvents
   to remove contaminants in soil
     483
   used in heap leaching   452
sonic logging devices   555
South Africa
   gold mining   416, 417–18
   mining of tabular stopes with
     substantial spans   416
   see also Johannesburg, South Africa
South Wales, assessment of degree of
   risk due to mining subsidence
   422
spalling   407
specific capacity, of a well   265
specific retention   251
specific risk   7
specific yield   251, **252**
spent fuel rods   475
spilling breakers   144
spills, accidental, treatment of
   278–9
spillways, marginal   219
spits, supplied by longshore drift
   149–50, *150*
splash erosion   174, 311
   dependent on vegetation cover
     310
spoil heaps   433
   acid generation   448
   rehabilitated   415, 432, *432*
   solids lost from areas of current
     tipping   451
   and their restoration   441–4
     need for additional land   443
     surface restoration after
       regrading   443–4
spontaneous combustion
   in coal seams   453–5
   colliery spoil heaps   442–3, *443*
spontaneous potential (SP) logs   555
spray irrigation, care over   272
spring lines   249
springs   249–50
   location of   249
Sri Lanka, dental and skeletal fluorosis
   507
stabilization, of mobile sand   178–9
   artificial stabilization   178–9
stacks   146, *148*
standing waves   140
   oscillation of, and pressure along
     discontinuities   140
steady supply *see* safe yield
steam, in volcanic eruptions   40
steel dowels, use of   107
step-performance test   258–9
stone, for revetments   156
stone pavements   170, 175
stone pitching   156

stone polygons, and frost heave   226
stone stripes   226
stopes, plugging with concrete for
   stabilization   417
storage capacity, estimation of for new
   reservoir site   235
storage coefficient
   defined   251–2
   unconfined aquifers   252
storm beach ridges   145
storm centres, large waves from   140
storm intensity, relationship to
   duration and magnitude of
     precipitation   130–1
storm surges
   influential factors   151–2
   major, types of   152
   and marine inundation   151–3
   risk often seasonal   151
   severity of   152
storm tide warning services   152
storm waves, remove beach material
   145
storms, tracked by satellite   153
strain release, sudden, along fault
   segments   72
strandlines (marine terraces)   220
strato-volcanoes
   polygenetic   36–7
   shape changes   37
stratum springs   249, *249*
stream action, arid/semi-arid regions
   169–74
stream classification, Horton   114
stream energy, lost by friction from
   turbulent mixing   121
stream flow records, for potential
   reservoir site investigations
   235
stream sediment sampling   496
   analysis of trace element content
     496–7
stream velocity   121
streams
   active, on pediments   171
   ephemeral   171
   ephemeral channels   174
   see also meltwater streams; rivers
stress corrosion, limestone   400
stress relief, after ice melt   226
striations, glacial   200, *201*
strip cropping   133, 315, 316
strip mining and opencasting   431–3
stripping   411
stripping ratios
   opencast coal   433
   surface mining   431
structural stability, of collapsible soils
   346
structures
   attacked by salt weathering   185
   conventional, classification of

mining damage 428–9, **429**
on different foundations, varied
responses to earthquakes
63–4
establishment of risk to in
earthquake regions 60
foundations on strata beneath
shallow workings 423
high values, flood proofing 133
mining areas, consolidation
grouting 423
removal or conversion of, in
landslide areas 105
surface, damaged by mining-
associated earth tremors 80
see also buildings
stubble strips 316
subglacial moraine 208
submarine bars 145
submerged forests 220
subsidence
assessment of risk due to mine
subsidence 422
associated with fluid abstraction
437–41
differential
fault steps 416
longwall mining 426
in South African gold reefs
417–18
due to abstraction of oil 438
due to groundwater abstraction
7, 437–8, *437*
due to gypsum solution 404
due to pillar collapse 416, 419,
419
prediction of 419–20
due to salt extraction 438–9
from metalliferous mining
415–18
mining methods 415–17
stabilization of old workings
417
in limestone 400
and longwall mining 426–30
measures to mitigate effects of
430–1
in the South African gold fields
416
urban, from mining 415
see also settlement
subsidence susceptibility maps, using a
GIS 403
subsidence troughs, from pillar failure
419
subsurface information, from
geophysical investigations
546
subsurface movements, measurement
of 98
suction pressure
of argillaceous materials 396

of ice lenses 230
Suez, Egypt, flood hazard map *132*
sulphate attack, by salt weathering
185
sulphate minerals, in clays,
detrimental to brick making
327
sulphate reduction 272
in landfill 466
sulphate(s) 261, 263
in mine water 452
sulphide minerals, oxidation of
and acid mine drainage 447–50
formation of secondary minerals
449
sulphur compounds
argillaceous rocks, volume
expansion can cause damage
399
decomposition causing attack on
Portland cement 399
oxidized by weathering 387
sulphur dioxide 443
sulphuric acid 41
summit craters 37
superficial deposits
confining function 255
and diagnostic landforms
recognized from aerial
photographs 531
superglacial till **207**
surf zone 143, *145*
surface coatings, on gravel and sands
325
surface ground motion, influenced by
physical properties of soils and
rocks 63
surface mining 431–7
blasting and vibrations 434–5
dredge mining 435–7
open pit mining 433
strip mining and opencasting
431–3
surface movement, monitoring of
97–8
surface pollution, from leachates
468
surging breakers 144
suspended load
concentration 124
distribution 124
suspended solids, pollutants from coal
mining 450–1
suspension, of sediment particles
164
swallow holes 403
swash 144, *145*
drives material up the beach 144,
145
movement of material in traction or
suspension 144
swash zone 143

swell–shrink, in clay soils 335, 357
black cotton soils 355
controlled by soil suction 339
swelling pressure, hydrating anhydrite
405

Taal volcano, Philippines, 1965
eruption prediction accurate
49–50
Tacubaya Clay, saturated deposits,
response to low-frequency
ground motion 65
tailings
deposition of 445
described 444
geotechnical properties of 445
rehabilitation of impoundments
446
used as liners in seepage control
445–6
tailings lagoons/dams 433
construction of 444, *445*
controlling seepage loss 445–6,
446
failure of 444–5, *445*
waste disposal in 444–6
Taiwan, arsenic toxicity 499–500
talus (scree) creep 83
Tees Laminated Clay 217, 218–19,
218, *219*
average shear strength 218–19
illite and kaolinite 217
temperature
desert surface, effects on common
salts 184–5
of groundwater 261
significant diurnal change in 385
tensile strength
building stone 406
rock 374
for facing stone 408
tension cracks, on slope crests 98
tension scars 90
from wild brine pumping 439,
440
tephra, defined 41
terminal moraines 208–9, *208*
in lakes 216
terraces
in arid areas 174
river and lake 220
terracettes 83
terracing 315
terrain evaluation 317, 322, 531–2
described 531
Hong Kong 101
landscape classification 531
means of correlating known and
unknown areas 531–2
parametric land evaluation 531
texture, of rocks 367, 372–3

Thames Barrier, London   153
thermal anomalies, due to heat of
        rising magma   49
thermal contraction, and desiccation
        226
thermal inertia, and infrared line
        scanning (IRLS)   526
thermal springs   249–50
thermokarst   225
Thiobacillus ferrooxidans, work of
        449
Third World countries *see* developing
        countries
thixotropy   301, 328
threshold, fiords   203
thyroid gland, requires iodine   504
tidal currents   139
tidal range   139, 142
tide gauges, measuring lake water
        level, Rabaul caldera   49
tides   139, 142–3
till sheets   205
tillite *see* drift deposits, stratified
tills   203–6, **207**
    basic properties   209–12
    compactness   205
    compressibility and strength
        212–13
    dense, undrained shear strength
        212
    fissures in   206
        influence shear strength and
            stability   213
    Holderness   210–12
        strength of   213, **214**
    north Norfolk   209, *209*
        strength of   213, **214**
    Northumberland   **211**, 212, *212*,
        213
tiltmeters   49, 72
    for measurement of surface
        movement   98
Tokyo earthquake (1923),
        destructiveness of   66
tolerability, of risk   9
tombolos   150
tooth decay, prevention using calcium
        fluoride   506
top water or normal pool level   235
topographical maps   535
    for potential reservoir site
        investigation   235
topography   **529**
toppling failure   90, 96
torrent stage   116
total dissolved salts, in pore water
        342–3
total dissolved solids (TDS), in
        drinking and industrial water
        264
total energy, of a river   121
total risk   7

toxic waste
    containment preferred   275
    disposal of   5
        protection of groundwater from
            472
toxicity
    of cyanide and metal cyanide
        compounds   453
    defined   490
    of metals   518
toxins, cumulative   490
trace element disease patterns,
        changed by new soil and crop
        management   496
trace elements
    areas of deficiency or toxicity   496
    can be toxic/lethal   487, 490
    essential to animals   487
    and health   498–511
    and humans   487, **488–90**
    in irrigation water   193
    levels of and disease, use of
        geochemical maps   497
    mobility of   491
    plant tolerance to   495
    speciation and mobility   490, *491*
    toxic, in acidic water   491
    uptake of   487, **488**
tracers, for investigation of
        groundwater movement
        259–60
training walls, to protect inlets and
        harbours   153–7
transgranular fracture   367
transition elements, removal from
        solution   493
translation waves   139, 140
translational slides   91
transmissivity   252, 551
    and safe yield   271
transportation, effects on
        sediments/soils   285, **286**,
        366
trap efficiency   238
travertine terraces   41, *41*
trees   161
    as windbreaks   195
trench drains   110
trenches
    for rapid, economic information
        gathering   539
    short-term protection against
        moving sand   180
tripods, used for coastal protection
        155, *155*
tropical regions, potential for toxicity
        or deficiency conditions   493
tropical soils   352–7, 493
    behave as if overconsolidated   356
    classification of   354, **354**
    containing halloysite/metahalloysite
        357

some engineering properties of
        354, **354–5**
trough's end   203
true formation factor   551
truncated spurs   203
tsunami hazard   161
tsunami risk maps   161
tsunamis   53, 70, 159–62
    frequency of recurrence   161
    intensity scale   **160**
    prediction of arrival time possible
        159
    seismic   159
tuff   43
    chaotic   43
    weathering of   388
tuffites   43
turbidity
    associated with discharge of colliery
        waste   451
    caused by oxidation of ferrous iron
        468
    of groundwater   261
turbulent flow   120, 171, 252, 308
    in lahars   46
    shooting flow   121
    streaming flow   121, 124

U-shaped valleys   203, *204*
unconformities, springs along outcrop
        of   249
underdrainage systems, for landfills
        462–3
underliners, and covers,
        geomembranes   463
underseepage, prevention methods
        245
undisturbed sampling   541
undrained loading, of rear of
        mudflows   92–3
UNESCO, definitions of vulnerability
        and risk   7
Unified Soil Classification System
        222, 299
    coarse grained soils   **296–7**
    fine grained soils   **298–9**
unit hydrograph concept   122
unit weight   95–6
Universal Soil Loss Equation
        311–13
upconing   276, *276*
uplift, in shale excavations   398
uplift pressure, on heel of dam   242
uranium   512
urban development   524
urbanization
    in developing countries   1
    increases flood hazard   129, 131
USA
    boundary of one hundred year
        flood defines compensation

area 8, 134
collapse of cavernous gypsum 405
costs of expansive soils problems 338
the Dust Bowl 307, 308
land use in the floodway 132–3
longest fissure zone 439
National Environmental Policy Act (1969) 3
Ohio survey on relationship between geology, water chemistry and disease 516–17
results of overgrazing 190
southeast, sinkholes 403
subsidized flood insurance 134
use of levées to prevent flooding 134
useful storage, reservoirs 235

vacuum extraction 482
vadose water 246
vadose zone 247
valley bulging 83
valley creep, means spreading of earth dam load over a wider area 244
valley fills, mine waste 441
valley glaciers 199
ponding back lakes 219
valley slopes, weakened by sheet (flat-lying) joints 243–4
valley springs 249, 249
valley tract 116
valley trains 213, 215
valley widening 115
valleys
V-shaped 115
glaciated, stepped long profile 203
U-shaped 203, 204
vanadium 489
varved clays 217–19
vegetation
beneficial and adverse effects of 309, 310
effects on soils 285
identifiable from aerial photographs 255
importance in soil erosion 309–10
influencing evapotranspiration 310
influencing slope stability 86–7, 87
loss of at desert margins 189
protection against wind erosion 165
on spoil heaps 441
transpiration from a major cause of

moisture loss 339
used to stabilize dunes 178
as windbreaks 179
see also afforestation; reafforestation; trees
vertisols 354
black cotton soils 291–2
prone to erosion, dispersion and swell-shrink problems 354–5
very low-frequency (VLF) electromagnetic method 552
vesicles, left in lava by gas action 40
Vesuvius, Plinian eruption and destruction of Pompeii and Herculaneum 39
vibrations
from blasting 434–5
reduction of possible effects 435
subjective response to 434, 435
vibrographs 435
virgin compression curve 306
viruses
elimination from drinking water 265
in groundwater 261
prone to adsorption 273
visibility, reduced by sand and dust 175
vitrification
in bricks 327–8
in situ, of contaminated soil 483
void filling, of old mine workings 423
void migration 420, 420
void ratios 285–6, 292, 306
in collapsible soils 345, 347
peat 357, 359
shale, application of load 399
tropical soils 356
volatile organic acids 470
volatile organic chemicals (VOCs)
removal from soil 483
water supply wells 277–8
volatiles, in magma, affect eruptive energy 34
volcanic activity
dealing with 56
hazards associated with 53, 55
volcanic bombs 43, 55
volcanic breccia 43
volcanic domes 37
volcanic form and structure 34–8
volcanic gas output, from volcanoes, monitoring of 49
volcanic gases
emissions variable in composition 40
injurious to people 55
poisonous, given off by volcanoes 53
volcanic hazard maps
Merapi volcano 54
preliminary, Galunggung volcano

54
Volcanological Survey of Indonesia, zones on 55
volcanic hazard and prediction 48–56
assessment of volcanic hazard and risk 50–6
dealing with volcanic activity 56
hazard categories 51–3
prediction methods 48–50
volcanic hazard reduction 56
volcanic hazard zones/zoning
delineation of 48
use of mean recurrence intervals (MRIs) 56
volcanic products
lava flows 44–6
pyroclasts 41–4
volatiles 40–1
volcanic risk assessment 50
assessment of hazard 50
volcanic risk maps 55–6
volcanic risk reduction 50
volcanic zones, and active plate boundaries 34, 35
volcano-tectonic sinks 36
volcanoes
active 48
in areas of high population density 50
need for surveillance 56
changes in elevation precede eruption 49
composite 35, 36
dormant 34
extinct 48
few with intensive monitoring 49
fissure 35, 36
gas phase, significant pre-eruption information 50
large, structural collapse of 52
lava 37
monogenetic or polygenetic 35
originating over hot spots 34
prehistoric record assists assessment of eruption probability 48
pyroclastic 35
shield 35, 37
speed of most dangerous phenomena 50
strato-volcanoes 36–7
vulnerability
defined by UNESCO 7
of property in volcanic emergencies 50
vulnerability analyses 4
vulnerability maps 11

wash boring method 540, 540
waste
described 459

dredge mining, flexibility of disposal 437
hazardous and non-hazardous 459
in landfills, degradation of 466
water-absorbing capacity of 462
see also hazardous waste; landfill sites; mine wastes; radioactive waste; toxic waste
waste waters, and effluents from coal mining 450–2
water
acceptability for human consumption 265
aggressiveness of towards limestone 400
fluorine in 506
fossil 261
fresh, concentration of arsenic 499
hardness of 264–5
in magma bodies 40
meteoric 246
natural
contains traces of most elements 493–4, *494*
lead in 502
resistivity values 548, **549**
optimization of use 282
physical properties in relation to water supply 261
produced during surface mining 431
residence time of 260
and reaction with host rock 495
retention of in soils 246
and soils 292
solute content of 260
water budget studies 250–1
water content variations, in soils and rocks 255
water hardness, and cardiovascular disease 516–17
water injection, to maximise oil yields, causing seismicity 78–9
water inventory of the Earth 233, **234**
water level change maps 256–7, *257*
water meadows, temporary flood storage 137
water pollution, from acid mine drainage 447
water pressure monitoring, differences between rock and soil 98–9
water quality
design of monitoring network 280
for irrigation 193, 194–5
and uses 260–5

of water from abandoned mines 448
see also sodium hazard
water quality monitoring wells
design and construction 279
used to observe effects of new developments 280
water release, and volcanic lahars 46
water spreading, for artificial recharge 272
water storage, desert, in coarse grained alluvium 174
water table 246–7
important at landfill sites 461
mapping of, northern Bahrain 186, 188, *188*
and new reservoir 235
perched 247, *247*
permanent and intermittent 247
rising, affecting opencast backfill 433
underlying coastal sabkhas *181*, 351
see also groundwater
water table maps 255, *256*
water-bearing deposits, consolidated through overpumping *437*, 438
water-injection test 258
watergas, potentially explosive 442
waterlain till *207*
waterlogging, over permafrost 222
watershed management, use of forests 133
watersheds, with high drainage density, effects of 128
wave height *141*
wave lengths, open sea 140
wave patterns, in sheltered areas 142
wave period *141*
wave reflection, and sea walls 154
wave refraction 141–2
causes longshore drift from headlands to bays 151
of tsunamis 159
wave velocity 140
wave-cut platforms 146, *147*
waves 139–42
acting on beach material 139
breaking 143–4, *144*
constructive and destructive 144, 145, 147
force and height of 140–1
importance of fetch 140
see also tsunamis
weathering 384–8
chemical and biological 386–7
effects on mudstones 396
engineering classification of 387–8
honeycomb 407

mobilization of arsenic 498
physical (mechanical) 169, 170, 351, 385–6, 396
of pyrite in colliery spoil heaps 442
rate of 385
and soil formation 285
and strength reduction 394
weathering grades, identification and classification 388, *389–90*
well waters, radon increase in 71
wells 265, *268*, 270
collector wells 249
deep
abstraction of groundwater for irrigation 190
for injection of hazardous liquids below fresh water aquifers 473–4
to drain slopes 111
domestic, safe distance from pollution sources 261–2, **262**, 273
excessive water abstraction, effect of 188
extraction, forming hydraulic barriers 472
factors in long-term yield 270
hydraulic efficiency of 259
increasing yields from 270
perennial yield to 250
polluting aquifers through poor design and maintenance 273
for recharge 272
sealing of 265, 270
Wenner configuration, electrodes 549, *549*
whalebacks 167
Wilmington oilfield, California, spectacular subsidence 438
wind
an agent of abrasion 163, 166, 175
causing orbital motion of water particles 139, *140*
and the movement of dust and sand 174–80
soil erosion by 308–9, 313
wind action 163–6
wind duration 140, *141*
wind erodibility 313
wind erosion 163–5, 166, 177–9, 313
control of by windbreaks 179–80
stabilization of mobile sand 177–8
Wind Erosion Equation 313
wind velocity, reduced by rough surfaces 165
wind-stable surfaces 166
windbreaks 179–80, 195, 316
Withernsea Till **210**, 211

world population, increase in and
    increased geohazard risk   7–8
World-Wide Network of Seismic
    Stations   60
World-Wide Seismograph Network
    75

xerophytes   255
xerosols (aridisols)   291

yardangs   166

yermosols (aridisols)   291
yield drawdown graphs   259
yield stress, tropical soils   356
Young's modulus   242–3
    influence of fissility on   396
    large plate bearing tests to
        determine value of   560

zeolites, absorption of water causes
    basalt disintegration   394

zeugens   166
Zimbabwe
    zinc in sediment, soil and forage
        508
    zinc in stream sediments   497
zinc   **489**, 496
    and health   508
zinc deficiency, in soils   508
zinc sulphide scintillation counting
    514